ELECTRON PARAMAGNETIC RESONANCE

ELECTRON PARAMAGNETIC RESONANCE

A Practitioner's Toolkit

Edited by

Marina Brustolon
University of Padova

Elio Giamello
University of Torino

A JOHN WILEY & SONS, INC., PUBLICATION

Copyright © 2009 by John Wiley & Sons, Inc. All rights reserved

Published by John Wiley & Sons, Inc., Hoboken, New Jersey

Published simultaneously in Canada

No part of this publication may be reproduced, stored in a retrieval system, or transmitted in any form or by any means, electronic, mechanical, photocopying, recording, scanning, or otherwise, except as permitted under Section 107 or 108 of the 1976 United States Copyright Act, without either the prior written permission of the Publisher, or authorization through payment of the appropriate per-copy fee to the Copyright Clearance Center, Inc., 222 Rosewood Drive, Danvers, MA 01923, (978) 750-8400, fax (978) 750-4470, or on the web at www.copyright.com. Requests to the Publisher for permission should be addressed to the Permissions Department, John Wiley & Sons, Inc., 111 River Street, Hoboken, NJ 07030, (201) 748-6011, fax (201) 748-6008, or online at http://www.wiley.com/go/permission.

Limit of Liability/Disclaimer of Warranty: While the publisher and author have used their best efforts in preparing this book, they make no representations or warranties with respect to the accuracy or completeness of the contents of this book and specifically disclaim any implied warranties of merchantability or fitness for a particular purpose. No warranty may be created or extended by sales representatives or written sales materials. The advice and strategies contained herein may not be suitable for your situation. You should consult with a professional where appropriate. Neither the publisher nor author shall be liable for any loss of profit or any other commercial damages, including but not limited to special, incidental, consequential, or other damages.

For general information on our other products and services or for technical support, please contact our Customer Care Department within the United States at (800) 762-2974, outside the United States at (317) 572-3993 or fax (317) 572-4002.

Wiley also publishes its books in a variety of electronic formats. Some content that appears in print may not be available in electronic formats. For more information about Wiley products, visit our web site at www.wiley.com.

Library of Congress Cataloging-in-Publication Data:
Brustolon, Marina.
  Electron paramagnetic Resonance: A Practitioner's Toolkit/Marina Brustolon.
    p. cm.
  ISBN 978-0-470-25882-8 (cloth)
 1. Electron paramagnetic resonance. I. Title.
   QC763.B78 2009
   543'.67—dc22

2008029606

Printed in the United States of America

10 9 8 7 6 5 4 3 2 1

On Exactitude in Science

... *In that Empire, the Art of Cartography attained such Perfection that the map of a single Province occupied the entirety of a City, and the map of the Empire, the entirety of a Province. In time, those Unconscionable Maps no longer satisfied, and the Cartographers Guilds struck a Map of the Empire whose size was that of the Empire, and which coincided point for point with it. The following Generations, who were not so fond of the Study of Cartography as their Forebears had been, saw that vast Map was Useless, and not without some Pitilessness was it, that they delivered it up to the Inclemencies of Sun and Winters. In the Deserts of the West, still today, there are Tattered Ruins of that Map, inhabited by Animals and Beggars; in all the Land there is no other Relic of the Disciplines of Geography.*

Suarez Miranda, Viajes de varones prudentes, Libro IV, Cap. XLV, Lerida, 1658
From Jorge Luis Borges, *Collected Fictions*, Translated by Andrew Hurley
Copyright 1999 Penguin

CONTENTS

FOREWORD xiii

PREFACE xv

CONTRIBUTORS xvii

I PRINCIPLES 1

1 Introduction to Electron Paramagnetic Resonance 3
Carlo Corvaja

   1.1 Chapter Summary 3
   1.2 EPR Spectrum: What Is It? 4
   1.3 The Electron Spin 5
   1.4 Electron Spin in a Magnetic Field (Zeeman Effect) 6
   1.5 Effect of Electromagnetic Fields 8
   1.6 Macroscopic Collection of Electron Spins 8
   1.7 Observation of Magnetic Resonance 10
   1.8 Electron Spin in Atoms and Molecules 11
   1.9 Macroscopic Magnetization 14
   1.10 Spin Relaxation and Bloch Equations 16
   1.11 Nuclear Spins 18
   1.12 Anisotropy of the Hyperfine Interaction 22
   1.13 ENDOR 25
   1.14 Two Interacting Electron Spins 28
   1.15 Quantum Machinery 31
   1.16 Electron Spin in a Static Magnetic Field 32
   1.17 Electron Spin Coupled to a Nuclear Spin 32
   1.18 Electron Spin in a Zeeman Magnetic Field in the Presence of a Microwave Field 34

2 Basic Experimental Methods in Continuous Wave Electron Paramagnetic Resonance 37
Peter Höfer

   2.1 Instrumental Components of a Continuous Wave Electron Paramagnetic Resonance (CW-EPR) Spectrometer 37
   2.2 Experimental Techniques 52

Acknowledgment 80
References 80
Bibliography 82

3 What Can Be Studied with Electron Paramagnetic Resonance? 83
Marina Brustolon

3.1 Introduction 83
3.2 Organic Radicals 84
3.3 Organic Molecules with More than One Unpaired Electron 92
3.4 Inorganic Radicals, Small Paramagnetic Molecules, and Isolated Atoms 96
3.5 Transition Metal Ions 98
3.6 Natural Systems and Processes 100
3.7 Tailoring and Assembling PS for Magnetic Materials 102
3.8 Industrial Applications of EPR 104
References 105
Bibliography 107

4 Electron Paramagnetic Resonance Spectroscopy in the Liquid Phase 109
Georg Gescheidt

4.1 General Considerations 109
4.2 Generation of Radicals and Radical Ions 110
4.3 Basic Interactions and Principles 116
4.4 Patterns and Line Shapes of Fluid-Solution EPR Spectra 120
4.5 Transition-Metal Ions 132
4.6 Biradicals 133
4.7 Simulation Software 134
4.8 How Fluid-Solution Spectra are Analyzed 135
4.9 Calculation of EPR Parameters 138
4.10 Molecular Properties Mirrored by EPR Spectra in Fluid Solution 139
4.11 Chemically Induced Dynamic Electron Polarization (CIDEP) and CID Nuclear Polarization (CIDNP): Methods to Study Short-Lived Radicals 151
Acknowledgments 154
References 154
Further Reading 157

5 Pulsed Electron Paramagnetic Resonance 159
Michael K. Bowman

5.1 Introduction 159
5.2 Vector Model for Pulsed EPR 162
5.3 Pulse Sequences 172

5.4 Data Analysis 185
5.5 Spectrometer 187
References 193

6 Electron Paramagnetic Resonance Spectra in the Solid State 195
Marina Bennati and Damien M. Murphy

6.1 Introduction 195
6.2 Anisotropy of the Zeeman Interaction: The **g** Tensor 198
6.3 The Hyperfine Interaction in the Solid State 209
6.4 TMIs 229
6.5 EPR Spectra for $S > 1/2$: ZFS 233
References 237
Appendix A.6.1 Simple Matrix Manipulations 239
Appendix A.6.2 Pauli Matrices 241
Appendix A.6.3 Transformation of Tensor Coordinates Via Matrices 241
Appendix A.6.4 Euler Angles 243
Appendix A.6.5 Matrix Elements of Spin–Orbit Coupling 246
Appendix A.6.6 Origin of the g and A Values for simple TMIs 246
Appendix References 249

7 The Virtual Electron Paramagnetic Resonance Laboratory: A User Guide to *ab initio* Modeling 251
Vincenzo Barone and Antonino Polimeno

7.1 Introduction 251
7.2 Modeling Tools 255
7.3 Tutorial and Case Studies 262
7.4 Conclusions 281
References 283

II APPLICATIONS 285

8 Spin Trapping 287
Angelo Alberti and Dante Macciantelli

8.1 What Is Spin Trapping and Why Use It? 287
8.2 Spin Traps 288
8.3 Experimental Methods 317
8.4 Applications 317
8.5 Spin Trapping in the Gas Phase or in the Solid State 320
8.6 Availability of Spin Traps 321
8.7 FAQs 321
Further Readings 322

9 Radiation Produced Radicals 325

Einar Sagstuen and Eli Olaug Hole

9.1 Introduction 325
9.2 Interaction of Radiation with Matter 326
9.3 Qualitative Detection of DNA Radicals 327
9.4 Tools and Procedures for Radical Structure Determinations 341
9.5 Quantitative Detection of Radicals 361
9.6 Highlighted Reading 374
Acknowledgments 375
References 375

10 Electron Paramagnetic Resonance in Biochemistry and Biophysics 383

Part I: Spin Labels, Paramagnetic Ions, and Oximetry 383
Michael K. Bowman

10.1 Introduction 383
10.2 Experimental Considerations 386
10.3 Dynamics 390
10.4 Saturation Transfer 393
10.5 Two-Dimensional Pulsed EPR 393
10.6 Protein Topology and SDSL 394
10.7 Surface Potentials/Accessibility and SDSL 394
10.8 Oximetry 395
10.9 Nanoscale Distance Measurement 396
References 402

Part II: Photosynthesis 403
Donatella Carbonera

10.10 Introduction 403
10.11 Oxygenic Photosynthesis 405
Appendix A.10.1: Pulse EPR Experiments on Radical Pairs 419
Appendix A.10.2: Recombination Triplet States of the Primary Donors 421
References 423
Further Reading 425

11 Electron Paramagnetic Resonance Detection of Radicals in Biology and Medicine 427

Michael J. Davies

11.1 Free Radicals in Disease Processes 427
11.2 Nature of Free Radicals Involved in Disease Processes and Potential Catalysts for Radical Formation 428

11.3 Direct EPR Detection of Reactive Radicals *In Vivo*
      and *Ex Vivo*   428
11.4 Spin Trapping of Reactive Radicals *In Vivo* and *Ex Vivo*   430
11.5 Spin Scavenging of Reactive Radicals *In Vivo* and *Ex Vivo*   439
11.6 Spin Trapping of Nitric Oxide   440
11.7 Verification of the Occurrence of Radical-Mediated Processes   443
11.8 Conclusions   445
Acknowledgments   445
References   446

12 Electron Paramagnetic Resonance Applications to Catalytic and Porous Materials   451
Daniella Goldfarb

12.1 Introduction   451
12.2 Paramagnetic TMIs   453
12.3 Spin Probes   475
12.4 Reaction Intermediates and Trapped Radicals   482
12.5 Sample Preparation Considerations   483
12.6 Summary and Outlook   484
References   484

13 Electron Paramagnetic Resonance of Charge Carriers in Solids   489
Mario Chiesa and Elio Giamello

13.1 Introduction   489
13.2 Point Defects, Charge Carriers, and EPR   490
13.3 Localized Electrons: Color Centers in Ionic Solids   492
13.4 Aggregate Color Centers   499
13.5 Localized Holes in Ionic Solids   500
13.6 Charge Carriers in Semiconductors   505
13.7 CESR in Metals   510
References   517

Appendix   519

SUBJECT INDEX   527

CHEMICAL INDEX   537

FOREWORD

It is more than 50 years since I ran my first electron paramagnetic resonance (EPR) spectrum. At that time the newly emergent field was developing rapidly, but it was still small so that in many ways life was simpler than it is today. For example, it was quite easy to keep up with the literature, and seminal papers on the understanding of hyperfine coupling constants in free radicals were not totally impenetrable even for a naive graduate student. Again, there was no sophisticated choice to be made about the type of experimental technique to use: it was continuous wave EPR or nothing. Of course, there were things that some today might regard as the downside: for example, if you wanted to make EPR measurements, you had to use a homemade spectrometer that contained thermionic valves (tubes). Its functioning tended to be idiosyncratic, and its operation was an art in itself. Further, all of the calculations that I performed, diagonalization of matrices, for instance, were done manually on electromechanical machines. Life in EPR could be tedious and frustrating, but it was exciting and it was fun and I reckon it still is.

Nowadays the field is amazingly more complex, not to say labyrinthine. There are numerous different types of experiments one can do in seeking answers to questions about fields ranging from solid-state physics to medicine. The Editors of this book perceived the need for a volume that would cover the whole of modern EPR and that would be useful both to newcomers to the field and to established practitioners who wished to broaden their perspectives. However, they recognized that it would not be feasible for them to write a comprehensive treatment themselves and were thus led to a multiple-author book. This was a brave decision: it is all very well for two people to share a concept, but to persuade authors to grasp that concept and shape their contributions to it so that the assembled volume would have coherence must have been like trying to herd cats. Thus, in my view, this book is a triumph of indefatigability.

When you are going to visit somewhere to which you have never been previously, or at best with which you are unfamiliar, it is always a good idea to study a map of the area you plan to visit and to carry a guidebook. I hope very much that this book will come to be seen both as a definitive map and as an indispensable *vade mecum* for the EPR community.

NEIL ATHERTON

PREFACE

The family of electron paramagnetic resonance (EPR) methods has expanded in the last 20 years, with new methods ushered in by instrumental and conceptual achievements. The basic qualities of the progenitor X-band continuous wave EPR, namely, its selectivity, sensitivity, and resolution, have been enhanced with the new generations of spectroscopies, which take advantage of multifrequency approaches, time resolution, pulsed techniques, and multiple irradiations. These developments have extended the range of applications and amplified the heuristic power of EPR spectroscopies, which are now increasingly employed in research in different fields, such as biology, medicine, material science, chemistry, physics, and earth sciences.

As soon as we began working on the book, we realized that our plan was somehow too ambitious. We were thinking of a book as a tool for the newcomers in the field; at the same time, we wished to address the developments of the experimental and theoretical EPR methods in recent years while also describing their most important applications in distinct fields. Assembling a book like this may be too great a task for anybody. However, we were also aware that the complexity was unavoidable for a book seeking to give an idea of what EPR means for science today.

A few EPR textbooks, some of which have been re-edited, have been used by generations of students. We have leaned heavily on those by Atherton and by Wertz and Bolton (and Weil), and one of our objectives here has been to update these. In thinking about how to give a comprehensive account of EPR today, we realized that it is like the Empire in the celebrated novel by Jorge Luis Borges, and that we ought not to act like the unwise Cartographers, set to draw "a Map of the Empire whose size was that of the Empire, and which coincided point for point with it." A real map was indeed necessary, with the right scale for each region, like an aerial photo taken by a plane flying at different altitudes. Moreover, to map such a wide territory we needed to enlist many more cartographers than the authors of the previous books. We looked therefore for 18 fine explorers of the various regions, including ourselves, and we have done our best to collect and assemble the diverse contributions. We have endeavored to connect the contents of the different chapters as much as possible to provide a compass to the reader.

We hope that this book will help to gauge the extent of the magnificent territory of EPR while also providing some indications as to where the path is easy and where it is difficult and how to find the way to one's destination.

Then, the readers will be just at the beginning of their tasks: exploring the real territory.

We want to thank our sixteen colleagues and friends, co-authors of this book, for participating to a living debate during its preparation, and to some intense discussions during the main EPR conferences of the past few years (Madrid 2006, Oxford 2007).

Without their knowledge, efforts and patience this book would have never be published.

We are also grateful to the support and precious advice of those colleagues of the international EPR community who have followed the progress of this book with keen interest.

Many thanks are also due to Dr. Marco Ruzzi (Università di Padova) for his kind assistance in various steps of the book editing.

MARINA BRUSTOLON, ELIO GIAMELLO

CONTRIBUTORS

ANGELO ALBERTI, ISOF-CNR, Area della Ricerca di Bologna, Via P. Gobetti 101, 40129-Bologna, Italy

VINCENZO BARONE, Dipartimento di Chimica and INSTM-Village, Università di Napoli Federico II, Napoli, Italy

MARINA BENNATI, Max Planck Institute for Biophysical Chemistry, Am Fassberg 11, Göttingen, Germany

MICHAEL K. BOWMAN, Department of Chemistry, University of Alabama, Tuscaloosa, AL 35487-0336

MARINA BRUSTOLON, Dipartimento di Scienze Chimiche, Università di Padova, Via Marzolo, 1, 35131 Padova, Italy

DONATELLA CARBONERA, Dipartimento di Scienze Chimiche, Università di Padova, Via Marzolo 1, 35131, Padova, Italy

MARIO CHIESA, Department of Chemistry IFM, University of Torino, Via P. Giuria 7, 10125 Torino, Italy

CARLO CORVAJA, Dipartimento di Scienze Chimiche, Università di Padova, Via Marzolo 1, 35131 Padova, Italy

MICHAEL J. DAVIES, The Heart Research Institute, 114 Pyrmont Bridge Road, Camperdown, Sydney, NSW 2050, Australia

GEORG GESCHEIDT, Institute for Physical and Theoretical Chemistry, Graz University of Technology, Technikerstraße, 4/IA-8010 Graz, Germany

ELIO GIAMELLO, Department of Chemistry IFM, University of Torino, Via P. Giuria 7, 10125 Torino, Italy

DANIELLA GOLDFARB, Department of Chemical Physics, Weizmann Institute of Science, Rehovot, Israel

PETER HÖFER, EPR Division, Bruker Biospin GmbH, 76287 Rheinstetten, Germany

ELI OLAUG HOLE, Department of Physics, University of Oslo, Oslo, Norway

DANTE MACCIANTELLI, ISOF-CNR, Area della Ricerca di Bologna, Via P. Gobetti 101, 40129-Bologna, Italy

DAMIEN M. MURPHY, School of Chemistry, Cardiff University, Main Building, Park Place, Cardiff CF10 3AT, United Kingdom

ANTONINO POLIMENO, Dipartimento di Scienze Chimiche, Università degli Studi di Padovav, Via Marzolo 1, 35131 Padova, Italy

EINAR SAGSTUEN, Department of Physics, University of Oslo, Oslo, Norway

PART I
Principles

1 Introduction to Electron Paramagnetic Resonance

CARLO CORVAJA

Dipartimento di Scienze Chimiche, Università di Padova,
Via Marzolo 1, 35131 Padova, Italy

Electron paramagnetic resonance (EPR), which is also called electron spin resonance (ESR), is a technique based on the absorption of electromagnetic radiation, which is usually in the microwave frequency region, by a paramagnetic sample placed in a magnetic field. EPR and ESR are synonymous, but the acronym EPR is used in this book. The absorption takes place only for definite frequencies and magnetic field combinations, depending on the sample characteristics, which means that the absorption is resonant.

The first EPR experiment was performed more than 60 years ago in Kazan (Tatarstan), which is now in the Russian Federation, by E. K. Zavoisky, a physicist who used samples of $CuCl_2 \cdot 2H_2O$, a radiofrequency (RF) source operating at 133 MHz, and a variable magnetic field operating in the range of a few millitesla and provided by a solenoid. More than five decades from the first experiment the technique has progressed tremendously and EPR has a broad range of applications in the fields of physics, chemistry, biology, earth sciences, material sciences, and other branches of science. Modern EPR spectrometers are much more complex than those used for demonstrating the phenomenon; they have much higher sensitivity and resolution and can be used with a large number of samples (crystalline solids, liquid solutions, powders, etc.) in a broad range of temperatures.

1.1 CHAPTER SUMMARY

The aim of this chapter is to provide the reader with the basic information about the phenomenon of electron magnetic resonance and the ways to observe it and to record an EPR spectrum. EPR spectra of very simple molecular systems will be

---

Electron Paramagnetic Resonance. Edited by Brustolon and Giamello
Copyright © 2009 John Wiley & Sons, Inc.

described together with the properties that influence the shape of the spectra and the intensity of the spectral lines. Moreover, it will be anticipated how the parameters characterizing the spectrum are related to molecular structure and dynamics. The approach will be as simple and *intuitive* as possible within the constraints of a rigorous treatment. Details on instrumentation, types of paramagnetic species studied, specific characteristics of EPR in solids and in solution, and theory are the subjects of the ensuing chapters. The second part of the book will consider applications to the investigation of complex chemical and biological systems and the improvements of the technique suitable for them.

An illustration of the spin properties of a single electron and its behavior in a magnetic field will be presented first, followed by a short discussion about the behavior of an electron spin when it is confined in a molecule, as well as when it interacts with one or several nuclear spins.

The macroscopic observation of EPR requires a collection of many electron spins the properties for which will be treated in a semiclassical way, leaving to more advanced EPR descriptions the quantum mechanical *density matrix* method. (You can find, e.g., a short account of the density matrix method applied to ensembles of spins in appendix A9 in the Atherton book in the Further Reading Section.) However, a quantum mechanical description is necessary to a deeper understanding of complex experiments, in particular pulse EPR experiments. A short introduction to quantum mechanics formalism will be presented at the end of this chapter. The concepts of spin–lattice (longitudinal) and spin–spin (transverse) relaxation processes will be introduced, and how the rate of these processes influences the spectra will be anticipated. Chapter 5 describes how the relaxation rates can be measured by pulsed EPR methods.

The presence of a second electron spin in the investigated paramagnetic system will be considered briefly. A second electron spin introduces the electron dipolar interaction, which constitutes a new important term in the energy. Chapters 3 and 6 contain more information on paramagnetic species with two or more unpaired electrons.

Analogies and differences with respect to the related phenomena of nuclear magnetic resonance (NMR), involving nuclear spins, will be provided when appropriate.

1.2 EPR SPECTRUM: WHAT IS IT?

The EPR spectrum is a diagram in which the absorption of microwave frequency radiation is plotted against the magnetic field intensity. The reason why the magnetic field is the variable, instead of the radiation frequency as it occurs in other spectroscopic techniques (e.g., in recording optical spectra), will be explained in Chapter 2. There are two methods to record EPR spectra: in the first traditional method, which is called the continuous wave (CW) method, low intensity microwave radiation continuously irradiates the sample. In the second method, short pulses of high power microwave radiation are sent to the sample and the response is recorded in the absence of radiation (pulsed EPR). This chapter is mainly focused on the CW method, and pulsed EPR is treated in Chapter 5. In CW spectra, for technical reasons explained in Chapter 2 (§2.1.4), the derivative of the absorption curve is plotted instead of the absorption itself. Therefore, an EPR spectrum is the derivative of the absorption curve with respect to the magnetic field intensity.

Microwave absorption occurs by varying the magnetic field in a limited range around a central value B_0, and the EPR spectrum in most cases consists of many absorption lines. The following main parameters and features characterize the spectrum: the positions of the absorptions, which are the magnetic field values at which the absorptions take place; the number, separation, and relative intensity of the lines; and their widths and shapes. All of these parameters and features are related to the structure of the species responsible for the spectrum, to their interactions with the environment, and to the dynamic processes in which the species are involved. This chapter will address these issues.

1.3 THE ELECTRON SPIN

Elementary particles such as an electron are characterized by an intrinsic mechanical angular momentum called spin; that is, they behave like spinning tops. Angular momentum is a vector property that is defined by the magnitude or modulus (the length of the vector used to represent the angular momentum) and by the direction in space. However, because an electron is a quantum particle, the behavior of its spin is controlled by the rules of quantum mechanics. For a first approach to the magnetic resonance phenomenon, it is sufficient to know that the electron spin can be in two states, usually indicated by the first letters of the Greek alphabet α and β. These states differ in the orientation of the angular momentum in space but not in the magnitude of the angular momentum, which is the same in the α and β states.[1] The spin vector is indicated by S and the components along the x, y, z axes of a Cartesian frame by S_x, S_y, S_z, respectively. The angular momenta of quantum particles are of the order of $\hbar$ (Planck constant h divided by 2π). Magnetic moments are usually represented in $\hbar$ units, and in these units the magnitude or modulus of S is

$$|S| = \sqrt{S(S+1)} \quad (1.1)$$

where $S = 1/2$ is the electron spin quantum number; therefore, $|S| = \sqrt{3/4}$. The usual convention is to consider the α and β electron spin states as those having definite components S_z along the z axis of the Cartesian frame. For an electron spin, quantum mechanics requires that S_z be in $\hbar$ units of either $1/2$ (α state) or $-1/2$ (β state). The components along the axes perpendicular to z are not defined in the sense that they cannot be determined (another requirement of quantum mechanics), and in the α and β states they could assume any value in the range of $-1/2$ to $1/2$. In the absence

---

[1] We are dealing here with a free electron, which is an electron whose motion is not constrained by Coulomb interactions with nuclei or with other electrons. In real systems studied by EPR the electrons belong to atoms, molecules, polymers, defects in crystalline solids or metal ion complexes, and so forth. However, in most cases their properties are not strongly influenced by the environment and their magnetic properties can be viewed as if they were free spins, at least in a first approximation.

6 INTRODUCTION TO ELECTRON PARAMAGNETIC RESONANCE

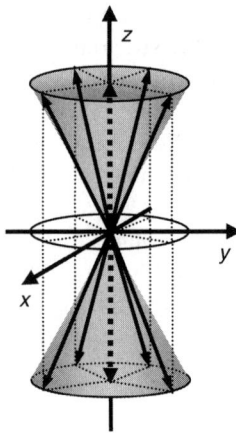

Fig. 1.1 The electron spin angular momentum is a vector represented in the figure as a solid arrow, whose length is $|S| = \sqrt{3/4}\,\hbar$. According to quantum mechanics, when a Cartesian frame x, y, z is chosen, only a component of the spin vector (usually assumed as the z component) has a definite value S_z of either $1/2\,\hbar$ or $-1/2\,\hbar$. The z component is shown in the figure as a dotted arrow pointing in the positive (α spins) or negative z direction (β spins). The components in the plane perpendicular to z are not defined: the α and β spins could point in any direction on the surface of a cone.

of any particular preferential direction in the space, connected with possible interactions of the electron spin with its environment, any choice for the direction in space of the z axis is allowed. As long as this space isotropy condition holds, the α and β electron spin states have the same energy (Fig. 1.1) and they are said to be *degenerate*. This is not the case if the electron spin is placed in a magnetic field.

1.4 ELECTRON SPIN IN A MAGNETIC FIELD (ZEEMAN EFFECT)

In EPR a crucial point to be considered is that a magnetic moment $\boldsymbol{\mu}_e$ is always associated with the electron spin angular momentum, where $\boldsymbol{\mu}_e$ is proportional to S, meaning that $\boldsymbol{\mu}_e$ and S are vectors parallel to each other. They have opposite directions because the proportionality constant is negative. The latter is written as the product of two factors g and μ_B:

$$\boldsymbol{\mu}_e = g\,\mu_B\,S \tag{1.2}$$

where g is a number called the Landé factor or simply the g factor. For a free electron $g = 2.002319$ and $\mu_B = -|e|h/4\pi m_e = -9.27410^{-24}\,\mathrm{J\,T^{-1}}$, where m_e is the electron mass; e is the electron charge; $h = 6.626 \times 10^{-34}$ Js is the Planck constant; and μ_B is the atomic unit of the magnetic moment, which is called the *Bohr magneton*. Because $\mu_B < 0$, to avoid confusion about the sign, the absolute value of μ_B will be used in several equations. The existence of a magnetic moment associated with the electron spin is the reason for having an energy separation between the α and β electron spin states when the electron is in the presence of a magnetic field. Suppose we apply a constant magnetic field $\boldsymbol{B}$ to an electron spin. Because the energy of a magnetic moment $\boldsymbol{\mu}_e$ is given by the scalar product between $\boldsymbol{\mu}_e$ and $\boldsymbol{B}$, the electron spin energy will depend on the orientation of $\boldsymbol{\mu}_e$ with respect to $\boldsymbol{B}$:

$$E = -\boldsymbol{\mu}_e \cdot \boldsymbol{B} = g|\mu_B|\boldsymbol{S} \cdot \boldsymbol{B} \tag{1.3}$$

The dot product reduces to a single term if the direction of $\boldsymbol{B}$ coincides with one of the axes respect to which the $\boldsymbol{B}$ and $\boldsymbol{S}$ are represented. The choice of the reference frame is arbitrary, and it can be chosen in such a way that the z axis is along the direction of $\boldsymbol{B}$. In this case the equation for the energy becomes

$$E = g|\mu_B|B_0 S_z \tag{1.4}$$

where B_0 is the magnetic field intensity.

If one takes into account that the electron spin can be in two states, either α or β, in which the z component of the spin is $1/2$ and $-1/2$, respectively, in the presence of a magnetic field the electron spin energy could assume only the two values,

$$E_\pm = \pm(1/2)g|\mu_B|B_0 \tag{1.5}$$

where the positive sign refers to the α state and the negative one to the β state.

The splitting of the electron spin energy level into two levels in the presence of a magnetic field is called the *Zeeman effect*, and the interaction of an electron magnetic moment with an external applied magnetic field is called the electron *Zeeman interaction*. The Zeeman effect is represented graphically in Fig. 1.2.

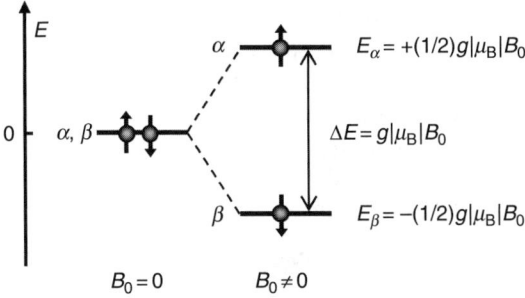

Fig. 1.2 The electron spin Zeeman effect. At zero field ($B_0 = 0$) the spin states α and β represented by up and down arrows have the same energy, which is zero in the energy scale. In the presence of a static magnetic field ($B_0 \neq 0$) the β spin state is shifted at low energy and the α one at high energy. The energy separation is proportional to the magnetic field intensity. It also linearly depends on the electron g factor.

1.5 EFFECT OF ELECTROMAGNETIC FIELDS

An electron spin in the β state, which is the low energy state, can absorb a quantum of electromagnetic radiation energy, provided that the energy quantum $h\nu$ coincides with the energy difference between the α and β states:

$$h\nu = E_\alpha - E_\beta = g|\mu_B|B_0 \quad (1.6)$$

where ν is the radiation frequency. Equation 1.6 is the fundamental equation of EPR spectroscopy.

In a 3.5-T magnetic field, which is the standard magnetic field intensity used in many EPR spectrometers, for $g = 2.0023$, Equation 1.6 gives $\nu = 9.5\,\text{GHz}$. This radiation frequency is in the microwave X-band region (8–12 GHz). The EPR spectrometers operating in this frequency range are called X-band spectrometers.

Other regions of higher microwave frequencies used in commercial EPR spectrometers are Q-band ($\sim 34\,\text{GHz}$) and W-band (95 GHz). Spectrometers operating at frequencies higher than 70 GHz are considered as high field/high frequency spectrometers. See Chapter 2 (§2.2.5) for a general introduction to multifrequency EPR. Applications of high field/high frequency EPR are described in Chapters 6 and 12.

For the spin system to absorb the radiation energy, the oscillating magnetic field $\boldsymbol{B_1}$ associated with the electromagnetic radiation should be in the plane xy, which is *perpendicular* to the static Zeeman field $\boldsymbol{B_0}$. In other words, the radiation should be polarized perpendicular to $\boldsymbol{B_0}$.

An electron spin in the α state cannot absorb energy because there are no allowed states at higher energy. However, the presence of an oscillating magnetic field of proper frequency corresponding to Equation 1.6 induces a transition from the α state to the β state with loss of energy and emission of a radiation quantum $h\nu$. This process is called *stimulated emission*, and it is just the opposite of the absorption. The spontaneous decay of an isolated spin to the lower energy state in the absence of radiation, with emission of microwave radiation (*spontaneous emission*), is a process occurring with negligible probability.

In conclusion, an *isolated* electron spin placed in a static magnetic field $\boldsymbol{B_0}$ and in the presence of a microwave oscillating magnetic field $\boldsymbol{B_1}$ perpendicular to $\boldsymbol{B_0}$ undergoes transitions from the low energy level state β to the upper one α, and vice versa. The net effect is zero because absorption and stimulated emission compensate each other. The next section will show that electron spins are never completely isolated, and the behavior of a collection of many electron spins is different.

1.6 MACROSCOPIC COLLECTION OF ELECTRON SPINS

In the usual experimental setup one considers samples of many electron spins, their number being on the order of 10^{10} or higher. Moreover, these electron spins are not independent, interacting with each other and with their environment. Furthermore, electron spins are not free; they are confined in atomic or molecular systems. The latter aspects will be considered later.

The electron spins of an ensemble are statistically distributed in the α and β states. Because these states are equivalent in the absence of a magnetic field, for $B_0 = 0$ half of the spins are α spins and half are β spins. In these conditions the z component of the total angular momentum is zero, as are also the components along any other direction. In fact, all directions in space are equivalent. The situation changes in the presence of a magnetic field B_0 *if the spin ensemble is allowed to interact with its environment (the "lattice")*. As learned in the previous section, if $B_0 \neq 0$, the α and β states do not have the same energy. In thermal equilibrium with the lattice the spins distribute between α and β states in such a way as to be in a small excess in the lower energy level (β state). The ratio between the number (N) of α spins and the number of β spins depends on the temperature. It is given by the Boltzmann distribution law:

$$N_\alpha/N_\beta = \exp(-g|\mu_B|B_0/k_B T) \qquad (1.7)$$

where k_B is the Boltzmann constant, which is equal to 1.3806×10^{-23} J K^{-1}; and T is the absolute temperature of the lattice.

At room temperature (300K) and for magnetic fields on the order of 0.3 T (X-band spectrometer), $g|\mu_B|B_0 \ll k_B T$ and the exponential can be expanded in series, retaining only the linear term. The approximate population ratio becomes

$$N_\alpha/N_\beta = 1 - g|\mu_B|B_0/k_B T \qquad (1.8)$$

This approximation is quite good, unless the spin system is at very high field or at very low temperature. According to Equation 1.8, at room temperature in the magnetic field of an X-band spectrometer there is an excess of β spins over the α spins of 1/1000. This small excess is enough for the microwave absorption to overcome the emission and to make possible the observation of an EPR absorption signal. In fact, a microwave field induces transitions from β to α, and the reverse one from α to β, in a number proportional to the number of spins in the initial state.

A further point should be considered regarding the interaction of the spin system with the lattice. If the spin system were not coupled with the lattice or weakly coupled to it, the microwave field acting continuously on the spin system (CW-EPR) would eventually equalize the level populations and after a short time the absorption EPR signal would disappear. Conversely, spin lattice interaction restores the thermal equilibrium, which is the excess spins in the low energy level, allowing the continued observation of the EPR absorption signal. Of course, because there is competition between the spin lattice interaction and microwave field, if the latter is strong enough with respect to the spin lattice interaction, the EPR signal *saturates*. The rate of the spin lattice process is usually reported as the inverse of a characteristic time T_1, which is called the *spin lattice relaxation time*. This is the time taken by a spin system forced out of equilibrium by an amount δ to reduce the deviation by a factor $1/e$.

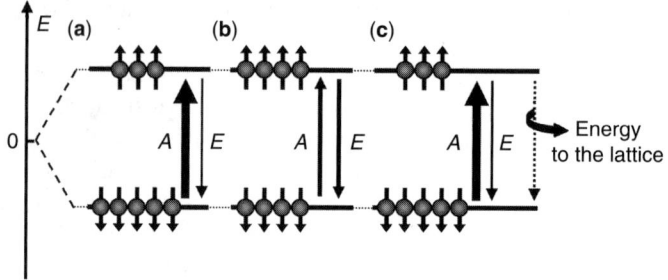

Fig. 1.3 The electron spin ensemble in a magnetic field. The populations of spin levels N_α and N_β are schematically indicated. (a) The A and E arrows indicate stimulated absorption and emission in the presence of resonant microwave radiation. The absorption is more efficient than the emission because of the difference in populations, corresponding to thermal equilibrium with the lattice. (b) The phenomenon of saturation with the two levels equally populated occurs when the energy transfer to the lattice is not efficient. No EPR signal is detectable in this case. (c) The energy transfer from the spin system to the lattice (spin lattice relaxation) indicated by the dotted arrow reestablishes a population difference.

Time T_1 is a measure of how strongly the spin system is coupled to the lattice. The experiments designed to determine T_1 are described in Chapter 5 (§5.3.2.2). The behavior of a spin ensemble in a magnetic field is shown schematically in Fig. 1.3.

1.7 OBSERVATION OF MAGNETIC RESONANCE

Equation 1.6 suggests two possible ways for performing an EPR experiment, which are illustrated in Fig. 1.4. The first one (Fig. 1.4a) consists of placing a spin ensemble in a constant magnetic field B_0 and irradiating it with microwave radiation of linearly variable frequency and constant intensity. When the frequency matches the resonance conditions for the magnetic field intensity B_0, microwaves are absorbed and the absorption is revealed by a microwave detector.

The alternative way (Fig. 1.4b) consists of irradiating the sample with microwave radiation of constant frequency ν_0 in a magnetic field of linearly variable intensity. In this second case EPR absorption is observed when the field intensity reaches the resonance conditions dictated by Equation 1.6 for the chosen frequency value ν_0.

For technical reasons discussed in Chapter 2, the preferred experimental procedure is the second one. However, in order to discuss the pattern of the EPR spectra, it will be more convenient to present energy level schemes for constant magnetic field conditions.

The resonance absorption line has a width, which means that the absorption of microwave radiation occurs in a range of magnetic field values, with a probability decreasing as the deviation from the value given by Equation 1.6 increases. The width is determined by dynamical processes and interactions described in a following section.

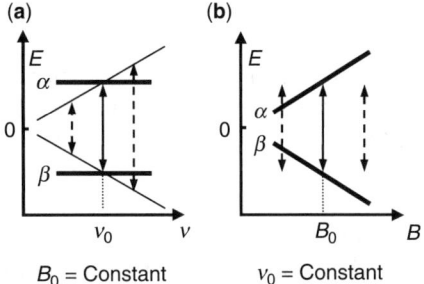

Fig. 1.4 Two alternative ways to record an EPR spectrum. (a) In the first one the spin system is placed in a constant magnetic field B_0 and irradiated with microwave radiation whose frequency is linearly changed. The double arrows represent the radiation quantum energies $h\nu$. Only when the frequency corresponds to the resonance condition given by Equation 1.6 the radiation is absorbed (solid double arrow). (b) In the second procedure, electron spins are irradiated with a microwave radiation of fixed frequency ν_0, and the magnetic field is swept. When the latter reaches the resonance condition, microwave radiation is absorbed by the sample.

1.8 ELECTRON SPIN IN ATOMS AND MOLECULES

Let us consider the simple case of an atom with a closed shell and an extra electron. In such an atom the electron spins are all coupled in pairs, except one. The electron angular momentum has two contributions: one arises from the electron spin, and another one arises from the orbital motion of the electron around the nucleus. The magnetic moment is the sum of two terms, referring to the two contributions,

$$\mu_e = \mu_B l + g\mu_B S \tag{1.9}$$

where l is the orbital angular momentum. The modulus of l is quantized and may assume only the values given by an equation analogous to Equation 1.1, which refer to the spin momentum:

$$|l| = \sqrt{l(l+1)} \tag{1.10}$$

where l is an integer, which depends on the electron spatial wavefunction. It can assume integer values or be zero, depending on the orbital occupied by the electron. Moreover, the component of l along z may assume only the $2l + 1$ quantized values:

$$l_z = -l, -l+1, \ldots, l \tag{1.11}$$

Equation 1.9 would be correct only if the spin motion and the orbital motion were independent of each other. In reality, they are not, because spin and orbit angular momenta are coupled by the spin–orbit coupling.

Contrary to the previous example of the electron in an atom with spherical symmetry, molecules are systems of low symmetry. For them the orbital angular momentum is quenched (its average is zero), and the electron angular momentum in the absence of spin–orbit coupling is only due to spin. The effect of spin–orbit coupling is to restore a small amount of orbit contribution, which results in a deviation $\Delta g = g - g_e$ of the g factor from the free electron value entering in Equation 1.2. The consequence is the shift of the resonance field intensity from the value corresponding to the free electron spin, as given by Equation 1.6.

Because Δg depends on the spin–orbit coupling, its value is large for metal complexes, where the electrons move in proximity to a heavy atom nucleus. For organic free radicals containing only light atoms, the spin–orbit interaction is small, and the deviation of g from the free electron value is also small ($\leq 1\%$). However, even such small deviations could give important structural information, as well as information about the radical environment. In any case, g is a parameter that characterizes a molecular system, and its measurement is also important as an analytical tool.

Note that the spin–orbit interaction is anisotropic, because it is related to the orbital motion, which means that the amount of orbital character in the angular momentum is different for the different directions in a molecule fixed frame. Therefore, the value of g to insert in Equation 1.6 will depend on the direction of the magnetic field with respect to the molecular axes. For example, for the popular nitroxyl radical 2,2,6,6 tetramethyl pyrrolidine-N-oxyl, g is 2.0090 if measured with the magnetic field directed along the N—O bond, 2.0027 if measured perpendicular to the plane formed by the N—O and N—C bonds, and 2.0060 in the direction perpendicular to the latter two (see Scheme 1.1).

The anisotropy of g can be measured by recording the EPR spectra of a single crystal, where the molecules are in fixed orientations, by rotating the crystal in the spectrometer's magnetic field. In liquids, because of rapid molecular tumbling, the g factor anisotropy is averaged out and a mean g value g_{iso} is measured.

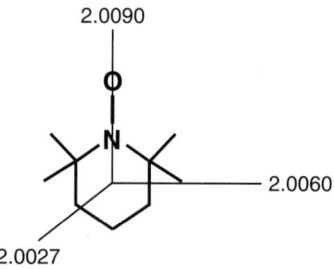

Scheme 1.1 The chemical structure of nitroxide radical 2,2,6,6 tetramethyl pyrrolidine-N-oxyl (TEMPO). Hydrogen atoms are not shown. The figures are the values of the g factor one would measure if the magnetic field were placed along the directions shown in the figure. If the same free radical is rapidly tumbling in solution, the average value $g_{iso} = (2.0090 + 2.0060 + 2.0027)/3$ is obtained.

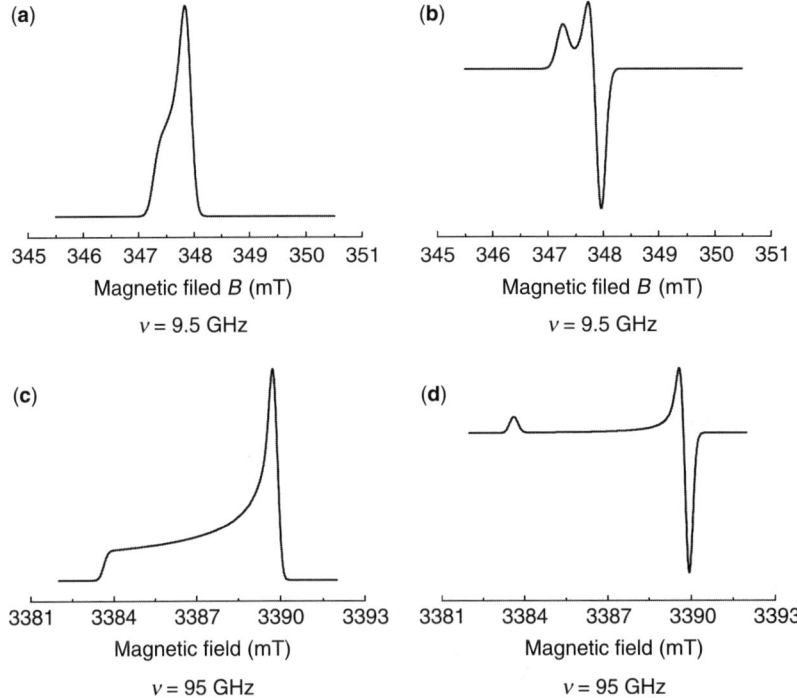

Fig. 1.5 EPR spectra of a sample of randomly oriented paramagnetic molecules with $S = 1/2$ and characterized by axial g anisotropy. In this example the values along three orthogonal axes fixed in the molecule are assumed to be $g_x = g_y = 2.0023$, $g_z = 2.0060$. (a, c) EPR absorption; (b,d) first derivative of the absorption with respect to the magnetic field. (a, b) spectra refer to a microwave frequency $\nu = 9.5$ GHz (X-band); (c, d) spectra obtained with the same g parameters refer to a 10 times higher microwave frequency $\nu = 95$ GHz (W-band). Note the different magnetic field scale and the higher resolution of the W-band spectra.

Conversely, the EPR spectrum of a powder sample (a collection of many microcrystals randomly oriented in space) is the superposition of the EPR lines of all of the microcrystals, each one corresponding to an individual orientation. The same is true for paramagnetic systems diluted in glassy matrices. However, the anisotropy of g can be measured even for such samples (see Chapter 6). Examples of powder EPR spectra of a simple ideal sample characterized by g anisotropy are shown in Fig. 1.5.

If the anisotropy is small, as in the case of organic free radicals, the effect on the EPR spectrum can be hidden in the resonance linewidth. In this case it is necessary to use high frequency/high field spectrometers to resolve the anisotropy, which is the spectrum spread proportional to the operating field frequency ν. High field EPR also presents other advantages, as illustrated in Chapter 2. The situation is similar to that encountered in NMR spectroscopy, where high frequency spectrometers were developed to increase the resolution of chemical shifts.

1.9 MACROSCOPIC MAGNETIZATION

In a real macroscopic system consisting of many electron spins that interact with the lattice and with each other, the spin properties are described by defining the total magnetization M as a vector resulting from the sum of the magnetic moments of the individual electron spins:

$$M = \sum \mu_i \qquad (1.12)$$

If such a system is in a magnetic field B_0 and it is in thermal equilibrium with the lattice, the resulting magnetization M is directed along the z axis, because of the excess of β spins with respect to α spins, as given by Equation 1.8. The components perpendicular to z are zero because the individual spins, while keeping a definite z component S_{zi} at either $1/2$ or $-1/2$, have random components along axes perpendicular to z (see Fig. 1.1)

$$\begin{aligned} M_z &= \sum \mu_{zi} = M_0 \\ M_x &= \sum \mu_{xi} = 0 \\ M_y &= \sum \mu_{yi} = 0 \end{aligned} \qquad (1.13)$$

M_0 is the thermal equilibrium magnetization and μ_{xi}, μ_{yi}, and μ_{zi} are the respective components of the magnetic moments of the individual spins. The sum is over all of the spins in the sample.

Suppose that the system is forced out of equilibrium. The time evolution of magnetization M is obtained by solving the equation of the motion of M. Before examining the proper equation, three points should be considered.

1. There are two ways to modify the equilibrium conditions. The first one consists of changing the z component and leaving the other two components equal to zero. This happens if the relative populations of the α and β spins are changed. The second one consists of tilting the magnetization with respect to z and generating a nonvanishing magnetization component in the plane perpendicular to z.
2. Even if the individual spins behave according to quantum mechanics, the motion of the macroscopic magnetization is accounted for by classical mechanics.
3. Magnetization M is associated with an angular momentum. In fact, M arises from the angular momenta associated with the spins of the individual electrons. The macroscopic angular momentum is $J = \sum S_i$. The relation between M and J is analogous to Equation 1.2: $M = g\mu_B J$.

1.9 MACROSCOPIC MAGNETIZATION

The time evolution of M in a magnetic field B is given by the vector product

$$dM/dt = M \times g\mu_B B \qquad (1.14)$$

This equation holds for any B field, static or time dependent.

Equation 1.14 indicates that the variation of the magnetization vector M is perpendicular to both M and B. For a constant magnetic field B, assumed to be along the z axis of a reference frame x, y, z, the z component of M is constant, because dM/dt is in the x, y plane. The motion, consisting of a rotation of M around the z axis (Fig. 1.6a) is called *Larmor precession*. The angular frequency is $\omega_L = g\mu_B B/\hbar$. It is called the *Larmor frequency*. When the magnetic field is not constant but changes with time, the motion of magnetization M is more complicated.

In magnetic resonance experiments (see Fig. 1.6b and c) the spin system is in the presence of both a large static magnetic field B_0 along z and a variable field oscillating at a microwave frequency in the plane perpendicular to z, for example, along the direction $x: B_{1x}(t) = B_1 \cos(\omega t)$. Note that an oscillating magnetic field can be decomposed in two mutually perpendicular components rotating at the same frequency, one clockwise and the other counterclockwise. Typical values used in EPR experiments are $B_0 = 0.3$ T and $B_1 = 1$ μT. For these conditions, a useful method to simplify the description of the motion of the magnetization is to use a *rotating* reference frame of axes, with x', y' rotating around z at *angular frequency* ω, $z' = z$, and x' is taken along B_1.

To understand the utility of this model of a rotating reference frame, let us discuss some cases. If B_1 is zero, in a reference system rotating around z at the

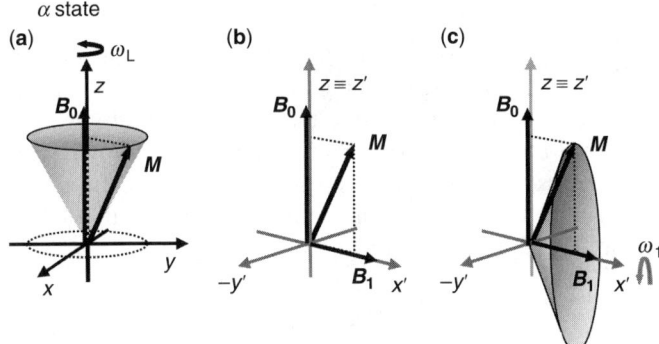

Fig. 1.6 (a) A classical magnetic moment M associated to an angular momentum when placed in a magnetic field $B_0 \parallel z$ undergoes a precession motion about the magnetic field direction. The precession angular frequency (Larmor frequency) is $\omega_L = \hbar^{-1} g\mu_B B_0$. (b) In a reference frame x', y', $z' = z$ rotating about z at the Larmor frequency, M is stationary. (c) If a small magnetic field B_1 also rotating at the Larmor frequency is superimposed to the static field B_0, the magnetic moment M for an observer in the x', y', z' precedes about B_1 at an angular frequency $\omega_1 = \hbar^{-1} g\mu_B B_1$. Note that $\omega_1 \ll \omega_L$ because $B_1 \ll B_0$.

Larmor frequency and in the same direction as vector M, the latter one is stationary. If $B_1 \neq 0$ and it is also rotating around z at the Larmor frequency, in the rotating frame B_1 is seen as constant by M and the motion of M becomes a rotation (precession) around B_1, with an angular frequency $\omega_1 = g\mu_B B_1/\hbar$. Note that three frequencies are considered here: the Larmor frequency ω_L, the rotation frequency of the microwave field B_1, ω (in this simple case assumed to coincide with the Larmor frequency $\omega = \omega_L$), and the precession frequency ω_1 of M around B_1 in the rotating frame. Moreover, note that $\omega_1 \ll \omega_L$.

In the general case, when $\omega \neq \omega_L$, magnetization vector M moves in a complex way by rotating around z and around the direction of B_1. The equation describing the motion of the magnetization vector in the frame rotating around z at angular frequency ω is the following:

$$(dM/dt)_R = M \times g\mu_B B_{\text{eff}} \qquad (1.15)$$

where

$$B_{\text{eff}} = (B_0 - \hbar\omega/g\mu_B)k + B_1 i \qquad (1.16)$$

and k and i are unit vectors directed along the z and x' axes, respectively.

Note that if frequency ω equals the Larmor frequency, B_{eff} coincides with B_1 and, as shown, M undergoes a rotation around x'. The same is true if ω deviates only slightly from the Larmor frequency. Conversely, if the deviation is large, B_{eff} is practically directed along axis z because $B_1 \ll B_0$.

1.10 SPIN RELAXATION AND BLOCH EQUATIONS

In real macroscopic systems the spin ensemble interacts with the lattice; and whenever the z component of the magnetization M_z deviates from the equilibrium value M_0, the spin–lattice interaction tends to restore it. Moreover, the spins interact with each other, and the M components perpendicular to z tend to become zero. To consider these processes, further terms should be included in the equations for the time evolution. These phenomenological terms transform Equation 1.15 into the following equations, called the *Bloch equations*:

$$\begin{aligned} dM_z/dt &= -M_y\omega_1 - (M_z - M_0)/T_1 \\ dM_x/dt &= M_y(\omega - \omega_0) - M_x/T_2 \\ dM_y/dt &= M_z\omega_1 - M_x(\omega - \omega_0) \end{aligned} \qquad (1.17)$$

Two time constants have been introduced. The spin–lattice or "longitudinal" relaxation time T_1 relates to the time for recovering the equilibrium value of the magnetization z component, which is also indicated as the longitudinal component. The

1.10 SPIN RELAXATION AND BLOCH EQUATIONS

second characteristic time is the transverse relaxation time T_2, which depends on how fast the M_x and M_y component tend to vanish. Note that these two times are generally different, because T_1 characterizes a process of energy transfer from the spin system to the lattice and vice versa, depending on how tight the connection is between them, whereas T_2 is related to processes of energy exchange within the spin system, which do not involve the lattice.

The EPR absorption signal is proportional to the component of the magnetization measured perpendicular to B_1, which is M_y, obtained by solving Equation 1.17. If the perturbation attributable to the microwave field is a continuous one and it is small, the Bloch equations are solved in the stationary regime by placing the time derivatives of the magnetization components equal to zero. In this CW regime the solution for M_y is

$$M_y = (M_0/B_0)B_1T_2/[1 + (\omega - \omega_0)^2 T_2^2] \qquad (1.18)$$

Equation 1.18 is a function of ω called the *Lorentzian function*, which represents the shape of the EPR resonance line. Figure 1.7 shows a Lorentzian line and its first derivative. The width of the line is inversely proportional to relaxation time T_2.

In solid or glassy samples each paramagnetic species is oriented in a specific way with respect to the magnetic field, and it is immobile. Therefore, the EPR lines are due to the superposition of many lines, each corresponding to a differently oriented species or to species having different interactions with their surroundings. The width of an EPR line in this case could be much broader than that expected from relaxation time T_2. In this case the linewidth is called *inhomogeneous* to distinguish it from the *homogeneous* width of the lines, which occur when the species are all magnetically equivalent. The shape of the EPR line in this instance is generally a Gaussian (see Chapter 2 and Fig. 2.10).

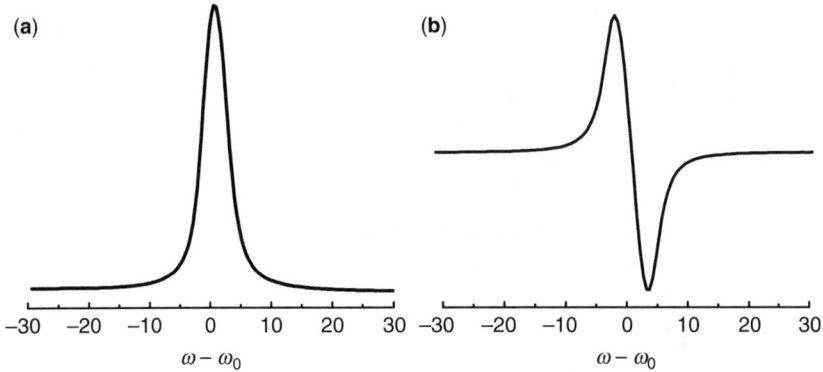

Fig. 1.7 The EPR Lorentzian (a) absorption line and (b) its first derivative. The abscissa shows the deviation of the microwave angular frequency from the resonance frequency ω_0.

1.11 NUCLEAR SPINS

Like the electrons, the nuclei are characterized by a spin angular momentum, usually indicated by I, and by the corresponding associated magnetic moment given by a relation analogous to Equation 1.2:

$$\mu_N = g_N \mu_N I \tag{1.19}$$

Note that the g_N factor depends on the isotope, and it can be positive (as for protons) or negative (e.g., in the case of ^{15}N).

Even for nuclei the magnitude (modulus) of the angular momentum and its component along a direction z are quantized. Depending on the type of nucleus, the nuclear spin quantum number I may have values other than $1/2$, including 0. A few examples are reported in Table 1.1. A comprehensive table of nuclear spin properties is found in the EPR-electron nuclear double resonance (ENDOR) Frequency Table.

The magnitude and z component of the nuclear spin moment in units of $\hbar$ are

$$|I| = \sqrt{I(I+1)} \tag{1.20}$$

and

$$I_z = -I, -I+1, \ldots, I \tag{1.21}$$

As it occurs for an electron spin, the energy of nuclear spins is influenced by a magnetic field according to the nuclear spin angular momentum component along the direction of magnetic field I_z. This effect is called the *nuclear Zeeman effect*.

$$E = -g_N \mu_N B_0 I_z \tag{1.22}$$

In the presence of a nuclear spin the electron spin experiences an additional magnetic field provided by the nuclear magnetic moment, which affects the resonance conditions. The electron–nucleus spin interaction is called *hyperfine interaction*. It gives rise to a splitting of the resonance EPR lines into several components: two

TABLE 1.1 Selection of Nuclear Spin

Isotope	I value
^{1}H, ^{13}C, ^{31}P, ^{15}N	1/2
^{2}H, ^{14}N	1
^{12}C, ^{16}O	0
^{55}Mn	5/2
^{23}Na	3/2

components for interaction with a nuclear spin $I = 1/2$, three components for an $I = 1$ nucleus, and in general $2I + 1$ components for the interaction with a spin I nucleus.

In the case of an electron spin interacting with a nuclear spin, Equation 1.4 for the energy in a magnetic field should be modified by adding two terms: the nuclear Zeeman interaction of the nuclear spin with the external magnetic field and a contribution that derives from the hyperfine field experienced by the electron spin. For the hyperfine field, two regions in the space should be distinguished: one inside the nuclear volume that is small but finite and another one outside the nuclear volume. Quantum mechanics does not exclude that the electron enters into the nucleus. In the region external to the nucleus, the hyperfine magnetic field is the classical field of a magnetic dipole. It decreases with the third power of the electron nucleus distance r, and it depends on the orientation of the vector connecting the electron and nucleus with respect to the directions of the dipoles. This dipole–dipole interaction is averaged out if the paramagnetic molecule is rapidly tumbling in an isotropic environment, as occurs in a normal liquid solution. In this case all orientations are covered with equal probability. Dipole–dipole interaction that is important for solid samples will be considered later.

Inside the nucleus the hyperfine field is constant, and it does not depend on the direction. The hyperfine energy contribution E_{hf} is called the *contact* (or *Fermi*) contribution, which is

$$E_{hf} = a\mathbf{S} \cdot \mathbf{I} \quad (1.23)$$

where a is a constant (*hyperfine coupling constant*) that depends on $|\Psi(0)|^2$, the square of the wavefunction that describes the electron motion calculated in the point where the nucleus is. $|\Psi(0)|^2$ gives a measure of how much electron spin enters the nucleus; a is given by the equation

$$a = (8\pi/3)g\mu_B g_N \mu_N |\Psi(0)|^2 \quad (1.24)$$

A detailed illustration of the hyperfine interaction, its characteristics in different molecular systems, and several examples of spectra dominated by hyperfine interactions will be given in Chapter 4.

Let us consider first the simple case of rapidly tumbling molecular systems in liquid solution. In such a system the energy terms dependent on the nuclear spin E_n are nuclear Zeeman interaction 1.22 and isotropic contact hyperfine interaction 1.23:

$$E_n = -g_N \mu_N B_0 I_z + a\mathbf{S} \cdot \mathbf{I} \quad (1.25)$$

The first term in Equation 1.25 is much smaller than the electron Zeeman interaction, by virtue of the much smaller nuclear magnetic moment compared with the electron magnetic moment ($\mu_N \ll \mu_B$). The hyperfine term is also much smaller than the electron Zeeman term, provided that magnetic field B_0 is large enough, as occurs for organic radicals and for most other systems studied in X-band spectrometers:

$$|a| \ll g|\mu_B|B_0 \tag{1.26}$$

Under these conditions (*high field approximation*) the energy term E_n, due to the nuclear spin, represents a small perturbation on the electron spin energy, which becomes

$$E_{tot} = g|\mu_B|B_0 S_z - g_N \mu_N B_0 I_z + a S_z I_z \tag{1.27}$$

For a free radical ($S = 1/2$) containing a single magnetic nucleus with $I = 1/2$, there are four possible values of the total energy, corresponding to the electron and nuclear spin components $S_z = \pm 1/2$ and $I_z = \pm 1/2$, as illustrated in Fig. 1.8. The corresponding energies are the following:

$$\begin{aligned} E_1 &= 1/2\, g|\mu_B|B_0 - 1/2\, g_N \mu_N B_0 + 1/4a \\ E_2 &= 1/2\, g|\mu_B|B_0 + 1/2\, g_N \mu_N B_0 - 1/4a \\ E_3 &= -1/2\, g|\mu_B|B_0 + 1/2\, g_N \mu_N B_0 + 1/4a \\ E_4 &= -1/2\, g|\mu_B|B_0 - 1/2\, g_N \mu_N B_0 - 1/4a \end{aligned} \tag{1.28}$$

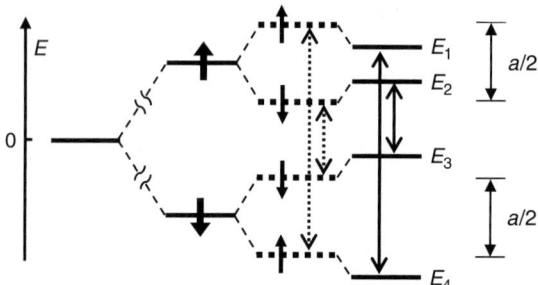

Fig. 1.8 The energy level scheme for an electron spin $S = 1/2$ coupled to a nuclear spin $I = 1/2$ (e.g., a proton) in the presence of a magnetic field B_0. The heavy arrows represent the $S_z = \pm 1/2$ electron spin components along the magnetic field direction. The light arrows represent the nuclear spin components. Each electron Zeeman spin level is split into two levels (dotted lines) by the hyperfine interaction with the nuclear spin. The hyperfine levels are then shifted by the nuclear Zeeman interaction term (solid lines). The shift is $\pm 1/2 g_N \mu_N$, and it has no effect on the energy difference between the levels connected by the allowed EPR transitions (vertical double arrows). Note that electron and nuclear Zeeman interactions have opposite sign. In the example there is a a positive sign for the hyperfine splitting constant a and $a > g_N \mu_N B_0$. The drawing is not to scale, because the hyperfine splitting and nuclear Zeeman interaction are exaggerated with respect to the electron Zeeman term for clarity.

EPR consists of transitions between pairs of energy levels characterized by *different values of S_z* but the *same I_z value*. In fact, when the electron and nuclear spin are weakly coupled, as we have assumed, the electromagnetic radiation acts by changing the component of a single spin: either the electron spin component S_z (in EPR) or the nuclear spin component I_z (in NMR) but not both at the same time. One says that the following selection rules apply, depending on the type of spectroscopic transition one is observing:

$$\Delta S_z = \pm 1; \quad \Delta I_z = 0 \quad (1.29)$$

for EPR transitions and

$$\Delta S_z = 0; \quad \Delta I_z = \pm 1 \quad (1.30)$$

for NMR transitions.

The NMR transitions would occur at much lower frequency than the EPR transitions (in the megahertz frequency range in the X-band spectrometer magnetic field).

It should be noted that EPR selection rule 1.29 holds for systems with hyperfine coupling such that the hyperfine field acting on the nucleus is much smaller than the external magnetic field, as pointed out. These transitions are the so-called EPR *allowed* transitions. When the high field approximation is no longer a good one, transitions with $\Delta S_z = \pm 1; \Delta I_z = \pm 1$ can also be observed. These are called *forbidden transitions*, and their intensity is lower than the allowed ones.

From Equation 1.28 it is derived that for a fixed field B_0 the EPR transitions allowed according to selection rules 1.29 take place at frequencies

$$\nu_\mathrm{I} = (E_2 - E_3)/h = (g\mu_\mathrm{B} B_0 - a/2)/h \quad (1.31)$$

and

$$\nu_\mathrm{II} = (E_1 - E_4)/h = (g\mu_\mathrm{B} B_0 + a/2)/h \quad (1.32)$$

Alternatively, using a fixed frequency ν_0 and a variable magnetic field B, EPR transition lines occur at magnetic field values

$$B_\mathrm{I} = h\nu_0/g\mu_\mathrm{B} + a/2 g\mu_\mathrm{B} \quad (1.33)$$

and

$$B_\mathrm{II} = h\nu_0/g\mu_\mathrm{B} - a/2 g\mu_\mathrm{B} \quad (1.34)$$

Because for EPR transitions we consider differences in energy between levels corresponding to the same nuclear spin state, the effect of the nuclear Zeeman interaction is canceled and the EPR spectrum of a system with an unpaired electron with hyperfine interaction with a single $I = 1/2$ nucleus consists of two lines separated by the *hyperfine splitting constant a*. This treatment is easily extended to the case of a nucleus with a generic nuclear spin I by taking into account that the allowed spin components are $2I + 1$. Therefore, each electron spin Zeeman level will be separated by the hyperfine interaction into $2I + 1$ levels. Because of the EPR selection rules (Equation 1.29), the spectrum will consist of $2I + 1$ lines.

In most cases, as in organic free radicals, the unpaired electron interacts with many magnetic nuclei and the total spin energy contains several hyperfine terms. For n nuclei,

$$\begin{aligned} E_{\text{tot}} &= g|\mu_B|B_0 S_z + a_1 S_z I_{z1} + a_2 S_z I_{z2} + \cdots + a_n S_z I_{zn} \\ &= g|\mu_B|B_0 S_z + \sum a_k S_z I_{zk} \end{aligned} \quad (1.35)$$

In Equation 1.35 the nuclear Zeeman terms are omitted because, even if present, they do not contribute to the EPR spectrum. Each electron Zeeman level is separated into a manifold of several sublevels. Their number is

$$N = (2I_1 + 1)(2I_2 + 1) \cdots (2I_n + 1) = \prod (2I_k + 1) \quad (1.36)$$

If all n nuclei have $I = 1/2$, the number of EPR lines is 2^n, which means that the number of EPR lines increases very rapidly with n.

In general, not all nuclei have distinct hyperfine splitting constants and several energy levels may coincide. They are said to be degenerate. For example, if two protons ($I = 1/2$) have the same splitting constant a, the hyperfine energy will cancel if one has component $I_z = 1/2$ and the other one has $I_z = -1/2$, independently of which one is which, and the energy level corresponding to this nuclear spin configuration is twofold degenerate. The population of degenerate levels corresponds to the equilibrium Boltzmann population multiplied by the degeneration factor. Consequently, the EPR transitions involving these levels have an intensity multiplied by the same factor. The relative intensity of the hyperfine components of a set of N nuclear spins having $I = 1/2$ follows the binomial distribution ratio (see Chapter 4).

1.12 ANISOTROPY OF THE HYPERFINE INTERACTION

The second contribution to the hyperfine interaction arises from the classical dipole–dipole interaction between the electron spin magnetic dipole and the nuclear magnetic dipole moment (see Fig. 1.9). The nuclear moment generates a magnetic field, which adds to the Zeeman field and is experienced by the electron spin. The dipolar hyperfine interaction is important for paramagnetic systems in single crystals, powder, glasses, and any other cases of molecules not tumbling in isotropic liquids.

The energy of the dipolar interaction between two magnetic dipoles depends on the inverse cube of their distance ($1/r^3$) and their orientation with respect to the

1.12 ANISOTROPY OF THE HYPERFINE INTERACTION

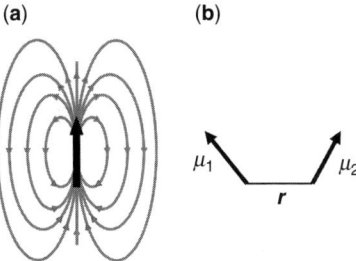

Fig. 1.9 (a) The magnetic field generated by a magnetic dipole. The figure represents a section of the three-dimensional surfaces obtained by a rotation along the dipole axis. The lines connect the points where the magnetic field intensity is the same, although the direction of the magnetic field is different. The latter one is always tangent to the curves. Note that in all of the points lying along the direction of the dipole and along the perpendicular direction the magnetic field is parallel to the dipole, but for all other points it is not. (b) Two magnetic dipoles μ_1 and μ_2 interact because each one feels the magnetic field generated by the other one. The interaction depends on the distance r between the dipoles and on their orientation with respect to the vector $\boldsymbol{r}$.

vector connecting them. It is given by

$$E_{\text{dip}} = (\mu_0/4\pi)[\boldsymbol{\mu}_1 \cdot \boldsymbol{\mu}_2/r^3 - 3(\boldsymbol{\mu}_1 \cdot \boldsymbol{r})(\boldsymbol{\mu}_2 \cdot \boldsymbol{r})/r^5] \tag{1.37}$$

where $\mu_0 = 4\pi \times 10^{-7}$ N A^{-2} is the magnetic permeability of the vacuum.

If μ_1 and μ_2 are substituted by the electron spin and nuclear spin magnetic dipoles $\boldsymbol{\mu}_e$ and $\boldsymbol{\mu}_N$ given by Equations 1.2 and 1.19, respectively, Equation 1.37 becomes

$$E_{\text{dip}} = (\mu_0/4\pi)g\mu_B g_N \mu_N [\boldsymbol{S} \cdot \boldsymbol{I}/r^3 - 3(\boldsymbol{S} \cdot \boldsymbol{r})(\boldsymbol{I} \cdot \boldsymbol{r})/r^5] \tag{1.38}$$

where $\boldsymbol{r}$ is a vector with components x, y, and z, which represents the position of the electron in a proper coordinate frame centered on the nucleus; and $r = \sqrt{x^2 + y^2 + z^2}$ is the electron nucleus distance.

When using Equation 1.38, one should take into account that the electron is not localized on a point but is distributed in space. Therefore, the terms contributing to the hyperfine dipolar interaction energy E_{dip} have to be averaged over the electron distribution. Because of the dot products, Equation 1.38 comprises nine terms that have the following form:

$$T_{ij}S_i I_j = (\mu_0/4\pi)g\mu_B g_N \mu_N \langle (\delta_{ij}/r^3 - 3ij/r^5) \rangle S_i I_j \tag{1.39}$$

where $i, j = x, y, z$ and the angle brackets $\langle \, \rangle$ are used to indicate the average over the electron spatial coordinates. Moreover, $\delta_{ij} = 0$ and $\delta_{ii} = 1$. The nine quantities T_{ij} can be arranged as the elements of a 3×3 symmetrical matrix ($T_{ij} = T_{ji}$) called the *hyperfine dipolar interaction tensor* or *hyperfine anisotropic tensor*, which is indicated by T:

$$T = \begin{pmatrix} T_{xx} & T_{xy} & T_{xz} \\ T_{yx} & T_{yy} & T_{yz} \\ T_{zx} & T_{zy} & T_{zz} \end{pmatrix} \quad (1.40)$$

The elements of T depend on the choice of the reference frame x, y, and z. However, whatever is the reference frame, because of the particular form of Equation 1.39 the sum of the diagonal elements (the *tensor trace*) is zero, because $x^2 + y^2 + z^2 = r^2$:

$$T_{xx} + T_{yy} + T_{zz} = 3/r^3 - 3(x^2 + y^2 + z^2)/r^5 = Tr(T) = 0 \quad (1.41)$$

This property of tensor T has an important consequence. Let us suppose that the electron spin distribution is spherical symmetric around the nucleus. In this case the three diagonal tensor elements are equal, and they are zero according to Equation 1.41. Moreover, all other tensor elements are also zero because for a spherical distribution the average over all directions of the products xy, xz, and yz vanishes. To have anisotropic hyperfine interaction, the electron spin should not be spherically distributed around the nucleus.

Because T is symmetric about the diagonal ($T_{ij} = T_{ji}$), it is always possible to find a particular reference frame X, Y, Z such that all tensor elements T_{ij} are zero if $i \neq j$. The X, Y, Z axes are called *principal axes* and the corresponding tensor elements T_{XX}, T_{YY}, and T_{ZZ} are called *principal values* of tensor T. Using the principal axes system, the dipolar energy (E_{dip}) assumes the simple form

$$E_{\text{dip}} = T_{XX} S_X I_X + T_{YY} S_Y I_Y + T_{ZZ} S_Z I_Z \quad (1.42)$$

The dipolar energy adds to the isotropic hyperfine term and to the Zeeman interaction to determine the total energy of the electron–nuclear spin system in the magnetic field. The dipolar term is anisotropic, because the electron and nuclear spin components depend on the direction of the Zeeman magnetic field B with respect to the X, Y, Z axes. Consequently, the hyperfine separation of the EPR transitions depends on the orientation of the paramagnetic molecule with respect to B.

If the paramagnetic molecule is rapidly tumbling in a normal liquid solution, all molecular orientations are explored and the dipolar contribution is averaged. Because the dipolar tensor has zero trace, the average is zero and the only contribution to the hyperfine interaction is the isotropic one.

This is not true if the liquid is partially oriented as occurs, for example, in liquid crystal solvents partially oriented in the magnetic field. In this case not all orientations occur with the same probability, and a residual dipolar interaction contribution remains.

1.13 ENDOR

With reference to Fig. 1.8, we note that the hyperfine coupling constant a is obtained from the difference $\Delta_{EPR} = \Delta_{14} - \Delta_{23} = (E_1 - E_4) - (E_2 - E_3) = a$ between the energy of the two allowed EPR transitions. Note that, because of the selection rule $\Delta I_z = 0$ (Equation 1.29), no information can be obtained in this way about μ_N, and therefore about the type of nucleus to which coupling a is referring. In general this is not too important because the type of nucleus is obtained from other information.

Hyperfine coupling a could also be obtained by taking the sum of the NMR transition energies:

$$\Delta_{34} + \Delta_{12} = (E_3 - E_4) + (E_1 - E_2) = a \tag{1.43}$$

In fact, the NMR transitions allowed according to the selection rules of Equation 1.29 have energies

$$\Delta_{34} = E_3 - E_4 = a/2 + g_N \mu_N B_0 \tag{1.44}$$

and

$$\Delta_{12} = E_1 - E_2 = a/2 - g_N \mu_N B_0 \tag{1.45}$$

These values are obtained by considering $a > 0$. The hyperfine coupling constants can be positive or negative, depending on the system (see Chapter 4). If $a < 0$, the two values of Δ_{34} and Δ_{12} reported in Equations 1.44 and 1.45 would be exchanged.

The two NMR transitions are centered at the frequency $a/2h$, separated by 2 times the nuclear Larmor frequency $\nu_N = g_N \mu_N B_0 / h$. These transitions occur in the RF range. However, the normal observation of the NMR absorption using a traditional NMR instrument would be prevented in this case by the low sensitivity and by the large width of the lines due to the fast electron spin relaxation. To overcome these difficulties, a different technique for observing the NMR transitions in paramagnetic species was introduced. It consists of the observation of the effect on the saturation properties of an EPR line, which is produced by an RF field that induces NMR transitions. ENDOR can be considered as NMR detected via EPR. Compared to normal NMR, it has the advantage of the much higher intensity of EPR transitions.

The principles of ENDOR are described in the following example referring to a simple system of an electron spin and a nuclear spin $I = 1/2$ with $g_N > 0$ and a positive hyperfine coupling a, as illustrated in Fig. 1.10. The energy levels are the same as in Fig. 1.8, with the two sets of levels corresponding to the two nuclear spin states, shifted one relative to the other for clarity.

In the absence of a microwave radiation field and at thermal equilibrium with the lattice, the populations of levels E_1, E_2, E_3, and E_4 are determined by the Boltzmann statistics. In particular, the populations of energy levels E_4 and E_3 are larger than those of levels E_2 and E_1. Low power microwave radiation at frequencies $(E_1 - E_4)/h$ or $(E_2 - E_3)/h$ would induce EPR transitions without perturbing the population distribution.

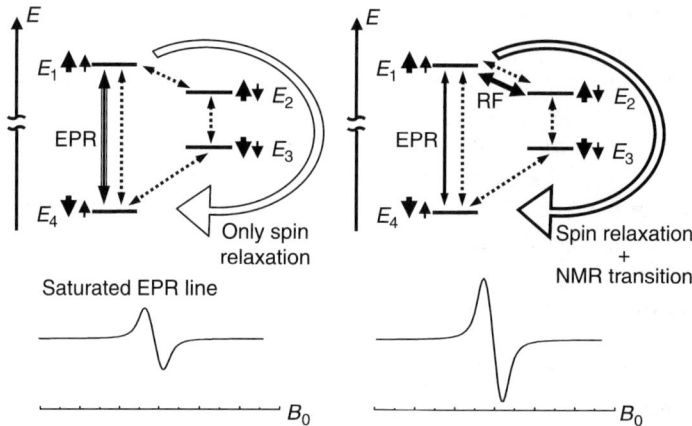

Fig. 1.10 Principles of ENDOR spectroscopy. (Left) The energy levels of an $S = 1/2, I = 1/2$ spin system in a magnetic field. One of the two EPR transitions ($1 \leftrightarrow 4$) is partially saturated as indicated by the vertical arrow, and the weakened EPR signal is shown below. The various spin relaxation steps are indicated by the dotted arrows, and the whole relaxation process between levels 1 and 4 is indicated by the circular arrow. It is too slow to recover the population difference. (Right) The NMR transition $1 \leftrightarrow 2$ induced by RF radiation at frequency $\nu_1 = a/2 - g_N\mu_N B_0$ is indicated by the solid bold arrow. Its effect is the speeding up of the relaxation path $1 \leftrightarrow 2 \leftrightarrow 3 \leftrightarrow 4$, increasing the rate of the population recovery between levels 1 and 4, with desaturation of the EPR line and increase of its intensity. The same desaturation effect would be produced by the NMR transition $3 \leftrightarrow 4$ at frequency $\nu_2 = a/2 + g_N\mu_N B_0$.

Let us suppose that microwave radiation is applied at a frequency corresponding to the EPR transition $E_1 \leftrightarrow E_4$ with a power high enough to partially *saturate* the EPR transition, as indicated by the solid arrow in the left part of Fig. 1.10. This means that the spin relaxation is not fast enough to compete with the rate of transitions induced by the microwaves, and the population difference between levels 1 and 4 is therefore reduced.

Note that various relaxation paths can be active in reestablishing the population difference between states 1 and 4, as the direct electron spin lattice relaxation between these latter states, as well as, for example, the path $1 \leftrightarrow 2 \leftrightarrow 3 \leftrightarrow 4$, with a sequence of steps consisting of a nuclear spin flip ($1 \leftrightarrow 2$), an electron spin flip ($2 \leftrightarrow 3$), and another nuclear spin flip ($3 \leftrightarrow 4$). In some systems, "cross relaxations" steps $1 \leftrightarrow 3$ and $2 \leftrightarrow 4$ (not indicated in Fig. 1.10) involving both electron and nuclear spin flips can also be very fast and important in determining the ENDOR effect. The curved arrow in Fig. 1.10 symbolically represents the combined efficiency of all of the paths.

In an ENDOR experiment the magnetic field is kept constant on an EPR line, and the RF frequency is swept. When the RF equals $(E_1 - E_2)/h$ (corresponding to $\nu_1 = a/2 - g_N\mu_N B_0$) or $(E_3 - E_4)/h$ ($\nu_2 = a/2 + g_N\mu_N B_0$), inducing one of the corresponding NMR transitions, the rate of the population recovery for state 4 increases, together with the EPR intensity. In fact, the $1 \leftrightarrow 2$ or the $3 \leftrightarrow 4$ nuclear spin flip steps become more efficient.

The ENDOR signal is the difference between the intensity of the EPR line with and without, respectively, the RF driving one of the two NMR transitions (ENDOR enhancement). If the intensity of the partially saturated EPR line is plotted as a function of the RF, the ENDOR spectrum schematically shown in Fig. 1.11 is obtained. In this example ($g_N > 0, a > 0$) the *low* frequency ENDOR line corresponds to the $E_1 - E_2$ NMR transition and the *high* frequency ENDOR line to the $E_3 - E_4$ NMR transition. Note that with a negative hyperfine coupling constant a different energy level scheme would be obtained, with the result of having the $E_3 - E_4$ transition at *low* frequency and the $E_1 - E_2$ transition at *high* frequency. However, the ENDOR spectrum looks the same, independently of the sign of the hyperfine constant a, and therefore it is not possible to get this sign from it. It is possible to get the relative signs of two hyperfine coupling constants by performing the so-called electron–nuclear–nuclear triple resonance or TRIPLE experiment, where two RF frequencies are used at the same time. This experiment is described in Chapter 2.

The CW-ENDOR effect heavily depends on the ratio of the various relaxation times. Sometimes it can be too small to be detected, in particular for radicals in solution, where often the ENDOR enhancement is detectable only with a convenient choice of solvent and temperature. Pulsed ENDOR, which is described in Chapter 5, is based on different principle. It can often be detected in cases where CW-ENDOR would be undetectable.

Two possible situations are encountered when the hyperfine coupling constant a is larger or smaller than the nuclear Zeeman interaction. In the first case the two ENDOR lines are centered at the frequency $a/2h$ and are separated by $2g_N\mu_N B_0/h$. In the second case the ENDOR lines are centered at $g_N\mu_N B_0/h$ and are separated by a/h (see Fig. 1.11). The same spectrum is obtained by saturating the second EPR line occurring between energy levels E_2 and E_3 of Fig. 1.10.

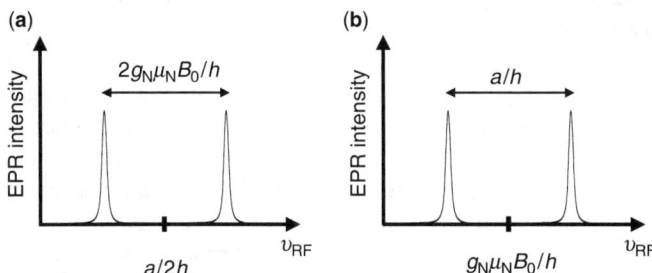

Fig. 1.11 The schematic ENDOR spectrum of an electron spin system coupled to a nuclear spin $I = 1/2$. The ENDOR lines represent the intensity variation of a partially saturated EPR line as the sample irradiated with RF radiation. An ENDOR line occurs when the RF radiation frequency matches the energy difference between energy levels 1 and 2 and between levels 3 and 4 (see Fig. 1.10). (a) If the hyperfine coupling a is larger then the nuclear Zeeman interaction ($|a| > g_N\mu_N B_0$), the ENDOR lines are centered at $a/2h$ and are separated by 2 times the nuclear Larmor frequency $g_N\mu_N B_0/h$. (b) If $|a| < g_N\mu_N B_0$, the ENDOR lines are centered at the nuclear Larmor frequency and are separated by a/h. In both cases information about hyperfine coupling and nuclear Larmor frequency is obtained.

If the unpaired electron system contains several hyperfine coupled nuclei, a pair of ENDOR lines is obtained for each set of nuclei having the same hyperfine coupling constant, independently of their number. This makes an ENDOR spectrum much simpler than the EPR spectrum, and it enhances the spectral resolution. For example, for an electron spin coupled to N spin $1/2$ nuclei one has $2N$ ENDOR lines instead of 2^N EPR lines. In contrast, the information about the number of nuclei responsible for the coupling is lost.

Note that in the case of small hyperfine couplings the ENDOR frequency could be very low, if the couplings refer to nuclei also having a small nuclear magnetic moment. Because at low frequencies the sensitivity is quite low, the use of high field/high frequency spectrometers is a big advantage. In fact, an increase of the nuclear Zeeman term permits the shifting of the frequency at higher values, but also its allows the discrimination of ENDOR lines corresponding to different nuclei with similar magnetic moments. This increase in sensitivity and resolution is parallel to the resolution increase obtained by high frequency NMR spectrometers. Examples of applications of high frequency ENDOR are provided in Chapter 12.

1.14 TWO INTERACTING ELECTRON SPINS

Some molecular systems have two unpaired electron spins. This could happen, for example, in some transition metal ion complexes or in high symmetrical molecules where electronic orbitals with the same energy (degenerate levels) are present. In fact, two electrons with unpaired spins may coexist if they are placed in different orbitals. Other systems in which two unpaired electrons are found are some excited states of organic molecules. In this case the two unpaired electron spins are placed on levels of different energies. A short list of systems with two unpaired electrons can be found in Chapter 3.

A system with two unpaired electron spins could be in a state characterized by the total spin quantum number S, either with value $S = 0$ or 1, corresponding to situations where the spins are *anti*parallel or respectively parallel to each other. The two electron spin configurations, called *singlet* and *triplet* states, respectively, correspond to different electron spatial distributions and have different energies, the separation of which is characterized by the exchange interaction parameter J. The names triplet and singlet derive from the fact that for a spin $S = 1$ three states are available with z components $S_z = 1, 0, -1$, whereas for $S = 0$ only $S_z = 0$ is possible. In the absence of a magnetic field the energy is independent of the z component, but in the presence of a magnetic field ($B_0 \neq 0$) the degeneration is removed and three separate energy levels are obtained for a triplet state.

If the electron distribution had spherical symmetry, the levels would be equally spaced. Consequently, transitions $-1 \leftrightarrow 0$ and $0 \leftrightarrow 1$, obeying the selection rule $\Delta S_z = \pm 1$, would coincide. This is not true if the symmetry is lower. In fact, for a two electron spin system one should take into account the dipolar interaction between the electron magnetic moments. This classical interaction is analogous to that of an electron spin magnetic moment with a nuclear magnetic moment discussed in Section 1.12.

Here, both magnetic moments are attributable to electron spins and the dipolar energy is

$$E_{\text{dip}} = (\mu_0/4\pi)g^2\mu_B^2[\mathbf{S_1}\cdot\mathbf{S_2}/r^3 - 3(\mathbf{S_1}\cdot\mathbf{r})(\mathbf{S_2}\cdot\mathbf{r})/r^5] \quad (1.46)$$

which comprises nine terms of the type

$$(\mu_0/4\pi)g^2\mu_B^2\langle\delta_{ij}/r^3 - 3ij/r^5\rangle S_i S_j \quad (1.47)$$

where $i, j = x, y, z$ are three orthogonal axes and S_i and S_j are the respective magnetic moment components along these axes of the individual electrons. Both electron spins are distributed in space orbitals, with a probability depending on the spatial coordinates x, y, and z. Therefore, the dipolar interaction should be averaged over the distribution of both electrons. This can be done because spatial and spin coordinates are separated. Upon integration the equation for the dipolar energy may be written as

$$E_{\text{dip}} = \sum_{i,j} D_{ij} S_i S_j \quad (1.48)$$

$$D_{ij} = (\mu_0/4\pi)g^2\mu_B^2\langle(\delta_{ij}/r^3 - 3ij/r^5)\rangle \quad (1.49)$$

The nine parameters D_{ij} constitute the elements of the *electron–electron dipolar interaction tensor*.

Note that parameters D_{ij} depend on the choice of the coordinate frame used for describing the electron distribution. As for the hyperfine dipolar tensor, there is a particular axes frame X, Y, Z (*principal axes*) such that all dipolar tensor elements are zero except the diagonal ones D_{XX}, D_{YY}, and D_{ZZ}, which are called the *principal values* of the dipolar tensor. Using the dipolar tensor principal axes frame, the electron dipolar interaction energy assumes the following simple form:

$$E_{\text{dip}} = [D_{XX}S_X^2 + D_{YY}S_Y^2 + D_{ZZ}S_Z^2] \quad (1.50)$$

Note that in Equation 1.50 the components of the total electron spin $S = S_1 + S_2$ are used instead of those of the individual spins. Moreover, because of the particular form of Equation 1.49, D_{XX}, D_{YY}, and D_{ZZ} are not independent:

$$D_{XX} + D_{YY} + D_{ZZ} = 0 \quad (1.51)$$

Taking into account Equation 1.51, the dipolar interaction energy can be written in terms of two independent parameters:

$$D = -3/2 D_{ZZ} \tag{1.52}$$

and

$$E = 1/2(D_{YY} - D_{XX}) \tag{1.53}$$

$$E_{\text{dip}} = DS_Z^2 + E(S_Y^2 - S_X^2) \tag{1.54}$$

Usually the label Z is used for the largest principal value while X and Y are taken in such a way that E has an opposite sign to that of D. The latter parameter offers information on the electron spin distribution and depends inversely on r^3, the cube of the average distance between the electron spins, whereas E is a measure of the deviation of the electron distribution from axial symmetry.

The effect of the electron dipolar interaction is to modify the energy spacing between the triplet sublevels in such a way that two distinct EPR lines occur. The line separation depends on the orientation of the magnetic field with respect to molecule fixed axes X, Y, Z.

In a frozen solution of molecules in a triplet state, all possible orientations are covered with equal probabilities and the EPR absorptions occur over a broad range of magnetic field values. However, the number of molecules responsible for the absorption at a particular field is not the same throughout the range. In fact, there are particular orientations where small changes of orientation have little effect on the resonant field value. At these field positions there is a piling up of EPR intensity. Consequently, the EPR spectrum assumes a particular shape, the analysis of which allows the determination of the characteristic parameters entering into Equations 1.53 and 1.54. Examples of EPR spectra of random orientation distributions of triplet state molecules are provided in Figs. 1.12 and 1.13.

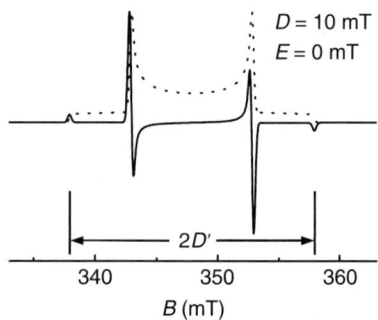

Fig. 1.12 The EPR spectrum expected for a random distribution of triplet state molecules ($S = 1$) characterized by an axial symmetry of the electron distribution ($E = 0$). The dotted line represents EPR absorption, and the solid line indicates the EPR first derivative spectrum. The linewidth is assumed as 0.3 mT, and $D' = D/g|\mu_B|$ is in magnetic field units.

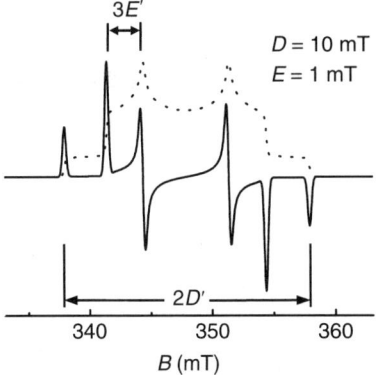

Fig. 1.13 The EPR spectrum of a random distribution of triplet state molecules ($S = 1$) with low symmetry ($E \neq 0$). The dotted line represents EPR absorption, and the solid line indicates the EPR first derivative spectrum. The linewidth is assumed as 0.3 mT, and $D' = D/g|\mu_B|$ and $E' = E/g|\mu_B|$ are in magnetic field units.

1.15 QUANTUM MACHINERY

This section deals with a more formal treatment of the general concepts of EPR outlined earlier. It is based on the quantum mechanical description of the interaction of electron and nuclear spins with a magnetic field and the quantum description of their hyperfine interaction. A basic outline of the quantum mechanics rules is now provided.

A quantum system, such as an electron or a nuclear spin, can be found in states described by a function that is usually represented by a Greek letter or by a symbol within a pair of brackets: a vertical line and an angular bracket as, for example, $|m\rangle$. In quantum mechanics, to each mechanical quantity is associated an *operator*, which is indicated by placing a circumflex (caret) over the quantity symbol. An operator acts on a state function by transforming it into another one. There are particular cases where an operator corresponding to a mechanical property acting upon a state function has the effect of multiplying the function by a constant. In such a case the state is said to be an *eigenstate* of that particular operator (property) and the constant is called its *eigenvalue*. In that state the property considered has a definite value equal to the eigenvalue.

In EPR one deals with angular momenta, which are described by vector operators $\hat{\mathbf{S}}$ (or $\hat{\mathbf{I}}$ for a nuclear spin) with three components $\hat{S}_x$, $\hat{S}_y$, $\hat{S}_z$ associated with the momentum components along the axes of an orthogonal frame.

Quantum mechanics does not allow us to know simultaneously the three components of an angular momentum $\mathbf{J}$, which means that there are no quantum states that are simultaneously eigenstates of the three operators $\hat{J}_x$, $\hat{J}_y$, $\hat{J}_z$. The maximum allowed information consists of knowing one component, usually set as the z component J_z and the square of the angular momentum $\mathbf{J}^2$.

The eigenvalues of the square of the angular moment operator $\hat{J}^2$ are limited to those given in $\hbar^2$ units by the products $j(j+1)$, where j is a characteristic quantum number that may be either an integer or a half-integer number. The eigenvalues of the $\hat{J}_z$ operator are the $2j+1$ values:

$$j_z = -j, \; -j+1, \ldots, j \qquad (1.55)$$

The state functions that are simultaneously eigenfunctions of both operators $\hat{J}^2$ and $\hat{J}_z$ can be indicated with the symbols that refer to the quantum number j, and the value of j_z placed within a vertical bar and angle bracket, as $|j, j_z\rangle$.

An electron spin angular momentum is characterized by the quantum number $1/2$ and the eigenvalue of the squared angular momentum operator $\hat{S}^2$ in $\hbar^2$ units is $(1/2)(1/2+1) = 3/4$. The eigenvalues of operator $\hat{S}_z$ corresponding to the spin component along z are $1/2$ and $-1/2$. The electron spin eigenfunctions of $\hat{S}^2$ and $\hat{S}_z$ can be written as $|1/2, 1/2\rangle$ and $|1/2, -1/2\rangle$. They correspond to those indicated in the previous paragraphs with letters α and β, respectively. The notation with Greek letters is used only for spin $1/2$ systems.

In quantum mechanics the total energy of a system is associated to an operator called the *Hamiltonian* operator (H), which contains all energy terms, and the eigenvalues of which are the quantum allowed energies of the system. Some examples of spin Hamiltonians, their eigenstates, and their eigenvalues are described in the following section.

1.16 ELECTRON SPIN IN A STATIC MAGNETIC FIELD

The simplest example of a spin Hamiltonian for EPR is that representing the energy of an electron spin placed in a static magnetic field B_0 directed along the z direction. It consists of the electron Zeeman term

$$H = g|\mu_B|B_0 \hat{S}_z \qquad (1.56)$$

Because g, μ_B, and B_0 are constants, the eigenfunctions of H are the same as $\hat{S}_z$; that is, $|1/2, 1/2\rangle$ and $|1/2, -1/2\rangle$ (or α and β). The eigenvalues of H are those of $\hat{S}_z$ multiplied by the factor $g|\mu_B|B_0$:

$$E_\pm = \pm 1/2 \, g|\mu_B|B_0 \qquad (1.57)$$

Here, $E_\pm$ is the allowed energies of the spin in the magnetic field.

1.17 ELECTRON SPIN COUPLED TO A NUCLEAR SPIN

Let us consider a single electron spin coupled to a nuclear spin with an isotropic hyperfine coupling a. We consider an $I = 1/2$ nucleus to make the example

1.17 ELECTRON SPIN COUPLED TO A NUCLEAR SPIN

as simple as possible. In this case Hamiltonian operator H comprises three terms: the electron Zeeman interaction, the nuclear Zeeman interaction, and the hyperfine interaction.

$$H = g|\mu_B|B_0\hat{S}_z - g_N\mu_N B_0\hat{I}_z + a\hat{\boldsymbol{S}} \cdot \hat{\boldsymbol{I}} \tag{1.58}$$

The spin functions of a system of *noninteracting* electron and nuclear spins ($a = 0$) can be written as the product of electron and nuclear spin functions represented by the symbols

$$\begin{aligned}\Psi_1 &= |1/2, 1/2\rangle \\ \Psi_2 &= |1/2, -1/2\rangle \\ \Psi_3 &= |-1/2, -1/2\rangle \\ \Psi_4 &= |-1/2, 1/2\rangle\end{aligned} \tag{1.59}$$

where the first number refers to the electron spin z component and the second one refers to the z component of the nuclear spin. The symbols referring to the square of the angular momenta S^2 and I^2 are omitted in this case to simplify the notation. They would be $1/2$ for both electron and nuclear spins. The above functions are eigenfunctions of the first two terms of the Hamiltonian with the corresponding eigenvalues:

$$\begin{aligned}E_1 &= 1/2\,g|\mu_B|B_0 - 1/2\,g_N\mu_N B_0 \\ E_2 &= 1/2\,g|\mu_B|B_0 + 1/2\,g_N\mu_N B_0 \\ E_3 &= -1/2\,g|\mu_B|B_0 + 1/2\,g_N\mu_N B_0 \\ E_4 &= -1/2\,g|\mu_B|B_0 - 1/2\,g_N\mu_N B_0\end{aligned} \tag{1.60}$$

For *interacting* electron and nuclear spins ($a \neq 0$) the same functions are not eigenfunctions of the total Hamiltonian because the hyperfine terms comprise the products $a\hat{S}_x\hat{I}_x$ and $a\hat{S}_y\hat{I}_y$. In fact, these operators transform the functions into different ones. For example, $\hat{S}_x$ and $\hat{S}_y$, acting only on the electron spin (the first number in the bracket) and leaving the nuclear spin unchanged, give the following:

$$\begin{aligned}\hat{S}_x|1/2, 1/2\rangle &= 1/2|-1/2, 1/2\rangle \\ \hat{S}_x|1/2, -1/2\rangle &= 1/2|-1/2, -1/2\rangle\end{aligned} \tag{1.61}$$

$$\begin{aligned}\hat{S}_x|-1/2, 1/2\rangle &= 1/2|1/2, 1/2\rangle \\ \hat{S}_x|-1/2, -1/2\rangle &= 1/2|1/2, -1/2\rangle\end{aligned} \tag{1.62}$$

$$\begin{aligned}\hat{S}_y|1/2, 1/2\rangle &= i/2|-1/2, 1/2\rangle \\ \hat{S}_y|1/2, -1/2\rangle &= i/2|-1/2, -1/2\rangle\end{aligned} \tag{1.63}$$

$$\hat{S}_y|-1/2, 1/2\rangle = -i/2|1/2, 1/2\rangle$$
$$\hat{S}_y|-1/2, -1/2\rangle = -i/2|1/2, -1/2\rangle \qquad (1.64)$$

where $i = \sqrt{-1}$. Analogous equations hold for $\hat{I}_x$ and $\hat{I}_y$, which act on the nuclear spin (the second number in the bracket).

According to quantum mechanics, if the energy associated with the hyperfine Hamiltonian $a\hat{\mathbf{S}} \cdot \hat{\mathbf{I}}$ is small compared to that corresponding to the Zeeman Hamiltonian, the terms $a\hat{S}_x\hat{I}_x$ and $a\hat{S}_y\hat{I}_y$ can be neglected in the computation of the energies and the operator $a\hat{\mathbf{S}} \cdot \hat{\mathbf{I}}$ can be replaced by $a\hat{S}_z\hat{I}_z$. This latter operator has functions in Equation 1.59 as eigenfunctions. The eigenvalues of the approximate hyperfine Hamiltonian are $\pm a/4$, which add to the Zeeman Hamiltonian energies given by Equation 1.60. The positive sign applies to the functions Ψ_1 and Ψ_4, corresponding to states with the same electron and nuclear spin z components (both $1/2$ or both $-1/2$), and the negative sign applies to the functions Ψ_2 and Ψ_3 relative to states having opposite electron and nuclear spin z components. These values correspond to the energy values reported in Equation 1.28.

1.18 ELECTRON SPIN IN A ZEEMAN MAGNETIC FIELD IN THE PRESENCE OF A MICROWAVE FIELD

This section considers the effect of the magnetic field associated to microwave radiation acting on the electron spin, considering first an isolated electron spin.

If an electron spin is placed in a magnetic field B_0 directed along the z axis and it is in the presence of a weak time dependent field $B_1 \cos(\omega t)$, perpendicular to B_0, say, directed along the x axis, the Hamiltonian becomes

$$H = g|\mu_B|B_0\hat{S}_z + g|\mu_B|B_1\cos(\omega t)\hat{S}_x \qquad (1.65)$$

The state functions $|\pm 1/2\rangle$ are not eigenfunctions of the complete Hamiltonian because the operator $\hat{S}_x$ acting on these spin functions gives as a result

$$\hat{S}_x|1/2\rangle = 1/2|-1/2\rangle$$
$$\hat{S}_x|-1/2\rangle = 1/2|1/2\rangle \qquad (1.66)$$

that is, it changes one function into the other one. The symbol referring to the spin squared moment is omitted here.

Because $B_1 \ll B_0$ and it changes with time, the effect of its presence can be considered in the frame of the time dependent perturbation theory. This theory states that a system initially in a state $|a\rangle$ will be found successively in the state $|b\rangle$, provided that the perturbation oscillates in time at an angular frequency $\omega = \hbar^{-1}(E_b - E_a)$, where $|a\rangle$,

$|b\rangle$ and E_b, E_a are the respective eigenfunctions and eigenvalues of the unperturbed system time independent Hamiltonian.

Moreover, to produce a transition from $|a\rangle$ to $|b\rangle$, it is necessary for the perturbing Hamiltonian H' to *connect* the two states in the sense that $H'|a\rangle = c|b\rangle$ with $c \neq 0$. This condition selects the allowed transitions between pairs of quantum states of a system. Furthermore, a perturbation that induces transitions from $|a\rangle$ to $|b\rangle$ also induces the opposite transition from $|b\rangle$ to $|a\rangle$.

When applied to the electron spin system, these rules describe the effect of the time variable field $\boldsymbol{B_1}$ on the electron spin. If the latter is initially in the state $|-1/2\rangle$, it will be found in the state $|+1/2\rangle$, provided that the perturbation varies in time at an angular frequency $\omega = \hbar^{-1} g \mu_B B_0$. Note that this is the Larmor frequency introduced earlier. For this case there are only two states, and the transition between them is allowed in virtue of the Equation 1.66, which ensures that the two states are connected by the S_x operator.

If a nucleus is also present, the time dependent Hamiltonian to be added to the Zeeman Hamiltonian contains two terms:

$$H(t) = g|\mu_B|B_1 \cos(\omega t)\hat{S}_x - g_N \mu_N B_1 \cos(\omega t)\hat{I}_x \qquad (1.67)$$

The first one acts only on the electron spin, inducing a change in its z component and leaving unchanged the nuclear z spin component. Conversely, the second term acts only on the nuclear spin, changing the nuclear spin z component. The Hamiltonian does not connect states differing by both components. This is the basis of the selection rules $\Delta S_z = \pm 1$; $\Delta I_z = 0$ and $\Delta S_z = 0$; $\Delta I_z = \pm 1$ (Equations 1.29 and 1.30).

To produce the electron spin flip transition, the oscillating field frequency should match the energy difference corresponding to states differing in S_z (those shown as allowed transitions in Fig. 1.8). Conversely, for producing nuclear spin transitions the oscillating field frequency should match the energy difference corresponding to states with the same S_z component and differing in I_z. These are NMR transitions observed by ENDOR.

2 Basic Experimental Methods in Continuous Wave Electron Paramagnetic Resonance

PETER HÖFER

EPR Division, Bruker Biospin GmbH, 76287 Rheinstetten, Germany

2.1 INSTRUMENTAL COMPONENTS OF A CONTINUOUS WAVE ELECTRON PARAMAGNETIC RESONANCE (CW-EPR) SPECTROMETER

This chapter discusses the hardware components of a CW-EPR spectrometer. The main components of an EPR spectrometer are the microwave bridge, the resonator, the magnet system, and the control electronics as depicted in Fig. 2.1.

Their individual properties and their significance for the experiments are described in this chapter. Although there has been enormous progress in instrumentation over the years, there are still technical limitations as well as fundamental physical properties that need to be understood in order to obtain high quality data and to make a correct analysis.

2.1.1 Microwave Resonators

The microwave resonator is the sample cell in EPR. Its two main purposes are, first, to generate a sufficient microwave magnetic field (B_1) at the sample position to drive the EPR transition and, second, to efficiently convert the sample response into a detectable microwave signal. Both of these requirements depend on the quality factor Q of the resonator that is therefore one of its most relevant characteristics.

The Q factor is defined as

$$Q = \nu_{\text{res}}/\Delta\nu \tag{2.1}$$

Electron Paramagnetic Resonance. Edited by Brustolon and Giamello
Copyright © 2009 John Wiley & Sons, Inc.

38 BASIC EXPERIMENTAL METHODS IN CW-EPR

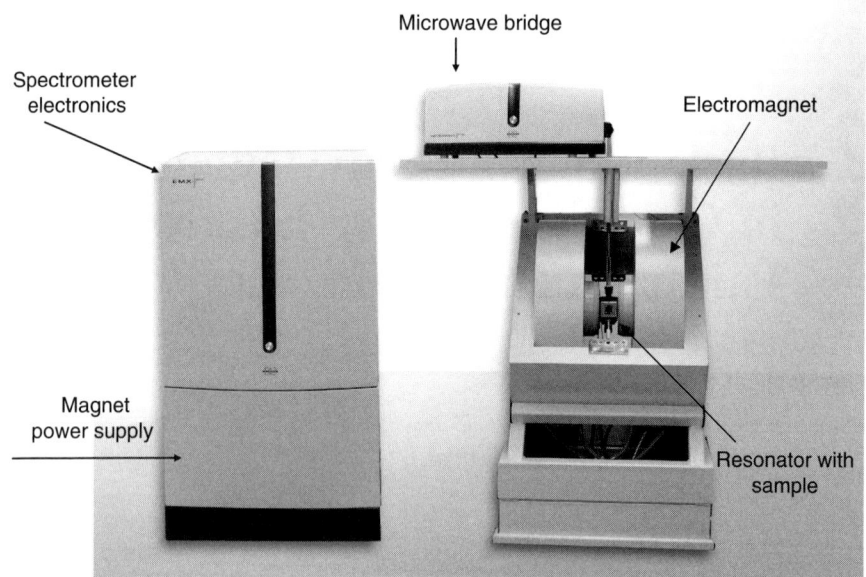

Fig. 2.1 The components of a Bruker EMX^{plus} CW-EPR spectrometer.

where ν_{res} is the resonance frequency of the resonator and $\Delta\nu$ is the width of the resonance curve (Fig. 2.2). A resonator with a frequency of 10 GHz and a bandwidth of 2 MHz therefore has a Q factor of 5000. Depending on the specific type of resonator, the Q factor values can range from $\sim$500 to 10,000.

Microwave energy is stored in the resonator leading to an enhancement of the microwave magnetic B_1 and electric E_1 field amplitudes relative to their free space values. The B_1 generated by microwave power P is given by [1] Equation 2.2:

$$B_1 = c \times (Q \times P)^{1/2} \qquad (2.2)$$

where c is the resonator conversion factor. The conversion factor is a specific number for a certain type of resonator and depends mainly on its size. As a general rule,

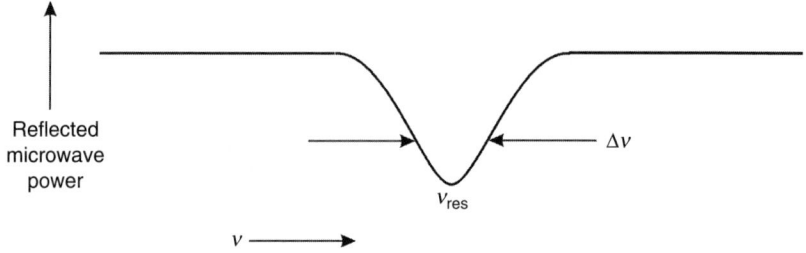

Fig. 2.2 Resonator Q factor definition. The resonator frequency is ν_{res} and the bandwidth is $\Delta\nu$.

a smaller resonator has a larger conversion factor. The unit of the conversion factor is gauss/$\sqrt{W}$.

The Q factor also enters into the EPR signal peak–peak amplitude (ΔS_{pp}) as

$$\Delta S_{pp} \propto \eta \times Q \tag{2.3}$$

which is consequently a central parameter for the detection sensitivity of the instrument. The filling factor η is defined as

$$\eta = \frac{\int_{sample} B_1^2 dV}{\int_{resonator} B_1^2 dV} \tag{2.4}$$

which is the fraction of the microwave energy stored in the resonator in interaction with the sample [1]. Therefore the signal amplitude increases with Q and with the sample volume. For a given resonator the Q factor is, however, lowered with increasing sample volume. Thus, to obtain the maximum signal amplitude the optimum sample volume has to be found for each resonator type. A good compromise for an X-band cavity is a 3–5 mm inner diameter tube for nonlossy samples.

The benefits of the resonator Q factor with respect to B_1 and the sensitivity are also the reasons why the EPR spectrum is usually recorded with a fixed microwave frequency and a variable magnetic field. It does not matter at which magnetic field value resonance occurs, the resonator always supplies the same B_1 field and the signal amplitudes across the spectrum can be compared and predicted. In the alternative approach where the magnetic field is constant and the microwave frequency is swept, a very low resonator Q factor, that is, a large bandwidth, would be required, resulting in a detrimental B_1 and sensitivity. In addition, there are technical difficulties arising from microwave reflections and varying B_1 during the sweep. Nevertheless, the frequency sweep technique has found use in the special area of zero field optically detected magnetic resonance (ODMR) spectroscopy [2]. Here, the requirements on the microwave part of the equipment are relaxed as the signal response is detected optically.

An important design property of the resonator is the spatial distribution of B_1 and E_1, the two field components of an electromagnetic wave. Only the magnetic component B_1 induces a magnetic resonance transition whereas the E_1 component leads to nonresonant microwave absorption through the electric dipole moments of the sample. This microwave absorption drops the Q factor and results in undesired sample heating (a very desirable effect in a microwave oven). EPR resonators are therefore designed to maximize B_1 at the sample position, which, by fundamental electrodynamics principles, concomitantly minimizes E_1 at that position. The field distribution of a rectangular TE_{102} cavity is illustrated in Fig. 2.3.

The field distribution in a resonator has an important consequence for positioning of the sample; this is especially true for so-called lossy samples. Aqueous solutions are the most prominent samples of this type (in general all highly polar solvents and conducting samples). If the aqueous sample extends sufficiently into the region of E_1, the resonator Q drops and the sample warms up, a condition that precludes

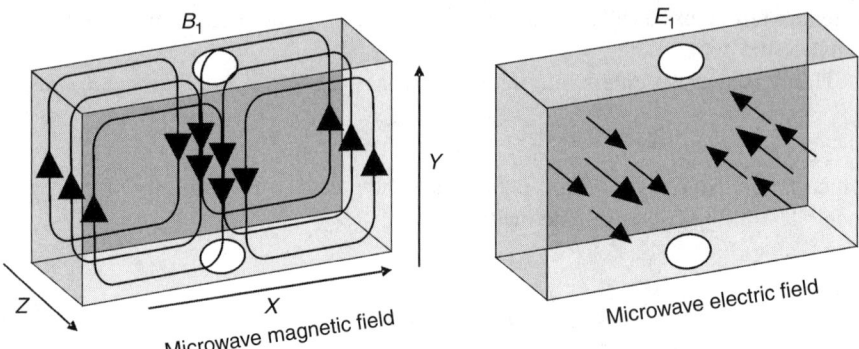

Fig. 2.3 The B_1 and E_1 distribution in an ER 4102ST rectangular resonator. Directions Y and Z correspond to the sample tube axis and the static magnetic field (B_0) direction, respectively.

a successful EPR measurement. To avoid this effect, the sample can be placed in a small capillary. Because the small capillary reduces the sample volume drastically, other sample containers with more volume have been developed. The most common of these are flat cells and multibore capillaries (Bruker AquaX). The flat cell is positioned in the resonator in such a way that it has minimum extension into the E_1 region. In the example in Fig. 2.3 the flat cell is aligned parallel to Z. The possible sample size depends strongly on the microwave frequency. In the X-band (9–10 GHz) the maximum inner diameter of a single tube of aqueous solution is <1 mm; it is only 0.1 mm in W-band (94 GHz) but goes up to several centimeters in L-band (1 GHz).

Many different EPR resonators have been developed to optimally deal with different sample classes and measurement requirements. We distinguish two main categories: waveguide cavity resonators [1] and lumped circuit resonators. A waveguide cavity has a size corresponding to the wavelength; that is, the inner dimension of a 10 GHz cavity is about 30 mm. In contrast, lumped circuit resonators at the same microwave frequency have dimensions in the range of 1–5 mm. As noted, the resonator size largely determines the conversion factor. Lumped circuit resonators therefore have a larger B_1 than cavity resonators but accommodate less sample volume. Representatives of the two resonator categories are rectangular and cylindrical waveguide cavities and split ring (Bruker ER 4118X-MS5), loop gap [3], and dielectric (Bruker ER 4118X-MD5) lumped circuit resonators. A typical value for the conversion factor of a cavity at 9.8 GHz is $1.4\,\text{G}/\sqrt{W}$ (Bruker ER 4102ST) and $8\,\text{G}/\sqrt{W}$ for a split ring resonator with a 2 mm inner diameter (Bruker ER 4118X-MS2).

The choice of the resonator depends on the sample properties and the type of experiment. For conventional CW-EPR with unlimited sample quantity the waveguide cavities (rectangular, cylindrical) are preferable. Whenever high B_1 values are required, for example, in saturation studies, transient EPR, and pulsed EPR, the small sized dielectric, split ring, and loop gap resonators are advantageous. If the sample quantity is limited, then the ideal solution would be a resonator adapted to the sample size, which cannot always be fulfilled.

A number of other common EPR techniques require very specialized resonators:

- Electron nuclear double resonance (ENDOR) resonator with an integrated radiofrequency (RF) coil (see §2.2.6)
- Double resonator for quantitative EPR (see §2.24)
- Dual mode resonator for the investigation of forbidden transitions
- Mixing resonator for the investigation of short-lived species in chemical reactions
- Resonator for optical excitation and detection
- High temperature resonator
- Electric field resonator for the investigation of conducting samples.

2.1.2 Microwave Bridge

Microwave frequencies are traditionally classified in terms of a band. The most universal and common band in EPR is called X-Band. This band covers the range of 8–12 GHz where the typical operating range of the spectrometer is 9–10 GHz. Other bands are used for more specific applications. Low frequency EPR at 1 GHz (L-Band) has its main application in the investigation of large biological samples, for example, living animals; high frequency EPR at 94 GHz (W-Band) and above [4] is mainly used in structural analysis because of the increase in resolution and absolute sensitivity. An increase in frequency also becomes necessary if the zero field splitting is larger than the microwave energy because otherwise the sample would be EPR silent. Other frequency bands like 3–4 GHz (S-band), 24 GHz (K-band), and 34 GHz (Q-band) are used to obtain additional information in the context of a multifrequency study (see §2.2.5).

The main components of a microwave bridge are sketched in Fig. 2.4. An oscillator supplies CW microwave power that is directed via a circulator to the microwave resonator where the EPR sample is placed (Table 2.1). Ideally, all microwave power is absorbed in the resonator. This condition is an important prerequisite for a successful EPR experiment and is called "matched" or "critically coupled." If not fulfilled, microwave power is reflected from the resonator and directed to the detector, resulting in a detector response that is not related to the EPR effect. This in turn tells how to check critical coupling: there should be no detector response when the microwave power is changed. When the magnetic field is swept through, resonance microwave energy is absorbed by the sample that causes a change of the resonator Q factor. It is this change that results in the reflection of microwaves from the resonator that in turn is registered as the EPR signal. Another essential element of the bridge is the reference arm consisting of a phase shifter and attenuator. To bring the microwave detector in its linear response regime a certain power (bias) is needed that is supplied by the reference arm and adjusted by the bias attenuator. In its linear response regime the detector converts large and small signals with their correct

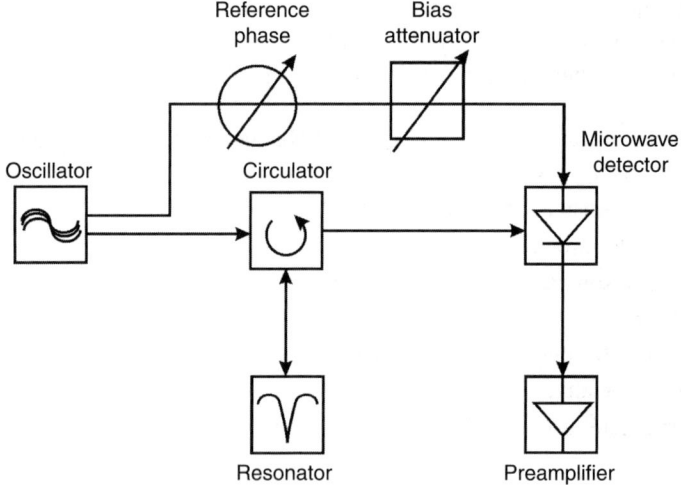

Fig. 2.4 Microwave bridge schematic.

amplitude ratios. Because EPR signals even within one spectrum can cover a large dynamic range, the linear detector response is an essential prerequisite for their truthful detection (in the spectrum of a Fig. 2.11 the ratio of the highest to the smallest line is 60). The microwave detector is phase sensitive, and the phase slider in the reference arm is used to set a defined phase relation between the reference and the reflected signal. The reflected signal and the reference signal are combined at the detector and converted to a zero frequency (DC) voltage output. Only when resonance

TABLE 2.1 Microwave Components and Their Essential Properties

Component	Function
Oscillator	Types: Synthesizer, Klystron, Gunn diode. Generates microwave power in the milliwatt range
Attenuator	Used to attenuate microwave power in well-defined manner. Attenuation range is typically 6 orders of magnitude, from milliwatt to nanowatt. The attenuation value is given in decibels (dB) relative to the maximum available microwave power.
Circulator	One-way street for microwaves
Detector	Converts microwave frequency into a zero frequency voltage, a process also called downconversion
Phase shifter	Changes the traveling pathlength of a microwave and thereby the phase relation of one wave with respect to another wave
Bias	Microwave power used to activate the detector
Resonator	Sample cell that stores microwave energy in the form of a standing wave; converts the sample response into a detectable microwave signal

occurs does the DC voltage have an additional AC component given by the modulation frequency that is the actual EPR signal. This AC signal is passed through the preamplifier for further processing (see §2.1.4).

A further important detail in the functioning of a microwave bridge is the so-called automatic frequency control (AFC). The AFC electronic modulates the microwave oscillator frequency with a rate of about 80 kHz and a small amplitude also in the kilohertz range. For the resonator to absorb all of the microwave power coming from the oscillator, the microwave frequency has to be concurrent with the center of the resonator absorption curve. Whenever the resonator changes its resonance frequency an error signal is generated at the AFC frequency that in turn is used to correct the oscillator frequency back to the actual resonator frequency. In this way the oscillator frequency is locked to the resonator frequency. The AFC is also required for the detection of the correct EPR line shape. While sweeping the magnetic field through resonance, not only the resonator matching but also its frequency are altered. The shift of the resonator frequency corresponds to the EPR dispersion signal; the change in the matching condition corresponds to the EPR absorption signal. Both effects always appear in parallel and are mathematically coupled to each other by the Kramers–Kronig relations [1]. Because the AFC takes care that the oscillator frequency follows the resonator shift, the dispersion component is eliminated and the pure absorption EPR signal is detected. For a line shape analysis it is important that the absorption and dispersion components are not mixed. From the spectroscopic point of view both signals are equivalent; that is, they have the same information content but show different saturation behavior. Therefore, special microwave bridges have been designed to detect the pure dispersion signal as well.

2.1.3 Magnet System

A magnet system for an EPR experiment comprises the magnet itself, a power supply, a field sensor, and a field regulator. To detect the EPR spectrum the magnetic field is swept over a certain range. Whenever the values of the microwave frequency and the magnetic field fulfill the resonance condition of the sample, a spin transition is induced and registered by the data system. The required magnetic field range is determined by the microwave frequency and the sample properties. An organic radical in solution, for example, has a g factor of about 2 and a spectrum width on the order of 100 G. To measure this radical in X-band (9.6 GHz) a magnetic field of about 3430 G with a sweep range of 100 G is required. This relatively simple task can be mastered by a permanent magnet equipped with a sweep coil. Table top EPR instruments for dedicated applications like quality control and dosimetry (Bruker e-scan) are actually designed in this way. For more general use and to reach higher fields, an electromagnet is the most commonly used field generator. With an air gap of 72 mm, electromagnets can reach maximum fields close to 15 kG (Bruker ER 073). The magnetic fields for high frequency EPR at 94 GHz (W-band) and above can no longer be reached by electromagnets, so one has to resort to sweepable superconducting magnets.

In addition to the maximum field, another important parameter for a magnet is the homogeneity of its field. The homogeneity over a certain volume determines the smallest detectable EPR linewidth if the sample fills this volume entirely. In general, commercial EPR magnets are designed in a way that a larger pole face results in a larger homogeneous volume. A magnet with a pole face of 25 cm has a homogeneity of 10 mG over a volume of $10 \times 10 \times 22$ mm^3 at a field of 3400 G (Bruker ER 073, 1800 kg weight). The longest axis of the homogeneous volume points along the sample tube axis. This definition takes into account that EPR samples are frequently filled in tubes or in flat cells (aqueous solutions).

The spectroscopic parameters characterizing the EPR spectrum on the field axis are the linewidth, line position, and line separation (Fig. 2.5). It is the task of the field measuring and control system to give the correct numbers of these parameters, but the spectroscopist has to be aware of the technical constraints involved. The various means to determine the magnetic field are illustrated in Fig. 2.6 and explained in the following. The setup used for electromagnets is usually a Hall sensor that is mounted in the magnet gap to measure and regulate the magnetic field. A control electronic reads the Hall voltage and feeds the power supply with this information either to generate the sweep or to stabilize the field at a certain value.

Especially in the determination of exact g factor values, one has to be cautious with respect to the magnetic field values. Although the microwave frequency can be determined with very high precision (10^{-7}) by a frequency counter, the magnetic field values are dependent not only on the absolute accuracy of the field sensor but also on its spatial position. In a standard setup the Hall sensor is mounted off center on the magnet pole face. The result is a field offset between the EPR sample position and the Hall sensor that amounts at 3400 G to about 3–5 G. This offset is usually taken care of by a calibration offset in the instrument software. The Hall sensor is mounted off center so that it is not influenced by the magnetic

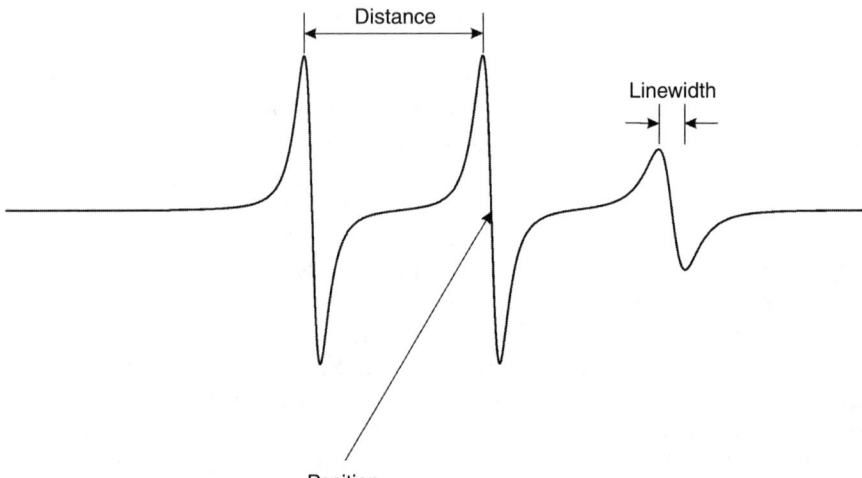

Fig. 2.5 The magnetic field parameters of an EPR spectrum. On the field axis the EPR spectrum is characterized by the line positions, distances between lines, and linewidth.

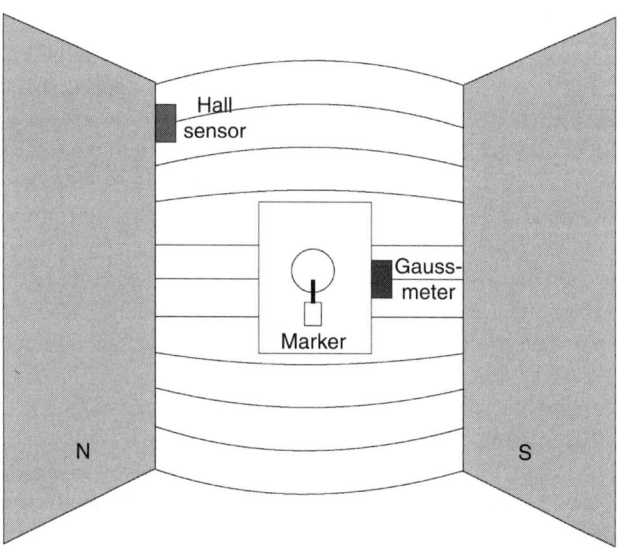

Fig. 2.6 Magnetic field sensors. The Hall sensor is used to measure and regulate the field, the gaussmeter only measures the field, and the marker serves as a *g* factor reference.

field modulation applied to the resonator. A Hall sensor is calibrated against another field measuring device and has an absolute accuracy on the order of 100 mG. A much higher absolute accuracy (10^{-6}–10^{-7}) is achieved by a NMR gaussmeter, a device that is a small and compact NMR spectrometer and measures the resonance frequency of a nuclear spin (usually ^{1}H) in the given field. The probe of the NMR gaussmeter should be as close as possible to the microwave resonator. Even then there is a small field offset between this probe and the center of the resonator that can again be calibrated out by the means of an in-cavity gaussmeter probe.

Another alternative way to determine the field at the sample position is a reference sample with a known *g* factor. Commonly the reference sample is inserted into the resonator together with the sample of interest and therefore experiences the same magnetic field. The achievable accuracy largely depends on the precision to which the *g* factor of the reference sample is known. Because both samples are measured simultaneously, it would be preferable to use a reference sample with a *g* factor different enough so that its spectrum does not overlap with the sample spectrum. Frequently used reference samples are powders of Cr^{3+} in MgO, Mn^{2+} in CaO, and 2,2-diphenyl-1-picrylhydrazyl (DPPH). The Cr^{3+} ion has a single line spectrum with a *g* factor of 1.97989; therefore, the resonance field is such that the line position is outside the spectrum of a typical free radical. The spectrum of Mn^{2+} consists of six lines with a separation in the range of 80–90 G and a *g* factor of 2.0011. DPPH also has a single line with a *g* factor of 2.0036, and it is likely to interfere with the spectrum of an organic radical. It should be noted that DPPH has a *g* factor anisotropy on the order of about 1.5×10^{-4}, which gives rise to a linewidth that varies, depending on the conditions of crystallization (see Ref. 1, p. 441).

The major problem in determining the g factor comes from the requirement to know the absolute field value at the sample position. The determination of distances on the field axis, for example, for linewidth or hyperfine coupling evaluation, is less problematic as constant field offsets are canceled. However, strictly speaking, the Hall offset determined at one field value is only valid at this field position and will vary over a sweep. The linearity of the sweep is usually very high so that the offset correction at the beginning and end of the sweep is sufficient. The difference in the Hall sensor offset at the low and high end of the sweep translates into a sweep range error. For a Bruker ER 073 magnet at a center field of 3480 G these errors are typically 100 mG for a 100 G sweep and 1 G for a 1000 G sweep. The error in determining distance parameters on the field axis is therefore on the order of 10^{-3}. By measuring the offset with a gaussmeter at the start and end of the sweep, the error can be reduced to a few milligauss, regardless of the sweep range.

A further concern with respect to the magnetic field is its temporal stability. The field stability is especially important in experiments where a static field is required, for example, in time resolved EPR, pulsed EPR, and ENDOR. Under proper environmental conditions a long-term stability of 10 mG/h can be achieved at the typical X-band field of 3400 G. To maintain good field stability it is important to have a cooling water supply with a constant flow rate and temperature and no drafts in the room.

For even higher long-term stability, more technical effort is required by means of a field–frequency (FF) lock. The FF lock is a device that registers at any moment the microwave frequency and the field value. It then regulates the field to keep a constant value for the ratio of the field and frequency. In this way field and frequency drifts are compensated and a g factor stable position in the spectrum is achieved. A long-term stability of a few milligauss per hour is realized with an FF lock.

2.1.4 Data System

The recording of the EPR spectrum requires a magnetic field sweep with a synchronous detection of the sample response. In the schematic setup shown in Fig. 2.7, the signal channel serves as the master of synchronization and drives the Hall controller. For a time period determined by the user, the signal channel accumulates the sample response coming from the microwave bridge at a given field position and then advances the Hall controller to the next field value.

The signal channel (Fig. 2.8, top) is a central part of the spectrometer; it comprises a lock-in detector and a modulation amplifier. A lock-in detector is a device that measures a signal of known frequency with high selectivity and sensitivity. To generate a signal of known frequency, CW-EPR uses the method of field modulation. Typically a 100 kHz frequency is generated by the lock-in and then amplified by the modulation amplifier. This modulation frequency (ν_{mod}) is fed to a modulation coil that then adds a magnetic field oscillation of certain amplitude (B_{mod}) parallel to the main magnetic field. When the main magnetic field is in resonance, the superimposed oscillating magnetic field modulates the resonance condition; thus, the sample response is also modulated at the rate of 100 kHz. The downconverted microwave signal coming from the bridge is therefore modulated with this 100 kHz frequency that is further processed by the lock-in detector to produce the EPR

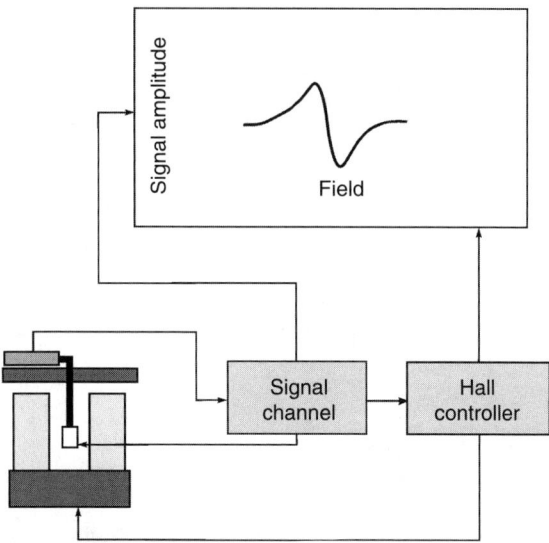

Fig. 2.7 A block diagram of an EPR spectrometer. The signal channel records the EPR signal and drives the Hall controller to advance the magnetic field.

signal as a function of the external field sweep. An important consequence of the field modulation is that the slope of the EPR absorption signal is measured, that is, its derivative. To obtain information about the number of spins in the sample, which is proportional to the integral of the absorption signal, now the double integral of the EPR spectrum has to be evaluated.

The basic principle of field modulation encoding is depicted in Fig. 2.8 (bottom). A modulation frequency with amplitude B_{mod} is applied at a certain point in the absorption line. For a small B_{mod} compared to the curvature of the absorption line, the response signal is an oscillation with ν_{mod} of amplitude A. Output amplitude A corresponds to the slope of the absorption line at the resonance position. Clearly, at the top of the absorption line the slope is zero; thus, the derivative signal has its zero crossing here. When increasing B_{mod} the effect of the curvature of the absorption line on the derivative signal increases, and higher harmonics of ν_{mod} are generated as well. If the modulation amplitude is large enough so that a sizable part of the absorption line is covered by B_{mod}, the derivative signal will no longer reflect the true line shape and linewidth (see §2.2.1). A full mathematical description of the field modulation technique can be found in Reference 1.

At the end of the chain of signal processing, an analog to digital converter (ADC) turns the EPR signal into a computer readable format. The ADC is characterized by its amplitude resolution that today is up to 24 bit, corresponding to $\pm 8 \times 10^6$ levels of digitization. A high amplitude resolution gives a large dynamic range in detectable signal variation. For example, we can follow an EPR signal time evolution over several orders of magnitude in amplitude change or we can record extremely weak signals in the presence of very large signals.

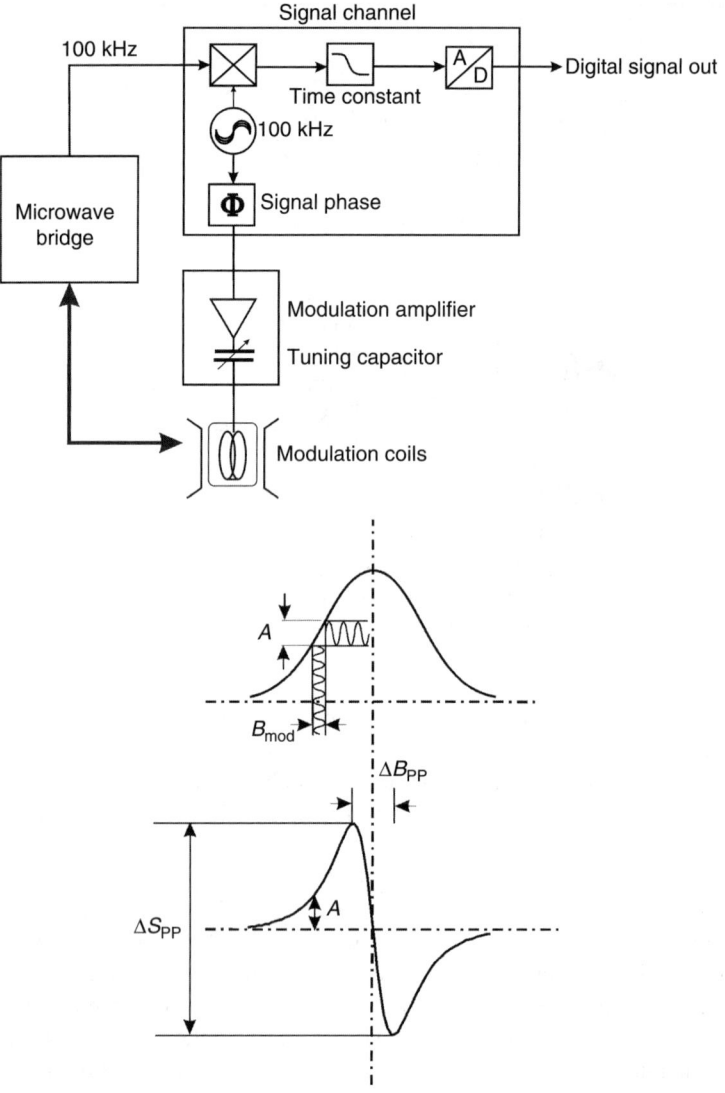

Fig. 2.8 The magnetic field modulation scheme and derivative line shape. The derivative line is characterized by the peak–peak amplitude (ΔS_{pp}) and the peak–peak linewidth (ΔB_{pp}).

As noted, if the lock-in detection frequency is equal to the modulation frequency the first derivative of the absorption line is registered. In general, the nth derivative is detected if the lock-in detection frequency is $n \cdot \nu_{mod}$. With increasing derivative order the resolution is increasing but the sensitivity is rapidly dropping. In practical work the second derivative can still be used to improve the resolution whereas for higher orders the loss in sensitivity is normally too severe.

2.1.5 Instrumental Parameters

To obtain high quality spectroscopic data, it is essential to be familiar with the instrumental parameters and how they influence the measurement. Depending on the aim of the measurement, not all parameters may be equally important and one can save considerable time by focusing on the relevant ones. However, regardless of what is to be measured, the signal/noise ratio (S/N) is the first and most important number to be considered because the accuracy of any numerical value inferred from the spectrum will depend on it. Signal and noise amplitudes have different instrumental parameter dependencies, so we consider them separately. The derivative peak–peak signal amplitude (ΔS_{pp}) in CW-EPR linearly depends on the following parameters:

$$\Delta S_{pp} \propto t \times B_{mod} \times \eta \times Q \times \sqrt{P} \tag{2.5}$$

This is correct as long the microwave power P is low enough so that no saturation occurs and the field modulation amplitude B_{mod} is much smaller than the peak–peak linewidth (ΔB_{pp}). These restrictions immediately tell us that optimization of S/N by P and B_{mod} has to be executed with care (see §2.2.1). During measuring time t, the signal amplitude is integrated (accumulated) and therefore grows linearly. Clearly, the filling factor η should be maximized within the constraints given by the resonator and the available sample quantity. The Q factor is a given value for a certain resonator, but it can be influenced by the sample geometry and is dependent on the dielectric loss characteristic of the sample.

Sensitivity is also a matter of the spectrometer operating frequency. In general, the higher the frequency is the smaller the minimum number of electron spins that can be detected. This means that if the sample is only available in a limited amount, a higher operating frequency is advantageous. In a broader view EPR samples have been classified in terms of available quantity, saturation behavior, and dielectric loss [5]. For most of these sample classes the sensitivity will increase with frequency, except for the combination of <*not limited, saturating, lossy*> and for <*not limited, not saturating, lossy*>. As reference numbers we note that the sensitivity limits for nonlossy samples today are 10^9 and 10^7 spins for 1 G linewidth and 1 Hz detection bandwidth in X- and W-bands, respectively [6].

Considering the noise, our first look is at white noise that has its origin mainly in the microwave components of the instrument and is independent of microwave power and modulation amplitude. This noise is purely statistical in nature and depends on the accumulation (or measuring) time and detection bandwidth (BW) as

$$N \propto \sqrt{t \times BW} \tag{2.6}$$

Because the signal amplitude is proportional to t and the noise to $\sqrt{t}$, the S/N $\propto \sqrt{t}$ as well. An important consequence is that the S/N can be improved by a longer measuring time, but to double the S/N the measurement takes 4 times longer. The noise amplitude is also proportional to the square root of the detection bandwidth. This fact is usually taken care of by the adjustable analog lock-in amplifier time constant (T_C) that controls the detection bandwidth (BW $\propto 1/T_C$). Thus, an increasing time constant causes the noise amplitude to decrease. Care has to be taken in the selection of T_C with respect to the sweep time. The time constant determines the instrument response time to any change of the signal amplitude. For example, if the time constant is larger than the sweep time across a signal, severe distortions of the line shape will occur. The analog time constant was an important control parameter at the time when the spectra were still registered with a pen recorder. Today the spectrum can be recorded with the lowest time constant (i.e., unfiltered) and then easily treated mathematically until the desired noise level is reached.

The counterpart to white noise is "correlated noise," a phenomenon that is much more difficult to address. Correlated noise has its origin in technical malfunctions, environmental conditions, and operator mistakes. It usually manifests itself in artificial signals, increased noise, and baseline drifts and tends to become worse with increasing microwave power and modulation amplitude.

Some common reasons for correlated noise that can be influenced by the operator are the following:

- Instabilities in cooling water pressure, temperature, and flow rate
- Mechanical vibration of the sample
- Mechanical vibration of the resonator
- Sample temperature instabilities in variable temperature experiments
- Excessive gas flow in variable temperature experiments
- Too large AFC amplitude
- Resonator instabilities due to excessive modulation amplitude

This chapter has shown that there are static and oscillating magnetic fields active in an EPR experiment. In order not to confuse them, they are summarized together with typical X-band values in Table 2.2.

TABLE 2.2 Magnetic Fields in EPR

Field	Property and Task	Typical Values
B_0	Static magnetic field, defines quantization axis of spins and is swept to achieve resonance	3450 G for $g = 2$ in X-band
B_{mod}	Oscillating magnetic modulation field parallel to B_0	100 kHz oscillation with typically 1 G amplitude
B_1	Oscillating magnetic field component of the microwave field, perpendicular to B_0, drives electron spin transitions	9.6 GHz oscillation (at X-band) with 10 mG amplitude in the rotating frame
B_2	Oscillating magnetic field component of RF field, perpendicular to B_0, drives nuclear spin transitions in an ENDOR experiment	15 MHz oscillation for ^{1}H nuclear spins in X-band with amplitude on the order of 10 G in the rotating frame
G	Static magnetic field gradient for spatial encoding in EPR imaging	40 G/cm

2.1.6 Variable Temperature

The temperature is a key parameter for many physical properties of a sample. The user can obtain a series of spectra at different temperatures to observe the effects on the EPR spectrum, for example how the viscosity of the solvent affects the EPR linewidth. In contrast, some EPR experiments require a temperature chosen in a narrow range of values. Some reasons for choosing a particular temperature include the following:

- Increase signal amplitude by the temperature dependent Boltzmann factor
- Slow down relaxation to allow spectrum detection, for example, for fast relaxing transition metal ions
- Slow down relaxation to achieve saturation, for example, in an ENDOR experiment
- Generate a frozen state from a liquid solution
- Hinder a chemical reaction
- Accelerate a chemical reaction (at high temperature)
- Establish physiological conditions

Low temperature equipment is usually based on either liquid nitrogen (77K) or liquid helium (4.2K). For variable temperature experiments the cryogenic liquid is supplied by a storage vessel from which it is taken either by overpressure or by a pump. It is passed through a heater and then supplied to the sample space. At the heater stage the liquid is evaporated and heated to the desired temperature. A thermocouple placed as close as possible to the sample is used to measure the temperature. The reading of the thermocouple is then supplied to a control unit to regulate heater power and flow rate to establish the desired temperature condition. Because the thermocouple and sample are in reality not at the same position, there is always a gradient between the thermocouple reading and the actual sample temperature. For a precise knowledge of the sample temperature this gradient has to be calibrated for different heater powers and flow conditions. To achieve a low temperature it is furthermore required to place the heater, thermocouple, and sample inside a dewar that has vacuum insulation to shield the inside from the room temperature environment. Two different designs are used for the dewar, depending on the size of the microwave resonator. For waveguide cavities at X-band and lower frequencies the dewar-in-cavity setup is normally used. These cavities are large enough so that a quartz dewar can be built that is inserted into the cavity. For lumped circuit or high frequency resonators a sufficiently small dewar can no longer be constructed. In this situation the setup is reversed and a cavity-in-dewar design is used. There is a fundamental difference in the functioning of the two designs. With the dewar-in-cavity setup the cavity stays at room temperature and only the sample experiences the set temperature. Conversely, the sample and the resonator are at the same temperature in the resonator-in-dewar setup. Especially in experiments where the EPR signal intensity is evaluated as a function of temperature, the effects due to the change of the resonator Q factor with temperature should

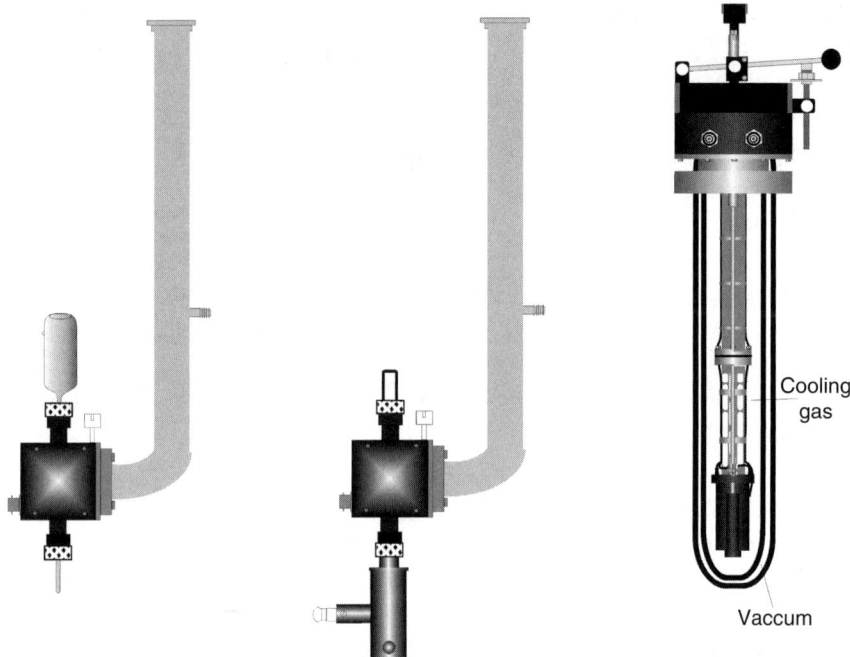

Fig. 2.9 (Left) A nitrogen temperature setup with a finger dewar for a 77K fixed temperature, and a variable temperature setup with (middle) dewar-in-cavity and (right) cavity-in-dewar.

not be overlooked. The resonator Q factor depends on the electrical conductivity of the resonator walls, and as the conductivity increases with lower temperature the Q factor also increases. A signal increase by lowering the temperature is then attributable to the increase of both the Boltzmann factor and the cavity Q factor.

Sometimes it is sufficient to keep the sample at a single fixed low temperature. This requirement can easily be mastered with a so-called finger dewar containing liquid nitrogen. The evacuated double walled dewar is inserted into an X-band waveguide cavity and filled with liquid nitrogen. With this setup the sample is at a constant temperature of 77K. The three types of temperature control setups are depicted in Fig. 2.9.

Although most variable temperature EPR work is done below room temperature, there are cases where high temperatures are required as well, for example, in catalysis. Temperatures up to 1300K can be reached with a dedicated resonator (Bruker ER 4114HT high temperature cavity).

2.2 EXPERIMENTAL TECHNIQUES

The basic experimental techniques in CW-EPR are now considered. One group of experiments can be distinguished based on the swept quantity, which is the field and time sweep. Another group deals with the use of additional parameters, for example, the microwave power in a saturation study, to obtain further information about the sample. The list of covered experiments is not meant to be exhaustive but encompasses the most frequently used techniques. There are still many other experimental techniques using special equipment. A few that are not treated here are high pressure [7], single crystal (see Chapters 6 and 9), parallel mode detection of half-field transitions [8], spin trapping (Chapter 8), electrochemistry [9], ODMR [2], and electrically detected magnetic resonance [10].

2.2.1 Field Sweep Experiments

Recording the EPR spectrum by a magnetic field sweep is either the only purpose of the measurement or the first step in a sequence of experiments. The EPR spectrum reflects the microscopic environment of the electron spin through its interaction with other electron spins and nuclear spins. In CW-EPR the spectrum is measured in the presence of a microwave field and a modulation field. In order to resolve the electron spin interactions in the EPR spectrum the influence of the measuring process itself has to be kept small compared to the interaction parameters. This is achieved by complying with the following conditions (for parameter definition, see Chapter 1):

$$B_1, B_{mod}, \nu_{mod} \ll a, \Delta g, D, 1/T_1, 1/T_2, \Delta B_{pp} \qquad (2.7)$$

Note that this requirement is completely reversed in pulsed EPR where B_1 should exceed all internal spin interactions (see Chapter 5).

In some cases the interaction parameters may be resolved in the EPR spectrum, but they can also be partially or completely hidden under the line. In the second case the EPR line is called inhomogeneous. An inhomogeneous line is composed of spin packets that have slightly different resonance fields. The resulting line shape frequently has a Gaussian character. Homogeneous broadening has its origin in the fluctuations of the EPR transition frequency. A homogeneous line has a Lorentzian shape, and its width is directly related to T_2 (see Chapter 1). Figure 2.10 shows homogeneous and inhomogeneous lines.

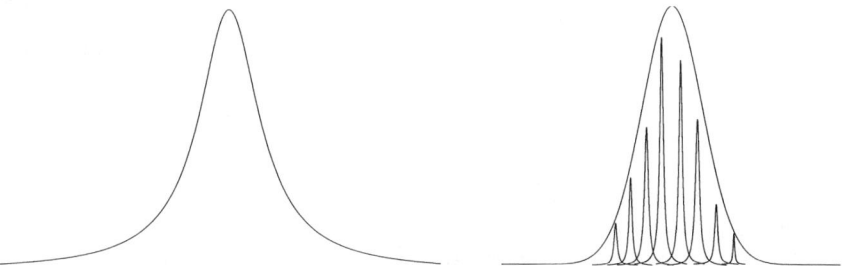

Fig. 2.10 (Left) A homogeneous line has a Lorentzian shape, and (right) an inhomogeneous line is built up by homogeneous lines and frequently has a Gaussian character.

Homogeneous lines are generally observed in liquid solutions, whereas inhomogeneous lines are typical for solids. An anisotropic powder line shape results from the random orientation of the orientation dependent interaction parameters with respect to the magnetic field, and it is always inhomogeneous. Crystals of strongly interacting radicals, such as DPPH, can show homogenous lines, because the hyperfine structure is washed out by the spin exchange (exchange narrowing).

Modulation Amplitude. For a sample with unknown interaction parameters the microwave power (B_1) and modulation conditions have to be optimized to measure the correct spectrum. Figure 2.11 shows the optimization of the modulation amplitude for the spectrum of the phenalenyl (PNT) radical in solution. The EPR spectrum is characterized by a linewidth of 100 mG and two hyperfine coupling constants of 1.85 and 6.3 G originating from coupling to three and six equivalent protons, respectively (how these coupling parameters lead to the spectrum is explained in Chapter 4). Successive reduction of the modulation amplitude until there is no further change in the linewidth results in the correct spectrum

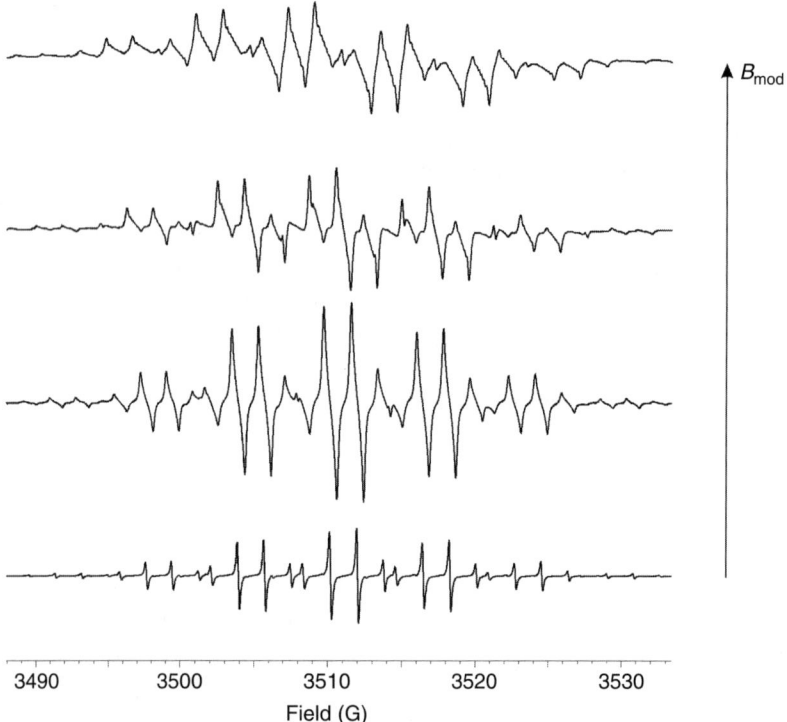

Fig. 2.11 The variation of B_{mod} to optimize the resolution of a spectrum with 100-mG linewidth; $B_{mod} = 6.0, 3.0, 1.0,$ and 0.1 G (top to bottom).

(assuming that there is no power saturation, see following, and the homogeneity of the field is high enough).

In this example the modulation amplitude for maximum resolution has been optimized. If resolution is not of major concern, higher signal amplitude can also be optimized. With increasing modulation amplitudes the derivative signal amplitude grows linearly initially, goes through a maximum, and then drops again. A good compromise is normally a modulation amplitude equal to the linewidth. Note that the double integral of the derivative line grows strictly linearly with increasing modulation amplitude, a fact that can be helpful in quantitative EPR of low S/N spectra. Technically, modulation amplitudes of several tens of gauss are possible, allowing broad lines to be sufficiently modulated. However, very high modulation amplitudes can lead to baseline instabilities and the overall quality of the spectrum may deteriorate.

Microwave Power. Like the modulation amplitude, the microwave power needs to be optimized. In the first step the microwave power is adjusted for maximum signal, and then checks for line shape distortions and saturation effects follow. Note that the spectrum depends on the value of B_1, and the value of B_1 corresponding to a given microwave power depends in turn on the type of resonator used. A quick check of the saturation condition is achieved by changing the microwave power by 6 dB. If not saturated, the signal amplitude will change by a factor of 2.[1] The PNT spectra in Fig. 2.12 were measured in a high Q cavity (Bruker ER 4122SHQE) that has a conversion factor of $2.2\,G/\sqrt{W}$. In the spectra measured with 14 and 0.9 mW this corresponds to B_1 values of 260 and 65 mG, respectively. Clearly, the B_1 of 260 mG exceeds the linewidth of 100 mG and leads to significant line broadening and consequently a loss in resolution and signal amplitude (the amplitude also drops due to saturation, see following).

Not all components of an EPR spectrum necessarily have the same microwave power dependence. This may be attributable to the presence of different species, the variances in the transition moment, or differences in the relaxation rates. In this situation the spectrum has to be recorded with different values of the microwave power to optimize or separate the various components. As an example, Fig. 2.13 (top) provides the spectrum of Mn^{2+} in CaO. The spectrum consists of the six main lines originating from the hyperfine coupling of the electron spin ($S = 5/2$) to the Mn nuclear spin with $I = 5/2$. In between the main Mn lines are the forbidden transitions, and in the spectrum center there is a radical signal.

The power dependence, plotted against $\sqrt{P}$, of the three spectrum components is shown in Fig. 2.13 (bottom). Clearly, the radical signal increases linearly over the full range of available microwave power while the allowed and forbidden transitions of the Mn^{2+} spectrum reach a maximum at different power levels. A typical property of forbidden transitions is that they require a higher power than the allowed ones to reach the same level of saturation. The different behavior of the allowed transitions

---

[1] An attenuation change of 6 dB is a factor of 4 in power and corresponds to a factor of 2 in B_1. In the absence of saturation the signal amplitude will then change by a factor of 2 as well.

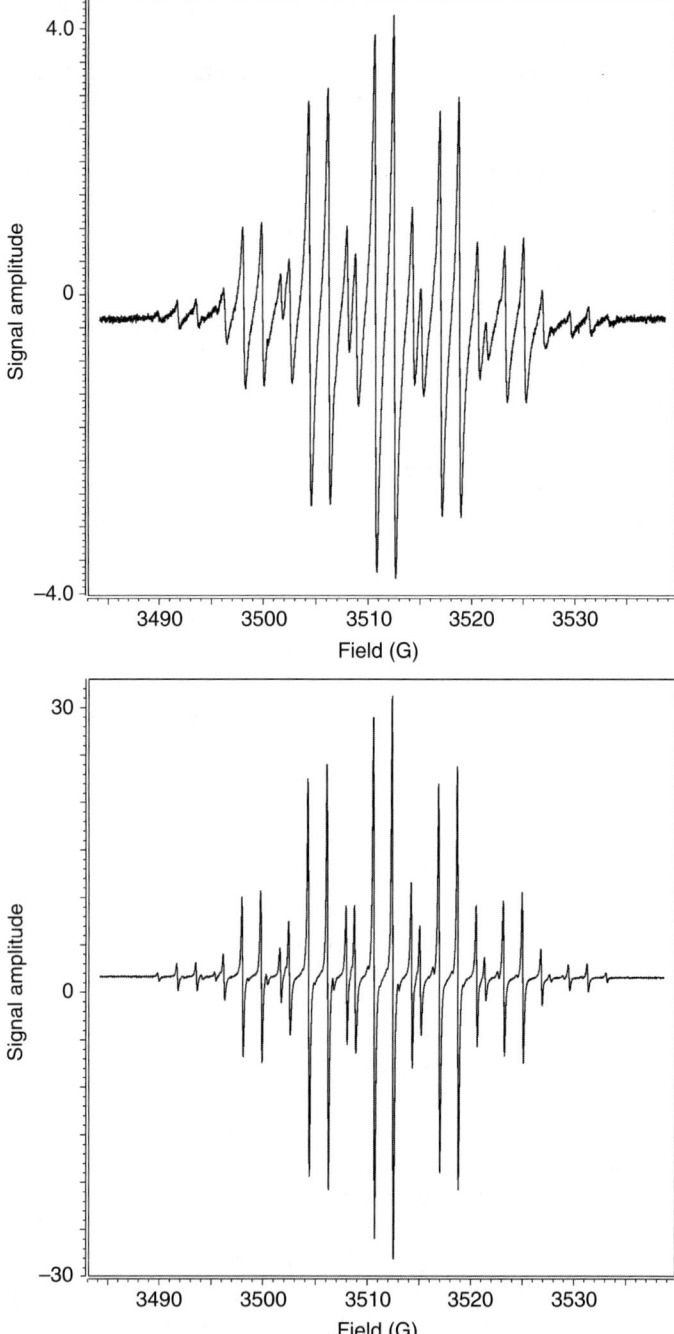

Fig. 2.12 The PNT spectra measured with B_1 values of (top) 260 mG and (bottom) 65 mG. The individual linewidth is 100 mG. If B_1 exceeds the linewidth, severe broadening occurs. The field modulation amplitude is 100 mG for both spectra.

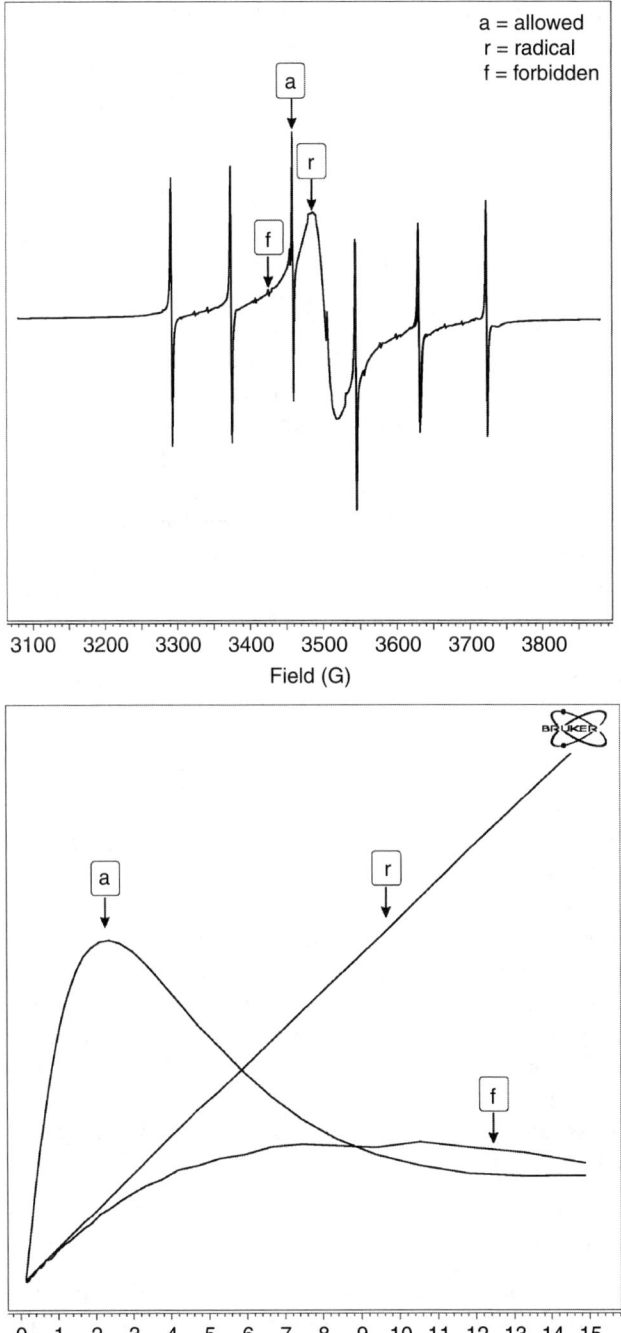

Fig. 2.13 (Top) The EPR spectrum on Mn^{2+} in CaO and (bottom) a saturation plot of the allowed (a), forbidden (f), and radical (r) spectral components.

of Mn and of those of the radical can originate from shorter relaxation times T_1 and T_2 of the radical and from the stronger transition moment of an $S = 5/2$ spin compared to an $S = 1/2$ spin (see Chapter 1). In a CW experiment these different contributions cannot be separated whereas in a pulse experiment each property (T_1, T_2, transition moment) can be measured separately (see Chapter 5).

Modulation Frequency. The range of available modulation frequencies is due to technical reasons, but the choice of a specific frequency is dependent on the sample properties (Equation 6). Because of the detector noise characteristics and susceptibility to low frequency mechanical vibration, a high modulation frequency is desired. In contrast, the resonator bandwidth sets an upper limit for the modulation frequency ($\nu_{mod} \ll \nu/Q$). As a good compromise 100 kHz has become well established as the maximum modulation frequency. Lower modulation frequencies can become important to fulfill the condition $\nu_{mod} \ll 1/T_{1,2}$ and are therefore entirely sample related. In the extreme case of $\nu_{mod} \gg 1/T_{1,2}$ (a situation that can easily happen at low temperatures), instead of the derivative of the absorption curve, the absorption signal itself is recorded [5].

Another subtle effect of modulation frequencies can be seen with extremely narrow EPR lines. As seen earlier, for maximum resolution the modulation amplitude should be less than the linewidth. However, this is no longer sufficient if the linewidth is considerably smaller than 100 mG. When the resonance condition is fulfilled, the EPR effect generates sidebands at $\pm \nu_{mod}$ around the carrier frequency, which is the microwave frequency [1]. This effect leads to a line broadening in the amount of $\pm \nu_{mod}$. The modulation frequency itself generates an uncertainty of the frequency and thus of the line position. A 100 kHz field modulation frequency corresponds to a field value of 35 mG,[2] and therefore it gives rise to a broadening of 70 mG. To resolve narrower spectrum structures, the field modulation frequency has to be reduced. A solution of a C_{60} fullerene with an encapsulated ^{14}N is a good example to demonstrate this effect (Fig. 2.14). The EPR spectrum consists of three lines with a width of $\approx$10 mG. At a 100 kHz modulation frequency the peak–peak linewidth of the center line shown in the figure is 70 mG, although the modulation amplitude used was only 10 mG. By reducing the modulation frequency to 60 kHz (± 20 mG uncertainty) the linewidth is reduced to 40 mG. With a further decrease to 10 kHz (± 3.5 mG uncertainty) the linewidth becomes 12 mG and approaches the correct value. As a general rule, whenever the linewidth is smaller than 100 mG a modulation frequency lower than 100 kHz should be used.

2.2.2 Saturation Study

An excessively high microwave power will lead to saturation of the spin transition and eventually to a loss of resolution and reduction in signal amplitude. However, because the occurrence of saturation is related to relaxation times T_1 and T_2,

---

[2]The conversion factor between the field and frequency is 2.8 MHz/G for a $g = 2$ signal.

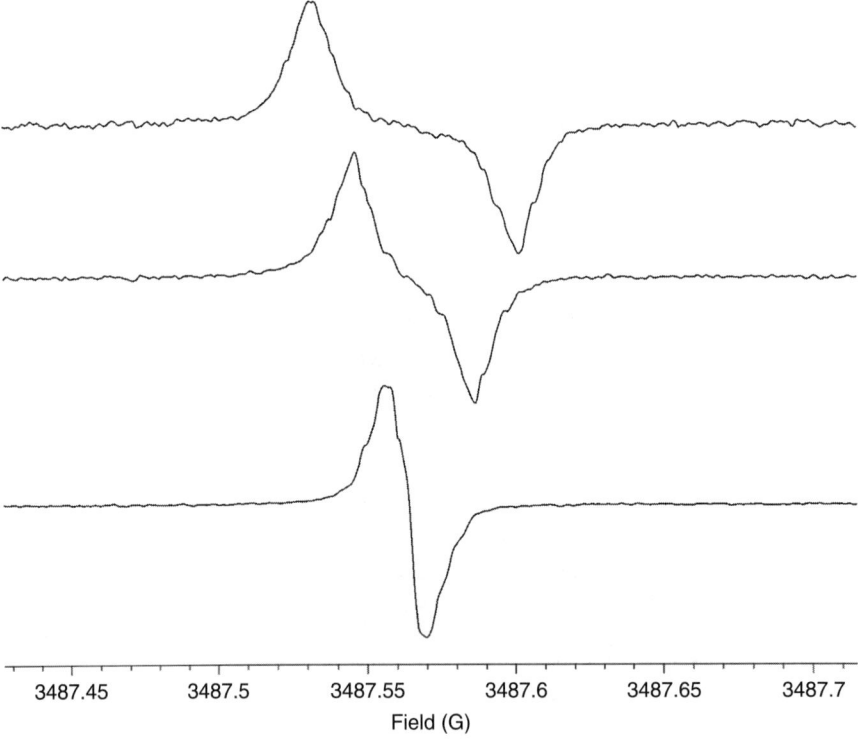

Fig. 2.14 The center line of N@C$_{60}$ measured with 100-, 60-, and 10-kHz modulation frequency (top to bottom) and 10 mG modulation amplitude. Sample courtesy of K. P. Dinse.

it gives valuable information about the local environment of the electron spin. In a systematic saturation study the EPR spectrum is recorded as a function of microwave power. The degree of saturation is related to the saturation factor s [1]:

$$s = \gamma^2 B_1^2 T_1 T_2 \tag{2.8}$$

where γ is the magnetogyric ratio ($\gamma = g_e/2m_e$). We see that the saturation factor not only depends on sample properties, and the product $\gamma^2 T_1 T_2$, but also on the instrumental parameter B_1. A saturation study should therefore be made in a comparative way using the same experimental setup.

It is convenient to analyze the saturation curve in terms of the so-called $P_{1/2}$ parameter, which is the microwave power level at which the signal has dropped to 50% of the signal amplitude in the absence of saturation.

Fitting a saturation function of the form

$$I = \frac{I_0 \sqrt{P}}{\left(1 + P/P_{1/2}\right)^{b/2}} \tag{2.9}$$

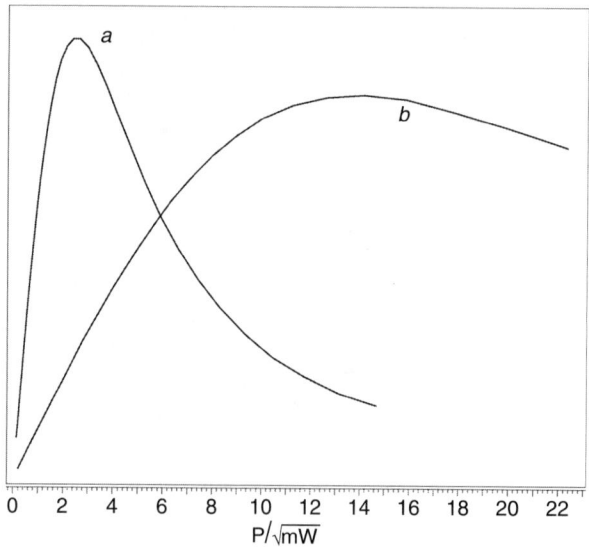

Fig. 2.15 The resonator dependence of saturation measured with a DPPH sample (homogeneous line, $T_{1,2} \approx 100$ ns) in ER 4123D (spectrum a) and ER 4102ST (spectrum b). The resulting $P_{1/2}$ values are 377 mW (ER 4102ST) and 11 mW (ER 4123D). In this example we have $b = 2.9$, indicative of the homogeneous character of the line.

allows us to typify the saturation curve in terms of $P_{1/2}$ and the parameter b. Here, b indicates a homogeneous (≈ 3) or inhomogeneous (≈ 1) character of the line, and I_0 is the initial slope of the saturation curve [11].

The influence on the saturation behavior due to the resonator conversion factor is shown in Fig. 2.15. There are two saturation curves of a small crystal of DPPH, obtained for the b curve in a standard rectangular X-band waveguide cavity (Bruker ER 4102ST) and for the a curve in a 5 mm dielectric resonator (Bruker ER 4123D).

Clearly, with the smaller resonator (see curve a, Fig. 2.15) the saturation is achieved at a much lower microwave power, thus allowing a wider range of samples to be studied. The saturation curve also gives an indication about the type of EPR line, which is homogeneous or inhomogeneous. In the case of a homogeneous line, like DPPH, the signal goes through a maximum and then drops with increasing microwave power. An ideal inhomogeneous line,[3] in contrast, levels off at a constant value. An example of the saturation of a predominantly inhomogeneous line is provided in Fig. 2.16 with a coal sample measured in the ER 4102ST cavity.

Because the saturation depends on the product $T_1 T_2$, all properties that influence T_1 or T_2 can be exploited in a saturation study, for example, temperature, viscosity, concentration, and oxygenation.

---

[3] An ideal inhomogeneous line has a much larger width than the underlying homogeneous lines.

2.2 EXPERIMENTAL TECHNIQUES 61

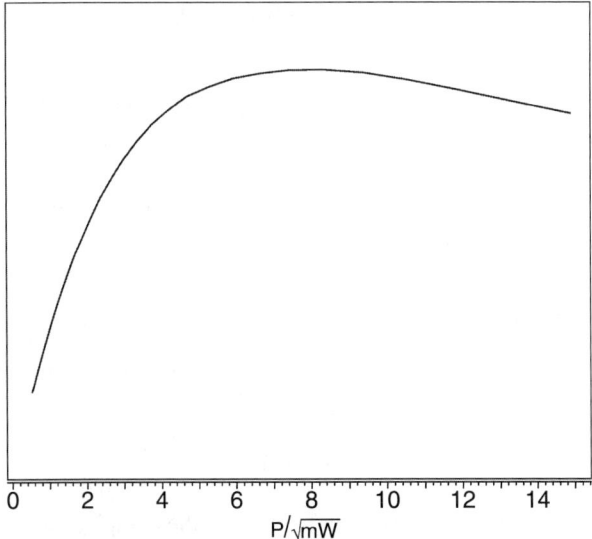

Fig. 2.16 The saturation behavior of a line with inhomogeneous character (coal sample); $P_{1/2} = 14$ mW, $b = 1.26$.

As an example let us consider how the presence of a second paramagnetic species changes the saturation behavior of the sample under investigation. A 10 μM solution of 4-hydroxy-2,2,6,6-tetramethylpiperidine-N-oxyl (TEMPOL) in water is investigated under ambient conditions (air) and in a nitrogen atmosphere (Fig. 2.17). If the sample is equilibrated in air the $P_{1/2}$ value is 16.3 mW, but it is only 4.92 mW

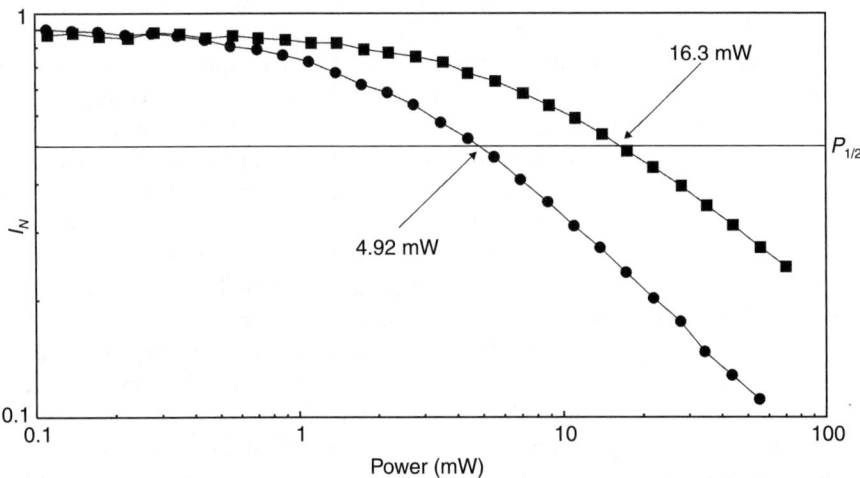

Fig. 2.17 Saturation curves of 10 μM TEMPOL in water measured with the ER 4123D resonator. An air or N_2 atmosphere result in $P_{1/2}$ values of 16.3 and 4.92 mW, respectively.

if the atmosphere is completely exchanged with nitrogen. The paramagnetic oxygen in air interacts with the TEMPOL electron spin and shortens its T_2, thus leading to a higher power for saturation [12].

The saturation analysis is today a widely used technique in the investigation of proteins. Site-directed spin labeling allows proteins to be marked with a paramagnetic probe that serves as a sensor for the local environment. By means of a gas exchange experiment the oxygen accessibility of this position is probed in the saturation study [13, 14].

2.2.3 Time Resolved EPR

In a time resolved experiment the EPR signal is observed as a function of time to study a kinetic process. The variation in time can be a decay or buildup of the signal driven, for example, by a chemical reaction.

How the experiment can be performed depends on the time scale of the process. If the time constant of the process is considerably longer than the time required for a field sweep, successive field sweeps can be run and the complete spectrum evolution can be monitored. When using an electromagnet the speed is limited by the time constant imposed by the magnet and the power supply. A field sweep over 200 G (sufficient for organic radicals) can be performed in a few seconds. Processes that are considerably longer than this can then be monitored by measuring the full spectrum by consecutive field sweeps. For even faster kinetics there are two alternatives: a rapid scan experiment and a time scan at constant field. In a rapid scan experiment the electromagnet only generates a static field (B_0) and an additional sweep coil supplies a field ramp centered at B_0. With this setup field ramps as fast as a few milliseconds over 200 G can be achieved. Of course, the faster the field ramp becomes, the less time the signal is accumulated and the more spins are needed for sufficient S/N. For the next step toward higher time resolution a field sweep is no longer feasible. Now the magnetic field is positioned at the EPR line, and the signal amplitude is recorded at a fixed field in a time scan. This approach has the drawback that any line shift or linewidth change will lead to an apparent signal amplitude change. However, if the kinetic process can be started again, the experiment can be repeated at another field value. In this way the complete spectrum kinetic can be reconstructed now by recording the time profile at successive field values. The achievable time resolution in this experiment is on the order of microseconds and no longer dependent on the field scan rate. At a 100 kHz modulation frequency the time resolution cannot be significantly lower than 10 μs. Consequently, for time resolution in the submicrosecond range the field modulation technique can no longer be used. Instead the direct detection mode (i.e., without field modulation and a lock-in detector) is utilized. Furthermore, because the inverse of the time resolution corresponds to the required detection bandwidth, more specialized hardware with a bandwidth in the range of 10–200 MHz is required as well.[4] Because of the large bandwidth, this technique inherently has low sensitivity.

---

[4]Time resolved EPR can also be performed in pulse mode (see Chapter 5).

2.2 EXPERIMENTAL TECHNIQUES

TABLE 2.3 Methods of Time Resolved EPR

Time Resolution	Experimental Technique
Greater than seconds	Electromagnet field sweep, lock-in detection
Greater than milliseconds	Rapid scan, lock-in detection
Greater than microseconds	Time scan at constant field, lock-in detection
Greater than nanoseconds	Transient EPR at constant field, direct detection, pulsed EPR

One of the most well-known systems studied by direct detection is optically excited triplet states (see Chapter 4 for transient EPR of chemically induced processes). Triplet states are usually highly polarized, so the low detection sensitivity is compensated to a large extent (Table 2.3).

An example of the first category with a time constant greater than seconds is illustrated in Fig. 2.18. Here a photochemical reaction is driven by continuous UV light irradiation and monitored by the time evolution of a phenylbutylnitrone spin trap signal. For baseline definition the UV is switched on only after the first field sweep is finished. The sweep time for one scan is 21 s, followed by a field fly back and settling time of 2 s. This corresponds to a time resolution of 23 s between slices. The signal evolution is monitored over 7 min until no further change is observed. An experiment with this time resolution and time scale can be conveniently performed with standard electromagnet and microwave equipment.

Time resolved (or transient) EPR with nanosecond resolution requires considerable changes in the hardware. The bandwidth of all elements in the signal detection path has to be large enough to pass and capture the fast transients. A microwave bridge with a special wideband receiver system (typically up to 200 MHz bandwidth) and a transient recorder with a high sampling rate (1 GHz) are employed. Furthermore, a resonator with low Q factor is essential. A high Q resonator with a bandwidth of 1 MHz (corresponding to $Q \approx 10{,}000$ in X-band) would superimpose a time constant of 320 ns to the transient signal and thus mask all faster time responses. Dedicated low Q resonators like split ring and loop gap are used instead. These resonators have Q factors of 1000 and lower (in X-band) and thus allow signal response times of a few tens of nanoseconds to be detected. As an example consider a transient EPR experiment with direct CW detection on a single crystal of pentacene in a p-terphenyl host matrix (Fig. 2.19). The triplet state is generated by a 10 ns laser pulse at a 532 nm wavelength in the presence of a microwave monitoring field. As the triplet state has a lifetime on the order of a 100 μs, it is regenerated by the laser pulses applied at a rate of a few hertz. The essential hardware components for the experiment are a split ring resonator (Bruker ER 4118X-MS5), a transient bridge (Bruker ER 046XT), a 250 MHz transient recorder (Bruker SpecJet), and a laser system (Spectra Physics, INDI). A common way to perform the experiment is to record the transient signal at successive values of the magnetic field (Fig. 2.19a). A slice parallel to the field axis (Fig. 2.19b) shows the EPR spectrum at a specific time after the laser pulse. This EPR spectrum exhibits positive and

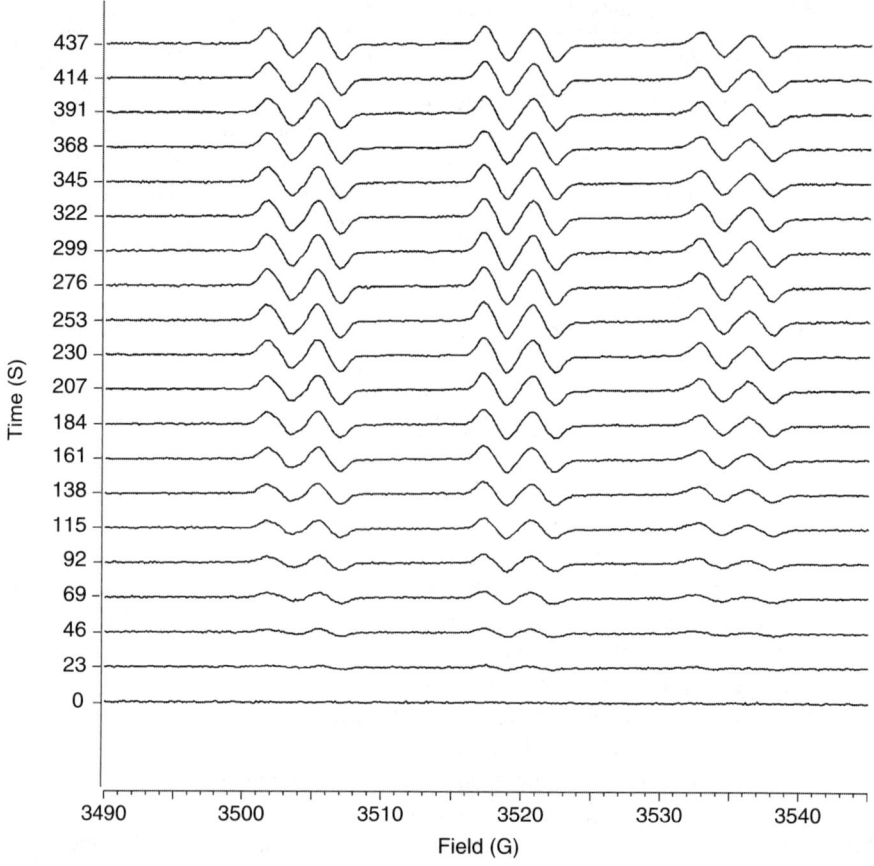

Fig. 2.18 The time evolution of a photochemical process in a suspension of TiO_2 in H_2O driven by UV light and monitored by a phenylbutylnitrone (PBN) spin trap signal. The measurement was performed by a conventional field sweep and lock-in detection.

negative amplitudes, and the shape of the spectrum is similar to that of a field modulated spectrum. However, no field modulation is employed for transient EPR; and positive or negative amplitudes are indicative of the absorption or emission character of the triplet transitions, which are due to the ratios between the transient populations of the spin states (see Chapters 3, 4, and 5, Part II, for examples). When the field is resonant with one of the triplet transitions, the microwave B_1 generates a so-called transient nutation signal (also called Torrey oscillation). The oscillation frequency of this pattern is given by the strength of B_1 and is thus resonator dependent (for details about the spin physics, see Refs. 15 and 16). Figure 2.19c shows the nutation signal driven by a microwave power of 200 mW. The signal has a period of $\approx$260 ns and an initial rise time of 40 ns, indicating the high time resolution achieved. The duration of the oscillation is determined by T_2, B_1 inhomogeneities, and eventually by the triplet state lifetime.

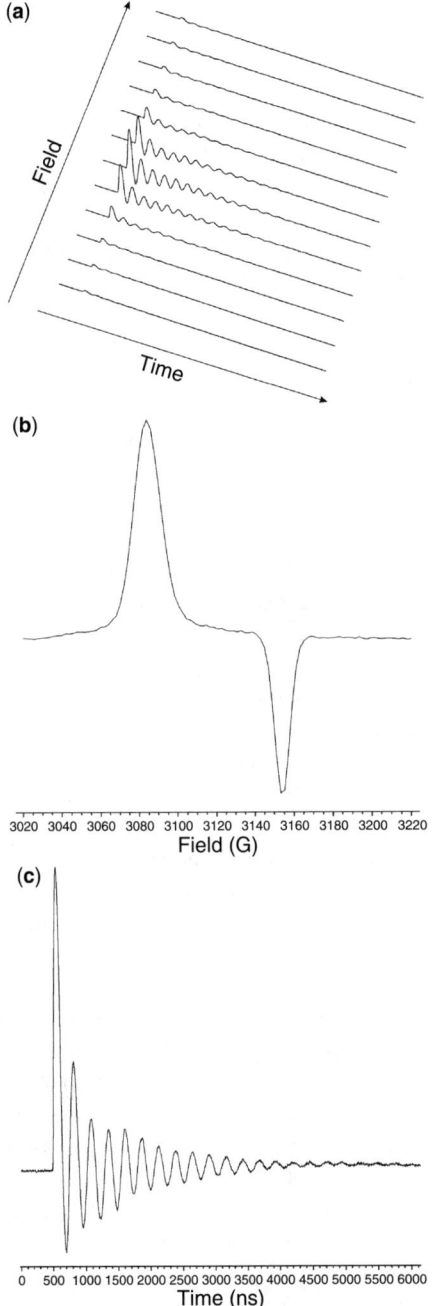

Fig. 2.19 The transient EPR of a single crystal of pentacene in a *p*-terphenyl host matrix. (a) The transient signal is recorded at successive field values. Slices parallel to the field and time axis results in the (b) EPR spectrum and (c) transient signal, respectively. Sample courtesy of R. Bittl.

2.2.4 Quantitative EPR

As noted, the analysis of the line positions and spectral profile of an EPR spectrum can be done on the basis of the magnetic interactions of the corresponding paramagnetic species. By contrast, the term "quantitative EPR" refers to the analysis of the signal amplitude to obtain information on the absolute number of unpaired electron spins in the sample or their changes induced by a certain sample treatment, for example, irradiation with different doses of γ rays (see Chapter 9).

Although this looks like a simple exercise, it is quite a difficult task (see also Chapter 9). However, as in any quantification, the required precision determines the degree of difficulty of the procedure.

Equation 2.5 outlines the instrumental parameters determining the signal amplitude. In a well-tuned and calibrated modern spectrometer the instrument related parameters are known and under control. A crucial parameter is the precise knowledge of the sample volume and mass. Care has to be taken that the instrument is properly warmed up and that the signal is not saturated by too much microwave power. Changes in the Q factor are negligible if samples of the same type in the same container are compared. However, it is wise to monitor the Q factor as it enters linearly into the signal amplitude. Alternatively, a reference sample can be used. The type of reference sample and its use depend on the nature of the quantitative analysis. It is necessary to distinguish between relative and absolute quantification. In relative quantification a series of samples is measured to determine the EPR signal as a function of some external parameter, for example, an annealing process. The reference sample is usually inserted into the resonator at a fixed position and is measured simultaneously with the unknown one. Typical reference samples are Cr^{3+} and Mn^{2+}, which are also employed as g factor standards (see §2.1.3). As long as reference and unknown signals are not power saturated, any Q change will cancel by calculating the ratio of the signal amplitudes. In case the spectral overlap of the unknown and reference signals is unacceptable, a double resonator is the best choice. The double resonator has two sample chambers that are microwave coupled and guarantees the most identical measuring condition for both samples. For the analysis of the signal variation the peak–peak amplitude can be used as long as the EPR linewidth and shape are identical for the whole series of unknown samples. If this is not the case the comparison has to be made based on the double integral of the spectra. The double integral has the advantage that it grows linearly with the modulation amplitude and thus allows high modulation amplitudes to improve the S/N. However, the evaluation of the double integral requires extra care because it can become a source of serious errors especially for an ill-defined baseline.

Another challenge in quantitative EPR is the determination of the absolute number of spins in the sample. Usually a double resonator and a reference sample with a known number of spins are employed. Ideally, the reference sample has EPR properties similar to the unknown, a requirement that cannot be met in many cases. A reference sample that is readily prepared with known concentration is a nitroxide spin label in solution. In general, the reference and unknown sample have different EPR spectra; thus, the double integrals over the full spectrum width have to be evaluated.

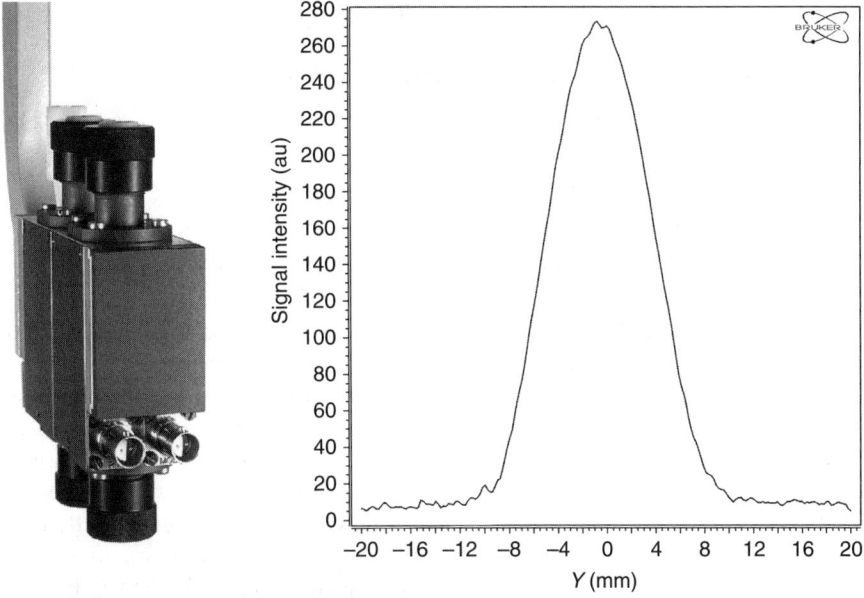

Fig. 2.20 A double resonator (Bruker ER 4105DR) and its sensitivity profile along the sample tube axis (Y direction).

Reproducible and known positioning of the samples in the resonator is essential as the resonator has a nonuniform volume sensitivity distribution. The sensitivity profile of a double resonator (Bruker ER 4105DR) along the sample tube axis is shown in Fig. 2.20. This clearly shows that positioning becomes crucial for samples that occupy only a part of the sensitive volume. Especially in the absolute quantification with extended samples, the resonator sensitivity profile has to be taken into account. Only if the reference and unknown have the same spatial dimensions and are at equivalent positions in the double resonator will the resonator sensitivity profile cancel in the ratio calculation.

2.2.5 Multifrequency EPR

The most universal frequency in EPR is X-band, which is well established. Aside from some restriction in the sample size, for example, no small animals, all sample classes are compatible with this frequency, all EPR techniques can be applied, and the sensitivity is good. Nevertheless, there are significant spectroscopic reasons to do EPR at more than one frequency. The spectral changes with frequency and the spectral features at other frequencies can be used to gain deeper insight or to enhance certain features. An example for the latter is the improvement in orientation selection with increasing frequency, a main motivation for high frequency EPR (see Chapters 1 and 12).

The main purposes of multifrequency EPR are the following:

- Identify field dependent and field independent interactions
- Simplify the spectrum
- Consistency check of the analysis
- Change the frequency window for motional sensitivity
- Improve orientation selection with high frequency (Chapters 1 and 12)
- Improve ENDOR resolution with high frequency (§2.2.6)

Chapter 1 introduced the energy terms of the various interactions of the electron spin. They can be grouped into field dependent and field independent components. The energy terms for hyperfine coupling, nuclear quadrupole coupling (see Chapter 6), and electron–electron dipolar coupling are independent of field whereas the electron and nuclear spin Zeeman terms are field dependent. Splittings arising from the field independent terms will not change if measured at different frequencies as long as the high field limit is fulfilled (see following). Splittings due to different g factors or g factor anisotropy are proportional to the microwave frequency. Measurements at two significantly different frequencies (e.g., X- and Q-band or X- and S-band) will help to differentiate these contributions.

As an example we consider the E' center in γ-irradiated quartz. Measured in X-band (9.6 GHz) an axial symmetric powder pattern with a width of 2.5 G is obtained (Fig. 2.21).

With a second measurement, now in S-band (3.6 GHz), we obtain a spectrum of the same shape but the width is only 1 G. As the spectrum width scales with the frequency ratio, the powder pattern is clearly dominated by an anisotropic g factor.

The field/frequency dependence of the interactions can also be used to change the spectrum signature. At low frequencies the spectra tend to be hyperfine dominated whereas at high frequencies the g factor features prevail.

Magnetic resonance is preferably performed under the condition that the Zeeman field is much larger than all other couplings (high field limit). The breakdown of the high field limit will lead to significant complications in the spectrum. Because the coupling constants in EPR can be rather large, second-order shifts and forbidden transitions are a common phenomenon and a signature for the breakdown of the high field limit [8]. In this situation an increase in frequency can help to approach the high field limit again and simplify the spectrum.

The X-band spectrum of $^{53}Cr^{3+}$ in $CsAl(SO_4)_2 \cdot 12H_2O$ is an example where the high field limit is no longer valid (Fig. 2.22, left). There is an electron spin $S = 3/2$, a zero field splitting of $D = -0.078$/cm, and an isotropic g factor of 1.975. Signatures of the high field limit breakdown are the forbidden transitions in the half-field region of the X-band spectrum ($\sim$1000 G). When going to higher frequencies (Q- and W-band), these features disappear (not shown) and the spectrum becomes increasingly symmetric. The multifrequency approach also allows the cross check of an analysis made at one frequency. The confidence in the analysis increases considerably

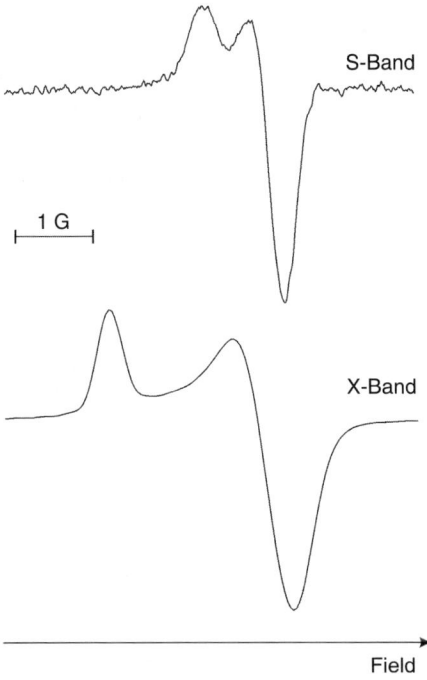

Fig. 2.21 The E' center in γ-irradiated quartz measured in (bottom) X-band and (top) S-band. In X-band the spectrum width is 2.5 G and in S-band it is 1 G, indicating the g factor nature of the anisotropy.

if one set of parameters in a simulation reproduces the spectra recorded at several frequencies.

The EPR spectra in solution are frequently characterized by narrow lines, a result of motional averaging of the anisotropies (for a full treatment, see Chapters 4 and 7). Complete averaging is achieved when the molecular rotations are isotropic and fast compared to one cycle of the microwave (100 ps in X-band, 10 ps in W-band). The linewidth can therefore change with the microwave frequency, and a fast motion spectrum in X-band can become a slow tumbling spectrum in W-band. In the example of a vanadyl acetylacetonate complex in toluene [17] with $S = 1/2$ and $I = 7/2$ in Fig. 2.23, in X-band there is already a linewidth variation across the spectrum, indicating incomplete motional averaging. This effect is strongly enhanced in W-band because the frequency is 10 times higher.

The large hyperfine coupling also results in a nonequidistant line separation across the spectrum in X-band. Because of second-order effects, the line separation increases from the low to the high field side of the spectrum. The first and second line are separated by 98 G, and the two high field lines have a distance of 119 G. The systematic change in line separation depends on the M_I of the observed EPR transition [8]. In W-band the high field limit is better fulfilled and the line separation

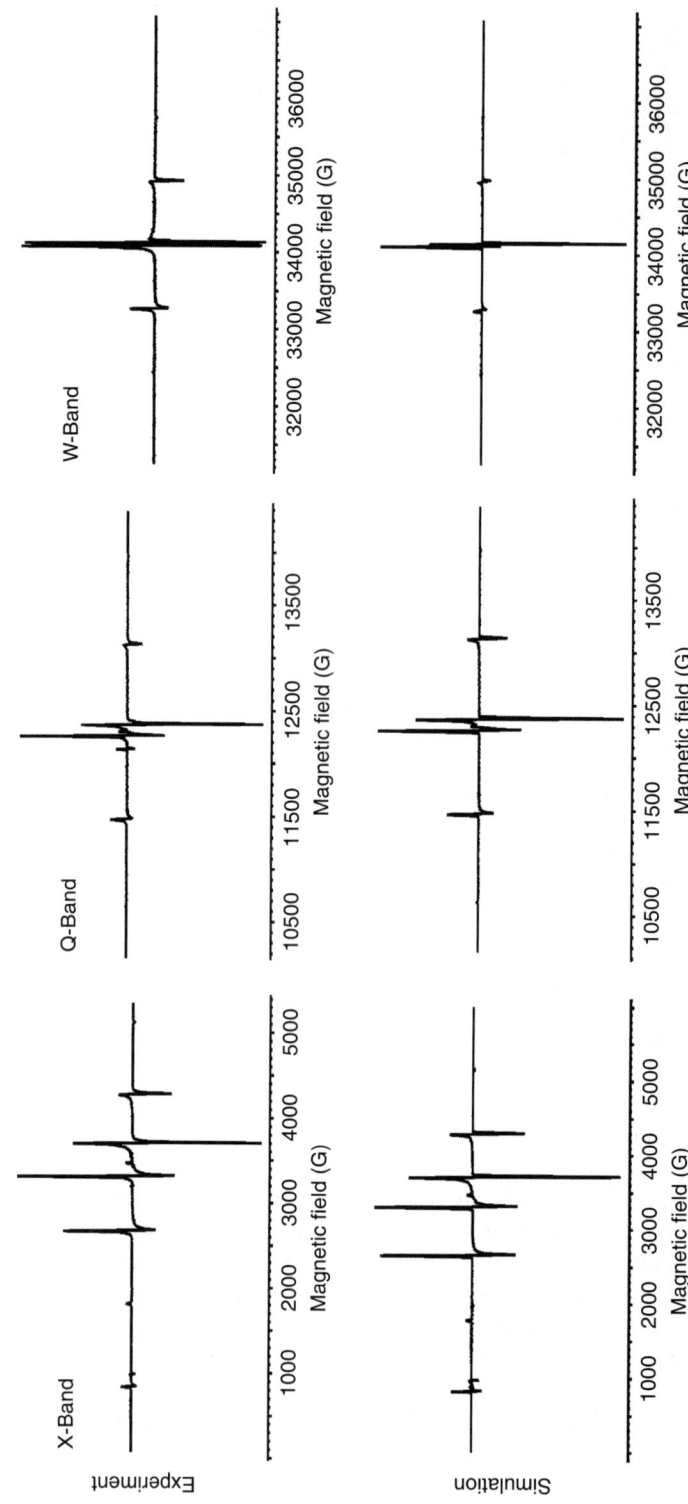

Fig. 2.22 Experimental (upper trace) multifrequency spectra and their simulations (lower trace). Sample: 1% $^{53}Cr^{3+}$ in $CsAl(SO_4)_2$, $S = 3/2$, $D = -0.078$/cm, $g = 1.975$. The simulation was performed with the Bruker XSophe program and reproduces all features at all frequencies with one set of parameters.

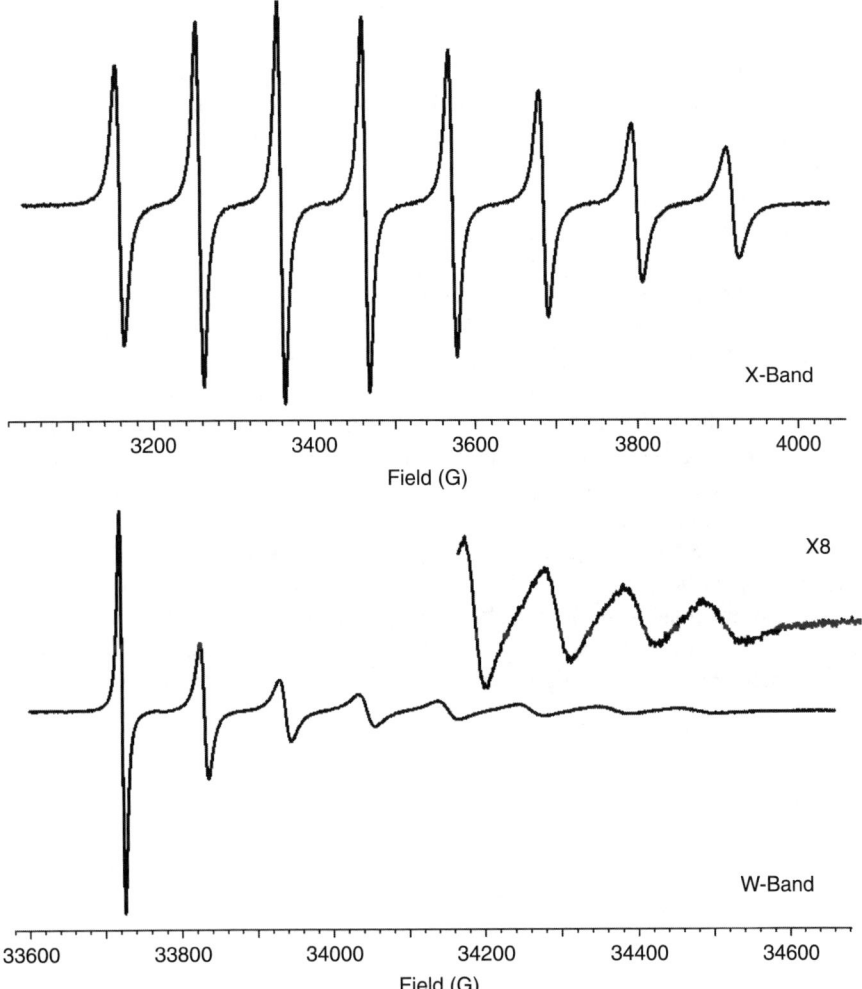

Fig. 2.23 The vanadyl complex in solution at room temperature measured in X-band and W-band. In W-band the motional averaging is less complete than in X-band, leading to significant line broadening.

becomes more equidistant: 106 G for the two low field lines and 110 G for the two high field lines.

2.2.6 CW-ENDOR

The interaction of the electron spin with nearby nuclear spins is a key element in structural analysis. It provides information about the local environment in a range of a few angströms around the electron spin. However, as the electron spin usually interacts with many nuclear spins, the EPR spectrum becomes very complicated (the number of EPR lines increases rapidly with the number of coupled nuclear spins, see Chapter 1). Furthermore, weak hyperfine coupling may just result in a broadening of the EPR line (one of the main reasons for inhomogeneous broadening), making it impossible to resolve the coupling constant in the EPR spectrum. The situation simplifies if considering the spectrum of the nuclear spins themselves. As each nuclear spin belonging to a group of equivalent nuclei couples only to a single electron spin (nuclear–nuclear spin coupling is too weak to be seen in EPR and ENDOR), the corresponding nuclear spin spectrum consists of just two lines separated by the hyperfine coupling constant (for $S = 1/2$). In fact, this is the NMR spectrum that cannot be detected with a standard NMR spectrometer because of the large coupling constants. Instead, the NMR spectrum is measured via the EPR transitions in an ENDOR experiment (see §1.1.3). In a double resonance experiment two resonance conditions have to be fulfilled simultaneously, thus reducing the number of observable transitions and thereby increasing the resolution. This increase in resolution, however, has to be paid by a loss in sensitivity. Only a few percent of the EPR signal contributes to the ENDOR spectrum (Fig. 2.24).

A prerequisite for a successful ENDOR experiment is the saturation of the EPR transition. A further necessity is that either the electron and nuclear spin relaxation rates are comparable or that cross relaxation via the forbidden transitions is effective [18]. Because all relaxation rates are strongly temperature dependent, an ENDOR setup usually requires variable temperature equipment. The critical balance of the

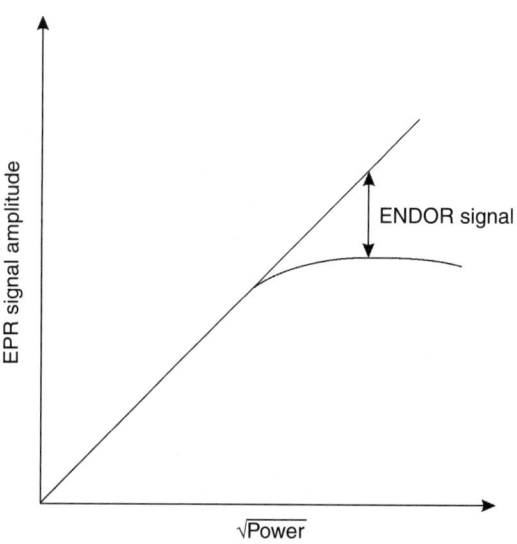

Fig. 2.24 The EPR signal amplitude dependence on microwave power. The maximum possible ENDOR amplitude corresponds to the difference between saturated and nonsaturated EPR signals.

relaxation rates is also the reason that the saturation level of the EPR transition necessary for successful ENDOR is unknown and has to be determined experimentally by means of the ENDOR signal itself.

Technically, the most demanding part of the ENDOR equipment is the resonator. The microwave structure has to be combined with an RF coil to generate the B_2 field that drives the ENDOR transition (Fig. 2.25). The required RF range depends on the nuclear spin studied and the coupling constants. In X-Band the ^{1}H Larmor frequency is about 14.5 MHz, and coupling constants can reach values of several tens of megahertz. However, even in X-band, there are cases with ENDOR transitions well above 100 MHz, for example, Mn^{2+}. Due to this large RF span, the coils are usually designed as a broadband structure allowing a sweep over the full range. A certain B_2 amplitude is required to drive the ENDOR transition. For an X-Band ^{1}H ENDOR transition at 14.5 MHz a typical B_2 amplitude in the rotating frame is around 10 G generated by a 200 W RF amplifier. A measure for the driving force experienced by the nuclear spin is the Larmor frequency in the rotating frame (also called the *Rabi frequency*), which is given by $\omega_R = \gamma_I B_2$. Thus, nuclear spins with lower γ_I require correspondingly higher B_2 to be driven at the same rate.

The local field generated by the electron spin at the nuclear spin, which is hyperfine coupling, leads to an effective Rabi frequency for the two ENDOR transitions:

$$\omega_{\text{eff}} = \gamma_I |1 \pm a/2\nu_I| B_2 \qquad (2.10)$$

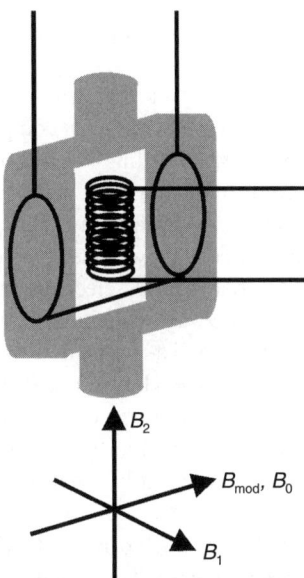

Fig. 2.25 The schematic of an ENDOR resonator. An RF coil is placed inside the microwave cavity as close as possible around the sample. In this setup the coil is mounted on a quartz dewar to allow variable temperature operation. Here, B_1 and B_2 are perpendicular to B_0.

TABLE 2.4 Selected Nuclear Spin Larmor Frequencies in Different Microwave Bands

	X-Band, 3.5 kG	Q-Band, 12.2 kG	W-Band, 33.6 kG
^{14}N (MHz)	1.07	3.73	10.3
^{2}H (MHz)	2.28	7.95	21.9
^{13}C (MHz)	3.74	13.0	35.9
^{31}P (MHz)	6.04	21.0	58.0
^{1}H (MHz)	14.9	51.9	143.0

Note that this equation is equivalent to Equations 1.44 and 1.45 in Chapter 1. The plus sign holds for the high frequency and the minus sign for the low frequency ENDOR line. Thus, the low frequency ENDOR transition requires a higher B_2 than the high frequency transition to be driven at the same rate. To some extent this requirement is fulfilled by the broadband RF coil that generates an increasing B_2 with decreasing frequency. However, this mechanism is not perfect and low frequency ENDOR lines are likely to be driven at a lower rate and therefore suffer from reduced amplitudes. An ENDOR line approaching zero frequency ($a = 2\nu_I$) would require infinite B_2 to be driven at the same rate as the corresponding high frequency line.

An approach to access even very low frequency nuclei is to perform ENDOR at higher microwave frequencies (an alternative pulsed method is electron spin echo envelope modulation, see Chapter 5). The high frequency has the further advantage that the spectra of different types of nuclear spins are better separated [19]. Several examples of ENDOR at high frequency are given in Chapter 12.

The Larmor frequencies of some common ENDOR nuclei at different microwave bands are summarized in Table 2.4.

The separation of different types of nuclear spins is illustrated in Fig. 2.26 by pulsed ENDOR spectra of ^{1}H, ^{31}P, and ^{195}Pt in a Cu complex at X-, Q-, and W-bands. In X-band all spectra overlap and the ^{31}P lines can easily be misinterpreted as ^{1}H lines. With the increase in Larmor frequency at Q- and W-bands the ENDOR spectra of the different nuclei are further apart from each other and become well separated.

To perform an ENDOR experiment the magnetic field is positioned on an EPR line of the spectrum (for a start one would select the most intense line), and then the EPR signal amplitude is recorded as a function of a swept RF. As in EPR, in CW-ENDOR a modulation frequency and a lock-in detector are used. However, instead of modulating the magnetic field, now the RF is modulated. One practical consequence is that the ENDOR effect is observed at the zero crossing of the EPR line because this is the position for the maximum EPR signal in the absence of field modulation.

Various modulation schemes are used in ENDOR today.

1. *Frequency modulation (FM)*: The RF driving the ENDOR transition is modulated in frequency, for example, an RF of 15 MHz oscillates between 14.9 and 15.1 MHz with a rate of 10 kHz. In this example the modulation depth is 200 kHz, the lock-in detection is at 10 kHz, and the derivative line shape is recorded. This method is advantageous for narrow ENDOR lines as in solution and single crystals.

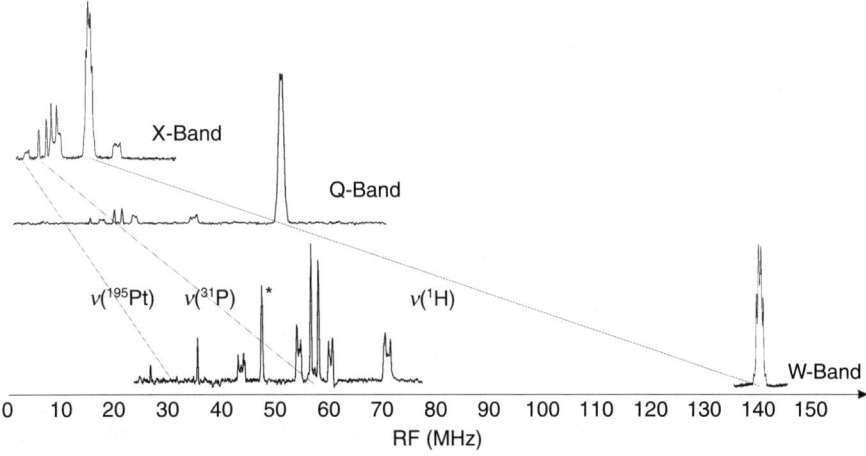

Fig. 2.26 Multifrequency pulsed ENDOR spectra in X-, Q-, and W-bands. The third harmonic of the ^{1}H line leads to an artifact in the W-band spectrum (H).

2. *Amplitude modulation (AM)*: The RF is switched ON and OFF at a rate corresponding to the lock-in frequency, typically ≤ 10 kHz. In this case the absorption line shape is recorded. The detection of AM is advantageous for broad lines like in powders (see Fig. 2.27).
3. *Stochastic ENDOR* [20]: From the modulation point of view this is also an AM technique, and the absorption line shape is recorded. However, instead of a linear RF sweep, the frequency between successive RF ON periods is now randomly selected within a given frequency band. A certain ENDOR transition is therefore hit at a much lower rate than the AM frequency. This technique is beneficial for extremely slowly relaxing spin systems.

The differences of the FM and AM techniques are illustrated in Fig. 2.27 with a powder sample of bisdiphenylene phenylallyl in polystyrene. The EPR spectrum is

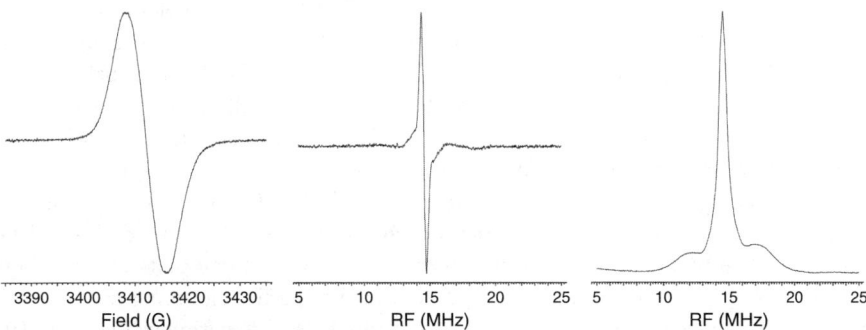

Fig. 2.27 (Left) EPR, (middle) FM-ENDOR, and (right) AM-ENDOR spectra of a bisdiphenylene phenylallyl (BDPA) powder sample.

76 BASIC EXPERIMENTAL METHODS IN CW-EPR

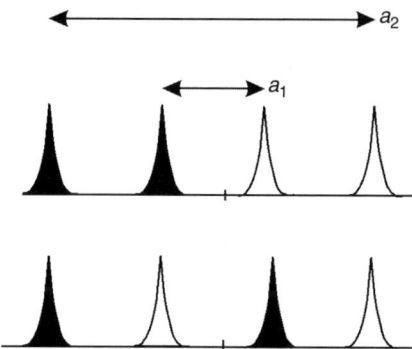

Fig. 2.28 A schematic representation of the ENDOR spectrum of two nuclear spins with hyperfine coupling of a_1 and a_2. The coupling constants have the (top) same sign or (bottom) different sign. Lines with the same shading belong to the same m_S state. If a_1 and a_2 have the same sign, both ENDOR lines from one m_S manifold are above (or below) the nuclear Larmor frequency. If a_1 and a_2 have opposite sign, one line is above and the other below the nuclear Larmor frequency.

a single inhomogeneous line with a width of about 8 G. In the FM recorded ENDOR spectrum the broad wings are hardly visible, but they are easily seen in the AM spectrum. Both spectra are dominated by the strong matrix line at the nuclear Larmor frequency.

In addition to the basic ENDOR experiment, there are two more ENDOR related techniques that further enhance the resolution. These are electron–nuclear–nuclear triple resonance (TRIPLE; special and general) and ENDOR induced EPR (EIE). In a general TRIPLE experiment a second, unmodulated static RF (called the pump frequency) is applied to one ENDOR transition and then the ENDOR spectrum is recorded. The additional information gained from a TRIPLE experiment is the relative sign of different hyperfine coupling constants as illustrated in Fig. 2.28. Further, the ENDOR spectra of different species can be separated.

In a general TRIPLE experiment the ENDOR lines are grouped according to the m_S state from which they originate. The pump RF set on one ENDOR line saturates this transition. Thus, when the other RF is swept over the pump line no further change is observed; that is, in the TRIPLE spectrum the pump line is suppressed. All other ENDOR lines belonging to the same m_S state as the pump line are reduced in amplitude as well. However, the ENDOR lines in the other m_S state grow in amplitude. Therefore, the common sign of the amplitude change assigns the ENDOR lines to the same m_S state.

As an example consider the ENDOR and TRIPLE spectra of the PNT radical (Fig. 2.29). The EPR spectrum of this radical consists of 28 lines (see Figs. 2.11 and 2.12). The ENDOR spectrum, however, consists only of two pairs of lines originating from three and six equivalent protons with hyperfine coupling constants of 5.2 and 17.6 MHz, respectively. For the general TRIPLE experiment the static RF was applied at the high frequency line (pump line). In the resulting TRIPLE spectrum the pump line and the line at 12 MHz are reduced in amplitude. As only lines within

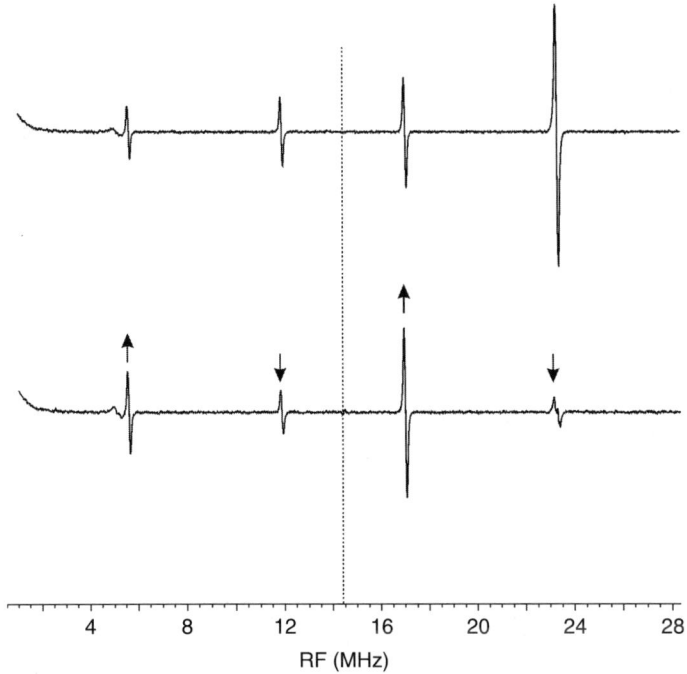

Fig. 2.29 (Top) ENDOR and (bottom) TRIPLE spectra of PNT. For the TRIPLE experiment the line at 23 MHz was pumped. The vertical bar indicates the ^{1}H Larmor frequency.

the same m_S manifold react in the same way and as these lines are above and below the nuclear Larmor frequency of 14.5 MHz, the two hyperfine coupling constants have opposite sign (the absolute sign cannot be established by TRIPLE and has to be found on the basis of theoretical considerations). The two lines at 5.5 and 17 MHz increase in amplitude and therefore belong to the other m_S state.

In special TRIPLE two RFs are swept simultaneously in such a way that both ENDOR lines of one coupling are hit at the same time. This experiment gives improved sensitivity and is less dependent on the balance of relaxation rates and therefore on temperature [18].

The EIE technique relates the ENDOR transitions to the EPR spectrum. It reveals the part of the EPR spectrum that is associated with one particular ENDOR line. A γ-irradiated single crystal of succinic acid serves as an example (Fig. 2.30). The ENDOR spectrum was recorded in the center of the EPR spectrum where many lines overlap. In the next step the RF is placed at one particular ENDOR line and the EPR spectrum is measured again through the detection of this ENDOR line. The EIE spectrum then only shows this part of the EPR spectrum that gives rise to the observed ENDOR line. In this way overlapping EPR spectra of different species and crystal sites can be disentangled (see also Chapter 9).

As the EIE spectrum is recorded without field modulation but by frequency modulating the ENDOR line, the absorption EPR spectrum is detected.

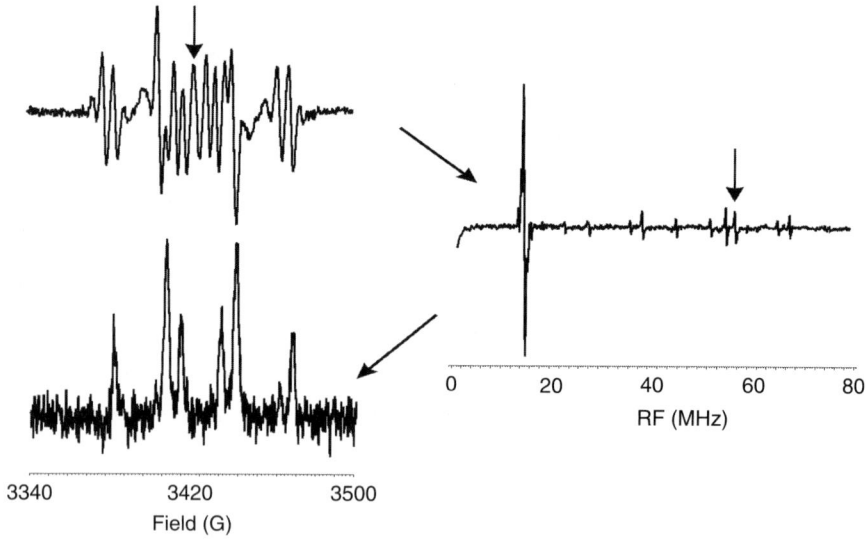

Fig. 2.30 (Left, top) EPR, (right) ENDOR, and (left, bottom) EIE spectra of a succinic acid single crystal. The ENDOR spectrum was recorded in the center of the EPR spectrum, and the EIE spectrum was measured by monitoring the ENDOR line at 58 MHz.

2.2.7 EPR Imaging (EPRI)

The majority of EPR deals with the microscopic environment of the electron spin. In contrast, EPRI observes sample heterogeneities on a macroscopic scale in the millimeter to micron range. Today EPRI is used in biomedical research of small animals and in material science. There are two main categories of imaging experiments: the detection of the spatial distribution of a paramagnetic species and the spatial resolution of the EPR spectrum properties, for example, the linewidth. The second category is also called spectral–spatial imaging or localized spectroscopy. How to achieve spatial resolution can be easily understood by means of the illustration in Fig. 2.31.

Consider an identical sample in two tubes in the resonator and the resonator being placed in the center of the magnet. In a conventional EPR experiment there is a homogeneous magnetic field and both samples, although separated in space, always experience the same field and their spectra will perfectly overlap. If we now generate a static magnetic field gradient G_x in the X direction the two samples will experience different field values. As a result of the gradient the spectra of the two tubes will show up at different positions on the field axis in a field sweep recorded spectrum. For a known gradient (G) the spectrum separation (ΔB) converts into a spatial distance of $d = \Delta B/G$. From this illustration one can also deduce an estimate about the possible spatial resolution that can be achieved. In order to distinguish two lines they should be separated by about their peak–peak linewidth. The minimum spatial distance (d_{min}) is resolved with the highest possible gradient and depends on the EPR

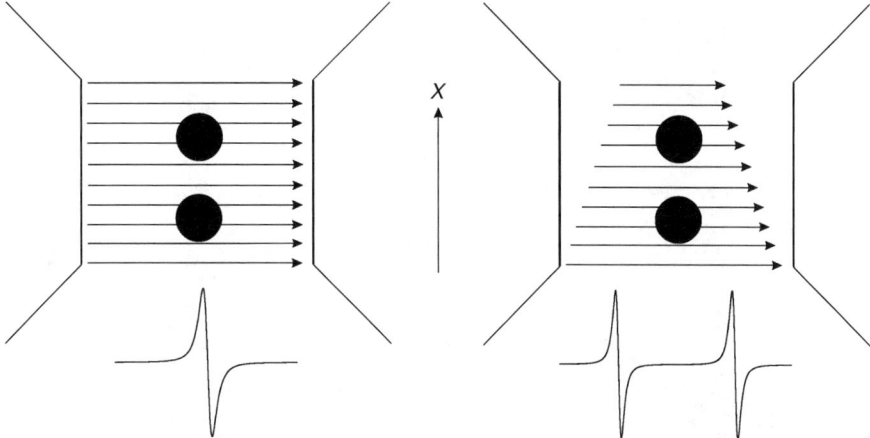

Fig. 2.31 Position encoding by static field gradient in the X direction. The arrows indicate the field strength and direction. (Left) In a homogeneous field the spectra of both samples perfectly overlap. (Right) A superimposed field gradient in the X direction separates the spectra on the field axis.

linewidth as $d_{min} \approx \Delta B_{pp}/G_{max}$. Gradients up to several hundred gauss/centimeter can be achieved with water cooled coils. We note that a gradient in the X direction is actually a change of B_z in the X direction, that is, $G_x = \delta B_z/\delta x$. In order to obtain a full three-dimensional picture, a series of spectrum profiles has to be recorded with different gradient directions. The spectrum profiles are then used to reconstruct the actual object by a backprojection algorithm [21].

A 19 bore AquaX tube filled with a Trityl solution is provided as an example (Fig. 2.32). The Trityl EPR linewidth is less than 100 mG, and a gradient strength of 40 G/cm is sufficient to resolve all tubes. The tube inner diameter is 0.5 mm, and the spacing is 0.1 mm.

As described earlier, the resolution depends on the gradient strength and linewidth. However, there is also an important dependence on sensitivity. If a certain volume element (voxel) should be resolved, then there must be at least that many spins in this voxel that correspond to the detection sensitivity of the instrument. As the absolute sensitivity increases with the microwave frequency, it is desirable to go to a high microwave frequency if resolution is a major concern. The application of EPRI in biomedical research, however, necessitates a low microwave frequency (≤ 1 GHz) to achieve sufficient penetration depth and because of the better compatibility of lossy samples with the microwave resonator. A major field of EPRI in the biomedical field is the measurement of oxygen in tissues. The linewidth of spin probes like Trityl or charcoal is sensitive to the partial oxygen pressure. These spin probes are introduced into a tissue and the spatial dependence of the linewidth is measured in a spectral–spatial imaging experiment. By means of a calibration, the linewidth is then converted into the oxygen partial pressure. A recent survey provided extensive coverage of the biomedical field [22]. For EPRI in material science a microwave

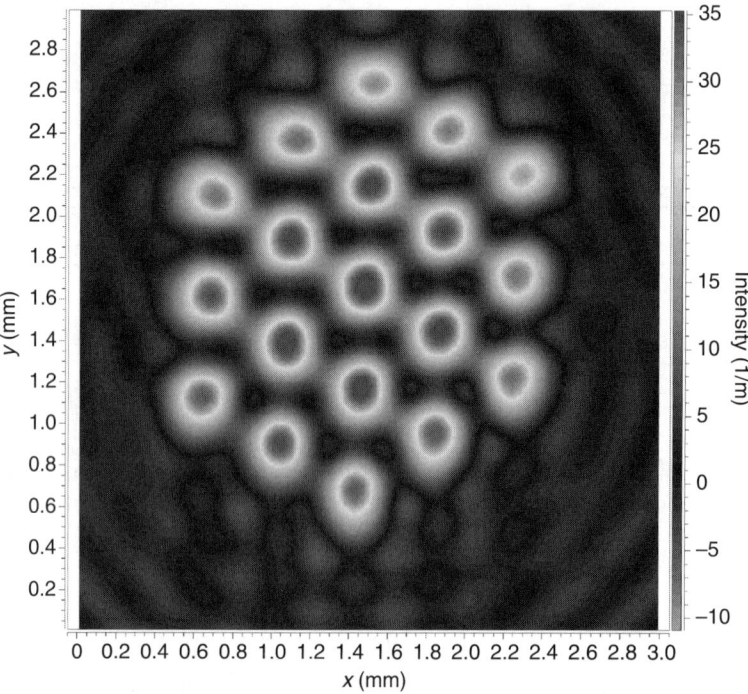

Fig. 2.32 A two-dimensional spatial image of a 19 bore AquaX filled with a Trityl solution.

frequency of X-band and higher can be used. Typical examples in this research area are diffusion processes, the spatial distribution and temporal evolution of radicals in polymers, and EPR dating of fossils [23–25].

ACKNOWLEDGMENT

I thank my colleagues Drs. P. Carl, A. Kamlowski, and R. T. Weber and Ms. S. Kummer for help in generating the schematics and spectra presented in this chapter. These data were collected in the Bruker Biospin EPR application labs over many years and have largely not been published elsewhere.

REFERENCES

1. Poole, C. P. *Electron Spin Resonance*, 2nd ed.; Dover Publications: New York, **1983**.
2. Hoff, A. J., Ed. *Advanced EPR*; Elsevier: Amsterdam, **1989**.
3. Hyde, J. S.; Froncisz, W. Loop gap resonators. In *Advanced EPR*; Hoff, A. J., Ed.; Elsevier: Amsterdam, **1989**.

4. Grinberg, O. Y.; Berliner, L. J. *Very High Frequency (VHF) ESR/EPR*; Biological Magnetic Resonance, Vol. 22; Kluwer Academic/Plenum: New York, **2004**.
5. Schneider, F.; Plato, M. *Elektronenspin-Resonanz*; Thiemig-Taschenbücher: München, **1971**.
6. Schmalbein, D.; Maresch, G. G.; Kamlowski, A.; Höfer, P. *Appl. Magn. Reson.* **1999**, *16*, 185.
7. Jaworski, M.; Checinski, K.; Bujnowski, W.; Porowski, S. *Rev. Sci. Instrum.* **1978**, *49*, 383.
8. Atherton, N. M. *Principles of Electron Spin Resonance*; Ellis Horwood PTR Prentice Hall: London, **1993**.
9. Bard, A. J.; McKinney, T. M.; Goldberg, I. B. EPR and Electrochemistry. In *Foundations of Modern EPR*; Eaton, G. R.; Eaton, S. S.; Salikhov, K. M., Eds.; World Scientific: Singapore, **1998**.
10. Greulich-Weber, S. *Mater. Sci. Forum* **1994**, *143–147*, 1337.
11. Rupp, H.; Rao, K. K.; Hall, D. O.; Cammack, R. *Biochim. Biophys. Acta* **1978**, *537*, 255.
12. Carl, P. *EPR Application Note*; Bruker BioSpin GmbH, Rheinstetten, Germany.
13. Columbus, L.; Hubbell, W. L. *Trends Biochem. Sci.* **2002**, *7*, 288.
14. Altenbach, C.; Hubbell, W. L. Site-directed spin labelling: A strategy for determination of structure and dynamics of proteins. In *Foundations of Modern EPR*; Eaton, G. R.; Eaton, S. S.; Salikhov, K. M., Eds.; World Scientific: Singapore, **1998**.
15. MacLauchlan, K. A. Time-resolved EPR. In *Advanced EPR*; Hoff, A. J., Ed.; Elsevier: Amsterdam, **1989**.
16. Stehlik, D.; Bock, C. H.; Thurnauer, M. C. Transient EPR-spectroscopy of photo induced electronic spin states in rigid matrices. In *Advanced EPR*; Hoff, A. J., Ed.; Elsevier: Amsterdam, **1989**.
17. Höfer, P.; Maresch, G. G.; Schmalbein, D.; Holczer, K. *Bruker Report 142/96*; Bruker Biospin GmbH, Rheinstetten, Germany, p.15.
18. Kurreck, H.; Kirste, B.; Lubitz, W. *Electron Nuclear Double Resonance Spectroscopy of Radicals in Solution*; VCH: Weinheim, **1988**.
19. Groenen, E. J. J.; Schmidt, J. High-frequency EPR, ESEEM and ENDOR studies of paramagnetic centers in single crystalline materials. In *Very High Frequency (VHF) ESR/EPR*; Grinberg, O. Y.; Berliner, L. J., Eds.; Biological Magnetic Resonance, Vol. 22; Kluwer Academic/Plenum: New York, **2004**.
20. Brueggemann, W.; Niklas, J. R. *J. Magn. Reson.* **1994**, *A108*, 25.
21. Eaton, G. R.; Eaton, S. S.; Ohno, K. *EPR Imaging and In Vivo EPR*; CRC Press: Boca Raton, FL, **1991**.
22. Gallez, B., Ed. *NMR Biomed.* **2004**, *17*(5) [Special Issue].
23. Watt, G. A.; Newton, M. E.; Baker, J. *Diamond Relat. Mater.* **2001**, *10*, 1681.
24. Lucarini, M.; Pedulli, G. F.; Motyakin, M. V.; Schlick, S. *Progr. Polym. Sci.* **2003**, *28*, 331.
25. Binet, L.; Gourier, D.; Skrzypczak-Bonduelle, A.; Delpou, O.; Derenne, S. *Lunar Planet. Sci.* **2007**, *38*.

BIBLIOGRAPHY

General Literature with Technical Background

Bruker BioSpin GmbH. *EMX and ELEXSYS User's Manuals*; Bruker BioSpin Corporation Billerica, MA, USA.

Eaton, G. R.; Eaton, S. S.; Salikhov, K. M., Eds. *Foundations of Modern EPR*; Singapore: World Scientific, **1998**.

Poole, C. P. *Electron Spin Resonance*, 2nd ed.; Dover Publications: New York, **1983**.

Schneider, F.; Plato, M. *Elektronenspin-Resonanz*; Thiemig-Taschenbücher: München, **1971**.

ENDOR

Dorio, M. M.; Freed, J. H. *Multiple Electron Resonance Spectroscopy*; Plenum Press: New York, **1979**.

Kevan, L.; Kispert, L. D. *Electron Spin Double Resonance Spectroscopy*; Wiley: New York, **1976**.

Kurreck, H.; Kirste, B.; Lubitz, W. *Electron Nuclear Double Resonance Spectroscopy of Radicals in Solution*; Weinheim: VCH, **1988**.

Imaging

Eaton, G. R.; Eaton, S. S.; Ohno, K. *EPR Imaging and In Vivo EPR*; CRC Press: Boca Raton, FL, **1991**.

Gallez, B., Ed. *NMR Biomed*. **2004**, *17*(5) [Special Issue].

3 What Can Be Studied with Electron Paramagnetic Resonance?

MARINA BRUSTOLON

Dipartimento di Scienze Chimiche, Università di Padova, Via Marzolo, 1, 35131 Padova, Italy

3.1 INTRODUCTION

Electron paramagnetic resonance (EPR) spectroscopy detects only paramagnetic species (PS), which are species with unpaired electrons. Because of its selectivity, an EPR study has the main advantage of providing insights into the nature of the paramagnetic center while also revealing detailed information on its environment and the dynamical processes in which it is involved. This chapter will describe and provide a few examples of the various PS that are studied using EPR methods for various experimental purposes.

Part of this chapter will be dedicated to the techniques and procedures used for production and stabilization of PS, which are often not persistent. It must be pointed out, however, that the very lability of PS can be exploited experimentally to study dynamic processes by means of a time resolved EPR study. Moreover, it should also be considered that in some systems, for example, in transition metal ions or radicals trapped in solids, PS are naturally present and stable. On the basis of these different properties, the types of the EPR investigations can be classified as follows: case 1, a PS (radical, triplet, etc.) is obtained from diamagnetic precursors to study its properties; case 2, a stable PS (radical, biradical) is introduced or generated in a system as a probe for testing the environment; case 3, a native PS (paramagnetic transition metal ions, radicals trapped in solids) is used to obtain information on the system; and case 4, a transient PS participating in a process is studied by a time resolved approach (paramagnetic states in photosynthesis, radicals produced in chemical reactions).

It may help the reader to consider EPR studies in a historical perspective. Briefly, from the first diffusion of EPR in the 1960s following the production of commercial spectrometers up to the 1980s, the emphasis was mainly on the production and

---

Electron Paramagnetic Resonance. Edited by Brustolon and Giamello
Copyright © 2009 John Wiley & Sons, Inc.

identification of new PS, and on the use of persistent radicals for testing the environment. The magnetic parameters obtained by CW-EPR or ENDOR (g and hyperfine tensors, fine tensors, quadrupolar tensors) were compared with theoretical models of electron distribution. The EPR spectral profiles of persistent radicals, used as spin labels attached to macromolecules or as spin probes, were simulated giving information on the intermolecular or intramolecular dynamics. The reader is referred to the two most authoritative textbooks of those years (Atherton and Weil, Bolton, and Wertz, in their various editions, see Bibliography) that illustrate all of the essential principles of the method and provide several examples of possible applications.

From the 1980s onward, the onset and diffusion of new EPR methods, such as time resolved, pulsed, and high field EPR, together with the development of suitable theoretical frameworks for the interpretation of the results, have fostered the entrance of EPR into the fields of biology, medicine, and material science. Indeed, the new tools, in conjunction with the knowledge accumulated in the previous 20–30 years, have paved the way to the investigation of systems of high structural or dynamic complexity. Due to its specificity for PS, EPR spectroscopy is particularly well suited for complex systems, because its selectivity gives detailed information on a specific part of the system regardless of its overall size.

The current applications of EPR have become far too complex and diverse to be exhaustively treated in a single textbook. Hence, only some of the most relevant aspects of EPR investigations will be described in detail in the next chapters. The main aim of the present chapter is to draw a blueprint of the EPR field, pointing out the relevance of the different branches and in some cases their foreseeable future development. The chemical nature and electron spin state of the PS is the first subject.

3.2 ORGANIC RADICALS

Organic radicals are generally classified following two criteria: the symmetry of the distribution of the unpaired electron (σ or π radicals) and their charge (neutral, anion, and cation radicals). These properties will be considered in Chapter 4. Why and how radicals are produced is explained here.

The lifetime of radicals is generally short compared to that of diamagnetic molecules. Many possible reactions can threaten their life, giving as a final product a more thermodynamically stable molecule, with a pair of electrons in a bonding orbital. A typical radical reaction is the combination of two radicals:

$$R\bullet + R'\bullet \longrightarrow R-R' \quad (3.1)$$

When two radicals with opposite spins collide in solution, this process can be extremely fast. However, the reaction rate is also strongly dependent on the conformation of the radicals. The kinetic stability, or persistence, is often more important than the thermodynamic stability in determining their lifetime. The persistence depends mainly on the steric protection of the unpaired electron. Moreover, we must distinguish between radicals obtained in a solid matrix (crystalline, glassy, or composite) and radicals obtained in liquid solution. In the former situation radicals are trapped and are therefore much more persistent. Radicals in solids are treated in the following section.

3.2.1 Radicals in Solution

Some of the methods for obtaining radicals in solution listed here will be described more extensively in Chapter 4.

The interactions between a solvent with a high dielectric constant, for example, water, and the electric component of microwaves results in losses of the available power and may make the detection of the spectrum impossible. As a consequence, the choice of solvent for an EPR experiment must always take its dielectric constant into account. Water and other polar solvents can be used with particular sample cells (see §2.1.1).

3.2.1.1 Persistent Radicals. The EPR spectra of radicals depend on the environment, and persistent radicals are typically used as probes for testing it. A typical example is that of nitroxide radicals, also called nitroxyl radicals (see compounds **1–3** in Scheme 3.1). Another type of persistent radicals similar to nitroxides are nitronyl nitroxides (see compound **4** in Scheme 3.1). The reader can see how the EPR line shape of a nitronyl nitroxide radical depends on the viscosity of the solvent in Chapter 7 (Fig. 7.12).

The names nitroxide and nitroxyl are not recommended by IUPAC, which instead recommends the general name aminoxyl. However, the traditional names are still used. Moreover, the most popular nitroxides have common names: 2,2,6,6-tetramethyl-piperidine-1-oxyl (**1**) is known as TEMPO; 2,2,6,6-tetramethyl-4-piperidone-1-oxyl

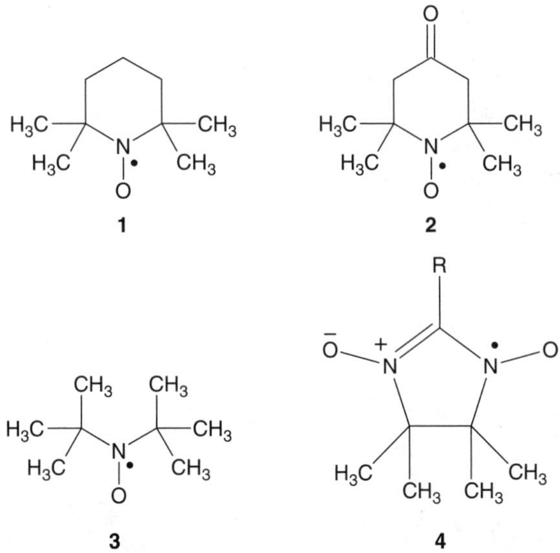

Scheme 3.1

(**2**) is known as TEMPONE; and di-*tert*-butylamine-*N*-oxyl (**3**) is known as di-*tert*-butyl-nitroxide.

These alkylnitroxides are remarkably persistent because the unpaired electron is nested between the four methyl groups. Therefore, they can be involved in various reactions without the unpaired electron being affected. Several hundred nitroxyl radicals have been synthesized to be used either as *spin probes* (when dispersed in an environment) or as *spin labels* (when chemically attached to a biological molecule as a protein). Because the EPR spectra of nitroxides depend on both the polarity of the medium and the mobility of the radicals, they can provide valuable information on the microenvironment of the radicals. A large part of EPR studies in biology (science of colloids and interfaces, soft matter) is based on nitroxide radicals. In the field of material science persistent nitronyl nitroxides are frequently used. They behave as bidentate ligands for various transition and rare earth metal ions, giving composite materials with new magnetic properties.

The field of EPR studies by spin probes is too wide to be more than outlined here; some examples and references are given in Chapters 4, 10, and 11.

The so-called spin trapping technique may also give rise to stable nitroxide radicals upon reaction of short-lived radicals with a suitable molecule called a spin trap. Details can be found in Chapter 8.

Other examples of persistent radicals are provided in Scheme 3.2. Radical **5**, 2,2-diphenyl-1-picrylhydrazyl, is traditionally used as a standard for *g* factor measurements. Radical **6** is a triarylmethyl radical, where T is a trityl substituent. Its EPR spectrum consists of a single narrow line, which makes it useful for a variety of applications.

3.2.1.2 Transient Radicals.

In the last 60 years, thousands of little persistent radicals in solution have been generated through different methods and studied by EPR. From their spectra, fully interpreted by means of simulation programs (see Chapters 4 and 7), two types of information can be obtained: the magnetic parameters of the spin Hamiltonian, which is informative about the radicals' spin distribution and conformation, can be derived from the resonance fields of the lines; and the spectral profile provides details of their motions in solution. The comparison between experimental and calculated spin Hamiltonian parameters provides extremely important and otherwise unachievable information on the properties of electronic distribution in molecules. Moreover, the success in calculating the *g* value and the

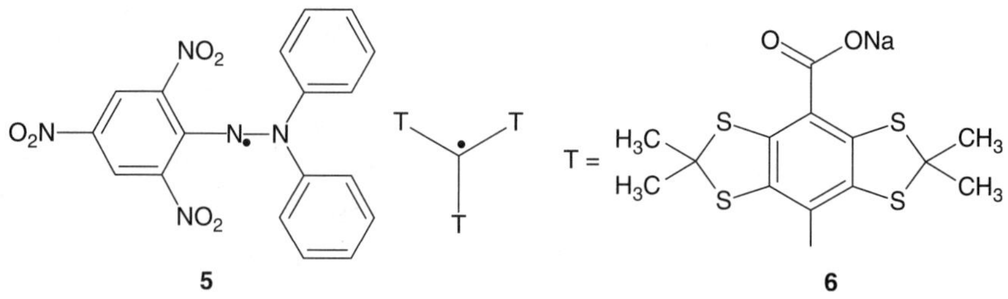

Scheme 3.2

hyperfine interaction tensors has proved a driving force for the development of computational methods (see Chapters 4 and 7). The strategies for the production of transient radicals and the method of radical formation depend on the type of EPR method that is used.

For the commonly used CW-EPR it is necessary to achieve a *steady-state concentration* of radicals, lasting at least several seconds, large enough to allow the recording of a spectrum without distortions. Some possible methods for obtaining the radicals from the parent compounds are described briefly.

In contrast, time resolved EPR methods allow the study of radicals with a much shorter lifetime down to hundreds of nanoseconds. Because they are *time resolved*, the spectrum is obtained after a fixed delay after the radical birth; therefore, it is necessary to know the time when the radical is born. This is commonly achieved by generating them photochemically with a laser pulse.

3.2.1.3 Strategies for CW-EPR.
In order to obtain a concentration of radicals constant in time, one must either find a continuous generation method so that newly formed radicals replace those disappearing or produce the radicals in an environment where radical reactions are inhibited as much as possible.

A lasting appropriate steady-state concentration of radicals can be produced directly in the microwave cavity, or near to it, through *in situ* photolysis and radiolysis, electrochemical methods, and flow systems.

3.2.1.3.1 Steady-State Photolysis and Radiolysis.
UV irradiation can break a relatively weak bond giving one or two neutral radicals. Photolysis of solutions of peroxides (frequently di-*tert*-butyl peroxide) has been widely used to initiate radical reactions. The success of the method depends on the efficiency of the photolysis coupled with the fact that alkoxyl radicals react very rapidly with many organic compounds via abstraction of a hydrogen atom. In most cases, a steady-state radical concentration greater than the EPR detection threshold ($\sim 10^{-7} M$) can be produced:

$$\begin{aligned} t\text{-BuOO}t\text{-Bu} &\xrightarrow{h\nu} 2t\text{-BuO}\bullet \\ \text{RH} + t\text{-BuO}\bullet &\longrightarrow \text{R}\bullet + t\text{-BuOH} \end{aligned} \qquad (3.2)$$

Irradiation of liquid hydrocarbons, alcohols, and other similar compounds with accelerated high energy electrons or with X or γ rays directly in the EPR cavity may result in the homolytic cleavage of C—H bonds with formation of the corresponding alkyl radical. Electron accelerators and sources of ionizing radiation are not normally available in EPR laboratories, so this technique is seldom used. However, in the early years of EPR it opened the route to many alkyl radicals, simple and fundamental such as methyl, ethyl, or allyl.

Primary and secondary alkyl radicals are very reactive and little persistent. If a steady-state concentration large enough to allow their detection cannot be attained, these species can be "trapped" in the form of a more stable radical adduct by reaction with a suitable diamagnetic reagent (spin trap). (Spin trapping is treated extensively in Chapter 8.)

3.2.1.3.2 Electrochemical Methods. The reduction or oxidation of molecules giving paramagnetic charged species can be performed by electrolysis either directly in the EPR cavity (*in situ*) or outside the cavity (external generation). In the first case a steady-state concentration is achieved, and in the second case the relatively persistent radicals produced in an external cell must be transferred to the EPR cavity with suitable methods (rapid flow, withdraw under inert atmosphere, etc.) [1].

An example of a typical electrochemical cell for EPR is that reported by Maki and Geske in the first reported example of electrochemical generation of radicals for EPR [2]. Various types of cells for *in situ* electrolysis are reported in Reference 3, together with experimental details.

The voltage to be applied to the cell to produce the desired charged species can be obtained by a preliminary study of the voltammetric behavior of the parent compound. The cell voltage V_c is varied, and the current i through the cell is measured. The current rises sharply at the half-wave potentials corresponding to transformations from one species to another. Cyclic voltammetry allows the testing of the reversibility of the reaction because for reversible reactions, if the potential is reversed after a redox step, the previous species is obtained again. A striking example is reported in Fig. 3.1 showing the cyclic and differential pulse voltammograms of fullerene C_{60}, with the six reduction steps from C_{60}^- to C_{60}^{6-}. This molecule has a high electron-accepting ability thanks to its three antibonding orbitals, which are degenerate, empty, and low lying in energy.

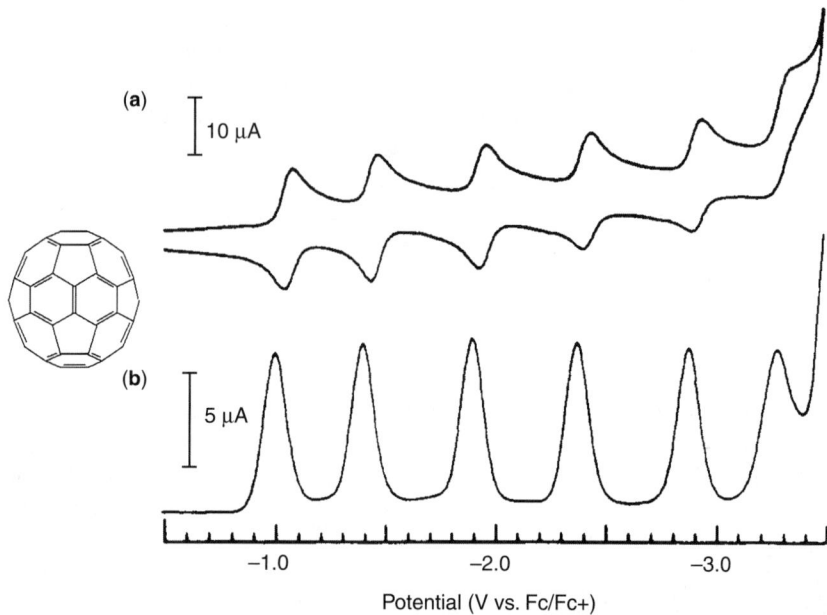

Fig. 3.1 The reduction of C_{60} in CH_3CN/toluene at $-10°C$ using (a) cyclic voltammetry, showing the reversibility of the reduction steps, and (b) differential pulse voltammetry. The six reduction steps from C_{60} to C_{60}^{6-} are evident. Reprinted from Xie, Q.; Perez-Cordero, E.; Echegoyen, L. *J. Am. Chem. Soc.* **1992**, *114*, 3978. Copyright 1992 American Chemical Society. Reprinted with permission.

The mono-, di-, and trianion were obtained by external generation and then transferred to the EPR spectrometer [4]. The EPR spectra showed that C_{60}^- is a radical as expected, C_{60}^{2-} is in a nonparamagnetic singlet state ($S = 0$), C_{60}^{3-} is again in a doublet state (a single unpaired electron), and C_{60}^{4-} is again in a singlet state.

3.2.1.3.3 Flow Systems. A variety of redox reactions has proved suitable, in conjunction with a relatively simple continuous flow and mixing system, for producing radicals in aqueous solutions. Typically, the two solutions of the redox reagents, at flow rates in the range of $1-5 \, \text{cm}^3 \, \text{s}^{-1}$, are mixed ~ 50 ms before passage of the combined solution through the cavity of the spectrometer.

$$\text{Ti(III)} + \text{H}_2\text{O}_2 \longrightarrow \text{Ti(IV)} + {}^\bullet\text{OH} + \text{OH}^-$$
$$^\bullet\text{OH} + \text{RH} \longrightarrow \text{H}_2\text{O} + \text{R}^\bullet \tag{3.3}$$

A number of other redox systems and primary radicals precursors have been used, such as Fe(II)/Fe(III) and NH$_2$OH [5].

Let us now move to the methods for increasing the lifetime of nonpersistent radicals by producing them in stabilizing environments: radical anions or radical cations.

3.2.1.3.4 Chemical Reduction in Ethereal Solvents. The most used method for producing radical anions of aromatic organic compounds is chemical reduction with an alkaline metal. A diluted solution of the parent compound dissolved in a deoxygenated and dehydrated ethereal solvent (frequently tetrahydrofuran or dimethoxyethane) is brought into contact with a mirror of an alkaline metal (more frequently sodium and potassium) under high vacuum (10^{-4} mbar). This reduction procedure and the properties of the π radical anions obtained, which are frequently detected by EPR as ion pairs with the alkaline cation, are described in detail in Chapter 4. The persistency of these radical anions varies, depending on the extension of their π distribution and on the conformation of the species. However, in any case it is generally on the order of days or months in samples sealed under vacuum, whereas traces of oxygen or water immediately destroy the radicals.

3.2.1.3.5 Chemical Oxidation. Several methods are used to oxidize a neutral compound to its radical cation. A simple method to obtain aromatic radical cations is that of dissolving the parent hydrocarbons in concentrated sulfuric acid. The detailed mechanism of this reaction is not fully understood. Other protic acids such as trifluroacetic acid can be used or nonprotic acids such as AlCl$_3$ or SbCl$_5$. More details on these methods are given in Section 4.2.3.1.

Note that EPR spectroscopists and theoretical chemists have done much work in the last 40 years on the electronic properties of radical anions and cations of aromatic compounds. The comparison of experimental and calculated hyperfine constants, mainly of protons and ^{13}C, has been an important driving force for a better understanding of the properties of aromatic compounds and for a progressive improvement of the methods of calculation (see, e.g., [6]).

3.2.1.4 Photolysis and Time Resolved EPR. Time resolved (§2.2.3) and pulsed EPR (Chapter 5) both allow the detection of a PS a few hundreds of nanoseconds after its birth following photoexcitation. The experiment consists of a repetitive process of production of the PS at time zero, generally with a laser pulse or a light flash, followed by a fast detection by EPR after a delay fixed by the operator. These two spectroscopies are both *time resolved* in this sense. However, time resolved EPR is characterized by a continuous microwave irradiation as in CW-EPR; in the second one the microwaves are pulsed, and during the detection of the signal the radiation is off. As far as the radicals in solution are concerned, time resolved EPR spectroscopy has been used the most. This spectroscopy is indicated also with the name of the effect it put in evidence, that is, chemically induced dynamic electron polarization (CIDEP). The CIDEP spectra show lines at the same resonance fields as the CW-EPR, whereas the spectral profile is generally very different. Indeed, it can show lines in enhanced absorption and in emission, as the spin populations of the new born radicals produced by the UV irradiation can be largely different with respect to that at thermal equilibrium (see §4.11.2). The CIDEP phenomenon has been utilized in numerous studies to investigate the chemical reactivity of radicals, especially concerning the radical–radical recombination, as well as addition reactions of the radicals to other chemical compounds [7] (see also §3.3.1.2).

3.2.2 Organic Radicals in Solids

Observing the EPR spectra of radicals in a solid phase is useful because they show anisotropic magnetic interactions (Chapter 6) differently than in liquid solution (Chapter 4).

Radicals can be produced directly in various solid phases, as detailed in the next paragraphs. Note that a liquid solution of PS can be transformed in a solid sample by lowering the temperature and obtaining a frozen solution, where the PS are immobilized. To obtain samples with constant and reproducible properties the frozen solution should be in a glassy state, and care must be taken to avoid the crystallization of the solvent. In fact some solvents, such as toluene, become easily polycrystalline on lowering the temperature. These phases are a collection of oriented microcrystals of different sizes, and they can give spectra that are not reproducible. Other types of solid samples used as matrixes of radicals are single crystals or polycrystalline samples obtained by grinding the crystals.

3.2.2.1 Ionizing Radiation. Radicals can be produced in solids by ionizing radiation. Any type of high energy radiation or particle bombardment of a solid generally gives rise to paramagnetic products. In fact, according to its name this class of radiation ionizes the atoms and molecules, with ejection of an electron and formation of radical cations on the one hand and formation of radical anions on the other hand when the lost electrons are captured by other molecules. These radicals are called "primary radicals."

As far as organic radicals are concerned, the most studied substances have been carboxylic acids and their salts, amides, amino acids, peptides, proteins, and nucleic acids constituents. Hundreds of these substances, irradiated generally with γ or X rays, have been studied by EPR, irradiated generally with γ or X rays. By varying the temperatures of the irradiation and of the EPR measurement it is possible to observe the PS produced in various reaction steps. The primary radicals produced by oxidation and reduction are unstable for $T > 4K$, and generally they react to give alkyl radicals ("secondary radicals") that are very persistent.

EPR and ENDOR studies of irradiated single molecular crystals or polycrystalline samples of the parent compounds give information on their radiation chemistry and on the conformation and electronic properties of radicals. In particular, hundreds of magnetic tensors of organic free radicals were derived from EPR studies of γ- or X-irradiated molecular single crystals in different orientations in the magnetic field.

Box 3.1 provides an example of the variety of radicals produced by irradiation from the same parent compound: zinc acetate.

BOX 3.1

As an example of the type of stable radicals obtained by ionizing irradiation, let us consider the radicals observed in zinc acetate irradiated with X or γ radiation.

1. Irradiation at 4K and EPR at the same temperature: detection of the primary radicals and the methyl radical by dissociation of the primary oxidized radical [57]:

$$CH_3COO^- \longrightarrow CH_3CO\dot{O} + e^- \quad (1)$$

$$CH_3COO^- + e^- \longrightarrow CH_3CO\dot{O}^{2-} \quad (2)$$

$$CH_3COO \longrightarrow \dot{C}H_3 + CO_2 \quad (3)$$

2. The same sample quickly warmed at room temperature and then cooled again shows another radical, CO_2^-.
3. Irradiation at 77K and EPR at the same temperature: detection of the methyl radical.
4. By warming above 130K the latter radical disappears and another radical is formed [58]:

$$\dot{C}H_3 + CH_3COO^- \longrightarrow CH_4 + \dot{C}H_2COO^- \quad (4)$$

5. Finally, by irradiation at room temperature, in addition to the radical in Reaction 4, another radical is formed [59]:

$$CH_4 + \dot{C}H_2COO^- \longrightarrow \dot{C}HCH_3COO^-$$

Therefore, six different radicals are formed just from this very simple molecule by varying the irradiation temperature and the subsequent thermal history.

Due to their biological relevance, the radicals obtained by irradiation of DNA and protein components are of particular interest. Chapter 9 is completely dedicated to EPR and ENDOR studies of the radicals produced by ionizing radiation in these compounds.

The radical concentration as determined by double integration of the EPR spectra can be used to obtain information on the administered radiation dose. Substances particularly convenient for recording the radiation doses by means of their EPR spectra are called EPR dosimeters, and they are used in medicine and in the food industry. In contrast, if the annual dose of ionizing radiation for a material in a particular place is known, a quantitative EPR measure of the trapped paramagnetic centers can be used for dating bones, teeth, rocks, clays, and so forth (§9.5.6).

3.2.2.2 Matrix Isolation: Freon Matrices. Freons are chemically inert fluorinated halocarbons that are widely used in many applications. Gamma radiolysis of freon matrices (e.g., CCl_3F) at cryogenic temperatures produces an electron ejection. In the presence of a solute, such as a hydrocarbon, the positive hole migrates to the solute molecule giving its radical cation. Identification of a large number of hydrocarbon radical cations was made in the 1980s [8]. This method is still used for studying properties and reactions of radical cations [9].

3.2.2.3 Matrix Isolation: Inert Gas Matrices. Inert gas (e.g., Ar, Ne, Xe, Kr, N_2) matrices at 4K can be used for trapping reactive species produced by different techniques, such as photolysis, pyrolysis, chemical reactions, and electron irradiation [10]. Knight et al. extensively studied radicals generated in neon matrices containing hydrocarbons by several independent methods including photoionization, electron bombardment, X irradiation, and the pulsed laser surface ionization technique [11]. A recent example of a study on PS produced in inert gas at 4.2K is a detailed EPR investigation on radical pairs between a hydrogen atom and a methyl radical observed in X-ray irradiated solid argon containing selectively deuterium-labeled methanes, CH_4, CH_2D_2, and CD_4 [12].

3.2.2.4 Zeolites and Porous Media. Inorganic matrices can be used to isolate and make persistent radicals obtained by ionizing radiation [13]. Zeolites are used as topologically defined matrices, but they are also chemically active matrices. An important and vast field of research involving EPR methods is the spontaneous oxidation giving radical cations of conjugated molecules absorbed in zeolite pores [14] (see Chapter 12).

3.3 ORGANIC MOLECULES WITH MORE THAN ONE UNPAIRED ELECTRON

Let us first consider the molecules with two unpaired electrons. A broad and useful classification of these molecules distinguishes them as *biradicals*, or molecules in a *triplet state*. Examples of these two types of systems are now shown, with the warning

3.3 ORGANIC MOLECULES WITH MORE THAN ONE UNPAIRED ELECTRON

that a precise attribution to one of these two categories can be done only on the basis of the hierarchy of the energy terms of their magnetic and exchange interactions. For example, see the Atherton book on this subject in the Bibliography. In a nutshell, the magnitude of the exchange interaction J determines the energy difference between singlet and triplet states. For a biradical this latter energy difference is of the same order of magnitude as the magnetic interactions, such as hyperfine couplings. In contrast, when the singlet–triplet separation is much larger than the magnetic interactions, the state corresponding to two unpaired electrons is called a *triplet* (see §1.14).

3.3.1 Biradicals and Correlated Radical Pairs

3.3.1.1 Biradicals. The type of biradical more commonly used can be defined as a molecule obtained by joining together with a spacer two persistent radicals, which are usually identical. In general, the segment linking the two paramagnetic centers is such that in a first approximation the two electrons can be considered as localized and weakly interacting (therefore with a small J value and of the same order of magnitude as the hyperfine coupling). Scheme 3.3 illustrates two biradicals: **7**, which has two nitroxides, and **8**, which has two nitronyl nitroxides.

The interaction between the two unpaired electrons depends on the length and chemical nature of the segment and on the conformation of the molecule. For very small electron–electron interactions the spectrum of a biradical in solution is indistinguishable from that of a radical, because the two electrons nearly "ignore" each other. On increasing the electron–electron interaction, for example, by shortening the tether between the two radicals, the spectrum changes (see Chapter 4).

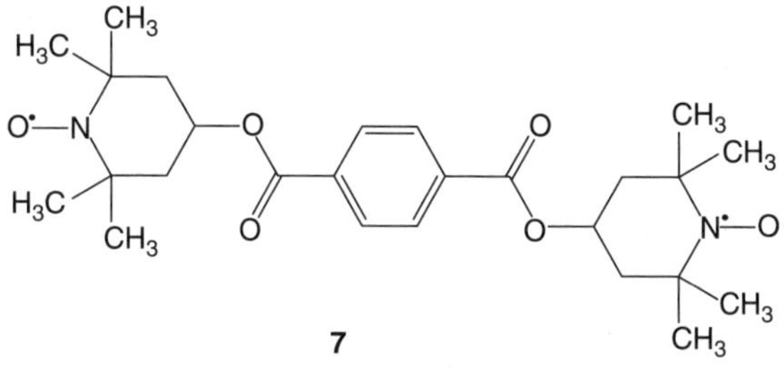

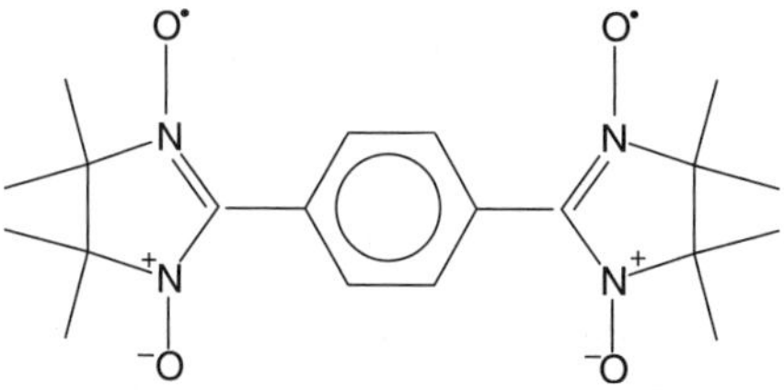

Scheme 3.3

Bisnitroxyls can be used as spin probes, because their EPR spectra, in addition to depending strongly on the distance between the two nitroxyl groups, also reflect the relative mobility of the two paramagnetic moieties. A comparison of the experimentally determined electron–electron interactions with the calculated parameters permits us to assess the reliability of calculation methods.

3.3.1.2 Correlated Radical Pairs. The electron–electron interaction in biradicals is similar to that of weakly interacting radicals not chemically bonded but only near to each other. The properties of these radical pairs are important in determining some photochemical processes in solution and in biological systems. Twin radicals, formed together by the breaking of a former bond between them, are called *correlated radical pairs*, because their spin state can be defined only for the pair. The two electrons, coming from their former singlet state when they were paired in the bond, will in fact also stay in a singlet state once separated, until a magnetic perturbation converts the singlet state to a triplet. The probability of this conversion process depends on the distance between the radicals and therefore on the strength of their interaction, on the dynamics of their motion and therefore on the time they spend feeling each other, and on the relative energies of the singlet and triplet levels. These latter energies depend on the presence of a magnetic field and on its intensity, and on the other magnetic interactions felt by each electron, as hyperfine couplings. Because two correlated radicals can recombine if they are still in a singlet state and they cannot if their state has been changed to a triplet, external magnetic fields and hyperfine interactions with the magnetic nuclei can affect the yield of a photochemical reaction. These effects (magnetic field effect and magnetic isotope effect) are being increasingly studied in a research field called spin chemistry (or dynamic spin chemistry) when taking into account the transient states of the mobile radicals [15, 16].

Correlated radical pairs can also be formed by electron transfer from a molecule in an excited state, acting as a donor, to an acceptor molecule. This process is of fundamental importance in the photosynthetic process, as the formation of a pair of charged radicals due to an electron transfer from the so-called primary donor excited by absorption of sunlight allows the conversion of radiation energy into chemical energy. More on this subject is provided in Section 3.6.1, and EPR on photosynthesis will be treated in Part II of Chapter 10.

Time resolved and pulsed EPR spectroscopy can give direct information on these processes, allowing the detection of short-lived radical pairs, their spin populations (CIDEP effect, §4.11.2) that depend on the photophysical pathway of the reaction, and the exchange interaction in the pair.

3.3.2 Triplet States ($S = 1$)

When the two unpaired electrons are interacting strongly (e.g., they belong to the same π conjugated system), a reliable description of the state of the molecule must take into account the total spin state of the two electrons together. The molecule therefore can be in a singlet or a triplet state (§1.14 and 6.5.1). We can distinguish between

3.3 ORGANIC MOLECULES WITH MORE THAN ONE UNPAIRED ELECTRON

molecules with a triplet ground state, with both triplet and singlet states thermally accessible, and with a triplet state that can be reached only by photoexcitation of the molecule. Note that generally triplet states are studied by EPR in solids, as the spectrum in liquid solution is severely broadened by spin relaxation.

3.3.2.1 Triplet Ground States. As a general rule, when the two interacting electrons in a neutral organic molecule belong to two half-filled molecular orthogonal orbitals, the triplet state is lower in energy with respect to the singlet. For example, in carbenes (**9**, Scheme 3.4) and nitrenes (**10**, Scheme 3.4) the two unpaired electrons are accommodated in a σ orbital and in a π orbital on a single C or N atom. These species can be obtained by photolysis of a suitable precursor in a glassy matrix or in a crystal, because in solution they would be nonpersistent.

Another interesting class of molecules with a triplet ground state is that of non-Kekulé hydrocarbons, trimethylenemethane being the simplest member of this family. The designation non-Kekulé implies that their conjugated system cannot be

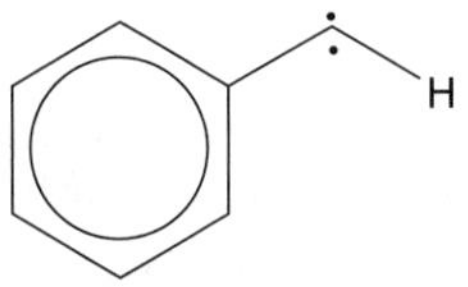

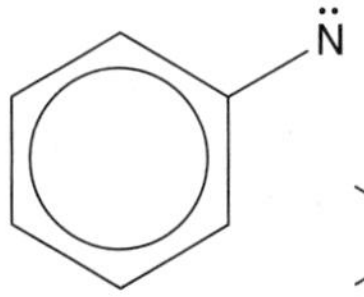

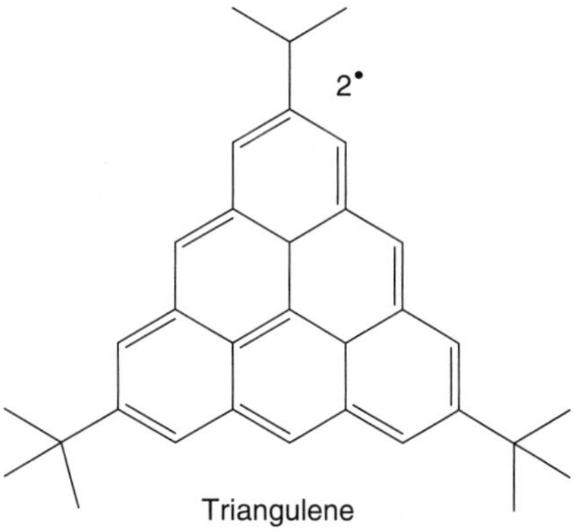

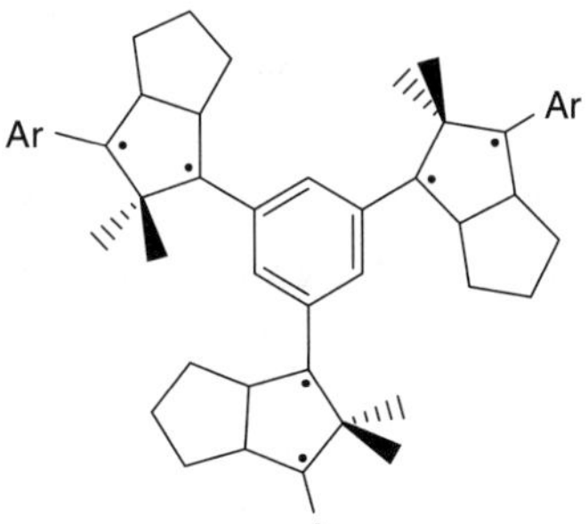

Scheme 3.4

represented by any resonance structure containing n double bonds derived from their $2n$ π electrons.

An example of this class of hydrocarbons in a triplet ground state is Triangulene (**4**) as provided in Scheme 3.4 [17]. Dianions obtained with the same methods illustrated in Section 3.2.1 by a two-electron electrochemical or chemical reduction of a molecule having a high symmetry, and therefore at least a pair of degenerate π orbitals, can have a triplet ground state.

3.3.2.2 Photoexcited Triplets. Hundreds of photoexcited triplets of organic molecules in solid matrices (glassy solutions, crystals, inclusion compounds) have been studied (see, e.g., [18]). Nowadays photoexcitation is generally done with a pulsed laser, and the EPR spectrum is obtained by time resolved EPR or pulsed EPR (as described earlier in this chapter). Magnetic spectroscopies with optical detection of the magnetic transitions as optically detected magnetic resonance are also used. These studies allow us to obtain information on the molecular electron distribution in the triplet state, the photophysical processes giving rise to the triplet state, and the energy transfer in crystals (triplet excitons) [19].

3.3.3 Systems with More Than Two Unpaired Electrons

3.3.3.1 Triplet–Radical Systems. The magnetic interactions between photoexcited triplets ($S=1$) and free radicals ($S=1/2$) can give excited quartet ($S=3/2$) states. This has been studied by photoexciting a fullerene derivative with an attached nitroxide radical [20].

3.3.3.2 Polyradicals. In the last few years organic molecules or polymers containing more than two interacting unpaired electrons have been studied quite intensively with the aim of preparing materials with interesting magnetic properties, such as ferromagnetism (see §3.7.1 and Lund and Shiotani, Chapter 12, Bibliography). Triradicals, depending on the strength of the interaction between electron spins, can be in a quartet ($S=3/2$) or in a doublet state ($S=1/2$). Tetraradicals and higher radicals (polyradicals) have also been synthesized and studied by EPR [21]. Compound **12** in Scheme 3.4, generated by photolysis of a triazoalkane precursor in a toluene matrix at 77K, is an example of a hexaradical [22].

3.4 INORGANIC RADICALS, SMALL PARAMAGNETIC MOLECULES, AND ISOLATED ATOMS

3.4.1 Inorganic Radicals

A profusion of inorganic radicals have been obtained and characterized with EPR, in particular in the 1970s and 1980s. Several examples are reported in the books by Weltner in 1983 and by Atkins and Symons in 1967 (see Bibliography).

The main methods of preparation are the high energy irradiation of solid parent compounds (γ or X irradiation, or radiolysis) or generation and trapping in inert gas matrices at 4K (see §3.2.2.3), whereby reactive species generated at ambient or high temperature are mixed with an inert gas of choice and then frozen into a solid upon the surface of a cold finger. Some examples of diatomic, triatomic, and tetra-atomic PS are metal hydrides (especially ZnH), carbides (especially RhC), metal–metal compounds (especially AgCa), oxygen compounds (especially ClO), anions (especially F_2^-, Cl_2^-), ScF_2, $Ag(CN)_2$, NO_2^{2-}, CO_2^-, CH_3, CF_3, and SiH_3.

3.4.2 Small Paramagnetic Molecules

Some small paramagnetic molecules are of paramount importance in many fields and were detected and characterized by EPR a long time ago. A short list follows.

1. Oxygen: As is well known, oxygen has a triplet ground state. An EPR study on O_2 in the gas phase at different pressures can be found in Reference 23. The presence of O_2 in samples broadens the EPR lines. This effect is normally avoided by degasing the samples. It can be exploited for measuring the oxygen content, as in oximetry (§10.8) [24, 25].
2. Oxygen Radicals O^-, O_2^-, O_3^-, OH ...: The oxygen radicals are studied by EPR both in biological systems, in this case called reactive oxygen species (ROS), and in solid matrices, in particular on metal oxide surfaces. Indeed, the study of the formation of ROS, their direct or spin trapping detection, and their scavenging may be seen as the most important application of EPR in the medical field (see Chapter 11).
3. Nitric oxide (NO): NO is a small, gaseous, inorganic radical that can also be formed in biological processes *in vivo*. EPR studies on this radical have relevance in pharmacology and toxicology.

The EPR studies on small inorganic paramagnetic systems are confining with the studies on defects and impurities in ionic and covalent solids (see Chapter 13).

3.4.3 Isolated Atoms

Many atoms have been isolated and stabilized in rare gas matrices at low temperature. The atom is obtained from a suitable parent compound by photolysis, laser ablation, vaporization, and so forth. Atoms possessing no orbital angular momentum in the inert matrices have magnetic parameters similar to those in the gas phase. For atoms with an orbital angular momentum different than zero, the interaction with the matrix is strongly perturbing and in some cases makes the detection of the atom difficult. This field of research is described in the book by Weltner and in the book edited by Lund and Shiotani (see Bibliography).

3.4.3.1 Atoms in Endohedral Fullerenes. A recent revival of EPR on "isolated" atoms has been due to the so-called endohedral fullerenes, which are fullerenes encasing an ion or neutral atom inside the carbon cage (see Fig. 3.2). Elements such as scandium, yttrium, and lanthanum have been encapsulated in C_{82} and studied

98 WHAT CAN BE STUDIED WITH EPR?

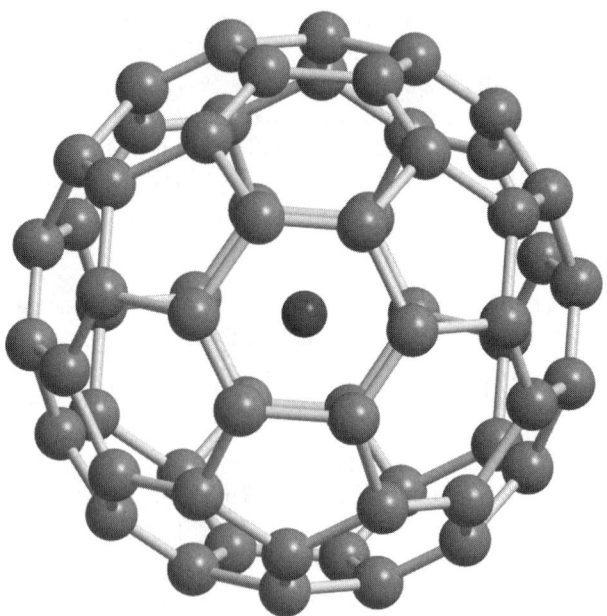

Fig. 3.2 The N atom is inside the carbon atoms cage. The symbol for this molecule is N@C$_{60}$.

with EPR [26a]. Particularly interesting is the field of studies on trapped nitrogen or phosphorus (N@C$_{60}$, N@C$_{70}$, P@C$_{60}$) that do not interact with the carbon cage and give an EPR spectrum typical of the three unpaired electrons ($S = 3/2$, quartet state) [26b]. The exceptional decoupling with the matrix makes the properties of these atoms identical to those in the gas phase. The shielding from outside gives rise to very long relaxation times and makes these systems interesting for applications such as quantum computing or EPR tomography.

3.5 TRANSITION METAL IONS

Transition metal ions, as well as rare earth and actinide ions, have been investigated intensively by EPR since the introduction of this technique. This field of EPR research is very wide, and the following gives merely a map for helping the reader's orientation. More detailed information can be found in Chapter 6. Note that the studies on paramagnetic transition metals have gained more than others from the onset of high frequency EPR (HF-EPR) [27, 28].

The number of unpaired electrons of transition metal complexes depends on the ion and its oxidation state, on the symmetry of the coordination, and on the strength of the crystal field. The EPR properties of each ion can be assessed starting from the electronic ground state of the free ion (spectroscopic atomic term) and then determining the symmetry of the electric fields coming from the ligands (octahedral, tetrahedral, square planar, trigonal, etc.) and how the degeneracy of the d orbitals is removed. The simplest cases are d^1 (V^{4+} mostly present as VO^{2+}, Cr^{5+}, Ti^{3+}) and d^9 ions (Cu^{2+}, Ni^+), which always have $S = 1/2$. For the other ions the spin state depends on the coordination, and the same ion can be in a low spin or high spin state in different compounds. For example, octahedral d^4–d^7 transition metal complexes can be either low spin, and in this case have spin ground states $S = 0$, $1/2$, or 1 (for d^6, $d^{5,7}$, and d^4, respectively), or high-spin, with $S = 3/2$, 2, or $5/2$ (for d^7, $d^{4,6}$, and d^5, respectively). The EPR spectra of transition metal ions, with the exception of $S = 1/2$ ions, are generally dominated by a so-called zero field splitting (ZFS) term, which is due to electron spin–spin interaction and to spin–orbit coupling. The ZFS interaction in a first approximation is equal to zero for Mn^{2+} ($3d^5$).

An introductory chapter can be found in chapter 8 of Weil, Bolton, and Wertz's book, which contains a general bibliography up to 1990. The classical textbook on magnetic properties of transition ions is by Abragam and Bleaney, whereas the most thorough and systematic analysis of the transition metal paramagnetic compounds is in the Mabbs and Collison book. Specific applications of EPR, ENDOR, and electron spin echo envelope modulation (ESEEM) to transition metal ions are described in the Pilbrow book, which also takes into account their biological relevance and coupled systems. The tome by Bencini and Gatteschi describes systems in which two or more spins are magnetically coupled, with particular attention to pairs of transition ions, clusters, low dimensional materials, and mixed systems with transition ions and organic radicals.

3.5.1 Metal Centers in Biology

The field of EPR studies on metal centers in biologically relevant molecules is developing thanks to the advanced EPR methods, in particular HF-EPR and ENDOR, and ESEEM. Reviews can be found in several of the books of the series Biological Magnetic Resonance in the Bibliography. Other relevant reviews on application of HF-EPR to transition metals, in particular in biological systems, can be found in References 27 and 28.

The EPR characterization of transition metal ions involves many research fields, and the following provides some nonexhaustive examples.

Paramagnetic ions are present in many enzymes and metalloproteins. They can also replace nonparamagnetic ions retaining the biological activity. Examples of very relevant subjects are Cu(II) proteins [29], Fe(III) heme–iron proteins [30], iron–sulfur proteins [31], Mn(II) proteins such as cytochrome c oxidase, and multinuclear Mn centers such as the oxygen evolving complex (OEC) in photosystem II.

Simple ion–ligand complexes are at present studied as models of bioinorganic systems. For example, the study of iron–porphyrin complexes is quite important in understanding the function and catalytic processes of naturally occurring heme proteins [32]. The study of manganese(imidazole)$_6$ in a single crystal can help in interpreting EPR and two-dimensional echo spectra (Hyscore spectra) of biomolecules containing manganese coordinated to protein residues [33a].

3.5.2 Electron Transfer Reactivity

The electron transfer reactivity of the coordination compounds has become an area of extensive research in recent years, for example, the multinuclear transition metal complexes containing second and third row transition ions, which are stable in different redox states {e.g., d^6 polypyridyl complexes with Re(I), Ru(II), Os(II) Rh(III), and Ir(III) [33b]}.

3.5.3 Paramagnetic Ions as Spin Probes

As native and stable spin probes present nearly everywhere, some of the transition ions are studied to characterize the material in which they are embedded. For example, Mn^{2+} is frequently used in this respect because of its ground-state electronic term with no ZFS. As a consequence, Mn^{2+} is only slightly affected by ligand/crystal fields, and its spectrum is well resolved and with rather sharp lines. A vast number of EPR studies have been performed giving important pieces of information on many materials (e.g., in minerals and rocks, where it substitutes Ca^{2+} and Mg^{2+}, see following; for zeolites, see Chapter 12).

3.6 NATURAL SYSTEMS AND PROCESSES

3.6.1 PS in Photosynthesis and Other Light Driven Biological Processes

In the last two decades one of the more important fields in which EPR has given an invaluable contribution, not achievable by any other method, has been the *photosynthesis* process. An overview of it and of the advanced EPR methods used for its study are given in Part II of Chapter 10. Here, the stress is only on attention to the type of PS produced in photosynthesis. This is a transformation of radiative energy (from the sun) into chemical energy (the growth of organisms, plants, or bacteria). Its initial steps, the light reactions, occur in two large membrane protein complexes, photosystems I and II. They catalyze the initial step of photosynthesis, which is the light induced charge separation across the photosynthetic membrane. The light gives rise to an excited singlet state of a *primary donor D*, a closely related pair of chlorophyll or bacteriochlorophyll molecules. Donor D^* ejects an electron that reduces a molecule acting as an *acceptor A*. Depending on the type of organism, A is a monomeric chlorophyll, a pheophytin, a quinone, or an iron–sulfur center. The electron is further transferred along a chain of acceptors. The transfer of the electron along the system creates a series of PS that can be studied by EPR. However, the transfer is very fast and thus for observing the various steps the transfer must be stopped (e.g., by chemical reduction of the following acceptor).

The electron transfer from D to A gives rise to a correlated radical pair $D^+ - A^-$, and the interaction between the two unpaired electrons has been studied by EPR because it gives structural and functional information. Cation radicals D^+ and anion radicals A^- have been trapped and characterized via their g factors and hyperfine structures with several EPR methods, such as pulsed and time resolved [34]. The

D^+–A^- pair, if further electron transfer is blocked, can recombine and give a triplet ^3D, which has also been studied by EPR (time resolved and pulsed ENDOR) [35].

Another PS of paramount importance is a cluster of four Mn ions, which are bound to a protein in PSII. This OEC has the ability to oxidize water to O_2 and $4H^+$.

The oxygen evolution occurs after a cycle of oxidized states, driven by the absorption of four photons. Some of these S states (S_2, S_0) exhibit a paramagnetic ground state with a total electron spin of $S = 1/2$. They are intensively studied by various EPR methods because their magnetic parameters will provide information about the electronic structure of the OEC, which is a prerequisite for the elucidation of the mechanism of water splitting [36].

3.6.2 EPR and Natural Materials

3.6.2.1 PS from the Degradation of Organic Matter. Natural organic matter is defined as the nonliving organic molecules found in the environment. It plays an essential role in most environmental and geochemical processes and contains free radicals and paramagnetic metal ions. Therefore, EPR is an invaluable method for its characterization. Persistent radicals (mainly semiquinones) are present, for example, in *humic substances*, a complex and ill-defined group of substances that results from the microbial degradation of organic material present in composts, soil, peat, soft brown coal, streams, and ocean water. EPR spectroscopy is an important tool to characterize these substances, and in particular their redox activity is important for *in situ* remediation of pollution [37]. Moreover, the presence of paramagnetic ions and their interactions with organic matter has been studied, for example, for Fe^{3+} (associated as colloidal iron oxides [38]) and Mn^{2+} [39].

3.6.2.2 Coal. The different types of coal are extremely complex mixtures of organic molecules varying in size and structure. They give an EPR spectrum due to superimposed lines attributed to persistent carbon centered PS with unpaired electrons delocalized on aromatic structures of various sizes. The EPR parameters have been correlated to important properties, such as the degree of cross-linking in the carbon network [40].

3.6.2.3 Graphite, Petroleum, Asphalt, Pitch, and Coal Tar. These substances give EPR signals used for assessing different properties. For example, EPR is used to detect radicals in cigarette tar.

3.6.2.4 Paramagnetic Ions and Defects in Minerals and Rocks. EPR spectroscopy is contributing more and more to geological and mineralogical studies. In geochemical studies, EPR has a main role in the characterization of diffusion of paramagnetic ions with environmental impact in soils and waters, and as a consequence decomposition of rocks [41].

Geochronology and environmental reconstruction exploits the method of dating by measuring the intensity of EPR signals produced by radioactivity. For example,

ESR dating has been applied to quartz grains from the fault gouge form assessing fault activity [42].

Insights into the structure, composition, and origin of rocks and minerals are provided by EPR studies on defects in crystals (e.g., in quartz and in zirconium silicate [43], see also Chapter 13); on identification of paramagnetic radicals as an indicator of geological events [44]; and on the study of paramagnetic ions, in particular Mn^{2+}, substituting Ca^{2+} in calcites, and Ca^{2+} and Mg^{2+} in dolomites such as in travertine and in marbles. The MnII EPR spectra of marbles are similar for samples coming from the same quarry, and this property is used in archaeological studies in assessing the origin of ancient marbles [45].

3.7 TAILORING AND ASSEMBLING PS FOR MAGNETIC MATERIALS

3.7.1 Magnetic Materials

Magnetic materials are those materials that show collective (bulk) magnetic properties of some interest. The most frequently encountered materials are composed of diamagnetic molecules and therefore show diamagnetic bulk behavior; that is, they practically do not interact with magnetic fields. Conversely, magnetic materials do interact. The magnetic materials known in the past were the naturally occurring materials such as bulk magnetic phases of some metals, for example, iron, nickel, and cobalt are classified as ferromagnetic. For the latter the building blocks of the magnetic phase are the metal atoms. The materials treated here are *molecule-based* magnetic materials that have *paramagnetic* building blocks such as organic radicals, transition ion complexes, or organometallic mixtures of the two. For any magnetic material the collective behavior of the magnetic moments can be *paramagnetic*, *antiferromagnetic*, *ferromagnetic*, or *ferrimagnetic*, depending on the type of interaction with an applied magnetic field. Provided here is a very short definition of the first three behaviors (see Carlin for a complete treatment, Bibliography).

The magnetic field creates a collective magnetization, according to the expression $\overline{M} = (\chi_m/\mu_0)\overline{B}$, where χ_m is the *magnetic susceptibility* and μ_0 is the vacuum permeability. If the atomic or molecular magnetic moments have magnetic interactions between them that are smaller than the Boltzmann energy (kT), the magnetization is simply given by the summing of the paramagnetic moments of the building blocks; this behavior is called *paramagnetism*. In contrast, under a sufficiently low temperature the magnetic interactions between the paramagnetic units will prevail on the Boltzmann energy, and a collective behavior will emerge. *Antiferromagnetism* is very common and corresponds to an antiparallel orientation of the magnetic moments in a magnetic field, leading to a vanishing of the magnetization on lowering the temperature. *Ferromagnetism* is much less common and very important for technology. Under a temperature called the *Curie temperature* the magnetic moments align in the magnetic field, giving rise to a large $\overline{M}$ (large value of χ_m). The switching off of the magnetic field for some ferromagnets called *hard* leaves a magnetization different than zero. This property is called *hysteresis*.

3.7.2 Molecule-Based Magnetic Materials

Note that the bulk magnetic properties are conveniently measured by specific methods different than EPR, such as the Gouy method or the Squid magnetometer. However, EPR spectroscopy has a major role in the research for molecule-based magnetic materials. These materials are built by assembling well-defined PS, such as organic radicals and transition metal ions, so the individual magnetic parameters of each magnetic unit can be determined by EPR. The magnetic and exchange interactions between the magnetic building blocks can be further studied by assembling them in materials of different dimensionality. Because the collective magnetic properties depend strongly on the dimensionality, this is one of the parameters for classifying magnetic materials, together with the nature of the molecular building blocks. The interest in these materials is due to the possibilities given by organic synthetic methods to modulate and combine properties of molecule-based materials, which is unparalleled in the world of atom-based solid-state chemistry.

3.7.2.1 Crystals of Organic Radicals. The design of organic ferromagnetic molecular crystals made of persistent radicals is being pursued by many research groups, particularly in Japan. The first report on a ferromagnetic phase of an organic stable radical was on *p*-nitrophenyl nitronyl nitroxide, and most known purely organic molecule-based magnets are members of the nitronyl nitroxide family (see **4** in Scheme 3.1) [46]. Many studies seek to establish a correlation between the crystal structure and the magnetic properties. Because it is the spin–spin interactions between electrons that determine the magnetic behavior, these studies start from the crystal packing and the bulk magnetism of an organic radical crystal and try to correlate these two properties with experimental (EPR) and calculated (density functional theory) magnetic parameters and spin–spin interactions between the radicals.

3.7.2.2 Mesoscopic Magnetic Clusters and Molecular Magnets. Magnetic materials of mesoscopic or nanoscopic size contain from thousands to tens of interacting magnetic units. These materials have particular properties of both fundamental and technological interest. They are composed by an interplay of organic radicals and paramagnetic transition metal ions [47].

3.7.2.2.1 Organic Polyradicals. In organic polyradicals many persistent radical fragments are connected to achieve the highest possible S value. Large polymers have been synthesized, for example, poly(*m*-phenylenecarbene) [48] with $S = 5$. However, the most promising magnetic materials are the organic–inorganic hybrid ones.

3.7.2.2.2 Organic Radicals: Transition Metal Ions. These materials are designed by linking transition metal ions with bidentate ligands bearing an unpaired electron, such as nitronyl nitroxide radicals [49]. Polyradicals can be used to obtain a polydimensional material [50].

3.7.2.2.3 Single Molecule Magnets (SMMs). These are small clusters of transition ions magnetically coupled between them where a permanent magnetization and

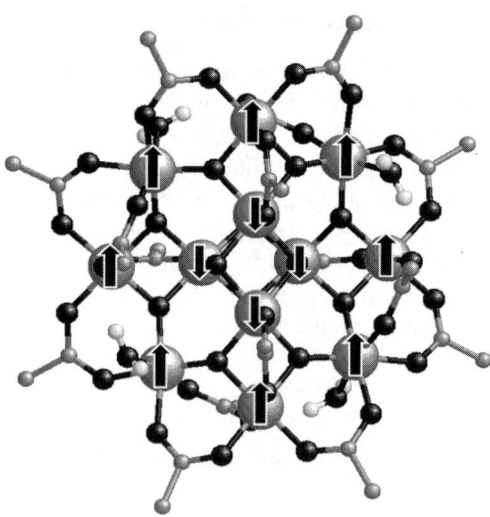

Fig. 3.3 The most studied SMMs: [Mn$_{12}$O$_{12}$(CH$_3$COO)$_{16}$(H$_2$O)$_4$] 2CH$_3$COOH·4H$_2$O. An external ring of eight manganese(III) ions ($S = 2$) and an internal tetrahedron of four manganese(IV) ions ($S = 3/2$). The cluster has a ground state with $S = 10$. Courtesy of R. Sessoli.

magnetic hysteresis can be achieved (although at extremely low temperatures) as a purely one-molecule phenomenon instead of through a three-dimensional magnetic ordering. The most studied one is the 12-ion manganese cluster in which 8 of the Mn ions are in the +3 oxidation state ($S = 2$) and 4 are in the +4 state ($S = 3/2$). These ions are magnetically coupled to yield an $S = 10$ ground state, giving rise to unusual magnetic relaxation properties [51] (see Fig. 3.3). EPR spectroscopy, in particular HF-EPR, has been a key tool for the characterization of SMMs, providing the spin Hamiltonian parameters that are responsible for the splitting in zero field of the large spin ground state and the consequent magnetic anisotropy [52].

3.8 INDUSTRIAL APPLICATIONS OF EPR

EPR is applied in many research fields interested in industrial production: medicine, food industry, polymeric materials, and degradation of materials. The application to medicine will be treated in Chapter 11.

3.8.1 Radical Polymerizations

These are chain reactions via short-lived and highly reactive propagating radicals. EPR can be applied to the study of the radicals involved when their steady-state concentration is sufficiently high. Spin trapping techniques can be applied where direct detection of the radical species fails. There are several examples, in particular bulk polymerizations of methylmethacrylate and styrene, in Reference 53.

3.8.2 Degradation of Materials

The degradation of several materials, such as the polymers themselves, occurs via free radicals. Polymeric materials undergo oxidative degradation triggered by oxygen and light. An example of an EPR test on the degradation of insulating materials used in the power cable industry, such as poly(vinyl chloride) and cross-linked polyethylene, is given in Reference 54.

In polymeric materials containing stabilizers as hindered amines, photo- and thermal degradation produce nitroxide radicals. By means of particular EPR methods such as space resolved, it is possible to study the degradation in the depth of the material [55].

3.8.3 Food Industry

3.8.3.1 Radiation Dosimetry. The irradiation of food for conservation requires an assessment of the dose of radiation. Radiation dosimetry by detection of radicals is described in Chapter 9 (see §9.5.2).

3.8.3.2 Antioxidant Properties of Food. EPR is used as one of the most important methods for assessing the antioxidant properties of foods and plant products, for example, beer, wine, and teas [56].

REFERENCES

1. Poole, C. P. *Electron Spin Resonance*, 2nd ed.; Dover: New York, 1983.
2. Maki, A. H.; Geske, D. H. *J. Chem. Phys.* **1959**, *30*, 1356.
3. Xie, Q.; Perez-Cordero, E.; Echegoyen, L. *J. Am. Chem. Soc.* **1992**, *114*, 3978.
4. Eaton, S. S.; Eaton, G. R. *Appl. Magn. Reson.* **1996**, *11*, 155.
5. Goldstein, S. *Free Rad. Biol. Med.* **1993**, *15*, 435.
6. Adamo, C.; Barone, V.; Subra, R. *Theor. Chem. Acc.* **2000**, *104*, 207.
7. McLauchlan, K. A. *Mol. Phys.* **1996**, *89*, 1423.
8. Shiotani, M. *Magn. Res. Rev.* **1987**, *12*, 333.
9. Bally, T.; Matzinger, S.; Bednarek, P. *J. Am. Chem. Soc.* **2006**, *128*, 7828.
10. Belevskii, V. N.; Feldman, V. I.; Tyurin, D. A. *High Energy Chem.* **2005**, *39*, 77.
11. (a) Knight, L. B., Jr.; Steadman, J.; Feller, D.; Davidson, E. R. *J. Am. Chem. Soc.* **1984**, *106*, 3007; (b) Knight, L. B., Jr.; King, G. M.; Petty, J. T.; Matsushita, M.; Momose, T.; Shida, T. *J. Chem. Phys.* **1995**, *103*, 3377.
12. Nomura, K.; Shiotani, M.; Komaguchi, K. *J. Phys. Chem. A* **2007**, *111*, 726.
13. Werst, D. W.; Trifunac, A. D. *Acc. Chem. Res.* **1998**, *31*, 65114.
14. Garcia, H.; Roth, H. D. *Chem. Rev.* **2002**, *102*, 3947.
15. Nagakura, S.; Hayashi, H.; Azumi, T., Eds. *Dynamic Spin Chemistry: Magnetic Controls and Spin Dynamics of Chemical Reactions*; Kodansha/Wiley: Tokyo/New York, 1998.
16. Brocklehurst, B. *Chem. Soc. Rev.* **2002**, *31*, 301.
17. Fukui, K. *Synth. Met.* **2001**, *121*, 1824.

18. Brustolon, M.; Barbon, A.; Bortolus, M.; Maniero, A. L.; Sozzani, P.; Comotti, A.; Simonutti, R. *J. Am. Chem. Soc.* **2004**, *126*, 15512.
19. Hirota, N.; Yamauchi, S. *J. Photochem. Photobiol. C* **2003**, *4*, 109.
20. Corvaja, C.; Maggini, M.; Prato, M.; Scorrano, G.; Venzin, M. *J. Am. Chem. Soc.* **1995**, *117*, 8857.
21. Rajca, A. *Chem. Eur. J.* **2004**, *10*, 3144.
22. Adam, W.; Baumgarten, M.; Maas, W. *J. Am. Chem. Soc.* **2000**, *122*, 6735.
23. Gardiner, W. C., Jr.; Pickett, H. M.; Proffitt, M. H. *J. Chem. Phys.* **1981**, *74*, 6037.
24. Hyde, J. S.; Subczynski, W. K. *Spin-Label Oxymetry*; Biological Magnetic Resonance, Vol. 8; Plenum Press, New York, 1989; Chapter 8.
25. Swartz, H. M.; Dunn, J. F. *Adv. Exp. Med. Biol.* **2003**, *530*, 1.
26. (a) Kato, T. *Dev. Fullerene Sci.* **2002**, *3*, 153; (b) Pietzak, B.; Weidinger, A.; Dinse, K.-P.; Hirsch, A. Group V endohedral fullerenes: N@C60, N@C70 and P@C60. In *Endofullerenes: A New Family of Carbon Clusters*; Akasaka, T.; Nagase, S., Eds.; Kluwer; New York, 2002.
27. Anderson, K. K.; Schmidt, P. P.; Katterle, B.; Strand, K. R.; Pallmer, A. E.; Lee, S. K.; Solomon, E. I.; Graslund, A.; Barra, A. L. *J. Biol. Inorg. Chem.* **2003**, *8*, 235.
28. Krzystek, J.; Ozarowski, A.; Telser, J. *Coord. Chem. Rev.* **2006**, *250*, 2308.
29. (a) Taiwo, F. A. *Spectroscopy* **2003**, *17*, 53; (b) Arieli, D.; Goldfarb, D. *Annu. Rev. Biophys. Biomol. Struct.* **2004**, *33*, 441.
30. van Doorslaer, S. *IUBMB Life* **2004**, *56*, 665.
31. Cammack, R.; Shergill, J. K. *Biochim. Biophys. Acta* **1994**, *1185*, 35.
32. Nakamura, M. *Coord. Chem. Rev.* **2006**, *250*, 2271.
33a. Angerhofer, A.; Schweiger, A.; Garcia-Rubio, I. *J. Magn. Reson.* **2007**, *184*, 130.
33b. Browne, W. R.; O'Boyle, N. M.; McGarvey, J. J.; Vos, J. G. *Chem. Soc. Rev.,* **2005**, *34*, 641.
34. Lubitz, W.; Lendzian, F.; Bittl, R. *Acc. Chem. Res.* **2002**, *35*, 313.
35. Lubitz, W. *Pure Appl. Chem.* **2003**, *75*, 1021.
36. (a) Lee, C.-I.; Lakshmi, K. V.; Brudvig, G. W. *Biochemistry* **2007**, *46*, 3211; (b) Kulik, L. V.; Lubitz, W.; Messinger, J. *Biochemistry* **2005**, *44*, 9368; (c) Kulik, L. V.; Epel, B.; Messinger, J.; Lubitz, W. *Photosynth. Res.* **2005**, *84*, 347.
37. (a) Scott, D. T.; McKnight, D. M.; Blunt-Harris, E. L.; Kolesar, S. E.; Lovley, D. R. *Environ. Sci. Technol.* **1998**, *32*, 2984; (b) Jezierski, A.; Siatecki, Z.; Chmielewski, P. *Spectrochim. Acta Part A* **2002**, *58*, 1293.
38. Benedetti, M. F.; Ranville, J. F.; Allard, T.; Bednar, A. J.; Menguy, N. *Colloids Surf. A* **2003**, *217*, 1.
39. Graham, M. C.; Gavina, K. G.; Farmer, J. G.; Kirika, A.; Britton, A. *Appl. Geochem.* **2002**, *17*, 1061.
40. Krzesinska, M.; Pilawa, B.; Pusz, S. *Energy Fuels* **2006**, *20*, 1103.
41. (a) Weber, T.; Allard, T.; Benedetti, M. F. *J. Geochem. Explor.* **2006**, *88*, 166; (b) Flogeac, K.; Guillon, E.; Aplincourt, M. *J. Colloid Interface Sci.* **2005**, *286*, 596.
42. Fukuchi, T. *Eng. Geol.* **1996**, *43*, 201.
43. (a) Lees, N. S.; Walsby, C. J.; Williams, J. A. S.; Weil, J. A.; Claridge, R. F. C. *Phys. Chem. Miner.* **2003**, *30*, 131; (b) Granwehr, J.; Weidler, P. G.; Gehring, A. U. *Phys. Chem. Miner.* **2004**, *31*, 203.

44. Franco, R. W. A.; Pelegrini, F.; Rossi, A. M. *Phys. Chem. Miner.* **2003**, *30*, 39.
45. Attanasio, D. *Ancient White Marbles: Analysis and Identification by Paramagnetic Resonance Spectroscopy*; L'Erma di Bretschneider: Roma, **2003**.
46. Takahashi, M.; Turek, P.; Nakazawa, Y.; Tamura, M.; Nozawa, K.; Shiomi, D.; Ishikawa, M.; Kinoshita, M. *Phys. Rev. Lett.* **1991**, *67*, 746.
47. Crayston, J. A.; Devine, J. N.; Walton, J. C. *Tetrahedron* **2000**, *56*, 7829.
48. Iwamura, H. *Proc. Jpn. Acad. B Phys. Biol. Sci.* **2005**, *81*, 233.
49. Caneschi, A.; Gatteschi, D.; Sessoli, R; Rey, P. *Acc. Chem. Res.* **1989**, *22*, 392.
50. Luneau, D.; Rey, P. *Coord. Chem. Rev.* **2005**, *249*, 2591.
51. Gatteschi, D.; Sessoli, R. *Angew. Chem. Int. Ed.* **2003**, *42*, 268.
52. Gatteschi, D.; Barra, A. L.; Caneschi, A.; Cornia, A.; Sessoli, R.; Sorace, L. *Coord. Chem. Rev.* **2006**, *250*, 1514.
53. Yamada, B.; Westmoreland, D. G.; Kobatake, S.; Konosu, O. *Prog. Polym. Sci.* **1999**, *24*, 565.
54. Morsy, M. A.; Shwehdi, M. H. *Spectrochim. Acta Part A* **2006**, *63*, 624.
55. Lucarini, M.; Pedulli, G. F.; Motyakin, M. V.; Schlick, S. *Prog. Polym. Sci.* **2003**, *28*, 331.
56. Polovka, M.; Brezová, V.; Staško, A. *Biophys. Chem.* **2003**, *106*, 39.
57. Box, H. C. *Radiation Effects*; Academic Press: New York, 1977.
58. Ohigashi, H.; Kurita, Y. *Bull. Chem. Soc. Jpn.* **1968**, *41*, 275.
59. Brustolon, M.; Maniero, A. L.; Segre, U. *Mol. Phys.* **1988**, *65*, 447.

BIBLIOGRAPHY

General EPR

Atherton, N. M. *Principles of Electron Spin Resonance*; Ellis Horwood PTR Prentice-Hall: London, 1993.

Weil, J. A.; Bolton, J. R.; Wertz, J. E. *Electron Paramagnetic Resonance*; Wiley: New York, 1994.

Organic Radicals

Gerson, F.; Huber, W. *Electron Spin Resonance Spectroscopy of Organic Radicals*; Wiley-VCH: Weinheim, 2003.

Radicals in Solids

Lund, A.; Shiotani, M., Eds. *EPR of Free Radicals in Solids*; Progress in Theoretical Chemistry and Physics, Vol. 10; Kluwer Academic: New York, 2003.

Transition Metal Ions

Abragam, A.; Bleaney, B. *Electron Paramagnetic Resonance of Transition Ions*; Dover: New York, 1986.

Bencini, A.; Gatteschi, D. *EPR of Exchange Coupled Systems*; Springer-Verlag: Berlin, 1990.

Mabbs, F. E.; Collison, D. *Electron Paramagnetic Resonance of d-Transition Metal Compounds*; Studies in Inorganic Chemistry, Vol. 16; Elsevier: New York, 1992.

Pilbrow, J. R. *Transition Ion Electron Paramagnetic Resonance*; Clarendon Press: New York, 1991.

Inorganic Radicals

Atkins, P. W.; Symons, M. C. R. *The Structure of Inorganic Radicals*; Elsevier: Amsterdam, 1967.

Weltner, W., Jr. *Magnetic Atoms and Molecules*; Van Nostrand Reinhold: New York, 1983.

Magnetic Materials

Carlin, R. L. *Magnetochemistry*; Springer-Verlag: Berlin, 1986.

Kahn, O.; Gatteschi, D.; Miller, J. S.; Palacio, F., Eds. *Magnetic Molecular Materials*; Kluwer: London, 1991.

4 Electron Paramagnetic Resonance Spectroscopy in the Liquid Phase

GEORG GESCHEIDT

Institute for Physical and Theoretical Chemistry, Graz University of Technology, Technikerstraße, 4/IA-8010 Graz, Austria

4.1 GENERAL CONSIDERATIONS

4.1.1 What Kind of Paramagnetic Species Can Be Detected by Electron Paramagnetic Resonance (EPR) in the Liquid Phase?

Electron transfer and (homolytic) bond cleavage are basic chemical transformations (Fig. 4.1). Starting with a generally occurring closed-shell diamagnetic molecule,

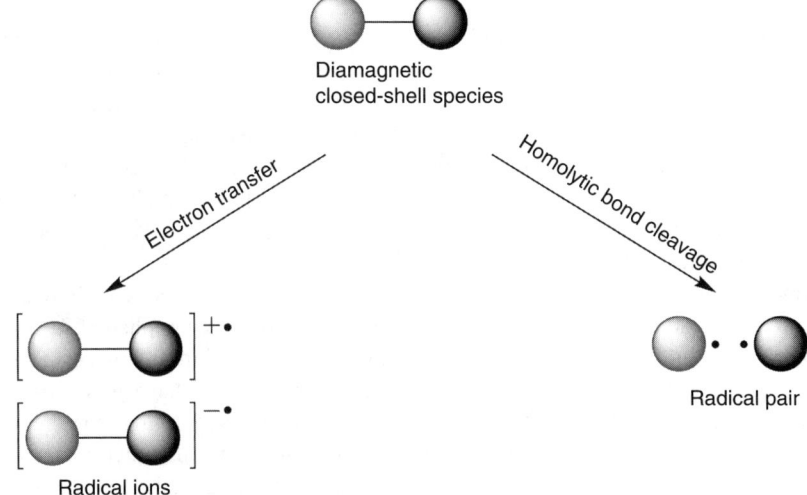

Fig. 4.1 Formation of radicals from closed-shell diamagnetic species.

Electron Paramagnetic Resonance. Edited by Brustolon and Giamello
Copyright © 2009 John Wiley & Sons, Inc.

both processes lead to the formation of paramagnetic species. Accordingly, a broad range of chemical reactions proceeds via radicals or radical ions. Such examples as Birch reduction, Kolbe electrolysis, and acyloin condensation can be found in organic chemistry textbooks (e.g., March's book [1]).

If such radicals are formed, their EPR spectra contain the information of several structural features like molecular symmetry and electron distribution. This will be represented here for aromatic radical ions and neutral carbon centered radicals.

EPR spectra of such radicals in solution may display dynamic phenomena if they appear at an appropriate time scale. Such phenomena offer insight into conformational or configurational changes, charge distribution, and electron-transfer processes.

In addition to organic radicals, salts and complexes of several paramagnetic transition-metal ions like Cu^{2+}, VO^{2+}, and Mn^{2+} exhibit characteristic EPR signals. Transition-metal ions are, in most cases, persistent in fluid solution, but radicals and radical ions are more often formed in the course of chemical reactions as intermediates with limited lifetimes.

This chapter will focus on the interpretation of essentially isotropic EPR spectra in liquid solutions of organic radicals and transition-metal ions. However, before describing the shape of the EPR signals and their analysis, some basic features of paramagnetic species will be introduced.

4.2 GENERATION OF RADICALS AND RADICAL IONS

4.2.1 General Considerations

Paramagnetic transition-metal salts are generally persistent. They are either paramagnetic in their usual oxidation states (Cu^{2+}, Mn^{2+}, Fe^{3+}, etc.) or a paramagnetic oxidation state can be conveniently generated by appropriate oxidants, reductants (mostly metal salts), or electrochemical methods in a broad range of solvents.

For radicals and radical ions, however, more specific techniques have to be applied because these species are rather unstable in most cases. Therefore, this section provides an exclusive description of how such short-lived species can be generated. For typical sample tubes utilized for X-band EPR experiments, ~ 0.7 mL of a millimolar solution of the diamagnetic precursor is necessary to perform the reaction that produces radicals. For apolar organic solvents, this is accomplished in quartz or pyrex cylindrical cells with an inner diameter of $\sim 3-4$ mm. For solvents of higher polarity, especially water, capillary tubes or flat cells have to be used to avoid dielectric losses (see §2.2.1). In some cases it is necessary to irradiate the reaction solutions. For this purpose, it is advantageous to use flat cells (quartz, suprasil); the solution is pumped through the cell by means of using peristaltic or syringe pumps to provide a sufficient steady-state concentration of the radicals. In the latter procedure, a sufficient amount of reaction solution is required (at least 10 mL) and the flow speed has to be adjusted. Several types of EPR tubes are illustrated in Fig. 4.2.

4.2 GENERATION OF RADICALS AND RADICAL IONS

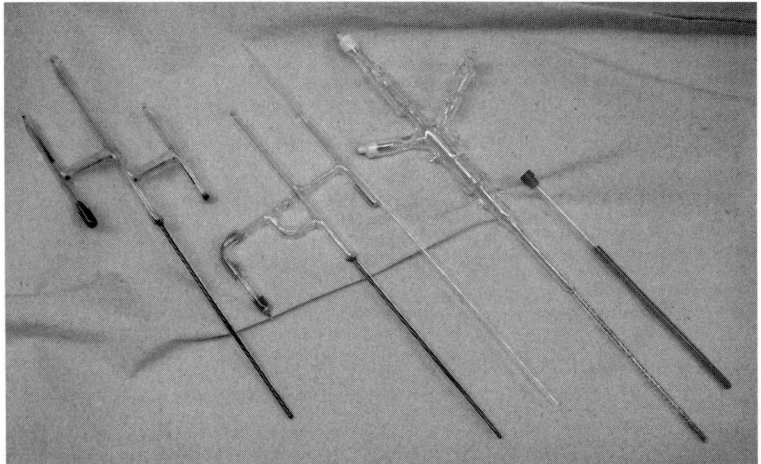

Fig. 4.2 Different EPR cells for the production of radicals and radical ions.

Care must be taken to use solvents that are inert toward the reagents, and oxygen-free solutions are mandatory (because in most cases, the presence of oxygen leads to line broadening that may obscure the EPR signal).

4.2.2 Neutral (Free) Radicals

Neutral free radicals can be formed via three conceivable pathways. The most straightforward one is homolytic bond cleavage, which can be achieved thermally or photochemically (Scheme 4.1). Alternatively, either oxidation of an anion or reduction of a cation lead to a radical (Scheme 4.2).

Although all of these methods have been widely used in practice, in most cases generation of radicals results from homolytic cleavage reactions. Historically, this

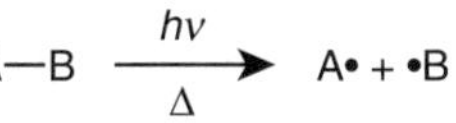

$$A\text{—}B \xrightarrow[\Delta]{h\nu} A\bullet + \bullet B$$

Scheme 4.1

$$X^{\ominus} \xrightarrow{-e^{\ominus}} X\bullet \quad \text{Oxidation}$$

$$X^{\oplus} \xrightarrow{+e^{\ominus}} X\bullet \quad \text{Reduction}$$

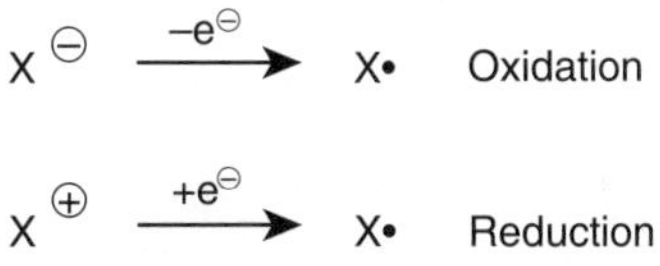

Scheme 4.2

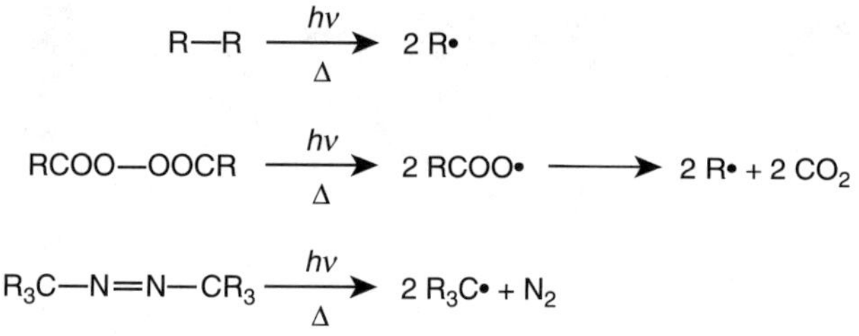

$$R-R \xrightarrow[\Delta]{h\nu} 2\,R\bullet$$

$$RCOO-OOCR \xrightarrow[\Delta]{h\nu} 2\,RCOO\bullet \longrightarrow 2\,R\bullet + 2\,CO_2$$

$$R_3C-N=N-CR_3 \xrightarrow[\Delta]{h\nu} 2\,R_3C\bullet + N_2$$

Scheme 4.3

was a method utilized in the early days of EPR of organic radicals. Hydrocarbons were irradiated by a van de Graaf accelerator (2.8 eV electrons) [2]. The simplest procedure is the (thermal or photochemical) cleavage of a weak bond [3, 4]. This is particularly efficient if the parent molecule consists of two identical moieties connected by a weak bond as in, for example, peroxides, peracids and peresters, or azo compounds (the latter rather efficient in the gas phase). The direct thermal cleavage of C—C bonds requires rather weak bonds as is found in sterically congested molecules (Scheme 4.3).

This procedure requires considerable synthetic effort. Therefore, more generally applicable procedures were developed. Their basis is the production of very reactive precursor radicals that act as (rather) specific abstracting agents.

Di(t-butyl)peroxide was shown to lead to hydrogen abstracting oxyl radicals [5] (Scheme 4.4). This method is particularly suited for the abstraction of H atoms from precursors, which possess one specifically weak X—H bond. Otherwise a mixture of different radicals is generated. The butoxyl radical itself does not give rise to a detectable EPR signal because of excessive line broadening.

A more specific procedure was developed by combining H-atom abstraction from trialkylsilanes or trialkylstannanes. The t-butoxyl radical abstracts an H atom from the silane (or stannane). Because the silicon and tin form particularly strong bonds with halogens, silyl and stannyl radicals selectively abstract halogens from alkyl halides [6–8]. The production of stannyl radicals can also be accomplished by direct photolysis of distannanes (Scheme 4.5).

Several Fenton-type reactions have been reported for the production of radicals. Here, hydrogen peroxide or organic hydroperoxides are reduced by Fe(II) or Ti(III) (Scheme 4.6).

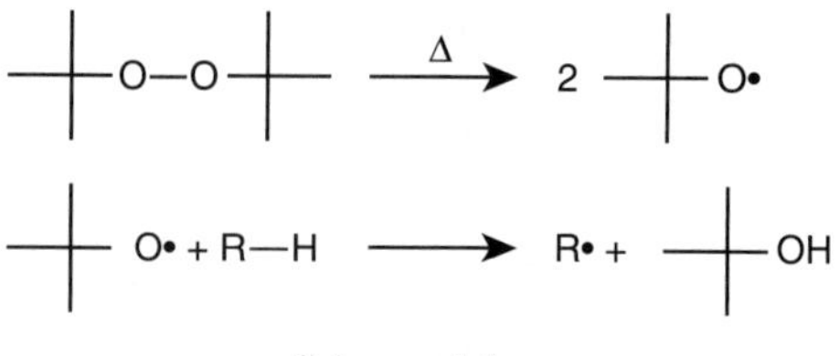

$$\text{t-BuO-OBu-t} \xrightarrow{\Delta} 2\,\text{t-BuO}\bullet$$

$$\text{t-BuO}\bullet + R-H \longrightarrow R\bullet + \text{t-BuOH}$$

Scheme 4.4

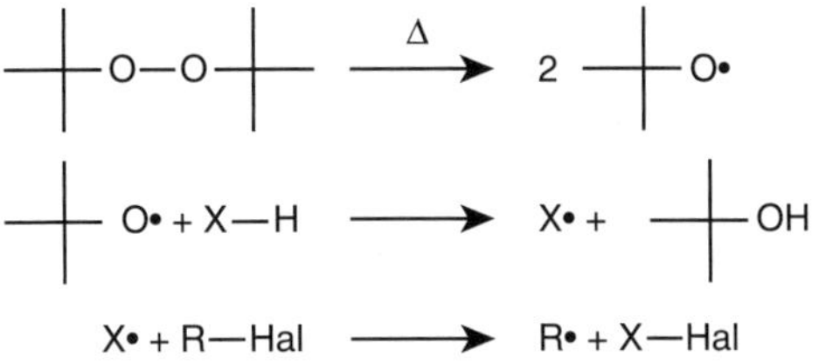

4.2 GENERATION OF RADICALS AND RADICAL IONS

$$\text{>C(O-O)C<} \xrightarrow{\Delta} 2 \text{ >C-O•}$$

$$\text{>C-O• + X-H} \longrightarrow \text{X• + >C-OH}$$

$$\text{X• + R-Hal} \longrightarrow \text{R• + X-Hal}$$

Scheme 4.5

$$\text{OH-OH + Fe(II)} \longrightarrow \text{HO• + OH}^- + \text{Fe(III)}$$

$$\text{RO-OH + Fe(II)} \longrightarrow \text{RO• + OH}^- + \text{Fe(III)}$$

Scheme 4.6

The primary, highly reactive HO• radicals undergo follow-up reactions (hydrogen abstraction or addition to unsaturated systems). The use of such methods is limited to aqueous solutions and mostly leads to short-lived paramagnetic products. Therefore, this approach is rather limited and generally requires rapid-slow flow mixing techniques.

Photolysis of ketones often leads to a Norrish I α cleavage of a C(O)—R bond with formation of an alkyl radical R• and an acyl radical R'(CO)•. The latter may further fragment to another alkyl radical R'• and carbon monoxide.

The electron-transfer methods introduced above make use of Zn powder, electrolysis cells for reductions, and oxygen in the case of oxidations. In alkaline solution, polyphenols are easily deprotonated and the anions are easily oxidized by O_2 to the corresponding phenoxyl radicals [9]. Similarly, one-electron oxidation of phenols, for example, by electron-transfer agents or UV light, is followed by rapid deprotonation of the initial radical cation.

Primarily formed radicals often undergo follow-up reactions, which in some cases may lead to other more persistent radicals. This can be observed when a radical adds to a double bond or to a nitrone or nitroso compound. These last reactions provide the basis of the so-called spin trapping (Chapter 8) used to scavenge short-lived radicals.

Aromatic ethers and halogenated alkanes or aromatics are solvents that are frequently used.

4.2.3 Charged Radicals

The production of charged radicals (radical ions) requires suitable electron-transfer agents. The detection of the EPR spectra of these species is substantially simplified if the follow-up product of the oxidant or reductant is EPR silent or has an EPR spectrum that does not overlap with that of the desired radical ion.

The simplest way of transferring electrons is the use of an electrode. There are commercially available electrolysis cells. However, a very efficient cell can be constructed in a rather simple way: a gold wire (working electrode) is wound into a helical form with a (straight) platinum wire in the center of the helix. This is inserted into an EPR tube (Fig. 4.2) [10]. This cell can be used with standard variable temperature accessories.

Depending on the polarization of the working electrode, such cells can be utilized in a broad range of potentials (reductions and oxidations), including multiple electron-transfer reactions.

If a chemical reaction is used to generate radical ions, distinct procedures have to be followed for radical cations and radical anions. Therefore, these two kinds of species are addressed separately.

4.2.3.1 Radical Cations.
A broad range of oxidants has been utilized to produce radical cations from parent, closed-shell neutral molecules. One of the first reagents was concentrated sulfuric acid that serves as the solvent and the reagent. Although the mechanism of this oxidation has not been solved, this method is rather efficient for a broad range of aromatic hydrocarbons. Alternatively, oxidating acids like CF_3COOH or FSO_3H can be used (nitromethane and dichlorobenzene solvents), which are often coupled with UV photolysis.

In addition, the Lewis acids $AlCl_3$ [11], $SbCl_3$, $SbCl_5$, and, rarely, $ZnCl_2$ were successfully utilized for the production of radical cations. The mechanisms underlying these procedures are not fully established.

One-electron transfer agents, which rely on well-defined mechanisms, are commercially available.

Tris(p-bromophenyl)ammoniumyl hexachloroantimonate is a paramagnetic salt with an oxidation potential of ~ 1 V versus a saturated calomel electrode (SCE). Upon electron transfer, the diamagnetic amine is formed together with the desired radical cation [12]. The 2,4-dibromo derivative has an even higher oxidation power ($E_{1/2} = \sim 1.6$ V vs. SCE). Bis(trifluoroacetoxy)iodo(III)benzene [13] is another reagent that is very efficient. 2,3-Dicyano-5,6-dichlorobenzoquinone [14] and tetrachlorobenzoquinone (chloranil) are strong electron acceptors (able to oxidize substrates with potentials up to ~ 1.6 V vs. SCE). Their ability for oxidation can be enhanced by UV irradiation. The primary follow-up products of these oxidants are the radical anions of these quinones. Accordingly, their EPR spectra may be superimposed over the EPR signal of the radical cation. Because these spectra are well described, the analysis of such signals is feasible. Furthermore, the addition of acids leads to the protonation of the primarily formed semiquinone anions followed by a fast quenching of the semiquinone.

Metal cations are also employed for oxidations. Here, Tl(III) [15], Hg(II) were shown to be particularly efficient (as trifluoroacetate salts). A somewhat weaker oxidant is $AgClO_4$.

Generally, precursors with (reversible) oxidation potentials up to ~ 2 V versus SCE can be produced by choosing an appropriate method. For species with oxidation potentials of >1.5 V, anodic oxidation is the most promising approach in most cases. The most efficient solvents are dichlormethane (CH_2Cl_2) and 1,1,1,3,3,3-hexafluoropropanol [16]. Both solvents must have the highest available purity and be free of moisture.

Radical cations can also be obtained via reduction of dications (or species with even more positive charges). This methodology corresponds to that utilized for the production of radical anions, which is discussed below.

An excellent review on the generation of radical cations is given by Davies [17].

4.2.3.2 Radical Anions. Radical anions are very sensitive against protic impurities, and almost any method for the generation of these species requires "superdry" reaction conditions. The most efficient procedure is the preparation of radical anions under high vacuum. The solvents of choice are ethers like dimethoxyethane (DME), tetrahydrofuran, and 2-methyltetrahydrofuran. The solvents should be dried with K/Na alloy, degassed, and stored under high vacuum. In early times, hexamethylphosphoric triamide was also used but should be omitted because it is highly toxic. For electrolysis and reductions with Zn metal, dry dimethylformamide can be chosen (for substrates with reduction potentials of less than ~ 1 V vs. SCE).

The standard reducing agents for the generation of radical anions are the alkali metals, Li, Na, K, Rb, and Cs (Scheme 4.7).

For the redox reactions, Li is used as a wire (or a mirror is produced by decomposition of *in situ* produced $LiHN_2$), Na and K are sublimed to form a metallic mirror (Fig 4.2), and Cs and Rb mirrors are often created by the initial decomposition of parent azides (which are commercially available).

Zinc mirrors (sublimation of the metal) are efficient for quinones with low reduction potentials. The parent molecule (R) is brought into contact with the metal surface at low temperatures (a dry ice based cooling mixture is suitable). Subsequently, the solution is inserted into an EPR spectrometer to check for a spectrum.

This procedure can be repeated several times and can be enhanced by UV irradiation, if necessary. Extended treatment of the reaction solution can lead to species with higher charges like dianions, (radical) trianions, and so forth.

Particularly for quinones, reduction with $NaBH_4$ (potential range down to ca. -1.2 V vs. SCE) has been shown to lead to highly intense and easily distinguishable EPR spectra [9].

The alkali metals, which are employed as reducing agents, possess magnetic nuclei (^7Li, ^{23}Na, $^{39}K : I = 3/2$; $^{85}Rb : I = 5/2$; $^{133}Cs : I = 7/2$). If spin is transferred from the radical anion to the metal cations, which are produced by the electron-transfer procedure (counterions), interactions between the radical anion and its counterion can be monitored in the EPR signal if this interaction allows an overlap between the orbitals of these two species. An example of this ion-pairing phenomenon will be given below. Tight ion pairs can be omitted by the addition of appropriate (dry) crown ethers.

$$R + M \longrightarrow R^{\bullet -} + M^+$$
$$M = Li, Na, K, Rb, Cs$$

Scheme 4.7

Ketyl or thioketyl radical anions ($R_2CO^{\bullet-}$ or $R_2CS^{\bullet-}$) can be formed by photolysing the corresponding alcohols (or thiols) in a mixture with di(*tert*-butyl) peroxide and potassium alkoxide [18].

4.3 BASIC INTERACTIONS AND PRINCIPLES

4.3.1 Fermi Contact

To find out what kind of information can be obtained from fluid-solution EPR spectra, it is helpful to first consider the principal magnetic interactions in radicals.

The EPR studies in liquid solution essentially address the study of isotropic interactions, which are phenomena that are independent of the spatial orientation of the paramagnetic centres. As indicated in Chapter 1 (§1.11), the Fermi contact term represents the isotropic interaction between the unpaired electron and the nucleus.

$$a_i = (8\pi/3) g \mu_B g_N \mu_N Q(0) \tag{4.1}$$

where a_i is the isotropic hyperfine coupling constant with nucleus i and $Q(0)$ represents the electron spin density on the nucleus, which gives rise to the interaction between the free electron and the nuclear spin. Because the s orbital is the only one with nonzero probability of the electron being present at the position of the nucleus, $Q(0)$ represents the s-orbital spin population; accordingly, a_i contains information about the s character of the singly occupied orbital. In Equation 4.1 the Bohr magneton μ_B and nuclear magneton μ_N are constants but the nuclear g factor g_N is specific for a given nucleus. Consequently, in the EPR spectra of paramagnetic atoms with 100% spin population in an s orbital, quite different a_i values (usually denoted a_0) would be observed for the various nuclei. For instance, the coupling constant for a hydrogen atom a_H is 50.8 mT whereas that of ^{31}P is 364 mT. This is illustrated in Table 4.1, which contains the a_0 values for various atoms.

TABLE 4.1 Size of Coupling Constants for 100% s-Spin Population

Atom	a_i (mT)	s-Spin Population (%) for $a = 0.1$ mT
^1H	50.8	0.20
^2H	7.8	1.28
^{13}C	11.3	0.09
^{14}N	55.2	0.18
^{19}F	1720.0	0.01
^{31}P	364.0	0.03
^7Li	13.0	0.76
^{23}Na	33.1	0.30
^{39}K	8.1	1.29
^{133}Cs	88.1	0.12

4.3 BASIC INTERACTIONS AND PRINCIPLES

The different values indicate that different s-spin populations are needed for different atoms to give the same coupling constant value. For example, to give a coupling constant of 0.1 mT (an isotropic hyperfine coupling constant easily detectable by EPR), a fluorine atom has to possess only 0.001% of s-spin whereas 1.28% s-spin population is necessary for ^{2}H. This also shows that replacing a proton (^{1}H) by a deuteron (^{2}H) leads to a decrease of the a value by a factor of 0.15. The s-spin populations for each nucleus needed for a coupling constant of 0.1 mT are reported in Table 4.1 (third column).

4.3.2 Spin Polarization

In most typical organic radicals the spin resides in π-type orbitals, and it is distributed within the entire molecule. The a_i values thus reflect the portion of spin population residing at nuclei with nuclear spin $\neq 0$ (magnetic nuclei). For example, have a look at the radical anion of benzene. Because the magnetic nucleus ^{13}C is only present at a low amount (^{12}C has no nuclear spin), only its six equivalent protons need to considered (^{1}H, nuclear spin $= 1/2$), for which an a_H value of -0.375 mT was experimentally determined [19]. This value indicates that only 0.74% of the spin is present at one H atom; that is, only $6 \times 0.74\% = 4.43\%$ of the overall spin reside at the H atoms. This is not much! Bear in mind that the benzene radical anion is a π radical. The unpaired electron is almost exclusively delocalized between the p$_z$ orbitals of the benzene sp^2 hybridized C atoms. From these orbitals the spin has to be transferred to the H nuclei to be detectable via the a_H. This is illustrated in Fig. 4.3: the adjacent hydrogen atoms (α hydrogens) reside in the nodal plane of the π system. Considering only the unpaired electron described by the π orbital, there should not be any spin density on these α protons. However, taking into account the interaction of the unpaired electron with the electron pair in the

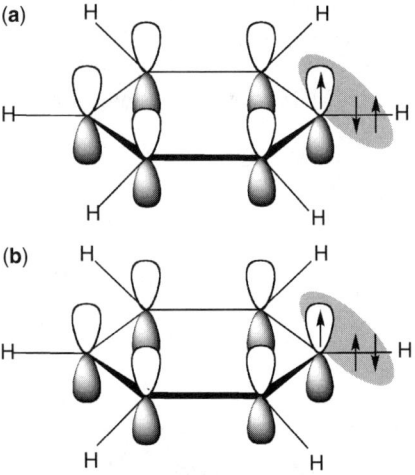

Fig. 4.3 Principle of the π–σ spin polarisation.

C—H bond, according to the model of π–σ *spin polarization*, a spin density is transferred to the α protons.

In a nutshell, the two "spin-up" and "spin-down" electrons of the C—H$_\alpha$ bond both have a *repulsive Coulomb interaction* with the unpaired electron. However, this repulsive interaction is slightly weaker with the electron of the C—H bond that has the *same* spin state as the unpaired electron than with the other one (an energy difference due to the quantum properties of the electron, in particular, due to the Pauli principle). Now let us suppose that the unpaired electron resides in the spin-up state (see Fig. 4.3). Because its repulsion interaction with the spin-up electron of the C—H bond is weaker than with the spin-down one, the former electron will have a slightly larger probability than the other one to be found nearer to the π distribution. This unbalanced spatial distribution depending on the electron spin state is called *spin polarization*. The two possible unbalanced distributions of the two electrons of the C—H bond are schematically indicated in Fig. 4.3a and b. The case in Fig. 4.3b is therefore the one favored energetically. Thus, on the H$_\alpha$ there is always a spin density with a sign opposite to that on the π orbital on the carbon atom. Therefore, the hyperfine coupling constant a_H is *negative*.

4.3.3 Hyperconjugation

Remarkably, the hyperfine coupling constant of the 12 methyl protons in the radical cation from 1,4,5,8-tetramethylnaphthalene is similar (in fact, even bigger than) to that exhibited by the four 1,4,5,8 protons in the radical cation of unsubstituted naphthalene (Fig. 4.4), even though the methyl protons are more distant from the π system carrying the unpaired electron and the electron distribution over the naphthalene skeleton is essentially unchanged.

Therefore, the spin transfer between the π orbitals of the naphthalene moiety and the methyl hydrogens must be rather efficient. This is explained by a *hyperconjugative* mechanism of spin transfer.

Figure 4.5a displays a methyl group attached to an sp^2 center. Suppose that the 2p orbital of the sp^2 center belongs to the π orbital bearing the unpaired electron. A simple molecular orbital (MO) treatment demonstrates that an amount of spin

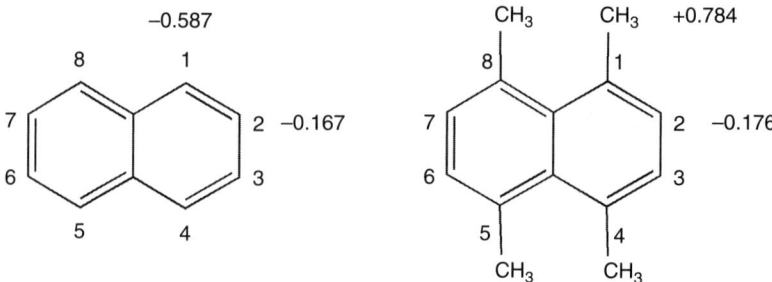

Fig. 4.4 The hyperfine coupling constants (mT) for the naphthalene and the 1,4,5,8-tetramethylnaphthalene radical cation.

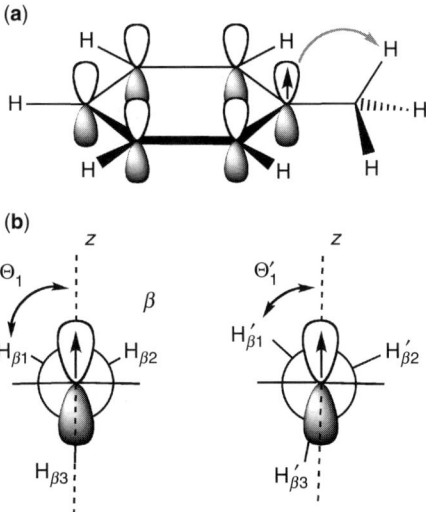

Fig. 4.5 Principle of hyperconjugation.

density is transferred from the π orbital to the H atoms of the CH_3 group in a "direct" way by the hyperconjugative mechanism. Therefore, the sign of the spin density on the H atoms (in β position with respect to the unpaired electron) is the same as that in the π orbital. Consequently, this through-space type of interaction leads to a *positive* sign of the coupling constant. This interaction is particularly efficient if the C—H bond is oriented parallel to the z axis of the π orbital, because the overlap between the C—H bond and the $2p_z$ orbital is maximum in this case, as is the spin density transfer.

Figure 4.5b shows the C—CH_3 group, looking along the C—C bond, with the —CH_3 group in two different orientations with respect to the $2p_z$ axis, corresponding to two values of dihedral angle Θ between the z axis of the spin carrying p_z orbital and the C_β—H_β bond.

A useful semiempirical expression correlates the values of $a_H(H_\beta)$ with dihedral angle Θ and the spin density $\rho(C_\alpha)$ on the π center C_α [20, 21]:

$$a_H(\beta) = \rho(C_\alpha) \cdot (A + B\cos^2(\Theta)) \tag{4.2}$$

where A and B are empirical parameters: A describes the spin polarization transferred through two bonds, and B reflects the transfer of spin by the hyperconjugative effect. In the semiempirical expression 4.2 the parameter A is frequently neglected, because $|A| \ll B$. Note that the B parameters are significantly different for radical anions (4–5 mT) and radical cations (8–9 mT).

According to this model, the value of $a_H(\beta)$ depends on the conformation of the radical and particularly on the orientations of the C_β—H_β bonds with respect to the π axis. If dihedral angle Θ is 0°, $a_H(\beta)$ reaches its maximum (see $H_{\beta 3}$ in Fig. 4.5b). The two virtually equivalent protons $H_{\beta 1}$ and $H_{\beta 2}$ with $\Theta_1 = 60°$ would have smaller

$a_H(\beta)$ values. In contrast, if the methyl group is freely rotating, the average value $\langle \cos^2 \Theta \rangle = 1/2$ must be introduced in Equation 4.2; therefore, the $a_H(\beta)$ values for all three methyl hydrogens are averaged and become identical.

In general, Equation 4.2 is very useful for the determination of the conformations of radicals. This will be demonstrated in an example below.

4.3.4 The g Factor

The center of the EPR signal of a single paramagnetic species is determined by its g factor (see Chapter 1). The g factors of organic radicals are generally similar to that of the free electron, as far as they are composed by "light" atoms. Spin–orbit coupling leads to characteristic shifts of the g factor, particularly when heavier elements are involved. Although these shifts for organic radicals are only on the order of 10^{-3}–10^{-4}, they can be established experimentally and are rather characteristic for types of radicals, depending on the atoms bearing the spin density. Therefore, they help in determining when the unpaired electron resides on atoms different from carbon, information that is particularly useful for nuclei (e.g., ^{16}O, ^{32}S), which do not give rise to hyperfine coupling.

For nitrogen-containing radical ions, g is in the range of 2.0035–2.0055. For nitroxyl radicals it can rise to 2.0065, and for radicals carrying a high amount of spin at sulfur the g factors are 2.0075 ± 0.001. In the case of a high σ character, they are close to the value of the free electron, $g_e = 2.0023$.

4.4 PATTERNS AND LINE SHAPES OF FLUID-SOLUTION EPR SPECTRA

The fundamental mechanisms of spin transfer detectable in a fluid-solution (isotropic) spectrum, spin polarization, and hyperconjugation have been introduced. The following questions will be answered in this section:

1. How do the hyperfine interactions affect the EPR spectra of paramagnetic species?
2. Which phenomena can be monitored by the EPR spectra?

4.4.1 Interactions and Multiplicities

Section 4.1.2 demonstrated that the interactions between an unpaired electron and nuclear spins are mainly responsible for the patterns of fluid-solution EPR spectra. Whereas the spin quantum number of a free electron is $S = 1/2$ with magnetic spin quantum number $m_s = \pm 1/2$ (thus producing two energy levels), the corresponding quantum number for a magnetic nucleus can amount to $I = 1/2$, 1, 3/2, 2, ..., with magnetic nuclear spin quantum numbers m_i ranging from $-I$, $-I+1$, ..., to I ($2 \cdot I + 1$ values). Figure 4.6 displays the energy level schemes

4.4 PATTERNS AND LINE SHAPES OF FLUID-SOLUTION EPR SPECTRA

produced by the interaction of an electron with a nucleus with $I = 1/2$ (e.g., ^{1}H, ^{31}P) and with $I = 1$ (e.g., ^{2}H, ^{14}N).

The arrows in Fig. 4.6 indicate the allowed EPR transitions with $\Delta m_s = 1$ and $\Delta m_i = 0$ (see Chapter 1). Accordingly, the two-spin systems, representative for the interaction of an unpaired electron with one $I = 1/2$ nucleus (e.g., ^{1}H or ^{31}P), leads to a two-line EPR spectrum in which the distance between the lines represents the isotropic hyperfine coupling constant a. In the same way, the interaction of an unpaired electron with an $I = 1$ nucleus leads to three equidistant EPR lines with a $1:1:1$ ratio. Generally, one nucleus with a nuclear spin $I \neq 0$ leads to $2 \cdot I + 1$ EPR lines.

What happens if several equivalent nuclei are present? This is displayed in Fig. 4.7 for two and three equivalent nuclei with $I = 1/2$. Evidently, the presence of two or

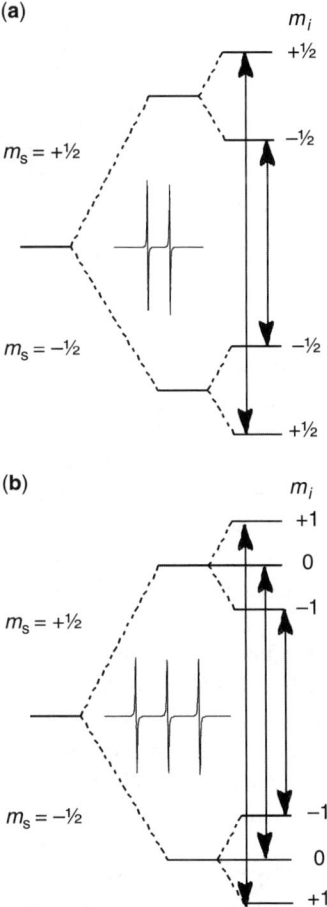

Fig. 4.6 Energy levels and EPR spectra for a two spin system (electron and nuclear spin) with (a) $S = 1/2$, $I = 1/2$, and (b) $S = 1/2$, $I = 1$.

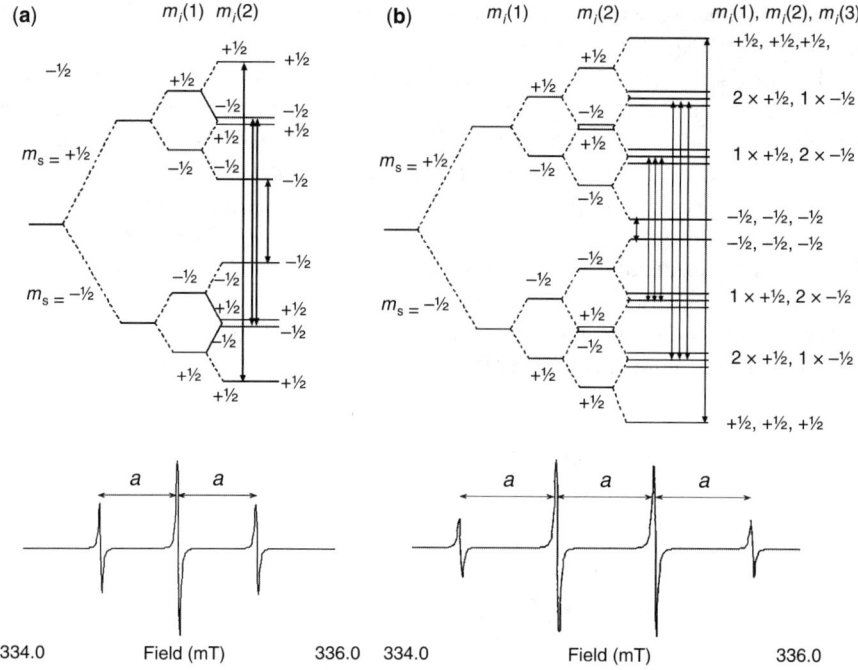

Fig. 4.7 Interaction of an unpaired electron with (a) two and (b) three equivalent nuclei with $I = 1/2$ and corresponding spectra.

more equivalent nuclei with identical a leads to sets of degenerate energy states. For two equivalent nuclei ($I = 1/2$), with (m_{I1}, m_{I2}) indicating their spin state, the two states (1/2, −1/2) and (−1/2, 1/2) are doubly degenerate because the nuclei are not distinguishable whereas (1/2, 1/2) and (−1/2, −1/2) are not. Therefore, some transitions possess a higher intensity than others: as displayed in Fig 4.7a, this leads to a triplet of EPR lines with an intensity ratio of 1 : 2 : 1. In an analogous way, in the presence of three equivalent $I = 1/2$ nuclei, a ratio of 1 : 3 : 3 : 1 is produced (Fig. 4.7b); again, the distance between the lines corresponds to the isotropic hyperfine coupling constant a.

Extending this concept, the interaction with a set of n nuclei with $I = 1/2$ will result in a multiplet of $n + 1$ lines with intensity ratios dictated by a binomial distribution. Accordingly, the outermost EPR lines always have the lowest intensity and, if the number of equivalent nuclei is high, it may become rather difficult to detect them.

If the nuclei possess a spin I, the number of lines for n equivalent nuclei is $2nI + 1$. The intensity ratios can be calculated in the same way as those for $I = 1/2$; for example, for a $I = 1$ nucleus, like ^{14}N, in the presence of two equivalent N atoms leads to five lines with a 1 : 2 : 3 : 2 : 1 ratio. This is illustrated in Fig. 4.8.

Using the above rules, it is simple to predict that the benzene radical anion with six equivalent H atoms gives rise to an EPR spectrum with seven equidistant lines with a 1 : 6 : 15 : 20 : 15 : 6 : 1 intensity ratio. Obviously, most molecules do not have such a

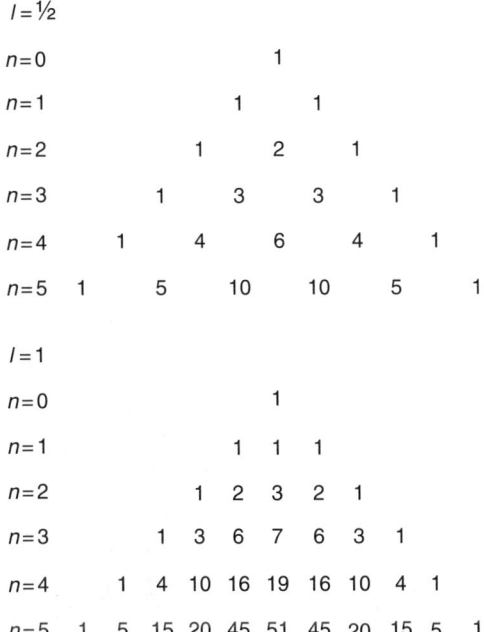

Fig. 4.8 Intensity ratio of the EPR lines for $S = 1/2$ radicals containing n equivalent nuclei having $I = 1/2$ (top) and $I = 1$ (bottom).

high symmetry. In naphthalene, two different sets of nuclei are present: four equivalent ones in the 1,4,5,8 position and four in the 2,3,6,7 position. Each of these sets has a different a_H value and leads to a splitting of the EPR resonance into five equidistant lines. This is shown in Fig. 4.9.

The signal is split into a $1:4:6:4:1$ quintet by the interaction of the electron spin with the four equivalent ^{1}H nuclei in the 1,4,5,8 position with an a_H of 0.495 mT. Each of these lines is again split into a quintet [H(2,3,6,7), $a_H = 0.183$ mT]. The $1:4:6:4:1$ ratio of these lines is related to the intensity of each component in the first quintet as indicated by the triangles in Fig 4.9. Accordingly, the spectrum contains $5 \times 5 = 25$ lines.

Independent of the number of nuclei, the distance between the two outermost lines always marks the smallest a. The overall width of the EPR spectrum corresponds to the sum $\sum(2 \times I_i \times n_i \times a_i)$, where n is the number of equivalent a. For the naphthalene radical anion the width of the EPR signal is $(2 \times 1/2 \times 4 \times 0.495 + 2 \times 1/2 \times 4 \times 0.183)$ mT $= 2.712$ mT.

From the scheme reported in Fig. 4.8 we see that in the case of an even number of nuclei with $I = 1/2$, a central line is produced. This line always possesses the highest intensity and can be easily distinguished in the spectrum displayed in Fig. 4.9. The overall number of lines is $\Pi(2 \times I_i \times n_i + 1)$. For example, for naphthalene, $(2 \times 1/2 \times 4 + 1) \times (2 \times 1/2 \times 4 + 1) = 25$.

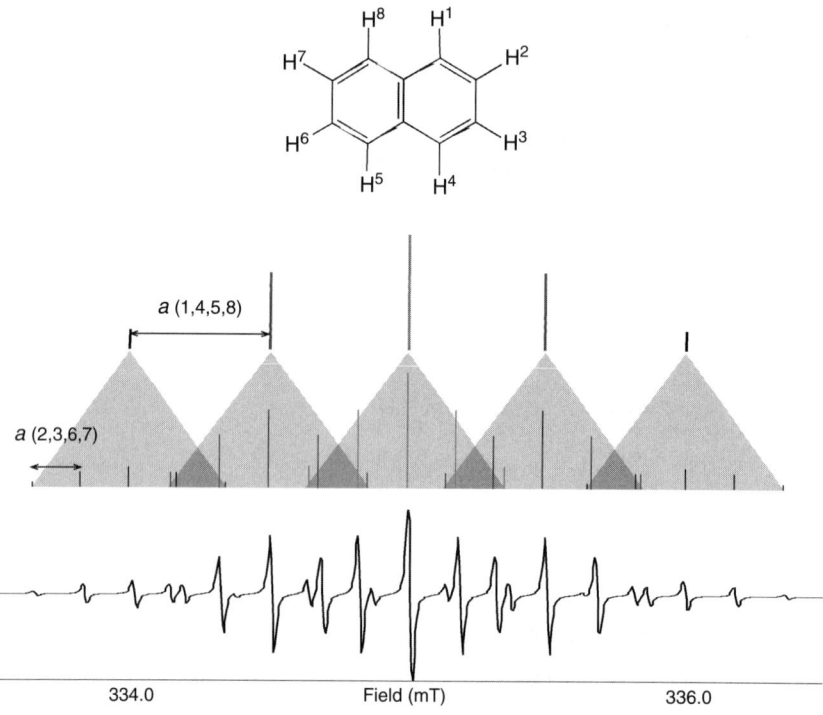

Fig. 4.9 EPR pattern of the naphthalene radical ion.

4.4.2 Dynamics

Consider the effect on the EPR spectral profile that is attributable to an intramolecular motion in the paramagnetic species. As in any spectroscopy, EPR has its own dynamic time scale.

Scheme 4.8 exemplifies a proton exchange between two distinguishable states A and B with different hyperfine coupling constants a_i^A and a_i^B. The frequency jump in the EPR spectrum due to the exchange Δv is given by the difference of the coupling constants between the two positions A and B:

$$\Delta v = 2\pi \gamma_e (|a_i^A| - |a_i^B|) \tag{4.3}$$

For example, A and B could be two positions of a proton in a rotating methyl group.

We can compare $\Delta \omega$ ($=2\pi\Delta v$) with the rate of rotation τ_R^{-1}. When $\tau_R^{-1} \gg \Delta \omega$, the rotation is so fast that during an EPR transition the hyperfine constant is averaged and

$$A \rightleftharpoons B$$

Scheme 4.8

4.4 PATTERNS AND LINE SHAPES OF FLUID-SOLUTION EPR SPECTRA

the protons are detected as being equivalent. If the rotation is severely hindered and $\tau_R^{-1} \ll \Delta\omega$, the protons are detected as not equivalent in the EPR spectrum.

In general, taking into account the hyperfine interactions, for an organic radical the $\Delta\nu$ values defined in Equation 4.3 range from ~100 to ~1 MHz. Therefore, for exchange motions between states with different hyperfine coupling constants, if the states are longer lived than ~10^{-6} s, they can be distinguished by EPR; if they are shorter lived then ~10^{-8} s, only averaged hyperfine data can be obtained. In the intermediate time regime, dynamic phenomena determine the shape of the EPR lines. The rate of interconversion between A and B defines the lifetime of the two states, and the simulation of the corresponding (e.g., temperature dependent) line shapes can be used to gain insight into the molecular dynamics (flexibility, rotational barriers, counterion migration).

Let us consider a radical containing a methylene (CH_2) group adjacent to a π system (Fig. 4.10). In conformation A in Fig. 4.10, one bond (C—H) forms a dihedral angle of 30° with the axis of the p_z orbital while the other bond (C—H') is in the nodal plane of the π system. Then, in conformation B the positions of the two protons are exchanged.

Conformations A and B correspond to identical EPR spectra. More precisely, assuming that the proton residing in the nodal plane of the p_z orbital has $a_H = 0$ whereas that close to the z axis of the p orbital has $a_H = 0.5$ mT, a doublet EPR spectrum results, with the latter undergoing hyperfine splitting (Fig. 4.11).

Now imagine what happens when the two protons exchange their positions in an intramolecular motion. The nuclear spin states of the two protons can be (1/2, 1/2), (1/2, −1/2), (−1/2, 1/2), and (−1/2, −1/2). When the two protons have the same

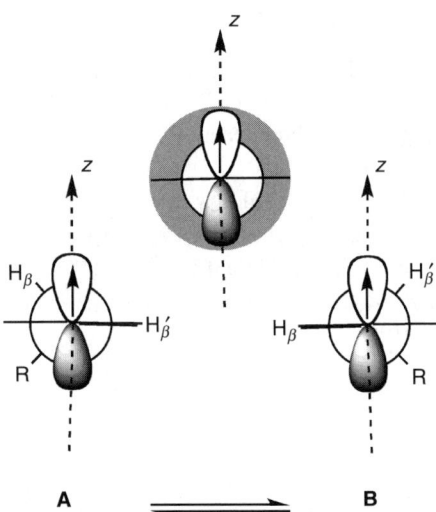

Fig. 4.10 Two hypothetical conformations, A and B, of a methylene group adjacent to an unpaired electron in a p orbital.

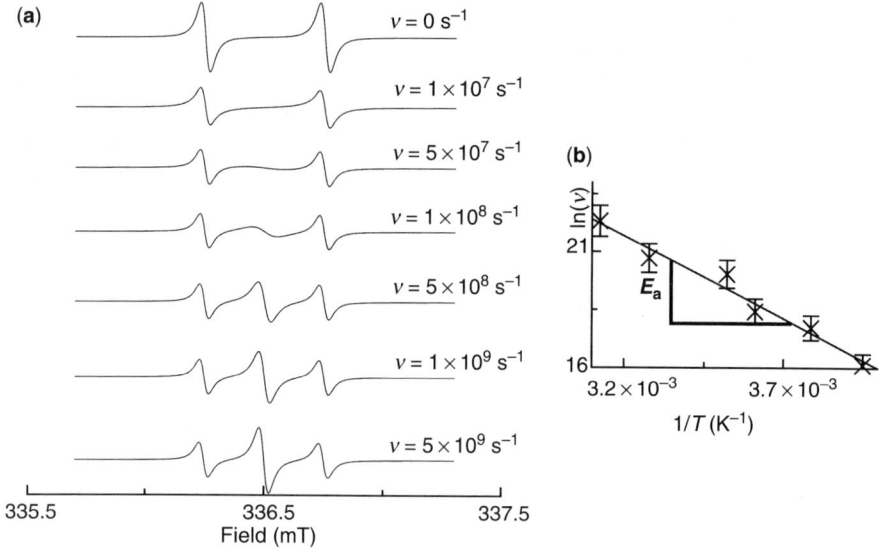

Fig. 4.11 (a) Line broadening calculated for the system shown in Fig. 4.10. (b) Arrhenius type plot for the determination of the activation energy.

spin, the conversion between A and B does not affect the spectrum; when they have different spin states, an exchange of the protons also exchanges the two lines of the doublet. When the rate of the exchange τ_R^{-1} becomes comparable with the hyperfine splitting (which corresponds to $\Delta\omega$ in this case), the latter components of the spectrum are broadened [22]. Figure 4.11a shows how the spectrum changes when increasing the rate of interconversion. Note that the two components at the positions of the initial doublet, corresponding to the EPR transitions between states with nuclear spins (1/2, 1/2) or (−1/2, −1/2), do not change at any rate of interconversion whereas the components corresponding to EPR transitions between states with different nuclear spins are progressively broadened. When the interconversion rate between A and B reaches the fast-exchange limit ($\tau_R^{-1} \gg \Delta\omega$), the two protons become equivalent and the spectrum becomes a triplet with 1 : 2 : 1 intensities, with an average $a_H = 0.25$ mT. The exchange frequency ($\nu = \tau_R^{-1}/2\pi$) can be determined from the simulation of the dynamic EPR spectra [23, 24] and plotting ln(ν) versus $1/T$ allows the determination of the activation energy E_a, according to an Arrhenius-type treatment (Fig. 4.11b):

$$\nu = \nu_0 \cdot e_a^{-E/RT} \qquad (4.4)$$

Line broadening due to dynamic phenomena depends on the exchange of EPR lines corresponding to specific magnetic quantum numbers, as shown above. Accordingly, in some cases the temperature-dependent linewidth variations may also provide information on the sign of the coupling constant of a given nucleus. An example is provided in Fig. 4.12a and b, which shows the spectra recorded and simulated for the adduct between benzo[2,1-b;3,4-b′]dithiophen-4,5-dione and the

4.4 PATTERNS AND LINE SHAPES OF FLUID-SOLUTION EPR SPECTRA

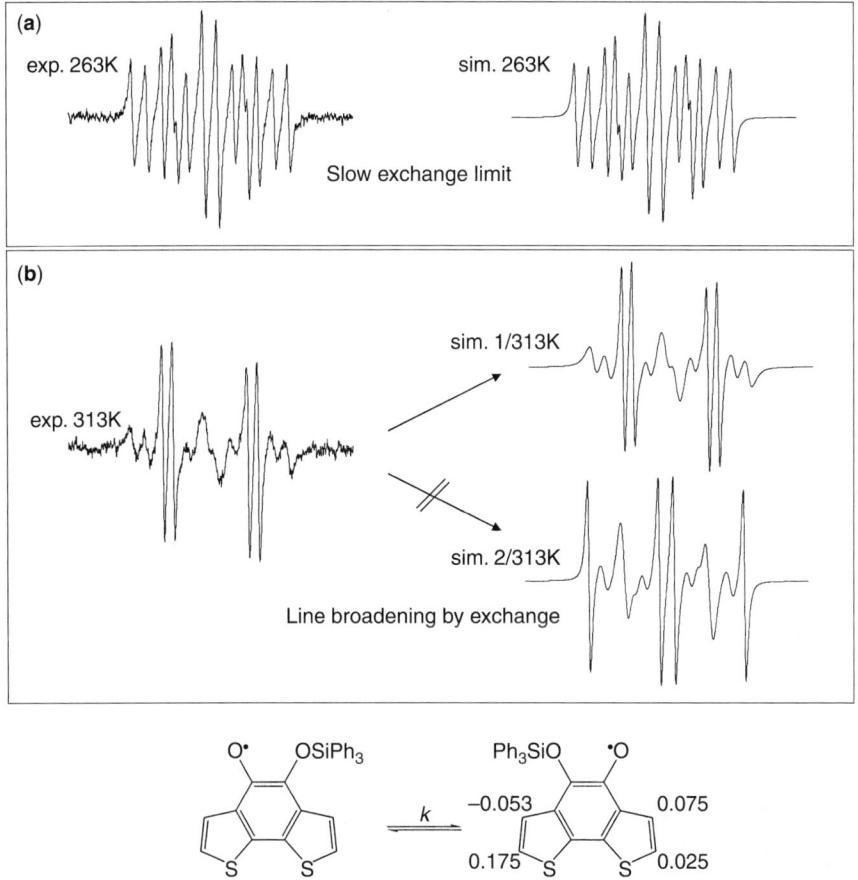

Fig. 4.12 EPR spectra and simulations of the triphenylsilyl adduct to TQ at two different temperatures (263K and 313K). Hyperfine values used in case (a) are indicated in the formula.

triphenylsilyl radical at 263 and 313K. Figure 4.12b shows two simulations: sim 1, obtained by using the values of the hyperfine coupling constants with the signs shown below, and sim 2 where the signs of the couplings are taken as all positive. The good agreement of sim 1 with the experimental spectrum confirms the choice of the relative signs of the hyperfine coupling constants.

4.4.3 Dynamics and Anisotropy

The tumbling of paramagnetic species in solution generally affects their EPR spectral profile. Generalizing the scheme discussed in the previous paragraph, the spectral profile will depend on the rate of the tumbling motion as compared to $\Delta\omega$ ($= 2\pi\Delta\nu$). In this case, the frequency jumps in the EPR spectrum are due to the varying orientations of the tumbling radical in the magnetic field and therefore to the anisotropy of the magnetic parameters of the spin Hamiltonian. If $\tau_c^{-1} \gg \Delta\omega$,

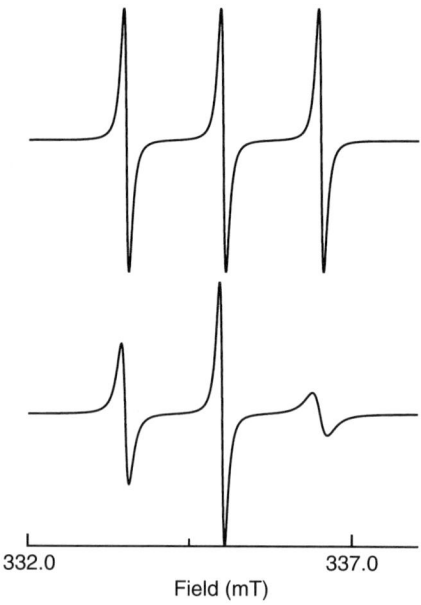

Fig. 4.13 Simulated spectra of a nitroxide radical in two solvents of different viscosity.

where τ_c^{-1} is the correlation rate of the motion in solution and $\Delta\omega$ is the anisotropy of the hyperfine and Zeeman interactions, the paramagnetic species tumble "freely" in fluid solution. However, if the (local) viscosity around the radical increases, not all anisotropic interactions are averaged out. This is particularly pronounced for atoms with a high amount of spin population in p-type orbitals (e.g., C, N, O, P, S) and therefore with high hyperfine anisotropies. This can lead to noticeable line broadening. The simulation of the spectral profile will be treated in detail in Chapter 7, but here a short account of the effects of relatively fast motions in solution (Redfield limit) are given. A detailed derivation of this case is given in the Atherton book in the Further Reading Section. Figure 4.13 reports an example of how a spectral profile for a nitroxide radical is modified by increasing the viscosity of the solvent (lower spectrum).

For a radical bearing a hyperfine interaction with a nucleus with m_I spin quantum number, the increase of the linewidth in magnetic field units (ΔB) depends on the nuclear magnetic spin quantum number associated to a specific EPR line [25]:

$$\Delta B = A + B \cdot m_I + C \cdot m_I^2 \tag{4.5}$$

Parameters A, B, and C depend linearly on the correlation time τ_c and on the anisotropic hyperfine and g tensors:

$$A = (2/15)\omega_0^2(g^0 \cdot g^0)\tau_c$$
$$B = (4/15\hbar)\omega_0(g^0 \cdot A^0)\tau_c \tag{4.6}$$
$$C = (2/15\hbar^2)(A^0 \cdot A^0)\tau_c$$

4.4 PATTERNS AND LINE SHAPES OF FLUID-SOLUTION EPR SPECTRA

where ω_0 is the Larmor frequency and g^0, A^0 are the **g** and hyperfine tensors, respectively, containing only the anisotropic part (after the trace of the tensors has been subtracted):

$$g^0_{ij} = g_{ij} - 1/3 \operatorname{Tr}(\tilde{g})\delta_{ij}, \qquad A^0_{ij} = A_{ij} - 1/3 \operatorname{Tr}(\tilde{A})\delta_{ij}$$

The inner products of the tensors are indicated in parentheses. (*In this case*, the inner product of the two tensors is the trace of their tensorial product. See Appendix A.6.1 in Chapter 6.)

Because A and C depend on the square of the anisotropies of the Zeeman and hyperfine interactions, respectively, the first and third terms in Equation 4.5 are always positive. In contrast, the term $B \cdot m_I$ can be positive or negative, and it critically determines which line is broadened more in the EPR spectrum. Note that because A and B depend on the Larmor frequency, the spectral profile will depend not only on the rate of the tumbling of the radical but also on the frequency band used in the EPR experiment (see Chapter 2).

Persistent nitroxide radicals (Chapter 3) have found manifold uses as spin labels or spin probes for the investigation of the local mobility in various contexts. The three-line EPR spectra of nitroxides offer information about the local environment, because their spectral profiles depend on the mobility of the >NO• moiety (see Fig. 4.13). Line broadening of the EPR lines of such spin probes has been applied in material science (e.g., phase transitions) as an indicator for local viscosities. Site-directed spin labeling [26] is a frequently utilized approach to gain insight into the properties of proteins or membranes (see Chapter 10).

Line broadening occurs for nitroxides when the rotational correlation time τ_c becomes longer then ~10 ns. A simple expression (Debye equation) gives the correlation time for the tumbling in solution:

$$\tau_c = \frac{\eta \cdot V}{k \cdot T} \qquad (4.7)$$

where η is the (local) viscosity, V is the rotational volume of the paramagnetic species, and k is the Boltzmann constant.

If the shapes of the EPR spectra are measured at different temperatures or under various conditions (pressure, pH value, etc.), the changes in the spectra reveal microscopic environments in terms of the local viscosity.

Frequently an empirical convenient formula [27, 28] is used to calculate τ_c from the intensities of the three lines:

$$\tau_c = C \cdot \Delta B_{+1} \left(\sqrt{\frac{h_{+1}}{h_{-1}}} - 1 \right) \qquad (4.8)$$

where ΔB_{+1} is the linewidth of the line at low field and h_{+1}, h_{-1} are the heights of the lines at low and high fields, respectively. Alternatively, the spectra can be simulated using a program developed by Freed et al. and the data extracted from the simulated spectra [29] or with the integrated approach described in Chapter 7.

The dynamic range of the spin probe technique in terms of rotational correlation time is 0.01–300 ns. This can be extended to lower processes (microseconds to milliseconds) by using more advanced saturation-transfer techniques [30].

The calculations of spectra that take dynamics into account will be treated in detail in Chapter 7.

4.4.4 Isotope Satellites

A wide range of elements is present as a mixture of isotopes in natural abundance. However, many dominating isotopes such as ^{12}C, ^{16}O, ^{32}S, or ^{28}Si are not magnetic ($I=0$), and their magnetic isotopes are present at low percentages (e.g., ^{13}C, ^{15}N, ^{29}Si, all with $I=1/2$). Taking ^{13}C as an example, the EPR spectrum of a radical with a carbon atom is given by the superposition of a spectrum corresponding to species with ^{12}C (main spectrum) and a much weaker one due to the radicals containing a ^{13}C isotope. This spectrum is wider than the main one, because there is further hyperfine coupling with ^{13}C. Therefore, it gives rise to the so-called isotope satellites outside the main spectrum. They are of much lower intensity (in the example above each satellite line has an intensity of 0.5% of that of the main spectrum) and consequently they become discernible only if the EPR signal possesses a very high intensity. The intensity of the satellite line is increased if the studied molecules are isotopically enriched or if the nuclei are present in several symmetrically equivalent positions.

The EPR spectrum of the radical anion of benzene can be used as an example. Its main EPR spectrum corresponds to radicals with six ^{12}C isotopes, and it is given by seven equidistant lines stemming from the six equivalent protons. For each equivalent position in the molecule there is a 1.1% probability of finding a ^{13}C isotope. Consequently, the probability of finding a benzene molecule with one ^{13}C nucleus is 6.6%. The EPR spectrum of the benzene radical anion is therefore given by the superposition of the main spectrum with 93.4% of the total intensity showing only ^1H a values and of the satellites' spectrum with 6.6% of the total intensity also showing a ^{13}C coupling constant. Two examples of EPR spectra in which satellite lines are clearly visible are shown in a later figure in this chapter and in Figs. 8.1d and 8.12 in Chapter 8.

4.4.5 Second-Order Effects

In very rare cases, if a values exceed ~ 1.2 mT, additional (unexpected) splittings can be detected in highly resolved EPR spectra. This observation can be traced back to a second-order effect [31] that becomes detectable only when the hyperfine energy is compatible with the Zeeman energy. An approximate formula for the line shift is

$$\Delta B = -\frac{a^2}{2 \cdot g \cdot \beta_e \cdot h \cdot \nu}[^\circ I(^\circ I + 1) - ^\circ m_i^2] \qquad (4.9)$$

where ΔB is the line shift in magnetic field units and $^\circ m_i$ is the magnetic quantum number of the overall nuclear spin $^\circ I$, which is the overall nuclear spin for a set of equivalent nuclei (e.g., 1 for two equivalent ^1H nuclei, $^\circ m_i$ then becoming +1, 0, −1).

Each EPR line thus has its individual second-order shift. An example is shown in Section 4.10.4.

4.4.6 Electron Nuclear Double Resonance (ENDOR) in Fluid Solution

Often in radicals and in radical ions several magnetic nuclei interact with the unpaired electron. Accordingly, the number of EPR lines becomes very high, leading to rather complicated patterns (Fig. 4.14a) or unresolved signals (owing to the high number of overlapping lines, Fig. 4.14b). Such EPR spectra generally cannot be analyzed in an unambiguous way. In such cases, it is very valuable to utilize ENDOR, which leads to much simpler spectra.

The ENDOR technique is treated in detail in Chapters 1 and 2 in this book. Remember that just a pair of ENDOR lines is obtained for each set of nuclei having the same hyperfine coupling constant, independently of their number. Accordingly, the number of allowed (or detectable) transitions (i.e., the number of lines) is generally

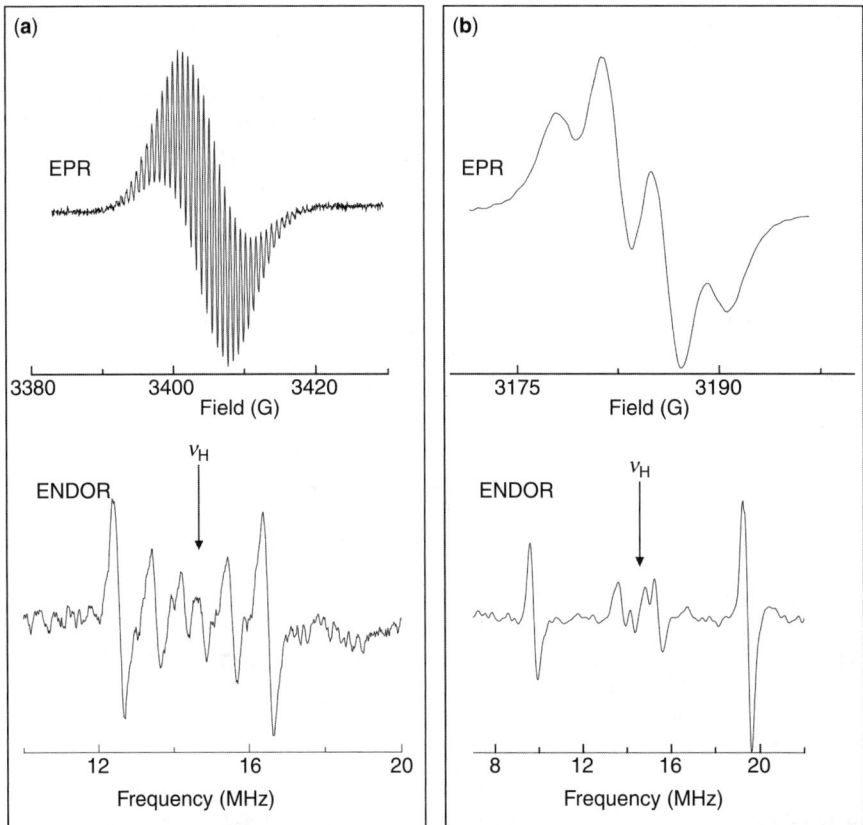

Fig. 4.14 Highly resolved multiline (a) and poorly resolved (b) EPR signals reported with the corresponding ENDOR spectra (bottom).

greatly reduced in ENDOR spectra with respect to EPR. For example, for a naphthalene radical anion the ENDOR spectrum is given by 4 lines compared to 25 EPR lines (see Fig. 4.9).

4.5 TRANSITION-METAL IONS

Fluid-solution EPR spectra of transition-metal ions with an $S = 1/2$ state are not reported as often as one might expect because there are some features that limit their detection. Obviously, any transition-metal ion in solution carries a ligand sphere. Depending on the character of the ligands and their arrangement (i.e., the shape of the first coordination sphere, e.g., octahedral, tetrahedral, tetragonal), the transversal relaxation times T_2 (Chapter 1) can vary considerably. They are usually quite short, causing broad EPR lines; it frequently happens that T_2 becomes so short that the lines are broadened beyond recognition. In these cases the samples have to be frozen to low temperatures (3–77K) to provide detectable EPR spectra. The analysis of such solid-state anisotropic spectra is discussed in Chapter 6.

The most typical $S = 1/2$ metal cations are Cu^{2+}, VO^{2+}, and Mn^{2+}.[1] The main isotopes of copper, ^{63}Cu and ^{65}Cu, have a nuclear spin $I = 3/2$, leading to EPR spectra with four lines with a $1:1:1:1$ intensity ratio (Fig. 4.15). The distance between these lines corresponds to the isotropic hyperfine coupling constant of the Cu nuclei, a_{63Cu} and a_{65Cu}. Typically, this value is in the range 4–6 mT. The vanadyl cation VO^{2+} gives EPR spectra exhibiting eight equidistant lines ($a_{51V} = 10-11$ mT)

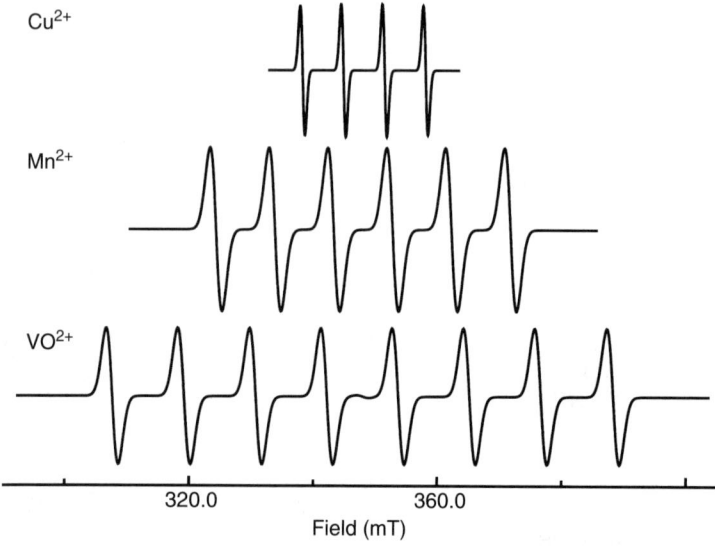

Fig. 4.15 Calculated EPR spectra of Cu^{2+}, VO^{2+}, and Mn^{2+} ions in low viscosity solutions.

[1]Mn^{2+} is a $S = 5/2$ ion that in solution can be treated as a $S = 1/2$ ion. See Chapter 6.

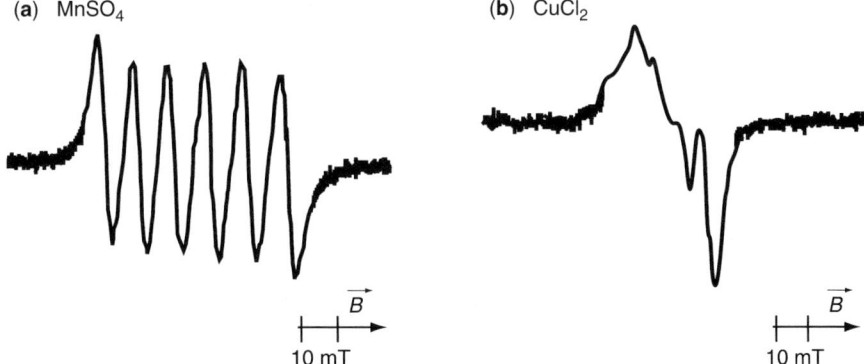

Fig. 4.16 Experimental spectra of Mn^{2+} in water and Cu^{2+} in DMSO.

corresponding to $I = 7/2$ for the ^{51}V nucleus. Analogously for Mn^{2+}, the ^{55}Mn nucleus with $I = 5/2$ leads to six equidistant lines.

In many cases the unpaired electron of the transition-metal cation (residing in a d-type orbital) interacts with magnetic ($I \neq 0$) nuclei of the adjacent ligands. Often, interactions with ^{14}N nuclei give rise to distinguishable splittings. They are ~1 order of magnitude smaller than those of the metal cations and are denominated as superhyperfine splittings. The shapes of these EPR spectra obey the same rules as discussed for the organic radicals.

The simulated spectra shown in Fig. 4.15 display "ideal cases" of isotropic solutions with complete averaging of the magnetic anisotropies. Real EPR spectra of samples containing transition-metal ions reveal lines selectively broadened by slow molecular tumbling and second-order effects. Figure 4.16 present the experimental spectra of Mn^{2+} and Cu^{2+} in solution.

For the spectrum of $MnSO_4$ (Fig. 4.16a), the separation between the two outermost lines on the right edge is by 1.1 mT less than that at the left edge. This is due to a second-order effect. The average value of 9.577 mT corresponds to the distance between the two central lines. In the case of the signal attributed to $CuCl_2$ (Fig. 4.16b), the strongly differing widths of the four lines stemming from Cu^{2+} can be predominately attributed to slow-tumbling phenomena.

4.6 BIRADICALS

Biradicals are species containing two unpaired electrons. In many cases they are constituted of two nitroxide moieties that are connected by a nonconjugated spacer (bisnitroxides)

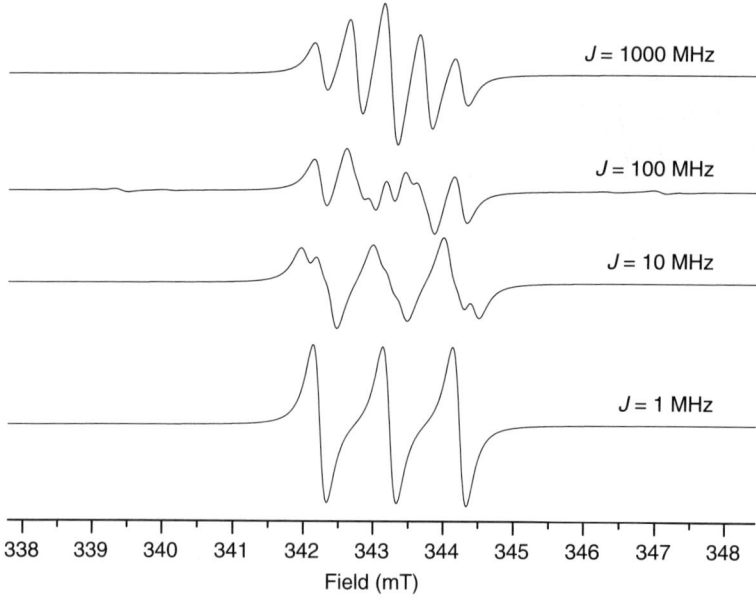

Fig. 4.17 Simulated EPR spectra of bisnitroxide radical having different exchange interaction energy.

The properties of biradicals are governed by the exchange interaction J. If $J = 0$, the two radical moieties behave as two independent radicals, giving rise to an EPR spectrum that is expected for "usual radicals" (i.e., three lines for a nitroxide). However, if $J \neq 0$, phenomena similar to the exchange processes introduced in Section 4.4.2 are observed. In the case of the isotropic hyperfine coupling constant a, which is substantially smaller than J, a pattern corresponding to two (equivalent) nuclei with $a/2$ is observed. In intermediate cases, rather complex spectra are obtained, which require specific simulations. (The reason for these complicated spectra is that off-diagonal elements appear in the Hamiltonian matrix, e.g., chapter 4.7 in Atherton's book, see Further Reading.) Some calculated examples are shown in Fig. 4.17.

4.7 SIMULATION SOFTWARE

There are several freely available programs that can be utilized to manipulate and simulate EPR spectra. Some programs are provided by the manufacturers of EPR spectrometers, for example, SIMFONIA and WINEPR from Bruker. Whereas WINEPR allows manipulation of EPR spectra imported from spectrometers, like determinations of line distances, baseline corrections, smoothing, comparisons between several spectra, and so forth, SIMFONIA allows the simulation of a variety of EPR spectral types.

The program WINSIM [32] is oriented toward EPR spectra in fluid solution. It can be downloaded from the homepage of the EPR Center of the NIH (http://epr.niehs.nih.gov/pest.html). This program can perform some basic manipulations on imported experimental EPR spectra, contains cursor functions, and allows simulations of isotropic (fluid-solution) signals in a very convenient way. This includes the superposition of several spectra, which is particularly valuable when EPR signals indicating satellite lines have to be evaluated. Older versions of this program contain several additional features but only run on the DOS level. This program also provides fitting of the experimental spectra.

A program allowing the simulation of almost any kind of EPR spectra and being almost unlimited on the computer platform is EasySpin (http://www.easyspin.ethz.ch) developed at ETH Zuerich [24]. This software package is based on MatLab. Its application is not as straightforward as that of WINSIM, but using the templates leads to a rather efficient advance in the simulation procedure.

Chapter 7 presents a new software package for an integrated approach to an *ab initio* calculation of EPR spectra of radicals and biradicals in any motion regime.

4.8 HOW FLUID-SOLUTION SPECTRA ARE ANALYZED

The analysis of fluid-solution spectra is closely connected to their construction principle shown in Fig. 4.9. Two features have to be regarded: the multiplicity (intensity ratios) of the signals or line pattern (doublet, triplet, etc.) and the separations between the lines.

However, it should first be checked if the EPR signal stems from one or more paramagnetic species. In the former case the spectrum should be symmetric, with a center of inversion in the middle point, which means that both wings of the spectrum should be identical. This is preferably checked by increasing the signal intensity. One should be confident to have found the outermost EPR lines on each side of the spectrum. If these are at symmetric distances from the center, their separation marks the overall width of the EPR spectrum representing the sum of all splittings caused by Fermi contact interaction.

How can the signal be further analyzed? This is illustrated in Fig. 4.18, showing how to interpret the EPR spectrum of the naphthalene radical anion introduced in Fig. 4.9.

The upper part of Fig. 4.18 represents the stick spectrum, which is constructed by the two quintet patterns displayed in the middle of the figure. The lower part shows the shape of the EPR signal. The distance between the two outermost lines on one side marks the smallest a [assigned to H(2,3,6,7)] and the intensity ratio between these two lines indicates the number of equivalent magnetic nuclei, here protons, having this a (the ratio is 1:4). If the same distance holds for the next line, this is just another line of the multiplet representing the smallest a. The third line is the third component in the 1:4:6:4:1 quintet. The fourth line reveals a distance that is not a multiple of $a(2,3,6,7)$ and has the same intensity as the second line. Accordingly, it has to belong to a different set of nuclei, representing the quintet with $a(1,4,5,8)$

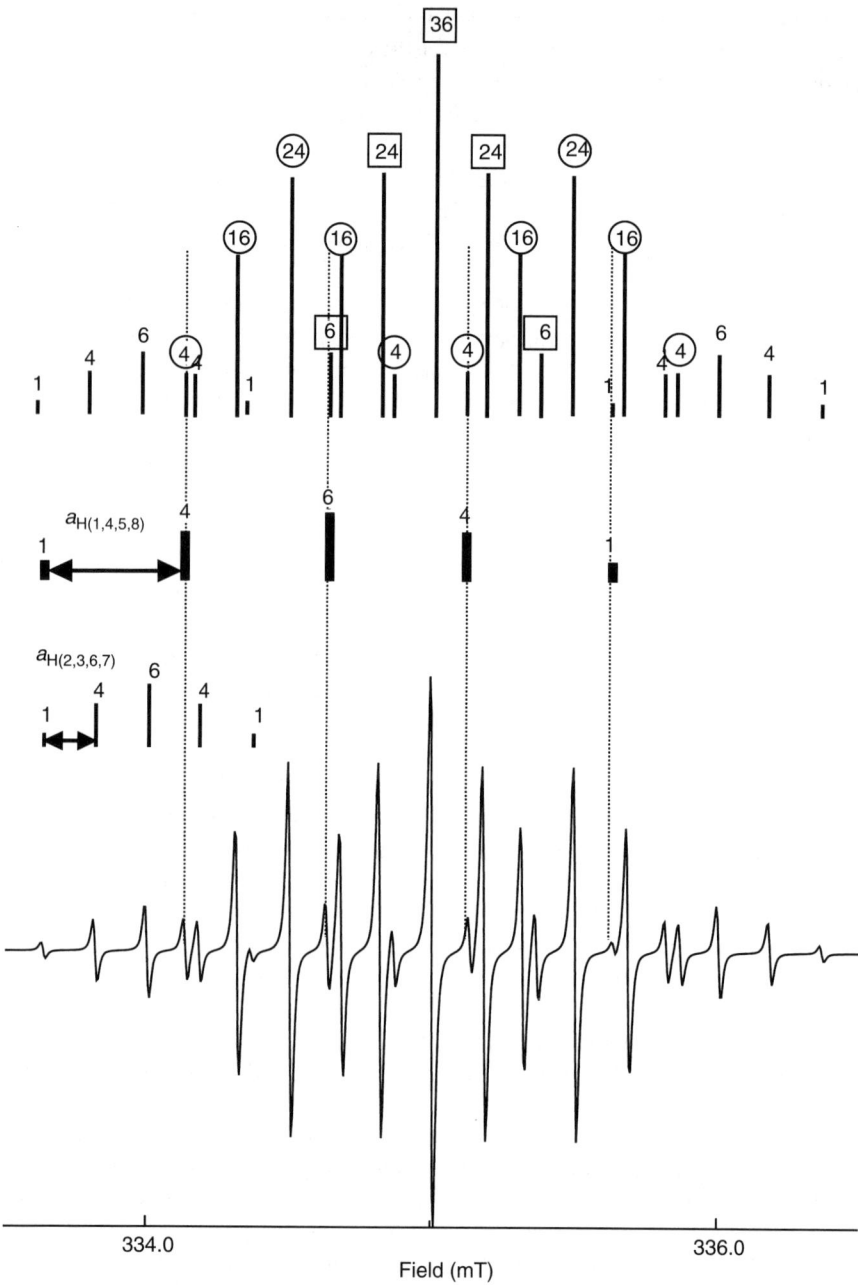

Fig. 4.18 Analysis of the EPR spectrum of the naphthalene radical anion.

4.8 HOW FLUID-SOLUTION SPECTRA ARE ANALYZED

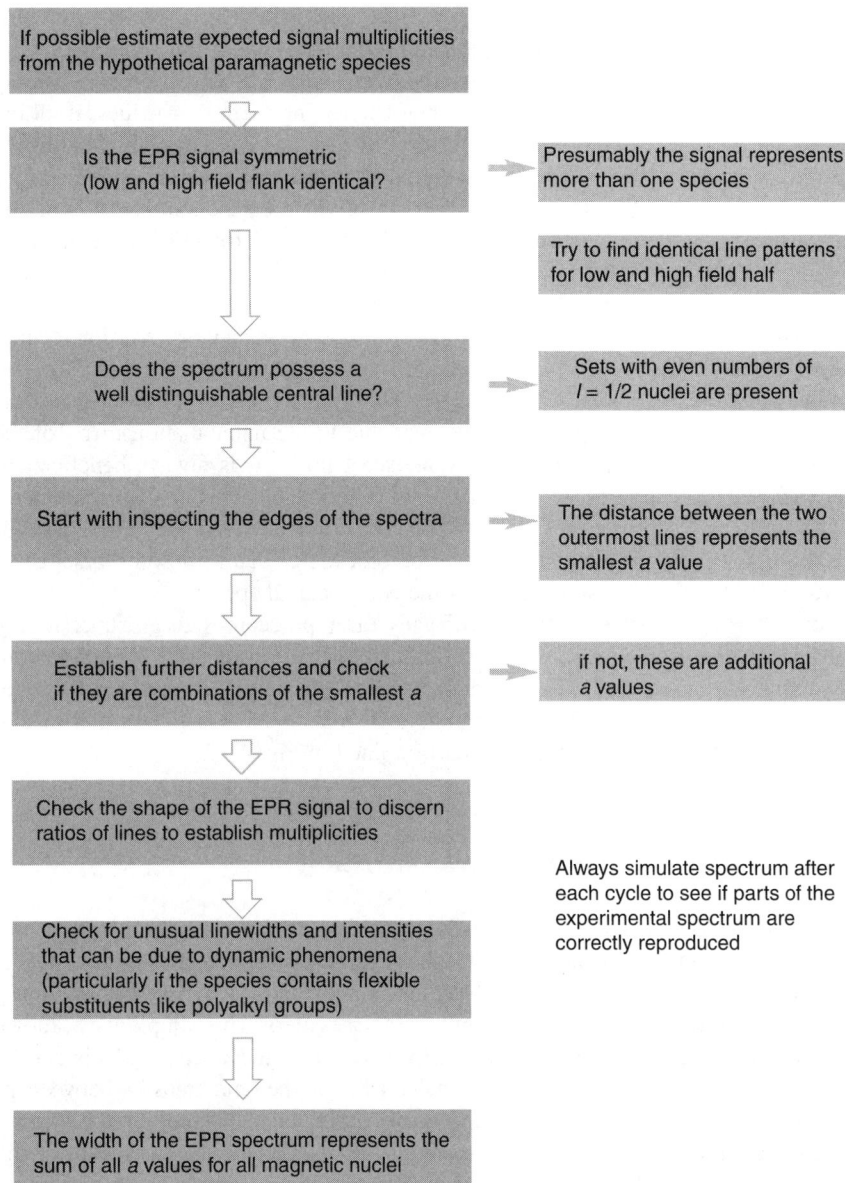

Fig. 4.19 The essential steps of the procedure for the determination of a values and multiplicity from an experimental spectrum.

marked with the bold lines in the center of Fig. 4.18. This line is now a new starting point with the quintet pattern with $a(2,3,6,7)$ and so on. This principle is replicated for all components of $a(1.4.5.8)$, which are marked by the dotted lines in Fig 4.18, and can be generalized for the treatment of any well-resolved EPR spectrum.

In most cases, the precursor of the radical responsible for the observed EPR spectrum is known and there are reasonable hypotheses about the expected sets of equivalent nuclei. Hence, one can check the overall signal shape in terms of finding the appropriate signal multiplicities and then connecting them to a values. If all but one a value have been found, it is rather simple to calculate the sum of the established a values times the number of expected signal distances and subtract these from the overall width of the EPR signal. This difference divided by the number of expected intervals (e.g., 1 for a doublet for one proton, 2 for a 1:2:1 triplet of two equivalent protons, or 2 for a 1:1:1 triplet of one ^{14}N nucleus, 4 for a 1:2:3:2:1 quintet of two equivalent ^{14}N nuclei) yields the final a value.

It may also be useful to check if the spectrum possesses a clearly visible central line with a high intensity, a feature that, in most cases, is indicative for an even number of equivalent nuclei with $I = 1/2$ (^{1}H, ^{31}P, etc.). All of these distances are preferentially checked using the cursors available in the manipulation menu of the spectral acquisition or simulation software (see below). It is always beneficial if, after each step, the appropriate EPR spectrum is simulated. This procedure immediately specifies if substructures of the EPR spectrum are simulated correctly. Of course, this is a step-by-step trial and error procedure; but with some practice, it rapidly leads to matching simulations of the experimental spectra.

Some of the available software packages offer procedures that extract highly probable a values from an imported experimental EPR spectrum. This is often helpful but generally not sufficient *per se* to obtain a reasonable simulation of the EPR signal.

The essential steps for the determination of EPR parameters from an experimental spectrum are summarized in the flowchart outlined in Fig. 4.19.

4.9 CALCULATION OF EPR PARAMETERS

The first organic radicals observed in fluid solution were derived from planar, delocalized π systems. The analysis of their EPR spectra was performed by the Hückel method, which was very popular at that time. The Hückel MO (HMO) model only considers π orbitals and the connectivity of the π centers. This simple view allows the estimation of a values for atoms directly attached to the π system. This can be traced back to the $\pi-\sigma$ polarization mechanism for the spin transfer between p_z orbitals and the adjacent σ bonded atom as described above. Therefore, the a_H of a hydrogen atom attached to a π system should be proportional to the square of the π orbital coefficient of the adjacent carbon atom c_i (Fig. 4.20):

$$a_H = Q \cdot c_i^2 \tag{4.10}$$

where Q is a proportionality factor with a value varying between about -2 and -3 mT for protons [33], depending also on the charge of the radical. For delocalized planar π systems, this procedure leads to fairly good agreement with the experimental values (Fig. 4.20).

Although several refinements have been added to Hückel-based procedures for the calculation of a values, this approach can be regarded as qualitative because no

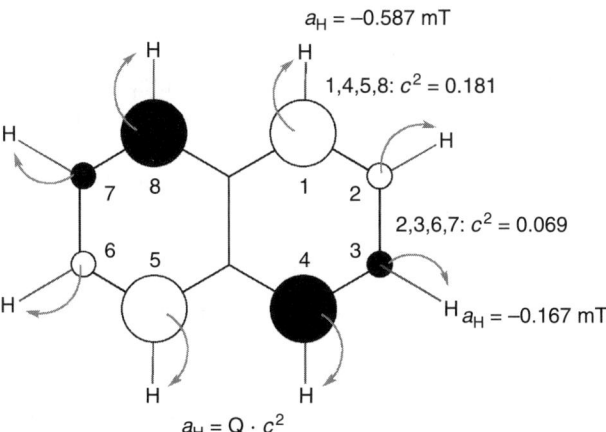

Fig. 4.20 HOMO of the naphthalene molecule and experimental a_H values of the radical cation.

structural optimizations are included in this method. Nevertheless, the use of this very fast and straightforward method for extended π systems is still useful for the estimation of the expected a values (is a big or small a expected for a specific position?).

Since the 1970s several approaches by semiempirical quantum chemical and *ab initio* methods have been followed. In recent years, computational protocols based on density functional theory (DFT) have been considered as the methods of choice for the determination of the geometry and the electronic structure of paramagnetic systems [34], because they offer reasonable results at an acceptable computational effort and provided hyperfine parameters of organic radicals with rather high accuracy [35–37]. In addition, the progress in the field of transition-metal complexes has also been considerable [38, 39].

A recent integrated approach for an *ab initio* calculation of the EPR spectrum combines calculations of magnetic parameters and of dynamic effects on the EPR spectral profile [40]. This approach is treated in detail in Chapter 7 with examples of applications.

For such calculations, frequently used functionals are B3LYP, PWP86, and PBE, combined with appropriate basis sets. These functionals and a number of additional ones are implemented in several program packages like GAUSSIAN (www.gaussian.com), ORCA (www.thch.uni-bonn.de/tc/orca), TURBOMOL (www.turbomole.com), ADF (www.scm.com), or GAMESS (www.msg.ameslab.gov/gamess/gamess.html). Applications of these methods are demonstrated in the examples below.

4.10 MOLECULAR PROPERTIES MIRRORED BY EPR SPECTRA IN FLUID SOLUTION

The previous sections outlined the background of the fluid-solution EPR parameters a and g. Whereas the g factor offers information on the contribution of heavier elements to the character of the radical, the isotropic hyperfine coupling constants a_i essentially map the electron density (Equation 4.1), allowing the characterization of the singly occupied MO (SOMO). Consequently, the multiplicities of the a values are indicative of the symmetry of a paramagnetic species and the size of a corresponds to MO coefficients.

The following section provides examples to demonstrate how EPR spectra can be analyzed and which molecular properties can be discerned.

4.10.1 σ Character of Radicals: A Direct View on the Fermi Contact Interaction

Amines can often be readily oxidized to their radical cations. Upon oxidation, an electron is formally removed from the nitrogen lone pair and the originally pyramidalized nitrogen atom becomes essentially planar (Fig. 4.21).

There are many examples of rather persistent trialkylamino radical cations in the literature [41, 42]. Most of them have rather similar a_N values of $\sim$2 mT for the ^{14}N nucleus (basically independent of the substituents). However, when the nitrogen atom is embedded in a rather rigid skeleton, the a_N values increase substantially. In 1-azaadamantane (**N1**), the a_N value is +2.16 mT; it rises to 2.51 mT in 1-azabicyclo[2.2.2.]octane (**N2**) and even to 3.02 mT in 1-azabicyclo[2.2.1.]heptane (**N3**) [43].

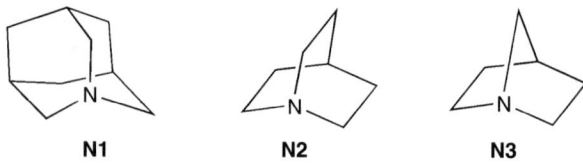

This reflects the increasing amount of pyramidalization at the nitrogen center as a result of the increasing molecular rigidity when going from **N1** to **N3**; that is, if the N center cannot become planar, the σ character (which corresponds to the Fermi contact) rises. Thus, the a_N values become higher (without a change of the spin population).

Remarkably, in radical cations of diamines, in which the two N atoms are oriented in such a way that their lone pairs are able to interact through space, the a_N values can possess even higher values. This has been observed for a series of bicyclic diamines (e.g., **N4** and **N5**) as well as for diiminoannulenes (**N6–N8**) [44].

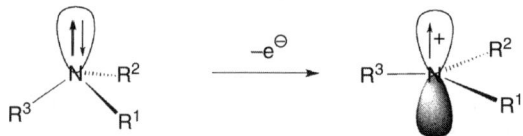

Fig. 4.21 Planarisation of N centres in amines upon oxidation.

Figure 4.22 shows the EPR spectrum of **N8** that consists of 15 groups of lines. The small splittings within these groups stem from couplings of the 10 protons of the 14-membered π system (a discussion of these a values can be found in Ref. 45). Here, the concentration is on the dominant splittings that are caused by the interaction of the unpaired electron with the two equivalent bridgehead nitrogen atoms and the

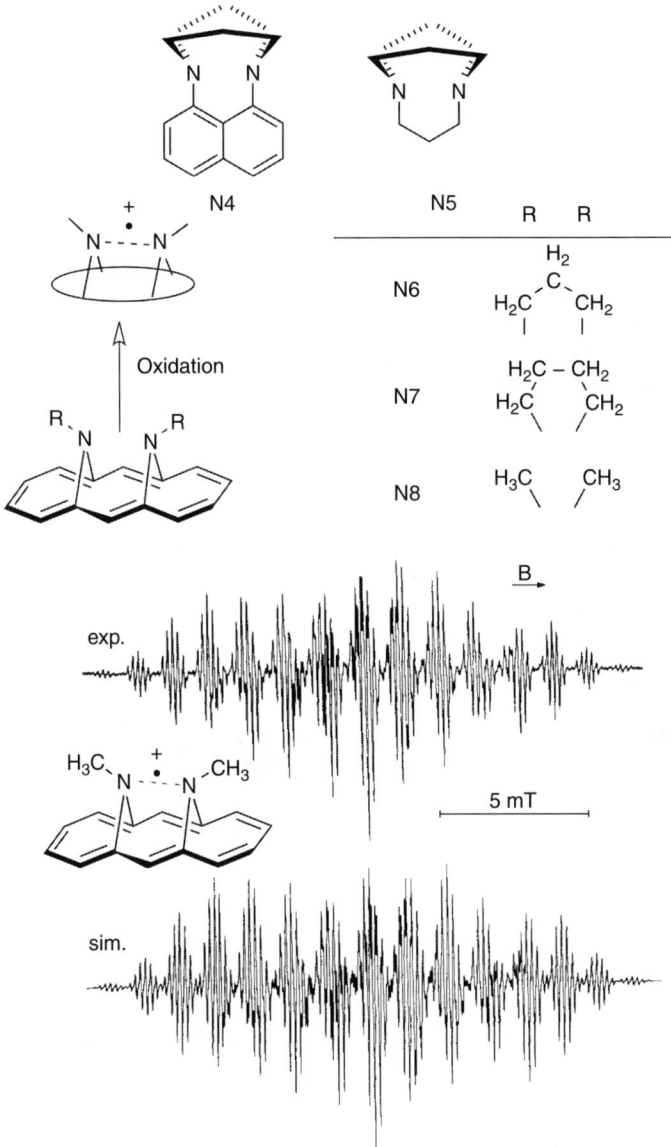

Fig. 4.22 EPR spectrum and simulation of **N8**.

six equivalent β protons of the two methyl groups. The first and second groups of lines, which are septets, exhibit a 1:6 intensity ratio, which corresponds to six equivalent protons ($a_H = +1.213$ mT). The third multiplet differs slightly from the second one. This is because lines attributed to the two nitrogen nuclei ($a_{14N} = +2.66$ mT) start to overlap with the hydrogen septets. These data clearly reveal that the spin population is basically confined to the two nitrogen atoms. This is explained by the formation of a σ^*-type bond between the two nitrogen atoms: the N⋯N bond is formed by the interaction of the two nitrogen lone pairs, which are "forced" to interact by the molecular skeleton. In the parent neutral stage of the diamines, each lone pair has two electrons and their interaction is unfavorable. However, if one electron is removed by oxidation, stabilization is achieved as shown in Fig. 4.23: a two-center three-electron bond is formed with the unpaired electron residing in a σ^*-type orbital. It is this σ character that causes the considerable size of the a_N.

The a_N value of **N6**$^{\bullet+}$ is the smallest (+1.70 mT); it rises to +2.57 mT in **N7**$^{\bullet+}$ and reaches +2.66 mT in **N8**$^{\bullet+}$. This is parallel to the amount of pyramidalization at the two identical N atoms: for **N6** the trimethylene bridge causes an almost planar arrangement around the N atoms causing an increase of the π character. The longer tetramethylene chain in **N7** allows a more pronounced bending; finally, the methyl groups in **N8** permit even further deviations from planarity. The more pyramidalized the N atoms are, the more the σ character. Consequently, the Fermi contact builds up, reflecting the increasing a_N values.

Such direct coherence between the a values and the orbital character was established for several (particularly N-containing) radicals [44–46].

Notably, the EPR spectrum of **N8**$^{\bullet+}$ shows some asymmetry. This can be traced to "second-order" effects that are observed when a_H values are higher than ~1.2 mT (see §4.4.5 and pp. 45–47 in Ref. 41).

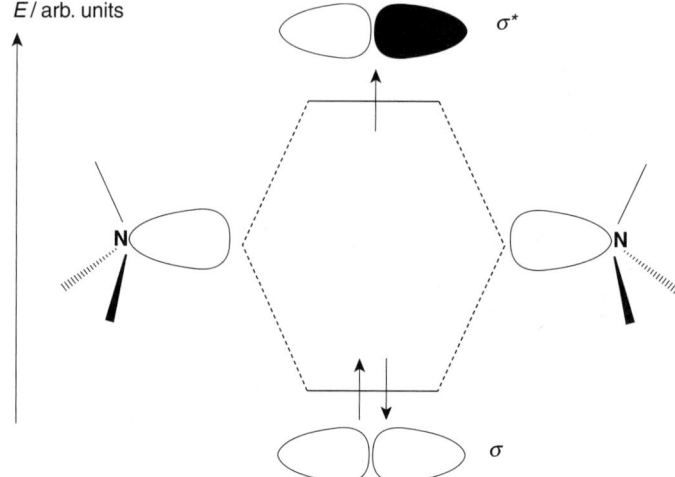

Fig. 4.23 A two-center three-electron bond between the lone pairs of two nearby nitrogen centers.

4.10.2 Radical Ions of Aromatic Molecules: The Case of Dibenzo[b,h]biphenylene

Historically, the first studies on organic radicals by EPR in solution were focused on molecules consisting of extended π systems (aromatic molecules). This is particularly because such molecules, in most cases, form rather persistent radical ions exhibiting well-defined EPR spectra (as shown previously for naphthalene). Moreover, the a_H values determined for aromatic hydrocarbons ideally map the character of the SOMO (see §4.9).

The following example shows how the orbital character can be mapped from the knowledge of the a_H values of α-hydrogen atoms (spin transfer via $\pi-\sigma$ polarization). It will also be indicated how helpful simple Hückel calculations can be for the calculation of the a values. However, the limitations of this method will also be presented simultaneously.

According to Hückel theory, an alternant hydrocarbon π system (having only even-numbered rings) possesses "paired" orbitals: each bonding orbital has its antibonding counterpart where each orbital has the same energy distance from the reference value α. Paired orbitals have identical absolute orbital coefficients but differing signs. Thus, the highest occupied MO (HOMO) of such a molecule has absolute values that are identical to the lowest unoccupied MO and so forth. This is shown for dibenzo[b,h]biphenylene (**D**) in Fig. 4.24.

Accordingly, the radical anion and the radical cation of **D** should, in principle, have identical EPR spectra. Because the MO coefficients are proportional to the a_H values, the largest values for four equivalent protons are expected at positions 3,4,9,10,

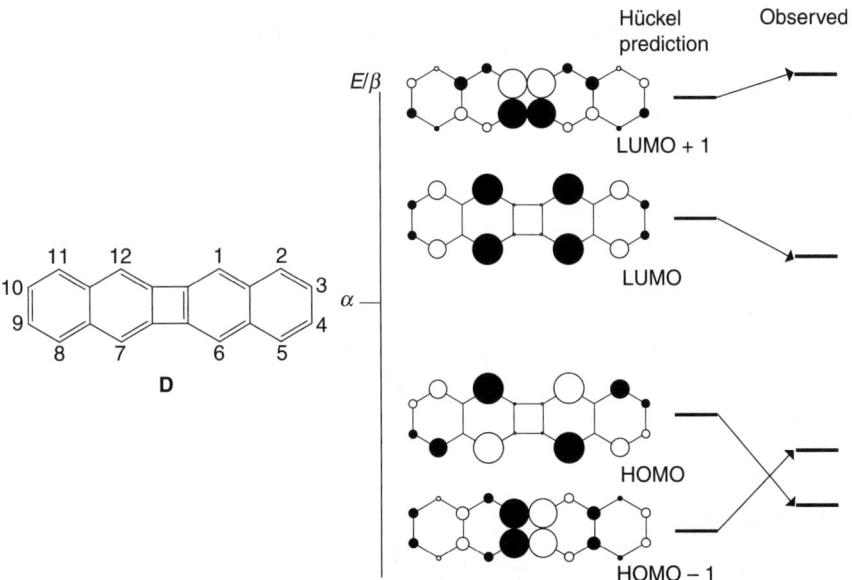

Fig. 4.24 Frontiers orbitals of dibenzo [b,h] biphenylene.

followed by those at 2,5,8,11 and finally rather small ones for 1,6,7,12. This is reflected in the EPR spectrum of **D**$^{\bullet-}$ (Fig. 4.25, top). The spectrum consists of a quintet of quintets of quintets (125 lines) with a_H values of 0.09, 0.157, and 0.423 mT, respectively (overall width = 2.68 mT). The signal detected after oxidation of **D** reveals a completely different pattern. It is substantially narrower (0.94 mT) and shows fewer lines. In fact, the splittings are due to two sets of four equivalent lines (the weak lines are due to ^{13}C satellites). The two easily distinguishable a_H values are 0.173 mT (four equivalent H) and 0.061 mT (four equivalent H); a third a_H can be detected via ENDOR spectroscopy and is beyond EPR resolution (<0.004 mT; Fig. 4.25) [47].

Evidently, the orbital sequence has been changed in the case of the SOMO for the radical cation (Fig. 4.24). The shape of the SOMO corresponds to the HOMO − 1 of the Hückel calculation, which possesses substantially smaller coefficients at H-carrying C atoms than the HOMO. This unexpected sequence of orbitals is based on the strain introduced by the central cyclobutane moiety and is well predicted by *ab initio* and DFT calculations, which obviously cannot be predicted by the Hückel method. This is in accord with the prediction of the Mills–Nixon effect [48] and illustrates the power of EPR in providing the shape of SOMOs.

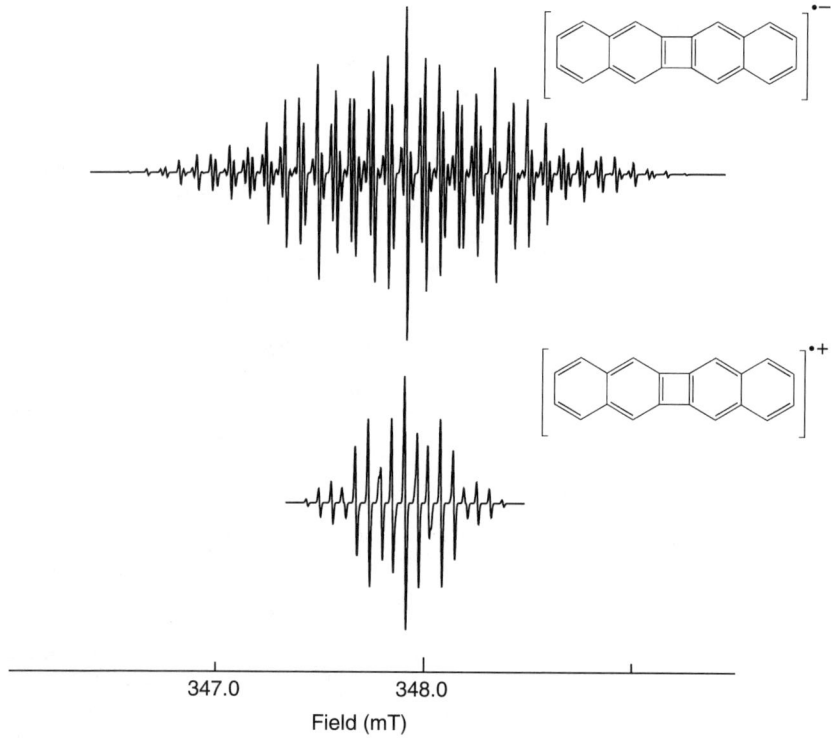

Fig. 4.25 EPR spectra of the radical anion (top) and radical cation (bottom) of **D**.

4.10.3 Conformations and Dynamics in Polymers

The following example shows how preferred conformations of radicals can be determined by analysing β-hydrogen splittings from experimental EPR spectra. Moreover, it will be indicated that the temperature dependence of these spectra can be evaluated in terms of conformational dynamics (cf. §4.4.2). The combination of EPR and theoretical calculations leads to a fairly precise description of intramolecular motions.

Many polymers are produced by processes involving radicals. In radical polymerization, the chain grows by the addition of an alkene to a radical, which is part of a growing polymer chain. The mode in which the chain grows is mainly governed by steric factors. To gain insight into the structure and dynamics of radicals involved in the mechanism of radical polymerization, the [D$_3$]methyl (2SR,3RS)-2-methyl-4,4,4-trichlorobutanoate-2-yl radical (**R**) was investigated (Scheme 4.9).

According to the hyperconjugation mechanism, the hyperfine coupling constants with β protons should be large enough to be detected. As no steric congestion is expected for the CH$_3$ group, the methyl group should rotate freely and the three protons should be equivalent. In contrast, a more pronounced overcrowding can be anticipated for the two β protons H$_{\beta'}$ and H$_{\beta'}$, which also acquire spin population by hyperconjugation, and dynamic phenomena can be observed in the temperature-dependent EPR spectra of **R**.

At relatively high temperatures above 273K, the EPR spectrum of **R** consists of a "quartet of triplets" (the construction principle is shown in Fig. 4.26). The outermost triplet stems from two equivalent protons H$_{\beta'}$ and H$_{\beta'}$ of the methylene group ($a_{H\beta}$ = 1.12 mT) whereas the quartet has to be ascribed to the three equivalent protons of the CH$_3$ moiety ($a_{H(CH_3)}$ = 2.36 mT). The spin population at the three deuterons of the acetoxy group is too small to be detected (which simplifies the spectral analysis).

Upon lowering the temperature to 213K, the overall width of the EPR spectrum is not modified, but the intensity ratios of the three outermost lines change from 1:2:1 to 1:1:1 (Fig. 4.26, bottom) with the central line being broadened. This is characteristic for a dynamic process introduced in Section 4.4.2.

The alteration of the line intensities and the broadening of the central line of the 1:2:1 triplet is characteristic of a slowed down movement of the methylene group: at low temperatures the two H atoms H$_{\beta'}$ and H$_{\beta''}$ become nonequivalent. Simulation of the experimental spectra (at least four temperatures, with EasySpin) and a plot analogous to that in Fig. 4.11 reveal an activation energy of 12 ± 1.8 kJ mol^{-1} [49].

Because hyperconjugation is the main mechanism for spin transfer for H$_{\beta'}$ and H$_{\beta''}$, it is straightforward that the temperature-dependent line intensities and widths reflect

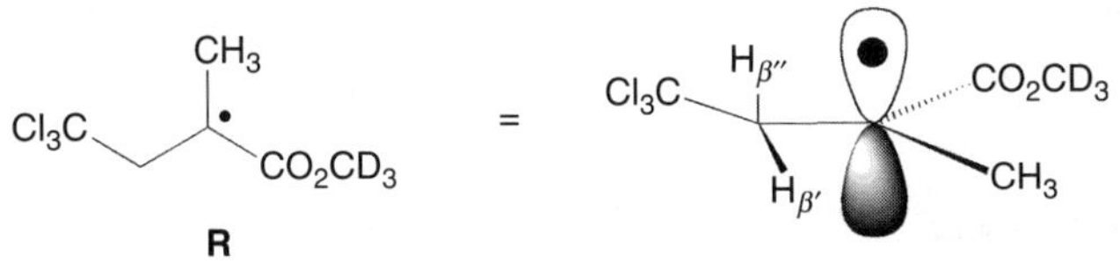

Scheme 4.9

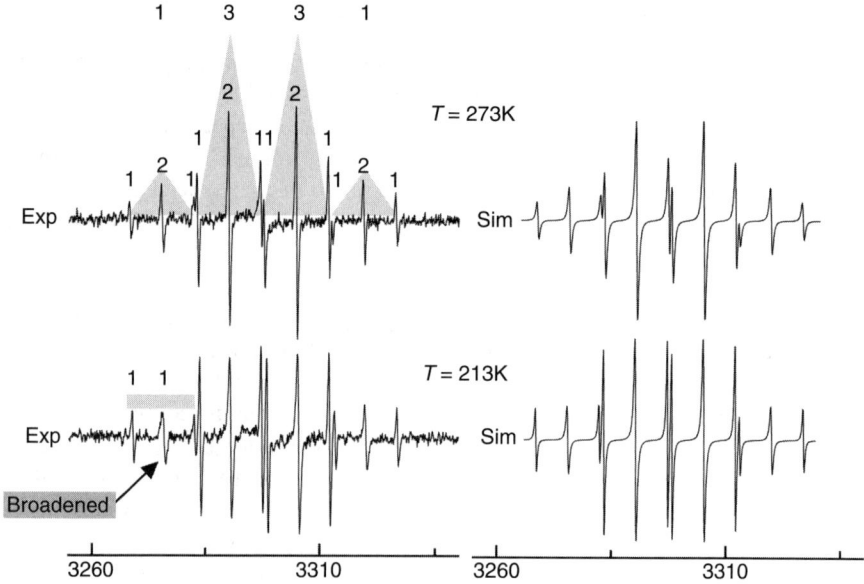

Fig. 4.26 Temperature-dependent EPR spectra of **R**.

a jump of the radical **R** between specific conformations. The rate of the jump decreases on decreasing the temperature, and the spectral profile is consequently affected.

This can be established very well by DFT calculations, which predict that conformations **R1** and **R2** are energy minima of **R**, being separated by a barrier of 11 kJ mol^{-1}, closely resembling the experimental value of 12 ± 1.8 kJ mol^{-1}. In the same way, the calculated a values match the experimental ones: for 273K, $a_{H\beta'} = a_{H\beta''}$ 1.16 mT (exp. = 1.12 mT) and $a_{H(CH_3)} = 2.22$ mT (exp. = 2.36 mT).

In cases where dynamic phenomena can be expected, it is always advisable to check the temperature dependence of the EPR spectra. It is also necessary to ascertain if the spectral changes are reversible, that is, if increasing or lowering the temperature always leads to identical spectra, to exclude that the altered spectra indicate the formation of follow-up products.

4.10.4 Ion Pairing

Ions of opposite charges tend to pair. This is of great importance for the course of many chemical reactions. If ion pairing occurs in charged radicals and the orbitals of the radical ion and its counterion interact, spin can be transferred to the counterion. Such interactions can be frequently established when a radical anion is produced by alkali–metal reduction. Evidence for such an ion pairing is provided by the appearance of hyperfine splittings from paramagnetic nuclei of ^{7}Li, ^{23}Na, ^{39}K, ^{85}Rb, ^{87}Rb, and ^{133}Cs. The corresponding hyperfine splittings can often be observed in the EPR spectra of such radical anions. The strength of association with the counterion depends on several factors, such as the solvating power of the solvent, the temperature, and the radius of the cation. As the solvating power decreases, ion pairs tighten and the coupling constants of the alkali–metal nucleus increase. This effect is promoted by raising the temperature.

The radical anion of 2,3-di(*tert*-butyl)buta-1,3-diene (**B**) is an illustrative example of several features of ion pairing detected via EPR [50]. Reduction of **B** with K metal in DME at 175K leads to the spectrum shown in Fig. 4.27a, consisting of two overlapping EPR spectra. Clearly the dominating one is a *quintet* with an intensity ratio of $1:4:6:4:1$ and a line spacing with $a_H = 0.71$ mT, due to four equivalent protons. This coupling can be assigned to the four virtually equivalent protons at C(1) and C(4). No a_H values can be established for the protons of the *tert*-butyl groups because the MO coefficients at the 2,3 positions are small and the H atoms are separated from the π system by three bonds. The weaker overlapping spectrum increases in intensity on increasing the temperature, and it can be analyzed as due to an ion pair B$^-$K$^+$, with an a_K of $+0.134$ mT (a $1:1:1:1$ quartet of ^{39}K, $I = 3/2$). The behavior described here is reversible and is characteristic for an ion-pairing phenomenon. It is attributable to an equilibrium between a loose ion pair, where the alkaline cation is solvated and therefore the interaction between radical and metal cation is undetectable, and a tight ion pair, where there is an overlap between the orbitals of **B**$^{\bullet-}$ and K$^+$, with a spin density transfer to the metal cation, and a detectable hyperfine coupling with the ^{39}K nucleus.

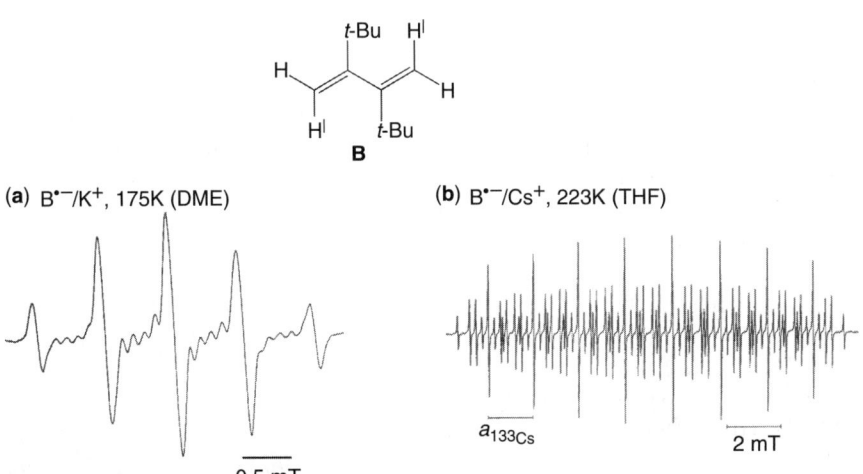

Fig. 4.27 EPR spectra of the **B** radical anion obtained by reduction with K (a) and Cs (b) respectively.

As an example of a tight ion pair, Fig. 4.27b provides the spectrum of **B**$^{•-}$/Cs$^+$ at 210K. This spectrum can be analyzed as due to a triplet of triplets, each in a 1:2:1 intensity ratio, due to the coupling with two *nonequivalent* pairs of protons [a_H = −0.69 mT (2 H) and −0.46 mT (2 H)], with each of the nine lines split further in an *octet*, with an intensity ratio 1:1:1,..., by the coupling with one ^{133}Cs nucleus ($I = 7/2$) with coupling constant a_{Cs} = +1.59 mT. It is interesting that the four equivalent methylene protons giving the quintet in the spectrum in Fig. 4.27a are now no longer equivalent. The symmetry of the spin distribution on the radical anion is lowered by the interaction with the ^{133}Cs$^+$ cation, and the spin distribution interacting with the two methylene groups is now different.

4.10.5 EPR Spectra Indicating Satellites

Kinetic stabilization by steric overcrowding often leads to astonishingly stable molecular stages. An example is the 1,2,3,4-tetra-*tert*-butyl-4-trimethylsilyl-4-sila-2-cyclobuten-1-yl radical **S**, which reveals unexpectedly high persistence, giving rise to very intense EPR spectra with a high signal/noise ratio [51]. This allows the detection of satellite lines as demonstrated below.

S

Figure 4.28 displays the EPR spectrum of **S**. The central multiplet (19 equidistant lines with an a_H of 0.037 mT) is straightforwardly assigned to the 18 equivalent protons of the *tert*-butyl groups at C(1) and C(3). The protons at the remaining *tert*-butyl groups carry nondetectable spin populations. This central signal is symmetrically sided by three additional groups of lines of substantially lower intensity. At increased gain, these groups reveal the same pattern as the central signal with the outermost signal group presenting ∼2.2% of the intensity relative to the main signal. This corresponds to a presence of ^{13}C in the radical that is the double of the natural abundance (1.1%), and therefore the ^{13}C isotopes are substituting ^{12}C in two equivalent positions. The distance from the center of the main spectrum to the center of each of the outermost groups represents $1/2\ a_{13C}$ (see the construction in Fig 4.28) because ^{13}C has $I = 1/2$. Therefore, $a_{13C} = 2.56$ mT.

The next group has an intensity ratio versus the main signal of ∼4.7%, which matches the natural abundance of ^{29}Si ($I = 1/2$) with an a_{29Si} of 1.495 mT. The determination of the nuclei giving rise to the line groups next to the central one is rather

4.10 MOLECULAR PROPERTIES MIRRORED BY EPR SPECTRA

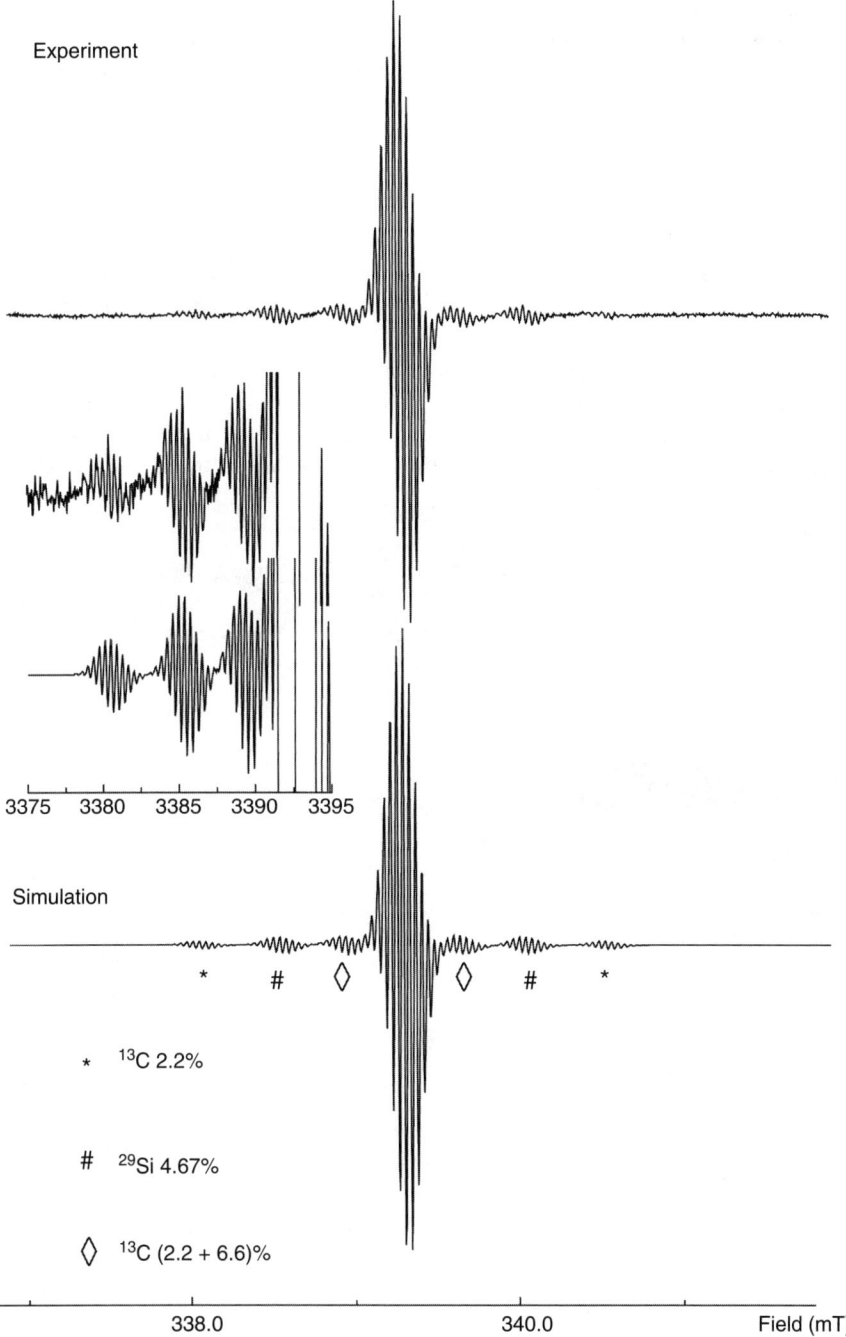

Fig. 4.28 The EPR spectrum of **S** and its simulation.

delicate: by simulation, the best fit was found for a a_{13C} splitting of 0.692 mT for six equivalent ^{13}C nuclei and a a_{13C} of 0.656 mT for two equivalent ^{13}C atoms. In summary, the whole spectrum can be composed by adding four components with the appropriate ratios as depicted in Fig. 4.28.

4.10.6 Electron Self-Exchange: A Special Case of Dynamics

Section 4.10.3 demonstrated how EPR spectroscopy provides knowledge about intramolecular dynamics in terms of conformational alterations. Under specific conditions EPR can also be used to obtain information on intermolecular electron exchange phenomena and to assess the kinetics of degenerate electron transfer.

Generally such experiments are performed with substrates that form exceptionally stable (kinetically and thermodynamically) radicals, because the concentrations of the diamagnetic precursors and the corresponding radicals have to be rigorously controlled. The equilibrium reaction in Scheme 4.10 is followed by EPR as a function

$$A + A^{red/ox} \underset{\text{Electron transfer}}{\overset{\text{Degenerate}}{\rightleftharpoons}} A^{red/ox} + A$$

Scheme 4.10

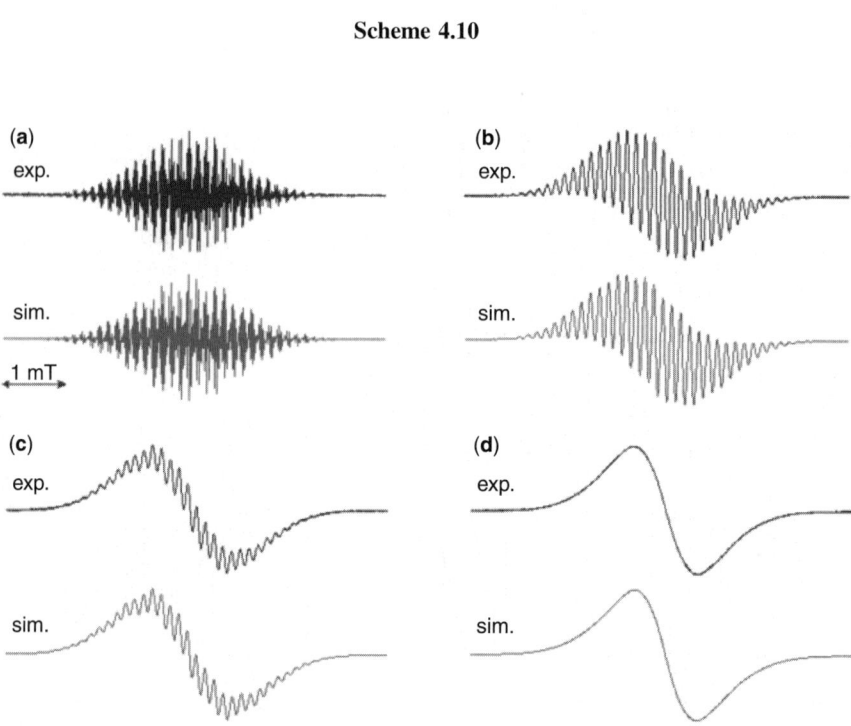

Fig. 4.29 EPR spectra and corresponding simulations of the methylviologen radical cation (MV^{2+}) in solutions at increasing concentration: (a) 0.02 mol/l, (b) 0.03 mol/l, (c) 0.04 mol/l, (d) 0.05 mol/l.

of the concentration ratio between **A** and $A^{red/ox}$. At a high concentration of **A**, the lifetime of $A^{red/ox}$ decreases. Because the linewidth of the EPR lines is proportional to the reciprocal value of the lifetime of the species, the corresponding EPR lines substantially broaden with an increasing concentration of **A**. Figure 4.29 shows the example of methylviologen (MV^{2+}).

MV²⁺

The simulations shown in Fig. 4.29 provide the rate constants for the degenerate electron transfer (range = $10^7 - 10^9$ M^{-1} s^{-1}). Performing these measurements at different temperatures provides activation energies and information on solvent dynamics [52].

4.11 CHEMICALLY INDUCED DYNAMIC ELECTRON POLARIZATION (CIDEP) AND CID NUCLEAR POLARIZATION (CIDNP): METHODS TO STUDY SHORT-LIVED RADICALS

4.11.1 General

All methods and examples described above require a paramagnetic species lifetime of longer than milliseconds. In several contexts, particularly when light-induced processes are the focus of interest, paramagnetic species are formed at the nanosecond (or even shorter) time scale. Such short-lived species cannot be observed with "classical" EPR. As illustrated in Chapter 2, in the short time domain, time-resolved spectroscopy is feasible at X-, Q- or W-band using modified spectrometers operating in either the continuous wave (CW) or the pulsed mode [53]. Such paramagnetic species can also be detected indirectly by a nuclear magnetic resonance (NMR) related technique: CIDNP.

In most cases irradiation produces only a low concentration of radical species. Nevertheless, it is possible to detect radicals at these very low concentrations. The reason is that the light pulse primarily creates a radical pair. Because of the excited states participating in the creation of the radical pair, it is formed in a non-Boltzmann populated state. An excited state can lead to a radical pair by α cleavage, an atom transfer reaction (e.g., hydrogen abstraction), or electron transfer (Scheme 4.11).

The follow-up reactions of the radical (ion) pairs depend on the relative orientation of the unpaired spins. This reactivity obeys different mechanisms that depend on the character of the radical pair and reaction dynamics (radical pair mechanism, triplet mechanism [54]). This leads to "polarized" signals with enhanced absorption or emission patterns and high intensity with the result that even radicals present at

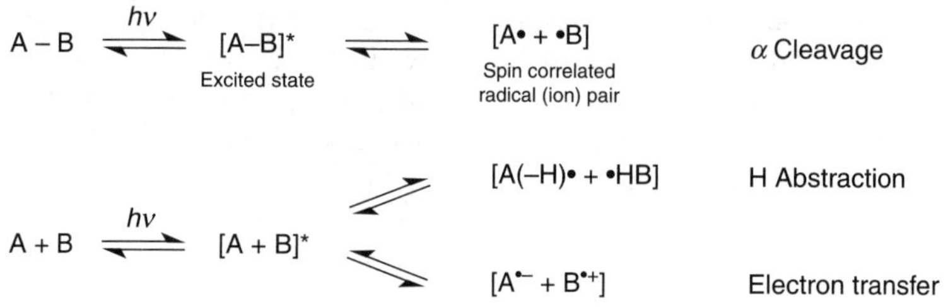

152 EPR SPECTROSCOPY IN THE LIQUID PHASE

$$A-B \xrightleftharpoons{h\nu} [A-B]^* \rightleftharpoons [A^\bullet + {}^\bullet B] \quad \alpha \text{ Cleavage}$$
Excited state   Spin correlated radical (ion) pair

$$A + B \xrightleftharpoons{h\nu} [A+B]^* \begin{array}{c} \nearrow [A(-H)^\bullet + {}^\bullet HB] \quad \text{H Abstraction} \\ \searrow [A^{\bullet -} + B^{\bullet +}] \quad \text{Electron transfer} \end{array}$$

Scheme 4.11

low concentrations become detectable. Such magnetic field effects can be found in many areas of chemistry and biology; it is very likely that they contribute to the navigation ability of birds [55].

4.11.2 CIDEP: Time-Resolved EPR in the 100-ns Regime

CIDEP experiments can be performed in both CW and pulsed fashions [56]. Both ways provide compatible information. Usually a laser pulse produces a radical (ion) pair. After a delay of ~50 ns, the EPR signal is recorded.

This has to be performed step-by-step at several magnetic field intervals for the CW experiment and the EPR spectrum is constructed slice by slice (see Fig. 4.30). Although the pulsed technique provides the entire EPR signal with one "shot," several spectra have to be summed up to obtain a reasonable signal/noise ratio. The time-resolved spectra are recorded with the field modulation switched off;

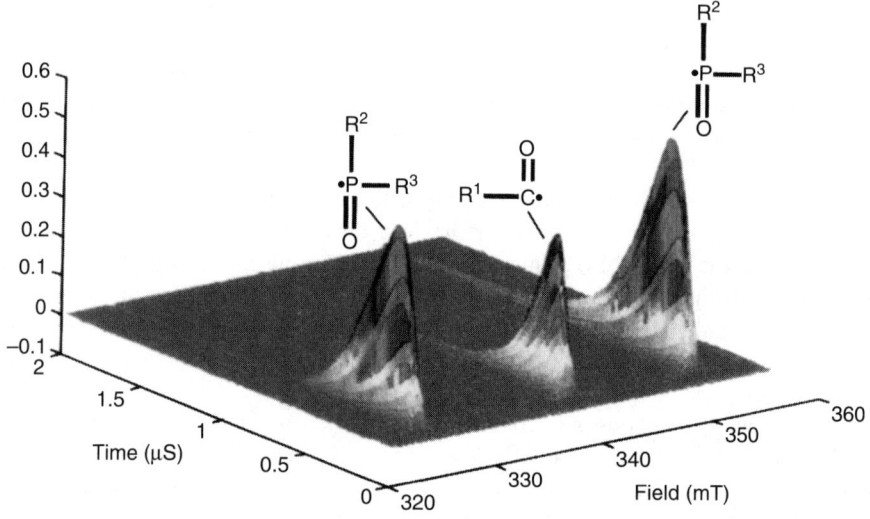

Fig. 4.30 CIDEP spectrum obtained upon photolysis of phosphinoxide **P**.

4.11 CHEMICALLY INDUCED DYNAMIC ELECTRON POLARIZATION

as a consequence, the EPR lines appear in absorption or emission (Fig. 4.30). Unfortunately, the decay of the EPR signals is not necessarily representative of kinetic phenomena because it contains contributions from magnetic field effects, which imply longer lifetimes [57]. The analysis of the CIDEP spectra follows similar rules as the analysis of usual fluid-solution EPR spectra. The signal, however, contains the spectra of both radicals. For example, in Fig. 4.30, the spectra of the benzoyl radical (the unresolved line, positioned roughly in the center) and the phosphinoyl radical (the outer doublet with a_{31P} of 34.3 mT) are shown, produced simultaneously by photolysis of phosphinoxide **P**.

P

CIDEP experiments in the fluid solution have to be performed with the use of flow systems because many repetitions have to be performed and each laser shot triggers a chemical reaction and a fresh solution must be available for the next one.

4.11.3 CIDNP

An alternative way to observe radicals is provided by CIDNP spectroscopy. This method is derived from NMR and provides indirect information about radicals formed in the course of chemical reactions. These reactions can happen in the ground state or can be induced by temperature jumps or light pulses. Similar to CIDEP, the principle underlying CIDNP is based on the occurrence of radical (ion) pairs and magnetic field effects. The method allows an indirect observation of radicals formed on the nanosecond time scale.

After presaturation of the NMR signals of the reaction mixture, a light pulse (in most cases) leads to the formation of radical (ion) pairs. Then, an NMR spectrum is taken (Fig. 4.31). It contains resonances from products formed by the geminate radical (ion) pair (so-called cage products) or by radicals that have diffused out of this pair and have undergone follow-up reactions ("escape" products). According to spin sorting mechanisms, it is possible to distinguish between these products because of the polarized signal intensities. From the intensity ratios of the CIDNP signals, the a values of the transient radicals and the photophysical course of the reaction can be derived [58–63].

The ^{31}P-CIDNP spectrum presented in Fig. 4.31, which was recorded after photolysis of molecule **X** shown in the figure, shows lines characteristic for three- and five-bonded phosphorus atoms. In addition to regenerated **X**, molecules containing $R_2P(O)$—$P(O)R_2$, and $R_2P(O)$—O—PR_2 moieties can be detected.

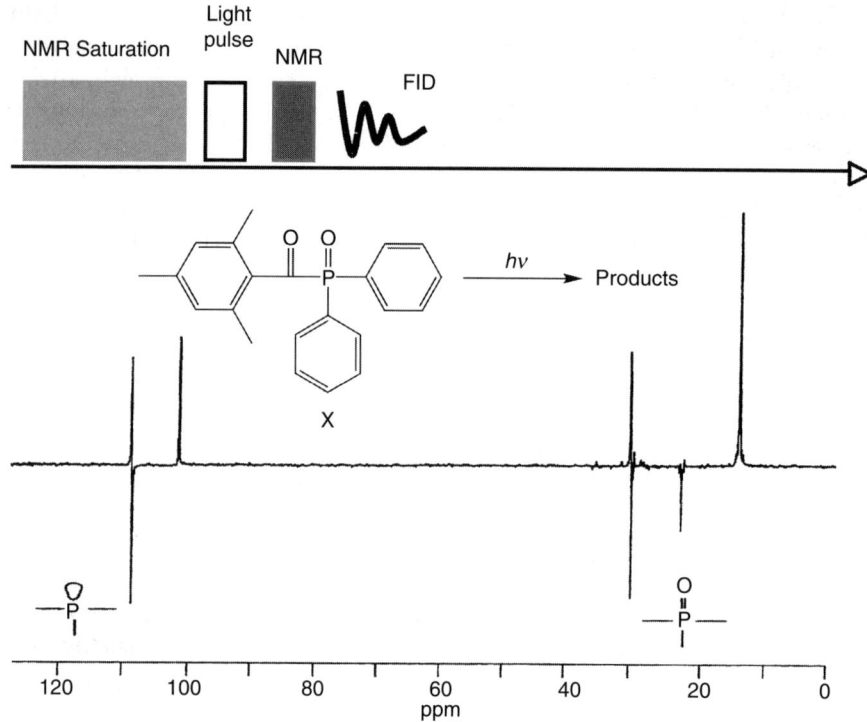

Fig. 4.31 CIDNP timing of the experiment (top) and spectrum, obtained upon photolysis of **X**.

ACKNOWLEDGMENTS

Thanks are due to Prof. A. G. Davies (University College London), Dr. H.-G. Korth (University of Essen, Germany), and M. Griesser (TU Graz) for fruitful discussions and suggestions.

REFERENCES

1. Smith, M. B.; March, J. *March's Advanced Organic Chemistry: Reactions, Mechanisms, and Structure*, 5th ed.; Wiley-Interscience: Chichester, 2001.
2. Fessenden, R. W.; Schuler, R. H. *J. Chem. Phys.* **1960**, *33*, 935.
3. Rüchardt, C.; Bechaus, H.-D. *Angew. Chem. Int. Ed.* **1980**, *19*, 429.
4. Krusic, P. J.; Kochi, J. K. *J. Am. Chem. Soc.* **1969**, *91*, 3940.
5. Krusic, P. J.; Kochi, J. K. *J. Am. Chem. Soc.* **1968**, *90*, 7155.
6. Hudson, A.; Jackson, R. A. *Chem. Commun.* **1969**, 1323.
7. Davies, A. G.; Griller, D.; Ingold, K. U.; Lindsay, D. A.; Walton, J. C. *J. Chem. Soc. Perkin Trans. 2* **1981**, 633.

8. Placucci, G.; Grossi, L. *Gazz. Chim. Ital.* **1982**, *112*, 375.
9. Pedersen, J. A. *CRC Handbook of EPR Spectra from Quinones and Quinols*; CRC Press: Boca Raton, FL, 1985.
10. Ohya-Nishiguchi, H. *Bull. Chem. Soc. Jpn.* **1979**, *52*, 2064.
11. Bock, H.; Lechner-Knoblauch, U. *J. Organomet. Chem.* **1985**, *294*, 295.
12. Bell, F. A.; Ledwith, A.; Sherrington, D. C. *J. Chem. Soc. C* **1969**, 2719.
13. Eberson, L.; Hartshorn, M. P.; Persson, O. *Acta Chem. Scand.* **1995**, *49*, 640.
14. Davies, A. G.; Ng, K. M. *Aust. J. Chem.* **1995**, *48*, 167.
15. Eberson, L.; Hartshorn, M. P.; Persson, O. *Chem. Commun.* **1995**, 1131.
16. Eberson, L., Hartshorn, M. P.; Persson, O. *J. Chem. Soc. Perkin Trans. 2* **1995**, 1735.
17. Davies, A. G. *J. Chem. Res. Synop.* **2001**, 253.
18. Davies, A. G.; Neville, A. G. *J. Chem. Soc. Perkin Trans. 2* **1992**, 163.
19. Bolton, J. R. *Mol. Phys.* **1963**, *5*, 219.
20. Heller, C.; McConnell, H. M. *J. Chem. Phys.* **1960**, *32*, 1535.
21. Horsfield, A.; Morton, J. R.; Whiffen, D. H. *Mol. Phys.* **1961**, *4*, 425.
22. Hudson, A.; Luckhurst, G. R. *Chem. Rev.* **1969**, *69*, 191.
23. Freed, J. H.; Fraenkel, G. K. *J. Chem. Phys.* **1962**, *37*, 1156.
24. Stoll, S.; Schweiger, A. *J. Magn. Reson.* **2006**, *178*, 42.
25. Kivelson, D. *J. Chem. Phys.* **1960**, *33*, 1094.
26. Berliner, L. J. *Spin Labeling: The Next Millennium*; Plenum: New York, **1998**.
27. Lebedev, Y. S. *Pure Appl. Chem.* **1990**, *62*, 261.
28. Kuznetsov, A. N.; Wasserman, A. M.; Volkov, A. U.; Korst, N. N. *Chem. Phys. Lett.* 1971, 12, 103.
29. Freed, J. H. Theory of slow tumbling ESR spectra of nitroxides. In *Spin Labeling, Theory and Applications*; Berliner, L. J., Ed.; Academic Press: New York, 1976, pp. 53–132.
30. Biological Magnetic Resonance, Vol. 8, *Spin labeling, Theory and Applications*. Berliner, L. J.; Reuben, J. Eds. Plenum Press: New York, 1989.
31. Weil, J. A.; Wertz, J. E.; Bolton, J. R. *Electron Paramagnetic Resonance*; Wiley: New York, 1994.
32. Duling, D. R. *J. Magn. Reson., B* **1994**, *104*, 105–110.
33. McLachlan, A. D.; Dearman, H. H.; Lefebvre, R. *J. Chem. Phys.* **1960**, *33*, 65.
34. Ban, F.; Gauld, J. W.; Boyd, R. J. Computation of Hyperfine Coupling Tensors to Complement EPR Experiments. In *Calculation of NMR and EPR Parameters: Theory and Applications*; Kaupp, M., Bühl, M., Malkin, V. G., Eds. Wiley-VCH: Weinheim, 2004.
35. Hermosilla, L.; Calle, P.; Garcia de la Vega, J. M.; Sieiro, C. *J. Phys. Chem. A* **2006**, *110*, 13600.
36. Hermosilla, L.; Calle, P.; Garcia de la Vega, J. M.; Sieiro, C. *J. Phys. Chem. A* **2005**, *109*, 1114.
37. Hermosilla, L.; Calle, P.; Garcia de la Vega, J. M.; Sieiro, C. *J. Phys. Chem. A* **2005**, *109*, 7626.
38. Neese, F. *J. Biol. Inorg. Chem.* **2006**, *11*, 702.

39. Munzarova, M. L. DFT Calculations of EPR Hyperfine Coupling Tensors. In *Calculation of NMR and EPR Parameters: Theory and Applications*; Kaupp, M., Bühl, M., Malkin, V. G., Eds. Wiley-VCH: Weinheim, 2004.

40. Barone, V.; Brustolon, M.; Cimino, P.; Polimeno, A.; Zerbetto, M.; Zoleo, A. *J. Am. Chem. Soc.* **2006**, *128*, 15865.

41. Gerson, F.; Huber, W. *Electron Spin Resonance Spectroscopy of Organic Radicals*; Wiley-VCH: Weinheim, 2003.

42. Nelsen, S. F. Cation radicals from nitrogen-containing compounds. In *Landold Börnstein, Magnetic Properties of Free Radicals*; **Vol II/9**, Chapt. 19, Fischer, H. and Hellwege, K.-H., Ed., Springer: Heidelberg, 1980.

43. Danen, W. C.; Rickard, R. C. *J. Am Chem. Soc.* **1975**, *97*, 2303.

44. Zwier, J. M.; Brouwer, A. M.; Keszthelyi, T.; Balakrishnan, G.; Offersgaard, J. F.; Wilbrandt, R.; Barbosa, F.; Buser, U.; Amaudrut, J.; Gescheidt, G.; Nelsen, S. F.; Little, C. D. *J. Am. Chem. Soc.* **2002**, *124*, 159.

45. Gerson, F.; Gescheidt, G.; Knoebel, J.; Martin, W. B., Jr.; Neumann, L.; Vogel, E. *J. Am. Chem. Soc.* **1992**, *114*, 7107.

46. Gerson, F.; Gescheidt, G.; Buser, U.; Vogel, E.; Lex, J.; Zehnder, M.; Riesen, A. *Angew. Chem.* **1989**, *101*, 938.

47. Davies, A. G.; Gescheidt, G.; Ng, K. M.; Shepherd, M. K. *J. Chem. Soc. Perkin Trans. 2* **1994**, 2423.

48. Finnegan, R. A. *J. Org. Chem.* **1965**, *30*, 1333.

49. Spichty, M.; Giese, B.; Matsumoto, A.; Fischer, H.; Gescheidt, G. *Macromolecules* **2001**, *34*, 723.

50. Gerson, F.; Hopf, H.; Merstetter, P.; Mlynek, C.; Fischer, D. *J. Am Chem. Soc.* **1998**, *120*, 4815.

51. Maier, G.; Kratt, A.; Schick, A.; Reisenauer, H. P.; Barbosa, F.; Gescheidt, G. *Eur. J. Org. Chem.* **2000**, 1107.

52. Grampp, G.; Mladenova, B. Y.; Kattnig, D. R.; Landgraf, S. *Appl. Magn. Reson.* **2006**, *30*, 145.

53. Savitsky, A.; Möbius, K. *Helv. Chim. Acta* **2006**, *89*, 2544.

54. Steiner, U. E.; Ulrich, T. *Chem. Rev.* **1989**, *89*, 51.

55. Brocklehurst, B. *Chem. Soc. Rev.* **2002**, *31*, 301.

56. Forbes, M. D. E. *Photochem. Photobiol.* **1997**, *65*, 73.

57. Gatlik, I.; Rzadek, P.; Gescheidt, G.; Rist, G.; Hellrung, B.; Wirz, J.; Dietliker, K.; Hug, G.; Kunz, M.; Wolf, J.-P. *J. Am. Chem. Soc.* **1999**, *121*, 8332.

58. Goez, M. *Concepts Magn. Reson.* **1995**, *7*, 69.

59. Goez, M. *Concepts Magn. Reson.* **1995**, *7*, 137.

60. Goez, M. *Concepts Magn. Reson.* **1995**, *7*, 263.

61. Roth, H. D. *J. Photochem. Photobiol. C* **2001**, *2*, 93.

62. Closs, G. L.; Miller, R. J.; Redwine, O. D. *Acc. Chem. Res.* **1985**, *18*, 196.

63. Wan, J. K. S.; Elliot, A. J. *Acc. Chem. Res.* **1977**, *10*, 161.

FURTHER READING

Radicals

Gerson, F.; Huber, W. *Electron Spin Resonance Spectroscopy of Organic Radicals*; Wiley-VCH: Weinheim, 2003.

This book contains an introduction into EPR of mainly organic radical ions and comprises an extended database of isotropic hyperfine coupling constants.

Landolt–Börnstein Series on Numerical Data and Functional Relationships in Science and Technology: *Magnetic Properties of Coordination and Organometallic Transition Metal Compounds* and *Magnetic Properties of Free Radicals*; Springer: Berlin; 1965–2005.

This series comprises a huge variety of data obtained by EPR spectroscopy and serves as an ideal reference for a and g values of many paramagnetic compounds.

Fossey, J.; Lefort, D.; Sorba, J. *Free Radicals in Organic Chemistry*; Wiley: New York, 1997.

Linker, T.; Schmittel, M. *Radikale und Radikalionen in der Organischen Synthese*; Wiley-VCH: Weinheim, 1998.

The last two books describe the properties and the reactivity of organic radicals.

Calculations

Kaupp, M., Bühl, M., Malkin, V. G., Eds. *Calculation of NMR and EPR Parameters: Theory and Applications*; Wiley-VCH: Weinheim, 2004.

This book contains a series of articles about calculations of magnetic parameters. Several chapters are quite useful to obtain insight into computational procedures and the quality of the expected results. The computational methods are continuously being improved; therefore, this book is a useful starting point to find more recent developments.

5 Pulsed Electron Paramagnetic Resonance

MICHAEL K. BOWMAN

Department of Chemistry, University of Alabama, Tuscaloosa, AL 35487-0336

5.1 INTRODUCTION

5.1.1 What Is Pulsed Electron Paramagnetic Resonance (EPR)?

An important area of modern EPR is pulsed EPR. It is an approach to EPR measurement in which the electron spins are excited by a series of microwave pulses and then the signal induced by these pulses is measured while there are no microwaves applied. The initial pulsed EPR measurements were made in the 1950s shortly after the initial pulsed nuclear magnetic resonance (NMR) experiments. Some of the early pulsed EPR measurements were made for very practical applications: using electron spins as computer memory and using EPR for remote detection of submarines. By the early 1960s, pulsed EPR was used in a few physics laboratories to measure relaxation in materials for quantum optics.

In the mid-1960s, V. V. Voevodsky of Novosibirsk, Bill Mims of Bell Laboratories, and Ian Brown of McDonnell–Douglas Research Laboratories realized that pulsed EPR had broad potential for important problems in chemistry and biology. Mims, Brown, and Yuri Tsvetkov (who led the Novosibirsk effort after the untimely death of Voevodsky) established a technological and theoretical approach for pulsed EPR that paved the way for early chemical and biological applications by Blumberg, Brown, Freed, Kevan, Mims, Norris, the Orme-Johnsons, Peisach, Trifunac, and Tsvetkov. Two reviews from those early beginnings make fascinating reading and contain a lot of practical information and ideas that are still being reinvented today: a chapter by Mims [1] and a much more detailed and comprehensive book by Salikhov, Semenov, and Tsvetkov [2], which unfortunately was never translated from Russian.

Electron Paramagnetic Resonance. Edited by Brustolon and Giamello
Copyright © 2009 John Wiley & Sons, Inc.

The original pulsed EPR spectrometers did not offer the flexibility of modern spectrometers. Each pulse sequence was implemented in hardware and required changing an array of cables and switches to change a two-pulse into a three-pulse sequence. Data collection was equally difficult, with signals either photographed from an oscilloscope or written on chart paper and then manually digitized using a ruler. It was not until 1980 that a pulsed EPR spectrometer was described [3, 4] in which the pulse sequence and the data acquisition are controlled by computer (Box 5.1).

The development of commercial pulsed EPR instrumentation lagged about a decade behind that of pulsed NMR instrumentation because of the nature of the spins involved. The unpaired electron spin has a magnetic moment that is roughly

---

BOX 5.1 CW VERSUS PULSED EPR

The prototypical CW experiment consists of measuring the absorption or phase shift (dispersion) of a constant microwave field by the sample as the external magnetic field is swept slowly (compared to any spin relaxation time) through resonance for each possible transition. This CW measurement yields the EPR spectrum, which is the splittings between energy levels of the electron spin system, but provides no direct information about the dynamics of the spin system.

The prototypical pulsed EPR experiment consists of measuring the signal emitted by the spin system following a series of one or more microwave pulses. The resulting signal is a consequence of the dynamics of whatever state of the spin system was prepared by the microwave pulses and modified by instrumental characteristics. The pulsed EPR signal contains both spectral information and spin relaxation information that can be recovered by further signal processing.

In practice, most EPR measurements fall somewhere in between these two extremes. CW measurements can use field modulation that is faster than their relaxation rates and are performed on transient or slowly decaying radicals or following a perturbation such as a long microwave pulse. The signals from such CW measurements are strongly influenced by the dynamic properties of the spin system. Similarly, pulsed EPR measurements are often performed with instrumental bandwidths that are much narrower than the EPR spectrum, and the signals are strongly dependent on both the magnetic field and microwave frequency.

A useful, but very arbitrary, classification of EPR measurements into CW (which may be time resolved) and pulsed types is based on whether microwaves are applied to the sample simultaneously with signal acquisition. CW EPR has microwaves applied during signal acquisition, so that resonator match and frequency stability are critical experimental issues. Pulsed EPR has no microwaves applied during signal acquisition, but pulse timing and resonator bandwidth are critical.

a 1000-fold greater than the nuclear spin moments of protons or carbon-13. Relaxation times scale roughly as the square of the magnetic moment, so that electrons relax $\sim 10^6$ times faster than nuclei, requiring pulse widths, pulse delays, and repetition times that are 10^6 times shorter for pulsed EPR than for pulsed NMR

BOX 5.2 PITFALLS AND HOW TO EXTRACT MORE INFORMATION WITH PULSED EPR

A common pitfall for beginners in EPR, as in many other forms of spectroscopy, is to overanalyze an experimental measurement and attempt to extract more information than is possible.

For instance, the EPR spectrum of many free radicals in frozen solution is, to within the experimental noise, a Gaussian line that is completely described by only three parameters: the intensity or area under the curve, the position of the center of the peak, and the width or second moment. Yet, the EPR signal is determined by the number of spins, the resonator Q factor, the field modulation amplitude, the average g factor, the g factor anisotropy, the magnetic field inhomogeneity, the spin relaxation times, hyperfine interactions with many nuclei, and dipolar interactions with all other spins in the sample, to name just a few.

It is quite possible to calculate a spectrum for particular values of all of those parameters that matches the experimental signal to within the experimental noise level. However, those values of the parameters can be totally unreliable for two reasons. (1) The parameters may be highly correlated with each other, so that very different values produce virtually identical calculated spectra. In other words, the fit is not unique. (2) The spectral model may be incomplete so that a different, preferably more accurate model, say one that includes radical motion, could also reproduce the spectrum but with quite different parameter values. In other words, the model used is incorrect.

The situation might be helped by independently determining some of the parameters, for instance, calibrating the magnetic field modulation amplitude. Reducing the number of unknown parameter values typically improves the simulation of the experimental data, but it does have drawbacks. One major drawback is that it may require extensive measurements unrelated to the phenomenon of interest. For instance, determining the dipolar interactions with other spins in the sample accurately by spectral simulation may require detailed field mapping of the magnet, precise measurement of all hyperfine couplings, and determination of the isotopic composition of the free radical. All of these are possible but not the main focus of the measurement.

A better strategy is to use a signal that is very sensitive to the few parameters of interest and insensitive to all others. For instance, if spin relaxation is the object of the measurement, then a spin echo or inversion recovery measurement of spin relaxation is a much better approach; or if unresolved hyperfine interactions are of interest, then ENDOR or ESEEM measurements would be more effective.

and easily exceeded the capabilities of the best available digital integrated circuits of that period.

5.1.2 Why Utilize Pulsed EPR?

In view of the technological challenges facing pulsed EPR, why not stay with the continuous wave (CW) EPR methodology? It is certainly a mature, well-developed methodology providing spectra that contain a lot of information about the spin system. However, that information is convolved with all of the other information in the spectrum and can be impossible to reliably extract (Box 5.2).

Pulsed EPR provides much better control over the production of the signal and the information it contains. A common pitfall for beginners in EPR, as in many other forms of spectroscopy, is to overanalyze an experimental measurement and attempt to extract more information than is possible. The goal in pulsed EPR is to provide a set of signals that contain only the desired information. There should be at least one pulse sequence for each parameter that one wishes to measure. Developing a new, useful pulse sequence can be a challenge because it requires considerable insight into how the spins evolve, how to make the measurement insensitive to most properties of the spin system, and how to compensate for practical shortcomings of the measurement equipment so that the result is independent of the spectrometer.

The number of signals generated by a pulse sequence increases very rapidly as the number of pulses increases [4], so the challenge is to recognize which sequences and signals will provide the desired information. Fortunately, dozens of pulse sequences have been described, and there are approaches for designing sequences for certain classes of measurements [5]. Each pulse measurement shows how the spin system responds to that pulse sequence, but it does not necessarily deliver an abstract parameter such as T_2 and does not consistently deliver the same parameter for all types of samples. This will be evident in the discussion of different pulse sequences.

5.2 VECTOR MODEL FOR PULSED EPR

An unpaired electron has spin angular momentum with $S = 1/2$ that can be described in terms of two states defined by its projection in some arbitrary direction in space. The behavior of a single spin is subject to the arcane laws of quantum mechanics. However, an ensemble of many identical, noninteracting electron spins, as in the prototypical EPR sample, can be described by the sum or ensemble average of the electron spin moments, often called the magnetization. The magnetization, in contrast, follows the laws of classical mechanics and actually behaves as a classical vector with a length and a direction in space. This property of the magnetization is fortunate because we all have a practical, intuitive feel for classical mechanics honed by living in a "classical" rather than a "quantum" world, and that feel will help us understand much of pulsed EPR. This classical vector is an exact description of the quantum mechanical density matrix describing a single unpaired electron, but in a more intuitive fashion.

Of course, there are atoms, ions, and molecules with spin angular momentum $S > 1/2$. They have more than two energy levels and require the full quantum treatment. However, even then, if transitions involve only two levels of those complicated spin systems, those two levels can be treated exactly as an "effective" $S = 1/2$ system, although sometimes with properties, such as the g factor, that are quite different from that of an isolated, unpaired electron. Thus, pulsed EPR can be usefully described in terms of the magnetization from effective $S = 1/2$ electron spins, except in cases where transitions involve more than two levels of the electron spin. Each transition has an effective g or magnetogyric ratio that is simply a measure of how it responds to magnetic fields. The effective g can vary with the magnetic field and across a spectrum.

In considering pulsed EPR it is very useful to follow the magnetization of a "spin packet" of identical spins (also called isochromats because they all have the same resonant frequency). The pulsed EPR signal and the total magnetization from all the different spin packets in the sample are just the sum of the results for the individual spin packets if the spins behave independently of each other and do not interact. This assumption of noninteracting spin packets yields a very simple and intuitive magnetization vector model for pulsed EPR whose simplicity far outweighs its shortcomings.

The behavior of the spin packet magnetization vector is usually described by the Bloch equations [6] (introduced in Chapter 1) in a coordinate system whose z direction is defined by the magnetic field B_0 or B_z of the EPR spectrometer. The magnetization vector **M**, like any vector, can be written in terms of components along the x, y, and z axes as M_x, M_y, and M_z. The behavior of these three components is quite simple in the absence of any applied microwaves: M_z simply relaxes with a time constant T_1 (spin lattice relaxation) toward its thermal equilibrium value determined by the temperature and the splitting between the two energy levels of the spin; M_z is directly proportional to the population difference between these two levels, but M_x and M_y oscillate or precess around B_z at the EPR frequency γB_0 of that particular isochromat. At the same time, they also shrink or decay toward zero with a time constant T_2 (spin–spin relaxation). Resonant or nearly resonant microwave pulses convert M_z into M_x and M_y and vice versa. This motion is much easier to follow in what is called a rotating frame, similar to the interaction representation of quantum mechanics.

By using a coordinate system rotating at the frequency of the microwaves ω_0 around B_z, the microwave magnetic field has a constant direction during the pulse, often labeled the X direction in the rotating frame. In addition, the amplitude of B_z is reduced to a small field along the z axis equal to the difference ΔB between B_z and the resonant field for that spin packet, ω_0/γ. The spin packet magnetization, when expressed in terms of the X, Y, and Z axes of the rotating coordinate system, is even simpler. In the absence of microwaves, **M** precesses about ΔB. Magnetization M_Z remains directed along the Z axis and its length relaxes toward the equilibrium value with a time constant T_1, whereas M_X and M_Y now precess around Z at a rate determined by ΔB and decay toward zero with a time constant T_2. This behavior is the same as in the laboratory axis system except that the speed of precession is much slower, proportional to $\Delta B = B_0 - \omega_0/\gamma$ rather than B_0 (Fig. 5.1).

Things are much simpler during the microwave pulse. The microwave magnetic field no longer appears to rotate because the coordinate system is rotating at the same speed. The effective magnetic field B_{eff}, affecting **M** in the rotating frame, is then the vector sum of ΔB and the microwave magnetic field B_1, which is assumed to lie along the X axis. It is generally assumed that the microwave pulses are short and intense so that T_1 and T_2 relaxation can be neglected during the pulses and $B_{\text{eff}} \sim B_1 \gg \Delta B$. In this limit, the microwave pulse makes **M** rotate around X from Z toward Y.

The motion of **M** during the microwave pulse and the motion and relaxation in the absence of microwaves is all that is needed to start to understand the basic pulse sequences used in pulsed EPR and the types of information that those sequences

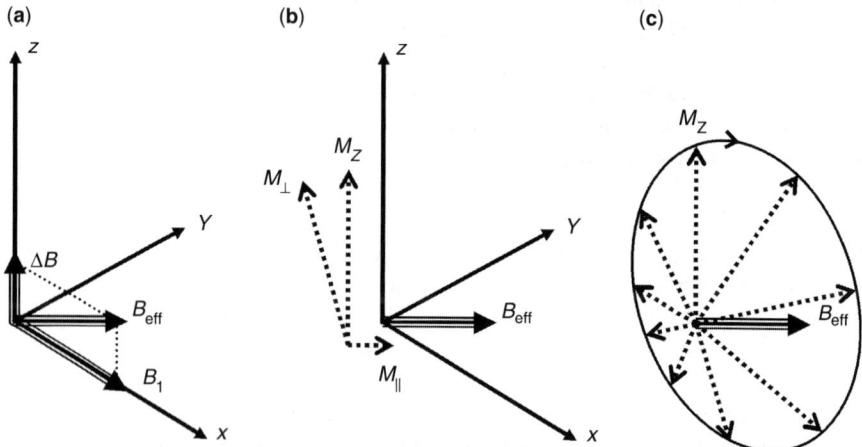

Fig. 5.1 Illustration showing the fields and magnetization of the vector model in the rotating frame. (a) The effective magnetic field B_{eff} that the individual spins experience is the vector sum of the offset field ΔB and the microwave magnetic field B_1. The B_{eff} is different for each spin packet because each spin packet has a different offset ΔB from resonance. During the periods between microwave pulses, $B_1 = 0$ and $B_{\text{eff}} = \Delta B$. (b) The equilibrium magnetization M_Z of a spin packet and the B_{eff} is shown. Here, M_Z lies along the applied magnetic field B_0 and is not perpendicular to B_{eff} if ΔB is not zero. M_Z, or any magnetization, can be represented as the sum of two components: $M_\parallel$, which is parallel to B_{eff}, and $M_\perp$, which is perpendicular to B_{eff}. These two components have the useful property that $M_\parallel$ is unaffected by B_{eff} whereas $M_\perp$ precesses in a plane perpendicular to B_{eff}. (c) Breaking the magnetization into two components can be helpful in tracking the response of a spin packet during a pulsed EPR measurement because the motion of the individual vectors is simpler than the motion of the total magnetization that generally traces out a cone around B_{eff}. That cone is the vector sum of a stationary $M_\parallel$ and the $M_\perp$ precessing around B_{eff}. For each pulse and delay in a particular sequence, the direction and magnitude of B_{eff} changes and consequently the two magnetization components are redefined at each transition between microwave pulse and delay.

provide. However, it must be kept in mind that this vector model, based on the Bloch equations, is only approximate. The Bloch equations have been modified, extended, and replaced numerous times; for example, see the discussion in Chapters 1 and 2 of Reference 7. This vector model violates the dictum attributed to Albert Einstein: "Everything should be made as simple as possible, but not one bit simpler." The vector model's great utility comes from its use of classical, three-dimensional (3-D) mechanics that most people have a highly developed, intuitive feel for as a result of a lifetime's experience in a classical, 3-D world. The vector model is a great tool and a great mathematical first approximation that is easily extended and adjusted. The vector model in 3-D space is an exact description of the quantum mechanical density matrix for a two-level system. However, a four-level system, such as an unpaired electron interacting with the spin of a single proton, can be completely described only by a vector in 15-D space where intuition fails most of us.

The density matrix formalism of Chapter 7 provides a consistent method for approaching the dynamics of spin systems. Yet, the density matrix is just an idealized model of the spin system and is always simplified in order to make calculations feasible. These models and calculations must never be confused with reality. As Einstein noted: "As far as the laws of mathematics refer to reality, they are not certain; and as far as they are certain, they do not refer to reality." The electron spin is a quantum mechanical phenomenon and, on occasion, it will exhibit counterintuitive behavior. In the words of Sir Arthur Eddington, "Not only is the universe stranger than we imagine, it is stranger than we can imagine."

5.2.1 One Pulse: Free Induction Decay (FID)

The simplest pulsed EPR sequence consists of a single microwave pulse. The pulse intensity B_1 and width t_p are usually adjusted to a "turning angle" of $\gamma B_1 t_p = \pi/2$, although virtually any pulse that is shorter than the spin relaxation times and with a nonzero turning angle will produce a signal. The pulse will excite spins over a bandwidth of approximately $\pm 0.68/t_p$. Resonator Q will stretch the pulse and, to some extent, make it longer and will also act as a filter for the signal as will the spectrometer receiver. The actual bandwidth will always be somewhat less than expected from the pulse width alone. At the end of such a pulse, the magnetization of each spin packet with a resonant frequency near ω_0 lies along the Y axis in the rotating frame. The M_X and M_Y components of **M** at the end of the pulse then precess around Z at a rate $\gamma \Delta B$ while shrinking with a time constant T_2. During this precession, the M_X and M_Y induce a microwave signal in the resonator of the EPR spectrometer that is amplified and detected. The frequency of the signal induced by this spin packet is $\gamma \Delta B + \omega_0$, which is the precession rate plus the frequency of the rotating frame (because the resonator is in the laboratory frame and not rotating). When this induced microwave signal is demodulated in the receiver, it becomes a "video" frequency signal at $\gamma \Delta B$. Because this signal is "induced" in the receiver during the "free" precession of **M** (no applied microwaves), it is commonly called a free induction signal or free induction decay (FID) (Fig. 5.2).

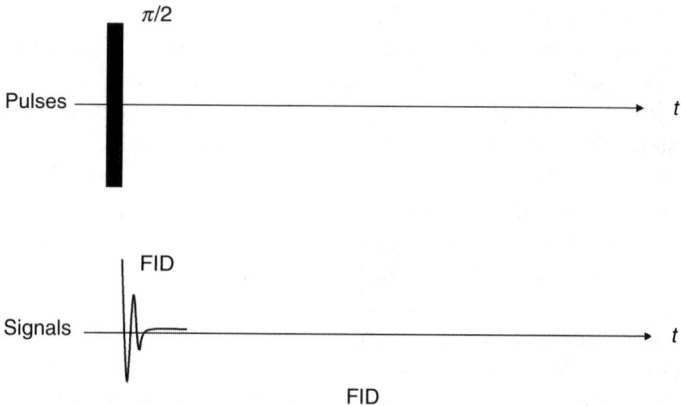

Fig. 5.2 The FID pulse sequence. A single microwave pulse, indicated by the black rectangle, with an optimal turning angle of $\pi/2$ is immediately followed by the FID.

The free induction signal has four very important properties:

1. It occurs at precisely the resonant frequency of the spin packet.
2. It decays with the T_2 of that spin packet (if the resonator bandwidth is large).
3. If the pulse deviates from the ideal $\pi/2$, the amplitude or phase of the FID may be affected but not its frequency or lifetime.
4. If the phase of the excitation pulse is shifted (B_1 along some other direction in the rotating frame), the phase of the FID shifts by exactly the same amount.

Thus, the FID is a good way of measuring the frequency and T_2 of a spin packet.

Most samples are composed of many spin packets, and the FID is composed of the sum of the FIDs of all of the individual spin packets. Because the spin packets have different resonant frequencies, the signals from the individual spin packets can beat against each other or interfere with each other to produce a highly modulated FID from the sample.

Nevertheless, it is often possible to reconstruct the EPR spectrum of the sample from such an FID using the same signal processing methods employed in 1-D Fourier transform NMR spectroscopy. The main challenge comes from the "dead time" following the microwave pulse before the receiver and resonator have recovered from the intense ($\sim$1 kW) pulse and are able to detect the weak (nanowatt) FID. If the distribution of spin packet resonant frequencies is broad, the FID from the individual spin packets can interfere and leave no detectable signal extending beyond the dead time (Fig. 5.3).

5.2.2 Two Pulses

As one might expect, the response of a spin packet to a pair of microwave pulses is a bit more complex than the response to a single pulse, but still understandable. There

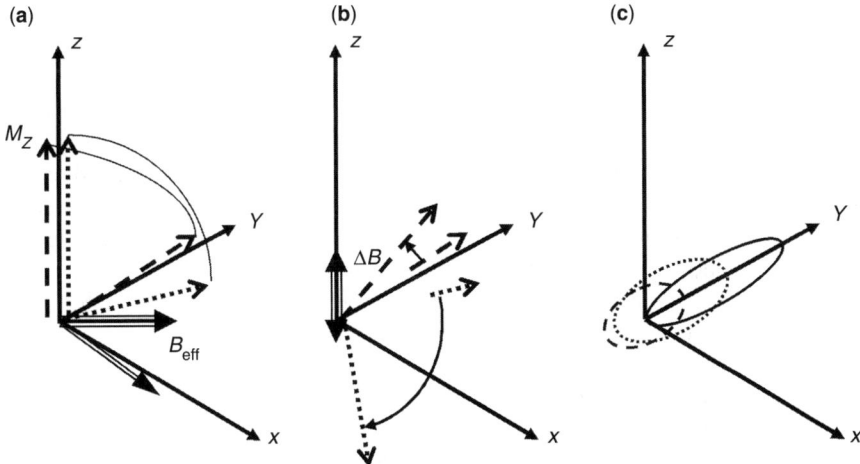

Fig. 5.3 Illustration showing the production of the FID from a set of spin packets. (a) The M_Z of two spin packets (dotted and dashed lines) are driven by their respective B_{eff} to near the Y axis by a pulse that has a nominal $\pi/2$ turning angle. Because the spin packets move in a cone around B_{eff}, there will be a slight spread of magnetization. (b) After the pulse, the B_{eff} is equal to the ΔB for each spin packet. The dotted spin packet has the larger, positive ΔB whereas the dashed spin packet has the small, negative ΔB. During the period following the pulse, the spin packets precess around ΔB as illustrated. (c) The net effect for an inhomogeneously broadened EPR line is shown, where the initial distribution of magnetization from all spin packets is concentrated along the Y axis giving a net magnetization indicated by the solid curve. Somewhat later, the distribution of ΔB has caused the spin packets to precess and the distribution of magnetization is less peaked as shown in the dotted curve until finally in the dashed curve there is no net magnetization remaining because the spin packet magnetizations point in all directions.

are three experimentally distinguishable EPR signals that can be generated by a pair of pulses, and each will be discussed in turn. Most discussions of pulsed EPR describe the pulse sequence in terms of the turning angles and pulse phases that maximize a particular signal. This is done partly for simplicity and partly because it is the condition that one generally tries to achieve experimentally for maximum sensitivity. However, pulsed EPR signals are rather robust and each type of signal is observable for a wide range of turning angles. It is usually only for a very specific set of turning angles that a signal vanishes. The discussion in this chapter will refer to arbitrary pulses except where specific pulses greatly simplify matters.

The first signal is simply the FID produced by the first microwave pulse and discussed above. The first pulse leaves the spin packet with an altered M_Z and with magnetization precessing in the X, Y plane. The second signal to consider is the FID produced by the second pulse from the M_Z left over by the first pulse or regenerated by spin–lattice relaxation. This second signal provides a very nice way to directly measure the T_1 of the spin packet in the context of the Bloch equations—by measuring the amplitude of the FID as the delay between the first and second pulses is varied. In a sample with many spin packets, recording the FID and reconstructing the EPR signal as discussed above for each delay allows measurement of T_1 across the EPR spectrum. This T_1-sensitive signal is most intense when the first pulse is a π or 180° pulse followed by a $\pi/2$ or 90° pulse. This is often called a 180–90 pulse sequence or an inversion recovery sequence because the first pulse conceptually inverts **M** to −**M** and then measures its recovery. The excitation bandwidth of the inversion recovery sequence is a bit difficult to define because the typical inverting pulse inverts spins over a much narrower bandwidth than the second, detecting pulse. In addition, the microwave field is usually very inhomogeneous along the length of the sample so that parts of the sample that are not in the center of the resonator are far from being inverted even when they are at the same frequency as the microwave pulse (Fig. 5.4).

Inversion recovery (or saturation recovery if the FID is initially partially saturated rather than fully inverted) is probably the most accurate description of what is measured. In the Bloch equations, the only mechanism for recovery of the signal is spin–lattice relaxation, but in the real world other processes that have little to do with transfer of spin energy to the lattice can make the signal recover. One possibility is cross relaxation, also called Heisenberg exchange, spin exchange, or spin diffusion in various limiting cases, which occurs when spins from two different spin packets interact and exchange or swap electron spins. If one of the spin packets is farther from its equilibrium M_Z (more saturated) than the other, the result of exchanging spins is the approach of both spin packets to a common degree of saturation. Thus, the signal from one spin packet recovers intensity at the expense of another. Such cross relaxation can be observed when the two radicals have different T_1 values or when the first pulse saturates the two spin packets to different extents.

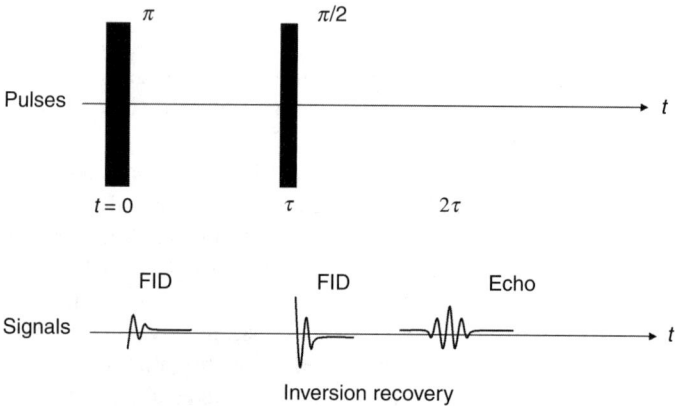

Fig. 5.4 The inversion recovery pulse sequence. The nominal sequence consists of an inverting π pulse followed by a $\pi/2$ pulse to generate the signal, which is the FID. The delay t is varied. This sequence often produces an FID following the first pulse and a spin echo. The unwanted signals can be avoided in recorded data by cycling the phase of the microwave pulses.

Another way for the signal to recover is through a process known as spectral diffusion, when the resonant frequency of a spin packet actually changes with time so those spins appear to diffuse across the spectrum. One example is the rotation of a radical with an anisotropic spectrum, as in saturation transfer EPR. If spectral diffusion replaces a well-saturated spin packet with a less saturated one, the result is the recovery of the FID signal from saturation unrelated to relaxation of the electron spins.

5.2.2.1 The Spin Echo.
The final signal produced by a pair of pulses is known as the spin echo, and it results from the action of the second pulse on the M_X and M_Y produced by the first pulse. It is easier to understand how the signal arises using very specific pulses, but almost any pair of pulses will produce a spin echo (Fig. 5.5).

For the sake of this description, the pulses are assumed to have the same phase, with B_1 along the X axis of the rotating frame; the pulses have turning angles of 90° and 180°, respectively; and ΔB for all spin packets can be neglected relative to B_1. The excitation bandwidth for the echo when the 90° and 180° pulses have the same power ranges between $\pm 0.27/t_p$ and $\pm 0.21/t_p$.

In this ideal case, the first pulse converts all the M_{Z0} into M_Y that starts precessing and decaying in the rotating frame after the first pulse. After a delay time usually called τ, the magnetization in the X, Y plane has become $M_Y[\tau] = M_{Z0} \exp[-\tau/T_2]\cos[\gamma\Delta B\tau]$ and $M_X[\tau] = M_{Z0} \exp[-\tau/T_2]\sin[-\gamma\Delta B\tau]$ through the combined effects of precession in the effective field of ΔB and T_2 relaxation. It is usually more convenient to combine these two equations and define a precessing magnetization

$$M_+[\tau] = M_X[\tau] + i\,M_Y[\tau] = M_{Z0}\,e^{-i\gamma\Delta B\tau - \tau/T_2 - i\pi/2} \qquad (5.1)$$

which shows both the phase and amplitude of the precessing magnetization and of the demodulated, induced signal in the spectrometer receiver. The terms in the exponential

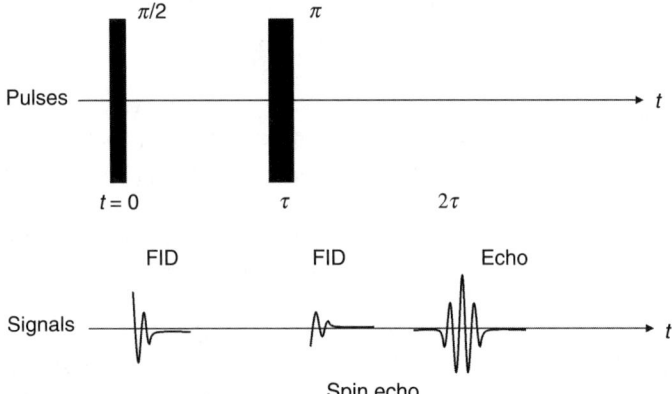

Fig. 5.5 The primary or two-pulse electron spin echo sequence. The nominal sequence consists of a $\pi/2$ pulse followed by a π pulse to generate the signal, which is the echo appearing at time 2τ. This sequence often produces FIDs following each pulse. The unwanted signals can be avoided in recorded data by cycling the phase of the microwave pulses.

function are the precession of the magnetization, the T_2 decay, and an overall phase shift of $-\pi/2$ or $-90°$ between $M_+[0]$ and the microwave magnetic field. It is precisely the FID of the first pulse!

At time τ the 180° pulse is applied with B_1 along the X axis of the rotating frame. This pulse converts M_Y into $-M_Y$ and M_Z into $-M_Z$, which we are not interested in and will ignore. The 180° pulse converts M_+ into its complex conjugate so that at a time τ after the second pulse, the magnetization is

$$M_+[\tau+t] = M_+^*[\tau]\, e^{-i\gamma\Delta Bt - t/T_2} = M_{Z0}\, e^{-i\gamma\Delta B(t-\tau)-(t+\tau)/T_2 + i\pi/2} \quad (5.2)$$

which has a strong resemblance to the FID in the previous equation. The magnetization of the spin packet after the second pulse is precessing with the same frequency, decaying with T_2, and is inducing a signal in the receiver, although with a different phase. In fact, the common reaction of people when this sequence is applied on a sample with very narrow lines approaching a single spin packet is that the signal is still just the FID and nothing special has happened.

So why is this called a spin echo? It is because something quite unexpected happens when there are many spin packets and a broad distribution of ΔB, as a result of the phase conjugation caused by the second pulse. The FID following the first pulse from each spin packet is proportional to $\exp[-i\gamma\Delta B\tau - (\tau/T_2) - i\pi/2]$, so that for a broad distribution of $\Delta B\tau$ values, the signals from the individual spin packets interfere with each other and the sum of the individual FID goes to zero before the magnetization has decayed from T_2 relaxation as mentioned earlier. The second pulse reverses the M_Y of every spin packet, so that the destructive interference of the FID is just as complete immediately after the pulse as immediately before. However, because of the phase conjugation, something amazing takes place. The destructive interference that attenuated the FID from the collection of spin packets now is reversed because the $\Delta B(t-\tau)$ term gets smaller for all spin packets. At the instant that $t = \tau$, the magnetization from each spin packet is

$$M_+[2\tau] = M_{Z0}\, e^{-2\tau/T_2 + i\pi/2} \quad (5.3)$$

and the signal from each spin packet is independent of ΔB. All of the signals are in phase and add together, so that from little or no signal immediately after the second pulse, a signal reappears at time 2τ. This signal is an "echo" of the FID from the first pulse.

At 2τ the magnetization is the same as after the first pulse except for some T_2 decay and a 180° phase shift, so the evolution and interference of the original FID is reproduced (Fig. 5.6).

The formation of the echo just before 2τ also reproduces the FID, but in reverse (and with the effect of T_2 on the FID reversed). Thus, the echo is sometimes used as a way of capturing the entire FID in samples where the dead time causes severe distortions of the FID following a single pulse.

A major use of the spin echo is based on Equation 5.3. At time 2τ the echo amplitude is independent of ΔB and the underlying inhomogeneous broadening of spin packets. It is commonly said that the second pulse "refocuses" the static,

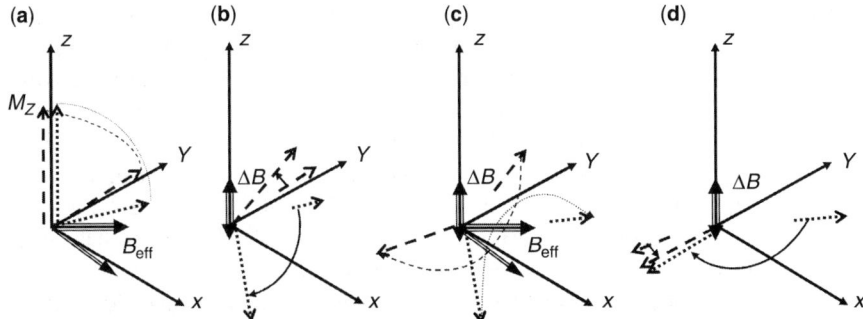

Fig. 5.6 Illustration of the formation of a spin echo by the refocusing of magnetization. (a) The M_Z of two spin packets (dotted and dashed lines) are driven by their respective B_{eff} to near the Y axis by a pulse that has a nominal $\pi/2$ turning angle. Because the spin packets move in a cone around B_{eff}, there will be a slight spread of magnetization. (b) After the pulse, the B_{eff} is equal to the ΔB for each spin packet. The dotted spin packet has the larger, positive ΔB whereas the dashed spin packet has the small, negative ΔB. During the period following the pulse, the spin packets precess around ΔB as illustrated. (c) The second, or "refocusing," pulse after a delay τ rotates the magnetizations around their B_{eff} by approximately an angle of π. This has the effect of performing a "phase conjugation" on each of the spin packets so that the M_X and M_Y of each spin packet is transformed into M_X and $-M_Y$. (d) The magnetization of each of the spin packets continues to precess in the same direction and the same rate determined by the respective ΔB following the refocusing pulse. The result of this continued precession is that after an additional delay τ, the magnetization of all of the spin packets, regardless of their individual ΔB, align along the $-Y$ axis. This net magnetization produces the spin echo signal.

inhomogeneous broadening. This means that the echo amplitude decays as a function of 2τ with a time constant T_2, unaffected by inhomogeneous broadening that can dominate the FID and the CW EPR line shape.

What are real spin echo decays like? In liquid solutions of free radicals, the echo decay is usually exponential, as expected from Equation 5.3 with a time constant of T_2 with allowance made for collisions of the radical with other paramagnetic species in the solution. A careful study of nitroxides [8] showed complete agreement for collision rates, temperature dependences, and concentration dependence between CW EPR line shapes and spin echoes. There was a minor numerical difference that arose because the CW analysis did not take into account that the deuteration of the nitroxide was not perfect. This study is a clear illustration that T_2 can be measured by CW and pulsed methods. Pulsed EPR gives a direct measurement whereas CW EPR requires additional careful determination of small hyperfine couplings to all protons in the radical as well as measurement of the exact isotopic composition and extensive line shape simulations. In short, the CW EPR approach required learning a lot more about the radical than was the goal of the study.

The decay of the echo in solids is often quite different from Equation 5.3. Rather than being exponential, the decay is often a stretched exponential of the form $\exp[-(2\tau/T_M)^n]$, where n can have a value from 0.5 to 2. The decay is often written with a time constant of T_M, the phase memory time, to distinguish it from the T_2 defined by the Bloch equations. Even stranger are the decays that are not even monotonic, but show oscillations in amplitude known as electron spin echo envelope modulation (ESEEM). Both types of decays can be understood by relaxing some of the assumptions implicit in the Bloch equations.

One of the assumptions is that ΔB is constant during the time between the pulses that generate the echo. This assumption allows complete refocusing of ΔB so that ΔB has no impact on the decay of the echo. However, processes such as spectral diffusion mean that ΔB can vary during the echo generation so that the refocusing following the second pulse does not completely duplicate and cancel the defocusing or "dephasing" of the magnetization between the first two pulses. The result is an additional loss of echo intensity whose dependence on 2τ depends in turn on the dynamics of the spectral diffusion.

An interesting form of spectral diffusion is the instantaneous diffusion described by Klauder and Anderson [9]. In this case, the second microwave pulse flips the spin of other radicals in the sample that then changes the dipolar interaction between those spins and the observed radical, shifting its ΔB instantaneously with the second pulse. This instantaneous spectral diffusion prevents complete refocusing, causing an additional exponential decay component that depends on $\Delta B_d \sin^2(\theta/2)$, where ΔB_d is the average dipolar field and θ is the turning angle of the second pulse. Instantaneous diffusion is very useful for determining dipolar interactions in solids by measuring echo decays with different turning angles for the second microwave pulse. Unfortunately, it also means that the T_M in solids with large radical concentrations is only obtained after correcting for instantaneous diffusion.

Finally, in many solid samples the second microwave does not cleanly cause a phase inversion. The traditional selection rules of EPR break down and the microwave pulse flips the electron spin and a nuclear spin coupled to it. This nuclear spin flip causes a change in ΔB related to the electron nuclear double resonance (ENDOR) frequencies of the nucleus so that the magnetization does not refocus properly as the shifted and unshifted spin packets beat against each other, modulating the decay of the echo with ENDOR frequencies. This ESEEM effect is a very powerful method to measure the hyperfine and quadrupole interactions of nuclei coupled to the unpaired electron spin. Strong ESEEM is generally observed when the isotropic hyperfine interaction is small and the anisotropic hyperfine interaction is comparable to the nuclear Zeeman interaction or when the hyperfine and quadrupole interactions are comparable. The ESEEM effect and its applications have been extensively reviewed [10].

5.3 PULSE SEQUENCES

Most EPR pulse sequences are based on strategies seen in the one- and two-pulse sequences just described. One strategy is the measurement of relaxation times or frequencies in the spectrum from the free evolution of the spin system between or following microwave pulses. This aspect of pulsed EPR is seen in the discussion of the

FID and provides access to the spectrum and ultimately to the energy levels and splittings of the free radical or paramagnetic center. The two-pulse sequence shows that relaxation times can be measured from the recovery of the spin system following perturbation.

The spin echo illustrates the other strategy, that is, of refocusing interactions. In this case, static inhomogeneous broadening is removed from the measurement, in this example, of the decay of the echo. The interactions that are refocused still have an effect on the measurement; inhomogeneous broadening has a major role in determining the width and shape of the echo itself. However, there is that one point in the echo at $(t - \tau) = 0$ where the signal amplitude becomes independent of static inhomogeneous broadening and allows "clean" measurement of T_2, for example.

Fortunately for the experimentalist, it is not necessary to measure at exactly the point in the echo where $(t - \tau) = 0$. All that is necessary in practice is to keep $(t - \tau)$ constant, which is to measure the same part of the echo throughout the measurement, a much easier task than to keep $(t - \tau)$ precisely equal to zero. The signal then does depend, to some extent, on the static inhomogeneous broadening. However, measuring at constant $(t - \tau)$ keeps the impact of the inhomogeneous broadening constant for all values of τ, thus making a clean measurement of T_2 possible.

5.3.1 Preparation, Evolution, and Detection Paradigm

It can be very useful to think of a pulse sequence in terms of three periods of time where different things take place: preparation of the spin system; evolution of the spin system; and detection of the signal. This paradigm of preparation, evolution, and detection periods during a pulse sequence applies naturally to most, but not all, of the common pulse sequences and pulsed EPR measurements, as well as to NMR. It also aids in the design of new pulse sequences and has led to the idea of standard "modules" that can be used to prepare the spin system in certain ways [5].

Typically the preparation period consists of a series of pulses and delays that remains constant for a measurement. Its function is to convert the initial equilibrium **M** into some initial nonequilibrium magnetization at the start of the evolution period. The evolution period consists of delays and/or pulses that usually vary from point to point in a measurement. The detection period usually starts with one or more microwave pulses that convert some of the evolved magnetization into M_X and M_Y for detection and continues throughout the measurement of the signal.

A simple example of this paradigm is the two-pulse inversion recovery sequence (Fig. 5.4). The preparation period consists of the initial pulse that converts $\mathbf{M_0}$ into $-\mathbf{M_0}$; the evolution period is the delay between the two pulses during which the magnetization recovers through spin–lattice relaxation. The second pulse of the sequence converts the recovering **M** into M_X and M_Y. This ends the evolution period and starts the detection period during which the signal is measured. During a typical inversion recovery measurement, the preparation and detection periods do not change but the evolution period is systematically lengthened.

EPR measurements and pulse sequences can be classified on the basis of the number of dimensions that their data occupies or the number of independent variables that uniquely identify each point in the spectrum. Most measurements produce either 1-D or 2-D data. The standard CW EPR spectrum is a 1-D measurement, as is the analogous spectrum resulting from the Fourier transform of an FID. A series of ENDOR spectra measured at each point in the EPR spectrum is an example of a 2-D measurement. Each point in that spectrum is characterized by the magnetic field at which it was measured and by the ENDOR frequency.

The one-pulse FID measurement described above is strictly a 1-D pulsed EPR measurement. Most other measurements and pulse sequences are potentially multidimensional measurements if the full shape of that FID or echo were to be recorded. This is seldom done except in time-resolved pulsed EPR measurements of photochemical reactions. Current commercial pulsed EPR spectrometers are not designed to take such multidimensional data efficiently and there is no readily available, simple-to-use software to process EPR data of greater than two dimensions. Modern pulsed NMR suggests that multidimensional pulsed EPR is inevitable.

Several pulse sequences will be described briefly. They are divided into 1-D and 2-D sequences, although the sequences could be 2-D and 3-D if the shape of the echo or FID were recorded as described in Reference 11. The sequences here are chosen because they are in wide use, because they appear as building blocks for more complicated sequences, or because they reveal an important concept of pulsed EPR. There are specialized variations of each of these pulse sequences that perform much better in some situations but are beyond the scope of this chapter.

5.3.2 1-D Sequences

5.3.2.1 FID. The FID or one-pulse sequence (Fig. 5.2) consists of a single pulse producing maximal signal for a turning angle of $\pi/2$. It is used to measure the Fourier transform EPR spectra of samples with very narrow, resolved lines. This sequence was described in more detail in 5.2.1. It is also used to detect free radicals in time-resolved measurements of photochemical reactions in liquids.

5.3.2.2 Inversion Recovery. The nominal inversion recovery sequence (Fig. 5.4) is $\pi-t-\pi/2$–detect, where the first pulse serves as the preparation period, the variable time t as the evolution period, and a FID-like signal appears during the detection period. This sequence was also discussed in detail in 5.2.2. Signal amplitude is recorded as a function of t in order to measure T_1, although the evolution can be affected by spin diffusion or spectral diffusion.

A variation of this sequence is the saturation recovery sequence, nominally $(t'-\pi/2m)_n-t-\pi/2$–detect. This sequence uses the initial series of pulses as the preparation period to prepare a state with $M=0$ at the beginning of the evolution period (consisting of time t). An FID-like signal is generated during the detection period. The purpose of this sequence is to produce uniform magnetization over a very broad region around ω_0 in order to minimize the effects of spectral diffusion. For a single preparation pulse, the turning angle is set to approximately $\pi/2$. For multiple preparation pulses, t' should be larger than T_M and m should be between 1 and n, so that subsequent pulses produce no echoes and only continue to diminish the magnetization. With large n, the magnetization can be driven to zero over a range much wider than the inverse of the pulse width. Such a set of preparation pulses are also known as a "picket fence" because they resemble one on the oscilloscope screen as n approaches 100.

For samples with broad lines where the FID is lost in the dead time, the last pulse is often replaced by a pair of pulses to produce a two-pulse spin echo.

5.3.2.3 Two-Pulse Echo. The two-pulse spin echo is also known as the primary or the Hahn echo. (Erwin Hahn discovered the spin echo in NMR and did early work on the electron spin echo [12, 13].) The nominal primary echo sequence is $\pi/2-\tau-\pi-\tau$–echo (see Figs. 5.5 and 5.6). The echo amplitude is measured as a function of time τ. As discussed in detail in 5.2.2.1, several types of information appear in the decay of the echo, including T_2, T_M, chemical and Heisenberg exchange, instantaneous diffusion, and ESEEM.

The virtue of the electron spin echo is that it refocuses interactions that remain static during the generation of the echo. It is useful in removing much of the effect of inhomogeneous broadening on the evolution of the spins. Because of its ability to effectively shift the signal out of the dead time in inhomogeneously broadened systems, it is often used at the end of the evolution period to generate a signal for the detection period, for example, in Davies ENDOR.

5.3.2.4 Echo Induced EPR. One common use of the primary echo is to measure the EPR spectrum by recording the echo amplitude as the magnetic field is swept. Such a spectrum is often called the echo induced EPR spectrum. It is an invaluable means of getting an idea of what the EPR spectrum is. However, it must never be confused with the true EPR spectrum because the generation and detection of the echo distorts the resulting spectrum. Some of these "distortions" can be quite useful. For instance, the echo intensity is affected by T_M, so it is possible to obtain a T_M-filtered spectrum showing only the species with large values of T_M by recording the echo induced EPR spectrum at a large value of τ. This can separate fast relaxing metal impurities from slowly relaxing free radicals (Box 5.3).

5.3.2.5 Stimulated Echo. Another commonly used pulse sequence is the three-pulse or stimulated echo (Fig. 5.7). The nominal stimulated echo sequence is $\pi/2-\tau-\pi/2-T-\pi/2-\tau$–echo. The echo amplitude is typically measured as a function of time T at constant τ. The stimulated echo, as well as several other echoes produced by this pulse sequence, was reported by Hahn [12] at the same time as the discovery of the primary echo. The stimulated echo is usually presented as being sensitive to T_1 during the time period T, but it is even more sensitive to spectral diffusion and chemical and spin exchange. If there is any appreciable spectral diffusion, the decay rate for the stimulated echo as a function of T also depends on τ, making their interpretation in the context of the Bloch equations problematic. The major application of the stimulated echo is in ESEEM where it provides simpler selection rules and higher resolution than the ESEEM from the primary echo.

The first microwave pulse starts spin packets precessing in the X, Y plane of the rotating frame. The second pulse converts M_Y, assuming all pulses are phased with their magnetic fields along the X axis of the rotating frame, into M_Z. The M_Z forms

BOX 5.3 PITFALLS IN ECHO INDUCED EPR

- On large field sweeps, the change in nuclear Zeeman frequency can cause the ESEEM pattern to modulate the amplitude of a broad spectrum, giving the appearance of a hyperfine structure when there is none. This is fairly easy to diagnose because the number and position of such lines will change as τ is changed.
- A more insidious artifact has been reported a number of times when there is substantial instantaneous diffusion with inhomogeneous broadening that is somewhat larger than B_1. At small τ, instantaneous diffusion has little impact on the echo intensity and the echo induced EPR spectrum gives a reasonable profile for each line in the spectrum. At large τ, instantaneous diffusion is significant at the center of each line but not in the wings. Thus, the echo signal at the center of each line decays to zero rapidly and it not observed, whereas at the wings some of the echo survives and gives a signal. The first impression is that a signal with a short T_M has decayed, revealing a new species with a very different spectrum and a much longer T_M. It is easy to identify the true cause of the change in the spectrum. Simply decrease the turning angle of the second pulse by at least a factor of 2, either by attenuating both pulses or by shortening the second pulse. If instantaneous diffusion is the cause, the spectrum shape should change very noticeably.
- The echo is usually integrated over a narrow window to produce the amplitude that is plotted as the echo induced EPR signal. The window position and width can affect the resulting spectrum. An integration window that covers the entire complete echo will give a good approximation of the EPR line shapes but at the expense of a noisier spectrum. A window that is narrower than the echo tends to broaden the line shapes but reduce random noise. An off-center, narrow integration window tends to put oscillations and negative going excursions at the sides of each line. These are artifacts caused by truncating or missing part of the signal during the detection period.

a pattern of polarization across the EPR spectrum proportional to $-\cos[\gamma \Delta B \tau]$. After a delay T, the third pulse converts the M_Z into $-M_Y$ and the echo forms a time t after the third pulse, just as after the second pulse in the primary echo sequence. In effect, the π pulse of the primary echo sequence has been split in half with the magnetization "stored" along the Z axis during time T (Fig. 5.7).

What can happen during the delay T to affect the echo? Anything that affects the pattern of M_Z polarization will affect the echo. The T_1 relaxation of the Bloch equations will destroy the polarization pattern and replace it with M_{Z0}, destroying the echo and replacing it with an FID from the third pulse. The polarization pattern can also "smear" out and disappear if the resonant field ΔB of a spin packet changes as a result of spectral or spin diffusion, chemical or spin exchange, molecular motion,

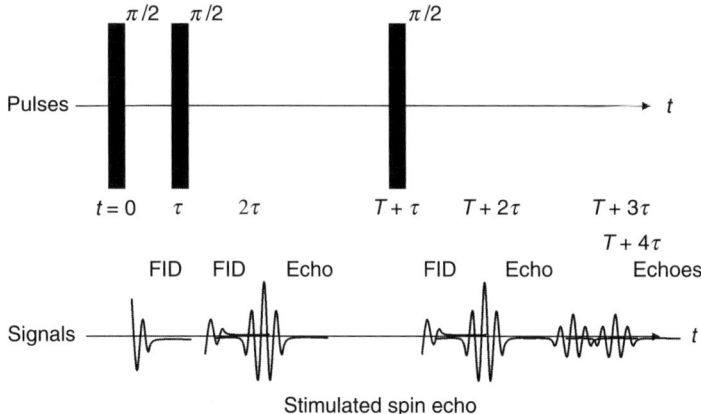

Fig. 5.7 The stimulated or three-pulse electron spin echo sequence. The nominal sequence consists of three $\pi/2$ pulses to generate the signal that is an echo. This sequence often produces FIDs following each pulse and other spin echoes as indicated. The unwanted signals can be avoided in recorded data by cycling the phase of the microwave pulses. Phase cycling is important when measurements are made varying either the τ or T delay because the unwanted echoes are prone to overlap the desired echo at some stage of the measurement.

or basically anything that changes the resonant field of the spin packet. Even a radio-frequency (RF) pulse that flips a nuclear spin with a hyperfine coupling to the electron can alter the polarization pattern and reduce the echo amplitude. This is the basis of the popular Mims ENDOR sequence that will be discussed later.

The most common application of the stimulated echo is to measure ESEEM at X-band. This is an effect in solids caused by the interaction between nuclear Zeeman and anisotropic hyperfine interactions or between hyperfine and nuclear quadrupole interactions. The ESEEM effect has been described in many ways and in many places. Each person who is interested in the details of ESEEM should be able to find a treatment with which they are comfortable.

ESEEM occurs when the usual selection rules of EPR, $\Delta m_S = \pm 1$, $\Delta m_I = 0$, break down and the selection rule is $\Delta m_S = \pm 1$, $\Delta m_I = 0, \pm 1, \ldots, \pm 2I$. (This breakdown leads to the appearance of the "spin-flip satellite" lines from protons flanking many narrow EPR lines in solids.) For a pair of pulses such as the first two in the stimulated echo sequence, the selection rules are time dependent. For different values of τ, the formally "forbidden" transitions in which a nuclear spin flips have varying probabilities. One result of flipping a nuclear spin along with the electron spin is a precessing nuclear magnetization that then evolves during time T at the ENDOR frequency of that nucleus. The total state of the spin system is not the same at the time of the third pulse as it was just after the second pulse. The ability of the third pulse to convert the spin system back into M_X and M_Y depends on the state of the spin system including the precessing nuclear magnetization. The result of all of this is that the echo intensity as a function of T is modulated at the ENDOR frequencies, but the amplitude of that modulation depends on τ. That modulation can be Fourier transformed to yield the ENDOR frequencies. The selection rules for which frequencies can appear in the ESEEM spectrum can become very complex. Notably, for $I = 1/2$ nuclei, the ENDOR frequencies corresponding to the principal values of the hyperfine interaction are absent from the modulation. As a result, ESEEM spectra from randomly oriented radicals lack the sharp features found in conventional ENDOR spectra and may be more difficult to interpret.

A number of descriptions of ESEEM have been published over the years. The simplest case for $S = 1/2$, $I = 1/2$ was first described in a very geometric fashion by Rowan, Hahn, and Mims [13] and recently in a very quantum fashion in section 11.4 in Slichter's book [7]. A completely general equation for ESEEM for an effective $S = 1/2$ with any combination of nuclear spins was worked out by Mims [14, 15]. The two-pulse ESEEM is the product of the ESEEM from each of the nuclei coupled to the electron spin with a similar product formula for the three-pulse ESEEM [10]. The two-pulse ESEEM from each coupled nucleus is

$$\varepsilon(\tau) \alpha \sum_{i,k} |M_{ik}|^4 + 2 \sum_{(i<j),k} |M_{jk}|^2 |M_{ik}|^2 \cos(\omega_{ij}\tau) + 2 \sum_{i,(k<n)} |M_{in}|^2 |M_{ik}|^2 \cos(\omega_{kn}\tau)$$

$$+ 2 \sum_{(i<j),(k<n)} \mathrm{Re}\left\{ M_{jk}^* M_{jn} M_{in}^* M_{ik} \cos[(\omega_{ij} + \omega_{kn})\tau] + M_{ik}^* M_{in} M_{jn}^* M_{jk} \right.$$

$$\left. \times \cos[(\omega_{ij} - \omega_{kn})\tau] \right\} \quad (5.4)$$

where the indices i and j run over the m_I for one value of m_S and the indices k and n for the other [16]; M is a nuclear spin overlap matrix, analogous to the Franck–Condon factors in electronic spectra; and the ω are differences in nuclear eigenfrequencies. The ESEEM from the stimulated echo is similar [16]:

$$\varepsilon(\tau, T) \alpha \sum_{i,k} |M_{ik}|^4 + 2 \sum_{(i<j),k} |M_{jk}|^2 |M_{ik}|^2 [\cos(\omega_{ij}\tau) + \cos(\omega_{ij}(\tau+T))]$$

$$+ 2 \sum_{i,(k<n)} |M_{in}|^2 |M_{ik}|^2 [\cos(\omega_{kn}\tau) + \cos(\omega_{kn}(\tau+T))]$$

$$+ 2 \sum_{(i<j),(k<n)} \mathrm{Re}\left\{ M_{jk}^* M_{jn} M_{in}^* M_{ik} \right\} [\cos(\omega_{nk}\tau + \omega_{ij}(\tau+T)) + \cos(\omega_{nk}\tau - \omega_{ij}(\tau+T))$$

$$+ \cos(\omega_{ij}\tau + \omega_{nk}(\tau+T)) + \cos(\omega_{ij}\tau - \omega_{nk}(\tau+T))] \quad (5.5)$$

5.3.2.6 ENDOR.

ENDOR is a family of methods that measures the magnetic resonance frequencies of nuclei interacting with an electron spin by exciting the electron spins with a microwave field and the nuclei with an RF field. When this is done using a CW EPR approach, the intensity of the spectrum depends on a balance between the relaxation times of the electron and the nuclei. CW-ENDOR was treated in Chapters 1 and 2. In pulsed EPR the relaxation times are not critical, but the spectral shapes depend on the pulse timings. The use of pulsed ENDOR is now wide spread, and it allows the study of systems with short relaxation times. Examples of pulsed ENDOR applications are found in Chapters 6 and 12. There are two basic approaches to pulsed ENDOR: the original Mims ENDOR based on the stimulated echo and the Davies ENDOR based on the inversion recovery measurement.

Both approaches use one or more microwave pulses to prepare a nonequilibrium spin state and one or more pulses to detect it. Between these preparation and detection periods is an evolution period during which the RF pulse is applied. When the RF pulse is more or less coincident with an ENDOR frequency of a nucleus, its spin is flipped, changing the frequency of the electron spin via the hyperfine coupling and resulting in a change in the detected signal. A good description of pulsed ENDOR and its variations can be found in Reference 17.

5.3.2.6.1 Mims ENDOR. The first pulsed ENDOR method was described by Bill Mims in 1965 [18]. It consists of a stimulated echo sequence into which an RF pulse is inserted during delay T. The preparation period consists of the first two microwave pulses, the evolution period is delay T during which the RF pulse is applied, and the detection period is the time following the third microwave pulse.

If the RF pulse flips a nuclear spin coupled to an electron spin, that electron spin changes resonant frequency by the hyperfine splitting A. At a time τ after the third microwave pulse when the stimulated echo appears, that spin will have a phase $A\tau$ different than without the RF pulse. The refocusing of M_Y to form the echo will be incomplete (unless $A\tau$ is a multiple of 2π) and the ENDOR signal appears as a decrease in the intensity of the echo (Fig. 5.8).

The Mims ENDOR spectrum has a periodic modulation with the amplitude oscillating as $\sin^2(A\tau/2)$, producing "blind spots" where no ENDOR signal can be seen. A common pitfall is to mistake these blind spots for real structure in the ENDOR spectrum or to use too large a value of τ so that the edges of the spectrum fall in the blind spots.

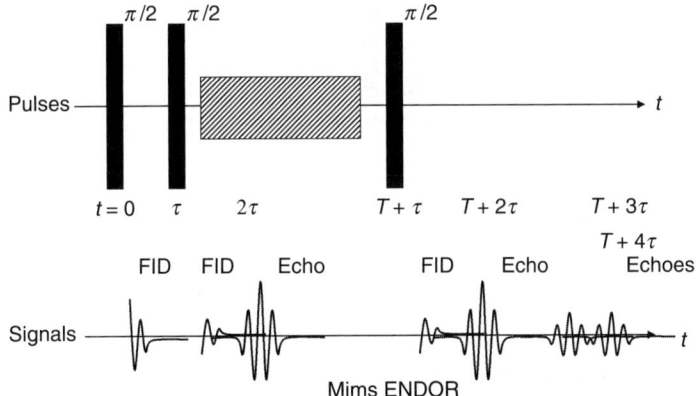

Fig. 5.8 The Mims ENDOR sequence. This is a stimulated echo sequence in which an RF pulse (pulse with diagonal banding) is inserted into the delay T.

Unfortunately, it is not possible to simply make τ very small and measure just one spectrum. The reason is that there is effectively a blind spot at the nuclear Zeeman frequency (for negligible quadrupole couplings) where A approaches zero. Use of a small value of τ suppresses the ENDOR signals from small hyperfine couplings, although there is often some residual intensity from the so-called matrix ENDOR line. It generally is necessary to measure spectra at several values of τ, small delays for large couplings and large delays for small couplings, and explicitly consider the blind spots in the spectral analysis and simulation of the ENDOR spectra. When hyperfine couplings are larger than the EPR spectral width excited by the preparation pulses, Mims ENDOR no longer has blind spots and becomes similar to Davies ENDOR.

5.3.2.6.2 Davies ENDOR. The second approach to ENDOR was described by Davies in 1974 [19] and is similar to an inversion recovery measurement. The preparation period consists of a single long microwave pulse that inverts the magnetization in a narrow whole near the resonant frequency ω_0. The evolution period consists of a delay T during which the RF pulse is applied. The detection period follows either a single pulse that produces an FID from the inverted hole or a pair of pulses that produces a primary echo from the spins in and around the hole.

The ENDOR signal appears as a loss of the inverted hole as the RF is swept through a nuclear resonance frequency. The hole is lost, or filled in, by shifting an off-resonance spin packet that has a hyperfine coupling into resonance by changing the spin of the hyperfine coupled nucleus. Obviously, shifting a spin packet into resonance will not change the detected EPR signal if that spin packet was also inverted by the preparation pulse. The ENDOR spectrum does depend on the preparation pulse to the extent that small hyperfine couplings are lost for narrow preparation pulses that produce broad inverted holes. Other processes that can fill in the inverted hole can

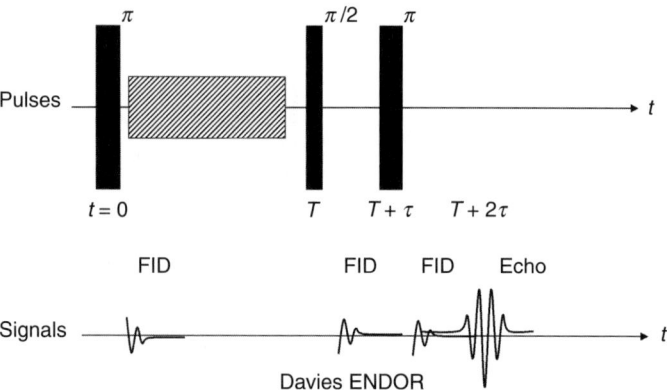

Fig. 5.9 The Davies ENDOR sequence. This is a variant of the inversion recovery sequence in which the final $\pi/2$ pulse that produces the detected FID has been replaced by a primary echo sequence to produce a spin echo. The RF pulse for ENDOR is inserted into the recovery period.

> **BOX 5.4 PITFALLS IN PULSED ENDOR**
>
> Both Mims and Davies ENDOR are affected by the same forbidden spin flip transitions that are responsible for ESEEM. Both types of ENDOR rely on the ability of the preparation pulses to produce a pattern of magnetization or polarization across the EPR spectrum that is then altered by the RF pulse and measured during the detection period. The forbidden transitions interfere with the sinusoidal pattern of magnetization that generates the stimulated echo in Mims ENDOR. They also produce additional holes ("discrete saturation") that distort the Davies ENDOR spectrum. This effect from forbidden transitions seems to reduce pulsed ENDOR sensitivity and to distort ENDOR spectra at X-band for samples with strong ESEEM, particularly from nitrogen nuclei. This effect may underlie the apparent increase in performance of ENDOR at frequencies above X-band where such transitions are more strongly forbidden.

interfere with Davies ENDOR, including spectra and spin diffusion, molecular motion, and chemical exchange.

The placement of the detection window with respect to the beginning of the FID or the center of the echo affects both the sensitivity and resolution in the ENDOR spectrum (Fig. 5.9). Like Mims ENDOR, the Davies ENDOR spectrum depends on the measurement conditions, but in its own peculiar ways (Box 5.4).

5.3.2.7 Electron–Electron Double Resonance (ELDOR). ELDOR is known variously as pulsed ELDOR, PELDOR, or double electron–electron resonance (DEER). One form is the analog of the classic CW-ELDOR where two microwave frequencies excite transitions of the same radical, perhaps at different times. The other form, commonly called DEER, excites two different radicals to detect an interaction between those radicals.

5.3.2.7.1 Pulsed ELDOR. The first type of pulsed ELDOR (Fig. 5.10) is based on the inversion recovery sequence described earlier. The preparation pulse however is at a different microwave frequency, called the pumping frequency ω_p, than the pair of pulses at ω_0 starting the detection period. The initial pumping pulse inverts or saturates spin packets near ω_p. The resonant frequency of that spin packet then changes during the evolution period via spectral or spin diffusion, chemical exchange, nuclear relaxation, or molecular motion and eventually is resonant at ω_0, where it appears as a decrease in the signal intensity if it has not undergone T_1 relaxation. A combination of this type of ELDOR together with inversion recovery measurements allows a fairly complete unraveling of the dynamics of the spin system and has been applied to slow motions of nitroxides and spin-labeled proteins.

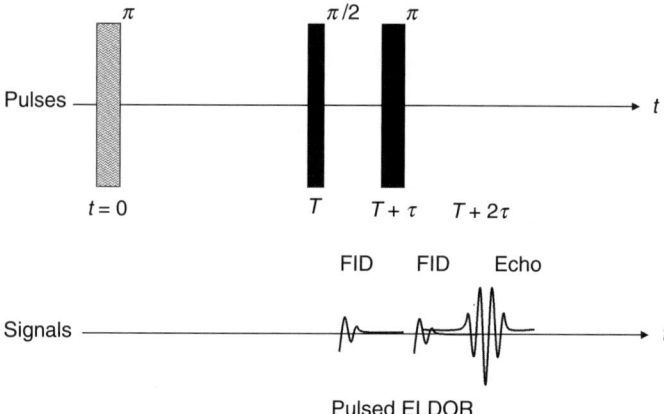

Fig. 5.10 The classic pulsed ELDOR sequence. This is a primary echo sequence preceded by a microwave pulse at ω_p.

This sequence is also used to probe transitions of a multilevel system sharing a common level, either with $S > 1/2$ or when there are forbidden spin flip transitions. In this case the evolution period is largely irrelevant; the ELDOR effect is the change in population of one of the energy levels involved in the detected EPR transition caused by pumping another transition involving that level. For instance, if the transition between the $m_S = +1/2$ and $-1/2$ levels is detected, the amplitude of that signal will change when the transition between the $m_S = +3/2$ and $+1/2$ levels is pumped by the preparation pulse because the two transitions share the $m_S = +1/2$ level.

5.3.2.7.2 DEER or PELDOR. The ELDOR measurement of interradical distances, particularly in proteins, has undergone strong growth in the last few years. The ability to place pairs of spin labels at almost any desired position in a biological macromolecule and then measure their separation to within a couple tenths of a nanometer over a range of 1.5–8.0 nm or more has really opened up the study of motion and conformation of large, noncrystalline biomolecules in membranes, micelles, or solutions. The measurement of distance is based on the dipolar interaction between two spins that can be calculated from first principles without any calibration. For the observed spin, the resonant frequency of a spin packet shifts by the dipolar interaction, depending on the spin state of the second radical.

The original DEER sequence used three microwave pulses and was based on a primary echo sequence. This has largely been replaced by the four-pulse sequence described here (see Fig. 5.11).

The value of τ producing an echo near 2τ whose peak defines the time $t' = 0$ when all spin packets are "focused". The evolution period of fixed length T starts at the end of the second preparation pulse and ends with a π pulse at ω_0 that initiates the detection period. With this definition, the evolution time t' has values ranging from $-\tau$ to $T-\tau$. The M_Y at the peak of the echo dephases during the evolution period, undergoes a phase conjugation at the π pulse and rephases to form a "reflected" echo at $2T$ that

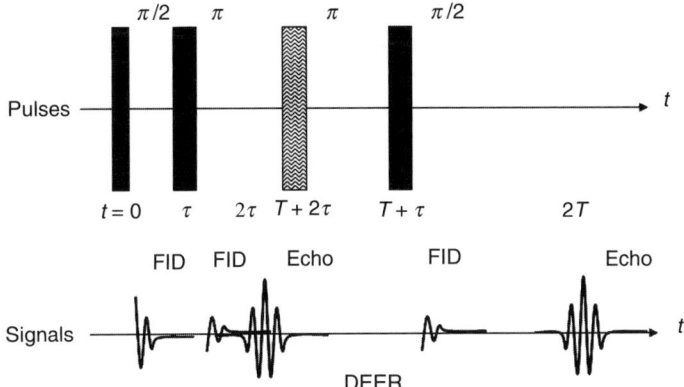

Fig. 5.11 The four-pulse DEER sequence for measuring distances in doubly spin-labeled proteins. The first two pulses produce a spin echo that is then refocused by the final pulse. The pumping pulse at ω_p is varied in time relative to the initial primary echo that defines the relative time of the pump pulse.

is detected. During the evolution period, a π pulse at frequency ω_p is applied that flips electron spins in the spectrum near ω_p. If one of those spins is dipolar coupled to an observed spin, the observed spin undergoes a frequency shift equal to the dipolar coupling ω_d. Thus, during the evolution period, the observed spin accumulates a phase of $\gamma\Delta Bt$ before the ω_p pulse and $(\gamma\Delta B \pm \omega_d)(T - t')$ until the final pulse. During the detection period the spin packets refocus at a rate of $\gamma\Delta B \pm \omega_d$ until they reform into a reflected echo at $2T$. At the peak of the reflected echo, there is a net remaining phase of $\pm\omega_d t'$. As t is varied from the peak of the primary echo to the final microwave pulse of the sequence, the echo amplitude is modulated by a frequency ω_d. In a sample of randomly oriented pairs of radicals, the distribution of ω_d forms the well-known Pake pattern and the echo amplitude traces out its Fourier transform.

One of the difficult aspects of DEER distance measurement is the extraction of a distance or a distance distribution function from the experimental data. This is one of those ill-conditioned mathematical problems in which the calculation of the DEER time-domain spectrum from a distance distribution is easy, but the recovery of the distance distribution from a DEER spectrum is next to impossible. The problem is that many distance distributions can reproduce (within the noise level) the experimental data. Fortunately, often only one distribution is physically reasonable and can be found using mathematical techniques such as Tikhonov regularization or fitting to a Gaussian distribution of distances. Data analysis is still in a state of flux, particularly at high fields and for intrinsic radicals where orientation selection can have a major impact on the experimental data.

5.3.3 2-D Sequences

There are basically two types of 2-D pulse sequences. One is simply a 1-D sequence measured at many different values of another parameter, for instance, a series of 1-D primary echo decays measured at different positions across an EPR spectrum with no further processing. Such spectra make it possible to see how a feature in the 1-D spectrum depends on a second variable such as the magnetic field, delay time after a laser, or orientation of a crystal. This type of 2-D spectroscopy is a very natural and obvious extension of the underlying 1-D spectrum.

The second type of 2-D sequences largely falls into the class of correlation spectroscopies. The signal is recorded as a function of two different delay times and then processed, usually by 2-D Fourier transformation into a 2-D spectrum with peaks correlating one spectral parameter, usually a frequency, with another. Correlation spectroscopy has a few important aspects relative to 1-D spectroscopy.

1. It provides additional information about the two frequencies connected by a cross-peak.
2. Resolution can be much greater, because roughly the same number of peaks are spread over a plane rather than piled up on a line.
3. The sensitivity is comparable to a 1-D spectrum taken in the same amount of time, because the signal power is put in roughly the same number of peaks while the noise power is spread over the entire spectral plane.

5.3.3.1 ESEEM. An important class of 2-D pulse sequences concerns the measurement of ESEEM. The measurements are usually based on the stimulated echo sequence discussed earlier. The stimulated echo sequence has two delay times that affect the echo in different ways. The echo intensity is modulated at ENDOR frequencies of nuclei with suitable couplings to the electron spin. The amplitude of the modulation at each of the ENDOR frequencies is a sinusoidal function of the product of delay τ and an ENDOR frequency from that nucleus but with a different value of m_S (Eqn. 5.5). Thus, the simplest 2-D ESEEM experiment is to record the stimulated echo intensity as a function of τ and T. The resulting data look like a pattern of waves and ripples on a plane. A 2-D Fourier transform converts the waves and ripples into a set of peaks whose coordinates are $(\omega_\alpha, \omega_\beta)$ and $(\omega_\beta, \omega_\alpha)$, where ω_α and ω_β are the ENDOR frequencies of the same nucleus in the α and β electron spin manifolds, respectively. This correlation is very useful for assigning ENDOR frequencies in a radical with several interacting nuclei, particularly if there are nuclear quadrupole couplings involved.

This sequence is seldom used because of one major drawback: the echo decays rapidly with τ and slowly with T. This is to be expected because decay along one direction is a rather rapid T_2 or T_M decay whereas the other is a T_1 or analogous decay. The disparity in decay rates produces a 2-D spectrum with good resolution in the T dimension and poor resolution in the τ dimension.

5.3.3.2 Hyperfine Sublevel Correlation Spectroscopy (HYSCORE). This disparity in resolution was solved very cleverly by adding another pulse to the stimulated echo sequence. A π pulse was inserted into delay T, breaking it into delays t_1 and t_2. Delay τ is kept fixed while t_1 and t_2 are varied independently. The sequence starts like a stimulated echo ESEEM measurement through delay t_1. At that point,

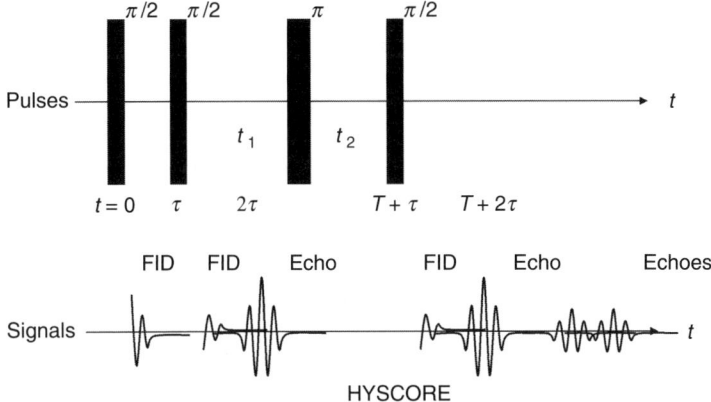

Fig. 5.12 The HYSCORE pulse sequence. This sequence consists of a stimulated echo sequence with an additional π pulse inserted during period T, defining the two delay periods t_1 and t_2 that serve to index the 2-D data set. Phase cycling is required to eliminate artifacts from unwanted echoes moving across the echo signal. Unfortunately, the stimulated echo from the $\pi/2$ pulses cannot be removed by phase cycling, so the measured data nearly always consist of the HYSCORE signal and a residual stimulated echo signal. The HYSCORE spectrum contains mainly cross-peaks correlating frequencies and is easy to distinguish from the peaks from the stimulated echo that appear along the diagonal.

there is nuclear spin magnetization precessing at an ENDOR frequency. The π pulse flips all of the electron spins but leaves the nuclear magnetization precessing—not at the same ENDOR frequency, but at an ENDOR frequency of that nucleus in the other electron spin manifold. After a 2-D Fourier transform, the spectrum has the same correlation of ENDOR frequencies as in the 2-D stimulated echo just discussed, but the spectral resolution is the same in each dimension because the signal decay during t_1 and t_2 is the slow decay of the stimulated echo (Fig. 5.12).

This four-pulse 2-D ESEEM measurement is known as HYSCORE. It was developed in 1986 [20] but was not applied extensively until graphical methods were developed to analyze HYSCORE spectra from randomly oriented radicals [21–24]. For $I = 1/2$ nuclei or in the absence of quadrupole interactions, the HYSCORE cross-peaks form ridges that have some curvature to them because of second-order hyperfine shifts. When the ridges are plotted as a function of frequency squared rather than frequency, they become straight lines that directly give the principal values of the hyperfine tensor. Even under strong orientation selection conditions where only a fraction of the ENDOR spectrum can be seen, the hyperfine tensor can be determined.

5.4 DATA ANALYSIS

The analysis of data has become much less tedious because of the widespread availability of computer packages designed for pulsed EPR that are supplied with commercial spectrometers or available as shareware. The choice of a particular package is largely a matter of personal taste and availability. Once the data are in a computer, the data can, in principle, be transferred for analysis anywhere.

5.4.1 Time Versus Frequency Domain

A major choice in the analysis of data is whether to perform the analysis in the time domain or in the frequency domain. A simple Fourier transform losslessly converts data back and forth between the time domain (as the data are usually acquired) and the frequency domain (where it is usually easier to view). There is no concern that information is lost by making the wrong choice. That is not to say that the tools available in each domain are equivalent.

5.4.2 Data Fitting

For relaxation measurements where decay curves are recorded, it is faster and easier to fit the decays in the time domain with exponential functions than to convert decays to the frequency domain and fit with the appropriate Laplacian conjugate function. However, there are some specialized analysis tools for multicomponent decays that do their work in the frequency domain.

In contrast, analysis of ESEEM is now done mostly in the frequency domain where the connection to other forms of spectroscopy is more apparent. In addition, ESEEM data analysis usually only requires a subset of the information available in the spectrum. The analysis of ESEEM was originally based on fitting the intensity of the time-domain ESEEM signal because the intensity was felt to provide additional information that might be lost by considering only the frequencies. However, it was finally realized that the ENDOR frequencies alone contained all of the information present in the spin Hamiltonian.

ESEEM analysis should work equally well using the frequencies from the frequency-domain spectrum or fitting the amplitudes, phases, and frequencies in the time-domain data. However, there are some practical details that impact the amplitudes and phases, but not the frequencies.

1. The phases in the time-domain data are affected by the point taken to be $t = 0$. This means that the time origin of the data must be carefully calibrated or the phases will not be consistent with the frequencies.
2. The amplitudes in ESEEM are affected by the bandwidth of the resonator and pulse width. This requires calibration of the resonator response as a function of the ESEEM frequency for each sample and pulse length or the amplitudes will not be consistent with the frequencies.

Fortunately, as long as the time base for the excitation and detection is accurate, the ESEEM frequencies are accurate and can be used to determine the hyperfine and quadrupole interactions if transitions can be resolved and assigned. The requirements for an accurate time base for the pulsed EPR are readily met by available digital circuitry. Most analyses of ESEEM, particularly of HYSCORE spectra, are currently based on reproducing the frequencies in the spectrum not their amplitudes or phases. The analysis of ESEEM is well developed for $I = 1/2$ nuclei, such as ^{1}H or ^{31}P, or nuclei with negligible quadrupole couplings such as ^{2}H. The analysis often starts with a rough estimation of isotropic and anisotropic hyperfine couplings from primary ESEEM followed by refinement from intelligently planned ENDOR, HYSCORE, or stimulated ESEEM measurements. Conditions are not as good for nuclei with significant quadrupolar couplings, such as ^{14}N. Spectral patterns are somewhat understood in the limits of a completely isotropic hyperfine interaction or of "complete cancellation" where $A/2 \sim \omega_N$. In the latter case, the ESEEM spectrum contains two or three sharp and intense peaks corresponding to the zero-field nuclear quadrupole resonance frequencies of the nitrogen. At X-band, this occurs for the distant nitrogen of histidine coordinated to Cu(II) and has come to be diagnostic of histidine ligation by Cu(II) in biological systems.

5.5 SPECTROMETER

Pulsed EPR instrumentation can have a quite daunting appearance. At a minimum, there is the complete spectrometer with a magnet, power supply, coolers, and computer and microwave equipment for performing the pulsed EPR measurements. Often a cryostat, liquid helium dewars, transfer lines, temperature controllers, pumps, and gas lines are needed to control the sample temperature. For studies of photochemistry and excited states, there is usually a laser system or high powered lamp. All of these systems need to operate together, preferably under control of the spectrometer's computer. The lasers and cryogen systems are used with CW EPR systems as well. Their description goes beyond the scope of this chapter that now concludes with a brief discussion of the spectrometer and some operational details.

Modern pulsed EPR spectrometers are composed of a microwave bridge to induce the pulsed EPR signals, a resonator where the sample is excited by the microwave pulses and subsequently emits its signals, a digitizer to capture and record the signals and spectra, a magnet, and a computer to control everything else. There are several ways to design and construct each of these components, and there is no clearly superior approach. In fact, different EPR frequencies require radically different designs. One of the benefits of extensive computer control of the operation of the spectrometer is that the technical details do not really matter to the user. The computer interface allows critical operations such as tuning and pulse amplitude adjustment to be performed in a manner independent of the underlying hardware. For the beginner the functional operations are more important than their implementation in hardware. In many respects, the considerations that are important in CW EPR are irrelevant for pulsed EPR and vice versa. A spectrometer designed for both CW and pulsed EPR measurements is much more complicated and compromised than one designed solely for CW or for pulsed measurements. The major systems and the requirements relevant to pulsed EPR operation are described in this final section.

5.5.1 Microwave Bridge

The microwave bridge has two major sections: the transmitter and the receiver. The transmitter generates microwave pulses with the desired amplitude, width, and phase at the desired times for each pulse sequence and measurement. These pulses are sent or transmitted to the sample in its resonator, producing a pulsed EPR signal. The microwaves from the sample return to the receiver in the bridge where they are amplified, processed, detected, and finally converted into the signal. The "signal" at this point may be either digital or analog, but it does represent, as accurately as possible, the magnitude and phase of the microwaves from the sample. The signal is generally represented as in-phase and out-of-phase components, or real and imaginary or I and Q.

For pulsed EPR operation, no microwaves are applied to the sample while the signal is being detected. This means that the microwave bridge is no longer a "bridge" in the sense of a carefully balanced electrical circuit. Phase noise and maintaining the spectrometer frequency precisely at the resonator frequency are no longer significant concerns. A free running microwave oscillator without automatic frequency control is more than sufficient, but fair long-term stability should be ensured by stabilizing its temperature and operating voltages. The pulses are typically generated using very nonlinear devices operated near saturation so that amplitude stability is likewise not a significant concern. Microwave pulse power and relative pulse phases need to be accurately adjusted, so accurate attenuators and well-matched switches and phase shifters need to be used. Important considerations in the bridge for pulsed EPR are the stability and reproducibility of pulse widths and relative pulse phases as pulse positions and duty cycles are changed.

The receiver needs to survive the intense microwave pulses that are returned from the resonator yet faithfully amplify and reproduce signals only a few nanoseconds later. This often requires power limiting or gating devices in the signal path before the low noise microwave amplifier. These are critical devices that must go from strong attenuation of reflected pulses to minimal, stable attenuation almost instantly. In particular, there must be no transient phase shifts during the signal. These are typically not parameters specified by microwave component manufacturers, but they are vital.

Most pulsed EPR bridges have some sort of phase adjustment in the receiver, so that the signal can be adjusted to have only a nonzero in-phase component and zero out of phase. This is a remnant from CW spectrometers where a phase adjustment was necessary for good sensitivity. Much time can be wasted on pulsed EPR spectrometers trying to obtain good phasing of weak, noisy signals. A more effective approach is to ignore the spectrometer phase adjustment and record both I and Q components of the signal as the spectrum. That spectrum can be phased much better later without loss of information. Unfortunately, the software operating the pulsed EPR spectrometers is not designed to support routine handling of unphased data.

5.5.2 Magnet System

The performance of the magnet system is not a major concern for pulsed EPR applications. The field should be stable, calibrated, and easily changed with homogeneity and stability less than the EPR linewidth of the sample. Access to and the size of the sample space are usually more important factors. The sweep width, sweep rate, and cryogen consumption of superconducting magnets are as important for pulsed EPR as for CW EPR and nearly identical.

5.5.3 Data System

5.5.3.1 Computer. The modern pulsed EPR spectrometer needs to be controlled by a computer system that also stores the recorded data. The computer has to take care of several functions at a time, such as magnet and bridge control, data acquisition, signal display, and user interface. This is typically done with a PC or workstation running a multitasking or threaded operating system. Several functions, such as data acquisition, are time critical and are best handled by offloading that operation to a dedicated microprocessor that will carry out that function efficiently and return the result to the computer upon completion. The design of a fast, reliable, and efficient system is very complex and demanding; but its operation should be transparent to the user, who should see a consistent interface that hides the details of the implementation.

5.5.3.2 Pulse Programmer. The pulse programmer is another important subsystem in the data system. It has the responsibility for generating the pulses and timings for the bridge and for the digitizer. It needs to have a timing resolution of a few nanoseconds, which no longer challenges digital technology. It also needs to generate delays of several minutes with this precision, for instance, to measure T_1 at very low temperatures. Probably the greatest challenge for the pulse programmer is that the delays and widths of the many channels or output lines need to track precisely. That is, the data acquisition gate must exactly track the changes in τ in a primary echo measurement: when the phase of a pulse is changed, the pulse must not change in position or width; the pulse width must not change when the pulse separation or pulse repetition rate is changed; and when pulse widths are changed, the change must be accurate.

These pulse timing issues are not the sole concern of the pulse programmer, but they include the interface to the bridge and the bridge itself. A systemwide approach must be taken in the design. Careful matching of impedances in digital signal lines, attention to toggle rate effects in circuits and switches, and compensation for capacitive loading are technical issues that have been handled well, if not perfectly, in commercial spectrometers and rarely need to concern the user unless modifications are made to the spectrometer system.

Another vital aspect of the pulse programmer is the user interface for specifying and modifying the pulse sequence. The current EPR user interface lags far behind what is available for NMR. The user interface should allow the user to enter a pulse sequence with delays and sweeps without having to understand how it will be implemented, and it should be possible to view the sequence graphically.

5.5.3.3 Digitizer. At some point, the signal in the receiver has to be converted into digital form and then stored as part of a spectrum. It is desirable to capture the entire shape of the signal (FID or echo) and store it at each point in a measurement. This is done in a very few measurements, for example, when the FID is measured at a series of delays after a laser pulse. Usually, the signal is integrated or summed over a window of some duration. This integration reduces the amount of data to be stored, but that is not an issue with modern disk drives. It is also easier to write the software to process and analyze the data if there are less data.

This integration or summation within a finite acquisition window does some significant filtering of the information content of the spectrum. It is very simple to show that use of a gate or window about as wide as one cycle of the proton Zeeman frequency noticeably attenuates the frequencies in the proton ESEEM spectrum. However, use of a narrow window decreases the signal to noise ratio because the recorded signal contains noise from a broader bandwidth than when a broad window is used. A better approach would be to acquire the complete shape of the time-domain signal and then during signal processing apply a suitable filter to reject as much noise as possible while preserving the desired signal.

5.5.3.4 Phase Cycling. One point mentioned in the discussion of the different pulse sequences is that the number of distinguishable responses grows rapidly with the number of pulses. For example, the three-pulse sequence is often used for inversion recovery (with detection through a primary echo) and three-pulse ESEEM measurements; different information and different signals are desired from the same sequence. In addition, there are other primary echoes and FIDs that potentially overlap with the echoes used in these measurements. All of these signals have different dependences on the turning angle of the different pulses, but B_1 inhomogeneity in the resonator and the off-resonant field ΔB make the turning angle a poor method for separating signals. Instead, the dependence of the phase of each signal on the phase of the pulses is a practical method.

5.5.3.4.1 Signal Selection. It is simple, but tedious, to show that the phase of the echo used for inversion recovery depends on the phase of the three microwave pulses as $\varphi_{IR} = 2\varphi_3 - \varphi_2$, but the stimulated echo depends on the pulse phase as $\varphi_{SE} = \varphi_2 - \varphi_1 + \varphi_3$. All of the other signals generated by three pulses have still other phase dependences. As a result, it is possible to systematically vary the phase of all of the pulses and add or subtract the signals together in a precise pattern so that whichever signal is desired is recovered and all other signals vanish. As an added benefit, it is often possible with such phase cycling to eliminate ringing from the resonator and correct for any DC offsets in the signal.

Unfortunately, most pulse spectrometers are not optimized for phase cycling and there is appreciable overhead associated with it. Some trade-offs are usually made by the user, sacrificing the effectiveness of the phase cycle for efficiency in data collection.

5.5.3.4.2 Cyclically Ordered Phase Sequence Phase Cycle (CYCLOPS). There is a second purpose for phase cycling in pulsed EPR, and that is for ensuring that the in-phase and out-of-phase signal channels are really in quadrature with each other. The transmitters and receivers in pulsed EPR spectrometers do not have perfect phase shifts or perfect amplitudes in either the pulses or in the receiver channels. As a result, when the FID or the echo shape are Fourier transformed to obtain an EPR-like spectrum, there appear what are known as image peaks in the spectrum. That is, a spin packet $+\Delta B$ from resonance will also appear as a minor response or "image" at $-\Delta B$. The phase and amplitude imbalances that produced this image can be eliminated to a high degree by yet another phase cycle known as CYCLOPS [25], which needs to be superimposed on top of the phase cycling used for signal selection.

5.5.4 Resonators

The sample is placed in some sort of structure designed to contain microwaves without letting them leak into the laboratory and to funnel the emitted microwaves back to the receiver. There are many resonator designs, some for general purpose operation with a wide range of samples and others designed for the specific properties of a single type of sample. The structure holding the sample is generally resonant near ω_0 and is therefore commonly referred to as a resonator or a resonant structure. The resonator must hold the sample, allow change of samples without undue hardship, and must be fairly efficient at converting microwaves from the bridge into microwave magnetic fields at the sample and conversely at converting the microwave signals induced by the sample in the resonator into microwaves going back to the bridge.

The resonators used for CW EPR need to allow high frequency field modulation to reach the sample, maintain a high Q factor for optimum sensitivity, and accommodate aqueous and other lossy samples without severe degradation of their Q factor. These requirements are almost irrelevant for pulsed EPR, at least at X-band and lower frequencies. Field modulation is not used, the Q factor is limited to rather low values by the need for resonator bandwidth, and lossy samples have much less impact on the Q factor. Consequently, different compromises are appropriate in the design of pulsed EPR resonators.

Many different types of resonators are in use for pulsed EPR and have their passionate advocates. However, there are a few general rules that govern resonators. The smaller the resonator, the higher resulting B_1 field from a given microwave pulse. Similarly, the higher the B_1 field, the larger the signal from a given sample. The smaller the bandwidth or the greater the resonator Q, the larger the signal and the larger the energy density. Unfortunately, these place conflicting demands on the resonator designer.

One solid point is that pulsed EPR measurements require a certain bandwidth and the resonator Q is therefore limited by that requirement to values on the order of 100 at X-band. It is certainly possible to make resonators at X-band with Q values of several thousand. A common question is then how should one limit Q. The important value of Q is the loaded Q_L of the resonator that determines the actual bandwidth in the spectrometer. Another parameter is the match or coupling of the resonator that determines the efficiency of energy transmission from the bridge to the resonator. When the resonator is matched or critically coupled, all of the energy coming toward it is absorbed by the resonator and sample (and dissipated as heat).

5.5.4.1 Overcoupling.
A common emotion among pulsed EPR users is that after spending a lot of money on high power microwave amplifiers, they ought to send all of that power into the resonator by using a matched resonator with the desired Q_L. But this desire treats the resonator as a microwave oven where the object is to heat the sample as quickly as possible.

However, the object in a pulsed EPR spectrometer is to generate the strongest signal for a given Q_L and available microwave power. That is achieved with an overcoupled resonator where much of the microwave power is reflected by the resonator and never enters. In fact, thorough circuit analysis [26] and empirical observation show that for a constant Q_L and microwave power, both the EPR signal and the B_1 fields are significantly more intense with the resonator overcoupled than with it matched.

5.5.4.2 Dead Time.
At X-band and lower frequencies, the Q_L has a major impact on the dead time of the spectrometer. That is because a significant fraction of the full microwave power of each pulse enters and is stored in the resonator during the pulse. At the end of the pulse, the stored energy in the resonator radiates toward the receiver as ringing of the resonator. This ringing is a significant fraction of the full microwave power of each pulse that is typically 1 W to 1 kW while the receiver is trying to measure pulsed EPR signals of 10^{-9} W or less. Such powers can overload if not destroy critical parts of the receiver, so a limiter or receiver protection switch is placed in the signal path to attenuate the pulses and ringing until they have decayed to a safe level.

The ringing of the resonator is a particular problem because even after it decays to a safe level of ~ 100 mW, it is still many orders of magnitude more intense than the signal and completely masks it. The ringing from the resonator decays with a time constant equal to Q_L/ω_0, which is only a few nanoseconds. To reach the level of the EPR signal and thermal noise takes 20–30 time constants, so the spectrometer dead time caused by resonator ringing is much longer (60–200 ns) than might be initially expected.

5.5.5 Variable Temperature Accessories

Pulsed EPR often requires lower temperatures for measurements than CW EPR. Molecular motion even at 77K can be very effective at shortening T_1 and/or T_M, which can prevent detection of an FID or echo but not the CW EPR signal. Yet, many free radicals and excited states are readily measured by pulsed EPR at room temperature or above; it all depends on the system. As a result, many pulsed EPR systems include variable temperature accessories that operate down to 4K. These variable temperature accessories typically cool the entire resonator as well as the sample, so-called immersion cryostats. The small resonators often used in pulsed EPR systems prevent dewars from being inserted into the resonator as is common in CW EPR systems.

For operation between 100 and 350K, the temperature is usually controlled by the flow of nitrogen gas around the resonator and sample in a simple evacuated jacket.

Such systems are prone to large temperature oscillations when changing the temperature because of the large heat capacity of the system and the low heat capacity of the nitrogen gas.

One of the advantages of an immersion cryostat with flowing helium is that its large heat capacity damps out temperature fluctuations so that temperature stability can be excellent. In addition, they can be well insulated and quite efficient. The cost for this is that it takes a long time to change the temperature and reestablish equilibrium. They sometimes operate below atmospheric pressure to improve performance, which can cause problems if air enters the cold sample chamber through a leak or during sample change. High efficiency also presents problems when heat needs to be dissipated at low temperatures. The heat produced by optical irradiation of samples and by long, intense RF pulses in ENDOR can produce large relative increases in temperature.

There are a few cryostat systems that actually immerse the resonator and sample in liquid helium. The temperature stability and cooling power are excellent, but their working range is extremely limited, usually between 2.1 and 4K. In the long term, variable temperature cryogenic systems will change radically as the current worldwide shortage of helium spawns new designs for sample temperature control in EPR systems.

REFERENCES

1. Mims, W. B. Electron spin echoes. In *Electron Paramagnetic Resonance*; Geschwind, S., Ed.; Plenum: New York, 1972; pp 263–352.
2. Salikhov, K. M.; Semenov, A. G.; Tsvetkov, Y. D. *Electron Spin Echoes and Their Applications*; Nauka: Novosibirsk, 1976.
3. Norris, J. R.; Thurnauer, M. C.; Bowman, M. K. *Adv. Biol. Med. Phys.* **1980**, *17*, 365.
4. Bowman, M. K. Fourier transform electron spin resonance. In *Modern Pulsed and Continuous Electron Spin Resonance*; Kevan, L., Bowman, M. K., Eds.; Wiley: New York, 1990; Vol. 1, pp 1–42.
5. Jeschke, G. Dissertation for the degree of Doctor of Natural Sciences, New Concepts in Solid-State Pulse Electron Spin Resonance, Swiss Federal Institute of Technology, Zürich.
6. Bloch, F. *Phys. Rev.* **1946**, *70*, 460.
7. Slichter, C. P. *Principles of Magnetic Resonance*; Springer: Berlin, 1996.
8. Schwartz, R. N.; Jones, L. L.; Bowman, M. K. *J. Phys. Chem.* **1979**, *83*, 3429.
9. Klauder, J. R.; Anderson, P. W. *Phys. Rev.* **1962**, *125*, 912.
10. Dikanov, S. A.; Tsvetkov, Y. D. *Electron Spin Echo Envelope Modulation (ESEEM) Spectroscopy*; CRC Press: Boca Raton, FL, 1992.
11. Gorcester, J.; Millhauser, G. L.; Freed, J. H. Two-dimensional electron spin resonance. In *Modern Pulsed and Continuous-Wave Electron Spin Resonance*; Kevan, L., Bowman, M. K., Eds.; Wiley: New York, 1990; Vol. 1, pp 119–194.
12. Hahn, E. L. *Phys. Rev.* **1950**, *80*, 580.
13. Rowan, L. G.; Hahn, E. L.; Mims, W. B. *Phys. Rev. A* **1965**, *137*, 61.

14. Mims, W. B. *Phys. Rev. B Condens. Matter* **1972**, *5*, 2409.
15. Mims, W. B. *Phys. Rev. B Condens. Matter* **1972**, *6*, 3543.
16. Bowman, M. K.; Massoth, R. J. Nuclear spin eigenvalues and eigenvectors in electron spin echo modulation. In *Electronic Magnetic Resonance of the Solid State;* Weil, J. A., Bowman, M. K., Morton, J. R., Preston, K. F., Eds.; Canadian Society for Chemistry: Ottawa, 1987; pp 99–110.
17. Grupp, A.; Mehring, M. Pulsed ENDOR spectroscopy in solids. In *Modern Pulsed and Continuous-Wave Electron Spin Resonance*; Kevan, L., Bowman, M. K., Eds.; Wiley: New York, 1990; Vol. 1, pp 195–230.
18. Mims, W. B. *Proc. R. Soc. Lond. Ser. A Math. Phys. Sci.* **1965**, *283*, 452.
19. Davies, E. R. *Phys. Lett. A* **1974**, *47*, 1.
20. Höfer, P.; Grupp, A.; Nebenfuehr, H.; Mehring, M. *Chem. Phys. Lett.* **1986**, *132*, 279.
21. Dikanov, S. A.; Tyryshkin, A. M.; Bowman, M. K. *J. Magn. Res.* **2000**, *144*, 228.
22. Dikanov, S. A.; Bowman, M. K. *J. Magn. Res.* **1995**, *116*, 125.
23. Maryasov, A. G.; Bowman, M. K. *J. Magn. Res.* **2006**, *179*, 120.
24. Maryasov, A. G.; Bowman, M. K. *J. Phys. Chem. B* **2004**, *108*, 9412.
25. Hoult, D. I.; Richards, R. E. *Proc. R. Soc. Lond. Ser. A Math. Phys. Sci.* **1974**, *344*, 311.
26. Rinard, G. A.; Quine, R. W.; Eaton, S. S.; Eaton, G. R.; Froncisz, W. *J. Magn. Res. A* **1994**, *108*, 71.

6 Electron Paramagnetic Resonance Spectra in the Solid State

MARINA BENNATI

Max Planck Institute for Biophysical Chemistry, Am Fassberg 11, Göttingen, Germany

DAMIEN M. MURPHY

School of Chemistry, Cardiff University, Main Building, Park Place, Cardiff CF10 3AT, United Kingdom

6.1 INTRODUCTION

Chapter 4 examined in detail the form and analysis of the electron paramagnetic resonance (EPR) spectra in the liquid phase. However, paramagnetic species can also be conveniently studied in the solid state as magnetically dilute systems, in doped single crystals, or in frozen solutions (briefly treated in Chapter 3). In the solid state, the EPR spectra become far more complex compared to the liquid phase. This added complexity is however also advantageous because of the additional information now obtainable from the spectrum, via the spin Hamiltonian, including details on the site symmetry, electronic configuration, and nature of the neighboring atoms. The reason for the added complexity in the spectrum comes from *anisotropies* in the general spin Hamiltonian introduced earlier (Chapter 1). In this chapter, the form and shape of the EPR spectra in the solid state arising from these anisotropies will therefore be treated in more detail.

To illustrate the main differences between a solution phase EPR spectrum compared to a solid-state EPR spectrum, consider the representative example shown in Fig. 6.1. The figure illustrates the fluid solution, frozen solution (also called polycrystalline), and single crystal EPR spectra of a transition-metal complex containing a Cu^{2+} ion. At room temperature, only the average isotropic g_{iso} and a_{iso} values can be determined from the spectrum (Fig. 6.1a), as described in Chapter 4. That is not to say that the solution spectrum is uninformative; on the contrary, it can greatly assist the analysis of the solid-state spectrum and in many cases provide informative

---

Electron Paramagnetic Resonance. Edited by Brustolon and Giamello
Copyright © 2009 John Wiley & Sons, Inc.

196   EPR SPECTRA IN THE SOLID STATE

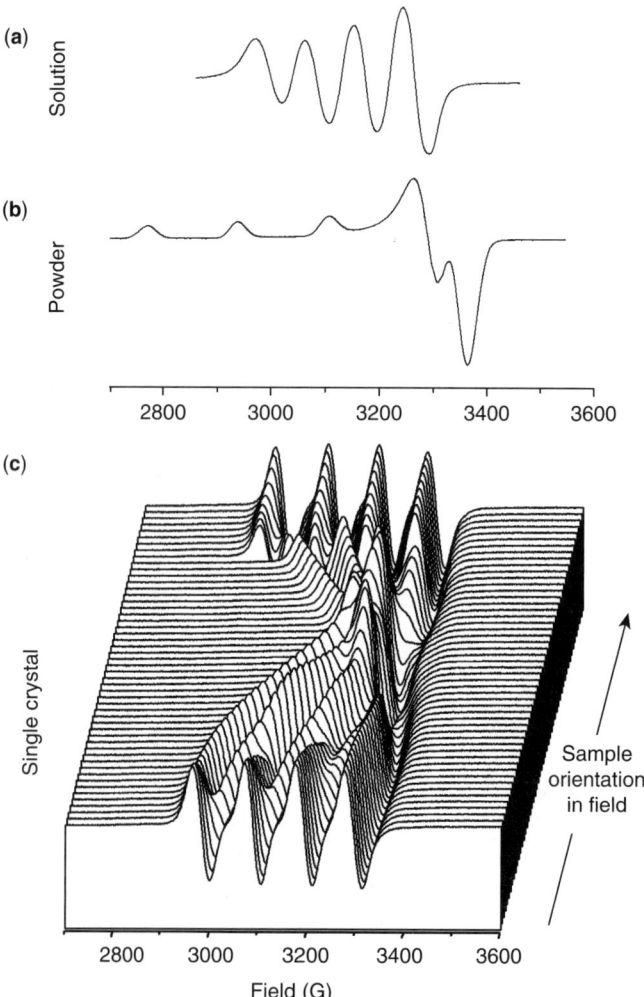

Fig. 6.1 The X-band CW EPR spectra of a Cu^{2+} cycloclam complex (labeled [DCM Cu][ClO$_4$]$_2$) dissolved in toluene and recorded at (a) 300K, (b) 100K, and (c) as a 1% doped single crystal in the Ni^{2+} DCM lattice at various orientations of the crystal relative to the magnetic field. Sample kindly synthesized by I. A. Fallis, Cardiff.

data on the structure and dynamics of the paramagnetic species. Upon freezing the solution the paramagnetic species become immobilized and oriented in a completely disordered way in the magnetic field (Fig. 6.1b). The spectrum now becomes broad and splits into distinct regions arising from the anisotropies in the **g** and hyperfine tensors, as all orientations of the paramagnetic species, randomly frozen in solution, are observed by EPR. In the single crystal (Fig. 6.1c), the spectra are very well resolved and the resonance position of the peaks is now seen to depend on the relative orientation of the sample in the laboratory magnetic field (or viewed from the viewpoint of the local paramagnetic center, in the direction of the magnetic field with respect to the principal axes of the paramagnetic species). Single crystal samples can only be recorded under magnetically dilute conditions (i.e., each paramagnetic species acts independently of all others), because at high concentrations, the electron spins will interact significantly with each other altering the nature of the spectrum (specifically the linewidths). One of the primary aims of this chapter is to explain why these powder and single crystal spectra take this form (discussed in §6.2).

In the next stage of spectral interpretation, one must know how to extract and analyze the principal axes of a tensorial interaction from an EPR spectrum. This will be discussed later in this chapter with some representative examples. In simple terms, the isotropic g_{iso} and a_{iso} values previously given in Chapters 1 and 4 are now replaced by 3×3 matrices (called the **g** and **A** tensors) that arise from the anisotropic electron Zeeman (§6.2) and hyperfine (§6.3) interactions. In the case of the single crystal example shown above, the paramagnetic system can be considered as having three mutually perpendicular inherent directions (*principal axes*) such that these, together with the resulting observed values (known as the *principal values*) measured along these directions, completely describe the anisotropic properties in **g** and **A** (discussed in §6.2.1). If the system contains interacting nuclei with $I > 1/2$, then anisotropic nuclear quadrupolar (**Q**) interactions must also be treated and included (§6.3.8) in our description of the spectrum. Generally the quadrupolar interactions will only be observed in EPR spectra for nuclei with small magnetic moments and large quadrupole moments, but they are readily observable by other techniques such as electron nuclear double resonance (ENDOR; Chapters 1 and 2) or hyperfine sublevel correlation spectroscopy (HYSCORE, Chapter 5). Thus, it is important to know how to recognize and interpret such interactions. Furthermore, if the paramagnetic system contains two or more unpaired electrons, an anisotropic fine term (**D**) introduced in Chapter 1 must also be considered (§6.5).

All of these terms (**g**, **A**, **D**, **Q**) will generally be anisotropic, complicating the solid-state spectrum, and such anisotropic interactions can be expressed in terms of a tensor. The concept of the tensor was introduced in Chapter 1, but it will be elaborated in greater detail in this chapter, showing how it can be derived and utilized. This analysis will be performed specifically for the **g** and **A** tensors (described in §6.2 and §6.3, respectively), because the same methodology can then be easily established to encompass the quadrupolar term (§6.3.8) and zero-field splitting (ZFS; §6.5). Attention will also be focused on powder or polycrystalline spectra (§6.2.2), as opposed to single crystals (§6.2.1), because the former are most commonly encountered in the laboratory.

Most of the illustrative examples given in Section 6.3 will be based on organic type radical species. However, a frequently encountered group of paramagnetic species include the d-block transition-metal ions (TMIs). In very broad terms, the main differences between the EPR spectra of TMIs compared to free radicals is that the spin–orbit coupling is unusually strong and exhibits large anisotropy, so the g factors are no longer close to g_e and may in fact vary over a wide range. The analysis of the spectra now requires some knowledge about the splitting of the d orbitals in various point symmetries. For this reason, the spectra of simple TMI systems, with an $S = 1/2$ effective ground state, will be treated separately in this chapter (§6.4).

6.2 ANISOTROPY OF THE ZEEMAN INTERACTION: THE g TENSOR

The shape of the solid-state EPR spectrum depends on the relative orientation of the applied magnetic field (**B**) with respect to the orientation of the paramagnetic species itself. The symmetry of the local site at the paramagnetic center is therefore directly reflected in the final profile of the resulting EPR spectrum.

For simplicity, consider three limiting cases that are used to define the local symmetry of the electronic wavefunction (introduced in Chapter 1) in a paramagnetic site: *cubic*, *axial*, and *rhombic*. In the first case, three subsystems called cubal, octahedral, and tetrahedral exist (Fig. 6.2a–c). Because of the high symmetry of the site, anisotropy in the EPR spectrum is absent. In simple terms, the g factor is independent of the sample orientation. In the second case (Fig. 6.2d), however, a rotational symmetry about a unique axis is present. Anisotropy will now be observed in the EPR spectrum because the resonance field will be different when **B** is aligned along the unique axis (Z) compared to the observed case when **B** is aligned in the XY plane perpendicular to the applied field (Fig. 6.2d). In other words, different g factors will be observed in the parallel and perpendicular orientations of **B** with respect to the axes. In the third case, anisotropy occurs along all three directions and three unequal values of the g factor will occur (Fig. 6.2e).

Because of these differences in symmetry, the basic EPR resonance equation presented in Chapter 1 must now take into consideration the importance of sample orientation (or angle) and therefore has the form

$$h\nu = \mu_B g(\theta, \phi) \mathbf{B} \tag{6.1}$$

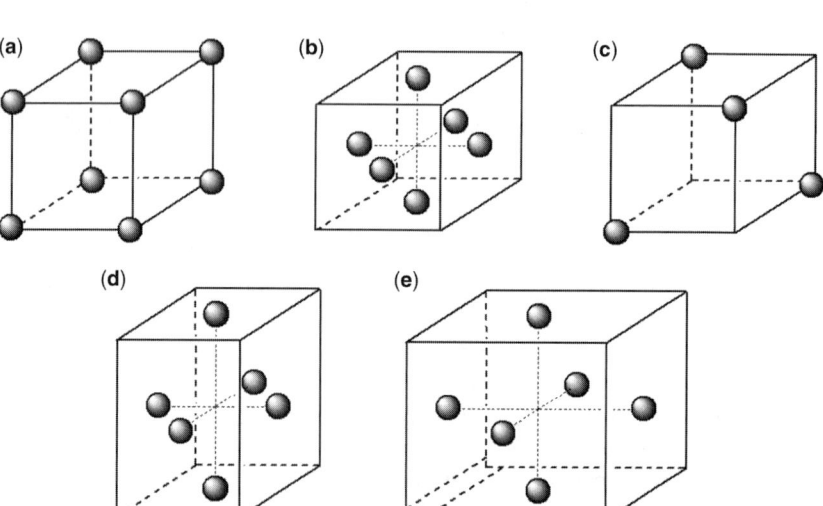

Fig. 6.2 An illustration of the cubic symmetry classes: (a) cubal, (b) octahedral, and (c) tetrahedral. Elongation along the (d) Z direction or (e) X direction causes a lowering of the site symmetry. One can imagine, for example, a paramagnetic complex with the metal surrounded by four or six ligands with the different geometries shown in this figure.

6.2 ANISOTROPY OF THE ZEEMAN INTERACTION: THE g TENSOR

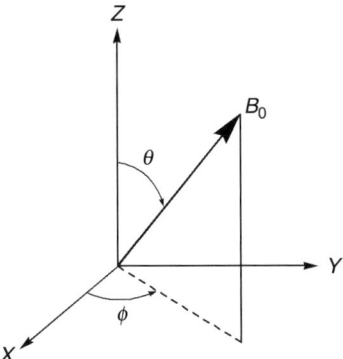

Fig. 6.3 The orientation of the magnetic field vector with respect to the **g** tensor principal axes system denoted by X, Y, and Z.

where θ and ϕ refer to the polar angles of the applied magnetic field direction within the molecular **g** tensor principal axes system (see Fig. 6.3 and a later figure for an explanation of the axes systems). Because the observed g value depends on the angles (θ, ϕ), one must derive a more complete spin Hamiltonian that accounts for the g factor at any arbitrary orientation. Such an equation for the electron Zeeman Hamiltonian (in the absence of any hyperfine interaction) would take the form

$$\hat{H} = \mu_B \cdot \mathbf{S} \cdot \mathbf{g} \cdot \mathbf{B} \tag{6.2}$$

where μ_B is the Bohr magneton, $\mathbf{S} = (S_x\ S_y\ S_z)$ is the Pauli spin operator vector, and **g** is the so-called 3×3 **g** matrix (no longer written as a simple scalar quantity independent of orientation as in Chapter 1). The principal values of the **g** tensor can be simply denoted as g_1, g_2, and g_3.

An explicit expression for $g(\theta, \phi)$ in Equation 6.1 can be obtained by first writing the components of the applied magnetic field in the **g** tensor frame using the polar angles as defined in Fig. 6.3, such that

$$B_0^x = B_0 \sin\theta \cos\phi;\quad B_0^y = B_0 \sin\theta \sin\phi;\quad B_0^z = B_0 \cos\theta \tag{6.3}$$

where B_0^x is the component of the applied magnetic field in the x direction and so forth. The **g** tensor can be more easily represented by a 3×3 matrix, so that the Zeeman Hamiltonian in Equation 6.2 is more conveniently written in matrix form as

$$\hat{H} = \mu_B \begin{pmatrix} S_x \\ S_y \\ S_z \end{pmatrix} \cdot \begin{pmatrix} g_1 & 0 & 0 \\ 0 & g_2 & 0 \\ 0 & 0 & g_3 \end{pmatrix} \cdot \begin{pmatrix} B_0^x \\ B_0^y \\ B_0^z \end{pmatrix} \tag{6.4}$$

Inserting Equation 6.3 into Equation 6.4, after manipulation of the matrix (see Appendix A.6.1 for manipulation of these and similar matrices), the following equation is obtained:

$$\hat{H} = \mu_B \left(g_1 B_0 \sin\theta \cos\phi \cdot S_x + g_2 B_0 \sin\theta \sin\phi \cdot S_y + g_3 B_0 \cos\theta \cdot S_z \right) \tag{6.5}$$

It is important to realize at this point that the actual field experienced by the spin arises not only from the applied magnetic field (**B**) but also from internal local fields (**B**$_{local}$), which add together and produce the total effective magnetic field (**B**$_{eff}$). Formally, the **B** in Equation 6.1 should therefore be replaced by **B**$_{eff}$ (although in practice **B** is usually used). The anisotropy of **g** ensures that **B**$_{eff}$ is not parallel to the applied field **B**, but it is **B**$_{eff}$ that defines the quantization axis for the spin system. The result for the g value is that the spin **S** is quantized along **B**$_{eff}$ (not **B**), and this is the reason why the three components of S_x, S_y, and S_z were considered in the previous equations. A final manipulation and diagonalization of Equation 6.5 after inserting the Pauli matrices (see Appendix A.6.2) leads to the following effective g value:

$$g(\theta, \phi) = \sqrt{\sin^2\theta \cdot \cos^2\phi \cdot g_1^2 + \sin^2\theta \cdot \sin^2\phi \cdot g_2^2 + \cos^2\theta \cdot g_3^2} \qquad (6.6)$$

This is the expected g value (hence the expected resonant field) for a paramagnetic center characterized by g_1, g_2, and g_3 principal values of the **g** tensor and having a generic orientation (θ, ϕ) with respect to the magnetic field.

6.2.1 The Single Crystal Case

To first understand the position of the EPR resonance (i.e., why the peak appears where it does in a single crystal EPR spectrum), it is instructive to consider a single paramagnetic $S = 1/2$ spin center, with no hyperfine interaction, in the solid state possessing cubic symmetry. In this case g remains a scalar constant (isotropic), and its value is the same regardless of the orientation of the applied magnetic field. The spin Hamiltonian is therefore analogous to that described for an isotropic liquid solution.

If the symmetry of the site is now reduced from cubic to uniaxial (Fig. 6.2d), the g value will vary, depending on the magnetic field direction. Two limiting cases for the magnetic field direction can be identified in which (1) **B** is parallel to the unique Z axis (labeled $B_\parallel$) and (2) **B** lies in the XY plane (labeled $B_\perp$) and so is perpendicular to the Z axis. An EPR resonance (or peak) will appear for the field position of $B_\parallel$ and a different peak will appear at the field position $B_\perp$. As the orientation of the field moves from $B_\parallel$ to $B_\perp$, the EPR resonance will also move. The orientation dependence of this peak with respect to the variation in applied field direction can be seen in the lower trace of Fig. 6.4. The g values corresponding to $B_\parallel$ and $B_\perp$ (i.e., when the field is parallel and perpendicular to the unique axis) are labeled $g_\parallel$ and $g_\perp$.

The variation in the resonance position for uniaxial symmetry (Fig. 6.4) can be represented by an effective g factor of the form

$$g^2 = g_\perp^2 \sin^2\theta + g_\parallel^2 \cos^2\theta \qquad (6.7)$$

where θ represents the angle between **B** and the unique Z axis (Fig. 6.3). Equation 6.7 is a special case of Equation 6.6 for the situation where $\phi = 0$, $g_1 = g_2 = g_\perp$, and $g_3 = g_\parallel$. It is important to realize here that the g value is actually determined by the square of **g**, as nicely described by Atherton [1].

6.2 ANISOTROPY OF THE ZEEMAN INTERACTION: THE g TENSOR 201

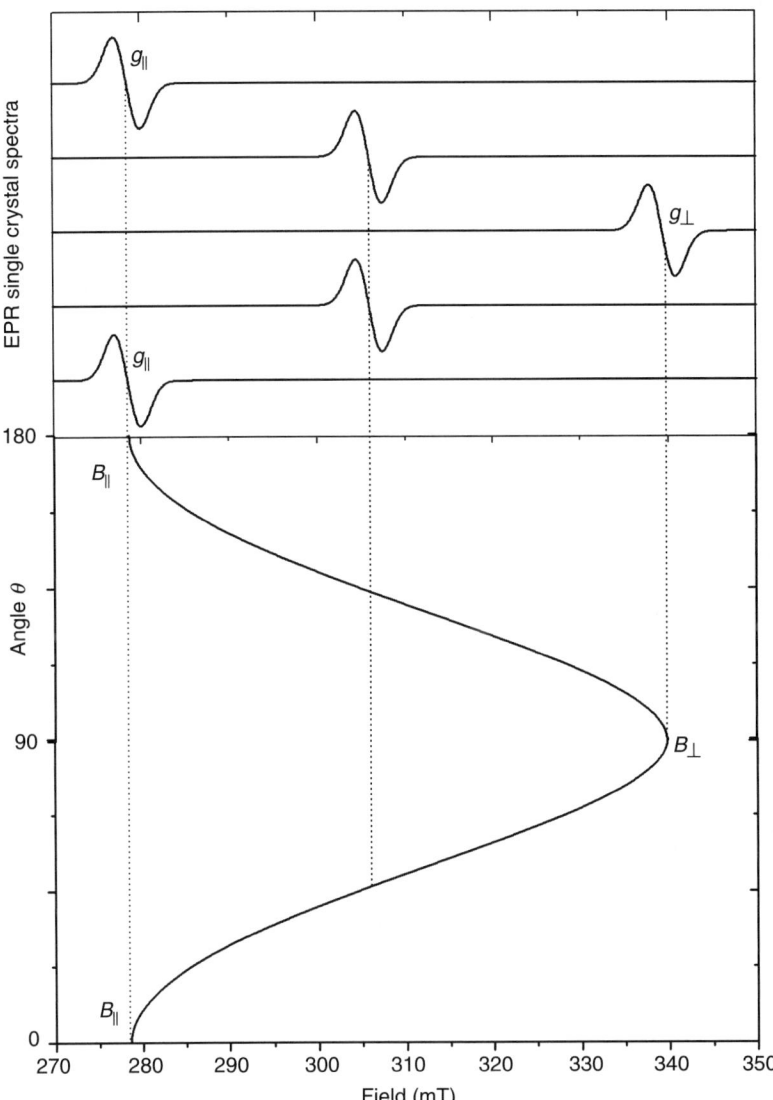

Fig. 6.4 Simulated EPR profiles for a single crystal $S = 1/2$ spin system possessing axial symmetry. As the crystal is rotated from $\theta = 0°$ to $180°$ within the field, the peak positions change. The angular dependency profile of these peaks for various angles of θ are shown in the lower curve.

Measurement of the resonance position, as discussed above, corresponding to the effective g value in single crystal samples is generally made for different orientations of the magnetic field with respect to three orthogonal axes in the sample (labeled $X'\,Y'\,Z'$ in Fig. 6.5). The fixed laboratory reference axes are labeled x, y, z. It is important to realize that $X'\,Y'\,Z'$ axes represent an arbitrary frame for the single crystal

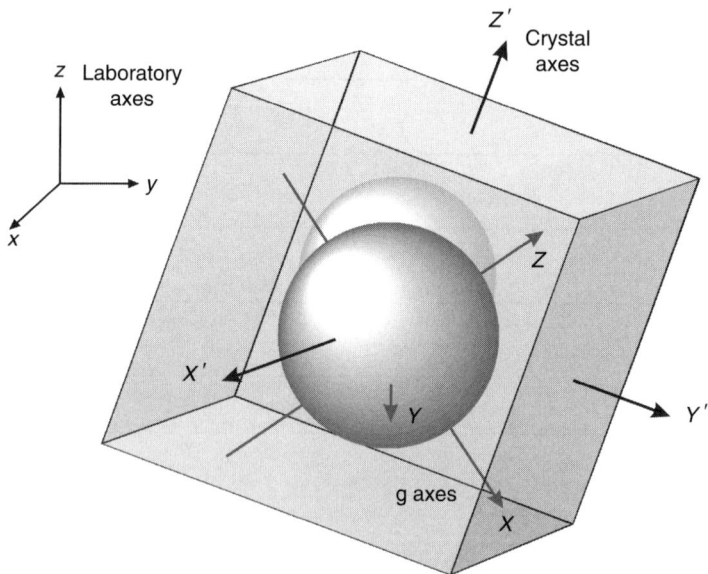

Fig. 6.5 A schematic illustration of the different sets of coordinate axes including the laboratory fixed axes (x, y, z), the crystal axes (X', Y', Z'), and the principal axes of the paramagnetic species within the crystal (X, Y, Z).

mounted within the cavity, which are often chosen for practical purposes, and the EPR spectra are obtained by systematically rotating the crystal (using a goniometer) about them. One set of such spectra was shown in Fig. 6.1c after several rotations about one crystal axes. In this arbitrary frame $(X' Y' Z')$, the resulting matrices (**g** or **A**) are not generally diagonal, because this axes system does not coincide with the principal axes one (labeled X, Y, Z in Fig. 6.5). The experimental determination of the **g** tensor is done by measuring the g^2 value and obtaining the **g · g** matrix. In the arbitrary frame this matrix is nondiagonal with six independent elements $(gg)_{ij}$. Because the **g · g** matrix is symmetric, with $(gg)_{ij} = (gg)_{ji}$, these six independent elements provide enough information to diagonalize the matrix. Diagonalizing the matrix is equivalent to performing a rotation of $X' Y' Z'$ axes to coincide with the reference frame of the principal axes. This manipulation is achieved by transformation of the matrices relative to **g** or **A** tensors, and the steps involved in this manipulation will be systematically explained later in Section 6.3.2 and in Appendix A.6.3.

Measurements of the g^2 value at different orientations gives the elements of the **g · g** tensor $(gg)_{ij}$ that can then be transformed into its principal axes system. We could begin by rotating the field in the $X' Z'$ plane ($\phi = 0°$), where the three components in Equation 6.5 are $(\sin \theta \ 0 \ \cos \theta)$. After inserting these field components and the 3×3 **g · g** matrix into the spin Hamiltonian given in Equation 6.4 and diagonalizing [1], the effective g^2 value can be written in a compact form as

$$g^2 = [\sin \theta \ \ 0 \ \ \cos \theta] \cdot \begin{bmatrix} (gg)_{xx} & (gg)_{xy} & (gg)_{xz} \\ & (gg)_{yy} & (gg)_{yz} \\ & & (gg)_{zz} \end{bmatrix} \cdot \begin{bmatrix} \sin \theta \\ 0 \\ \cos \theta \end{bmatrix} \quad (6.8)$$

6.2 ANISOTROPY OF THE ZEEMAN INTERACTION: THE g TENSOR

Further matrix multiplication (see Appendix A.6.1) gives

$$g^2 = (gg)_{xx} \sin^2 \theta + 2(gg)_{xz} \sin \theta \cos \theta + (gg)_{zz} \cos^2 \theta \quad (6.9a)$$

so that three tensor elements are measured in this experiment. Measurements in the $X'Z'$ ($\phi = 90°$, field components of 0 sin θ cos θ) and $X'Y'$ ($\theta = 90°$, field components of cos ϕ sin ϕ 0) planes gives the remaining values (Equations 6.9b and 6.9c), so that eventually enough information is obtained enabling the $\mathbf{g} \cdot \mathbf{g}$ tensor to be diagonalized. In this way, the squares of the principal values and the orientation of the $\mathbf{g}$ tensor axes within the crystal axes can be extracted from the single crystal spectra.

The other two rotations in the respective $Y'Z'$ and $X'Y'$ planes give

$$g^2 = (gg)_{yy} \sin^2 \theta + 2(gg)_{yz} \sin \theta \cos \theta + (gg)_{zz} \cos^2 \theta \quad (6.9b)$$

$$g^2 = (gg)_{xx} \sin^2 \phi + 2(gg)_{xy} \sin \phi \cos \phi + (gg)_{yy} \sin^2 \phi \quad (6.9c)$$

6.2.2 Powder Patterns

In many situations, especially in biological systems, it is not always possible to grow and obtain magnetically dilute single crystals. Frequently, the paramagnetic species may only be formed and stabilized in frozen solution, polycrystalline, or glassy media. The powder spectrum of an EPR line that is dominated by **g** anisotropy is given by the contributions of all molecules in their specific orientations with respect to the applied magnetic field. Each paramagnetic center exhibits its own resonance position depending on its orientation, and therefore the resultant powder spectrum is usually broad because the resonance is an envelope representing a weighted distribution of all possible resonance fields. The orientation of the applied magnetic field with respect to the **g** principal axes was illustrated in Fig. 6.3. In a powder, all orientations have the same probability and the total intensity of the EPR spectrum is given by the sum of the contributions of each single molecular orientation in a sphere. Powder patterns are usually computed numerically (using simulation techniques as discussed in Chapter 7) by systematic variation of the angles θ and ϕ between 0 and π and 0 to 2π, respectively, and weighting the spectral contributions with sin θ.

As discussed in Section 6.2.1, the effective g value can be expressed in terms of three principal values directed along three axes or directions (Equation 6.6). The powder spectra can give these principal values directly but not the principal directions of the tensor with respect to the molecular axes. These can be determined only in single crystal EPR experiments or predicted by quantum chemical calculations (see Chapter 7). For example, let us first again consider the case of a paramagnetic species ($S = 1/2$) with uniaxial symmetry, which can be characterized by two principal axes g values of $g_\perp$ and $g_\parallel$. As given in Equation 6.7, the variation in the g value will depend solely on the angle θ between **B** and the unique axis:

$$B(\theta) = \frac{h\nu}{\mu_B} \left(\sqrt{g_\parallel^2 \cos^2 \theta + g_\perp^2 \sin^2 \theta} \right)^{-1} \quad (6.10)$$

The values of $g_\perp$ and $g_\parallel$ therefore set the range of **B** over which absorption occurs. When $g_\parallel > g_\perp$, then no absorption occurs at fields lower than

$$B_\parallel = \frac{h\nu}{\mu_B} \frac{1}{g_\parallel} \tag{6.11}$$

and none at fields higher than

$$B_\perp = \frac{h\nu}{\mu_B} \frac{1}{g_\perp} \tag{6.12}$$

When the field is aligned along the unique axis ($B_\parallel$), absorption occurs for those paramagnets whose field lies along the symmetry axis, corresponding to an angle of $\theta = 0°$ (see Fig. 6.6). At this orientation only very few spins contribute to the pattern, and the spectral intensity has a minimum (at the edge of the powder pattern) close to 280 mT in this specific example. As the field moves progressively from $B_\parallel$ to $B_\perp$, more spins come into resonance and correspondingly the intensity of the absorption line increases. This can be represented by the shading of the spheres in Fig. 6.6 for $\theta = 0°$ and $90°$. At $B_\perp$, the absorption reaches a maximum because there is now a large plane of orientations with the field perpendicular to the symmetry axis [see sphere (**2**) in Fig. 6.6]. As discussed in earlier chapters, continuous wave (CW) EPR spectra are recorded as the first derivative of the absorption (Fig. 6.6b); nevertheless, as seen from Fig. 6.6, it is still possible to extract the values of $g_\parallel$ and $g_\perp$ even from the powder spectrum. The variation in the resonance absorption (analogous to the variation in g) as a function of the angle θ can be seen as a smooth curve with two prominent resonances at $B_\parallel$ ($g_\parallel$) and $B_\perp$ ($g_\perp$) (Fig. 6.6c).

In the second example, let us consider the case of a paramagnetic species with rhombic symmetry (represented earlier in Fig. 6.2e), which is now characterized by the three g values g_1, g_2, and g_3. The variation in the g values now depends on the two polar angles θ and ϕ in Equation 6.9, and a typical example of the absorption and first derivative profiles for such a system with distinct g values is provided in Fig. 6.7a and b.

Three special cases (called singularities) now occur for the resonant field positions corresponding to $\theta = 0°$, $\theta = 90° = \phi$, and $\theta = 90°$ and $\phi = 0°$. At $\theta = 0°$, the spins that come into resonance are those for which the applied field lies along the Z axis and an absorption edge occurs producing the derivative peak corresponding to g_1. As the field moves away from the Z axis, in the ZY plane (such that $\phi = 90°$ and only the angle of θ varies) the resonance field position will also vary and a maximum in intensity of the absorption occurs when $\theta = 90° = \phi$ (g_2 in Fig. 6.7). A similar situation occurs when the field moves from the Z axis but now in the ZX plane, such that all intermediate values of θ contribute to the intensity of the absorption line (because $\phi = 0°$ in this plane). The limiting point for this trend is reached when $\theta = 90°$ and $\phi = 0°$ (g_3 in Fig. 6.7).

The angular dependency plots illustrating the variation in the resonant field positions are thus shown in Fig. 6.6c. The most intrinsic feature in the powder

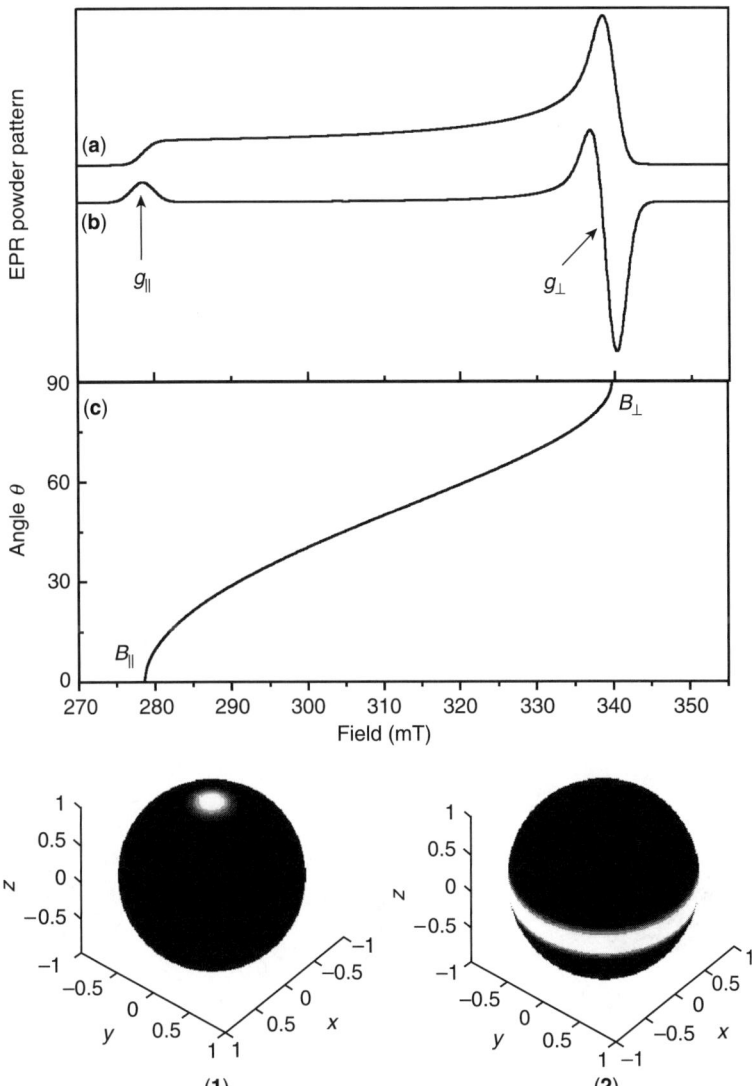

Fig. 6.6 The (a) absorption and (b) first derivative EPR line shape for a randomly oriented $S = 1/2$ spin system with axial symmetry ($g_\| = 2.437$, $g_\perp = 2.000$, $\nu = 9.5$ GHz). (c) The angular dependency curve (θ vs. field) is shown. The orientation selection on the unit spheres are also provided for the observer position in the $g_\|$ plane (**1**) and the $g_\perp$ plane (**2**).

pattern of a rhombic **g** tensor consists of the fact that, although only very few orientations contribute to the spectrum at $B\|g_1$ and $B\|g_3$ (single crystal like case), several intermediate orientations reveal the same resonance as $B\|g_2$, resulting in a maximal absorption at this field. The situation is best illustrated in the sphere (**2**) in Fig. 6.7.

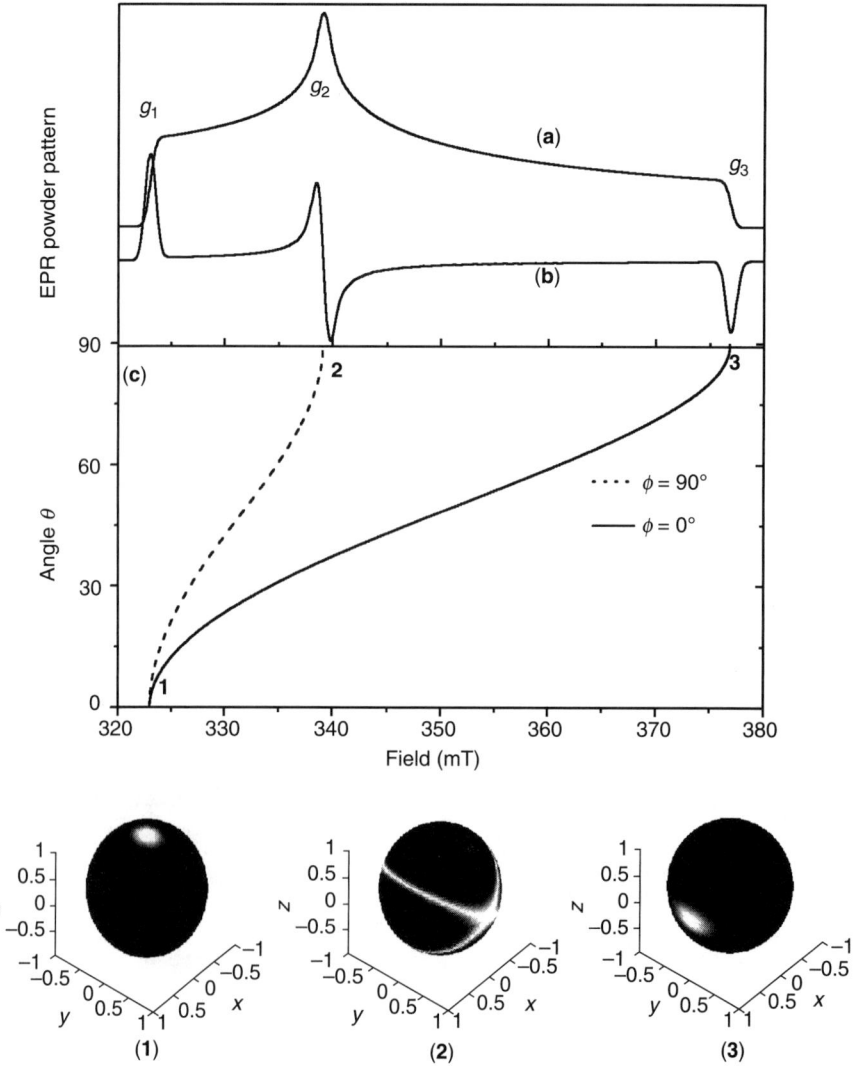

Fig. 6.7 The (a) absorption and (b) first derivative EPR line shape for a randomly oriented $S = 1/2$ spin system with rhombic symmetry ($g_1 = 2.101$, $g_2 = 2.000$, $g_3 = 1.800$, $\nu = 9.5$ GHz). (c) The angular dependency curve (θ vs. field) is shown for two angles of $\phi = 0°$ and $90°$. The orientation selections on the unit spheres are also shown for the observer position in the $g_\parallel$ plane (**1**), $g_\perp$ plane (**2**), and g_3 plane (**3**).

These peaks in the rhombic powder spectrum correspond to turning points in the angular variation of the resonant field and can be represented in another way for clarity using the contour plot shown in Fig. 6.8. Point 3 on the contour plot corresponds to the g_3 peak in Fig. 6.7 (the maximum in the contour plot close to 380 mT).

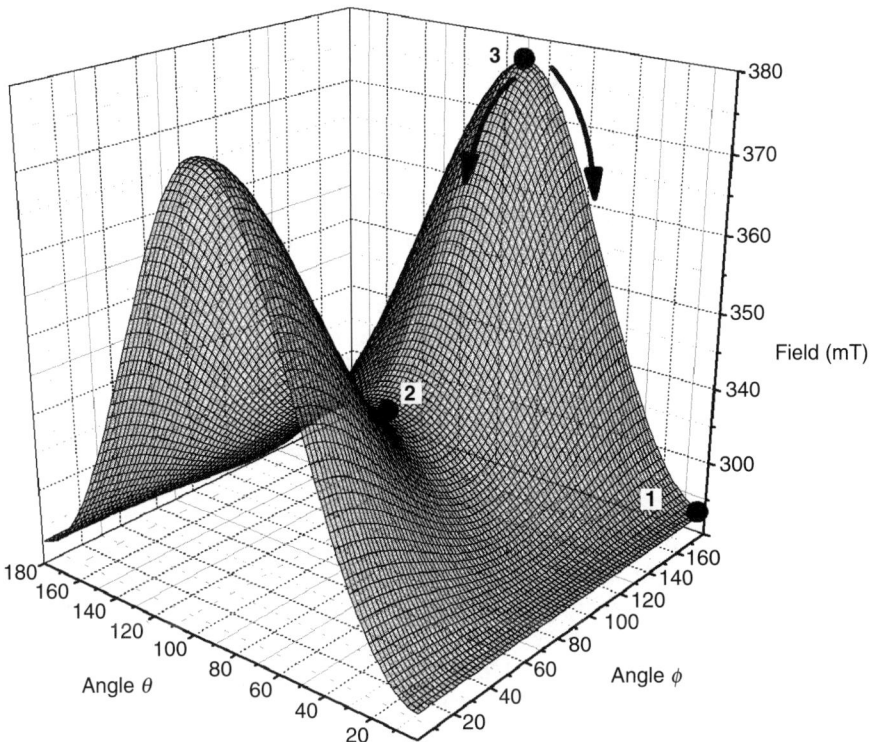

Fig. 6.8 A contour plot of the resonant field positions versus the polar angles of θ and ϕ (0° to 180°) for the randomly oriented $S = 1/2$ spin system with rhombic symmetry shown in Fig. 6.7 ($g_1 = 2.101$, $g_2 = 2.000$, $g_3 = 1.800$, $\nu = 9.5$ GHz).

As θ varies (but ϕ remains constant at 0°) the resonant field moves from point 3 to point 1 on the curve. As ϕ varies (but θ remains constant at 90°) the resonant field moves down the slope from point 3 to point 2 at the bottom of the saddle. For intermediate angles of θ and ϕ, all orientations of the applied field ultimately contribute to the absorption pattern. Nevertheless, it can be easily seen that the turning points can still be identified in the plot and hence can be extracted from the powder spectrum.

6.2.3 Interpretation of the g Tensor and the Role of Spin–Orbit Coupling

In the previous section, anisotropies in the **g** tensor were qualitatively shown to be responsible for the anisotropic profile of the spectrum as a function of site symmetry and the relative orientation with respect to the applied magnetic field. In all of these discussions, only the electron spin (S) interaction with the applied field was considered. However, as introduced in Chapter 1 (§1.8), the electron angular momentum has two contributions in real paramagnetic systems, namely, spin and orbital angular momentum (l), that are coupled by spin–orbit coupling (λ). Although l is usually suppressed, the influence of λ restores some of the orbital momentum and this is the primary reason for the observed deviation of g from g_e ($\Delta g = g - g_e$). The spin–orbit interaction admixes the ground state with certain excited states, causing a small amount of orbital angular momentum to appear in the ground state. This produces a small local magnetic field ($\mathbf{B}_{local}$) that adds to the external field ($\mathbf{B}$), producing Δg shifts. Clearly, the extent of this admixing depends on which orbital (p, d, or f) contributes to the spin ground state and on the energy difference between ground and excited states. All of these considerations (λ, energy difference, coupled orbitals) are responsible for the shift in the g values and can therefore be accounted for by the following equation:

$$g_{ij} = g_e \delta_{ij} + 2\lambda \sum_{m \neq 0} \frac{\langle m|l_i|n\rangle\langle n|l_j|m\rangle}{E_n - E_m} \qquad (6.13)$$

where ij are the molecular coordinate axes, λ is the spin–orbit coupling constant, E_n is the energy of the singly occupied molecular orbital (SOMO), m denotes the filled and empty orbitals with energy E_m, and l_i is the component of the orbital angular momentum operator [1]. Let us suppose that the molecular orbitals are written as linear combinations of p or d orbitals, then the integrals in Equation 6.13 (e.g., $\langle m|l_i|n\rangle$) can be calculated. The results of the operation by l_i on these integral functions are given in Appendix A.6.4.

The important point to note from this discussion is simply that λ affects the g values, so it is therefore instructive to make some generalized comments at this stage on the role of the orbital angular momentum in the g value. First, spin–orbit coupling to empty molecular orbitals produces a negative contribution to g_{ij} whereas coupling to filled molecular orbitals produces a positive effect on g_{ij} [2]. This is best illustrated with respect to d^1 TMIs that have negative g shifts (less than g_e) compared to d^9 metal ions for which positive g shifts are found due to admixture of empty versus filled molecular orbitals (§6.4). An analogous case is frequently encountered with inorganic radicals, such that negative g shifts occur for 11e$^-$ π^* diatomic radicals (such as NO, N_2^-, CO$^-$) compared to the expected positive g shifts for 13e$^-$ π^* radicals (such as O_2^-). Second, because l_z does not couple the d_{z^2} orbital to any other orbital, then according to Equation 6.13, a g_{zz} value close to free spin is predicted for a SOMO based primarily on d_{z^2}. Therefore, as a first and crude approximation, a g value close to g_e for a TMI complex can be indicative of a d_{z^2} based SOMO.

The **g** tensor of organic radicals can also be interpreted in a somewhat similar fashion. For example, in the simple ·C–H fragment (with the axis along the singly occupied 2p orbital being defined as the Z axis), no mixing with other orbitals about the Z axis occurs via spin–orbit coupling. As a result, the g_{zz} value for the radical should be close to g_e. However, a shift in the g values will occur by admixing p_x and p_y in the XY plane. The orbitals involved are the σ-bonding and σ^*-antibonding orbitals that lie below and above the $2p_z$ orbital, and they are respectively filled and empty in the ground state. Because the σ^*-antibonding orbital lies further above 2p than the σ-bonding orbital lies below, the contributions from the latter would be greater so there should be an overall net positive contribution to the g values. It is worth noting that in general the spin–orbit coupling constants for organic radicals are small, and for this reason the **g** tensor values are close to g_e. This can be compared to TMIs, where much larger spin–orbit couplings produce much larger deviations in the g values (§6.4).

To summarize, deviations of the main g values from the free electron g_e value arise from residual orbital momentum of the unpaired electron. They can be used as a simple fingerprint to identify the radical species or in comparison with quantum chemical calculations to determine the electrostatic surroundings of the paramagnetic molecule in the matrix. If supplementary UV visible data are available, a value of $E_n - E_m$ can be obtained, which can give a first approximation on the molecular orbitals involved in the **g** matrix.

6.3 THE HYPERFINE INTERACTION IN THE SOLID STATE

The hyperfine interaction describes the interaction of a paramagnetic species with nuclear spins in the close surroundings ($r \leq 1$ nm). This is the most common interaction observed in EPR spectroscopy, producing splittings in the EPR line in liquid solution (see Chapters 3 and 4) but rather inhomogeneous line broadening in the solid state. As reported in Chapters 1 and 4, the isotropic part of the hyperfine interaction can be extracted from the EPR spectra in the liquid phase and gives insight into the spin density distribution of the unpaired electron spin. However, the anisotropic part arises from the dipolar interaction between the electron and the nuclei and contains important information about the distance and the orientation of the two interacting spins with respect to the applied magnetic field. This interaction is generally averaged out in liquids and must be obtained by analysis of the spectra in the solid state. Furthermore, for many commonly studied systems, in particular biological systems and inorganic materials, the physical nature or reactivity of the samples often precludes study in liquid solution so that the isotropic hyperfine interaction must also be extracted from the solid-state spectrum. Under these conditions, it is very important to develop experimental protocols to resolve the complete hyperfine tensor in the solid state.

The spectroscopic approach of choice will depend on the expected size of the hyperfine interaction with respect to the intrinsic line broadening in the EPR spectrum at the observed frequency. However, the number of coupled nuclei will also substantially influence the complexity of the spectra and eventually several methods have to be combined to obtain unambiguous results. In general, two limiting cases are readily distinguished: (1) the EPR case for $A_{1,2,3} > \Delta B_{1/2}^{\text{EPR}}$, where $\Delta B_{1/2}^{\text{EPR}}$ is the EPR linewidth of a spin packet and $A_{1,2,3}$ are the principal values of the hyperfine tensor and (2) the ENDOR or electron spin echo envelope modulation (ESEEM) case for $A_{1,2,3} < \Delta B_{1/2}^{EPR}$. In case 1 the hyperfine tensor can be extracted directly from the EPR spectrum, whereas in case 2 the so-called hyperfine spectrum has to be recorded by ENDOR or ESEEM methods. The powder line shapes generated by the hyperfine interaction in EPR (§6.3.1 and §6.3.2) as well as in the hyperfine spectra are described first, and then these cases are illustrated with examples (§6.3.3–6.3.5).

6.3.1 EPR and Hyperfine Powder Spectra in the $S = 1/2$ and $I = 1/2$ Case

The spin Hamiltonian that describes an electron spin coupled to one or more nuclei i is given by

$$\hat{H} = \mu_B \mathbf{S} \cdot \mathbf{g} \cdot \mathbf{B} + \sum_i (g_N^i \mu_N^i \mathbf{I} \cdot \mathbf{B} + \mathbf{I} \cdot \mathbf{A}^i \cdot \mathbf{S}) \quad (6.14)$$

The first and second terms in the equation describe the Zeeman splitting of the electron and nuclear spins, respectively. The third term describes the hyperfine interaction, and $\mathbf{A}^i$ represents the hyperfine tensor of nucleus i and is composed of the isotropic (a_{iso}) or scalar part and the anisotropic ($\mathbf{T}$) or dipolar part, respectively:

$$\mathbf{A} = a_{iso} + \mathbf{T} = \begin{pmatrix} A_{xx} & A_{xy} & A_{xz} \\ A_{yx} & A_{yy} & A_{yz} \\ A_{zx} & A_{zy} & A_{zz} \end{pmatrix} \quad (6.15)$$

As introduced in Chapter 1, $\mathbf{T}$ is a traceless tensor. For an axially symmetric tensor, the principal values would be $T_\parallel$ and $T_\perp$. Because the tensor is traceless, it results in $\text{Tr}(\mathbf{T}) = 0$ and $T_\parallel = -2 \cdot T_\perp$. In contrast, the tensor as written in Equation 6.15 is not diagonal because it is expressed in the laboratory frame.

There is something very important to note here about the spin Hamiltonian in Equation 6.14. Because of the much smaller gyromagnetic ratio, the nuclear Zeeman splittings are 3–4 orders of magnitude smaller compared to the electron Zeeman splitting. Thus, in almost all cases encountered at EPR frequencies in the gigahertz range, the electron Zeeman interaction in Equation 6.14 will dominate, which means that the electron spin is quantized along the applied magnetic field direction. In this case, the S_x and S_y components of the electron spin vector (see also Chapter 1) can be neglected. However, the same does not apply for the nuclei, because the hyperfine coupling is often larger than the nuclear Zeeman frequency. The nucleus will therefore experience the $\mathbf{B}_{eff}$ generated by the applied field in addition to the hyperfine field and is not necessarily quantized along the applied field. This causes the energy eigenvalues to depend on the relative size of the nuclear Zeeman versus the hyperfine interaction.

To understand the shape of EPR and hyperfine spectra, one must first consider the simplest case of an electron spin $S = 1/2$ coupled to a nucleus with $I = 1/2$

(i.e., a simple two-spin system). The Hamiltonian Equation 6.14 can be rewritten after vector and matrix multiplication (Appendix A.6.1) as

$$\hat{H} = \mu_B \cdot g \cdot S_z \cdot B_0 + g_N \mu_N I_z \cdot B_0 + A_{zz} I_z S_z + A_{yz} I_y S_z + A_{xz} I_x S_z \qquad (6.16)$$

where we have neglected all terms containing S_x and S_y. To find the explicit matrix form of this Hamiltonian one must first expand the spin matrices according to the total spin space with dimension $(2S+1)(2I+1)$, which is equal to four in this case as $S = 1/2$ and $I = 1/2$. This procedure is excellently described in specific literature [3]. After inserting the spin matrices and carrying out matrix multiplication, the Hamiltonian in Equation 6.16 achieves the following matrix form:

	$\|++\rangle$	$\|+-\rangle$	$\|-+\rangle$	$\|--\rangle$
$\|++\rangle$	$\frac{1}{2}\mu_B g B_0 - \frac{1}{2}\mu_N g_N B_0 + \frac{1}{4}A_{zz}$	$\frac{1}{4}(A_{xz} - iA_{yz})$		
$\|+-\rangle$	$\frac{1}{4}(A_{xz} + iA_{yz})$	$\frac{1}{2}\mu_B g B_0 + \frac{1}{2}\mu_N g_N B_0 - \frac{1}{4}A_{zz}$		
$\|-+\rangle$			$-\frac{1}{2}\mu_B g B_0 - \frac{1}{2}\mu_N g_N B_0 - \frac{1}{4}A_{zz}$	$-\frac{1}{4}(A_{xz} - iA_{yz})$
$\|--\rangle$			$-\frac{1}{4}(A_{xz} + iA_{yz})$	$-\frac{1}{2}\mu_B g B_0 + \frac{1}{2}\mu_N g_N B_0 + \frac{1}{4}A_{zz}$

(6.17)

Notice that the diagonal elements contain A_{zz} only, whereas the four off-diagonal terms contain contributions from A_{xz} and A_{yz}. The diagonal elements are called "secular" terms, and the off-diagonal elements are called "pseudosecular" terms [4]. The Hamiltonian has the form of a block diagonal matrix, which can be easily diagonalized to the following eigenvalues:

$$E_{1,2} = \frac{\mu_B g B_0}{2} \pm \frac{1}{2} K(+) \quad \text{(for the upper block, } m_S = +1/2) \qquad (6.18a)$$

$$E_{3,4} = \frac{\mu_B g B_0}{2} \pm \frac{1}{2} K(-) \quad \text{(for the lower block, } m_S = -1/2) \qquad (6.18b)$$

where

$$K(\pm) = \sqrt{\frac{1}{4}\left(A_{xz}^2 + A_{yz}^2\right) + \left(-\mu_N g_N B_0 \pm \frac{1}{2}A_{zz}\right)^2} \qquad (6.18c)$$

The EPR transitions energies are obtained from the selection rules $\Delta m_S = \pm 1$, $\Delta m_I = 0$, which hold in the introduced approximation if the off-diagonal elements in the spin Hamiltonian are small. The transition energies correspond to

$$E_1 - E_3 = \mu_B g B_0 - \frac{1}{2}K(+) + \frac{1}{2}K(-) \tag{6.19a}$$

$$E_2 - E_4 = \mu_B g B_0 + \frac{1}{2}K(+) - \frac{1}{2}K(-) \tag{6.19b}$$

The nuclear magnetic resonance (NMR) or hyperfine transitions result from the selection rules $\Delta m_I = \pm 1$, $\Delta m_S = 0$ and correspond to

$$\begin{aligned} E_1 - E_2 &= K(+) \\ E_3 - E_4 &= K(-) \end{aligned} \tag{6.20}$$

The simplest case to discuss is the so-called high field limit for both electron and nuclei, when the nuclear Zeeman frequency is larger than the hyperfine interaction and the x and y components of the nuclear spin vector can also be neglected in Equation 6.16. In this limit, matrix Equation 6.17 is diagonal and the EPR and hyperfine transition energies reduce to

$$\Delta E_{\text{EPR}} \approx \mu_B g B_0 \pm \frac{1}{2} A_{zz} \tag{6.21a}$$

$$\Delta E_{\text{hyperfine}} \approx \mu_N g_N B_0 \pm \frac{1}{2} A_{zz} \tag{6.21b}$$

This approximation is usually valid for EPR and hyperfine spectra at high frequencies ($\nu \geq 90$ GHz) [5], because the nuclear Zeeman frequency scales linearly with the magnetic field whereas the hyperfine interaction is not field dependent. The energy level diagram of the resulting two-spin system is represented schematically in Fig. 6.9.

The actual shape of the EPR spectrum with a hyperfine powder pattern is determined by the matrix elements A_{zz}, A_{xz}, and A_{yz}. Remember Fig. 6.5, where the laboratory axes were different from the molecular axes (in that case specified as the **g** tensor axes). It is desirable to express these matrix elements of **A** as a function of its tensor principal axes values, which is possible via a coordinate transformation

$$\begin{pmatrix} A_1 & 0 & 0 \\ 0 & A_2 & 0 \\ 0 & 0 & A_3 \end{pmatrix} \longrightarrow \begin{pmatrix} A_{xx} & A_{xy} & A_{xz} \\ A_{yx} & A_{yy} & A_{yz} \\ A_{zx} & A_{zy} & A_{zz} \end{pmatrix} \tag{6.22}$$

where $A_1 > A_2 > A_3$ denotes the principal axes values of the **A** tensor. As shown in Appendix A.6.3, such a transformation is performed by the following rotation: $\mathbf{A}_{\text{Lab}} = R(\theta, \varphi) \cdot \mathbf{A}_{\text{Diag}} \cdot R^{-1}(\theta, \phi)$.

6.3 THE HYPERFINE INTERACTION IN THE SOLID STATE

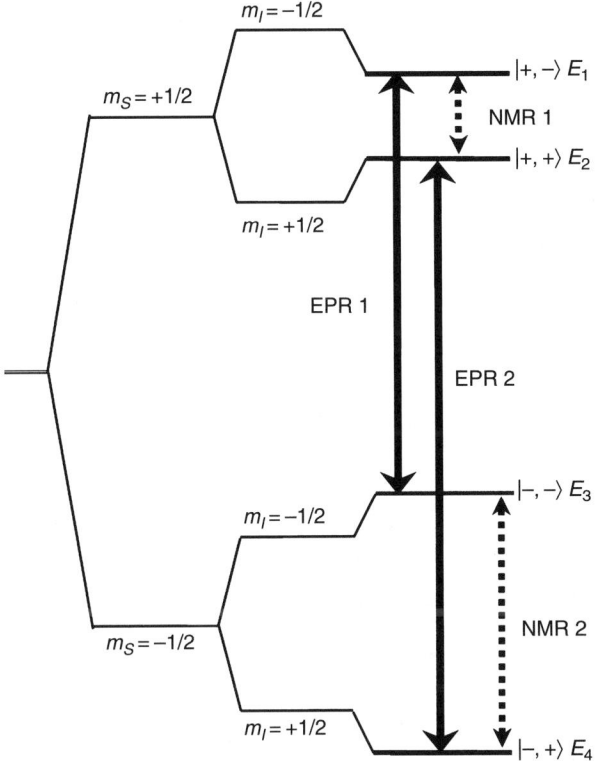

Fig. 6.9 A schematic energy level diagram for an electron $S = 1/2$ coupled to a nucleus $I = 1/2$, according to the Hamiltonian in Equation 6.17.

In Equation 6.21, only the orientation dependence of A_{zz} is required and it is given by [6]

$$A_{zz} = A_1 \sin^2\theta \cdot \cos^2\phi + A_2 \sin^2\theta \cdot \sin^2\phi + A_3 \cos^2\theta \quad (6.23)$$

We note also from Equation 6.21 that in the high field limit the EPR and hyperfine powder patterns for one single nucleus are identical but centered at different frequencies. However, EPR spectra are much less resolved when several nuclei contribute because of the multiplicity of the number of lines (Chapters 1 and 4), whereas equivalent nuclei all give the same hyperfine spectrum. Using Equation 6.21 and the strategy to compute powder patterns (Chapter 7), the absorption line shapes of hyperfine powder patterns for three different cases have been calculated: $a_{iso} \gg T_\|$, $a_{iso} \ll T_\|$, and $a_{iso} = 0$, **T** only. The two transitions could be either EPR or hyperfine and are therefore centered either at ν_e or ν_n, respectively (i.e., in frequency units rather than field units). The resulting spectra for the three cases are illustrated in Fig. 6.10.

In the first case, the transitions are well separated by a_{iso} and broadened by the anisotropic part of the interaction. Each powder pattern has a typical shape for a uniaxial

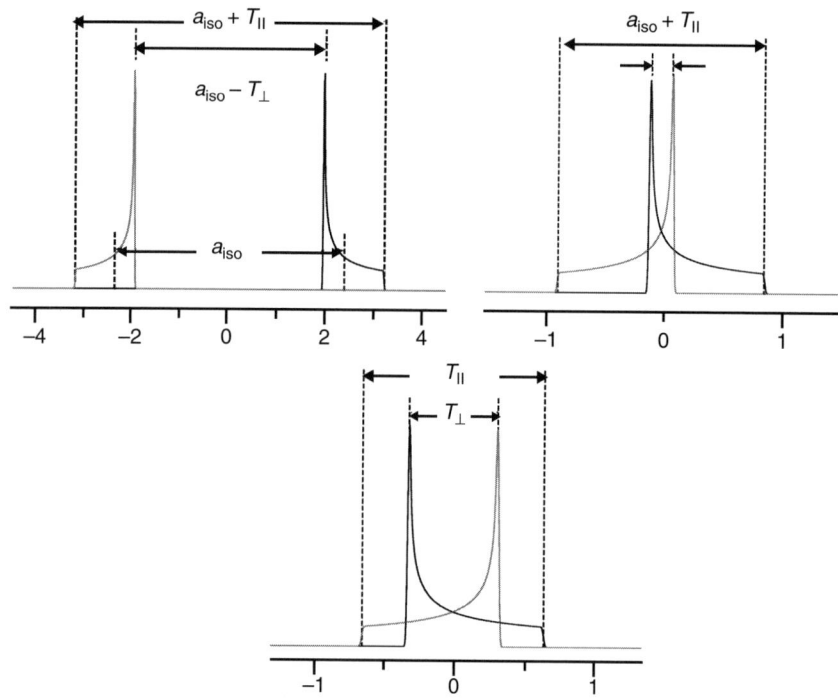

Fig. 6.10 Powder patterns for a hyperfine tensor **A** in three different cases: $a_{iso} > T$; $a_{iso} < T$; and $a_{iso} = 0$, T only. The two transitions represents either the EPR or the hyperfine transitions for an $S = 1/2$ spin coupled to an $I = 1/2$ spin.

tensor (§6.2) with two singularities. The **A** tensor components can be readily extracted from the spectrum. In the second case, an unusual spectral form exists, because the two maxima almost collapse. The hyperfine tensor components might not be determined very precisely because only the spectral edges are resolved. The third case is analogous to an NMR Pake pattern [7], with maxima at $\nu \pm T_\perp/2$ and edges at $\nu \pm T_\parallel/2$.

Note that in all cases the shape of the spectrum does not provide any information about the absolute sign of the hyperfine coupling. To determine the sign of the coupling, more sophisticated experiments such as triple ENDOR or high field ENDOR with variable mixing times must be performed. Furthermore, for a calculation of an EPR and hyperfine spectrum at low EPR frequencies (such as X- or Q-band), one needs to evaluate transition energies and probabilities with higher precision than first order approximation and the resulting spectra might differ slightly from the patterns shown in Fig. 6.10.

6.3.2 Spectra with Anisotropic g and A Interactions

In an EPR powder pattern, the resonance frequencies depend on the orientation of the paramagnetic center with respect to the applied magnetic field. Consider, for

6.3 THE HYPERFINE INTERACTION IN THE SOLID STATE

example, an organic radical with an anisotropic **g** tensor and one anisotropic hyperfine coupling to a nucleus. Both interaction frames (**g** and **A** tensor frames) are fixed to each other by the molecular frame. To compute resonance frequencies and a powder pattern, these interactions need to be expressed in the laboratory frame. However, because the two tensor frames are fixed to each other, one must first define the rotation of the **g** tensor from its principal axes system into the laboratory frame with $R(\theta, \phi)$:

$$g_{\text{diag}} \xrightarrow{R(\theta, \phi)} g_{\text{lab}} \quad \text{or} \quad g_{\text{Lab}} = R(\theta, \phi) \cdot g_{\text{Diag}} \cdot R^{-1}(\theta, \phi) \tag{6.24}$$

Hyperfine tensor **A** in the laboratory frame is then expressed via two consecutive rotations from its principal axis frame ($\mathbf{A}_{\text{diag}}$): first into the **g** tensor principal axes system and second into the laboratory frame:

$$A_{\text{diag}} \xrightarrow{R(\alpha, \beta, \gamma)} A_g \xrightarrow{R(\theta, \phi)} A_{\text{lab}} \tag{6.25}$$

or more explicitly,

$$A_{\text{Lab}} = R(\theta, \phi) \cdot R(\alpha, \beta, \gamma) \cdot A_{\text{Diag}} \cdot R^{-1}(\alpha, \beta, \gamma) \cdot R^{-1}(\theta, \phi) \tag{6.26}$$

The matrix $R(\alpha, \beta, \gamma)$ establishes the mutual orientation of the **g** and **A** tensors. After the whole transformation, the matrix elements of $\mathbf{A}_{\text{lab}}$ become a complex function of all angles $\alpha, \beta, \gamma, \theta$, and ϕ. Their evaluation can be performed analytically using more sophisticated mathematical tools, like spherical tensor theory [6]. However, in spectral simulations they are readily computed numerically.

To illustrate the features of an EPR spectrum with substantial g and A anisotropy, let us consider the EPR transition energies in Equation 6.21. For anisotropic g and A values, Equation 6.21a becomes

$$\Delta E_{\text{EPR}} \approx g_{zz}(\theta, \phi) \cdot \mu_B B_0 \pm \frac{1}{2} A_{zz}(\alpha, \beta, \gamma, \theta, \gamma) \tag{6.27}$$

When the g anisotropy is well-resolved, the singularities in the powder pattern are found at $B\|g_1$ ($\theta = 90°$, $\phi = 0°$), $B\|g_2$ ($\theta = 90°$, $\phi = 90°$), and $B\|g_3$ ($\theta = 0°$) (see Fig. 6.7). For these molecular orientations, it becomes easy to visualize the effect of the hyperfine splitting if the two tensors are collinear. For instance, if $\alpha = \beta = \gamma = 0$, then $A_{zz} = A_1$ at $B\|g_1$ and correspondingly $A_{zz} = A_2$ at $B\|g_2$ and $A_{zz} = A_3$ at $B\|g_3$. This can be further generalized for any principal axis of g, which is collinear to any principal axis of A. For instance, if the principal axis of g_3 is collinear with the principal axis of A_1, one will observe a hyperfine splitting in the magnitude of A_1 at the field position $B\|g_3$. In high field EPR, when the **g** tensor usually dominates the EPR spectrum, hyperfine splittings larger than the intrinsic EPR line broadening become visible at the singularities (turning points) of the g powder pattern and are given by the effective value of A_{zz} at this molecular orientation. At first glance this gives insight into the size of the anisotropy of the hyperfine interaction.

In contrast, the situation is different at low frequencies (X-band) when hyperfine interaction dominates the EPR spectrum. Spectral simulations and a multifrequency approach are the ideal method to disentangle the g and A values. Because the orientation of the **g** and **A** tensors within a paramagnetic center are usually consistent with the molecular symmetry axes of the center itself, the case of coincident axes is encountered quite often in nature and simplifies the interpretation of the spectra.

6.3.3 EPR Powder Spectra of Nitroxide Spin Labels

A first example is an illustration of the case of a nitroxide "static" EPR spectrum. Nitroxide radicals were introduced in Chapters 3 and 4 and are frequently employed as spin probes. The nitroxide radical is characterized by an unpaired electron mostly localized over an N–O moiety and therefore strongly coupled to a nitrogen (^{14}N) $I = 1$ nucleus. Although some variety exists in the chemical structure of spin labels that are commonly used, the solid-state EPR spectra at low and high fields all have a similar appearance.

Reported here are the EPR absorption spectra of the methanethiosulfonate spin label at 9 and 140 GHz (Fig. 6.11). At 9 GHz the spectrum is dominated by the three hyperfine lines and by the anisotropy of the hyperfine interaction. The principal axes values of the hyperfine tensor to the ^{14}N nucleus are $A_1 = 6$ G, $A_2 = 6$ G, and $A_3 = 36$ G with $a_{iso} = 16$ G [8]. The central EPR transition ($m_I = 0$) is not broadened by the anisotropic hyperfine interaction because $m_I = 0$ states are not shifted by hyperfine coupling (Equation 6.14). The two outer EPR transitions connecting electronic states with $m_I = \pm 1$ are orientation dependent as explained in the previous section. This leads to hyperfine powder patterns as illustrated in Fig. 6.10 and results in a total spectral width at 9 GHz of about $2 \cdot A_3 = 72$ G.

By comparison, at 140 GHz the g anisotropy is well resolved with principal axes g values of $g_1 = 2.00848$, $g_2 = 2.00610$, and $g_3 = 2.00217$ [9]. The largest hyperfine splitting is observed at $B \| g_3$, corresponding to the A_3 value of the hyperfine tensor. The two smaller hyperfine components (A_1 and A_2) are not resolved because the intrinsic EPR line broadening at this frequency amounts to 6–8 G. The high field EPR spectra indicate that the principal axes of the largest hyperfine tensor component A_3 is coincident with g_3, and thus all tensor axes are collinear because of the equivalency of A_1 and A_2.

6.3.4 EPR Spectra of Tyrosyl Radicals

An important example of a frequently encountered radical in biology is the tyrosyl radical. Because this amino acid radical will also be treated in other chapters of this book, the focus here will be on the basic features of the EPR and hyperfine spectra. The chemical structure of the radical is illustrated in Fig. 6.12: the spin density on the aromatic ring alternates [10, 11], with large values at C1, C3, and C5, and is

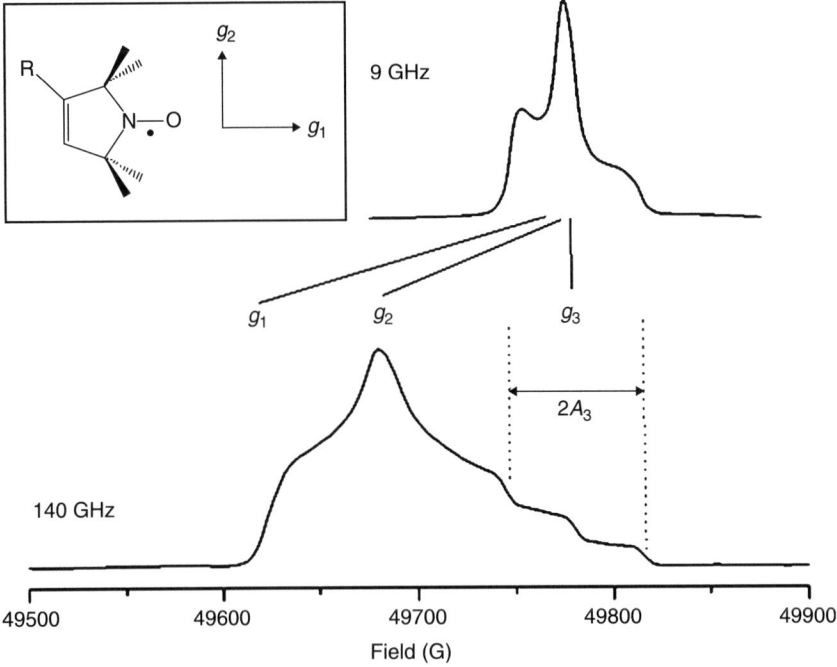

Fig. 6.11 The 9- and 140-GHz EPR absorption spectra of the methanethiosulfonate spin label in frozen solution. For a comparison of the spectral width, spectra are superimposed in a way that the field values corresponding to $g \approx 2.0022$ overlay. (Inset) The chemical structure of the methanethiosulfonate spin label and orientation of the **g** tensor principal axes within the molecule. Adapted with permission from Reference 9. Copyright 1999 Elsevier.

partially delocalized on the oxygen bound to C4. This results in hyperfine couplings to four ring protons and to two side-chain protons (β-methylene protons). The couplings are large ($A \geq 5$ G) for the protons at C3 and C5 and for one β-methylene proton and rather small for the protons at C2, C6, and the second β-methylene proton.

Representative 9- and 180-GHz EPR spectra of the tyrosyl radical in *Escherichia coli* ribonucleotide reductase are provided in Fig. 6.12. The 9-GHz EPR spectrum is dominated by the largest hyperfine splitting to the β-methylene proton. This interaction is almost isotropic, because of hyperconjugation (Chapter 4), and leads to a resolved "two-line" spectrum. However, 9-GHz spectra of various tyrosyl radicals in proteins differ in their appearance because this coupling strongly varies as a function of the side-chain orientation in the protein [10, 11]. In contrast, the high field EPR spectrum is dominated by the **g** tensor anisotropy and displays three well-resolved g values at $g_1 = 2.00912$, $g_2 = 2.00457$, and $g_3 = 2.00225$ [12]. In addition, one large splitting is visible at $B \| g_2$ and a quintet structure attributable to the protons at C3 and C5 appears at $B \| g_1$ and $B \| g_3$. This anisotropy in the hyperfine splittings at high fields indicates that the hyperfine tensors of the protons at C3 and C5 have two large values at orientations close to $B \| g_1$ and $B \| g_3$ and a small unresolved

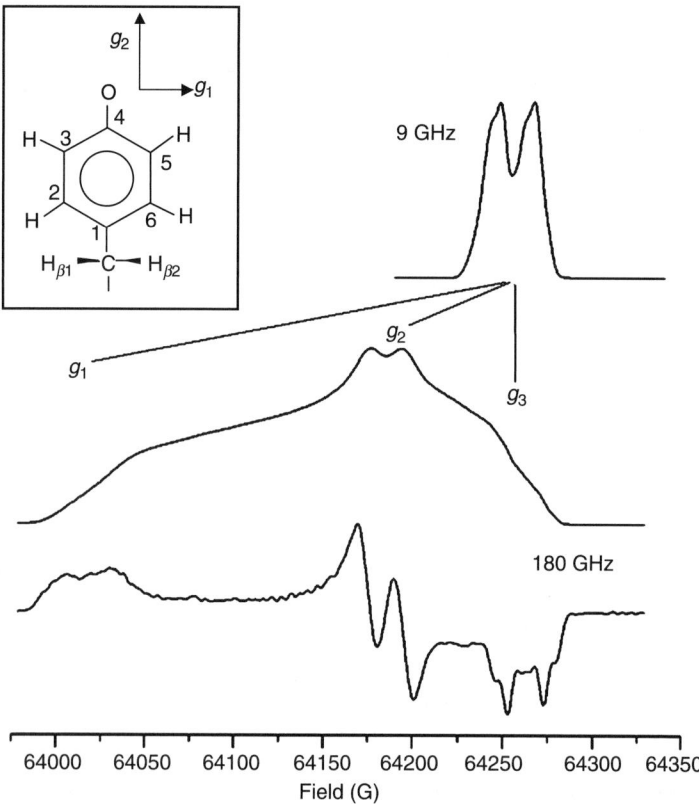

Fig. 6.12 The 9- and 180-GHz EPR spectra of the tyrosyl radical in *Escherichia coli* ribonucleotide reductase in frozen solution. For a comparison of the spectral width, spectra are superimposed in a way that the field values corresponding to $g \approx 2.0002$ overlay. (Inset) The chemical structure of the tyrosyl radical showing the numbering of the carbon atoms and the orientation of the **g** tensor principal axes within the molecule.

value at $B \| g_2$, revealing a tensor with structure $A_1 > A_2 < A_3$. The mutual orientation of the **g** and **A** tensors for the protons at C3 and C5 was determined by simulations of the high field EPR spectra [12]. Note also that tyrosyl radicals have been the subject of numerous high field EPR [13] and quantum chemical investigations [14] since it was established that the shift of the g_1 value is very sensitive to the electrostatic environment and, in particular, is indicative of hydrogen bonding to the protein.

The small hyperfine couplings arising from the protons at C2, C6, and the second β-methylene proton are not resolved in any EPR spectra because they are smaller than the intrinsic EPR linewidth. These couplings were entirely resolved with high field ENDOR [9].

6.3.5 EPR Spectra of Dithiazolyl Radicals

A final example in this section concerns a series of main group nitrogen–sulfur based radicals known as dithiazolyl radicals. These radicals constitute one member of a

6.3 THE HYPERFINE INTERACTION IN THE SOLID STATE 219

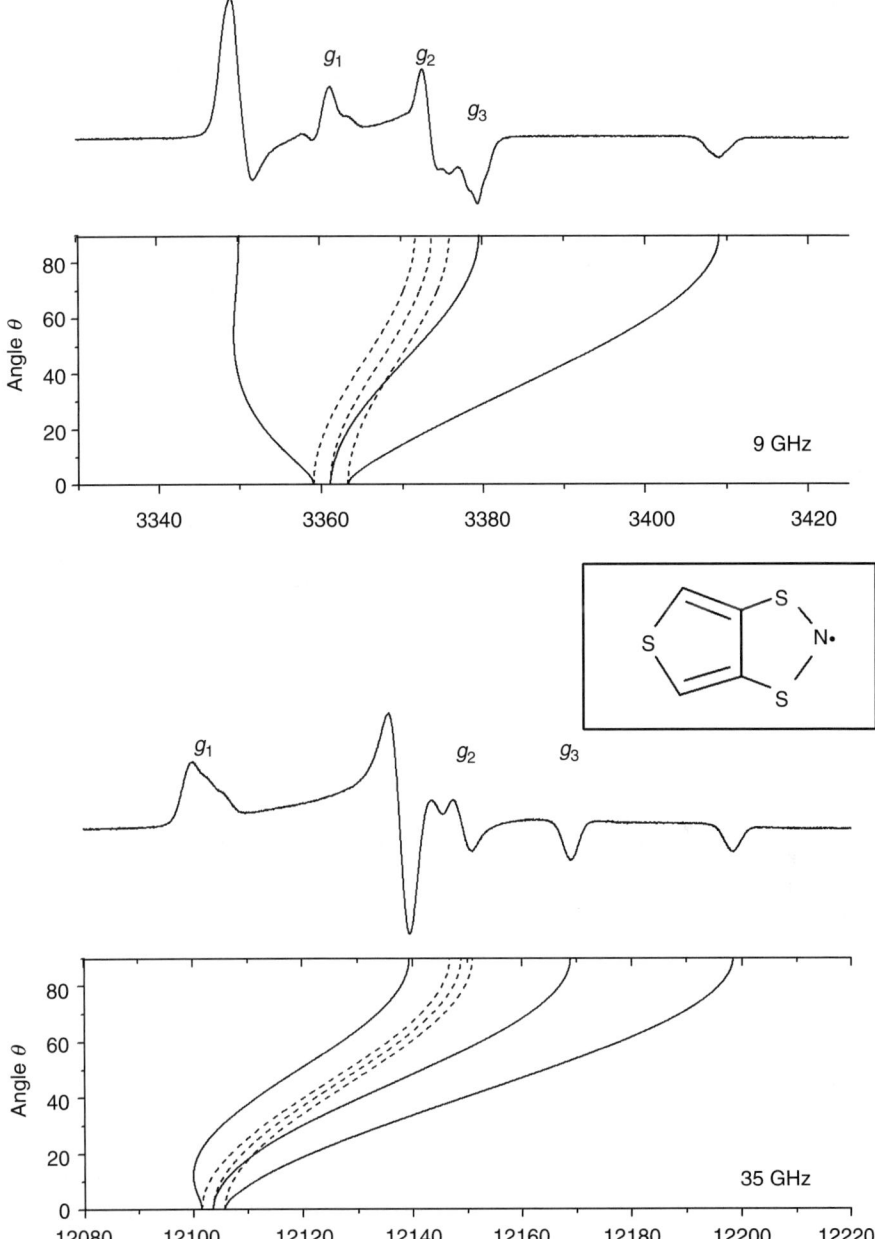

Fig. 6.13 The 9- and 35-GHz EPR spectra of the dithiazolyl radical $C_4H_2S_3N\cdot$ recorded in CD_2Cl_2 at 100K. Distortions to the low field resonance, as seen in the angular dependency plot as a function of θ, are due to off-axis extrema known as "undershoot" features. (Inset) The chemical structure of the dithiazolyl radical. Adapted with permission from Reference 15. Copyright 2005 RSC Publishing.

family of thermally stable thiazyl-based π radicals whose properties have been investigated extensively as building blocks for the construction of both conducting and magnetic materials [15]. Crucial to the optimization of the magnetic properties of these materials is the information on the spin density distribution within the radicals.

The 9- and 35-GHz EPR spectra of the thiophene derivative TDTA· reveal a rhombic **g** tensor with a pronounced anisotropic hyperfine coupling to the dithiazolyl ring ^{14}N but no visible hyperfine to the two chemically equivalent thiophenic H atoms (although these protons are resolved in the ENDOR spectrum) [15]. The spectra can be simulated with the g and A values $g_1 = 2.0126$, $g_2 = 2.0051$, $g_3 = 2.0017$, $^NA_1 = 2.1$, $^NA_2 = 2.1$, and $^NA_3 = 29.6$ G. It is common that the principal ^{14}N tensor axes are collinear with the molecular symmetry axes in these dithiazolyl radicals with one large ^{14}N value and two small values [16].

This example illustrates the importance of simulations when interpreting EPR spectra. In many systems, the features observed in the EPR powder pattern are closely associated with the principal directions of the **g** tensor, because these directions usually give rise to extremes in the magnetic field of a particular resonance (i.e., the turning points discussed in §6.2.2). However, in some cases, additional features referred to as "overshoot" or "undershoot" lines may also be observed, arising from the angular anomalies associated with the field dependencies of the **g** and **A** tensors as a function of orientation [17]. Such angular anomalies do not correspond to resonances from principal directions and, as stated by Mabbs and Collison [18], should not be interpreted as such. Having their origin at angles in between the principal axes, they are also known as "off-axis turning points." It is noteworthy that in Fig. 6.13 the lowest field features (in both the X- and Q-band spectra) are off-axis features, or undershoot resonances, in this case arising from the highly anisotropic **A** tensor because $^NA_3 \gg {^NA_2}$ and $^NA_1$. Although most commonly observed in TMIs, as this example illustrates, these additional spectral features can also occur in organic radicals, so care must be exercised during interpretation (Box 6.1).

BOX 6.1 ANGULAR ANOMALIES (OVERSHOOTS)

For the copper powder EPR pattern discussed in Section 6.4.1, $2I + 1 = 4$ overlapping lines occur in the parallel and perpendicular directions. Occasionally additional lines can appear arising from angular anomalies. They occur because, to first order, the hyperfine spacings remain equal at any orientation: the relative anisotropy of **g** and **A** in any plane determines whether additional turning directions occur for orientations away from the principal or canonical directions. They are usually found in systems with large anisotropy in the principal g values combined with substantial hyperfine splittings.

6.3 THE HYPERFINE INTERACTION IN THE SOLID STATE

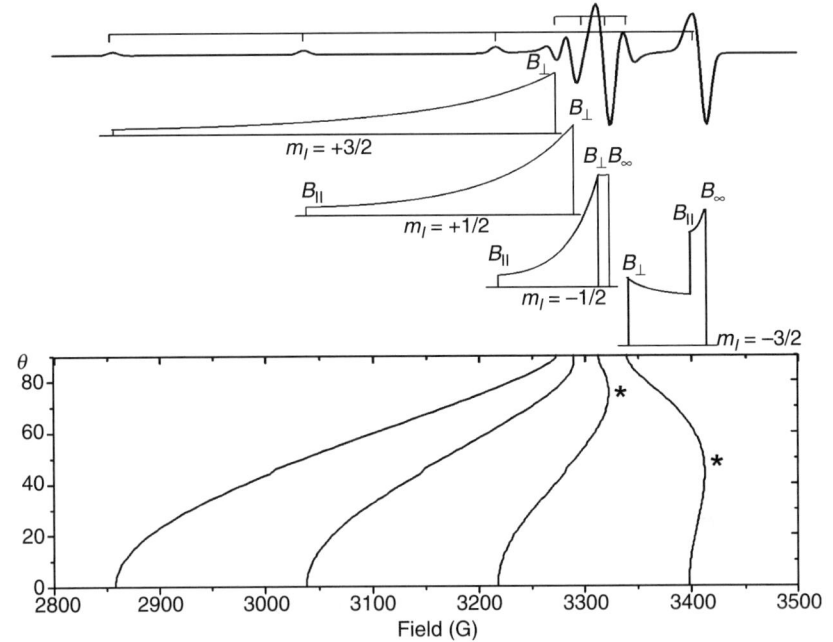

Fig. B.6.1 A simulated X-band spectrum of a Cu^{2+} ion in frozen solution. The absorption profile for the individual m_I lines is plotted assuming infinitely sharp linewidths. The angular dependency profile, highlighting the overshoot features (∗), is also shown.

In the Cu^{2+} example (Section 6.4.1) with axial symmetry (D_{4h} with $Z \neq Y = X$), two resonances are expected for each value of $m_I = \pm 3/2, \pm 1/2$, corresponding to the orientation of **B** along the principal axes (i.e., $B_\parallel$ along the Z direction and $B_\perp$ along the X, Y direction). These turning points are shown in the Fig. B.6.1. However, extra lines also appear that are due to off-axis orientations of **B**, most noticeable for the $m_I = -3/2$ transition (and to a lesser extent, the $m_I = -1/2$ transition). For $m_I = -3/2$, when the field is parallel to the Z direction ($\theta = 0°$), absorption occurs producing the peak at 3398 G ($B_\parallel$); when B is oriented along the X, Y axis ($\theta = 90°$), a peak is observed at 3340 G ($B_\perp$). However, another maximum or turning point also appears in between the Z and X, Y axes, producing the peak at 3413 G (B_∞); this is referred to as an off-axis field orientation. The angular dependency profile of this Cu hyperfine pattern is also shown in Fig. B.6.1, where these "overshoot" features are labeled with an asterisk (∗). They are not confined to TMIs and can be manifested in any spectrum displaying relatively large **g** and **A** anisotropy. In these circumstances, a higher frequency (Q-band, W-band) should be utilized if any ambiguities exist in the interpretation of the low frequency spectra.

6.3.6 Orientation Selection in EPR and ENDOR Spectra

The EPR spectra of amorphous samples or polycrystalline systems usually reflect an average of all molecular orientations with respect to the external field. If the **g** and **A** tensors are known, one can easily associate distinct sets of molecular orientations with a given resonant field value. This was described in Section 6.2.2, where distinct resonant fields ($B_\parallel$ and $B_\perp$) can be associated with specific orientations and g values (Equations 6.11 and 6.12) for a system with no hyperfine interaction. This set of molecular orientations can however also be selected for nuclear resonances by the fixed field settings with hyperfine techniques such as ENDOR, HYSCORE, and ESEEM. The result is a hyperfine spectrum that contains only one part of the powder pattern or at least very few molecular orientations. This selection can substantially simplify hyperfine spectra and help in determining hyperfine coupling parameters.

The concept of orientation selection in ENDOR spectroscopy was first described by Rist and Hyde [19]. Later Hoffman et al. [20] and Hurst et al. [21] provided further theoretical insights into the analysis of polycrystalline ENDOR spectra, and these seminal papers give an excellent overview of orientation selection in general [19–21].

The EPR spectrum is first simulated by searching for all of the molecular orientations that contribute to a resonance position within the field interval. The ENDOR spectra can then be calculated by diagonalization of the spin Hamiltonian given in the **g** tensor principal axes system, as described by Schweiger [22]:

$$\hat{H} = \mu_B B(0, 0, 1) R^T(\phi_i, \theta_i, 0) g \hat{S} - \mu_n g_n B(0, 0, 1) R^T(\phi_i, \theta_i, 0) \hat{I}^1$$
$$+ \hat{S} R^T(\alpha, \beta, \gamma) A^1 R(\alpha, \beta, \gamma) \hat{I}^1 \qquad (6.28)$$

where A^1 is the hyperfine interaction of the interacting nucleus given in the **g** tensor principal axes system. Equation 6.28 is a more generalized form of Equation 6.27, which was only valid in the high field case. The quantity R is the Euler matrix (see Appendix A.6.4), and the Euler angles (α, β, γ) define the transformation of the diagonal **A** tensor into the **g** tensor coordinate frame (as described in §6.3.2). The ENDOR transition frequencies are then calculated from the differences between the obtained eigenvalues within the two electron spin manifolds, and finally a function is applied to determine the relative ENDOR intensities. This procedure is repeated for each set of contributing **g** tensor orientations, and the final powder ENDOR pattern is obtained by adding all of the individual ENDOR spectra.

This general approach can be illustrated for a simple two-spin system ($S = 1/2$, $I = 1/2$) with coaxial **g** and hyperfine tensors [$A_1 = A_2 (A_\perp) = 4$ MHz and $A_3 (A_\parallel) = 8$ MHz] as illustrated in Fig. 6.14. Both **g** and **A** are treated as axial with $A_\parallel$ perpendicular to $g_\parallel$. When $B \parallel g_\parallel$ (i.e., $g_\parallel = 2.437$ in our example), only two resonance peaks are observed, corresponding to $\pm A_\perp/2$ (2 MHz). At other field positions, the simulated ENDOR pattern reveals that the intensity occurs over a frequency range, with singularities occurring at the two frequencies. One remains fixed at $\nu = \pm A_\perp/2$ (line a, Fig. 6.14), and the other moves smoothly from $A_\perp/2$ to $A_\parallel/2$ at $B \parallel g_\perp$ (i.e., $g_\perp = 2.000$; line b, Fig. 6.14. When the components of the hyperfine tensor have opposite sign, as in a pure dipolar interaction, the peaks or singularities will cross

6.3 THE HYPERFINE INTERACTION IN THE SOLID STATE

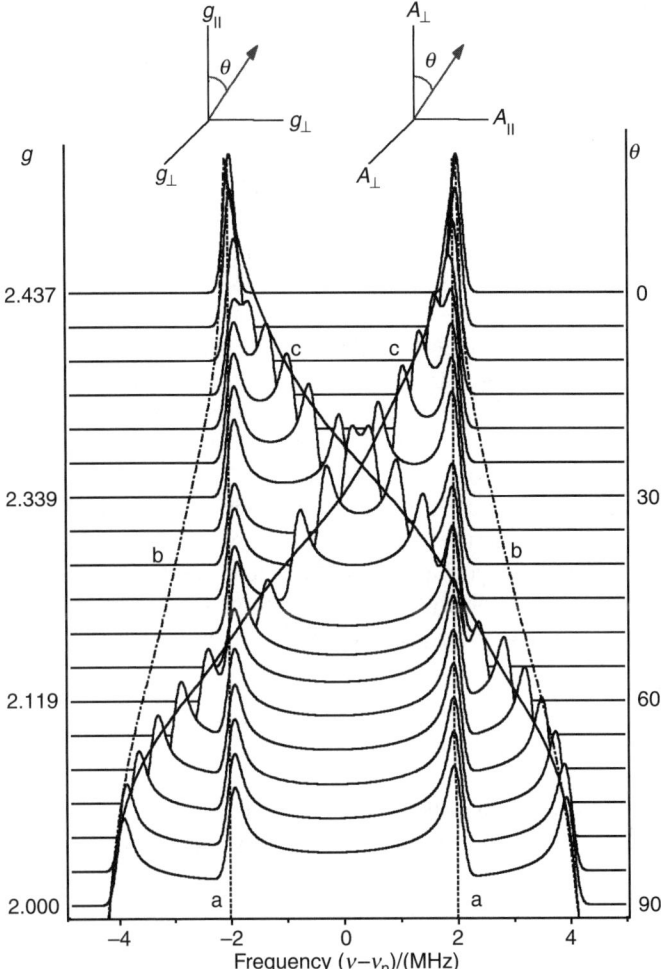

Fig. 6.14 Simulated ENDOR profiles for an axial $S = 1/2$ spin system (based on the EPR spectrum shown in Fig. 6.6; $g_{\|} = 2.437$, $g_{\perp} = 2.000$) interacting with a single $I = 1/2$ nucleus with hyperfine coupling values of $A_1 = A_2$ $(A_{\perp}) = -4$ MHz and A_3 $(A_{\|}) = 8$ MHz. The variation in resonance frequency as a function of θ is illustrated by solid lines a and c. In the case where $A_1 = A_2$ $(A_{\perp}) = 4$ MHz and A_3 $(A_{\|}) = 8$ MHz, the variation in resonance frequency as a function of θ is represented by dashed line b (peaks not shown for this line).

in the spectrum as shown by line c in Fig. 6.14. In theory both principal axes values for the hyperfine tensor can be obtained from a single spectrum at $g_{\perp}$ (in this simple case where **g** and **A** are coaxial and $A_{\|} \perp g_{\|}$) but not the relative sign of the values. In practice, such simple cases are not commonly observed. Only by analyzing the powder type spectra at intermediate fields, in addition to the single crystal-like spectra, can the true values of the hyperfine tensor be confidently assigned.

6.3.7 Examples of Hyperfine Spectra for $I = 1/2$ Spin Nuclei

Examples for simple hyperfine spectra of $I = 1/2$ nuclei are the 94-GHz ENDOR spectra of ^{31}P and ^{13}C detected in the ligand sphere of Mn^{2+} in a protein [23]. At 94 GHz (W-band) the ENDOR spectra of most nuclei are well resolved from each other and ENDOR line shapes often approach the high field limit of Equation 6.21b. The EPR spectrum of Mn^{2+} at 94 GHz is characterized by six sharp isotropic transitions (see also Chapter 4) resulting from the hyperfine splitting of the $m_S \pm 1/2$ electron spin manifolds with $I = 5/2$ of the ^{55}Mn nucleus. Excitation of the ENDOR spectrum on one of the six lines leads to a full hyperfine powder pattern.

The first illustration is of the ENDOR spectra of ^{31}P, arising from the nucleotide guanosine-5′-diphosphate (GDP) bound to Mn^{2+} via an oxygen of the β-phosphate group. Two phosphorus atoms are coupled to the ion. From the structure displayed in Fig. 6.15, two different ^{31}P ENDOR spectra are expected because of the different distances of the ^{31}P to the Mn^{2+} center. The Davies ENDOR spectrum (Fig. 6.15a) represents the coupling to the closest phosphorus nucleus: the spectrum is symmetric around the nuclear Zeeman frequency of ^{31}P and two transitions are observed with the characteristic shapes of the powder patterns due to an axially symmetric tensor (see case i, Fig. 6.10). From this spectrum, the isotropic ($a_{iso} = 4.7$ MHz) and anisotropic part ($T_{\parallel} = -1.6$ MHz) of the hyperfine tensor could be readily obtained. Using $T_{\parallel} = 2T_{\perp}$ and the point–dipole model,

$$T_{\perp} = \frac{g_e \cdot g_n \cdot \mu_B \cdot \mu_N}{h \cdot r^3} \quad (6.29)$$

where r is the dipole–dipole distance, a Mn^{2+}–^{31}P distance of $r = 3.4$ Å was determined and found to be in agreement with the X-ray structure and density functional theory calculations. The coupling to the second nucleus is not visible in the Davies ENDOR because it lies in the wide "spectral hole" of the Davies ENDOR spectrum. Instead, this coupling is easily visible with the Mims ENDOR technique (Chapter 5). Again, a powder pattern similar to the Davies spectrum (Fig. 6.15a) is visible but the singularities at $\nu = \pm 1/2 T_{\perp}$ are more attenuated compared to the edges of the spectrum because of the spectral hole at zero frequency. Note that in the latter case, the hyperfine coupling is entirely anisotropic (dipolar) because of the larger distance to the ion ($r = 4.9$ Å).

Within the same study, the ^{13}C ENDOR spectra of uniformly labeled Ras protein bound to GDP and guanosine-5′-triphosphate (GTP) were also recorded. In the GDP-bound state only one amino acid binds to the Mn^{2+} ion, giving rise to a single strongly coupled ^{13}C atom. In contrast, in the GTP-bound state a conformational change leads to binding of an additional amino acid residue to the paramagnetic center, a situation that can be best detected in the ENDOR spectra. The ^{13}C Mims ENDOR spectra are displayed in Fig. 6.16. The ^{13}C hyperfine is mostly dipolar, leading to a powder pattern line shape similar to case ii in Fig. 6.10. A spectral simulation, which correctly reproduced the intensities, gave the component of the dipolar tensor as $T_{\parallel} = 1.3$ MHz and a small isotropic contribution of $a_{iso} = 0.46$ MHz. The

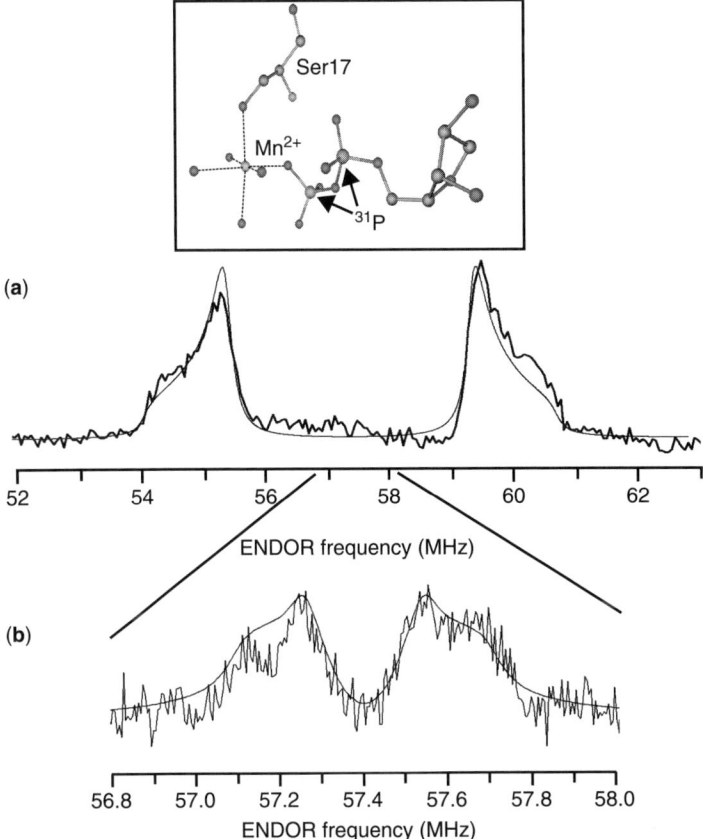

Fig. 6.15 The 94-GHz ^{31}P (a) Davies and (b) Mims ENDOR spectra of Ras(wt) · Mn^{2+} · GDP measured at 4K. (Inset) The structure of the binding site of Mn^{2+} in Ras(wt) · Mn^{2+} · GDP. In addition to the two coordinating residues, the amino acid serine and GDP, the remaining ligands are water molecules. Adapted with permission from Reference 23. Copyright 1982 Springer-Verlag.

simulation also showed that the powder pattern can be interpreted with the coupling to a single ^{13}C nucleus. When GTP binds to the center, an intrinsic change in the ^{13}C spectrum is observed: two more peaks appear within the spectrum and the total spectral width increases (Fig. 6.16d). This new spectrum could be easily deconvoluted into two contributions, which are the same spectrum as observed for GDP and a new one arising from an additional ^{13}C nucleus.

6.3.8 Hyperfine Spectra for $I > 1/2$ Spin Nuclei: The Quadrupole Interaction

If the nuclear spin is $>1/2$, the electric quadrupole moment (Q) of the nucleus interacts strongly with the electric field gradients generated by the surrounding electron

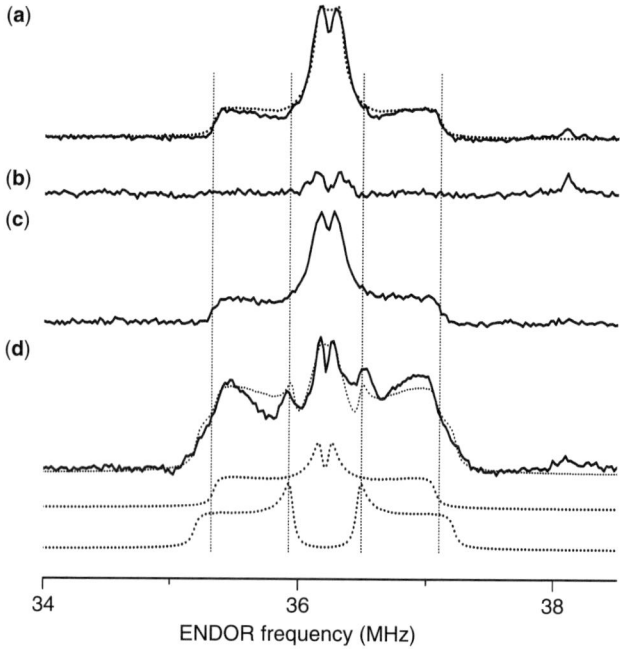

Fig. 6.16 The 94-GHz ^{13}C Mims ENDOR spectra of (a) uniformly ^{13}C-labeled Ras(wt) · Mn^{2+} · GDP, (b) selectively ^{13}C-1,4 Asp57 labeled Ras(wt) · Mn^{2+} · GDP, (c) uniformly ^{13}C-labeled Ras(G12V) · Mn^{2+} · GDP, and (d) uniformly ^{13}C-labeled Ras(wt) · Mn^{2+} · GppNHp · Raf-RBD. Adapted with permission from Reference 23. Copyright 1982 Springer-Verlag.

clouds. The coupling of Q (a property of the nucleus) with an electric field gradient (a property of the sample) is called the quadrupole interaction. This quadrupole interaction, when smaller than the nuclear Zeeman and hyperfine interactions (first order), only shifts the energy levels according to the nuclear states m_I. The quadrupole Hamiltonian ($\widehat{H}_Q$) can be written as [7]

$$\widehat{H}_Q = \mathbf{I} \cdot \mathbf{Q} \cdot \mathbf{I} \tag{6.30}$$

where $\mathbf{I}$ is the nuclear spin vector operator and $\mathbf{Q}$ is the quadrupole tensor that is traceless and symmetric. Within the high field approximation (first order), the spin Hamiltonian in the laboratory frame becomes

$$\widehat{H}_Q \approx 1/2 Q_{zz}\left(3I_z^2 - I(I+1)\right) \tag{6.31}$$

Because the selection rules for EPR transitions are $\Delta m_S = \pm 1$ and $\Delta m_I = 0$, quadrupole effects in first order are usually not observed in EPR spectra but are readily observed in hyperfine spectra such as ENDOR, HYSCORE, or ESEEM ($\Delta m_I = \pm 1$, $\Delta m_S = 0$).

The principal values of the nuclear quadrupole tensor are $Q_{xx} = [-(e^2qQ/4h)](1-\eta)$, $Q_{yy} = [-(e^2qQ/4h)](1+\eta)$, and $Q_{zz} = e^2qQ/2h$, where asymmetry parameter η is equal to $(Q_{xx}-Q_{yy})/Q_{zz}$ with $|Q_{zz}| > |Q_{yy}| > |Q_{xx}|$, and Q is the coupling constant. Because the quadrupole tensor $\mathbf{Q}$ is traceless, it is determined, apart from its orientation, by only two parameters. The literature usually reports the two quantities e^2qQ/h and η.

The quadrupole interaction can substantially affect the line shape of hyperfine spectra, and therefore needs to be considered explicitly in the analysis. To determine Q_{zz} in Equation 6.28 in addition to A_{zz} and g_{zz}, the principal axes of the quadrupole tensor $\mathbf{Q}$ need to be correlated with the principal axes of the $\mathbf{g}$ and $\mathbf{A}$ tensors via another Euler transformation $R(\alpha', \beta', \gamma')$:

$$Q_{\text{diag}} \xrightarrow{R(\alpha', \beta', \gamma')} Q_A \xrightarrow{R(\alpha, \beta, \gamma)} Q_g \xrightarrow{R(\theta, \phi)} Q_{\text{lab}} \qquad (6.32)$$

where Q_{zz} and the matrix elements of $R(\alpha', \beta', \gamma')$ are usually determined by spectral simulations, possibly in combination with orientation selected ENDOR experiments.

As a typical example of quadrupole effects in hyperfine spectra, the ENDOR spectra of a single deuterium nucleus (^{2}H, $I=1$) coupled to an unpaired electron spin $S=1/2$ are shown in Fig. 6.17. The example is reported from a study of the hydrogen bond interaction in a protein containing endogenous tyrosyl radicals, and the deuterium spectrum was obtained by exchanging the protein buffer with ^{2}H$_2$O enriched water [24]. In the particular system under study, the oxygen of the tyrosyl radical binds to one deuterium atom of the deuterated water. The structure of the complex is provided in Fig. 6.17. The ^{2}H Mims ENDOR spectra were recorded across the EPR line at 140 GHz, following the method of orientation selection (§6.3.6).

All spectra are characterized by a doublet of doublets. Indeed, for a spin $I=1$ system, the spin Hamiltonian in Equation 6.31 generates four ENDOR transitions at frequencies

$$v(+) = |-v_n + 1/2A_{zz} \pm 3/2Q_{zz}| \quad \text{and} \quad v(-) = |-v_n - 1/2A_{zz} \pm 3/2Q_{zz}| \quad (6.33)$$

It is evident that the largest splittings observed in Fig. 6.17 are close to $B \| g_1$ and the smallest one at $B \| g_3$. Because the quadrupole tensor is traceless, and in this particular case the hyperfine tensor that is pure dipolar is also traceless, the number of free parameters is reduced and a quantitative analysis of the spectra becomes straightforward. The simulations of the spectra indicated that the hyperfine and quadrupole tensor are collinear $[R(\alpha', \beta', \gamma') = 0]$ and the direction of the hydrogen bond was determined from the Euler matrix $R(\alpha, \beta, \gamma)$ with the angles $\alpha = 30°$, $\beta = 15°$, and $\gamma = 0°$.

Another typical nucleus producing characteristic quadrupole effects in hyperfine spectra is nitrogen (^{14}N, $I=1$), particularly in complexes bearing TMIs. The ^{14}N hyperfine couplings are much larger than the nuclear Zeeman interaction and the spectra are centered at $A/2$ and split by $2v_n$. For example, the ^{14}N ENDOR spectra for a Cu^{2+}–Schiff base complex in Fig. 6.18 were recorded at the field positions

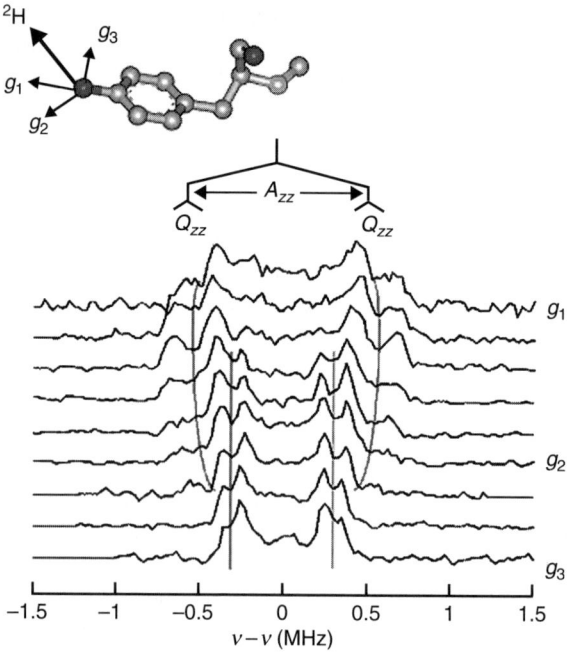

Fig. 6.17 The ^{2}H Mims orientation selected ENDOR spectra of the tyrosyl radical in ribonucleotide reductase from yeast in ^{2}H$_2$O buffer recorded across the EPR line. (Inset) The structure of the radical and direction of the detected H bond obtained from the mutual orientation of **A** and **g** tensors. Adapted with permission from Reference 24. Copyright 2001 American Chemical Society.

of $B\|g_\|$ and $B\|g_\perp$. Both the hyperfine and quadrupole tensors of the two equivalent nitrogen nuclei deviate slightly from axial symmetry with the largest principal axes oriented approximately along the Cu–N bond direction (to within 7° based on single crystal studies) [25]. The hyperfine and quadrupole values were determined as $A_1 = 50.5$, $A_2 = 36.6$, $A_3 = 38.2$, $Q_1 = -1.12$, $Q_2 = 0.64$, and $Q_3 = 0.48$ (all values in megahertz) with $e^2qQ/h = -2.24$ MHz and $\eta = 0.14$, respectively, in line with similar square planar copper complexes. At low frequencies (9 GHz) overlap of ^{1}H signals can complicate the interpretation of the ^{14}N spectrum for strongly coupled Cu–N systems; but this problem is eliminated at higher frequencies (because ν_n is field dependent), which provides the added advantage of improved orientation selection.

It is most common for the quadrupolar term to be very small compared to the hyperfine term, as discussed above, and therefore the quadrupolar interaction can only be observed in hyperfine spectra. However, occasionally quadrupolar effects can be directly observed in EPR spectra when the $I > 1/2$ nucleus combines a small magnetogyric ratio with a relatively large quadrupolar moment. This occurs, for example, with ^{191}Ir and ^{193}Ir for which $I = 3/2$ with a small magnetic moment to quadrupolar moment ratio [26]. Such cases are rare, however, and hyperfine spectra are usually required to observe and extract the **Q** tensor.

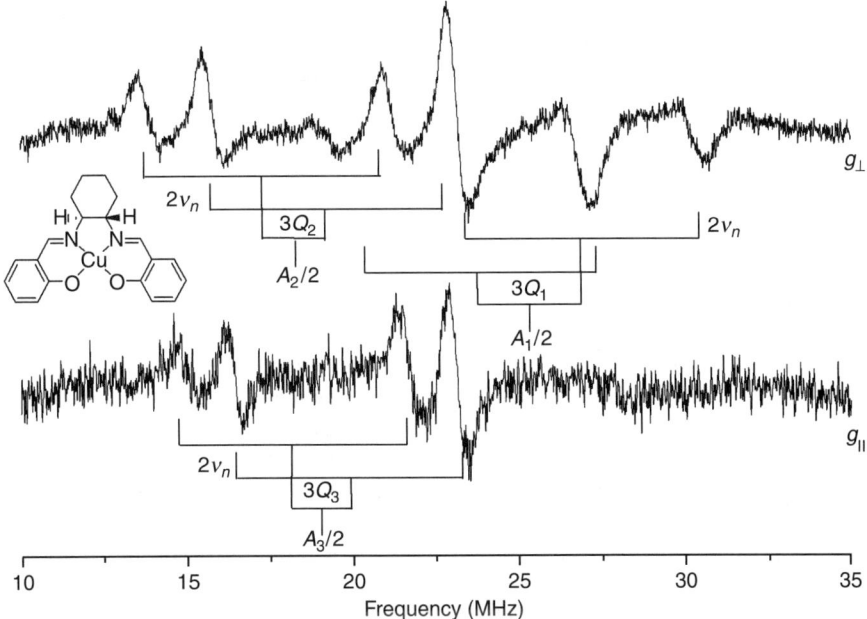

Fig. 6.18 The ^{14}N CW ENDOR spectra of a Cu^{II}–Schiff base complex recorded at parallel and perpendicular field positions. The spectra were recorded in CD$_2$Cl$_2$/Tol-d^8 frozen solution at 35 GHz. For illustration, the stick diagram shows how the values of A and Q are extracted (via simulation) from the experimental spectrum. In CW ENDOR, the peaks appear as a derivative line shape.

6.4 TMIs

To derive the g and A parameters for TMI, good knowledge of crystal field theory and some understanding of group theory are required. However, it is still possible to explain in simple terms the methodology used to derive the g values by reference to effective $S = 1/2$ spin states, such as d^1 or d^9 species, in common stereochemistries. Derivation of the A values can be achieved by a similar approach (not considered here but briefly covered in Appendix A.6.6). Only an outline of the steps involved are given here, but more details of the origin of the g values with reference to group theory can be found in the Appendix A.6.6. Note that the widths of the EPR lines are often larger compared to organic radicals because of the short relaxation times. In many cases, low temperature measurements are frequently required to lengthen the relaxation time sufficiently for the EPR spectrum to be observable.

6.4.1 Origin of the g Values

The g values of TMIs frequently deviate considerably from free spin ($g \neq g_e$) because of orbital angular momentum. Although the electron orbital angular momentum is usually quenched in the ground state, excited states can mix with the ground state such that some orbital angular momentum is reintroduced. This creates a field dependent local magnetic field that will be manifested in the g values, as discussed in Section 6.2.3. For many TMIs the g values can then be estimated using an equation analogous to Equation 6.13.

The most important contributions in Equation 6.13 for the d block elements are (1) the magnitude of the spin–orbit coupling (λ), (2) the separation between the ground state and the excited states ($E_n - E_m$) equivalent to the crystal field splitting term (Δ), (3) the components of the orbital angular momentum operator, and (4) the admixture of excited states with ground states [27]. For ions in which the d shell is less than half full, $\lambda > 0$ and therefore $g < g_e$; for ions in which the d shell is greater than half full, $g > g_e$.

To illustrate how the g values vary as a function of the d orbital splittings and the ground state of the ion, consider a d^9 ion in octahedral symmetry (O_h), such as [Cu(H$_2$O)$_6$]$^{2+}$. In this case, the five d orbitals are filled with the nine electrons and are split into a lower t_{2g} set (d_{xy}, d_{yz}, d_{xz}) and a higher e_g set ($d_{x^2-y^2}$ and d_{z^2}). This orbital splitting situation is not stable and, due to the Jahn–Teller effect, will undergo a tetragonal distortion to remove the degeneracy in the t_{2g} and e_g sets. Elongation of the metal ligand bonds in the z direction (Fig. 6.19a) creates this tetragonal distortion and thus lowers the symmetry from O_h to D_{4h}.

Simple crystal field arguments will predict that the unpaired electron exists in the $d_{x^2-y^2}$ orbital rather than d_{z^2} for this situation (Fig. 6.19a). Because the missing electron can be regarded as a positive "hole," the result of this orbital ground state is that the hole is well separated in energy from all other orbitals. Orbital motion can contribute to the energy differences measured by EPR, only if the hole can move from

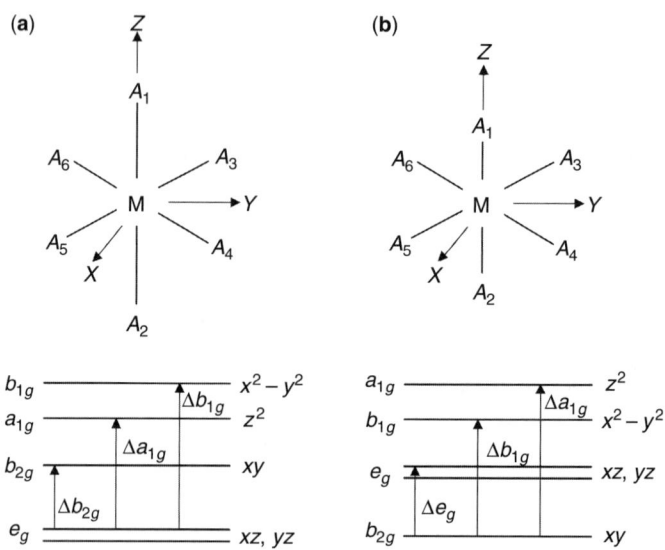

Fig. 6.19 Geometry and d orbital splittings of the antibonding metal orbitals for a d^9 ion in D_{4h} point symmetry undergoing a tetragonal distortion by (a) elongation and (b) compression along the Z axis.

one orbital to another. Because this is not possible, the orbital contribution to the magnetism is quenched and the g values are predicted to be close to g_e.

However, deviations from g_e can arise by spin–orbit coupling causing the unpaired hole to spend some of its time in orbitals other than $d_{x^2-y^2}$, and this effectively reintroduces some orbital motion. The extent to which this occurs depends on the magnitude of λ and Δ and on the symmetry elements of the orbitals involved (see Appendix A.6.6 for more details). In simple terms, when the field lies along the Z axis, $d_{x^2-y^2}$ can mix with d_{xy} and the magnitude of g_{zz} (equivalent to $g_\parallel$) will thus be given by

$$g_{zz}(g_\parallel) \propto g_e + \frac{8\lambda}{E_{xy} - E_{x^2-y^2}} \tag{6.34}$$

When the field lies along the X or Y axis, $d_{x^2-y^2}$ can mix with d_{yz} or d_{xz} and the magnitude of g_{xx}, g_{yy} (equivalent to $g_\perp$) will thus be given by

$$g_{xx}, g_{yy}(g_\perp) \propto g_e + \frac{2\lambda}{E_{yz,xz} - E_{x^2-y^2}} \tag{6.35}$$

Because the values of Δ in Equations 6.34 and 6.35 are different, where $E_{yz,xz} - E_{x^2-y^2} > E_{xy} - E_{x^2-y^2}$, the g values will also be different and this is the reason why $g_{zz} > g_{xx}, g_{yy} > g_e$. The origin of numerical coefficients 8 and 2 in the above equations is explained in Appendix A.6.6.

A typical frozen solution EPR spectrum for this situation is shown in Fig. 6.20a. In simple terms, a hole in $d_{x^2-y^2}$ can acquire angular motion about the Z axis by spending part of its time in the d_{xy} orbital. To induce orbital motion around the axes perpendicular to Z requires "mixing" of $d_{x^2-y^2}$ with d_{yz} or d_{xz}, which is energetically more difficult compared to d_{xy}; this is why the g_{zz} value is different compared to g_{xx} and g_{yy}.

One can also visualize this admixture of orbitals as a consequence of an induced current around the direction of the applied field. For example, in a d^1 system, if the external field is applied along the Z axis, the electron can weakly commute from $d_{x^2-y^2}$ to d_{xy}. The effect of the applied magnetic field is thus to make the electron "rotate" about the Z axis, producing a small current and therefore a small local magnetic field. This small local field will oppose the applied laboratory magnetic field; to achieve the original resonance field at the electron, the field must be increased by a small amount. For this reason, $g_z < g_e$ for d^1 systems. For the d^9 Cu^{2+} system, the opposite effect occurs as the positive hole commutes from $d_{x^2-y^2}$ to d_{xy}, and in this case the field decreases by a small amount (so that $g_{zz} > g_e$).

Next consider the situation of the tetragonally distorted Cu^{2+} ion that experiences compression along the Z axis (Fig. 6.19b). In practice this situation occurs when the ion possesses some unusual copper geometries such as tetragonally compressed octahedron, cis-distorted octahedron, trigonally compressed tetrahedron, or trigonal bipyramidal configurations. For the compressed octahedron case, the d orbital splittings are as shown in Fig. 6.19b, and the unpaired electron now resides in the d_{z^2} orbital (formally $3_{z^2-r^2}$).

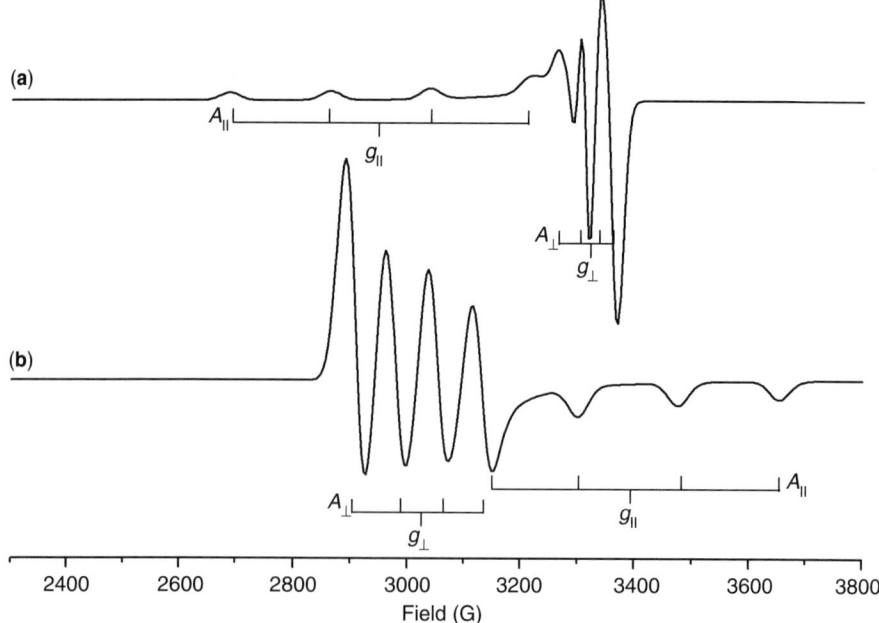

Fig. 6.20 Simulated EPR frozen solution spectra of a Cu^{2+} ion with D_{4h} symmetry in the case of (a) elongation [$g_z = g_\parallel = 2.30$, $g_x(g_y) = g_\perp = 2.05$, $A_\parallel = 175$ G, $A_\perp = 20$ G] shown in Fig. 6.19a and (b) compression [$g_z = g_\parallel = 2.001$, $g_x(g_y) = g_\perp = 2.25$, $A_\parallel = 175$ G, $A_\perp = 75$ G] shown in Fig. 6.19b, along the Z axis.

The predicted g values in this case become

$$g_{zz} = g_e \quad \text{and} \quad g_{xx}, g_{yy}(g_\perp) \propto g_e + \frac{6\lambda}{E_{yz,xz} - E_{z^2}} \tag{6.36}$$

It is important to note here the reversed trend in g values with $g_x, g_y > g_z \approx g_e$. A typical profile of such a spectrum is provided in Fig. 6.20b.

As one moves from a simple isolated TMI (i.e., where the unpaired electron is confined to the metal d orbitals) to a transition-metal complex with strongly coordinated σ- or π-bonded ligands, the unpaired electron will spend more time associated with ligand-based orbitals. With a slightly reduced spin density at the metal center, less orbital angular momentum occurs about the metal ion so the deviation of the g values from the free electron value are usually reduced. To correctly account for this, appropriate molecular orbital coefficients must be included in the above equations in order to derive a more accurate value of g [28].

The d^1 configuration is similar to the d^9 configuration discussed above and, assuming that the unpaired electron is confined to the metal d orbitals, the **g** tensors may be derived using the same expressions as for the d^9 case. In the d^1 case, the appropriate value of λ for the ion in question must be included, and this will have opposite sign compared to the d⁹ case. The derived g values for the d¹ case are therefore

$$g_{zz}(g_{\parallel}) \propto g_e - \frac{8\lambda}{E_{xy} - E_{x^2-y^2}} \tag{6.37}$$

$$g_{xx}, g_{yy}(g_\perp) \propto g_e - \frac{2\lambda}{E_{yz,xz} - E_{x^2-y^2}} \tag{6.38}$$

With the exception of the d_{z^2} ground state, where $g_{zz} = 2.0023$, all principal g values are expected to be less than 2.0023 for any other ground state.

6.4.2 m_I Dependencies of the Linewidths

The fluid solution EPR spectra of TMIs were discussed in Chapter 4 (§4.5). Due to slow tumbling phenomena and second-order effects, the relative intensities and widths of the EPR lines (e.g., for Mn^{2+} and Cu^{2+}) varied considerably. Despite the partial averaging of the **g** and **A** tensors in solution, note that in some cases it is still possible to extract the sign of the isotropic coupling constant from such spectra. The partial averaging of the signals arises from the dependence of the linewidth (ΔB) on the value of m_I (see Chapter 4, Equation 4.5). Because the peak–peak linewidths depend on the value of m_I, the m_I^2 term in Equation 4.5 causes the outer lines to broaden more compared to the inner lines. An asymmetric variation in the intensity across the spectrum arises from the m_I term, and it depends on the relative signs of m_I and of the parameter B. For $B > 0$ the transitions with the largest negative value of m_I will be broadened the least, whereas transitions with the largest positive m_I value will be broadened the most. This correlation is extremely useful because, if the sign of B is known, it provides a means of determining the sign of a_{iso} from the spectrum if the term in m_I dominates that in m_I^2. For example, for $B > 0$, if a_{iso} is positive, then the resonance at lowest field must be due to $m_I = +I$ and that at high field due to $m_I = -I$, as occurs for Cu^{2+} in Fig. 4.17b (Chapter 4), indicating a positive $+a_{iso}$ value.

6.5 EPR SPECTRA FOR $S > 1/2$: ZFS

When two or more unpaired electrons are present in the system of study, and the coupling between them is strong enough that their magnetic moments (and thus their wavefunctions) cannot be considered separately, the magnetic moments will add up to build a total new magnetic moment corresponding to a total spin larger than 1/2. These systems exhibit an additional interaction called ZFS, the origin of which arises from both strong dipole–dipole and spin–orbit coupling between the electron spins and their orbital angular momentum. The two more representative classes of paramagnetic centers, which exhibit such interactions, are organic molecules in triplet or higher multiplicity states (Chapters 1 and 3) and TMIs other than the d¹ and d⁹ ions discussed in the Section 6.4. The fundamental feature of the ZFS compared to all other magnetic interactions treated is that it causes a splitting of the electronic energy states even in the absence of an external magnetic field (therefore the notation ZFS). This electron–electron spin interaction is called also *fine* interaction.

The physical origin of this interaction is mainly attributable to dipole–dipole interactions in the case of organic molecules and to spin–orbit coupling in the case of TMIs. In this latter case, the ZFS can reach values up to hundreds of gigahertz; consequently, the EPR spectra in the solid state can become much broader than spectra dominated by g and hyperfine interaction, and the overall intensity of the spectrum is correspondingly weak. The EPR spectra of $S > 1/2$ spin systems are characterized by more than one EPR transition, because the number n of electronic spin states increases according to $n = 2S + 1$ (see Chapters 3 and 4). Figure 6.21 illustrates the energy levels of a triplet state compared to those of a Mn^{2+} (d^5) ion. The triplet state originates from the interaction of two electrons and has a total spin $S = 1$. The possible spin states are built by the combinations of the spin-up ($+1/2$) and spin-down ($-1/2$) states of the single electrons and have (in the high field limit) quantum numbers $m_S = +1, 0$, and -1. This is at first glance the simplest case for an $S > 1/2$ system. In contrast, the Mn^{2+} ion has a total spin $S = 5/2$, with $2S + 1 = 6$ electron spin states ($m_S = 5/2, 3/2, 1/2, -1/2, -3/2$, and $-5/2$). Moreover, Mn^{2+} ions show a large hyperfine interaction with the ^{55}Mn ($I = 5/2$) nucleus, and the fine and hyperfine interactions give rise therefore to a considerably large number of energy levels ($6 \times 6 = 36$). This situation of several spin energy levels attributable

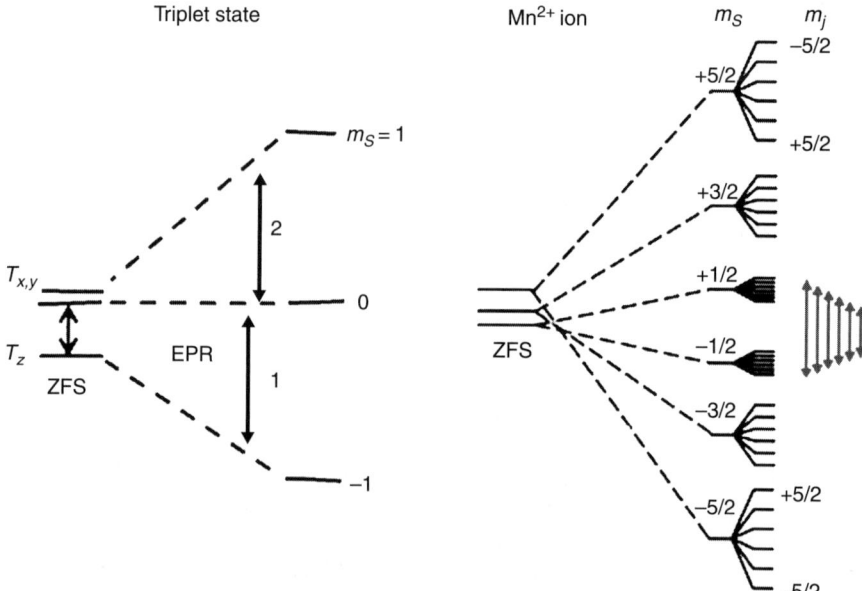

Fig. 6.21 A schematic energy level diagram for a triplet $S = 1$ (left) and a Mn^{2+} ion with $S = 5/2$. The ZFS in the absence of a magnetic field and the subsequent additional Zeeman splitting for an applied external field are also illustrated.

6.5.1 Integer versus Half-Integer Spins

In TMIs, the spin–orbit coupling is mainly responsible for ZFS and it also gives rise to a **g** tensor that is strongly anisotropic and largely shifted from g_e [28]. Thus, their EPR spectra are generally very broad and in most cases difficult to detect. In this respect, it is necessary to distinguish between TMIs with an even or odd number of unpaired electrons. Systems of the latter type must always have a twofold degenerate electron spin state in zero field (Kramer theorem). These $m_S = \pm 1/2$ states are called "Kramer doublets," and the EPR transition from $m_S = -1/2$ to $1/2$ is detectable around the $g = 2$ resonance at any EPR spectrometer frequency. In simple terms, in the case of half-integer spins the ZFS contributes to the separation of the complex line multiplets from the central EPR transition. Metal ions with half-integer spins are also called Kramer ions and are usually amenable to any type of more complex EPR spectroscopy, because of the strong intensity of the central transition. However, one should take into account that for these cases, if the spin–orbit coupling is strong, the relaxation rates can be so fast that the resonances can be detected only at very low temperatures.

A particular case is that of TMIs with $S = 5/2$. This is the case for Mn^{2+} ions, with a symmetric $3d^5$ configuration. For this ion one expects a g factor near to the free electron and no ZFS. In addition, if higher order effects of spin–orbit coupling give rise to a small ZFS, the EPR transition from $m_S = -1/2$ to $1/2$ is easily detectable at any temperature and in any environment. As a consequence, its EPR spectrum is detected and studied in a number of different systems, so that Mn^{2+} can be considered as a true inorganic spin probe (see Chapter 3).

6.5.2 Example of EPR Spectra Dominated by ZFS: The Triplet State

EPR spectra of triplet states are characterized by two EPR transitions according the energy level scheme shown in Fig. 6.21. In the single crystal, these two transitions give rise to a two-line spectrum. For triplet states of photoexcited organic molecules (Chapter 3), the contribution of spin–orbit coupling to ZFS is small (but not negligible in general). As a consequence, the **g** anisotropy is small. Moreover, because of the spin distribution the hyperfine anisotropy is also small; therefore, in general, only the ZFS is resolved in the organic molecules' triplet EPR spectra. Their powder pattern in frozen solutions is determined by the principal values of the ZFS tensor. These values represent the points at which the singularities of the powder pattern occur. As the ZFS tensor is traceless, the transition energies are expressed through two quantities D and E, or zero-field parameters, defined as (see also Chapter 1) $D = -3/2 D_{ZZ}$ and $E = 1/2(D_{YY} - D_{XX})$, where D_{ii} are the principal values of the fine tensor. The D and E values provide a means for the deviation from spherical and axial symmetry, respectively, of the two-electron wavefunction.

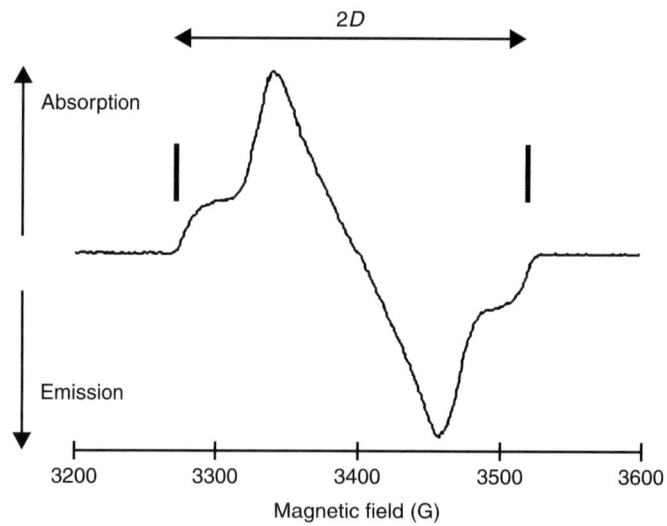

Fig. 6.22 An experimental EPR spectrum of the triplet state of C_{60} in frozen solution. The line shape displays absorptive and emissive features, as explained in the text. The value of the ZFS parameter D can be extracted from the powder pattern as illustrated. Adapted with permission from Reference 29. Copyright 1992 Elsevier.

6.5.2.1 Polarization of Triplet Spectra. One intrinsic feature of photoexcited triplet states is that they are generated after intersystem crossing (ISC) of an excited molecule in the singlet state or by other photophysical paths and detected by time-resolved EPR (described in Chapter 2). Therefore, the populations of the energy levels of the triplet state are due to the competition between population and depopulation rates of the spin levels, so that their resulting populations are quite different compared to the Boltzmann equilibrium. This considerably affects the intensity and appearance of the EPR spectra, which are said to be *spin polarized*. The population differences are much larger than the Boltzmann ones, and the lines are either in *emission* or in *enhanced absorption*. The large population differences allow detection of spin polarized triplet spectra that could never be detected for equilibrium spin populations.

For triplets produced by ISC a simple rule can be kept in mind to understand the appearance of the triplet line shape: in the high field limit, the $m_S = \pm 1$ states are populated equally as required by the symmetry of the spin wavefunctions. Consequently, the two EPR transitions have equal intensity but opposite polarization (i.e., one line is absorptive and the other emissive or vice versa). The powder pattern acquires a symmetric appearance around the Zeeman resonance frequency but with opposite polarization. Deviation from the high field limit leads to a breakdown of this symmetry but to an overall similar line shape. Figure 6.22 displays the triplet powder spectrum of the photoexcited C_{60} molecule. Because of the spherical shape of the molecule, the ZFS interaction is small ($D = 120$ G, $E \approx 0$) [29].

Examples of spin polarized spectra of triplets are given in Chapter 10, Part II, which is devoted to applications of EPR to the study of photosynthesis.

REFERENCES

1. Atherton, N. M. *Principles of Electron Spin Resonance;* Ellis Horwood PTR Prentice-Hall: New York, 1993.
2. Weil, J. A.; Bolton, J. R. *Electron Paramagnetic Resonance: Elementary Theory and Practical Applications*, 2nd ed.; Wiley: New York, 2007.
3. Poole, C. P.; Farach, H. A. *The Theory of Magnetic Resonance;* Wiley-Interscience: New York, 1972.
4. Schweiger, A.; Jeschke, G. *Principles of Pulse Electron Paramagnetic Resonance Spectroscopy;* Oxford University Press: Oxford, U.K., 2001.
5. Bennati, M.; Farrar, C.; Bryant, J.; Inati, S.; Weis, V.; Gerfen, G.; Riggs-Gelasco, P.; Stubbe, J.; Griffin, R. G. *J. Magn. Reson.* **1999**, *138*, 232.
6. Mehring, M. *Principles of High Resolution NMR in Solids;* Springer: Berlin, 1983.
7. Levitt, M. H. *Spin Dynamic: Basics of Nuclear Magnetic Resonance;* Wiley: New York, 2001.
8. Todd, A. P.; Millhauser, G. L. *Biochemistry* **1991**, *30*, 5515.
9. Bennati, M.; Gerfen, G. J.; Martinez, G. V.; Griffin, R. G.; Singel, D. J.; Millhauser, G. L. *J. Magn. Reson.* **1999**, *139*, 281.
10. Bender, C. J.; Sahlin, M.; Babcock, G. T.; Barry, B. A.; Chandrashekar, T. K.; Salowe, S. P.; Stubbe, J.; Lindström, B.; Petersson, L.; Ehrenberg, A.; Sjöberg, B.-M. *J. Am. Chem. Soc.* **1989**, *111*, 8076.
11. Honganson, C. W.; Sahlin, M.; Sjöberg, B.-M.; Babcock, G. T. *J. Am. Chem. Soc.* **1996**, *118*, 4672.
12. Gerfen, G.; Bellew, B. F.; Un, S.; Bollinger, J. M.; Stubbe, J.; Griffin, R. G.; Singel, D. *J. Am. Chem. Soc.* **1993**, *115*, 6420.
13. Bennati, M.; Prisner, T. F. *Rep. Prog. Phys.* **2005**, *68*, 411.
14. Un, S. *Magn. Reson. Chem.* **2005**, *43*, S229.
15. Alberola, A.; Farley, R. D.; Humphrey, S. M.; McManus, G. D.; Murphy, D. M.; Rawson, J. M. *Dalton Trans.* **2005**, 3838.
16. Mattar, S. M. *Chem. Phys. Lett.* **1999**, *300*, 545.
17. Ovchinikov, I. V.; Konstantinov, V. N. *J. Magn. Reson.* **1978**, *32*, 179.
18. Mabbs, F. E.; Collison, D. *Studies in Inorganic Chemistry 16: Electron Paramagnetic Resonance of d-Transition Metal Compounds;* Elsevier: New York, 1992.
19. Rist, G. H.; Hyde, J. S. *J. Chem. Phys.* **1970**, *52*, 4633.
20. (a) Hoffman, B. M.; Martinsen, J.; Venters, J. *J. Magn. Reson.* **1984**, *59*, 110; (b) Hoffman, B. M.; Venters, J.; Martinsen, J. *J. Magn. Reson.* **1985**, *62*, 537.
21. (a) Hurst, G. C.; Henderson, T. A.; Kreilick, R. W. *J. Am. Chem. Soc.* **1985**, *107*, 7294; (b) Henderson, T. A.; Hurst, G. C.; Kreilick, R. W. *J. Am. Chem. Soc.* **1985**, *107*, 7299.
22. Schweiger, A. *Struct. Bond. (Berl.)* **1982**, *51*, 1.
23. Bennati, M.; Hertel, M.; Fritscher, J.; Prisner, T. F.; Weiden, N.; Spörner, M.; Hofweber, R.; Horn, G.; Kalbitzer, H. R. *Biochemistry* **2006**, *45*, 42.
24. Bar, G.; Bennati, M.; Nguyen, H.-H.; Stubbe, J.; Griffin, R. G. *J. Am. Chem. Soc.* **2001**, *123*, 3569.
25. Kitao, S.; Hashimoto, M.; Iwaizumi, M. *Inorg. Chem.* **1979**, *18*, 3.
26. Vugman, N. V.; Caride, A. O.; Danon, J. *J. Chem. Phys.* **1973**, *59*, 4418.

27. Pilbrow, J. R. *Transition Ion Electron Paramagnetic Resonance*; Oxford Science Publications: Oxford, U.K., 1990.
28. Abragam, A.; Bleaney, B. *Electron Paramagnetic Resonance of Transition Metal Ions*; Oxford University Press: London, 1970.
29. Bennati, M.; Grupp, A.; Dinse, P.; Fink, J.; Mehring, M. *Chem. Phys. Lett.* **1992**, *200*, 440.

APPENDIX A.6.1 SIMPLE MATRIX MANIPULATIONS

Various manipulations of matrices are performed throughout this chapter, and some basic rules for these are briefly given here. In mathematics, a matrix is a rectangular table of $n \times m$ elements (usually numbers or any abstract quantities) that can be added, multiplied, or decomposed in various ways. The horizontal lines n refer to the number of rows, and the vertical lines m are the number of columns.

A.6.1.1 Addition of Matrices

Two or more matrices of identical dimensions m and n can be added together by combining the corresponding elements of each matrix. For example,

$$\begin{bmatrix} 1 & 2 & 3 \\ 4 & 5 & 6 \\ 7 & 8 & 9 \end{bmatrix} + \begin{bmatrix} 10 & 11 & 12 \\ 13 & 14 & 15 \\ 16 & 17 & 18 \end{bmatrix} = \begin{bmatrix} \underline{11} & \underline{13} & \underline{15} \\ 17 & 19 & 21 \\ 23 & 25 & 27 \end{bmatrix}$$

In the first row of each matrix, $1 + 10 = \underline{11}$, $2 + 11 = \underline{13}$, $3 + 12 = \underline{15}$, and so forth.

A.6.1.2 Multiplication of Matrices

Multiplication depends on the form of the individual matrices (e.g., whether a row or column matrix are being multiplied). Multiplication of a matrix by a scalar is achieved by multiplying each matrix element by the scalar:

$$2 \cdot \begin{bmatrix} 1 & 2 & -3 \\ -4 & 5 & 6 \\ 7 & 8 & -9 \end{bmatrix} = \begin{bmatrix} 2 & 4 & -6 \\ -8 & 10 & 12 \\ 14 & 16 & -18 \end{bmatrix}$$

A column and row matrix can be multiplied in a similar way, but the result is a scalar quantity (also called scalar product):

$$[1 \quad 2 \quad 3] \cdot \begin{bmatrix} 1 \\ -2 \\ 3 \end{bmatrix} = (1 \cdot 1 = 1) + (2 \cdot -2) + (3 \cdot 3) = 6$$

For example, the Zeeman energy of a spin is a scalar quantity that is computed as

$$E = g \cdot \mu_B \cdot S = g \cdot \mu_B \cdot [B_x \quad B_y \quad B_z] \cdot \begin{bmatrix} S_x \\ S_y \\ S_z \end{bmatrix}$$

$$= g \cdot \mu_B \cdot (B_x S_x + B_y S_y + B_z S_z)$$

where S and B are the vectors representing the spin and the magnetic field and g is in this particular case isotropic.

Multiplication of two matrices is *only* possible if the number of columns of the matrix on the left side is equal to the number of rows of the matrix on the right side. For instance, in the product **C = A B**, each element c_{mn} of **C** is given by a scalar product of the m row of A and the n column of B:

$$\left(\begin{array}{c} m \\ n \end{array} \right) = \left(m \right) \cdot \left(\begin{array}{c} n \\ \end{array} \right)$$

For instance,

$$[1 \quad 2 \quad 3] \cdot \begin{bmatrix} 1 & 2 & 3 \\ 4 & -5 & 6 \\ -7 & 8 & 9 \end{bmatrix} = [-12 \quad 16 \quad 42]$$

From this procedure it becomes clear that a product of matrices is not generally commutative: **A B ≠ B A**. When dealing with coordinate transformations (§A.6.3), it becomes *imperative* to establish the correct sequence of multiplication.

If multiplication of several matrices is required, then multiplication is done sequentially and always under consideration of the fundamental rule, which is that the number of columns and rows have to match (see above). For example, in the multiplication of three matrices **A B C** = product, then you can build **A B = D** and **D C** = product or **B C = D** and then **A D** = product. For instance, for the Hamiltonian in Equation 6.4 of Section 6.2,

$$\widehat{H} = \mu_B \cdot \begin{pmatrix} S_x \\ S_y \\ S_z \end{pmatrix} \cdot \begin{pmatrix} g_1 & 0 & 0 \\ 0 & g_2 & 0 \\ 0 & 0 & g_3 \end{pmatrix} \cdot \begin{pmatrix} B_0^x \\ B_0^y \\ B_0^z \end{pmatrix}$$

the multiplication can be carried out from the *right* to the *left* (or vice versa):

$$\begin{pmatrix} g_1 & 0 & 0 \\ 0 & g_2 & 0 \\ 0 & 0 & g_3 \end{pmatrix} \cdot \begin{pmatrix} B_0^x \\ B_0^y \\ B_0^z \end{pmatrix} = \begin{pmatrix} g_1 B_0^x \\ g_2 B_0^y \\ g_3 B_0^z \end{pmatrix}$$

Further,

$$(S_x \quad S_y \quad S_z) \cdot \begin{pmatrix} g_1 B_0^x \\ g_2 B_0^y \\ g_3 B_0^z \end{pmatrix} = (S_x g_1 B_0^x + S_y g_2 B_0^y + S_z g_3 B_0^z).$$

where the values of B_0^x, B_0^y, B_0^z were written in Equation 6.3 as $B_0^x = \sin\theta \cos\phi$ and so forth, and the final Hamiltonian becomes (written as Equation 6.5 in §6.2)

$$\widehat{H} = \mu_B \cdot (g_1 B_0 \sin\theta \cos\phi \cdot S_x + g_2 B_0 \sin\theta \sin\phi \cdot S_y + g_3 B_0 \cos\theta \cdot S_z)$$

APPENDIX A.6.2 PAULI MATRICES

Quantum mechanical spin operators cannot be generated like orbital angular momentum operators by the operators of the position and momentum, because the spin angular momentum has no classical analogue. Nevertheless, spin operators, as any momentum operator, have to satisfy the Heisenberg commutation rules. Pauli matrices fulfill these requirements in the simplest mathematical form. For a spin $S = 1/2$, these matrices are

$$S_x = \hbar \frac{1}{2}\begin{pmatrix} 0 & 1 \\ 1 & 0 \end{pmatrix} \quad S_y = \hbar \frac{1}{2}\begin{pmatrix} 0 & -i \\ i & 0 \end{pmatrix} \quad S_z = \hbar \frac{1}{2}\begin{pmatrix} 1 & 0 \\ 0 & -1 \end{pmatrix} \quad (A6.1)$$

The eigenvalues of all three matrices are $\pm 1/2\hbar$. Pauli matrices are simple square matrices with complex entries that are equal to their own conjugate transpose (i.e., the element in the ith row and jth column is equal to the complex conjugate of the element in the jth row and ith column, $a_{ij} = a_{ji}^*$ or $A = A^T$). The dimension of these matrices can be easily expanded to represent spin operators with $S > 1/2$. This method is described in detail in Reference 3 in the text.

APPENDIX A.6.3 TRANSFORMATION OF TENSOR COORDINATES VIA MATRICES

An anisotropic physical quantity can be described by a tensor using a 3×3 matrix (introduced in Chapter 1 and §6.2 and §6.3.1). Each tensor defines a set of spatial coordinates (known as the *principal axes* system) along which this physical quantity possesses three *principal values*. When the tensor is expressed in its principal axes system, then its matrix representation is diagonal with elements given by the principal axes values. However, in other frames, for instance, the laboratory frame, the tensor matrix contains off-diagonal elements.

In EPR spectroscopy the diagonal (or principal values) form of a tensor matrix is mostly used when reporting anisotropic coupling constants like the A or g values. Furthermore, the orientation of the principal axes system with respect to the magnetic field is specified via the polar angles θ/ϕ or the direction cosines (which is easier to handle). However, for manipulation purposes we must be able to rotate a tensor matrix into any chosen reference frame.

The mathematical operation that permits the transformation of an orthogonal Cartesian coordinate system into another one with the same origin is given by a

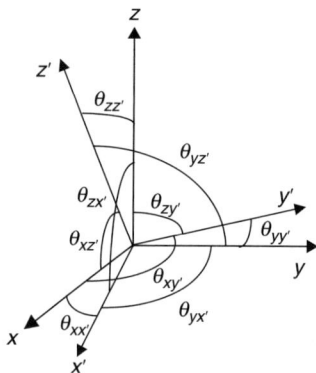

Fig. A.6.1 Direction cosines $\cos \theta_{i,j}$ between the primary coordinate system (x, y, z) and the new coordinate system (x', y', z').

rotation using a 3×3 orthogonal matrix. The most general form of this matrix is found when considering the two coordinate systems illustrated in Fig. A.6.1.

The direction of each basis vector in the new rotated system is related to the primary system by the three direction cosines between the basis vectors, $\cos \theta_{x',x} = \vec{x}' \cdot \vec{x}$, and so forth, where the expression on the right-hand side represents a scalar product. The full rotation matrix has the form

$$R = \begin{pmatrix} \cos \theta_{x',x} & \cos \theta_{x',y} & \cos \theta_{x',z} \\ \cos \theta_{y',x} & \cos \theta_{y',y} & \cos \theta_{y',z} \\ \cos \theta_{z',x} & \cos \theta_{z',y} & \cos \theta_{z',z} \end{pmatrix} \qquad (A6.2)$$

This matrix rotates any vector from the original system into the new system as follows: $\vec{a}' = R \cdot \vec{a}$ and vice versa $\vec{a} = R^T \cdot \vec{a}'$ for the backrotation with the transposed matrix R^T. Such a rotation matrix is orthogonal with $R^T = R^{-1}$. In a similar manner we can rotate a 3×3 tensor as $A' = R \cdot A \cdot R^T$ and backwards as $A = R \cdot A' \cdot R^T$. Another fundamental property is that consecutive rotations like $A \xrightarrow{R} A' \xrightarrow{R'} A''$ are simply generated as $A'' = R' \cdot R \cdot A \cdot R^T \cdot R'^T$.

A simple case will be described here to illustrate how these transformations occur.

A.6.3.1 Example: Rotation of Axially Symmetric Tensors

Consider a simple case of a paramagnetic species with an axially symmetric **g** tensor and a dipolar interaction with a remote nucleus. This could represent a TMI, interacting with the proton of a coordinated ligand. The coordinate axis system for this case is shown in Fig. A.6.2. The metal ion will have its principal axes **g** tensor frame (g_x, g_y, g_z), and the hyperfine interaction to the remote proton will also have its own **A** tensor frame (A_x, A_y, A_z).

Before constructing the spin Hamiltonian, it is necessary that either the **A** tensor be transformed into the principal g axes system or vice versa. Normally the **g** tensor is

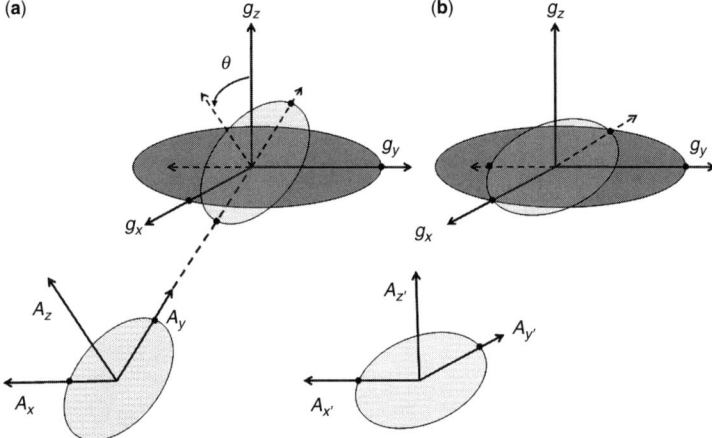

Fig. A.6.2 The general orientation of the principal axes systems for the **g** tensor and hyperfine **A** tensor. If both tensors are axially symmetric ($X = Y$), then one rotation through θ is sufficient to align both frames: (a) before and (b) after rotation.

the reference frame. In this case, because both tensors are axially symmetric (the X and Y directions are equivalent), it is necessary to only rotate the **A** frame through one angle θ to align both z axes.

$$R = \begin{bmatrix} 1 & 0 & 0 \\ 0 & \cos\theta & \sin\theta \\ 0 & -\sin\theta & \cos\theta \end{bmatrix} \quad (A6.3)$$

The transpose of a matrix is obtained by replacing the nth column with the nth row so that a_{ij} becomes a_{ji}. In other words,

$$R^T = \begin{bmatrix} 1 & 0 & 0 \\ 0 & \cos\theta & -\sin\theta \\ 0 & \sin\theta & \cos\theta \end{bmatrix} \quad (A6.4)$$

This rotation step is illustrated in Fig. A.6.2. In this case, one of the **g** and **A** tensor axes was parallel (i.e., g_y and A_x).

APPENDIX A.6.4 EULER ANGLES

Every orthogonal coordinate transformation can be decomposed into three transformations around a principal axes direction [A.1]:

$$R(\alpha, \beta, \gamma) = R_z(\alpha) \cdot R_y(\beta) \cdot R_z(\gamma) \quad (A.6.5)$$

Each transformation on the right-hand side of Equation A.6.5 is characterized by an angle called the Euler angle. The lower index indicates the principal axis around which the rotation is performed. Note that some text books prefer to perform the second rotation about the x axis rather than the y axis and attention must be paid by comparing Euler matrices. The rotations are carried out counterclockwise. The explicit matrix elements of $R(\alpha, \beta, \gamma)$ are [A.1]

$$R(\alpha, \beta, \gamma) = \begin{pmatrix} \cos\beta\cos\alpha\cos\gamma - \sin\alpha\sin\gamma & \cos\beta\sin\alpha\cos\gamma + \cos\alpha\sin\gamma & -\sin\beta\cos\gamma \\ -\cos\beta\cos\alpha\sin\gamma - \sin\alpha\cos\alpha & -\cos\beta\sin\alpha\sin\gamma + \cos\alpha\cos\gamma & \sin\beta\sin\gamma \\ \cos\alpha\sin\beta & \sin\alpha\sin\beta & \cos\beta \end{pmatrix}$$

(A6.6)

Unfortunately, Euler angles cannot easily represent geometrical or structural parameters as direction cosines do. However, their introduction is necessary to correctly reduce the number of free parameters. In the computation of EPR spectra, Euler angles are usually required as input parameters. As mentioned above, if the direction of the considered interaction tensors is known within the molecular frame, the Euler angles can be calculated from the direction cosine matrix using $R(\cos\theta_{i,j}) = R(\alpha, \beta, \gamma)$. Further, the transformation of any interaction tensor into the laboratory frame, usually denoted as $R(\theta, \phi)$, is also an Euler rotation and only two angles are required because no distinction is made in the laboratory frame between axes x and y. Note that within this nomenclature, θ denotes the Euler rotation around the y axis, which is $R_y(\theta)$, and corresponds to angle β in the above formulas.

A.6.4.1 Example: Low Symmetry Case

In low symmetry cases, we transform the three **A** tensor axes into the three **g** tensor axes by the Euler angles. To do this transformation, we construct a rotation matrix R from three Euler rotations as given in Equation A.6.5. Broken down into individual steps, the three individual rotations that are occurring are systematically shown in Fig. A.6.3 [A.2]:

1. The first rotation (from Fig. A.6.3a to b) occurs *anticlockwise* about the Z axis through an angle α, giving rise to the new axes system (A_{x1}, A_{y1}, A_z) (see Fig. A.6.3b):

$$R_\alpha = \begin{bmatrix} \cos\alpha & \sin\alpha & 0 \\ -\sin\alpha & \cos\alpha & 0 \\ 0 & 0 & 1 \end{bmatrix}$$ (A6.7a)

2. The second rotation (from Fig. A.6.3b to c) occurs *anticlockwise* through an angle β about the NEW x_1 axis, giving rise to the new axes system

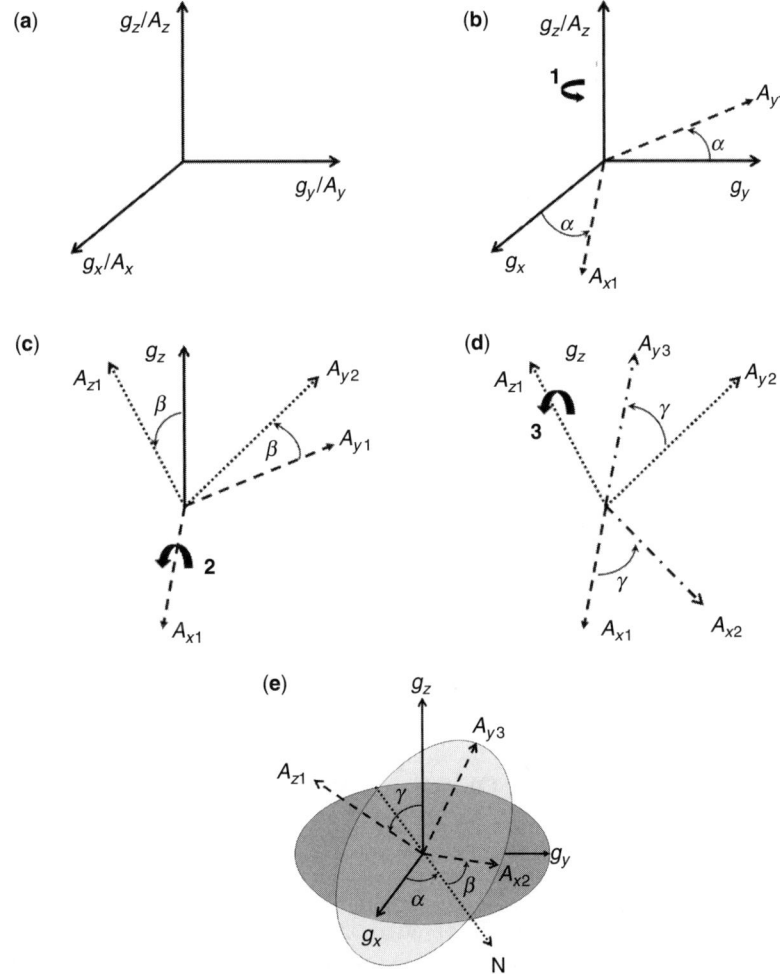

Fig. A.6.3 An illustration of the sequential transformation steps involved during the rotation of the **g** and **A** tensors from a position of (a) coincidence to (d) complete noncoincidence.

(A_{x1}, A_{y2}, A_{z1}) (see Fig. A.6.3c):

$$R_\beta = \begin{bmatrix} 1 & 0 & 0 \\ 0 & \cos\beta & \sin\beta \\ 0 & -\sin\beta & \cos\beta \end{bmatrix} \quad (A6.7b)$$

3. Finally, the third rotation (from Fig. A.6.3c to d) occurs *anticlockwise* through an angle γ about the NEW z_1 axis, giving rise to the new axes system (A_{x2}, A_{y3}, A_{z1})

(see Fig. A.6.3d):

$$R_\gamma = \begin{bmatrix} \cos\gamma & \sin\gamma & 0 \\ -\sin\gamma & \cos\gamma & 0 \\ 0 & 0 & 1 \end{bmatrix} \quad (A6.7c)$$

The final position of the **A** tensor axes after rotation, from the **g** tensor axes, is shown in Fig. A.6.3e for clarity. The intersection of the **A** tensor and the **g** tensor coordinate planes is represented by the *line of nodes* (N) in this figure.

APPENDIX A.6.5 MATRIX ELEMENTS OF SPIN–ORBIT COUPLING

The modes of angular momentum operation on the real forms of the p and d orbitals are given in the Table A.6.1. The table can be read by looking at the operator at the top of a column operating on the function at the left of a row, producing the result entered at the intersection of the row and column. For example,

$$\langle z | l_x | y \rangle = i$$

$$\langle z^2 | l_x | yz \rangle = i\sqrt{3}$$

$$\langle z^2 | l_y | xz \rangle = -i\sqrt{3}$$

APPENDIX A.6.6 ORIGIN OF THE g AND A VALUES FOR SIMPLE TMIs

A.6.6.1 The g Values

The symmetry elements of the TMI affect the g values. To explain this, one requires some basic understanding of group theory. In simple terms, if the applied

TABLE A.6.1 Modes of Angular Momentum Operation on Real Forms of p and d Orbitals

	l_x	l_y	l_z
$\lvert p_x \rangle \ldots \lvert x \rangle$	0	$-i\lvert z\rangle$	$i\lvert y\rangle$
$\lvert p_y \rangle \ldots \lvert y \rangle$	$i\lvert z\rangle$	0	$-i\lvert x\rangle$
$\lvert p_z \rangle \ldots \lvert z \rangle$	$-i\lvert y\rangle$	$i\lvert x\rangle$	0
$\lvert d_{x^2-y^2} \rangle \ldots \lvert x^2 - y^2 \rangle$	$-i\lvert yz\rangle$	$-i\lvert xz\rangle$	$2i\lvert xy\rangle$
$\lvert d_{xy} \rangle \ldots \lvert xy \rangle$	$-i\lvert xz\rangle$	$-i\lvert yz\rangle$	$-2i\lvert x^2 - y^2\rangle$
$\lvert d_{yz} \rangle \ldots \lvert yz \rangle$	$i\lvert x^2 - y^2\rangle + i\sqrt{3}\lvert z^2\rangle$	$i\lvert xy\rangle$	$-i\lvert xz\rangle$
$\lvert d_{xz} \rangle \ldots \lvert xz \rangle$	$-i\lvert xy\rangle$	$i\lvert x^2 - y^2\rangle - i\sqrt{3}\lvert z^2\rangle$	$i\lvert yz\rangle$
$\lvert d_{z^2} \rangle \ldots \lvert z^2 \rangle$	$-i\sqrt{3}\lvert yz\rangle$	$i\sqrt{3}\lvert xz\rangle$	0

APPENDIX A.6.6 ORIGIN OF THE g AND A VALUES FOR SIMPLE TMIs

magnetic field lies along the x, y, or z axis, this is equivalent to a rotation about that axis. In group theory language, this has the symmetry representation of R_x, R_y, and R_z. For D_{4h} symmetry (as discussed in §6.4, Fig. 6.19a), the $d_{x^2-y^2}$ orbital transforms as B_{1g} (see the point group character tables for D_{4h} in any general chemistry textbook) and when the field is parallel to the Z axis, this will transform as A_{2g}. The coupling of B_{1g} and A_{2g} gives B_{2g} symmetry according to the character tables that is analogous to d_{xy}. In other words, when the field lies along the Z axis, $d_{x^2-y^2}$ can mix with d_{xy} and the magnitude of g_{zz} (equivalent to $g_\parallel$) will thus be given by Equation A.6.8:

$$g_{zz}(g_\parallel) \propto g_e + \frac{8\lambda}{E_{xy} - E_{x^2-y^2}} \qquad (A6.8)$$

When the field lies in the X or Y plane, the orbitals transform as E_g. The coupling of B_{1g} and E_g gives E_g symmetry that is analogous to the degenerate d_{yz}, d_{xz}. In other words, when the field lies along the X or Y axis, $d_{x^2-y^2}$ can mix with d_{yz} or d_{xz} and the magnitude of g_{xx}, g_{yy} (equivalent to $g_\perp$) will thus be given by

$$g_{xx}, g_{zz}(g_\perp) \propto g_e + \frac{2\lambda}{E_{yz,xz} - E_{x^2-y^2}} \qquad (A6.9)$$

Note that the functions obtained when the operators l_x, l_y, and l_z are applied to the various d orbitals were given in Table A.6.1, and the same results as described above are obtained for the mixing of the orbitals. From this table we also find that the B_{1g} state (in this case $|d_{x^2-y^2}\rangle$) is connected to the B_{2g} ($|d_{xy}\rangle$) and E_g states ($|d_{xz}\rangle$, $|d_{xz}\rangle$) by the spin–orbit operators.

In the second case considered in Section 6.5.1 (D_{4h} symmetry with compression along the Z axis; Fig. 6.19b), it can be shown by simple crystal field arguments that the unpaired electron will now reside in the d_{z^2} orbital. In group theory the d_{z^2} orbital transforms as A_{1g}; when the applied magnetic field is parallel to the z axis, this state does not mix with any other ground state by spin–orbit interaction. In other words, in the D_{4h} character tables $A_{1g} \cdot A_{2g} = A_{2g}$, so that $g_z = g_e$. When the field is parallel to the x and y axes, the A_{1g} state couples with the E_g state, and this results in an E_g state (i.e., d_{yz} or d_{xz}). As mentioned in Section 6.2.3 (see also Appendix A.6.3), l_z does not couple to any other orbital whereas l_x and l_y will couple d_{z^2} with d_{yz} or d_{xz}. The predicted g values in this case become

$$g_{zz} = g_e \qquad (A6.10a)$$

$$g_{xx}, g_{yy}(g_\perp) \propto g_e + \frac{6\lambda}{E_{yz,xz} - E_{z^2}} \qquad (A6.10b)$$

For other ground state configurations, these equations will obviously have different values, depending on which orbitals are now coupled. In the above examples, the singly occupied molecular orbital was based on $d_{x^2-y^2}$ or d_{z^2} for the d^9 ion in D_{4h} symmetry. If the metal contribution to the SOMO was purely d_{xy}, spin–orbit coupling will mix in d_{xz}, d_{yz}, and $d_{x^2-y^2}$ character such that the **g** matrix components are as

TABLE A.6.2 Expressions for the g Matrices for Single d Electron SOMO

SOMO	Δg_{xx}	Δg_{yy}	Δg_{zz}
d_{z^2}	$6\lambda/E_{xy} - E_{z^2}$	$6\lambda/E_{xz} - E_{z^2}$	0
$d_{x^2-y^2}$	$2\lambda/E_{xz} - E_{x^2-y^2}$	$2\lambda/E_{yz} - E_{x^2-y^2}$	$8\lambda/E_{xy} - E_{x^2-y^2}$
d_{xz}	$2\lambda/E_{xy} - E_{xz}$	$2\lambda/E_{x^2-y^2} - E_{xz}$ $+ 6\lambda/E_{z^2} - E_{xz}$	$2\lambda/E_{yz} - E_{xz}$
d_{yz}	$2\lambda/E_{x^2-y^2} - E_{yz}$ $+ 6\lambda/E_{z^2} - E_{yz}$	$2\lambda/E_{xy} - E_{yz}$	$2\lambda/E_{xz} - E_{yz}$
d_{xy}	$2\lambda/E_{yz} - E_{xy}$	$2\lambda/E_{xz} - E_{xy}$	$8\lambda/E_{x^2-y^2} - E_{xy}$

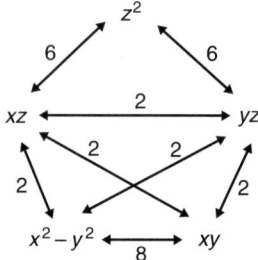

Fig. A.6.4 The values of n for the differently coupled d orbitals in the d^9 state. The diagram is often referred to as the "magic pentagon."

given in Table A.6.1. Nevertheless, one can see that a simple analysis of the trends in the g values can often give one an initial estimate about the ground state of the TMI.

The values of n given in Table A.6.2 are sometimes summarized in a figure referred to as the "magic pentagon" (Fig. A.6.4), and this value varies depending on which excited state d orbitals are mixed into the appropriate ground state. For example, in the first case discussed earlier for the $d_{x^2-y^2}$ orbital coupled to the d_{xy} orbital, the coefficient n has the value 8 (Equation A.6.8) whereas the coupling of $d_{x^2-y^2}$ with d_{yz} or d_{xz} produces a value of 2 (Equation A.6.9).

A.6.6.2 The A Values

The simplified theoretical expressions for the copper hyperfine couplings A_z ($A_{\parallel}$) and $A_x = (A_{\perp}) = A_y$ in the d^9 configuration for the $d_{x^2-y^2}$ ground state are given by [A.3]

$$A_{zz}(A_{\parallel}) = P\left[-\kappa + \Delta g_z - \frac{4}{7}\alpha^2 + \frac{3}{7}\Delta g_x\right] \quad (A6.11)$$

$$A_{xx}, A_{yy}(A_{\perp}) = P\left[-\kappa + \frac{2}{7}\alpha^2 - \frac{11}{14}\Delta g_x\right] \quad (A6.12)$$

where $\Delta g_z = g_{\parallel} - 2.0023$, $\Delta g_x = g_{\perp} - 2.0023$, P is the dipolar hyperfine coupling and depends on the metal ion; for copper $P = (\mu_0/4\pi h)g_e^{Cu}\beta_e\beta_n\langle r^{-3}\rangle_{3d} = 1171$ MHz.

Without going into any detail about how these equations are derived, the most important point to note is that, just like the g values, the A values also depend on the orbital ground state. The terms containing α^2 arise from dipole–dipole interactions between the magnetic moments associated with the spin motion of the electron and nucleus. As the unpaired electron is more delocalized onto the surrounding ligands, this contribution to α^2 is reduced. The terms containing g factors are now included because of the similar coupling between the orbital motion and the nucleus. Finally, κ arises from the Fermi contact interaction that has its origin in a nonvanishing probability of finding the unpaired electron at the site of the nucleus. For the d_{z^2} ground state, a similar set of equations can be derived for the Cu hyperfine, with different values of the coefficients, where

$$A_{zz}(A_\parallel) = P\left[-\kappa + \frac{4}{7}\alpha^2 - \frac{1}{7}\Delta g_x\right] \tag{A6.13}$$

$$A_{xx}, A_{yy}(A_\perp) = P\left[-\kappa + \frac{2}{7}\alpha^2 + \frac{15}{14}\Delta g_x\right] \tag{A6.14}$$

APPENDIX REFERENCES

A.1. Rose, M. E. *Elementary Theory of Angular Momentum;* Wiley: New York, 1967.

A.2. Mabbs, F. E.; Collison,, D. *Studies in Inorganic Chemistry 16: Electron Paramagnetic Resonance of d-Transition Metal Compounds;* Elsevier: New York, 1992.

A.3. Pilbrow, J. R. *Transition Ion Electron Paramagnetic Resonance*; Oxford Science Publications: Oxford, U.K., 1990.

7 The Virtual Electron Paramagnetic Resonance Laboratory: A User Guide to *ab initio* Modeling

VINCENZO BARONE

Dipartimento di Chimica and INSTM-Village, Università di Napoli Federico II, Napoli, Italy

ANTONINO POLIMENO

Dipartimento di Scienze Chimiche, Università degli Studi di Padovav, Via Marzolo 1, 35131 Padova, Italy

7.1 INTRODUCTION

7.1.1 Modeling Continuous Wave Electron Paramagnetic Resonance (CW EPR) Spectra

The information buried in an EPR spectrum can be extracted by two types of calculations, based on suitable models: the magnetic parameters of its spin Hamiltonian, and its spectral profile. Previous chapters have provided some of the possible approaches for calculations in both of these fields, based on the approximate methods developed during the life of EPR spectroscopy. For example, the magnetic parameters of the spin Hamiltonian (Chapter 4) in the early days of EPR used to be calculated by semiempirical methods (see, e.g., the well-known "McConnell equation"). Today, for medium to large sized molecules, computer implemented methods based on the density functional theory (DFT) are paving the route toward an accurate and effective computation of structural, electric, and magnetic properties. This has proved to be a turning point for the calculations of the spin Hamiltonian parameters [1]. Reliable methods for the evaluation of hyperfine tensors are available for several cases and, particularly for radicals in solution, the agreement between experimental and calculated parameters of the spin Hamiltonian by DFT is outstanding [1–3].

Because of its favorable time scale, EPR experiments can be highly sensitive to the details of the rotational and internal dynamics. In the so-called fast motional regime

---

Electron Paramagnetic Resonance. Edited by Brustolon and Giamello
Copyright © 2009 John Wiley & Sons, Inc.

(as, e.g., a radical tumbling in a liquid solution that is not too viscous), simple Lorentzian line shapes are observed, the widths of which can be predicted by using a perturbative approach (Redfield method). A well-known result of this approximate theory is the expression for the linewidths of the three Lorentzian lines for a nitroxide radical in a solution of known viscosity, taking into account the magnetic anisotropies, such as the hyperfine and **g** tensors. What can be obtained from this model are only estimates of molecular parameters (e.g., diffusion tensor values), regardless of the details of the molecular dynamics.

By contrast, in the so-called slow motional regime the spectral line shapes take on a complex form that is sensitive to the microscopic details of the motional process. To extract useful dynamic information from EPR experiments, a slow motional theory based on the stochastic Liouville equation (SLE) has been developed. Remember that the Liouville equation derives from the time-dependent Schrödinger equation for a quantum system. When the quantum system is exchanging energy with a reservoir, an irreversible term is introduced in the equation, taking into account the thermodynamic redistribution of energy. The equation is then called SLE. In our case the quantum system is the spin system, the energy reservoir is given by the degrees of freedom of the motions of the molecular species carrying the spin system, and the redistribution of energy produces the spin relaxation.

This more general approach models relaxation processes, like molecular tumbling, local solvent relaxation, internal conformational motions, and so on in terms of suitable classic time evolution operators that depend upon phenomenological dissipative parameters, such as the diffusion tensor.

The new approach presented in this chapter is given by the link between these two traditionally independent worlds: quantum computation of physicochemical properties of complex molecules in solution, and structural and dynamic properties of solvated probes. The resulting unified interpretative tool paves the route toward a true "virtual" spectroscopy.

It is interesting that virtual spectroscopic investigations are popular in fiction and movies. In the 1981 science fiction movie "Outland," Space Marshal O'Neil (Sean Connery) investigates the mysterious deaths of a number of mine workers on one of Jupiter's moons. Soon he discovers that the mine boss has been giving his workers an amphetamine-like, work-enhancing drug that keeps them productive for months, until they finally snap, go berserk, and die. One of the topic moments of the movie, at least for a computational chemist, is when Marshal O'Neil asks Dr. Lazarus, the station resident physician, to identify the mysterious drug. She activates a wonderful panoramic screen, starts punching buttons on a complex console, no doubt attached to a gigantic computer, and from a tiny dry sample she extracts the killer molecule, visualizes it on the screen, and *calculates* all of its properties—reactivity, spectroscopic fingerprints, toxicology, and so on—all of this in less than 30 s [4].

Now, this is Hollywood computational chemistry of the 1980s. However, could it become real in a foreseeable future? Or perhaps, with a number of limitations and approximations that are inherent to real science, could it be already currently available? Perhaps so. In particular, today we can state that, if Marshal O'Neil's molecule is a free radical in solution, a full prediction of its CW EPR spectrum is available to us. Granted, it will probably take a little more than 30 s. However, if the molecule is made of, say, less than 50 atoms, and basic information on the solvent nature is available, there is a fair chance that Dr. Lazarus' trick is doable.

This is not surprising because the link between theoretical *predictive* methodologies and EPR spectroscopy dates back several decades, and it is attributable to a happy coincidence between experimental needs and available interpretative tools. On the one hand, the intrinsic resolution of the EPR spectra, together with the unique role played by paramagnetic probes in providing information about their environment, make EPR in principle one of the most powerful methods of investigation on the electron distribution in molecules and on the properties of their environments. On the other hand, EPR spectroscopy is intrinsically amenable to an advanced theoretical interpretation in terms of quantum chemistry and statistical thermodynamics. Today these tools have reached a degree of refinement, and a power given by efficient computer implementation, that promises, after effective integration, a leap from fitting simulation to prediction.

7.1.2 Fitting and Predicting

The relationship between EPR spectroscopic measurements and molecular properties can be gathered by the systematic application of modeling and numerical simulations to interpret experimental observables. The traditional way to achieve this goal rests on the definition of a phenomenological model including a number of molecular, mesoscopic, and macroscopic parameters to be fitted in order to minimize deviations between experimental and simulated spectra.

This strategy, based on the idea of a general fitting approach, can be very helpful and safe when the fitting is based on few parameters, such as the magnetic parameters when the diffusion tensor is known or vice versa. However, in general this approach suffers from some intrinsic limitations related to the presence of multiple minima, which is particularly dangerous when there are several fitting parameters, and related to the difficulty of justifying, in several cases, best-fit parameters for a single spectrum with more general approaches or known physical trends (e.g., temperature dependence in a series of spectra).

Some of these problems can be solved by a new *integrated computational strategy* (ICS), which is the combination of (1) quantum mechanical (QM) calculations of structural parameters and magnetic tensors, possibly including average interactions with the environment (by discrete-continuum solvent models) [5, 6] (cf. Box 7.1) and short-time dynamical effects [7, 8], and (2) direct feeding of calculated molecular parameters into dynamic models, possibly involving stochastic effects. EPR measurements are particularly amenable to this integrated strategy, thanks to the availability of an increasing number of correlated experimental data, as in the multifrequency approach, to the advancement in computational methods, and to the refinement of available dynamic models [9–11].

An *ab initio* interpretation of EPR spectroscopy needs to take into account different aspects regarding the *structural*, *dynamical*, and *magnetic* properties of the molecular system under investigation; and it requires as input parameters the known basic molecular information and solvent macroscopic parameters. The application of the SLE formalism integrates the structural and dynamic ingredients to directly give the spectrum.

> **BOX 7.1 SOLVENT EFFECTS**
>
> The most promising general approach to the problem of environmental (e.g., solvent) effects can be based on a system–bath decomposition. The system includes the part of the solute where the essential process to be investigated is localized together with possibly the few solvent molecules strongly (and specifically) interacting with it. This part is treated at the electronic level of resolution and is immersed in a polarizable continuum, mimicking the macroscopic properties of the solvent. The solution process can then be dissected into the creation of a cavity in the solute (spending energy E_{cav}) and the successive switching on of dispersion–repulsion (with energy $E_{dis-rep}$) and electrostatic (with energy E_{el}) interactions with surrounding solvent molecules.

Minimal additional fitting procedures can then be applied for the fine-tuning of a limited set of molecular or mesoscopic parameters [12–16].

The most sophisticated model of this kind is probably the so-called polarizable continuum model. By coupling physical soundness and computational efficiency, it allows the treatment of conventional, isotropic solutions, ionic strengths, and anisotropic media like liquid crystals (Box 7.1).

7.1.3 Chapter Overview

Our main objectives in this chapter are (1) to apply integrated theoretical tools to the modeling of CW EPR, (2) to shed light on methodological aspects [17, 18], and (3) to underline the applicability and user friendliness of ICS if a careful implementation is made available in the form of purposely tailored software. For several reasons, we concentrate on CW EPR of organic mono- and biradicals in solution, chiefly for the need of readdressing a relatively well-studied field of EPR spectroscopy with modern theoretical and computational tools. This approach is the starting platform for an extension of the methodologies presented here to other EPR techniques, such as electron nuclear double resonance (ENDOR) and Fourier transform EPR (FT-EPR), and to other classes of systems (e.g., metalloproteins). Some considerations on these topics are presented in the last section of this chapter.

Therefore, our work plan is the following. Section 2 provides a summary of the theoretical techniques for the ICS interpretation of CW EPR spectra of radicals in solutions. Formal derivations will be kept at a bare minimum. Section 3 is devoted to the actual application of the methodology to test cases, which are used as introductory tutorials to a general computational software tool implementing the overall theoretical procedure. A summary is given in Section 4. Throughout the entire chapter, a two-level teaching strategy is adopted: ideally, the reader will be able to follow the main presentation and discussion without being distracted by too many formal details, but *insets* (in the form of boxes) devoted to advanced methodological aspects will be available if the reader needs further clarification of technical points.

7.2 MODELING TOOLS

7.2.1 Theory

Some qualitative considerations on the foundation of an *ab initio* ICS to the interpretation of CW EPR spectra of free radicals are presented here. The calculation of EPR observables can be in principle based on the complete solution of the Schrödinger equation for the system made of a paramagnetic probe plus explicit solvent molecules. The system can be described by a "complete" Hamiltonian that contains electronic coordinates of the paramagnetic probe, nuclear coordinates, and all degrees of freedom of all solvent molecules. The basic object of study to which any spectroscopic observable can be linked is given by the density matrix $\hat{\rho}$, which in turn is obtained from the Liouville equation.

Solving for $\hat{\rho}$ in time, for instance, via an *ab initio* molecular dynamics scheme allows the direct evaluation of any molecular property in principle. However, significant approximations are possible, which are basically rooted in time-scale separation arguments (Box 7.2).

The usual implicit procedure for obtaining a spin Hamiltonian from a complete Hamiltonian of the paramagnetic species is the averaging of the latter on the femtoseconds and subpicoseconds dynamics of the electronic coordinates. In the ICS procedure a further averaging of the Hamiltonian is also done (in the frame of the Born–Oppenheimer approximation) on the vibrational picosecond dynamics of nuclear coordinates. This averaging allows us to introduce in the calculation of magnetic parameters the effect of vibrational motions, which can be very relevant in some cases.

Then taken into account are the slower nuclear coordinates Q, that is, *intermolecular rotation* degrees of freedom and, if intramolecular motions are present (e.g., a methyl group rotation), a further degree of freedom as an *intramolecular* "soft" torsional coordinate. The dependence upon solvent or bath coordinates can be treated at a classical mechanical level either by solving explicitly the Newtonian dynamics of the explicit set or by adopting a standard statistical thermodynamics argument. This is formally equivalent to averaging the density matrix with respect to solvent variables (Box 7.3 [19, 20]).

BOX 7.2 SYSTEM–BATH HAMILTONIANS

$$\hat{H}(\{\mathbf{r}_i\}, \{\mathbf{R}_k\}, \{\mathbf{q}_\alpha\}) = \hat{H}_{\text{probe}}(\{\mathbf{r}_i\}, \{\mathbf{R}_k\})$$

$$+ \hat{H}_{\text{probe-solvent}}(\{\mathbf{r}_i\}, \{\mathbf{R}_k\}, \{\mathbf{q}_\alpha\}) + \hat{H}_{\text{solvent}}(\{\mathbf{q}_\alpha\})$$

where probe and solvent terms are separated. Hamiltonian $\hat{H}(\{\mathbf{r}_i\}, \{\mathbf{R}_k\}, \{\mathbf{q}_\alpha\})$ contains (1) electronic coordinates $\{\mathbf{r}_i\}$ of the paramagnetic probe (where index i runs on all probe electrons); (2) nuclear coordinates $\{\mathbf{R}_k\}$ (where index k runs on all rovibrational nuclear coordinates); and (3) *coordinates* $\{\mathbf{q}_\alpha\}$, in which all degrees of freedom of all solvent molecules are included, each labeled by index α.

BOX 7.3 SAMPLING SHORT-TIME DYNAMICS

The computation of reliable magnetic properties in solution calls for the consideration of true dynamic effects connected to the proper sampling of the solvent configurational space. Discussed here briefly are the short-time effects leading essentially to averaged values.

As an illustration let us consider a prototypical nitroxide spin probe molecule di-*tert*-butyl nitroxide (DTBN) in aqueous solution: to overcome the limitation of currently available empirical force field parameterizations, we explicitly refer to first-principle molecular dynamics simulations of the DTBN aqueous solution and, for comparison, in the gas phase [19, 20]. The results can be summarized in three main points: the effect of the solvent on the internal dynamics of the solute, the very flexible structure of the DTBN–water hydrogen bonding network, and the rationalization of the solvent effects on the magnetic parameters. Magnetic parameters are quite sensitive to the configuration of the nitroxide backbone, and in the particular case of DTBN, the out-of-plane motion of the nitroxide moiety is strongly affected by the solvent medium. Although the average structure in the gas phase is pyramidal, the behavior of DTBN in solution presents a maximum probability of finding a planar configuration: this does not mean that the DTBN minimum in solution is planar, but that there is a significant flattening of the potential energy governing the out-of-plane motion and that the solute repeatedly undergoes an interconversion among pyramidal positions.

The vibrational averaging effects of this large amplitude internal motion can be taken into account by computing the EPR parameters along the trajectories. The hydrogen-bonding network embedding the nitroxide moiety in aqueous solution presents a very interesting result: the dynamics of the system points out the presence of a variable number of hydrogen bonds (0–2) with a highest probability of only one genuine hydrogen bond. Such features of DTBN–water interaction is actually system dependent; the high flexibility of the NO moiety and the steric repulsion of the *tert*-butyl groups decreases the energetically accessible space around the nitroxide oxygen. Analogous simulations for a more rigid five-ring nitroxide (proxyl) in aqueous solution provided a different picture with an average of two nitroxide water H bonds. In this case the substituents embedding the NO moiety are constrained in a configuration where methyl groups are never close to the nitroxide oxygen, and the backbone of the nitroxide presents an average value of the CNC angle that is lower than in the case of the DTBN, thus evidencing a better exposition of the NO moiety to the solvent molecules in the case of the proxyl radical. Nevertheless, the behavior of the closed ring nitroxide in water could not be generalized to all of the protic solvents: a similar simulation of the proxyl molecule in methanol solutions presents, on average, only one genuine solute–solvent H bond, possibly because the more crammed H-bonded methanol molecule prevents easy access to the NO moiety for other solvent molecules. Once again the reliable description of solvent dynamics plays a crucial role in an accurate prediction of spectroscopic data. Remarkably, a discrete-continuum solvent approach allows the decoupling of the different contributions and the quantification of their effect on each of the molecular parameters: the hydrogen-bonding interaction and the dielectric contribution of the solution bulk, taken independently, have a roughly comparable effect, with the dielectric contribution decreasing when going from DTBN to DTBN–water adducts.

In this way an *effective probe Hamiltonian* is obtained that is characterized by *magnetic tensors*. By taking into account only the electron Zeeman and the hyperfine interactions, for a probe with one unpaired electron and N nuclei we can define an averaged magnetic Hamiltonian:

$$\hat{H}(\mathbf{Q}) = \frac{\mu_B}{\hbar}\mathbf{B}_0 \cdot \mathbf{g}(\mathbf{Q}) \cdot \hat{\mathbf{S}} + \gamma_e \sum_n \hat{\mathbf{I}}_n \cdot \mathbf{A}_n(\mathbf{Q}) \cdot \hat{\mathbf{S}} \qquad (7.1)$$

where $\hat{H}(\mathbf{Q})$ is the spin Hamiltonian. The time evolution equation for $\hat{\rho}(\mathbf{Q}, t)$ can then be obtained by the so-called SLE formalism:

$$\frac{\partial}{\partial t}\hat{\rho}(\mathbf{Q}, t) = -i[\hat{H}(\mathbf{Q}), \hat{\rho}(\mathbf{Q}, t)] - \hat{\Gamma}\hat{\rho}(\mathbf{Q}, t) = -\hat{\mathcal{L}}(\mathbf{Q})\hat{\rho}(\mathbf{Q}, t) \qquad (7.2)$$

The Liouvillean operator $\hat{\mathcal{L}}$, governing the time evolution of the system, includes the motional dynamics in the form of the stochastic operator Γ. This operator describes the diffusive motion of degrees of freedom Q and it has the general form of a so-called Fokker–Planck operator, which is a parabolic operator governing the time evolution of the probability density function for a set of Markovian stochastic variable q. In other words, given an initial distribution, one gets the distribution at time t by solving analytically or numerically the corresponding Fokker–Planck equation (Box 7.4).

Let us consider a "rigid" paramagnetic probe, remembering that at this stage internal vibrational picosecond dynamics have been already averaged out in order to define the effective spin Hamiltonian. Computationally, these averages can be carried out by explicit molecular dynamics simulations (see Box 7.3). If our rigid paramagnetic probe is freely rotating, the set of stochastic relevant coordinates is restricted to the set of orientational coordinates $\mathbf{Q}\int\Omega$; these are described in terms of a simple formulation for a diffusive rotator, characterized by a diffusion tensor **D**. The diffusion tensor is determined by the shape of the molecule, deriving from the minimum energy conformations obtained from the QM calculations. This choice is formalized by adopting the following simple form for $\hat{\Gamma}$:

$$\hat{\Gamma} = \hat{\mathbf{J}}(\Omega) \cdot \mathbf{D} \cdot \hat{\mathbf{J}}(\Omega) \qquad (7.3)$$

where $\hat{\mathbf{J}}(\Omega)$ is the angular momentum operator for body rotation [21].

BOX 7.4 STOCHASTIC MODELING: AN EXAMPLE

In the case of a rotating molecule with 1 conformational degree of freedom, as the internal rotation around a single bond, the internal dynamics can be described by an extended stochastic model that explicitly includes torsional angle α. The torsional potential and diffusion properties for the internal rotation are obtained straightforwardly from QM and hydrodynamic estimates, respectively, whereas the modified stochastic operator is

$$\hat{\Gamma} = \hat{\mathbf{J}}(\Omega) \cdot \mathbf{D} \cdot \hat{\mathbf{J}}(\Omega) + D\frac{\partial}{\partial \alpha}P_{eq}(\alpha)\frac{\partial}{\partial \alpha}P_{eq}^{-1}(\alpha)$$

in its simplest form neglecting coupling in the diffusion tensor and assuming a constant diffusion coefficient D for conformational dynamics.

Complex solvent environments like highly viscous fluids can be described by an augmented set of stochastic coordinates to be included in $\mathbf{Q}$, which describe *slow relaxing local solvent structures*. Once the effective Liouvillean is defined, the direct calculation of the CW EPR signal is possible by evaluating the spectral density from the expression

$$I(\omega - \omega_0) = \frac{1}{\pi}\text{Re}\left\langle v|[i(\omega - \omega_0) + i\hat{\mathcal{L}}]^{-1}|vP_{eq}\right\rangle \quad (7.4)$$

where the Liouvillean $\hat{\mathcal{L}}$ acts on a starting vector proportional to the x component of the electron spin operator $\hat{S}_x$ related to the EPR signal (Box 7.5).

In summary, the basic parameters for the direct evaluation of Equation 7.4, for example, for a radical are the principal values and orientation of hyperfine tensors $\mathbf{A}_n$, the principal values and orientation of Zeeman tensor $\mathbf{g}$, and the rotational diffusion tensor $\mathbf{D}$.

BOX 7.5 STARTING VECTOR: AN EXAMPLE

If only one nucleus (e.g., nitrogen) is coupled to the electron, one has

$$\left|vP_{eq}^{1/2}\right\rangle = [I]^{-1/2}\left|\hat{S}_x \otimes 1_I P_{eq}^{1/2}\right\rangle$$

where $I = 1$; P_{eq} is the Boltzmann distribution in Ω space; ω is the sweep frequency; and $\omega_0 = g_0\mu_B B_0/\hbar = \gamma_e B_0$, where $g_0 = \text{Tr}(\mathbf{g})/3$.

A summary of the ICS is as follows. QM calculations allow the determination of the minimum energy molecular geometry and the evaluation of magnetic tensor parameters (hyperfine tensors, Zeeman tensors, fine tensors, etc.), with corrections based on averaging of fast motions, and the determination of the minimum energy molecular geometry. Then slow motional processes are explicitly modeled via stochastic treatment, taking advantage of the calculated geometry of the molecule giving the shape-dependent dissipative parameters (e.g., rotational diffusion tensor) obtained via a simple but effective hydrodynamic model. The overall strategy is sketched in Fig. 7.1.

7.2.2 Implementation

Without delving too deeply, at least at the first level, into the actual implementation of this scheme, let us make the following considerations: steps 1–4 of the ICS scheme described in Fig. 7.1 can be considered, in a utilitarian philosophical frame of mind, as black boxes in which suitable input should be inserted and from which suitable output should be obtained. Which kind of black boxes do we have around to be employed? How reliable are they? How can they be customized and adapted to the case of experimental interest? Some of these questions are addressed in the final part of this section and especially in the next one, which is dedicated to tutorials and test cases.

Basically, steps 1 and 2 are nowadays answered (partially or totally, depending on the specific cases) by a number of up-to-date quantum chemistry programs. The optimized structure of the free radical obtained by DFT calculations in a solvated

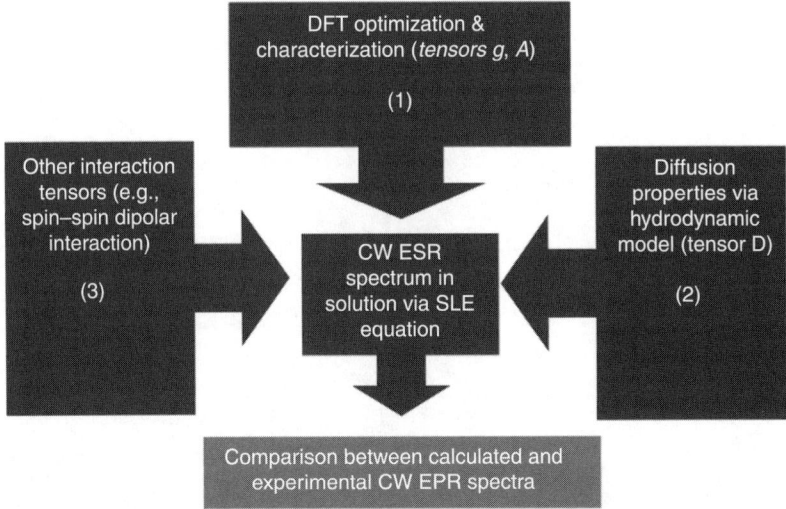

Fig. 7.1 A chart of the integrated computational approach to simulation of the CW EPR spectra in solution. Steps 2 and 3 are based on the optimized geometry and electronic structure obtained in step 1.

environment and hyperfine and **g** tensors can been computed *directly*. Notice that although dipolar hyperfine terms and **g** tensors are negligibly affected by local vibrational averaging effects, this is not the case for the isotropic hyperfine term, especially concerning those large amplitude vibrations that modify hybridization at the radical center. Of course, despite ongoing progress, the quantitative agreement between computed and experimental values is not always sufficient for a fully satisfactory interpretation of the spectrum, especially concerning isotropic hyperfine splittings (one-third of the trace of the corresponding hyperfine tensors). A minimal adjustment of this term from the computed value in the simulation of the EPR spectrum can therefore be allowed. All of these magnetic terms are local in nature, so they are scarcely dependent on conformational modifications. Other magnetic terms, which are relevant for biradicals (J and spin–spin dipolar interaction), have a long-range character and provide a signature of different molecular structures. Although the computation of J is, in principle, quite straightforward by, for example, the so-called broken symmetry approach [22], currently available density functionals are not always sufficiently reliable for the distances characterizing the systems under investigation (>5–6 Å) [23]. Thus, transfer of J values available for related systems is still widely employed. The situation is different for the spin–spin dipolar term, which is the most critical long-range contribution. For a nitroxide biradical this tensor is often calculated by assuming that the two electrons are localized at the center of the N–O bonds and considered just as two point magnetic dipoles. A more accurate computation, starting from the computed spin densities, is sketched in Box 7.10. Figure 7.2 shows the relevant frames or reference for a typical system: the laboratory frame (LF), the molecular frame (MF), the **g** tensor frame (GF) and the **A** tensor frame (A_nF).

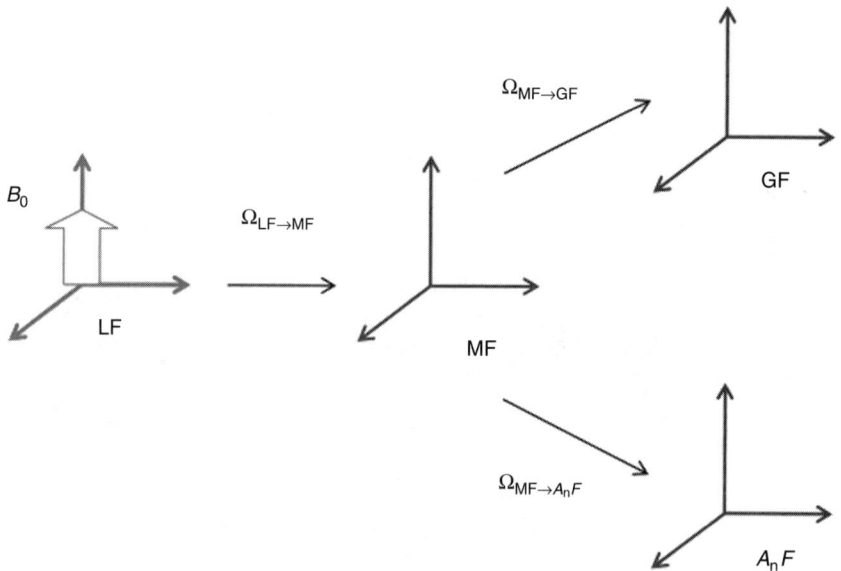

Fig. 7.2 Frames of reference: laboratory frame (LF), molecular frame (MF), **g** tensor frame (GF), and **A** tensor frame (A_nF).

BOX 7.6 FRICTION

Matrix ξ has only diagonal blocks of the form $\xi(T)\mathbf{1}_3$, where $\xi(T)$ is the translational friction of a sphere of radius R_0 at temperature T and is given by the Stokes law $\xi(T) = CR\eta(T)\pi$, where $\eta(T)$ is the solvent viscosity at a given temperature T and C depends on hydrodynamic boundary conditions. The system friction is then given as $\Xi = \xi(T)\mathbf{B}^{\text{tr}}\mathbf{B}$, where $\mathbf{B}$ is a rectangular matrix depending on atomic coordinates only.

The evaluation of the diffusion properties, which is step 3, can be based on a hydrodynamic approach in which the molecule under investigation is modeled as an ensemble of N fragments, each formed by spheres representing atoms or groups of atoms, immersed in a homogeneous isotropic fluid of known viscosity. By assuming a form for the friction tensor of nonconstrained atoms ξ, one can calculate the friction for the constrained atoms Ξ. Let us employ, for simplicity, the basic model for noninteracting or weakly interacting spheres in a fluid (Box 7.6).

The diffusion tensor (which can be conveniently partitioned into translation, rotational, internal, and mixed blocks) can now be obtained as the inverse of the friction tensor

$$\mathbf{D} = \begin{pmatrix} \mathbf{D}_{\text{TT}} & \mathbf{D}_{\text{TR}} & \mathbf{D}_{\text{TI}} \\ \mathbf{D}_{\text{TR}}^{\text{tr}} & \mathbf{D}_{\text{RR}} & \mathbf{D}_{\text{RI}} \\ \mathbf{D}_{\text{TI}}^{\text{tr}} & \mathbf{D}_{\text{RI}}^{\text{tr}} & \mathbf{D}_{\text{II}} \end{pmatrix} = k_{\text{B}} T \Xi^{-1} \quad (7.5)$$

Neglecting off-diagonal couplings, an estimate of the rotational diffusion tensor is given by $\mathbf{D}_{\text{RR}} \equiv \mathbf{D}$, which directly depends on the atomic coordinates, temperature, and solvent viscosity.

Finally, the numerical implementation of the SLE is usually based on a symmetrized equivalent of Equation 7.4, in the form

$$I(\omega - \omega_0) = \frac{1}{\pi} \text{Re} \left\langle v P_{\text{eq}}^{1/2} \left| [i(\omega - \omega_0) + \tilde{\mathcal{L}}]^{-1} \right| v P_{\text{eq}}^{1/2} \right\rangle$$

where $\tilde{\mathcal{L}} = P_{\text{eq}}^{-1/2} \hat{\mathcal{L}} P_{\text{eq}}^{1/2}$ is the symmetrized stochastic operator. The CW EPR spectrum is obtained by numerically evaluating the spectral density by adopting iterative algorithms, like Lanczos or conjugate gradients [8]. In particular, the Lanczos algorithm is a recursive procedure to generate orthonormal functions, which allows a tridiagonal matrix representation of the Liouvillean system. Several software packages implementing numerical solutions to the SLE for CW EPR are available. The first is the open-source ACERT package [24] by Freed and coworkers for simulation and analysis of EPR spectra, which includes the basic simulation program EPRLL and nonlinear least-squares fitting code NLSL [10]. Another example of implementation of (partial) solutions of the SLE is EasySpin, a free simulation toolbox for the MatLab package [25].

Other codes are based on a simplified description in terms of Lorentzian or Lorentzian–Gaussian function. Among others, there are the following commercial codes: Xsophe [26], which is intended for simulating CW EPR spectra in a user-friendly way thanks to an intuitive graphical user interface; Molecular Sophe [26] for the simulation of other spectra in addition to CW, such as FT-EPR, two- and three-pulse electron spin echo envelope modulation, hyperfine sublevel correlation spectroscopy, and pulsed ENDOR spectra; and SimFonia [26], which includes several kinds of magnetic interactions: electronic Zeeman, nuclear quadrupole, nuclear hyperfine, nuclear Zeeman, and zero-field splitting. Finally, EPRFIT [27] is available at the Institute für Chemie und Biochemie of the University of Berlin, and EWSim [28] is intended for organic anion radicals in solution and in the absence of anisotropy effects.

7.3 TUTORIAL AND CASE STUDIES

This section is essentially devoted to the description of some paradigmatic cases chosen to illustrate the potentialities as well as some current limitations of the ICS sketched in the preceding section. As a tutorial, a basic example is considered: tempone in aqueous solution. Several case studies are discussed afterward.

The software employed throughout this section is a novel outcome of the ICS approach, because it is based on the idea of full integration between QM tools (for structural and magnetic properties evaluation in the presence of solvent effects and fast motion averaging), and the stochastic approach to line shape evaluation (including a full evaluation of diffusion tensor properties via the hydrodynamic approximation). Although still in a preliminary version, the new software protocol Electron Spin Resonance Simulation (E-SpiReS) has some attractive features and can be thought of as a reasonable realization of a user-friendly, broad purpose virtual CW EPR spectrometer, at least for mono- and biradicals in solution, with the inclusion of rotational and internal (conformational) dynamics:

1. The code is highly modular, built (in C language) according to an object-oriented structure; changes, additions, and inclusions of new features are easy and transparent.
2. A fully integrated graphical interface (written in Java) allows the user to set up the numerical experiment in a natural way, setting up a *project* and adding information as needed (probe molecular structure, solvent physicochemical properties, etc.).
3. The code calls up independent software for quantum chemistry calculation (currently Gaussian), creating automatically customized input files and reading output files without any manual intervention from the user.
4. The code is parallelized and can be used on multiprocessor systems (cluster) under Linux OS.

As a general rule, the user can proceed by leaving most technical parameters (e.g., matrix dimensions, Lanczos step for generating a CW EPR spectrum) at their default value, which is chosen by the program; but access to refined choices is always available.

The reader can peruse through the tutorial without bothering to consider the boxes that will be used liberally throughout this entire section: detailed theoretical and computational information are confined there.

7.3.1 Tutorial: Tempone in Aqueous Solution

In the tutorial the CW EPR spectrum of Tempone in water at 298.15K is calculated. The numerical experiment is performed on a single node of a quadriprocessor Linux cluster.

As in all computational chemistry programs, initially the user must define the molecular structure of the probe (Fig. 7.3). This is currently done by defining the so-called **Z** matrix, which defines the topology and initial structure of the paramagnetic molecule of interest. This is usually generated via existing standard software tools, like Molden, GaussView, and so forth, or manually. In the case under investigation, assume that a suitable **Z** matrix file has been already generated when starting to use E-SpiReS. Most chemists are familiar nowadays with standard molecular drawing tools, which can easily generate a **Z** matrix file.

Now the data are uploaded step-by-step, generating the necessary structural and magnetic information and finally calculating the spectrum (Fig. 7.4).

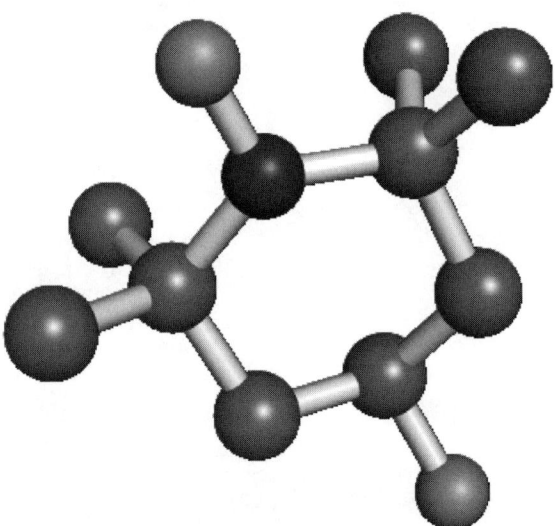

Fig. 7.3 The starting point is the molecular structure of tempone.

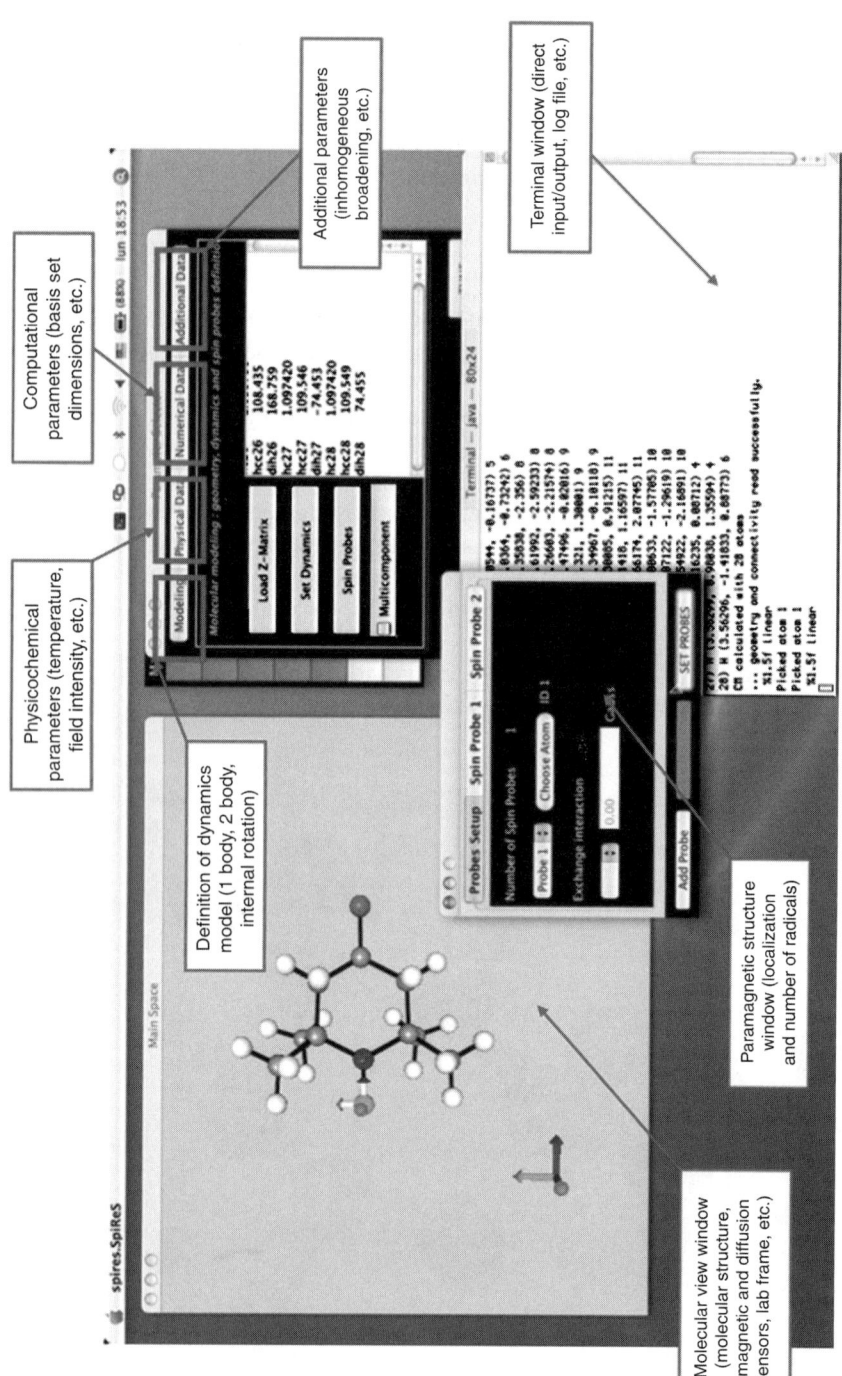

Fig. 7.4 A schematic view of the E-SpiReS main windows.

Steps 1–4. The user calls up the program (step 1). Technically, the E-SpiReS java graphical interface is activated; a main control panel and a three-dimensional (3-D) space is opened where molecules are drawn (step 2): by clicking the SetProject button a tagged window called Parameter Selector appears, where all the physical properties of the system under study can be set. The user first clicks on the Load Z-Matrix button to load the molecule (step 3). In the tutorials/tempone directory the user selects the tempone.zmt file that contains the **Z** matrix, generated in this case by the Molden package (step 4). (Fig. 7.5).

Step 5. The molecule is loaded. A 3-D representation appears in 3-D space; a reference frame is drawn. This is the inertial laboratory frame (LF). In the Parameter Selector window the **Z** matrix is written in the white area. This area is reactive to mouse clicking: when a row is clicked, the corresponding atom is highlighted.

Step 6. Clicking on the Set Dynamics button causes a new window to appear. Here the user chooses the form of the diffusive operator $\hat{\Gamma}$. Tempone is a small molecule that rapidly interconverts at ambient temperature between two twisted-crossover structures. Thus, a first average of magnetic tensors must be performed over the effective large amplitude motion connecting these two structures. Then the One Rigid Body model can be chosen for further rotational averaging (Fig. 7.6).

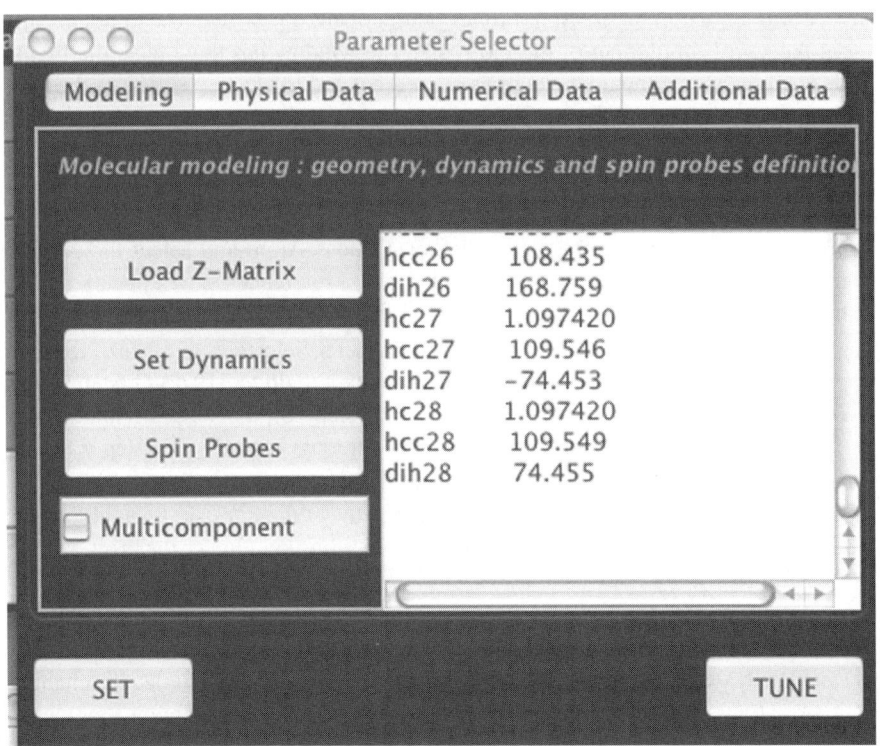

Fig. 7.5 The **Z** matrix for tampone is loaded in E-SpiReS.

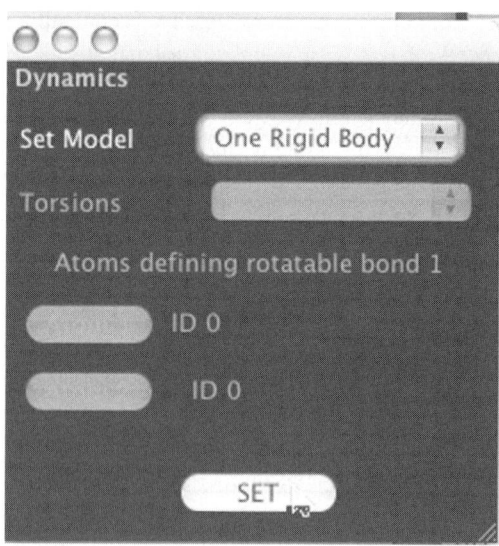

Fig. 7.6 Step 5: the user chooses the model dynamics.

Step 7. Mouse action on the Spin Probes button opens a new window to graphically set the spin Hamiltonian of the molecule. The window has three tags, the first to decide how many unpaired electrons are present and in which position of the molecule they are located via the Choose Atom button. Choose the O–N bond to set the probe of tempone. The O atom becomes green and a frame appears for the **g** tensor (Fig. 7.7).

Step 8. The second and third tags allow the adding of spin active nuclei to the probe(s). In the Spin Probe 1 tag click on Choose Atom and select the N atom in the 3-D space. The atom is highlighted and a reference frame appears for the hyperfine **A** tensor; set the spin number of the nucleus to 1.

Steps 9–12. In the Physical Data tag of the Parameter Selector a number of relevant parameters are set. The B_0 button sets the magnetic field to 3197.3 G (step 9); the B Sweep button sets the field sweep to 75.7 G (step 10); the Viscosity button sets the viscosity of the solvent (water in this tutorial) to 0.89 cP (step 11); the Temperature button sets the temperature to 298.15K (step 12) (Fig. 7.8).

Step 13. The user selects the Additional Data tag to set the intrinsic linewidth to 2.4 G, to take into account the unresolved superhyperfine coupling of the electron with the 12 surrounding hydrogen atoms.

Step 14. The user clicks on the Diffusion button in the Main Control Panel to enter in the Diffusion Environment. The diffusion tensor of the molecule is automatically calculated and a new frame appears in the 3-D space. The molecule changes color: atoms assume different colors if they belong to different fragments. In this case, there is only one fragment and so all of the atoms look the same (Box 7.7).

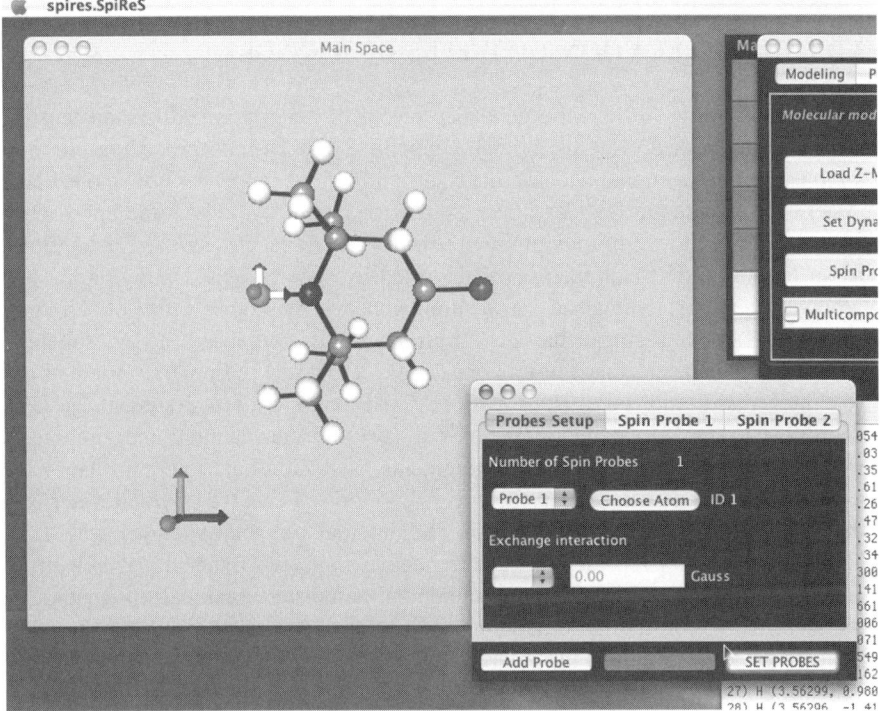

Fig. 7.7 The location of the magnetic tensors is chosen.

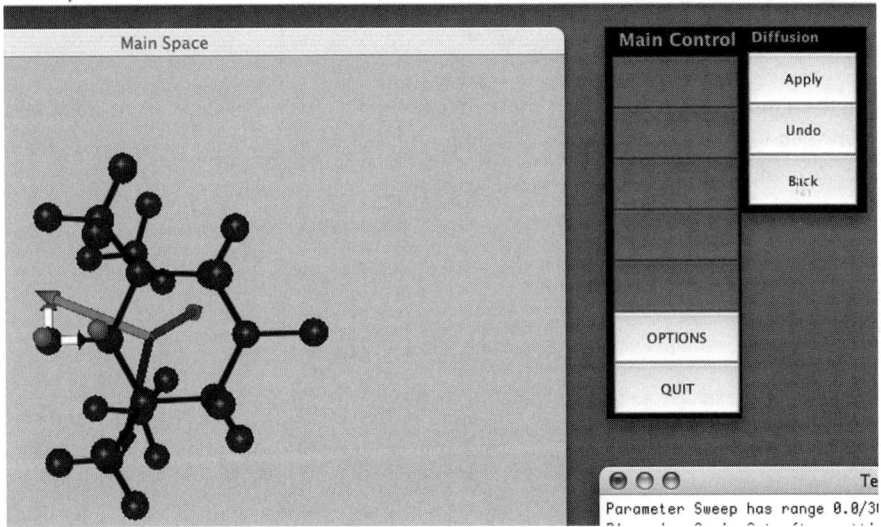

Fig. 7.8 The molecular diffusion tensor is evaluated.

BOX 7.7 EVALUATION OF DIFFUSION TENSORS

Before proceeding further, the general problem of determining diffusion tensor values is addressed. Let us briefly summarize the overall procedure, in its simplest implementation, to estimate the diffusion properties of molecular systems, with internal degrees of freedom, based on a hydrodynamic approach. The starting point is from a simplified view of the molecule under investigation as an ensemble of N fragments, each formed by spheres representing atoms or groups of atoms, immersed in a homogeneous isotropic fluid of known viscosity. Let us assume that the ith fragment is composed by N_i spheres (extended atoms) and that the torsional angle θ_i defines the relative orientation of fragments i and $i+1$. Our MF is fixed on a chosen fragment v. By definition, in the MF, atoms of fragment v have only translational and rotational motions whereas atoms of all other fragments have additional internal rotational motions. Let us now associate the set of coordinates $(\mathbf{r}, \mathbf{\Omega}, \mathbf{\theta})$, which describe the respective translational, rotational, and internal torsional motions with the velocities $(\mathbf{V}, \mathbf{\omega}, \dot{\mathbf{\theta}})$ representing the respective molecule translational velocity, angular velocity around an inertial frame, and associated torsional momenta. In the presence of constraints in and among fragments, the generalized force, made of force $\mathbf{F}$, torque $\mathbf{N}$, and the internal torque $\mathbf{N}_{\text{int}}$, is related to the generalized velocity by the relation

$$\mathcal{F} = -\Xi \mathcal{V}$$

$$\mathcal{F} = \begin{pmatrix} \mathbf{F} \\ \mathbf{N} \\ \mathbf{N}^{\text{int}} \end{pmatrix}, \quad \mathcal{V} = \begin{pmatrix} \mathbf{V} \\ \mathbf{\omega} \\ \dot{\mathbf{\theta}} \end{pmatrix}$$

In the absence of constraints, a similar relation holds for each single extended atom between its velocity and the force acting on it, which in compact matrix form can be written

$$f = -\xi v$$

$$f = \begin{pmatrix} \mathbf{f}_1^1 \\ \dots \\ \mathbf{f}_{n_N}^N \end{pmatrix}, \quad v = \begin{pmatrix} \mathbf{v}_1^1 \\ \dots \\ \mathbf{v}_{n_N}^N \end{pmatrix}$$

where $\mathbf{f}_j^i$ is the force acting on the jth atom of the ith fragment ($1 \leq j \leq n_i$), and so forth. Constrained and unconstrained forces and velocities can be related via geometric considerations as $\mathcal{F} = \mathbf{B}^{\text{tr}} f$, $\mathcal{V} = \mathbf{B} v$, where $\mathbf{B}$ is a rectangular matrix depending upon atomic coordinates. It follows that $\Xi = \mathbf{B}^{\text{tr}} \xi \mathbf{B}$. By assuming a form for the friction tensor of nonconstrained atoms ξ, one can calculate the friction for the constrained atoms Ξ. Assume for simplicity the simplest model for noninteracting spheres in a fluid, namely, that matrix ξ has only diagonal blocks of the from $\xi_0 I_3$, where ξ_0 is the translational friction of a sphere of radius R_0 given by the Stokes law $\xi_0 = CR_0 \eta \pi$, where η is the solvent viscosity and C depends on hydrodynamic boundary conditions. The system friction is then given as $\Xi = \xi_0 \mathbf{B}^{tr}\mathbf{B}$. The diffusion tensor (which can be conveniently partitioned in translation, rotational, internal, and mixed blocks) can now be obtained as the inverse of the friction tensor (cf. Equation 5).

Steps 15–17. To evaluate the magnetic tensors via QM calculations, the user enters in the Gaussian Environment. Here the user has the possibility to edit the Gaussian input file generated by E-SpiReS, launch Gaussian, or load a precalculated Gaussian output file (step 15); clicking Edit Input loads a simple editor, which one can use to personalize the Gaussian input file. In this tutorial no changes are introduced (step 16); clicking the Launch button, the input file is submitted to Gaussian (step 17). The user can choose in the OPTIONS to run Gaussian interactively (on a local computer) or to start the job on a remote multiprocessor system.

Step 18. The user needs to inform the program that magnetic tensors and structural information is superseded by output from Gaussian by checking the Use Output checkbox. When the Gaussian output is loaded, all of the tensors are updated (Box 7.8).

Step 19. Tensors can be modified manually in the Diffusion Environment. To modify a tensor, just click on it and change the desired quantity. In this tutorial, the Gaussian output (based on a slightly inefficient basis set) gives a value of the trace of the hyperfine constant about 2 G less than the experimental one. Thus, one needs to edit the hyperfine tensor by setting the isotropic value (Iso slide bar in the setting window) to 16.14 G. WARNING: changes are effective only by pressing the Apply button (Fig. 7.9).

Step 20. Further adjustments can be obtained by fitting (although, as a general rule, only small corrections should be necessary), in the Refine Environment. In this case, after entering the Refine Environment, the user adjusts the traces of g, A, and the intrinsic linewidth, checking the proper boxes.

Step 21. Next the user loads a reference experimental spectrum by clicking the Load Spectrum button and chooses the file tutorials/tempone/experimental_spectra/exp.dat (Fig. 7.10).

Steps 22–25. Now the user is ready to enter the electron spin resonance (ESR) Environment, where spectra can be calculated with or without fitting and then plotted. To refine three parameters, the user checks the Fit Mode checkbox (step 22); by clicking on the Calculate button the spectrum is obtained by solving the SLE (step 23). Note that it is possible to run the calculation interactively (choosing the number of dedicated processors) or via Portable Batch System (PBS). In this

BOX 7.8 EVALUATION OF MAGNETIC TENSORS

Let us first consider electron-field interactions. It is convenient, as far as the **g** tensor is concerned, to refer absolute values to shifts with respect to the free-electron value ($g_e = 2.002319$), namely, consider $\Delta \mathbf{g} = \mathbf{g} - g_e \mathbf{1}_3$, where $\mathbf{1}_3$ is the 3×3 unit matrix. Let us dissect $\Delta \mathbf{g}$ into three main contributions $\Delta \mathbf{g} = \Delta \mathbf{g}^{\text{RMC}} + \Delta \mathbf{g}^{\text{GC}} + \Delta \mathbf{g}^{\text{OZ/SOC}}$, where the first two terms are first-order contributions, which take into account relativistic mass (RMC) and gauge (GC) corrections, respectively. The first term can be expressed as

$$\Delta \mathbf{g}^{\text{RMC}} = -\frac{\alpha^2}{S} \sum_{\mu\nu} P_{\mu\nu}^{\alpha-\beta} \langle \varphi_\mu | \hat{T} | \varphi_\mu \rangle$$

where α is the fine structure constants, S is the total spin of the ground state, $P_{\mu\nu}^{\alpha-\beta}$ is the spin density matrix, $\{\varphi\}$ is the basis set, and $\hat{T}$ is the kinetic energy operator. The second term is given by

$$\Delta \mathbf{g}^{\text{GC}} = \frac{1}{2S} \sum_{\mu\nu} P_{\mu\nu}^{\alpha-\beta} \left\langle \varphi_\mu \left| \sum_n \xi(\mathbf{r}_n)(\mathbf{r}_n \mathbf{r}_0 - \mathbf{r}_{n,r} \mathbf{r}_{0,s}) \hat{T} \right| \varphi_\nu \right\rangle$$

where $\mathbf{r}_n$ is the position vector of the electron relative to the nucleus n; $\mathbf{r}_0$ is the position vector relative to the gauge origin; and $\xi(r_n)$, which depends on the effective charge of the nuclei, is defined below. These two terms are usually small and have opposite signs so that their contributions tend to cancel out. The last term is a second-order contribution arising from the coupling of the orbital Zeeman (OZ) and the spin–orbit coupling (SOC) operators. The OZ contribution in the system Hamiltonian is

$$\hat{H}_{\text{OZ}} = \beta \sum_i \mathbf{B} \cdot \hat{\mathbf{l}}(i)$$

It shows a gauge origin dependence, arising from the angular momentum of the ith electron $\hat{\mathbf{l}}(i)$. In our calculations a gauge including the atomic orbital approach is used to solve this dependence.

Finally, the SOC term is a true two-electron operator, but here it will be approximated by a one-electron operator involving adjusted effective nuclear charges. The one-electron approximate SOC operator reads

$$\hat{H}_{\text{SOC}} = \sum_{n,i} \xi(\mathbf{r}_{i,n}) \hat{\mathbf{l}}_n(i) \cdot \hat{\mathbf{s}}(i)$$

where $\hat{\mathbf{l}}_n(i)$ is the angular momentum operator of the ith electron relative to the nucleus n and $\hat{\mathbf{s}}(i)$ is its spin operator. The function $\xi(\mathbf{r}_{i,n})$ is defined as

$$\xi(\mathbf{r}_{i,n}) = \frac{\alpha^2}{2} \frac{Z_{\text{eff}}^n}{|\mathbf{r}_i - \mathbf{R}_n|^3}$$

where Z_{eff}^n is the effective nuclear charge of atom n at position $\mathbf{R}_n$.

The second term is the hyperfine interaction contribution, which, in turn, contains the so-called Fermi contact interaction (an isotropic term), which is related to the spin density at the corresponding nucleus n by

$$A_{n,0} = \frac{8\pi}{3} \frac{g_e}{g_0} g_n \beta_n \sum_{\mu,\nu} P_{\mu,\nu}^{\alpha-\beta} \left\langle \varphi_\mu \left| \delta(r_{kn}) \right| \varphi_\nu \right\rangle$$

and an anisotropic contribution, which can be derived from the classical expression of interacting dipoles

$$A_{n,ij} = \frac{g_e}{g_0} g_n \beta_n \sum_{\mu,\nu} P_{\mu,\nu}^{\alpha-\beta} \left\langle \varphi_\mu \left| r_{kn}^{-5}(r_{kn}^2 \delta_{i,j} - 3r_{kn,i}r_{kn,j}) \right| \varphi_\nu \right\rangle$$

Tensor components are usually given in Gauss (1 G = 0.1 mT); to convert data to megahertz, one has to multiply by 2.8025. From a computational point of view, evaluation of the **A** tensor (a first-order property) should be simpler than that of the **g** tensor. This is true for the anisotropic term, but evaluation of the Fermi contact contribution involves a number of difficulties related to the local quality of basis functions at the nuclei. The following examples use the purposely tailored NO7D basis sets together with B3LYP or PBE0 hybrid functionals, which have been proven to be very effective, especially for non hydrogen atoms.

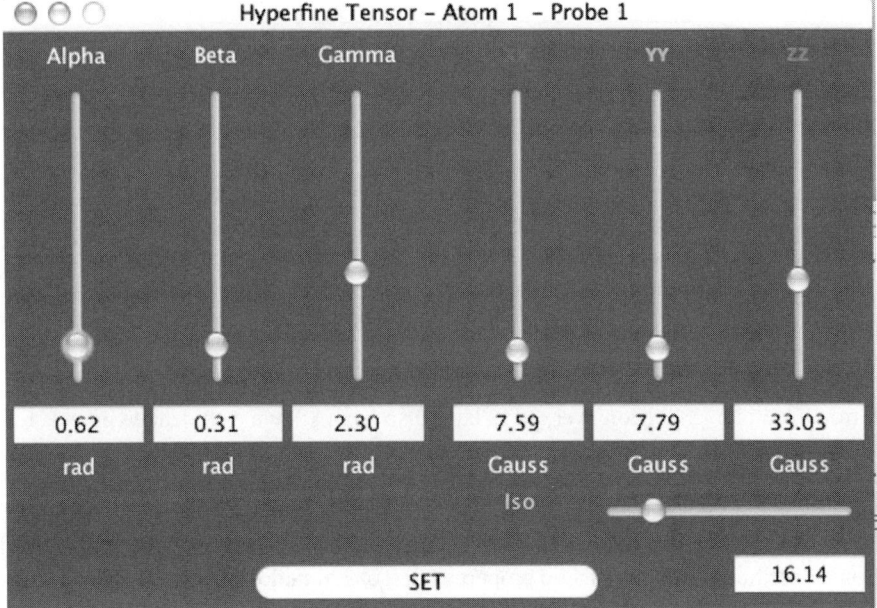

Fig. 7.9 Refining the starting data.

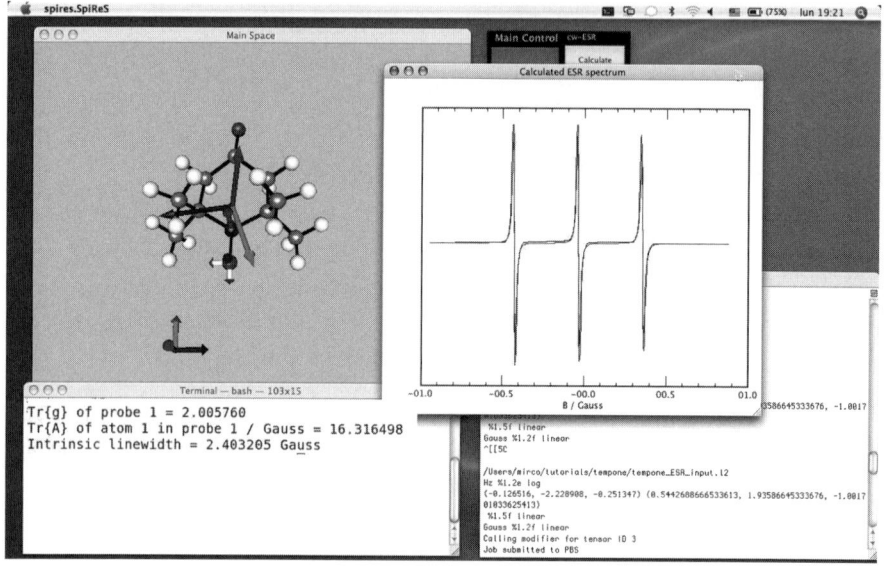

Fig. 7.10 The E-SpiReS evaluation of the CW EPR spectrum.

case the calculation was performed parallelizing the job on a four CPU nodes (four processors); after 15 s the calculation ends (step 24). In the present case, very small corrections (<0.1%) to the refined parameters are obtained. The theoretical and experimental spectra can be visualized by clicking the Plot button (step 25) (Box 7.9).

7.3.2 Case Study 1: p-(Methylthio)phenyl Nitronyl Nitroxide (MTPNN) in Toluene

The first case study addresses the interpretation, via an *ab initio* integrated computational approach, of CW EPR spectra of MTPNN dissolved in toluene for a wide range of temperatures (155–292K) with minimal resorting to fitting procedures, proving that the combination of sensitive EPR spectroscopy and sophisticated modeling can be highly helpful in providing structural and dynamic information on molecular systems. In particular, the search for new materials with tailored magnetic properties has intensified in recent years. In this field the most popular stable radicals are nitronyl nitroxide (NIT) free radicals. They exhibit a variety of magnetic behaviors: paramagnetism down to very low temperature, ferromagnetism, and antiferromagnetism [29]. Moreover, NITs have also been known as bidentate ligands for various transition and rare-earth metal ions. Ferromagnetic ground states have also been observed in these complexes [30]. For these particular magnetic properties, NIT radicals are particularly appealing as molecular units for composite new materials. In the path toward new magnetic materials, the characterization of the electronic distributions and magnetic properties of isolated radicals is of primary interest. Theoretical predictions of the spin distribution on the radicals by DFT calculations

BOX 7.9 SOLVING THE SLE

The SLE is usually solved by projecting the symmetrized time evolution operator $\tilde{\Gamma} + i\hat{H}^\times = i\tilde{\mathcal{L}}$ and the starting vector $\left|vP_{eq}^{1/2}\right\rangle$ on a suitable basis set that in our case can be initially defined as

$$|\Sigma\rangle = |p^S p^I\rangle \otimes |l\rangle = |\sigma, l\rangle$$

The basis set is given by the direct product of spin operators of the nitroxide, defined by electron and nuclear spin quantum numbers p^S, q^S, p^I, q^I, and of a complete (usually orthonormal for the sake of simplicity) basis set in the functional space in the generic set of stochastic coordinates Q, which is indicated here generically by $|l\rangle$. One needs to define the matrix operator and starting vector elements

$$(\mathbf{L})_{\Sigma,\Sigma'} = \left\langle \Sigma \left| i\tilde{\mathcal{L}} \right| \Sigma' \right\rangle, \quad (\mathbf{v})_\Sigma = \langle \Sigma | 1 \rangle$$

and the matrix–vector counterparts of the tridiagonal coefficient are

$$\beta_{n+1}\mathbf{v}_{n+1} = (\mathbf{L} - \alpha_n \mathbf{1})\mathbf{v}_n - \beta_n \mathbf{v}_{n-1}, \, \alpha_n = \mathbf{v}_n \cdot \mathbf{v}_n, \, \beta_n = \mathbf{v}_n \cdot \mathbf{v}_{n-1}$$

As a first example, let us consider the description of a rigid molecule in solution. No conformational degrees are included and only rotational motion is taken into account. A convenient definition of the stochastic coordinates is $Q = \Omega_{LF \to MF}$, where $\Omega_{LF \to MF}$ is the set of Euler angles defining the instantaneous orientation of frame MF, which is the principal frame of reference for the rotational diffusion tensor $\mathbf{D}$. In isotropic solvents one may write

$$\hat{\Gamma} = \hat{\mathbf{J}}(\Omega_{LF \to MF}) \cdot \mathbf{D} \cdot \hat{\mathbf{J}}(\Omega_{LF \to MF})$$

where $\hat{\mathbf{J}}(\Omega_{LF \to MF})$ is the angular momentum operator for body rotation. The Boltzmann distribution (equilibrium solution) is simply $P_{eq} = 1/8\pi^2$. By defining $\hat{\mathbf{J}}(\Omega_{LF \to MF})$ and $\mathbf{D}$ in the MF, a convenient form is obtained that is directly written in terms of the diffusion tensor principal values

$$\hat{\Gamma} = D_1 \hat{J}_1^2(\Omega_{LF \to MF}) + D_2 \hat{J}_2^2(\Omega_{LF \to MF}) + D_3 \hat{J}_3^2(\Omega_{LF \to MF})$$

The molecule-fixed properties are evidenced by frames MF, GF, and A_nF; the magnetic tensors' orientation is given now with respect to the MF, which is defined with respect to the laboratory frame. The parameters of the SLE for the case of a rigid paramagnetic probe dissolved in an isotropic medium are then defined as the principal values of the diffusion tensor $\mathbf{D}$; the principal values of the $\mathbf{g}$ and $\mathbf{A}_n$ tensors; and Euler angles $\Omega_{MF \to GF}$, $\Omega_{MF \to A_nF}$, which give the relative orientation of the magnetic tensors with respect to the diffusion tensor.

A basis set for the specific ensemble of stochastic coordinates is chosen, that is, $\Omega_{LF \to MF}$; usually normalized Wigner matrix functions are employed

$$|l\rangle \int \frac{1}{(8\pi)^{1/2}} \mathcal{D}_{MK}^{J*}(\Omega_{LF \to MF}) = |JMK\rangle$$

Symmetry is also typically employed by adopting a linearly transformed basis set that accounts for invariance of the Liouvillean for a rotation around the y molecular axis

$$|\Sigma\rangle_K = |\sigma, j^K LMK\rangle = [2(1+\delta_{K,0})]^{-1/2} e^{-i\pi(j^K-1)/4} \left[|+\rangle + j^K s^K |-\rangle\right]$$

where $s^K = (-1)^{L+K}$, with $K \geq 0$, and $j^K = \pm 1$ for $K > 0$, $(-1)^L$ for $K = 0$; and ket symbols $|+\rangle$, $|-\rangle$ stand for $|\Sigma\rangle$ with positive K and $|\Sigma\rangle$ with corresponding opposite K, respectively.

To evaluate explicitly symmetrized or unsymmetrized matrix elements, one needs to make explicit the dependence of the super Hamiltonian $i\hat{H}^\times$ from magnetic and orientational parameters. Following the established route, we adopt a spherical irreducible tensorial representation

$$\hat{H}^\times = \sum_\mu \sum_{l=0,2} \sum_{m,m'=-l}^{l} \mathcal{D}_{mm'}^l(\Omega_{LF \to MF}) F_{\mu,MF}^{(l,m')*} \hat{A}_{\mu,LF}^{(l,m)}$$

where μ runs over all possible interactions, $F_{\mu,MF}^{(l,m')*}$ is built from elements of g and A in the MF, and $\hat{A}_{\mu,LF}^{(l,m)}$ is obtained from spin operators.

are necessary in order to understand the radical–radical interactions in bulk and composite materials. In contrast, the spin density depends strongly on the interaction with the environment that can be very complex in a composite material. It is shown here that for a prototypical NIT like MTPNN [31] in a simple environment as a toluene solution, starting simply from the formula of the radical and the physical parameters of the solvent, it is possible to calculate EPR spectra that afterward demonstrate exceptionally good agreement with the experimental ones, from room temperature to a temperature very near to the glassy transition.

The system geometry is summarized in Fig. 7.11. A set of Euler angles Ω defines the relative orientation of a molecular frame (MF), fixed rigidly on the nitroxide ring, with respect to the LF; the local magnetic frames are in turn defined with respect to MF by proper sets of Euler angles.

The effective Hamiltonian for the system is given by Equation 7.1: two nuclei are explicitly coupled with the paramagnetic center. Within E-SpiReS, the users will modify steps 7 and 8 accordingly to identify the two nuclei. The spectra are then calculated without further adjustments of the temperature-dependent fitted parameters. Figure 7.12 compares the experimental and simulated CW EPR spectra of

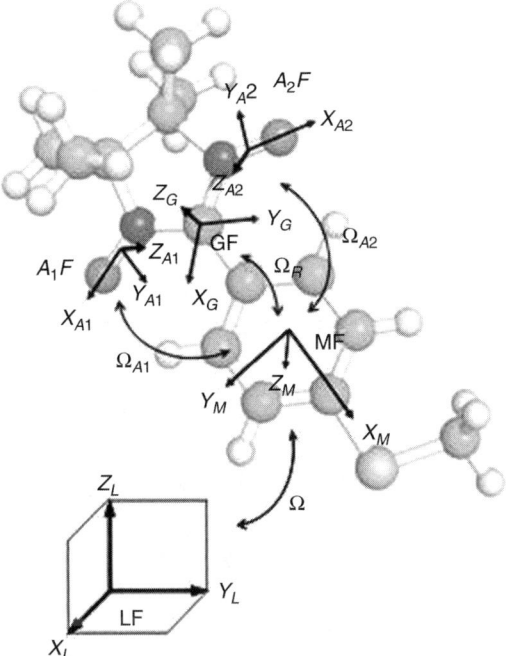

Fig. 7.11 Reference frames and geometry of MTPNN.

MTPNN in toluene in the temperature range of 155–292K. The measured value of g_0 at room temperature ($g_0 = 2.00681$) perfectly matches the predicted theoretical value, obtained as one-third of the trace of the **g** tensor, $g_0^{\text{calc}} = 2.00686$ [17].

7.3.3 Case Study 2: Fmoc-(Aib-Aib-TOAC)2-Aib-OMe in Acetonitrile

The CW EPR spectra of the double spin labeled, 3,10-helical peptide Fmoc-(Aib-Aib-TOAC)2-Aib-OMe dissolved in acetonitrile is considered next. The system is now described by a SLE for the two electron spins interacting with each other and with two ^{14}N nuclear spins, in the presence of diffusive rotational dynamics. Parameterization of diffusion rotational tensor is provided again by a hydrodynamic model. The system Hamiltonian is defined as

$$\hat{H} = \frac{\mu_B}{\hbar} \sum_i \vec{B}_0 \cdot \mathbf{g}_i \cdot \hat{S}_i + \gamma_e \sum_i \hat{I}_i \cdot \mathbf{A}_i \cdot \hat{S}_i - 2\gamma_e J \hat{S}_1 \cdot \hat{S}_2 + \hat{S}_1 \cdot \mathbf{T} \cdot \hat{S}_2 \quad (7.6)$$

where the two radicals are explicitly accounted for by the first term and J and $\mathbf{T}$ terms are included.

The system geometry is summarized in Fig. 7.13. Again, using E-SpiReS for determining the CW EPR of this biradical is comparatively simple: proper definition of paramagnetic centers and coupled nuclei proceeds as in previous examples (Box 7.10).

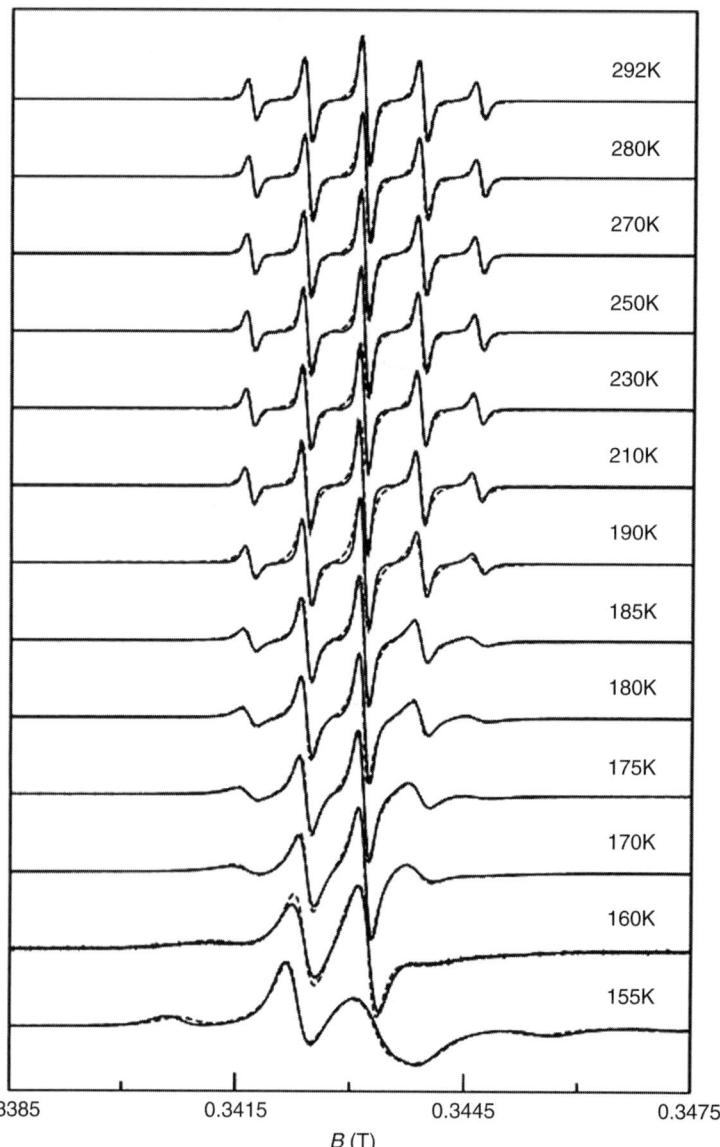

Fig. 7.12 (—) Experimental and (- - -) and simulated CW EPR spectra of MTPNN in toluene in a temperature range of 155–292K.

The spectra of the peptide dissolved in MeCN in a temperature range of 270–330K were simulated and compared. Figure 7.14 shows four theoretical spectra and their relative experimental counterparts.

An allowance was made for a limited adjustment δA of the scalar component $\text{Tr}(\mathbf{A})/3$ of the theoretical hyperfine tensor $\mathbf{A}$. The best agreement is obtained for

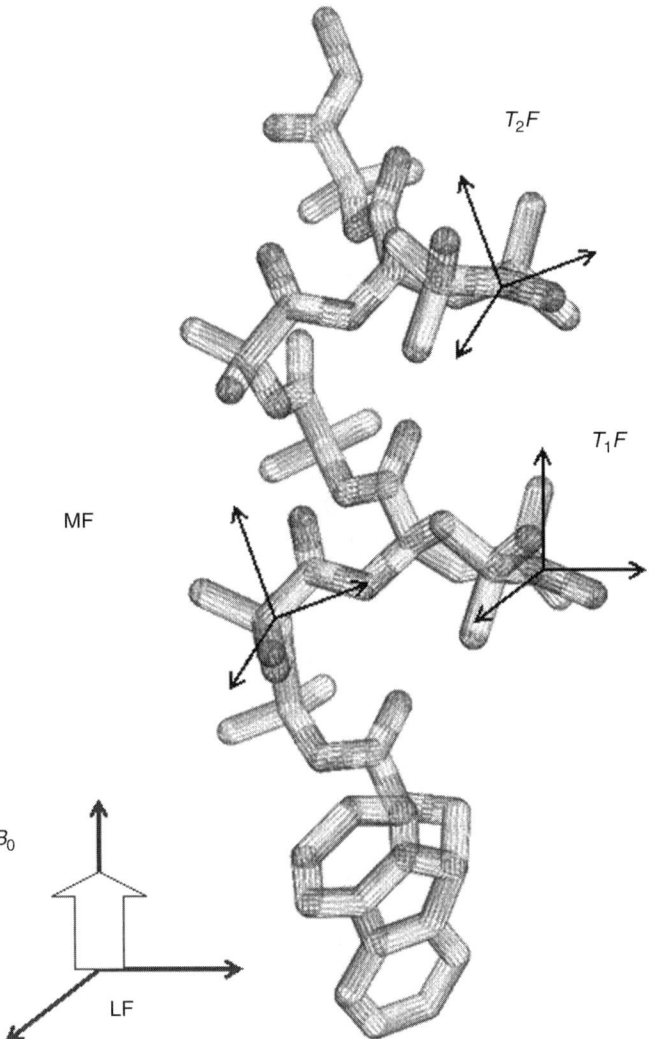

Fig. 7.13 Reference frames and geometry of Fmoc-(Aib-Aib-TOAC)2-Aib-OMe: LF, MF, T_iF (coincident g and A magnetic frame for ith radical, with $i = 1, 2$).

$\delta A = 0.3$ G, which is well within the estimated uncertainty of 0.5 G. The overall agreement between the theoretical and experimental spectra in the considered range of temperatures is good. The set of parameters employed in the simulations seems to be slightly less effective only at the lowest temperature (270K). It should be stressed that no internal dynamics model has been employed to describe collective motions in the heptapeptide, which has been treated again as a simple Brownian rotator, with diffusive properties predicted only on the basis of fixed molecular shape and solvent viscosity. Nevertheless, a reasonable prediction of the change in linewidth and change of intensity is observed in the whole range of temperatures considered, thus confirming that the molecular structure is essentially rigid in solution [18].

BOX 7.10 SPIN–SPIN DIPOLAR INTERACTION

The spin–spin dipolar term is the most critical long-range contribution. Usually, this tensor is calculated by assuming that the two electrons are localized and placed at the center of the N–O bond. In this view, the two electrons are considered just as two point magnetic dipoles and the interaction term is given simply by

$$\mathbf{T} = \frac{\mu_0}{4\pi} \frac{g_e^2 \mu_B^2}{\hbar r^3} \left[\mathbf{1}_3 - \frac{3}{r^2} \begin{pmatrix} r_x^2 & r_x r_y & r_x r_z \\ r_y r_x & r_y^2 & r_y r_z \\ r_z r_x & r_z r_y & r_z^2 \end{pmatrix} \right]$$

where r is the distance between the two localized electrons, which is the distance between the centers of the N–O bonds of the two TOAC nitroxides. Obviously, this is only an approximation because the electrons are not fixed in one point of space but delocalized in a molecular orbital. A complete QM computation starting from the computed spin density is still lacking for large molecules. Thus, we resorted to the following computational strategy based on the well-known localization of nitroxide singly occupied molecular orbitals (π^* orbitals) on the NO moiety (see Fig. 7.6). As a consequence, the corresponding electron density can be fitted by linear combinations (with equal contributions) of effective $2p_z$ atomic orbitals of nitrogen and oxygen:

$$\Psi' = N' \left[\phi_{210}^{N_1}(\mathbf{r} - \mathbf{R}_{N_1}) - \phi_{210}^{O_1}(\mathbf{r} - \mathbf{R}_{O_1}) \right]$$

$$\Psi'' = N'' \left[\phi_{210}^{N_2}(\mathbf{r} - \mathbf{R}_{N_2}) - \phi_{210}^{O_2}(\mathbf{r} - \mathbf{R}_{O_2}) \right]$$

Next the atomic orbitals are represented by Slater type orbitals of the form

$$\phi_{210}(\mathbf{r}) = \sqrt{\frac{4}{3} \alpha^5} r e^{-\alpha r} Y_{1,0}(\theta, \phi)$$

where $\alpha = Z_{\text{eff}}/2$ Hartree-1 and Z_{eff} is the effective nuclear charge; standard Clementi–Raimondi values of $Z_{\text{eff}} = 3.83$ for nitrogen and $Z_{\text{eff}} = 4.45$ for oxygen were used. The molecular geometry allows us to conclude that only the $T^{(2,0)}$ component contributes significantly to the dipolar tensor. As expected, at high distances the point approximation converges to the exact approach, but increasing differences are found when the distance is <7 Å.

7.3.4 Case Study 3: Tempo-Palmitate in 5-Cyanobiphenyl (5CB)

The last case study is an example of ICS applied to the case of nematic liquid crystalline environments by performing simulations of the EPR spectra of the prototypical

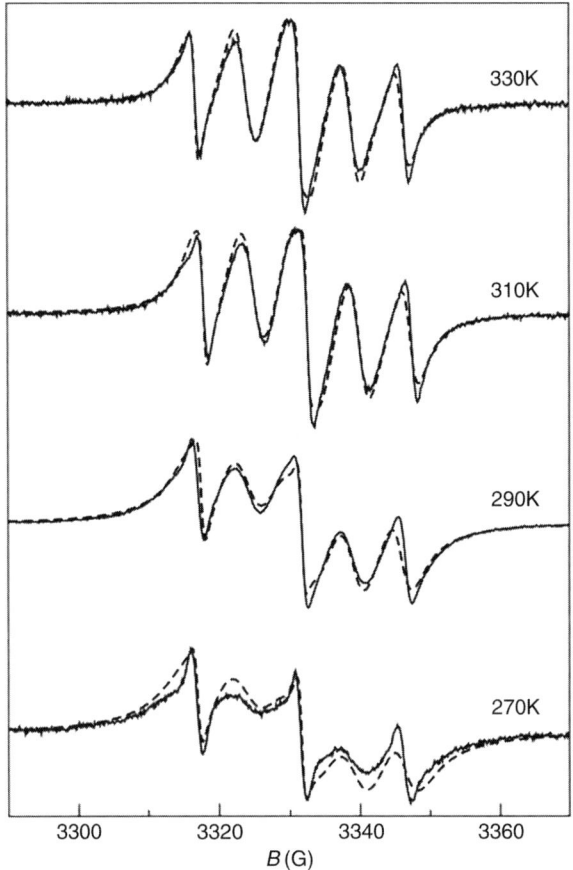

Fig. 7.14 Experimental (solid lines) and theoretical (dashed lines) CW EPR spectra of heptapeptide 1 in MeCN at 330, 310, 290, and 270K.

nitroxide probe 4-(hexadecanoyloxy)-2,2,6,6-tetramethylpiperidine-1-oxy in isotropic and nematic phases of 5CB. The procedure runs as (1) determination of geometric and local magnetic parameters by QM calculations taking into account the solvent and, when needed, vibrational averaging contributions; (2) numerical solution of a SLE in the presence of diffusive rotational dynamics, based on (3) parameterization of a diffusion rotational tensor provided by a hydrodynamic model. Note that an internal degree of freedom is explicitly taken into account (meaning that step 6 in E-SpiReS is modified) and that the necessary conformational potential is evaluated through the QM approach (Fig. 7.15).

The CW EPR spectra of tempo-palmitate in 5CB in a range of temperatures from 316.92K (isotropic phase) to 299.02K (nematic phase) was simulated. Figure 7.16 reports five simulated spectra, which are superimposed on experimental spectra

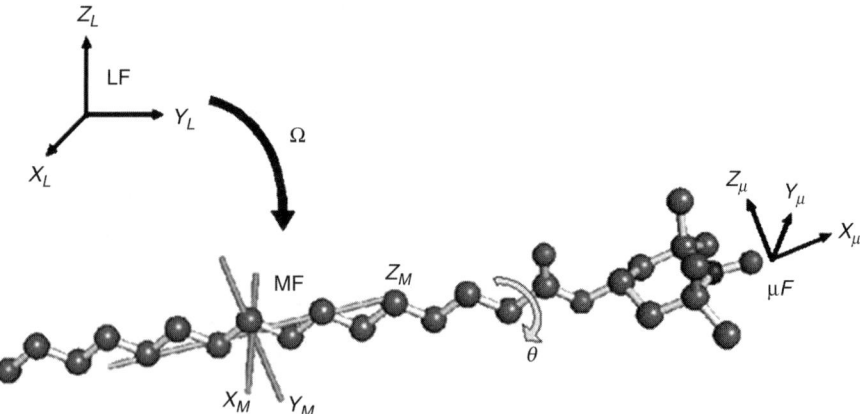

Fig. 7.15 Geometry, reference frames, and internal degrees of freedom for tempo-palmitate in 5CB.

taken from the literature. The results show that it is again possible to apply ICS even in the quite demanding situation represented by large nitroxides in nematic phases. In particular, the spectra at different temperatures and in different phases are reproduced with a very limited number of fitting parameters (ordering potential and isotropic parts of magnetic tensors), which could possibly be replaced by *a priori* computations in the near future. The computed value for the isotropic hyperfine splitting (15.3 G) nicely fits the experimental value in the isotropic phase. However, at a lower temperature local effects come into play that cannot be reproduced by the continuum solvent model employed in our computations.

7.3.5 E-SpiReS Demo Availability

A running demo of E-SpiReS is available on the website of the Italian Computational Network European Virtual Integrated Laboratory for Large-Scale Applications in a Geographically Distributed Environment or EU-VILLAGE (http://village.unina.it). The demo allows the user to perform most of the operations described above, including the evaluation and fitting of CW EPR spectra of mono- and biradicals at the desired temperature and/or solvent viscosity for rigid molecular systems only (no internal degrees of freedom) and in single-processor mode. The user should be aware that simulation of very slow motional regimes (freezing point) spectra are quite demanding computationally.

Instructions for downloading and setting up the code on a PC (with Linux OS or Windows Vista) are available online at the same location.

At the same address the fully functional form of E-SpiReS is also directly accessible as a web service with restricted access. At this stage of development, any interested user is asked to write directly to one of the authors of this chapter. Use of E-SpiReS will be provided via secure password-protected access and under proper agreements.

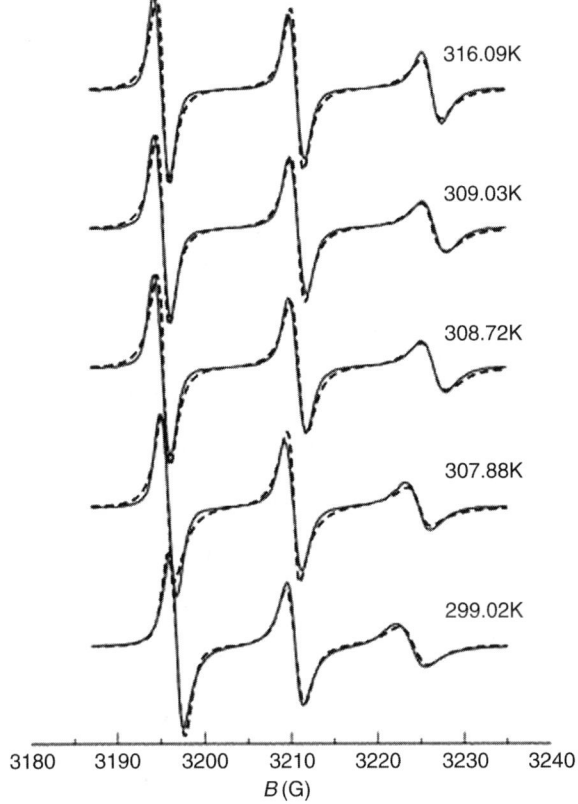

Fig. 7.16 Calculated and experimental spectra of tempo-palmitate in 5CB.

7.4 CONCLUSIONS

7.4.1 Perspectives

Understanding the influence of molecular dynamics and orientation on EPR spectroscopy of radicals can be seen in perspective as a logical process confined between two extremes. On the one hand, there is the case of freely rotating radicals in solutions of low viscosity: this is the so-called Redfield (or fast motion) limit that is understood as the dynamic regime for which average magnetic tensorial properties can be assumed. That is, because all molecular orientations are rapidly explored in the experimental time-scale window, magnetic tensors entering the spectrum evaluation are reduced to averaged scalar.

On the other hand, there is the completely dynamic-less situation of a radical embedded in a crystalline structure or in a glassy highly viscous fluid. In both cases inhomogeneous broadening, which is a superposition of spectra resulting from different spatial orientations of the paramagnetic probe, needs to be considered: averaging of the calculated spectra, one for each orientation, gives the final resulting line shape. In glassy fluids, powder, frozen liquid crystals, or single crystals local anisotropies are averaged out assuming a proper distribution of orientations, which can vary according to structural information: from a completely global isotropic function (glasses, powder) to a possible absolute orientation (liquid crystals) to a sharp function centered on a few preferential orientations (single crystals). In these cases, short-time dynamical effects can also be significant and can be studied by techniques analogous to those sketched for solutions.

However, computational experience in this field is just beginning and extension of our ICS to the spectroscopic techniques able to characterize dynamical effects in the solid state (e.g., echo-detected EPR) is one of the most exciting perspectives for the near future. In fact the stochastic Liouville approach, combined with the possibility of introducing additional variables to distinguish between local and global orientations, allows the simulation of spectra in any regime of motion and in any type of orienting potential.

In addition to this extension, the ICS approach can be further developed along three main lines of action of increasing complexity:

1. Setting up an online grid-oriented and user-friendly version of existing software.
2. Extension to advanced EPR spectroscopies and to paramagnetic metallic species. Both of these items involve either proper redefinitions of calculated observables (i.e., essentially an upgrade of Equation 7.4) or generalizations of already available code to multinuclei treatments (exact or approximate). Although relatively trivial, these upgrades impose significant additional computational burdens on the whole procedure, requiring, for instance, a complete diagonalization of the Liouvillean matrix instead of its reduction to tridiagonal form.
3. Inclusion of mixed dynamic approaches to account for multiscale processes in large biological molecules.

However, even at its present stage of development, the application of ICS to EPR allows a clear-cut explanation of EPR spectra in terms of structural and dynamic molecular characteristics. This is completely different from the conventional approach, which considers the spectrum as the "target" of a fitting procedure of molecular, mesoscopic, and macroscopic parameters entering the model. As observed above, this latter strategy can be very helpful in providing detailed characterization of molecular parameters. However, ICS, which is a combination of QM calculations of structural parameters and the direct feeding of calculated molecular parameters into dynamic models based on dynamic modeling, promises to couple prediction and interpretation in a very effective way.

7.4.2 Summary

The main objective in this chapter has been to discuss the degree of advancement of ICS to the interpretation of CW EPR of organic radicals and biradicals in solvated environments via the combination of advanced QM approaches and stochastic modeling of relaxation processes. The ICS *ab initio* prediction of CW EPR spectra is able to assess molecular characteristics entirely from computational models and a direct comparison with the experimental data. The sensitivity of the integrated methodology to the overall molecular geometry is proved, in all of the cases that are discussed, by the significant dependence of the calculated spectrum on arbitrary modifications of the molecular geometry or dynamic properties. For instance, in heptapetide biradicals, ICS is sensitive enough to distinguish between different helix conformations [18].

Some adjustment of computed magnetic tensors is probably unavoidable for a quantitative fitting of experimental spectra, particularly for large systems where only DFT approaches are feasible. However, the number of free parameters (if any) is limited enough that convergence to the true minimum can be granted. At the same time the allowed variation of parameters from their QM value is well within the differences between different structural models. Thus, pending further developments of DFT models, ICS is already able to predict CW EPR spectra of large molecular systems in solvents starting only from the chemical structure of the solute and some macroscopic solvent properties. Implementation in a user-friendly package finally could spread the systematic usage of ICS in current real-life EPR laboratories, as much as standard QM packages for structural molecular properties are already diffuse in most modern chemistry research facilities.

REFERENCES

1. (a) Barone, V. *J. Chem. Phys.* **1994**, *101*, 6834; (b) Barone, V. *J. Chem. Phys.* **1994**, *101*, 10666; (c) Barone, V. *Theor. Chem. Acc.* **1995**, *91*, 113; (d) Barone, V. Structure, Magnetic Properties and Reactivities of Open-Shell Species from Density Functional and Self-Consistent Hybrid Methods. In *Advances in Density Functional Theory, Part I;* Chong, D. P., Ed.; World Scientific: Singapore, 1995; p 287; (e) Improta, R.; Barone, V. *Chem. Rev.* **2004**, *104*, 1231.
2. Malkina, O. L.; Vaara, J.; Schimmelpfenning, J. B.; Munzarova, M. L.; Malkin, V. G.; Kaupp, M. J. *J. Am. Chem. Soc.* **2000**, *122*, 9206.
3. Neese, F. *J. Chem. Phys.* **2001**, *115*, 11080.
4. The Internet Movie Database, http://www.imdb.com/.
5. Tomasi, J.; Mennucci, B.; Cammi, R. *Chem. Rev.* **2005**, *105*, 2999.
6. Brancato, G.; Rega, N.; Barone, V. *J. Am. Chem. Soc.* **2007**, *129*, 15380.
7. Barone, V. *J. Chem. Phys.* **2005**, *122*, 014108.
8. Brancato, G.; Barone, V.; Rega, N. *Theor. Chem. Acc.* **2007**, *117*, 1001.
9. Van Doorslaer, S.; Vinck, E. *Phys. Chem. Chem. Phys.* **2007**, *9*, 4620.
10. White, G. F.; Ottignon, L.; Georgiou, T.; Kleanthous, C.; Moore, G. R.; Thomson, A. J.; Oganesyan, V. S. *J. Magn. Res.* **2007**, *185*, 191.
11. Jeschke, G.; Polyhach, Y. *Phys. Chem. Chem. Phys.* **2007**, *9*, 1895.
12. (a) Moro, G.; Freed, J. H. The Lanczos Algorithm in Molecular Dynamics: Calculation of Spectral Densities. In *Large-Scale Eigenvalue Problems;* Cullum, J., Willoughby, R., Eds.; Elsevier: New York, 1986; (b) Schneider, D. J.; Freed, J. H. *Adv. Chem. Phys.* **1989**, *73*, 487.

13. (a) Freed, J. H. *J. Chem. Phys.* **1977**, *66*, 4183; (b) Meirovitch, E.; Igner, D.; Igner, E.; Moro, G.; Freed, J. H. *J. Chem. Phys.* **1982**, *77*, 3915; (c) Vasavada, K. V.; Schneider, D. J.; Freed, J. H. *J. Chem. Phys.* **1987**, *86*, 647.
14. Schneider, D. J.; Freed, J. H. *Adv. Chem. Phys.* **1989**, *73*, 387.
15. (a) Polimeno, A.; Freed, J. H. *J. Phys. Chem.* **1995**, *99*, 10995; (b) Earle, K. A.; Moscicki, J. K.; Polimeno, A.; Freed, J. H. *J. Chem. Phys.* **1997**, *106*, 9996; (c) Sastry, V. S. S.; Polimeno, A.; Crepeau, R. H.; Freed, J. H. *J. Chem. Phys.* **1996**, *105*, 5753; (d) Sastry, V. S. S.; Polimeno, A.; Crepeau, R. H.; Freed, J. H. *J. Chem. Phys.* **1996**, *105*, 5773.
16. (a) Liang, Z.; Freed, J. H.; Keyes, R. S.; Bobst, A. M. *J. Phys. Chem. B* **2000**, *104*, 5372; (b) Liang, Z.; Lou, Y.; Freed, J. H.; Columbus, L.; Hubbel, W. L. *J. Phys. Chem. B* **2004**, *108*, 17649.
17. (a) Polimeno, A.; Zerbetto, M.; Franco, L.; Maggini, M.; Corvaja, C. *J. Am. Chem. Soc.* **2006**, *128*, 4734; (b) Barone, V.; Polimeno, A. *Phys. Chem. Chem. Phys*. **2006**, *8*, 4609; (c) Barone, V.; Brustolon, M.; Cimino, P.; Polimeno, A.; Zerbetto, M.; Zoleo, A. *J. Am. Chem. Soc.* **2006**, *128*, 15865.
18. (a) Zerbetto, M.; Carlotto, S.; Polimeno, A.; Corvaja, C.; Franco, L.; Toniolo, C.; Formaggio, F.; Barone, V.; Cimino, P. *J. Phys. Chem. B* **2007**, *111*, 2668; (b) Carlotto, S.; Cimino, P.; Zerbetto, M.; Franco, L.; Corvaja, C.; Crisma, M.; Formaggio, F.; Toniolo, C.; Polimeno, A.; Barone, V. *J. Am. Chem. Soc.* **2007**, *129*, 11248.
19. Rega, N.; Brancato, G.; Barone, V. *Chem. Phys. Lett.* **2006**, *422*, 367.
20. Pavone, M.; Cimino, P.; De Angelis, F.; Barone, V. *J. Am. Chem. Soc.* **2006**, *128*, 4338.
21. (a) Favro, L. D. *Phys. Rev.* **1960**, *119*, 53; (b) Hubbard, P. S. *Phys. Rev.* **1972**, *A6*, 2421; (c) Fixman, M.; Rider, K. *J. Chem. Phys.* **1969**, *51*, 2429.
22. (a) Bencini, A.; Totti, F.; Daul, C. A.; Doclo, K.; Fantucci, P.; Barone, V. *Inorg. Chem.* **1997**, *36*, 5022; (b) Improta, R.; Barone, V.; Kudin, K.; Scuseria, G. E. *J. Am. Chem. Soc.* **2002**, *124*, 113.
23. (a) Yawada, Y.; Tsuneda, T.; Yanagisawa, S.; Yanai, T.; Hirao, K. *J. Chem. Phys.* **2004**, *120*, 8425; (b) Kamiya, M.; Sekino, H.; Tsuneda, T.; Hirao, K. *J. Chem. Phys.* **2005**, *122*, 234111.
24. Cornell University. www.acert.cornell.edu/index_files/acert_ftp_links.php (accessed Oct. 2008).
25. EasySpin. www.easyspin.org (accessed Oct. 2008).
26. Bruker BioSpin. www.bruker-biospin.com/epr_software.html (accessed Sept. 2008).
27. University of Berlin. www.chemie.fu-berlin.de/chemistry/epr/eprft.html (accessed Sept. 2008).
28. Scientific Software Services. http://scientific-software.com/products.php (accessed Sept. 2008).
29. Osiecki, J. H.; Ullman, E. F. *J. Am. Chem. Soc.* **1968**, *90*, 1078.
30. (a) Awaga, K.; Inabe, T.; Okayama, T.; Maruyama, Y. *Mol. Cryst. Liq. Cryst.* **1993**, *232*, 79; (b) Caneschi, A.; Chiesi, P.; David, L.; Ferraro, F.; Gatteschi, D.; Sessoli, R. *Inorg. Chem.* **1993**, *32*, 1445; (c) Caneschi, A.; Ferraro, F.; Gatteschi, D.; Le Lirzin, A.; Novak, M.; Rentschler, E.; Sessoli, R. *Adv. Mater.* **1995**, *7*, 476; (d) Gorini, L.; Caneschi, A.; Menichetti, S. *SYNLETT* **2006**, *6*, 948; (e) Caneschi, A.; David, L.; Ferraro, F.; Gatteschi, D.; Fabretti, A. C. *Inorg. Chim. Acta* **1995**, *235*, 159.
31. Pillet, S.; Souhassou, M.; Pontillon, Y.; Caneschi, A.; Gatteschi, D.; Lecomte, C. *New J. Chem.* **2001**, *25*, 131.

PART II
Applications

8 Spin Trapping

ANGELO ALBERTI and DANTE MACCIANTELLI

ISOF-CNR, Area della Ricerca di Bologna, Via P. Gobetti 101, 40129-Bologna, Italy

8.1 WHAT IS SPIN TRAPPING AND WHY USE IT?

Because it is able to provide structural, quantitative, and mechanistic information on radical species present in a system, electron paramagnetic resonance (EPR) can be safely defined as the technique of choice for the study of free radicals. In contrast, in liquid systems free radicals are usually "transient species" that undergo very fast unimolecular (e.g., fragmentation and rearrangement) and/or bimolecular "termination" reactions (e.g., dimerization, coupling, addition). If the reaction leading to the formation of a given radical in solution has a rate comparable to those leading to its termination, a steady-state concentration can be reached that is large enough to allow its detection via EPR. Although this may often be the case when a certain radical is being generated for characterization and structural studies, which is when the radical is the reaction product, it is very rare that the steady-state concentration of transient radicals intervening in chemical reactions rises to exceed the detection threshold. Thus, in these cases EPR spectroscopy would be of no use. To overcome this impasse, the spin trapping technique has been devised, which basically consists of adding to the system under consideration a small amount of a diamagnetic molecule amenable to readily undergoing addition by the transient radical to yield a new, more persistent radical species, the concentration of which can reach detectable levels. The molecule undergoing radical addition is called a spin trap and the resulting radical is called a spin adduct. The spin trapping process can be simply schematized as in Reaction 8.1, where SP and SA stand for spin trap and spin adduct, respectively.

The introduction of spin trapping occurred in the late 1960s, and since then the technique has been applied to an enormous number of systems under a great variety of conditions. Although spin trapping can be carried out in liquid, gaseous, or solid phases, only spin trapping in liquid solution is dealt with here in detail, because it is the most exploited of the three possibilities.

Electron Paramagnetic Resonance. Edited by Brustolon and Giamello
Copyright © 2009 John Wiley & Sons, Inc.

$$R-N=O \; + \; \cdot X \;\longrightarrow\; \underset{X}{\overset{R}{>}}N-O\cdot$$

ST   Radical   SA

Reaction 8.1

8.2 SPIN TRAPS

Although any molecule inclined to undergo homolytic addition might in principle be regarded as a potential spin trap, to gain applicative interest the successful candidates should possess a number of features, some of which may depend on the actual purpose of the individual experiment to be carried out. Thus, a good spin trap has the following properties:

- should be stable under the actual reaction conditions,
- should not participate in any side reaction leading to paramagnetic species other than those involved in the process under study (to avoid misleading information),
- should undergo radical addition very quickly (to avoid further reactions of the transient radicals prior to their trapping),
- should lead to spin adducts as persistent as possible (obviously, more persistent than the trapped radical itself),
- should lead to spin adducts characterized by relatively simple EPR spectra (in many reactions, different radical species are present simultaneously and their trapping might lead to spectra that are too complex to be rationalized), and
- should provide the largest possible wealth of structural information on the trapped species. (Note that spin adducts can only provide indirect information on the trapped radical, but this "second hand" information is the only information available in most cases.)

In addition, the utmost care must be exercised in interpreting the results of a spin trapping experiment. Although the failure to detect a spin adduct does not necessarily mean that radical species were not present in the system under examination, the observation of a spin adduct does not necessarily imply the occurrence of a direct trapping process. As shown in Reaction 8.2, the spin trap might undergo a nucleophilic addition by a nucleophile to give an anionic species that in the presence of an oxidizing agent would afford the spin adduct SA· (*Forrester–Hepburn mechanism*). Under appropriated circumstances the spin trap might be initially oxidized to its radical cation, which by subsequent coupling with a nucleophile would again afford the

$$ST \xrightarrow{R^-} SA^- \xrightarrow{Ox.} \cdot SA$$

Reaction 8.2

$$\text{ST} \xrightarrow{\text{Ox.}} \cdot\text{ST}^+ \xrightarrow{\text{R}^-} \cdot\text{SA}$$

Reaction 8.3

spin adduct SA· (this is the so-called inverted spin trapping outlined in Reaction 8.3). Thus, whatever the spin trap used, its "chemistry" and characteristics, in particular its redox properties, must be well known beforehand.

A large variety of trapping agents have been developed since the introduction of the spin trapping technique. The most commonly used spin traps are aliphatic or aromatic nitroso compounds and open-chain or cyclic nitrones; in both cases the resulting spin adducts are nitroxides. The spin adducts from the two types of compounds differ in that in those from nitroso derivatives (nitroxides of type **1**) the trapped radical is directly bound to the nitrogen of the nitroxidic function, whereas in those from nitrones (nitroxides of type **2** or **3**) it is bound to the carbon atom adjacent to it (Reactions 8.4–8.6).

In nitroxides **1** information on the trapped radicals can be directly derived from the hyperfine splitting (hfs) of the nitrogen atom, which varies consistently depending on the nature of ·R′ (e.g., if it is a carbon, a nitrogen, an oxygen, a sulfur, or a phosphorus centered radical), and, in most cases, from the coupling constants of the magnetically active nuclei present in the residue R′. In the case of nitroxides **2**, instead the nitrogen hfs is not so dependent on the nature of the trapped radical and useful information can only be derived from the small but significant variations of the α-hydrogen coupling constant. Nitrogen splitting is slightly indicative also in nitroxides **3** from cyclic nitrones, which exhibit variations of the α-hydrogen coupling constant that are larger than those observed for nitroxides **2**.

$$\text{R—N=O} + \cdot\text{R}' \longrightarrow \begin{matrix} \text{R}' \\ \text{R} \end{matrix}\!\!\!>\!\!\text{N—O}\cdot \qquad (1)$$

Reaction 8.4

$$\begin{matrix}\text{R}\\\text{H}\end{matrix}\!\!\!>\!\!\text{C=N}^+\!\!\!<\!\!\begin{matrix}\text{O}^-\\\text{R}'\end{matrix} + \cdot\text{R}'' \longrightarrow \text{R}''\!-\!\underset{\underset{\text{H}}{|}}{\overset{\overset{\text{R}}{|}}{\text{C}}}\!-\!\text{N}\!\!\!<\!\!\begin{matrix}\text{O}\cdot\\\text{R}'\end{matrix} \qquad (2)$$

Reaction 8.5

(cyclic nitrone + ·R → cyclic nitroxide **3**)

Reaction 8.6

$$\underset{X}{\overset{Ph}{>}}C=S + \cdot R \longrightarrow \underset{X}{\overset{Ph}{>}}\overset{\cdot}{C}-SR$$

X = Ph, SiPh$_3$ (4)

Reaction 8.7

$$\underset{RS}{\overset{(R'O)_2\overset{O}{\overset{\|}{P}}}{>}}C=S + \cdot R'' \longrightarrow \underset{RS}{\overset{(R'O)_2\overset{O}{\overset{\|}{P}}}{>}}\overset{\cdot}{C}-SR''$$

(5)

Reaction 8.8

Although nitroso compounds and nitrones are by far the more widely used spin traps, other compounds have been successfully used in spin trapping experiments. Among these, a major role is played by aromatic thioketones and aryl-organometallic thioketones, which, following a thiophilic addition of the attacking radical to the C=S double bond, lead to adducts having a thioketyl-like general structure **4** (Reaction 8.7).

Here, the information on the nature of the trapped radical is derived from the values of the coupling constants of the magnetically active nuclei present in the R residue. Phosphoryldithioformates also act as spin trapping agents, undergoing a thiophilic addition to the C=S double bond and affording adducts **5** (Reaction 8.8).

Although the spectra of adducts **5** are dominated by a large phosphorus splitting, this is rather insensitive to the nature of the trapped radical and in this case the structural information is also derived from the coupling constants of the magnetically active nuclei present in R''.

Of course, many other molecules undergo radical addition and yield more or less persistent adducts, most of them specific for a particular family of radicals. Thus, although these molecules are not normally included among those considered as spin traps, they should not be disregarded *a priori* when designing a trapping experiment. Carbonyl compounds, in particular aromatic ketones and quinones, provide a typical example of such derivatives. These species are of no EPR use in the trapping of alkyl radicals because they undergo attack at the carbonyl carbon, leading to normally undetectable oxygen centered spin adducts. However, they are very efficient in trapping metal centered radicals $\cdot MR_n$ (M = Si, Ge, Sn, Pb, Mn, Re, etc.) to give ketyl-like spin adducts that may be particularly persistent in the case of *ortho*-quinones due to stabilization by metal chelation (spin adduct **6**, Reaction 8.9).

para-Quinones may also be seen as unconventional spin trapping agents, although, as demonstrated later, because of the presence of both carbon–carbon and carbon–oxygen double bonds, they might instead be included among polyfunctional spin traps.

[Reaction 8.9 scheme: phenanthrenequinone + •SnR$_3$ → cyclic adduct (6)]

Reaction 8.9

$$R-NO_2 + \cdot SiR'_3 \longrightarrow R-N(OSiR'_3)(O\cdot)$$

R = Phenyl, 2-thienyl, α/β-naphthyl,... (7)

Reaction 8.10

Nitrosubstituted aromatic compounds are particularly efficient spin traps for organometallic radicals centered at a group 14 element, and the trapping process leads to rather persistent spin adducts with general structure **7** (i.e., oxynitroxides; Reaction 8.10).

8.2.1 Conventional Spin Traps

8.2.1.1 Aliphatic Nitroso Compounds. 2-Methyl-2-nitrosopropane (**8**, MNP) is one of the first and certainly the most widely employed spin trapping agent. The reasons for its success are found in the high persistence of its adducts, in the simplicity of their EPR spectra, and, last but not least, in its commercial availability as a dimer in a pure form. In solution the dimer dissociates to give a light-blue colored monomer ($\lambda_{max} \sim 670$ nm, $\varepsilon_M = 21$ mol^{-1} cm^{-1} in cyclohexane).

As shown in Reaction 8.11, MNP readily undergoes radical addition to produce nitroxides, the spectral parameters of which provide a wealth of information on the nature of the trapped radical. Thus, the nitrogen hfs constant may vary significantly, assuming values of 0.7–0.8 mT when R is an acyl radical, 1.4–1.6 mT when R is an alkyl radical, 1.7–1.8 mT when R is a thiyl radical, and 2.6–2.9 mT when R is an alkoxy radical. When present, the coupling constants exhibited by the magnetically

$$^tBu-N=O + \cdot R \longrightarrow {}^tBu(R)N-O\cdot$$

(**8**, MNP)

Reaction 8.11

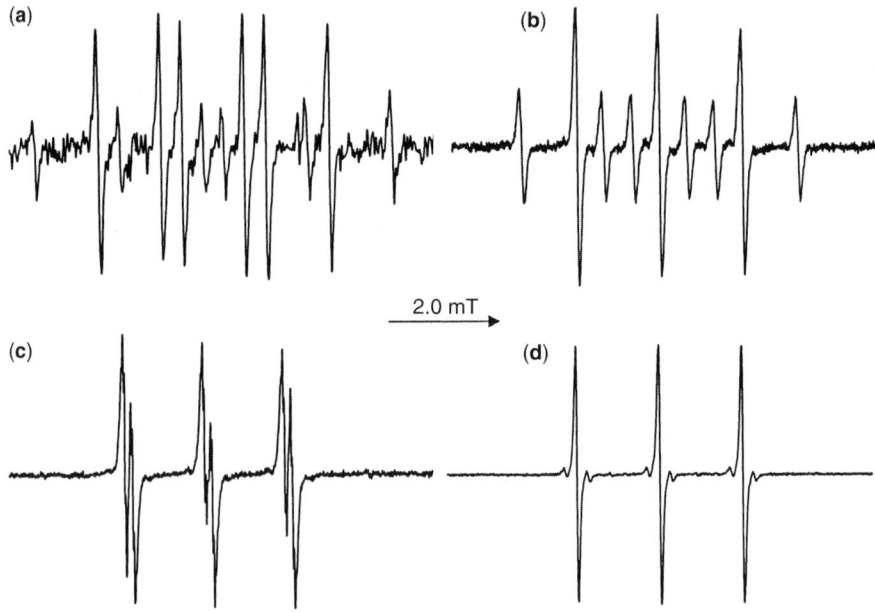

Fig. 8.1 The EPR spectra observed at room temperature upon photolysis of benzene solutions of MNP and $Re_2(CO)_{10}$ containing (a) CH_3I, (b) CH_3CH_2I, (c) $(CH_3)_2CHI$, and (d) $(CH_3)_3CI$.

active nuclei in the radical fragment are particularly useful for the identification of the trapped radicals. This is well evidenced in Fig. 8.1a–c, which displays the spectra of the nitroxides resulting from the trapping of four alkyl radicals.

Note that the coupling constants of the β-hydrogen atoms, which are those bound to the α carbon, decrease with the number of substituents carried by that carbon. This is because these hydrogen atoms acquire spin density through a hyperconjugative mechanism, the efficiency of which is strongly dependent on the average dihedral angle formed by the carbon–hydrogen bond and the p_z orbital on nitrogen, that is, on the conformational preference of the R residue. Thus, methyl-nitroxides are characterized by large hydrogen hfs constants, whereas small values are exhibited by secondary β hydrogens that tend to lie close to the nodal plane of the p_z nitrogen orbital, as is the case when R is an isopropyl group, and intermediate values are expected for primary β-hydrogen atoms. γ-Hydrogen coupling constants may also be detected, but this is not always the case. The small doublets appearing on the two sides of the main nitrogen lines in Fig. 8.1d are ^{13}C satellites and represent a very distinctive feature of di-*tert*-butyl nitroxide.

A feature clearly emerging from Fig. 8.1 is the evident improvement of the spectral signal to noise ratio with the degree of substitution of the α carbon of the R residue. Although nitroxides mainly decay bimolecularly, they do not normally dimerize. They instead tend to undergo a disproportionation process leading to the formation of a nitrone and a hydroxylamine, as outlined in Reaction 8.12.

$$2 \underset{RR'C}{\overset{^tBu}{\diagdown}} \underset{H}{\overset{|}{N}}{-}O\cdot \longrightarrow \underset{RR'C}{\overset{^tBu}{\diagdown}} \overset{+}{N}{-}O^- + \underset{RR'C}{\overset{^tBu}{\diagdown}} \underset{H}{\overset{|}{N}}{-}OH$$

A nitrone   A hydroxylamine

Reaction 8.12

TABLE 8.1 EPR Spectral Parameters for Selected Nitroxides $^tBuN(O\cdot)R$ Derived from MNP

Radical ·R	Solvent	a_N, mT	a_{other}, mT	g
Alkyls				
·CH$_3$	Benzene	1.535	1.165 (3H$_\beta$)	2.0060
·CH$_2$CH$_3$	Benzene	1.525	1.027 (2H$_\beta$)	
·CH$_2$CH$_3$	Water	1.710	1.130 (2H$_\beta$)	
·CH(CH$_3$)$_2$	Benzene	1.525	0.149 (H$_\beta$) 0.036 (6H$_\gamma$)	
·CH(CH$_3$)$_2$	Water	1.683	0.180 (H$_\beta$) 0.030 (6H$_\gamma$)	
·C(CH$_3$)$_3$	Benzene	1.536		2.0060
·C(CH$_3$)$_3$	Water	1.719		
Haloalkyls				
·CCl$_2$CHCl$_2$	Methylene chloride	1.280	0.45 (2Cl)	
·CCl$_3$	Chloroform	1.273	0.220 (3Cl)	2.0065
·CF$_2$CF$_3$	Methylene chloride	1.130	2.080 (2F$_\beta$) 0.39 (3F$_\gamma$)	2.0065
·CF$_3$	Methylene chloride	1.220	1.250 (3F)	
Aminoalkyls				
·N(CH$_3$)$_2$	Benzene	1.840	0.095 (N$_\alpha$)	2.0050
·N(CH$_2$CH$_3$)$_2$	Benzene	1.790		
Alkoxyls				
·OCH$_3$	Methanol	2.970	0.16 (3H$_\gamma$)	
·OCH$_2$CH$_3$	Ethanol	2.910	0.110 (2H$_\gamma$)	
·OCH(CH$_3$)$_2$	Isopropanol	2.840		
·OC(CH$_3$)$_3$	Benzene	2.717		2.0054
Alkythiyls				
·SCH$_3$	Benzene	1.880	0.080 (3H$_\gamma$)	2.0066
·SCH$_2$CH$_3$	Benzene	1.790		2.0070
·SCH(CH$_3$)$_2$	Benzene	1.714		2.0069
·SC(CH$_3$)$_3$	Benzene	1.542		2.0067
Acyls				
·C(O)CH$_3$	Toluene	0.725		2.0066
·C(O)CH$_2$CH$_3$	Benzene	0.820		
·C(O)C(CH$_3$)$_3$	Methyl acrylate	0.800		
Phosphorus radicals				
·P(CH$_2$CH$_3$)$_2$	Benzene	1.162	0.139 (P)	2.0068
·P(O)(CH$_2$CH$_3$)$_2$	Benzene	1.061	1.255 (P)	2.0067

$$>\!\!\ddot{N}\!\!-\!\!O\cdot \quad \longleftrightarrow \quad >\!\!\overset{\cdot}{\underset{+}{N}}\!\!-\!\!O^- \quad \quad >\!\!\overset{\cdot}{\underset{+}{N}}\!\!-\!\!O^-\!\!-\!\!H\!\!-\!\!O\!\!\diagup^{H}$$

    (9)   (10)   (11)

Block 8.1

The disproportionation process requires the presence of a β-hydrogen atom and is therefore easier the lower the degree of substitution of the α carbon. Thus, under the same experimental conditions the steady-state concentration of the nitroxides in Fig. 8.1a–d increases in the order N-methyl $<$ N-ethyl $<$ N-isopropyl $<$ N-tert-butyl as does the spectral signal to noise ratio.

The variations of the nitrogen hfs are almost irrelevant for MNP adducts resulting from the trapping of radicals of the same family (e.g., alkyl radicals), but they may be very relevant when radicals of different families are trapped (e.g., alkyls, alkoxyls, thiyls, etc.). As an example, Table 8.1 lists the spectral parameters of some MNP-derived nitroxides. The nitrogen hfs constants of nitroxides also exhibit a marked solvent dependence, because nitroxides exist as the result of the two mesomeric forms **9** and **10**; in apolar solvents, the two forms have the same weight and thus 50% of the spin density resides on the nitrogen atom and the remaining 50% on the oxygen. In contrast, polar solvents will favor form **10** with respect to **9**; as a result, an increase of the nitrogen splitting is observed, which is particularly evident for solvents where hydrogen bonding is possible (see structure **11**, Block 8.1).

An examination of Table 8.1 also reveals that the g factor values can contribute to some extent to the identification of the trapped radicals. Thus, dialkyl nitroxides exhibit a g value close to 2.0060; alkyl-alkoxy and alkyl-aminoalkyl nitroxides have g values from 2.0050 to 2.0055; and alkyl-acyl nitroxides, alkyl-thiyl nitroxides, and phosphorus nitroxides have g values from 2.0065 to 2.0070.

Table 8.2 shows that MNP is an efficient trap for nearly all sorts of radicals as indicated by the rate constants values. In general, the reactions of oxygen and sulfur

TABLE 8.2 Bimolecular Rate Constants for the Addition of Miscellaneous Radicals to MNP

Radical ·R	k, mol^{-1} dm^3 s^{-1}	Radical ·R	k, mol^{-1} dm^3 s^{-1}
·CH$_3^a$	1.7×10^7	·CCl$_2$CH$_2$CH$_2$Cl^b	1.7×10^5
·CH$_2$CH$_3^a$	5.3×10^7	·CCl$_3^b$	1.9×10^6
·CH(CH$_3$)$_2^a$	4.6×10^7	·C(O)OC(CH$_3$)$_3^c$	1.1×10^6
·C(CH$_3$)$_3^d$	3.3×10^6	·OCH$_3^e$	1.3×10^8
·CH$_2$OH^a	1.4×10^8	·OC(CH$_3$)$_3^d$	1.5×10^6
·CHOHCH$_3^a$	3.2×10^8	·SC$_6$H$_5^b$	1.7×10^8
·CHOH(CH$_3$)$_2^a$	6.9×10^8	·OH^f	2.5×10^9

^{a}Water, 291K.
^{b}Benzene, 293K.
^{c}tert-Butyl peroxide, 313K.
^{d}Benzene, 299K.
^{e}Methanol, 193K.
^{f}Borate buffer (pH 9.2).

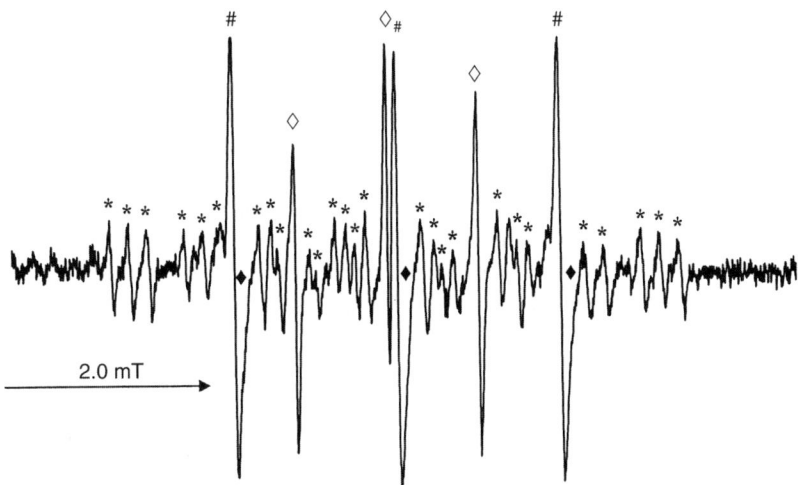

Fig. 8.2 The EPR spectrum observed at room temperature upon photolysis of benzene solutions of MNP and Irgacure 369, an industrial photoinitiator for industrial polymerization; (◇) **12**, (◆) **13**, (∗) **14**, and (#) **15**.

centered radicals with MNP are faster than those of alkyl radicals, but the rate of trapping of *tert*-butoxyl is an exception. It should also be noted that within a family of radicals the presence of electron donating substituents (e.g., α-hydroxyalkyls) enhances the rate of trapping whereas that of electron withdrawing substituents (e.g., α-haloalkyls) renders the trapping process more difficult. As indicated in Table 8.2, the hydroxyl radical exhibits the highest reactivity toward MNP, and the value of its addition rate constant is second only to that of the hydrated electron, which is 6.2×10^9 mol^{-1} dm^3 s^{-1}.

Although the spectra of the MNP adducts are fairly simple (see Fig. 8.1a–d) and hence easy to assign, complex spectra may be obtained when two or more transient radicals are simultaneously trapped, which is not unusual when trying to intercept transient radical intermediates in the course of an organic reaction. The spectrum shown in Fig. 8.2 was obtained by visible light irradiation of a benzene solution of 2-benzyl-2-(dimethylamino)-4′-morpholinobutyrophenone (Irgacure 369), in the presence of MNP, which is a typical example of such a situation.

Despite the complexity of the spectrum, it was possible to identify nitroxides **12**, **13**, **14**, and **15**, thus obtaining a wealth of information on the chemistry taking place in the system under consideration (Block 8.2).

Block 8.2

The major drawback of MNP is possibly its photolability. Thus, under visible light irradiation the monomeric ^{t}BuNO undergoes cleavage to nitric oxide and *tert*-butyl radicals that are readily trapped by MNP itself to give di-*tert*-butyl nitroxide (**15**), the signal of which may severely hamper the rationalization of the spin trapping experiment. MNP is also thermolabile to some extent, and as a result the triplet from **15** is often detected when working above room temperature in the dark. Another minor negative feature of MNP is the rather poor resolution of the spectra of its adducts that derives from the normally unresolved splitting from the nine *tert*-butylic hydrogen atoms. The spectral resolution can be drastically improved through the use of deuterated d_9-MNP, a compound that is not commercially available, however.

2-Diethoxyphosphoryl-2-nitrosopropano [$(EtO)_2(O)P(CH_3)_2CNO$, DEPNP] has recently entered the family of tertiary nitrosoalkanes, but it retains the main drawbacks of MNP, because it is photo- and thermolabile. Other tertiary aliphatic nitroso compounds can in principle be used as alternatives to MNP. Although some of these derivatives have actually been used in spin trapping experiments, most of them present inconveniences that make the use of MNP preferable.

8.2.1.2 Aromatic Nitroso Compounds. Nitrosobenzene (**16**, NB) is the archetype of the family of nitrosoarenes. Because it is able to intercept a variety of radicals of different natures, for example, alkyls, acyls, alkoxyls, thiyls, and phosphorus radicals (see Table 8.3), it has been widely employed in spin trapping experiments.

There are both advantages and disadvantages for NB with respect to MNP. The advantages are that it is not affected by visible light and only cleaves to phenyl radicals and nitric oxide if irradiated below 310 nm, it is more thermally stable than MNP, and it exists as a monomer also in relatively concentrated solutions. In addition, trapping with NB leads to more persistent nitroxides due to delocalization of the unpaired electron onto the aromatic ring (Scheme 8.1). A disadvantage is that electronic delocalization onto the aromatic ring leads to adducts **17** with more complex spectra (Fig. 8.3) that, although in principle are not difficult to interpret, may become intriguing because of additional hyperfine structures from the magnetically active nuclei present in the R fragment. Obviously, when two or more radicals are trapped simultaneously, the spectral pattern becomes so intricate that its rationalization is often impossible. The use of NB is therefore suggested for those systems where the formation of just one radical species is foreseen.

The delocalization of the unpaired electron onto the aromatic ring is also responsible for the lower values of the nitroxidic nitrogen splitting observed in the adducts of NB with respect to the corresponding adducts to MNP. Nevertheless, the nitrogen

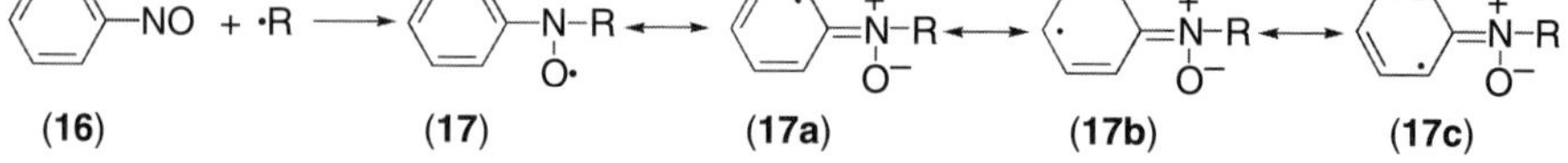

Scheme 8.1

TABLE 8.3 EPR Spectral Parameters for Selected Nitroxides PhN(O·)R Derived from NB

Radical R·	Solvent	a_N, mT	a_{other}, mT	g
Carbon radicals				
·CH$_3$	Benzene	1.045	0.273 (2H$_o$), 0.094 (2H$_m$), 0.282 (H$_p$), 0.975 (3H$_\beta$)	2.0056
·CH$_2$CH$_3$	Toluene	1.073	0.280 (2H$_o$), 0.095 (2H$_m$), 0.295 (H$_p$), 0.788 (2H$_\beta$)	
·C(CH$_3$)$_3$	Benzene	1.095	0.246 (2H$_o$), 0.086 (2H$_m$), 0.246 (H$_p$)	2.0057
·C$_6$H$_5$	Benzene	0.971	0.182 (4H$_o$), 0.080 (4H$_m$), 0.184 (2H$_p$)	2.0057
Acyls/aroyls				
·C(O)Ph	Benzene	0.770	0.160 (2H$_o$), 0.070 (2H$_m$), 0.160 (H$_p$)	2.0065
Alkoxyls				
·OCH$_3$	Methanol	1.510	0.310 (2H$_o$), 0.100 (2H$_m$), 0.310 (H$_p$), 0.100 (3H$_\gamma$)	2.0060
·OC(CH$_3$)$_3$	Benzene	1.470	0.303 (2H$_o$), 0.094 (2H$_m$), 0.303 (H$_p$)	2.0047
Thiyls				
·SC(CH$_3$)$_3$	Benzene	1.264	0.187 (2H$_o$), 0.085 (2H$_m$), 0.201 (H$_p$)	2.0061
·SC$_6$H$_5$	Benzene	1.160	0.265 (2H$_o$), 0.090 (2H$_m$), 0.265 (H$_p$)	2.0059
Phosphorus radicals				
·P(C$_6$H$_5$)$_2$[a]	Methylene chloride	1.000	1.000(P)	
·P(O)(C$_6$H$_5$)$_2$[a]	Methylene chloride	0.925	1.125(P)	
·P(S)(CH$_2$CH$_3$)$_2$	Benzene	0.911	0.288 (2H$_o$), 0.093 (2H$_m$), 0.288 (H$_p$), 1.206 (P)	2.0058

[a]Nitrosodurene.

hfs constants in nitroxides **17** varies with the nature of R in a fashion similar to that evidenced in the MNP adducts (see Table 8.3).

The presence of appropriate substituents in the aromatic ring reduces the number of hydrogen atoms and may largely simplify the spectral pattern of the spin adducts. Examples of substituted NBs frequently used in spin trapping experiments are compounds **18–20**. The adducts to nitrosodurene (**18**, ND) do not show any hyperfine pattern originating from the aromatic ring and normally consist of the nitrogen 1:1:1 triplet that may or may not be further split, depending on the nature of the trapped radical. The simplicity of these EPR spectra is attributable to steric hindrance that causes a partial rotation of the phenyl group out of the axial plane of the N-p$_z$ orbital, thus reducing the spin density on the ring. The linewidth is however rather large and the resolution poor (Block 8.3).

The adducts to compound **19** also have fairly simple spectra. This is well evidenced in Fig. 8.4b which reproduces the spectrum of nitroxide **22**.

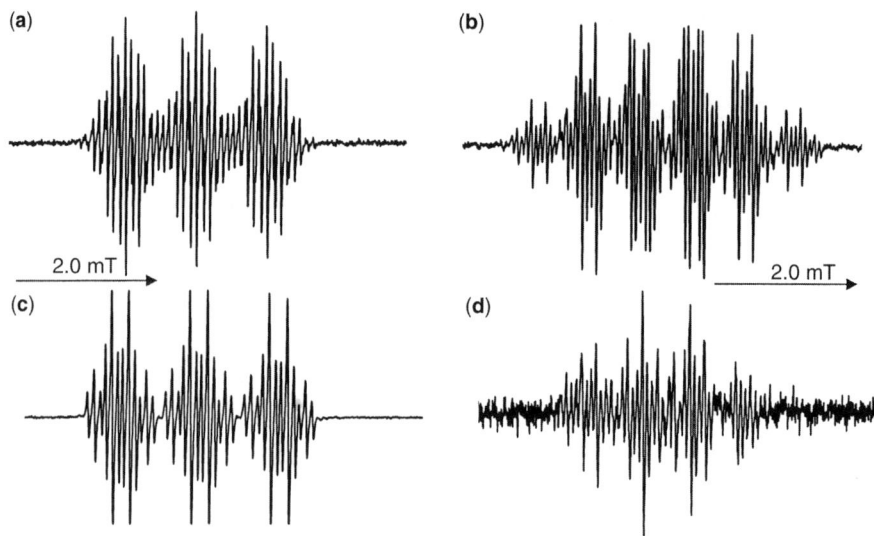

Fig. 8.3 The room temperature EPR spectra of (a) PhN(O·)Ph, (b) PhN(O·)Me, (c) PhN(O·)^{t}Bu, and (d) PhN(O·)P(S)Et$_2$ (see Table 8.3 for spectral parameters).

As can be readily seen, here the unpaired electron is coupled with the two meta hydrogens, the three hydrogens of the methyl group, and the nitrogen atom ($a_{2H_m} = 0.082$ mT, $a_{3H} = 1.235$ mT, $a_N = 1.288$ mT, $g = 2.0061$). Note that the spectrum of Fig. 8.4b was recorded after prolonged photolysis, whereas initially the spectrum of Fig. 8.4a was observed. This indicates the presence of another radical in addition to **22**. On the basis of its spectral parameter ($a_{2H_m} = 0.092$ mT, $a_{3H} = 0.187$ mT, $a_N = 2.379$ mT, $g = 2.0052$), this species was identified as methoxynitroxide (**21**), whose formation implies that the solution had not been carefully deoxygenated prior to UV irradiation. Thus, as soon as the methyl radicals are formed, they react with oxygen to give peroxy radicals that dimerize to a tetroxide that readily fragments to oxygen and the methoxy radicals that are trapped by 2,4,6-tri-t-butyl-nitrosobenzene (TBNB). As photolysis proceeds, oxygen is consumed and eventually only the spectrum from nitroxide **22** is detectable (Block 8.4).

The situation is clearly different when TBNB is reacted with a bulky radical, as may be the case of *tert*-butyl. Now the radical finds it easier to react with the

Block 8.3

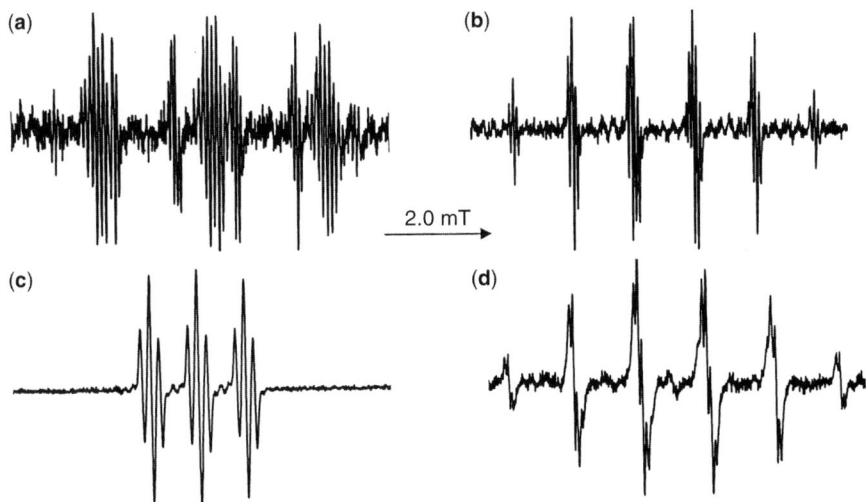

Fig. 8.4 The EPR spectra observed at room temperature upon photolysis of a benzene solution of (a) TBNB, MeI, and $Re_2(CO)_{10}$, beginning of irradiation; (b) TBNB, MeI, and $Re_2(CO)_{10}$, prolonged irradiation; (c) TBNB and ^{t}Bu-N=N-^{t}Bu; and (d) DBNBS and dimethyl sulfoxide in H_2O/H_2O_2.

oxygen atom of the nitroso function rather than with the nitrogen, and as a result the alkoxyaminyl radical (**23**, $a_{2H_m} = 0.189$ mT, $a_N = 0.997$ mT, $g = 2.0039$) is formed instead of the expected nitroxide. Note the lower nitrogen hfs constant exhibited by **23** as well as its low g factor value.

At variance with ND and TBNB, sodium 3,5-dibromo-4-nitrosobenzensulfonate (**20**, DBNBS) is a water-soluble spin trap that is frequently used in aqueous systems. It also leads to adducts with fairly simple EPR spectra as exemplified in Fig. 8.4d, which shows the spectrum of nitroxide **24** observed upon trapping of the methyl radical (Block 8.5).

Aromatic nitroso compounds are particularly reactive spin traps, with the possible exception of the sterically hindered TBNB. The rate constants for the addition of some typical radicals to NB, ND, TBNB, and DBNBS are collected in Table 8.4. Note that some of these values were obtained by means of EPR spectroscopy either directly or through competition experiments, whereas in some instances, such as DBNBS, they were determined by pulse radiolysis associated with optical detection of the transient adducts and might not be free of mechanistic ambiguity.

Block 8.4

$$Na^+ \ ^-O_3S-\underset{Br}{\overset{Br}{\underset{|}{\bigcirc}}}-\underset{O^\bullet}{\overset{|}{N}}-Me$$

(24)

Block 8.5

8.2.1.3 Open-Chain Nitrones. Methylene *t*-butyl nitrone (**25**, MBN) is the simplest and possibly the most reactive open-chain nitrone, but *t*-butyl-α-phenyl nitrone (**26**, PBN) is by far the most frequently employed, although in aqueous systems the use of nitrones **27** or **28** is preferable (Block 8.6).

Many other members have been recently added to the family of open-chain nitrones. Of particular interest are *N*-benzylidene-1-diethoxyphosphoryl-1-methyl-ethylamine *N*-oxide (PPN) (**29**), *N*-[pyridinium-4-yl]methylidene-1-diethoxyphosphoryl-1-methyl-ethylamine *N*-oxide (4PyPN) (**30**), and 1-diethoxyphosphoryl-1-methyl-*N*-[(1-oxidopyridin-1-ium-4-yl) methylidene]ethylamine *N*-oxide (PyOPN) (**31**) because of the presence of the phosphorus atom in the β position to the nitrogen atom and compounds **32** *N*-4-(lacto-bionamidomethylene)benzyliolene-*N*-butyl-amine *N*-oxide (LAMPBN), **33** diethyl-1-methyl-1[4-(lactobionamidomethylene) benzylidene]-zinoylethyl phosphonate (LAMPPN), and **34** for which the glycosyl or glycolipidic chain should allow specific localization of the trap within the system under consideration (Block 8.7).

Nitrones react with a wide variety of different radicals to give nitroxides of general structure **2**. In their spin adducts the trapped radical is bound to the carbon adjacent to the nitroxidic function rather than to the nitrogen atom itself. As a result of the large distance from the radical center, only rarely do the magnetically active atoms present in the trapped radical exhibit resolvable splittings. Thus, structural information about the trapped radical can only be derived from the β hydrogen, if present, and the nitroxidic nitrogen coupling constants, which vary in a rather small range of 0.00–0.30 mT for the former and 1.20–1.50 mT for the latter (see Fig. 8.5, Block 8.8).

TABLE 8.4 Room Temperature Bimolecular Rate Constants for the Addition of Miscellaneous Radicals to Some Nitrosoaromatics

Radical ·R	NB k, mol^{-1} dm^3 s^{-1}	ND k, mol^{-1} dm^3 s^{-1}	TBNB k, mol^{-1} dm^3 s^{-1}	DBNBS k, mol^{-1} dm^3 s^{-1}
·CH$_3$				1.55×10^9
·CH$_2$OH				2.55×10^8
·C(CH$_3$)$_3$	$>2 \times 10^8$	9.00×10^7	2.3×10^{5a}	
·cC$_6$H$_{11}$	9.00×10^7	1.60×10^{7b}	2.4×10^4	
·nC$_6$H$_{13}$		4.07×10^7	4.7×10^5	
·OH				3.95×10^9
·O$_2^-$				4.3×10^7
·SC$_6$H$_5$	2.2×10^8	3.2×10^8		

[a]100% Oxyaminyl.
[b]Pentamethylnitrosobenzene.

8.2 SPIN TRAPS   301

(25, MBN)   (26, PBN)   (27)   (28, POBN)

Block 8.6

(29, PPN)   (30, 4PyPN)   (31, PyOPN)

(32, LAMPBN)   (33, LAMPPN)   (34)

Gc = Glycosyl chain   Gl = Glycolipidic chain

Block 8.7

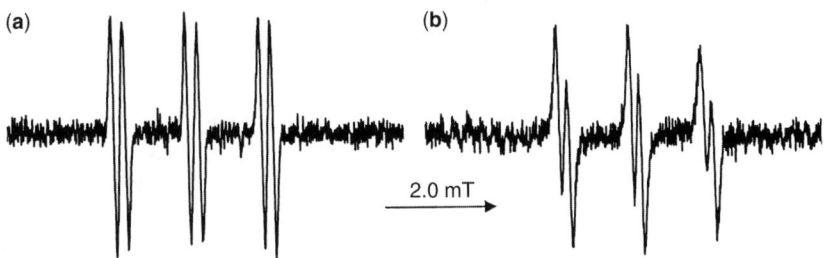

Fig. 8.5 The room temperature EPR spectra of (a) the methyl adduct to PBN (**35**) and (b) the *tert*-butoxyl adduct to PBN (**36**) (see Table 8.5 for spectral parameters).

(**35**, X = Me; **36**, X = O^{t}Bu)

Block 8.8

TABLE 8.5 EPR Spectral Parameters for Selected Nitroxides PhCHRN(O•)^{t}Bu Derived from PBN (26)

Radical •R	Solvent	a_N, mT	a_{other}, mT	g
Carbon radicals				
•CH$_3$	Benzene	1.496	0.375 (H$_\beta$)	2.0060
•CH$_2$CH$_3$	Toluene	1.458	0.326 (H$_\beta$)	
•CH(CH$_3$)$_2$	Toluene	1.466	0.258 (H$_\beta$)	
•C(CH$_3$)$_3$	Benzene	1.470	0.231 (H$_\beta$)	2.0061
•C$_6$H$_5$	n-Hexane	1.418	0.207 (H$_\beta$)	
Acyls/aroyls				
•C(O)CH$_3$	Methylene chloride	1.420	0.340 (H$_\beta$)	
•C(O)C$_6$H$_5$	Chloroform	1.460	0.455 (H$_\beta$)	
Nitrogen radicals				
•N$_3$	Acetonitrile	1.406	0.189 (H$_\beta$), 0.189 (N$_\beta$)	
•NH$_2$	Water	1.614	0.354 (H$_\beta$), 0.123 (N$_\beta$), 0.054 (2H$_\gamma$)	
Alkoxyls				
•OCH$_3$	Methanol/water	1.420	0.210 (H$_\beta$)	
•OCH$_2$CH$_3$	Ethanol	1.415	0.250 (H$_\beta$)	
•OC(CH$_3$)$_3$	Benzene	1.439	0.219 (H$_\beta$)	2.0069
Thiyls				
•SCH$_3$	Benzene	1.380	0.200 (H$_\beta$)	
•SC$_6$H$_5$	Benzene	1.400	0.161 (H$_\beta$)	2.0068
Phosphorus radicals				
•P(O)(OC$_2$H$_5$)$_2$	Methylene chloride	1.460	0.320 (H$_\beta$), 2.450 (^{31}P)	2.0059
•P(S)(OCH$_3$)$_2$	Methylene chloride	1.430	0.280 (H$_\beta$), 2.540 (^{31}P)	
Halogens				
•F	Benzene	1.220	0.118 (H$_\beta$), 4.56 (^{19}F)	
•Cl	Benzene	1.212	0.075 (H$_\beta$), 0.605 (^{35}Cl), 0.488 (^{37}Cl)	
Group 14 radicals				
•Si(C$_6$H$_5$)$_3$	Toluene	1.450	0.560 (H$_\beta$)	2.0056
•Ge(C$_6$H$_5$)$_3$	Toluene	1.481	0.558 (H$_\beta$)	2.0072

The data in Table 8.5 do not show any evident trend in relation to the trapped radicals. The identification of •R is therefore difficult, and the rationalization of the observed spectra requires that the spectral parameters are compared with those of the adducts of known •R obtained under identical experimental conditions. In this context it should be emphasized that the outcome of a spin trapping experiment may be strongly dependent on the relative concentration of the individual reactants. As an example, Fig. 8.6 shows the spectra obtained upon photolysis of the system PBN/Ph$_3$SiH/^{t}BuOO^{t}Bu.

With a ratio of peroxide to silane lower than 0.5, an almost clean spectrum of the PBN-SiPh$_3$ adduct is observed along with a weaker signal from the PBN-O^{t}Bu

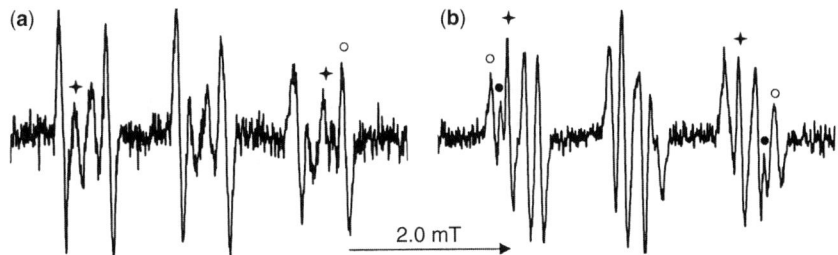

Fig. 8.6 The EPR spectra observed at room temperature upon photolysis of a benzene solution of PBN and ^{t}BuOO^{t}Bu/Ph$_3$SiH in ratios of (a) <0.5 and (b) >1.0 (the first and last lines of the different nitroxides are labeled); (○) PBN-SiPh$_3$ adduct, (+) PBN-O^{t}Bu adduct, and (●) PBN-Me adduct.

(37)   X = OMe   **(38)**

Block 8.9

adduct. In the presence of an excess of peroxide the spectrum becomes much more complex and is dominated by the signals from the PBN-O^{t}Bu and PBN-Me adducts.

The introduction of substituents onto the aromatic ring of PBN may result in a wider range of variations for the β-hydrogen splitting. Typical examples are the adducts **37** and **38** to t-butyl-α-(2,4,6-trimethoxyphenyl)nitrone, which exhibit a_{H_β} values of 0.235 and 0.927 mT, respectively (Block 8.9).

The opposite is true for α-(4-pyridyl-1-oxide)-N-t-butyl nitrone (POBN) (**28**), which, because of its greater solubility in water, may represent a valid substitute for PBN in aqueous media. Indeed, although in the adducts of this nitrone the nitrogen hfs constants are fairly similar to those measured for the same adducts to PBN, the range of the β-hydrogen atom splitting is definitely narrower (see Table 8.6).

TABLE 8.6 EPR Hyperfine Coupling Constants (mT) Measured in Water for Some Nitroxides Derived from PBN (26) and POBN (28)

Radical ·R	PBN-R	POBN-R
·CH$_3$	$a_N = 1.620$, $a_{H_\beta} = 0.335$	$a_N = 1.590$, $a_{H_\beta} = 0.275$
·CH$_2$OH	$a_N = 1.610$, $a_{H_\beta} = 0.375$	$a_N = 1.560$, $a_{H_\beta} = 0.200$
·CCl$_3$	$a_N = 1.554$, $a_{H_\beta} = 0.266$	$a_N = 1.480$, $a_{H_\beta} = 0.150$
·OH	$a_N = 1.553$, $a_{H_\beta} = 0.263$	$a_N = 1.500$, $a_{H_\beta} = 0.170$, $a_{H_\gamma} = 0.034$
·OOH	$a_N = 1.557$, $a_{H_\beta} = 0.403$	$a_N = 1.558$, $a_{H_\beta} = 0.262$

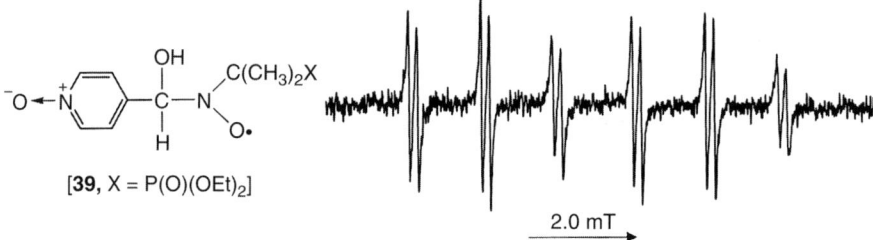

Fig. 8.7 The EPR spectrum of PyOPN-OH ($a_N = 1.380$ mT, $a_{H_\beta} = 0.170$ mT, $a_P = 4.281$ mT, $g = 2.0061$) observed at room temperature when carrying out a Fenton reaction in a pH 7 phosphate buffer in the presence of PyOPN (**31**).

The EPR spectra of the adducts to spin traps **29**, **30**, **31**, and **35** have twice as many lines as those to PBN and related compounds because of the presence of a phosphorus atom in the β position with respect to the nitroxidic nitrogen (see Fig. 8.7).

These phosphorylated nitrones are suitable for trapping carbon and oxygen centered radicals, although, once again, the value of the β-hydrogen hfs constant ($0.15 \leq a_{H_\beta} \leq 0.45$ mT) may not always be univocally informative as to the nature of the trapped species. Some additional information can be derived from the value of the hfs constant of the phosphorus atom ($3.80 \leq a_P \leq 5.00$ mT), which is also β to the nitroxidic nitrogen, but the most interesting feature of these nitrones is the significant persistence of their spin adducts, a persistence much greater (up to 20 times) than that of the adducts to PBN or other acyclic nonphosphorylated nitrones.

The addition of carbon centered radicals to structurally PBN-like acyclic nitrones is slower than that to MNP, and in general to nitroso spin traps, whereas that of radicals centered on such heteroatoms as oxygen or sulfur is comparable. As an example, the rate constants for the addition of some radicals to PBN are provided in Table 8.7.

TABLE 8.7 Bimolecular Rate Constants for the Addition of Miscellaneous Radicals to PBN

Radical ·R	k, mol^{-1} dm^3 s^{-1}	Radical ·R	k, mol^{-1} dm^3 s^{-1}
n-Alkyl[a]	1.3×10^5	·OCH$_3^b$	1.2×10^8
·C(CH$_3$)$_3^c$	< 10	·OC$_2$H$_5^b$	2.0×10^8
·CCl$_3^b$	7.0×10^4	·OC(CH$_3$)$_3$	$5.5 \times 10^{6c} - 9.0 \times 10^{7d}$
·CH$_2$OH[d]	4.3×10^7	·OH	$\sim 8.0 \times 10^{8e} - 8.5 \times 10^9$
·C$_6$H$_5^b$	1.2×10^7	·OC(O)C(CH$_3$)$_3^f$	1.0×10^6

[a]Benzene, 313K.
[b]Benzene, 293K.
[c]Benzene, 299K.
[d]Water, 298K.
[e]Water, 294K.
[f]Di-*tert*-butylperoxide, 313K.

TABLE 8.8 Room Temperature EPR Spectral Parameters for Selected Nitroxides Derived from DMPO

Radical R•	Solvent	a_N, mT	a_{H_β}, mT	a_{other}, mT	g
Carbon Radicals					
•CH$_3$	Water	1.610	2.300		2.0054
•CH$_2$CH$_3$	Water/benzene	1.620	2.360		2.0054
•CH$_2$OH	Water/methanol	1.595	2.269		2.0055
•CH(CH$_3$)$_2$	Benzene	1.430	2.180		2.0059
Acyls/Aroyls					
•C(O)CH$_3$	Acetone	1.330	1.750		
Alkoxyls					
•OCH$_3$	Benzene/methanol	1.360	0.750		
•OC(CH$_3$)$_3$	Benzene	1.315	0.745	0.195 (H$_\gamma$)	2.0059
Thiyls					
•SCH$_3$	Water	1.533	1.800		2.0061
•SC$_6$H$_5$	Benzene	1.309	1.403		2.0064
Phosphorus Radicals					
•P(O)(C$_2$H$_5$)$_2$	Benzene	1.350	1.860	2.960 (P)	2.0059
•P(O)(OC$_2$H$_5$)$_2$	Benzene	1.320	1.690	4.500 (P)	
•P(O)(C$_6$H$_5$)$_2$	Benzene	1.388	1.403	3.713 (P)	2.0059
Nitrogen Radicals					
•NH$_2$	Water	1.589	1.903	0.171 (N)	2.0057
•NHC$_4$H$_9$	Benzene	1.394	1.664	0.185 (N)	
Oxygen Radicals					
•OH	Hydrogen peroxide	1.509	1.509		2.0056
•OOH	Water	1.430	1.170	0.140 (H$_\gamma$)	

8.2.1.4 Cyclic Nitrones. A variety of cyclic nitrones have been successfully exploited in spin trapping experiments, and those belonging to the 1-pyrroline family are by far the most widely employed. Although 1-pyrroline *N*-oxide (**40**) is itself a spin trap, the real archetype of the family is 5,5-dimethyl-1-pyrrolyne *N*-oxide (**41**, DMPO) because the presence of the two β-methyl groups disfavor disproportionation of the resulting spin adducts, thus rendering them more persistent.

As is the case for the nitroxides from noncyclic nitrones, for those derived from DMPO the main information about the nature of the trapped radicals comes from the value of the hfs constant of the hydrogen atom in position 2 (Table 8.8, Block 8.10).

Block 8.10

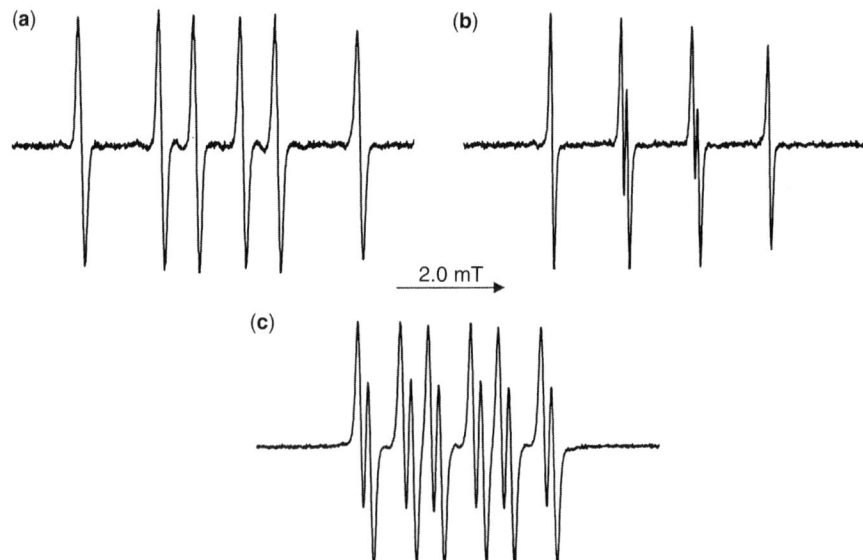

Fig. 8.8 The EPR spectra of (a) DMPO-Me (**48**) ($a_N = 1.571$ mT, $a_{H_\beta} = 2.546$ mT, $g = 2.0053$), (b) DMPO-SPh (**49**) ($a_N = 1.309$ mT, $a_{H_\beta} = 1.403$ mT, $g = 2.0064$), and (c) DMPO-O^{t}Bu (**50**) ($a_N = 1.308$ mT, $a_{H_\beta} = 0.801$ mT, $a_{H_\gamma} = 0.176$ mT, $g = 2.0059$).

The β-hydrogen splittings in nitroxides **42** vary with the nature of the trapped radical over a much wider range than that of the corresponding nitroxides from acyclic nitrones, although not so much within a single family. Thus, the β-hydrogen hfs constant is $a_{\beta H} \geq 2.0$ T for the adducts of alkyl radicals, $1.5 \leq a_{\beta H} \leq 2.0$ mT for those of thiyl radicals, and $0.6 \leq a_{\beta H} \leq 0.8$ mT for those of alkoxy radicals. The spectra of these last species generally exhibit an additional small splitting due to one of the γ-hydrogen atoms (Fig. 8.8, Block 8.11).

The nitroxides resulting from the trapping of radicals whose leading atom is magnetically active (e.g., phosphorus and nitrogen centered radicals or halogen atoms) are characterized by a larger number of lines. This is the case for butylaminyl-DMPO (**43**) and the diphenylphosphonyl-DMPO adducts (**44**), the spectra of which (see Fig. 8.9) exhibit either a small triplet due to the β-nitrogen atom or a large doublet due to the β-^{31}P nucleus.

In addition, the adducts of phosphinyl radicals exhibit a similarly large ^{31}P doublet. Because under normal experimental conditions phosphinyl radicals are frequently

Block 8.11

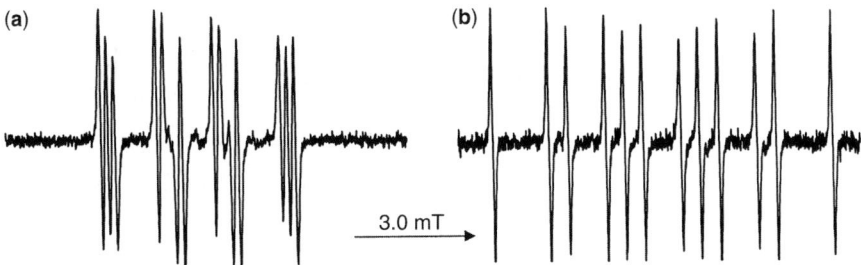

Fig. 8.9 The EPR spectra of (a) DMPO-NH^{n}Bu (**43**) and (b) DMPO-P(O)Ph$_2$ (**44**) (see Table 8.8 for spectral parameters).

oxidized to their phosphonyl analogues, it is often difficult to tell the nature of the trapped species simply on the basis of the value of the ^{31}P hfs constant.

The efficiency of DMPO in trapping alkyl radicals is comparable to that of MNP (see Table 8.9). However, the rate constant for the trapping of hydroxyalkyl radicals is about 1 order of magnitude smaller for the former spin trap, which instead reacts with alkoxy radicals much more readily than MNP.

DMPO is largely used for spin trapping experiments in biological systems, in particular for the detection of the hydroxyl radical ·OH, the hydroperoxyl radical ·OOH, or its conjugated base, the superoxide radical anion ·O$_2^-$, whose spin adducts have the very distinctive ESR spectra shown in Fig. 8.10.

Because the rate constant for the trapping of the superoxide radical anion is small and the decay of the adduct is fast, new compounds affording more persistent ·O$_2^-$/·OOH spin adducts have been introduced (Block 8.12).

Most of the new nitrones contain a phosphorus atom bound to a carbon adjacent to the heterocyclic nitrogen and give adducts with spectra more complex than those of the DMPO adducts (see Figs. 8.11 and 8.12).

TABLE 8.9 Bimolecular Rate Constants Measured at 298K in Aqueous Solution for the Addition of Miscellaneous Radicals to DMPO

Radical ·R	k, mol^{-1} dm^3 s^{-1}	Radical ·R	k, mol^{-1} dm^3 s^{-1}
·CH$_3$	1.4×10^7	·COH(CH$_3$)$_2$	6.8×10^7
·C$_2$H$_5$	1.6×10^7	·OC(CH$_3$)$_3$	9.0×10^{6a}–5.0×10^{8b}
·CH(CH$_3$)$_2$	5.8×10^6	·OH	2.8×10^9
·C(CH$_3$)$_3^b$	5.0×10^6	·OOH^c	6.6×10^3
·C(CH$_3$)$_2$C$_6$H$_5$	1.9×10^5	·OO^{-d}	1.0×10
·CH$_2$OH	2.2×10^7	·SO$_3^-$	1.6×10^7
·CHOHCH$_3$	4.1×10^7	·CO$_2^-$	6.6×10^7

^{a}Water.
^{b}Benzene.
^{c}pH ≤ 5.
^{d}pH ≥ 7.8.

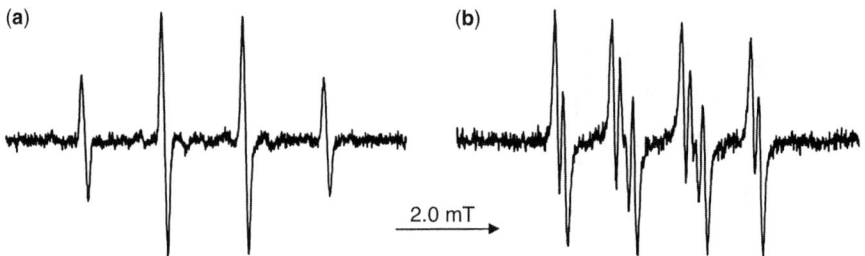

Fig. 8.10 The EPR spectra of (a) DMPO-OH in water (**45**) and (b) DMPO-O$_2^-$ in dimethyl sulfoxide (**46**) (see Table 8.8 for spectral parameters).

Block 8.12

Fig. 8.11 The half-life time for the ·O$_2^-$ adducts to eight DMPO-derived nitrones.

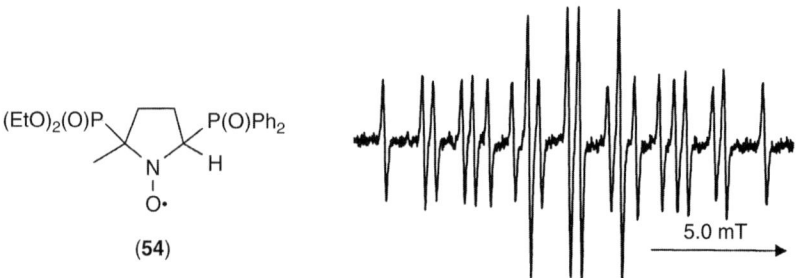

Fig. 8.12 The ESR spectrum of DEPMPO-P(O)Ph$_2$ in benzene (**54**) ($a_N = 1.343$ mT, $a_{H_\beta} = 1.728$ mT, $a_{31_P} = 3.579$ mT, $a_{31_P} = 4.900$ mT, $g = 2.0059$).

In addition, when the trapped species is a phosphorus centered radical (see **54**), the assignment of the two individual phosphorus splittings must be made via a comparison with the values observed in other DEPMPO (5-diethoxyphosphoryl-5-methyl-1-pyrroline *N*-oxide) adducts. Finally, yet importantly, because of the asymmetry of these nitrones, radical trapping may in principle afford different isomeric adducts leading to the observation of spectra the rationalization of which may become an intriguing task.

A favorable attribute is that the extra β-phosphorus coupling present in the spectra of these adducts may be of some additional help in the identification of the trapped radical.

8.2.2 Unconventional Spin Traps

8.2.2.1 Carbonyl Compounds. Aromatic carbonyl compounds and quinones are potential spin traps only for those species that react and attack the carbonyl oxygen, leading to ketyl-like adducts (see Reactions 8.13 and 8.14). Typically, these are phosphorus radicals (phosphinyls, phosphonyls, thiophosphonyls) and radicals centered at a metal atom such as silicon, germanium, tin, or lead.

The addition process is generally fast. The rate constant at 300K for the addition of triethylsilyl radical to **55** and to duroquinone (2,3,4,6-tetramethyl-*para*-benzoquinone) is 3.0×10^7 and 2.2×10^9 mol^{-1} dm^3 s^{-1}, respectively, although in the latter case addition to the ring takes place followed by a 1–3 carbon to oxygen migration of the organometallic moiety.

The identification of the trapped radical in these adducts is normally based on the hyperfine coupling constant of its leading atom, because all other splittings are not very sensitive to the nature of the attacking species. This is not a problem for phosphorus radicals (^{31}P has a natural abundance of 100%), whereas for silicon, germanium, tin, and lead, the most abundant isotopes of which are magnetically inactive, the intensity of the spectra of the adducts must be such to allow the detection of the satellites lines due to their less abundant but magnetically active isotopes. Benzophenone and related compounds give adducts with rather complex spectra

$$Ph_2C=O \;+\; HSiEt_3 \;\xrightarrow[C_6H_6]{^tBuOO^tBu/h\nu}\; Ph_2C\text{–}OSiEt_3$$

(55) (56)

Reaction 8.13

$$O=\!\!\bigcirc\!\!=O \;+\; HP(O)Ph_2 \;\xrightarrow[C_6H_6]{h\nu}\; \cdot O\text{–}\bigcirc\text{–}OP(O)Ph_2$$

(57) (58)

Reaction 8.14

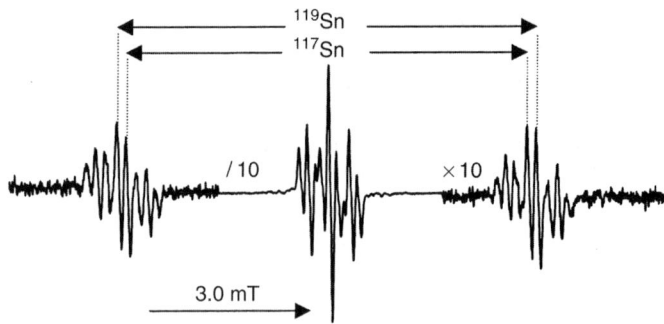

Fig. 8.13 The ESR spectrum of radical **60** observed at 373K by photolysis of **59** in neat hexabutylditin ($a_{2H_m} = 0.139$ mT, $a_{2H_o} = 0.394$ mT, $a_{117_{Sn}} = 7.568$ mT, $a_{119_{Sn}} = 7.880$ mT, $g = 2.0031$).

Reaction 8.15

due to the large number of coupled hydrogen atoms, and the situation becomes prohibitive if two or more radicals are trapped simultaneously (Fig. 8.13).

Things may be improved by replacing one of the aromatic rings with an organometallic moiety MR_3, where M is a silicon or a germanium atom [PhC(O)MR$_3$, M = Si, Ge]. Further simplified spectra can be obtained by inserting an appropriate substituent in the para position of the aromatic ring, as is case of the chloro derivative **59**, the spectrum of the tributylstannyl adduct of which is provided in Fig. 8.13 (Reaction 8.15).

Finally, aliphatic and aromatic α diketones as well as *ortho*-quinones are efficient spin traps for metal centered radicals because of the stabilization of their spin adducts via coordination of the metal atom by the oxygen of the carbonyl group adjacent to the one that has undergone the addition (see Reaction 8.9).

8.2.2.2 Aromatic and Aryl-Organometallic Thioketones. Radical addition to thiocarbonyl compounds is a more facile process than addition to carbonyl compounds. In addition, because of the different polarities of the oxygen and sulfur atoms, the vast majority of free radicals, whether nucleophilic or electrophilic, react with the thiocarbonyl function via a thiophilic attack that leads to the formation of thioketyl-like spin adducts (see Reaction 8.7, Block 8.13).

Block 8.13

On a spectral basis, the adducts of thiobenzophenone (**61**) are more informative than those of its aliphatic analogues. In addition to exhibiting the coupling of the unpaired electron with the 10 aromatic protons, additional splittings due to magnetically active atoms in the trapped radical are detected in most cases, thus providing useful information about the nature of the attacking species. As is the case for ketones, the spectra can be simplified by replacing an aromatic ring of thiobenzophenone with a silyl moiety, as in thiobenzoyl triphenylsilane (TBTPS, **62**), and replacing with appropriated substituents one or more hydrogen atoms of the aromatic ring. Indeed 3,5-di-*tert*-butylthiobenzoyl triphenylsilane (DBTBTPS, **63**) has proved to be the best spin trap of the thiocarbonyl family for both the persistence of its adducts and the simplicity of their spectra.

The structural information about alkyl radicals that can be derived from their adducts to thiocarbonyl derivatives (see Table 8.10) is equivalent to that obtainable with nitroso compounds, but the former traps are valuable substitutes when the use of conventional agents is prohibited by adverse reaction conditions. It is also possible to readily discriminate among $R_2P\cdot$, $R_2(O)P\cdot$, and $R_2(S)P\cdot$ species on the basis of the ^{31}P splitting that is typically ~ 2.0 mT for phosphinyl, 3.0–4.0 mT for phosphonyl, and ≥ 5.0 mT for thiophosponyl adducts. As for group 14 radicals, ^{29}Si, ^{73}Ge, $^{117/119}Sn$, and ^{207}Pb satellite lines can be easily observed by recording the spectra of the adducts at high gain.

Like nitrones and nitroso derivatives, thiocarbonyl compounds may also lead to the detection of radicals that originate through processes other than direct spin trapping. Aromatic thioketones are easily reducible compounds (their reduction potential is normally ≥ -1 V vs. normal hydrogen electrode), and when planning a spin trapping experiment one must therefore know the oxidation potentials of the other reactants present in the system. Thus, the reaction of Grignard reagents RMgX (R = alkyl, aryl; X = halogen) with compounds **61**–**63** at low temperature immediately leads to the detection of the intense spectra of their radical anions, whereas at or above room temperature R· adducts are observed instead. It has been suggested that these latter species originate via addition of radicals deriving from the fragmentation of the initially formed radical cations of the Grignard reagents to the thiocarbonyls, but the occurrence of a nucleophilic addition followed by adventitious oxidation has also been suggested.

TABLE 8.10 EPR Spectral Parameters for Selected Adducts of DBTBTPS (63) in Benzene at Room Temperature[a]

Radical R•	a_{H_o}, mT	a_{H_p}, mT	a_R, mT	g
Carbon Radicals				
•CH$_3$	0.318	0.383	0.174 (3H)	2.0043
•C$_2$H$_5$	0.316	0.334	0.095 (2H), 0.047 (3H)	2.0043
•C(CH$_3$)$_3$	0.308	0.350	0.049 (9H)	2.0039
•C$_6$H$_5$	0.345	0.386	0.039 (5H)	2.0040
Alkoxyls and Thiyls				
•OC(CH$_3$)$_3$	0.349	0.385	0.020 (9H)	2.0036
•SCH$_3$	0.352	0.394	0.103 (3H)	2.0040
•SCH(CH$_3$)$_2$	0.338	0.410	0.29 (6H), 0.57 (H)	2.0037
•SC(CH$_3$)$_3$	0.354	0.384	0.030 (9H)	2.0040
•SC$_6$H$_5$	0.343	0.386		2.0040
•SBtz[b]	0.354	0.404	0.066 (N)	2.0039
Phosphorus Radicals				
•P(C$_6$H$_5$)$_2$	0.380	0.410	2.040 (P)	2.0036
•P(O)(C$_6$H$_5$)$_2$	0.403	0.474	3.790 (P)	2.0033
•P(S)(C$_6$H$_5$)$_2$	0.397	0.462	5.380 (P)	2.0040
Metal Centered Radicals				
•Si(C$_6$H$_5$)$_3$	0.379	0.428	1.416 (Si$_\beta$)	2.0039
•Ge(C$_6$H$_5$)$_3$	0.364	0.408	0.720 (Ge$_\beta$)	2.0038
•Sn(CH$_3$)$_3$	0.352	0.376	11.36 (Sn$_\beta$)	2.0044
•Mn(CO)$_5$	0.245	0.245	0.632 (Mn)	2.0074
•Re(CO)$_5$	0.251	0.251	5.51 (Re)	2.0105

[a] The ^{29}Si$_\alpha$ and ^{13}C hfs constants are not reported.
[b] Btz, benzo-1,3-thiazol-2-yl.

The reactions of radicals with aromatic thiones have been the subject of few kinetic studies. On the basis of the available data, a value on the order of 10^7 mol^{-1} dm^3 s^{-1} has been estimated for the rate constant for the addition of alkoxy radicals whereas that for the addition of carbon centered radicals is believed to be 1 order of magnitude smaller. These values compare well with those determined for the addition of the same species to nitrones and nitroso derivatives and suggest a similar efficiency for thioketones.

The bad news comes from the availability and the stability of these compounds. Actually, thioketones are not commercially available, with the exception of dimethoxythiobenzophenone, which is of little use as a spin trap. Different synthetic procedures for thiobenzophenone (**61**) have been described that are fairly simple though delicate in the purification step. Conversely, the preparation of the more valuable compounds **62** and **63** is more painstaking and requires a good synthetic skill. In addition, thioketones need to be handled under an inert atmosphere to avoid degradation. Last but not least is the purity issue, because common impurities of thioketones are the corresponding ketones that can interfere in the spin trapping process.

8.2.2.3 Phosphoryl Dithioesters. Phosphoryl dithioformates are compounds with general structure **64**, where X is an alkyl group and Y may be either aliphatic or aromatic. Despite the presence of the PO and CS double bonds, they are not considered as polyfunctional spin traps because they exclusively undergo addition at the thiocarbonyl function (see Reaction 8.8 and Scheme 8.2) with the formation of unusually persistent spin adducts (**65**). The high persistence of these radicals is mostly attributed to capto-dative stabilization that is due to the presence of the strongly electron-withdrawing phosphoryl group and the electron-donating thiyl groups.

In all cases the EPR spectra of the spin adducts consist of a fairly large phosphorus doublet ($1.9 \leq a_{31_P} \leq 2.5$ mT), the value of which cannot however be directly related to the nature of the trapped radical except for the adducts of alkoxyls. Actually, these species exhibit an unusually low phosphorus coupling ($a_{31_P} < 1.7$ mT) and a g factor value ($g \sim 2.0055$) that is also lower than the average value of all other adducts (i.e., $g \sim 2006$). In contrast, all spectra also show additional hyperfine structures from the trapped •R radicals (or satellites for the •MR$_3$ species) that enables their unambiguous identification.

The versatility of compounds **64** as trapping agents equals that of thioketones **61–63**, and in some cases it is even greater. Although the nature of group Y has little effect on the trapping process, its results may be strongly affected by the experimental conditions and the nature of the X residue and the attacking radical •R. Indeed, the trapping process tends to become reversible with the increasing stability of the attacking radical and with increasing temperature. Thus, for a given X residue, the most persistent adducts are obtained when •R is either a methyl or a primary alkyl radical whereas with secondary and even more with tertiary radicals fragmentation of the adducts may take place (see left side of Scheme 8.2). By contrast, cleavage of the SX bond of radicals **65** may compete with that of the RS bond, which one will break depending on the relative stability of the •R and •X radicals. Decay of **65** through cleavage of the SX bond affords a new phosphoryl dithioformate (**66**), resulting in a homolytic substitution process that has proved of significant synthetic value. However, cleavage of the SR bond reinstitutes the starting trap **64** via a reversible addition–fragmentation process that has been successfully exploited to control radical polymerization of vinylic monomers.

The rate constant for the addition of either carbon or oxygen centered radicals to the CS double bond of the thioester function is generally greater than 10^7 mol^{-1} dm^3 s^{-1}, whereas that of the fragmentation of adducts **65** is on the order of 10^3 s^{-1}.

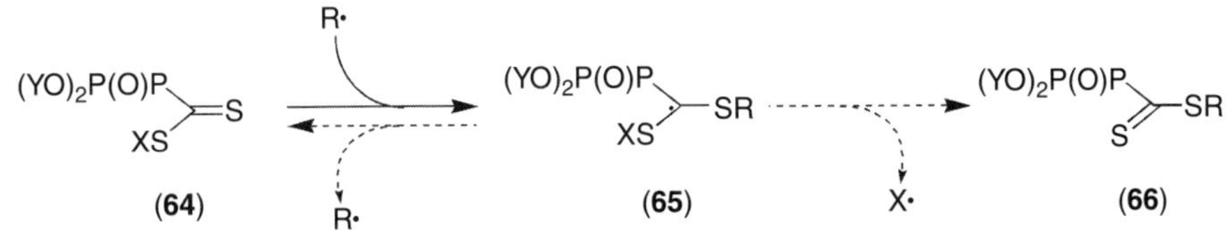

Scheme 8.2

Phosphoryldithioformates (**64**) are not commercially available, but they can be synthesized much more readily than thioketones. They are also fairly stable and can be handled in open air without any particular care.

8.2.2.4 Nitroderivatives.

Aromatic nitrocompounds are suitable to trap carbon centered and organometallic radicals centered at a group 14 element such as Si, Ge, and Sn. The trapping reaction (see Reaction 8.10) leads to oxynitroxides of general structure **67** that are characterized by nitrogen splittings in the range $1.0 \leq a_N \leq 1.6$ mT and g values in between 2.0040 and 2.0050. For 2,6-disubstituted aromatic nitrocompounds, the nitrogen coupling constant increases (up to 2.6–2.8 mT) because of rotation of the aromatic ring out of the ONO plane. In these aryl oxynitroxides the unpaired electron also interacts with the aromatic protons, leading to an additional hyperfine structure that may result in rather complex spectra, depending on the nature of the aromatic residue itself (Block 8.14).

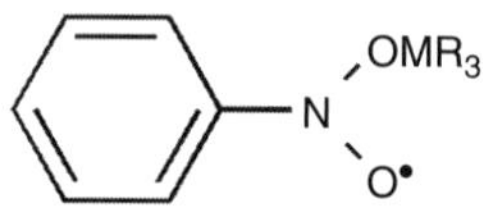

(**67**, M = C, Si, Ge, Sn)

Block 8.14

In nitroxides **67** the barrier to rotation of the aromatic ring about the CN bond is such that linewidth alternation effects are possible even at room temperature. If the aromatic ring is asymmetric or asymmetrically substituted, this hindered rotation may lead to geometrical isomers, therefore potentially increasing the complexity of the observed spectra.

Aliphatic nitrocompounds scavenge carbon centered and ·MR$_3$ radicals, leading to oxynitroxides with large nitrogen hfs constants ($2.5 \leq a_N \leq 3.0$ mT) and g values between 2.0045 and 2.0060. These nitroxides are more fragile than their aryl analogues. In particular, the reaction with tin centered radicals is used to generate alkyl radicals from nitroalkanes, whereas alkyl alkoxynitroxides mainly undergo cleavage of the N—OR bond, affording a nitrosoalkane and an alkoxy radical (see Scheme 8.3).

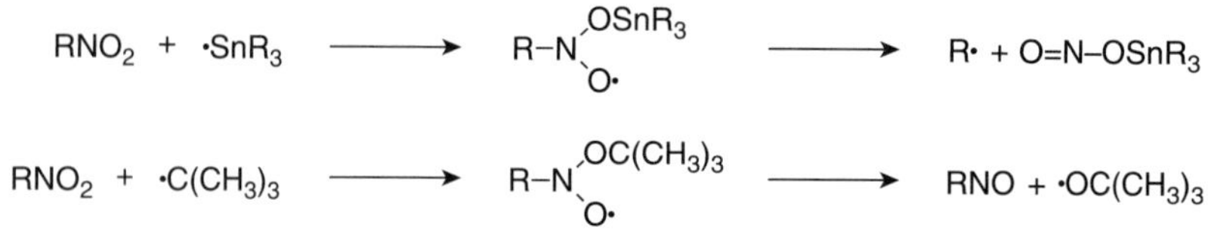

Scheme 8.3

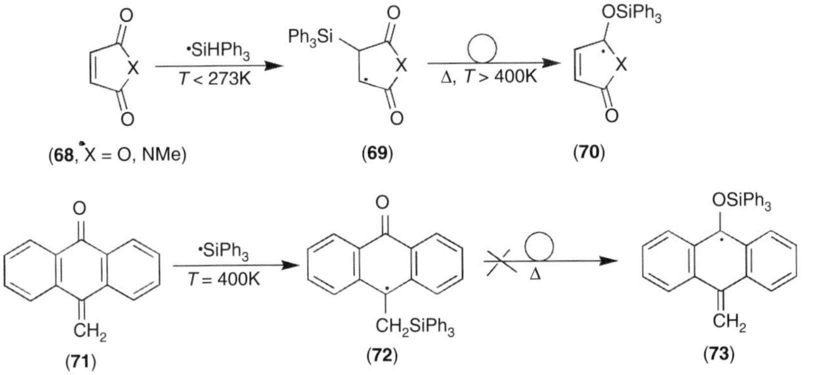

Scheme 8.4

8.2.2.5 Polyfunctional Spin Traps.
Polyfunctional spin traps are compounds that may potentially undergo radical addition at more than one molecular site.

As noted in a previous section, quinones and more generally unsaturated carbonyl compounds may undergo addition at either the CC or CO double bond. When the two functional groups are adjacent, migration from one site to the other may take place. This is the case of the addition of ·SiPh$_3$ radicals to cyclic anhydrides or imides (**68**) outlined in Scheme 8.4. At low temperature adduct **69** resulting from addition to the ring is detected; but if the temperature is raised while no new ·SiPh$_3$ radicals are generated, the spectrum of adduct **69** changes irreversibly to that of the oxygen adduct **70** because of a 1,3 migration of the silyl moiety.

Similar behavior (1,3 migration) is exhibited by *para*-quinones. Under the same conditions 9-methyleneanthrone (**71**) exclusively undergoes a selective addition to the exocyclic carbon–carbon double bond, thus behaving as a monofunctional spin trap.

Using *tert*-butyl-α-(3,5-di-*tert*-butyl-4-hydroxyphenyl)nitrone (**74**) allows the discrimination between oxygen and carbon centered radicals. The former (see Scheme 8.5) abstracts the hydroxylic hydrogen to give a vinylic nitroxide (**75**) characterized by a small nitrogen splitting ($a_N = 0.505$, $a_{H_\beta} = 0.26$, $a_{H_m} = 0.17$, $a_{H_m} = 0.15$ mT); the latter adds to the CN double bond, leading to nitroxide **76** ($a_N = 1.50$, $a_{H_\beta} = 0.36$ mT) with "normal" hfs constants.

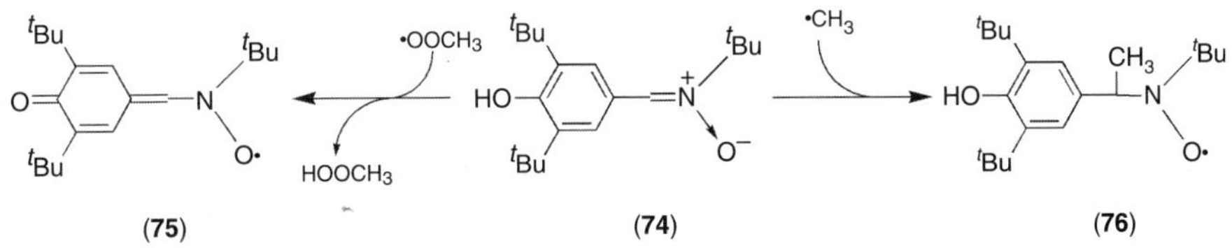

Scheme 8.5

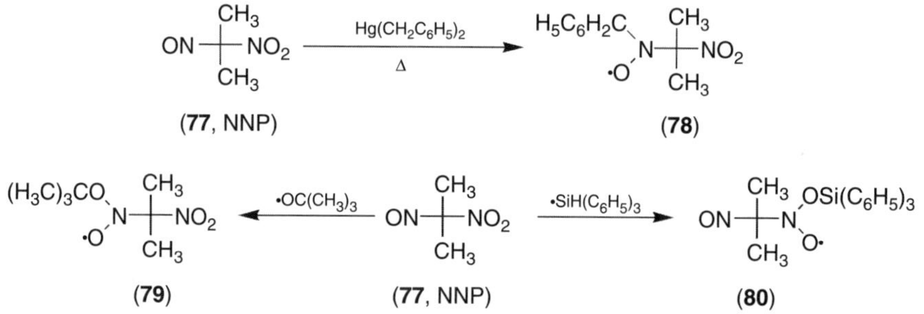

Scheme 8.6

As a last example of polyfunctional spin traps it is worth mentioning 2,nitroso-2,nitropropane (**77**, NNP), a compound where a nitroso and a nitro group are bound to the same carbon atom (Scheme 8.6).

Both carbon and oxygen centered radicals add to the nitroso function of NNP. In the former case dialkyl nitroxides are formed that exhibit a nitroxidic nitrogen splitting in the range of 1.1–1.5 mT. Oxygen centered radicals, in contrast, lead to alkyl alkoxynitroxides characterized by a much larger nitrogen splitting in the range of 2.5–3.0 mT. At variance with these species, organometallic radicals ·MR$_3$ react with NNP via attack to the nitro group, leading again to oxynitroxides with large nitrogen splittings ($2.5 \leq a_N \leq 3.0$ mT).

As it can be seen in Fig. 8.14, discriminating between carbon radicals adducts on one side and oxygen or organometallic radicals adducts on the other is easy. Actually, the EPR spectra of the latter adducts are fairly similar, both showing coupling of the unpaired electron with two different nitrogen nuclei; in those of carbon centered radical adducts, a hyperfine pattern deriving from the trapped species is observed in addition to that of the nitroxidic nitrogen.

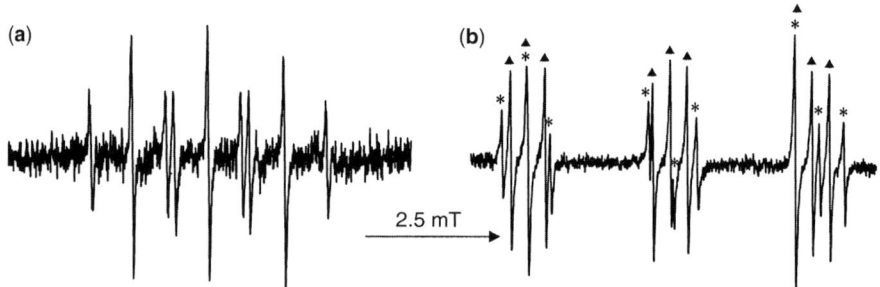

Fig. 8.14 The ESR spectra of (a) adduct **78** ($a_{2H} = 0.785$, $a_N = 1.418$ mT, $g = 2.0060$) observed upon heating a *tert*-butylbenzene solution of dibenzylmercury and NNP at 363K and (b) adducts (▲) **79** ($a_N = 0.326$, $a_N = 2.655$ mT, $g = 2.0053$) and (∗) **80** ($a_N = 0.450$, $a_N = 2.738$ mT, $g = 2.0051$) observed at 298K upon UV irradiation of a benzene solution of triphenylsilane, di-*tert*-butylperoxide, and NNP.

8.3 EXPERIMENTAL METHODS

It is impossible to define optimum conditions that are generally valid for all spin trapping studies, because each individual experiment may require special conditions related to the actual chemistry involved.

For screening purposes, relatively large modulation amplitudes ($\leq 0.05-0.075$ mT) can be used, whereas the use of 10-fold lower values is critical for resolving subtle spectral details that may be invaluable for an unambiguous characterization of the trapped radical. In general, it is advisable to eliminate oxygen from the system by either flushing the solution under study with an inert gas or via the freeze and thaw technique. This will lead to better quality spectra and help avoid the observation of "extra-adducts" resulting from the trapping of species derived from the reaction of oxygen and the radicals originated in the process under study. Of course, this does not apply to those cases when oxygen is actively involved in the reaction under investigation.

In order not to spoil the mode of the cavity, flat cells should be used when working with polar solvents, such as dimethylsulfoxide, acetonitrile, dimethylformamide, and especially water. Melting point capillary tubes (i.d. = $\sim$1 mm) may also be used, but flat cells have the advantage that larger amounts of solution can be used and they allow purging of the sample to eliminate oxygen. Standard sample tubes (i.d. = $\sim$4 mm) can be used for apolar organic solvents (benzene, toluene, etc.). Good quality fused silica (Suprasil) glassware must be employed when *in situ* UV irradiation is required.

Spin trapping experiments do not normally require large amounts of reactants. The volume of the samples may range from 20 µL (melting point tubes) to 200–300 µL (4-mm quartz tubes), although larger volumes are necessary if the flow system technique is used. As for the concentration of the spin trap, it varies from case to case and the optimum values are strictly related to the chemistry under investigation. Thus, if radicals are formed that may further evolve under the experimental conditions via fragmentation, addition, or atom-abstraction reactions, larger concentrations of the spin trap will favor the trapping of the initially formed species whereas a defective spin trap will lead to a greater chance of intercepting "secondary" radicals.

The choice of the right spin trap is normally dictated by the nature of the radicals to be trapped, but the solvent, temperature, other experimental conditions, and the possibility that a particular spin trap may lead to radicals through reactions other than spin trapping cannot be disregarded.

The commercial availability of the individual spin traps is also an important factor that may become critical for those research groups lacking access to synthetic facilities.

8.4 APPLICATIONS

Spin trapping is a powerful tool that extends the field of applicability of EPR spectroscopy. Its applications range from the elucidation of the mechanisms of radical-based reactions to kinetic studies aimed at determining the rate constants of radical processes, as well as to the interception of radicals in biological systems.

8.4.1 Mechanistic Studies

Spin trapping experiments may be very useful in the elucidation of the mechanisms of organic reactions involving radicals, because they allow the detection of paramagnetic species even if present in minute amounts. In general, there is no predefined protocol for spin trapping experiments aimed at elucidating the mechanism of an organic reaction. Most simply, the reaction must be carried out under usual conditions inside the cavity of the EPR spectrometer after having introduced in the reaction system a small amount of an appropriate spin trapping agent, the choice of which is based on the nature of the radicals that "are reasonably expected" to be involved in the process. A negative result (no detection of radicals) should not be taken as conclusive: actually, several attempts may be necessary before the right spin trap is found.

Spin trapping experiments using TBTPS were used to prove the mechanism outlined in Scheme 8.7 for the iododecarboxylation of carboxylic acids mediated by iodosobenzene diacetate (**81**).

DMPO spin trapping proved that the drug 5H-pyridophenoxazin-5-one (**85**, PPH) upon aerobic reduction leads to the superoxide radical anion (see Scheme 8.8) that

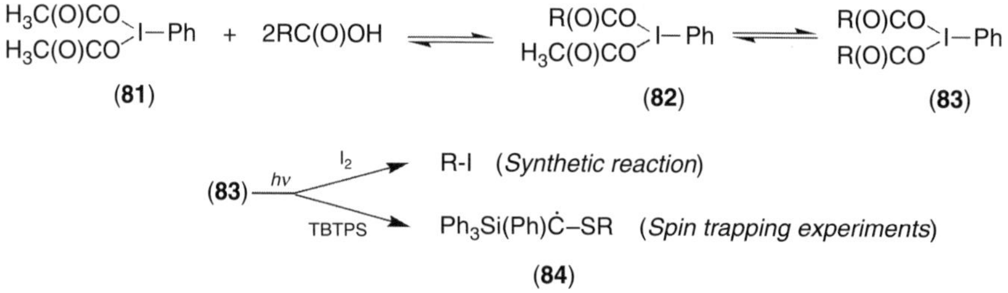

Scheme 8.7

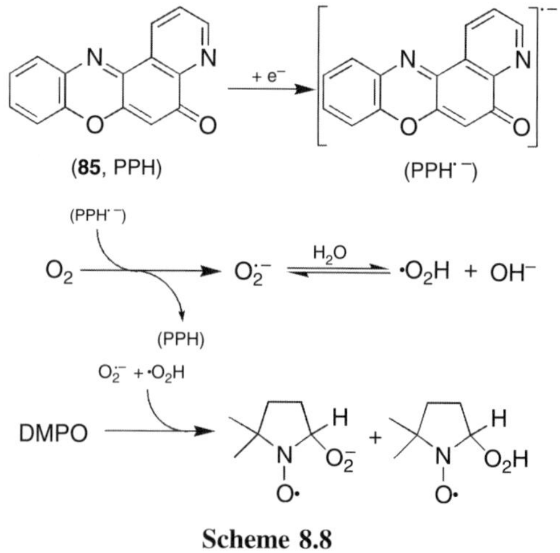

Scheme 8.8

under biological conditions triggers the formation of reactive oxygen species, in particular the hydroxyl radical, that are responsible for the activity of the drug.

In this particular context it is worthwhile emphasizing again that the observation of a radical adduct does not necessarily imply spin trapping. Although the observation of the alkoxy adducts of PBN, DMPO, or DEPMPO has been taken as evidence that alkoxy radicals are formed in the lead tetracetate or potassium persulfate oxidation of alcohols, one should be aware that the same adducts would be observed following a Forrester–Hepburn process, which is the nucleophilic addition of the alcohol followed by oxidation of the resulting anion.

Although the homolytic nature of many reaction mechanisms has been definitely proved through spin trapping experiments, it is generally recognized that the spectroscopic interception of radical species in the course of an organic reaction does not necessarily mean that these species lie on the main reaction coordinate, that is, that the reaction proceeds through a homolytic mechanism. It is in fact possible that, because of its high sensitivity, the EPR technique picks up species produced in relatively unimportant side reactions, the main process being essentially ionic. In many instances, however, this ambiguity can be overcome by combining spectroscopic results with product studies.

8.4.2 Kinetic Measurements

The direct measurement of spin trapping rate constants is based on the appearance or disappearance of a spectral line and requires special photolytic, electrochemical, or pulse-radiolytic facilities for the generation of the transient radicals. In particularly favorable cases it is possible to monitor both the decay of the signal of the primary radical and the growth of the signal of the spin adduct. The rate constants for the addition of a variety of alkyls and hydroxyalkyls to MNP (see Table 8.2) have been determined in this way through the use of time resolved EPR.

When appropriate generation techniques are not available, competitive experiments may be carried out by allowing the radical to undergo two simultaneous reactions, one of which with a known rate constant, and by determining the ratio of the different species from their EPR spectra.

For example, the rate constant for the addition of *tert*-butoxyl radicals to DMPO in cyclohexane can be determined from the intensity ratio of the signals of the adducts DMPO-O^{t}Bu and DMPO-cC_6H_{11}, provided the rate constants for hydrogen abstraction from cyclohexane by ^{t}BuO· and for the addition of cyclohexyl radicals to DMPO are known.

Competitive experiments may also involve the addition of a radical to two different spin traps as shown in Scheme 8.9 for the addition of *tert*-butoxyl radicals to DMPO and PBN.

Thus, the rate constants k_{ad2} to PBN can be determined according to Equation 8.1, where [(**86**)]/[(**87**)] is the ratio of the intensity of the corresponding EPR spectra.

$$k_{ad2} = k_{ad1} \times [\text{DMPO}][(\mathbf{86})]/[\text{PBN}][(\mathbf{87})] \qquad (8.1)$$

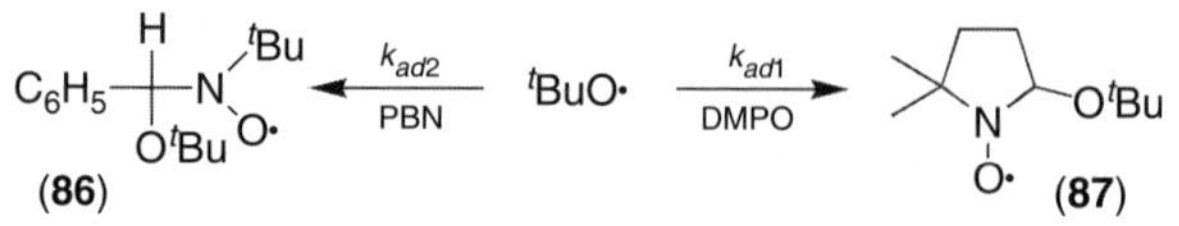

Scheme 8.9

$C_6H_5-\overset{H}{\underset{O^tBu}{C}}-N\overset{^tBu}{\underset{O^\bullet}{}}$ (86) $\xleftarrow{k_{ad2}}_{PBN}$ $^tBuO^\bullet$ $\xrightarrow{k_{ad1}}_{DMPO}$ [pyrrolidine N-oxide with O^tBu] (87)

Needless to say, in these experiments the intensities of the radical adducts must be determined by double integration of a "clean" line of each individual spectrum. In addition, if the spin adducts under consideration have a different persistence, corrections should be made accordingly.

Most spin adducts decay bimolecularly either via dimerization or via disproportionation. Because spin adducts are more stable than the corresponding transient radicals, it is normally simple to follow their decay by monitoring the decrease of the amplitude of a conveniently chosen spectral line. Special care must be exerted in those cases when the decay rate is of the same order of magnitude of its rate of formation.

8.4.3 Spin Trapping in Biological Systems

Spin trapping studies carried out in biological systems are mostly aimed at the detection of oxygen centered radicals (i.e., alkoxyl, hydroxyl, and hydroperoxyl radicals), the superoxide radical anion, and to a lesser extent sulfur and carbon centered radicals. For this purpose and taking into account that most studies are carried out in aqueous systems, the more widely used spin traps are the nitrones PBN (**26**), POBN (**28**), PPN (**29**), P$_y$OPN (**31**), DMPO (**41**), and DEPMPO (**51**) and the nitroso derivatives MNP (**8**) and DBNBS (**20**).

When planning spin trapping experiments in polar environments as biological systems, the possibility of the occurrence of electron transfer processes must be considered, because spin traps characterized by low oxidation potentials are amenable to undergoing single electron transfer (SET) processes leading to undesired radical species. Conversely, nitroxides obtained as a result of a trapping process *in vivo* can be converted to EPR silent hydroxylamines through a one-electron reduction process brought about by ascorbate, heme proteins, hepatocytes, or thiols in the presence of superoxide.

A detailed treatment of spin trapping in biological systems can be found in Chapter 9.

8.5 SPIN TRAPPING IN THE GAS PHASE OR IN THE SOLID STATE

Spin trapping in media other than fluid solutions normally ends up with conventional solution studies. In the gas-phase reactions, the stream of reacting gases (vapors) is allowed to pass over or through a solid trap (such as PBN or MNP) that is then dissolved in an appropriate solvent. The EPR spectra of the resulting solution are then recorded. In a similar fashion, crystals or frozen solutions of compounds that have undergone processes leading to radicals are added to a fluid solution of a spin trap that is then examined by EPR spectroscopy. Even when working under strictly anaerobic conditions, it is possible that the dissolving radicals decay before they are trapped and the failure to detect spin adducts does not exclude the presence of radicals in the original gaseous or solid samples.

8.6 AVAILABILITY OF SPIN TRAPS

Although ND (**18**) is not commercially available, most nitrosoalkanes and nitrosoarenes, including MNP (**8**), TBNB (**19**), and DBNBS (**20**), are commercially available. Acyclic nitrones PBN (**26**), (**27**), and POBN (**28**) are also commercially available, along with the cyclic derivatives DMPO (**41**), 5-(ethoxycarbonyl)-5-methyl-1-pyrroline *N*-oxide (EMPO) (**50**), and DEPMPO (**51**). More complex nitrones such as **29–34**, **47–49**, **52**, and **53** need to be prepared.

A large variety of ketones and quinones are present in the chemical catalogs, but organometallic derivative **59** is not included. Similarly, dithioformates **64** and **66** and thioketones **61–63** are not commercialized.

The Further Readings Section provides references about the synthetic procedures leading to most of the spin traps that have been dealt with in this chapter.

8.7 FAQs

Q1: Which spin trap should I use?

Answer: The choice of the spin trap to be used in a given experiment should take several factors into account. One should initially consider the experimental conditions, in particular, the solvent (whether the spin trap should be soluble in aqueous media or in organic solvents), temperature (some of the spin traps, e.g., DMPO, DEPMPO, and other pyrroline derivatives, are thermolabile and do not withstand high temperatures), and presence of UV irradiation (nitrosoalkanes and, to a lesser extent, nitrosoarenes are photolabile and tend to undergo loss of nitric oxide with formation of radicals not deriving from the system under study). Having tentatively individuated the spin trap to be used, one should then check for its commercial availability, a factor of vital importance if there is no access to synthetic facilities.

Q2: In my experiment I foresee the formation of alkyl radicals. Should I use a nitrosocompound or a thiocarbonyl derivative?

Answer: In the absence of UV irradiation and at normal temperatures, nitroso compounds are preferable because they can be bought in a pure form and are normally easier to handle. For photolytic reactions thiocarbonyl compounds are more convenient, but it should be born in mind that some phosphoryl dithioformates are photolabile.

Q3: When using DMPO or other pyrrolinic nitrones I trap more than one radical even in reactions where only one radical species should be formed. Why?

Answer: Pyrrolinic nitrones are fairly delicate compounds that, even if bought in the purest grade, may need purification prior to use especially if they have been stored for long time. Flash chromatography over an alumina column may however be sufficient.

Q4: Why in some cases do I observe EPR spectra upon simply mixing the reactants even when dealing with reactions that need being "initiated"?

Answer: Check for the redox potentials of the reactants. Spontaneous electron transfer reactions may occur if the potentials are compatible. Consider also the possibility of nucleophilic addition taking place followed by some oxidation process.

Q5: When using two spin traps in competitive experiments, does the ratio of the spectral intensity of the two spin adducts reflect the relative trapping efficiency of the two traps?

Answer: No, it does not. The ratio of the spectral intensity of the two spin adducts is dictated by both the rate of addition and the rate of decay of each individual adduct.

Q6: Is there an EPR database about spin adducts?

Answer: The most comprehensive database about the spectral parameters of radical species can be found in the *Landolt–Börnstein Series*, which includes four volumes on nitroxides and several other volumes concerning carbon centered radicals (where most of the adducts to nonconventional spin traps can be found).

FURTHER READINGS

Landolt–Börnstein Series: *Magnetic Properties of Free Radicals;* Springer-Verlag: Heidelberg, 1978, Vol. II9b-c; 1987, Vol. II17b-c-d1-d2; 2005, Vol. II26b-c-d.

Reviews

(a) Davies, M. J. In*Electron Spin Resonance. A Specialist Periodical Report;* Davies, M. J., Gilbert, B. C., Murphy, D. M., Eds.; Royal Society of Chemistry: Cambridge, U.K., 2002; Vol. 18, Chapter 2, pp 47–73, and references therein; (b) Alberti, A.; Benaglia, M.; Macciantelli, D. *Res. Chem. Intermed.* **2002**, *28*, 143, and references therein.

Synthesis of Spin Traps

Nitrosoarenes. (a) Terabe, S.; Kuruma, K.; Konaka, R. *J. Chem. Soc. Perkin Trans. 2* **1973**, 1252; (b) Smith, L. I.; Taylor, F. L. *J. Am. Chem. Soc.* **1935**, *57*, 2370; (c) Smith, L. I.; Taylor, F. L. *J. Am. Chem. Soc.* **1935**, *57*, 2460.

PBN Derivatives. (a) Kleiegel, W.; Lubkowitz, G.; Rettig, S.; Trotter, J. *Can. J. Chem.* **1993**, *71*, 2129; (b) Roubaud, V.; Tuccio, B.; Bouteiller, J.-C.; Tordo, P. *Res. Chem. Intermed.* **1996**, *22*, 405, and references therein; (c) Roubaud, V.; Rizzi, C.; Guérin, S.; Lauricella, R.; Bouteiller, J.-C.; Tuccio, B. *Free Radical Res.* **2001**, *34*, 237, and references therein.

DMPO Derivatives. (a) Frejaville, C.; Karoui, H.; Tuccio, B.; Le Moigne, F.; Culcasi, M.; Pietri, S.; Lauricella, R.; Tordo, P. *J. Med. Chem.* **1995**, *38*, 258; (b) Clement, J. L.; Barbati, S.; Frejaville, C.; Rockenbauer, A.; Tordo, P. *J. Chem. Soc. Perkin Trans. 2* **2001**, 1471, and references therein; (c) Tsai, P.; Elas, M.; Parasca, A. D.; Barth, E. D.; Mailer, C.; Halpern, H. J.; Rosen, G. M. *J. Chem. Soc. Perkin Trans. 2* **2001**, 875; (d) Zhao, H.; Joseph, J.; Zhang, H.; Karoui, H.; Kalyanaraman, B. *Free Radical Biol. Med.* **2001**, *31*, 599.

Carbonyl Compounds. (a) Brook, A. G.; LeGrow, G. E.; Kivisikk, R. *Can. J. Chem.* **1965**, *43*, 1175; (b) Alberti, A.; Seconi, G.; Pedulli, G. F.; Degl'Innocenti, A. *J. Organomet. Chem.* **1983**, *253*, 291; (c) Ricci, A.; Degl'Innocenti, A.; Chimichi, S.; Fiorenza, M.; Rossigni, G. *J. Org. Chem.* **1985**, *50*, 130.

Thiocarbonyl Compounds. (a) Alberti, A.; Colonna, F. P.; Guerra, M.; Bonini, B. F.; Mazzanti, G.; Dinya, Z.; Pedulli, G. F. *J. Organomet. Chem.* **1981**, *221*, 47; (b) Bonini, B. F.; Mazzanti, G.; Sarti, S.; Zanirato, P.; Maccagnani, G. *J. Chem. Soc. Chem. Commun.* **1982**, 1187; (c) Alberti, A.; Benaglia, M. *J. Organomet. Chem.* **1992**, *434*, 151, and references therein.

Phosphoryl Dithioformates. (a) Grisley, W. D. *J. Org. Chem.* **1965**, *26*, 2544; (b) Alberti, A.; Benaglia, M.; , Hapiot, P.; Hudson, A.; Le Coustumer, G.; Macciantelli, D.; Masson, S. *J. Chem. Soc. Perkin Trans. 2* **2000**, 1908, and references therein.

Kinetic Studies

(a) Zubaredv, V. E.; Belevskii, V. N.; Bugaenko, L. T. *Russ. Chem. Rev.* **1979**, *48*, 729; (b) Kemp, T. J. Prog. React. Kinet. Mech. **1999**, 24, 287, and references therein.

Biological Applications

(a) Burkitt, M. J. In *Electron Spin Resonance. A Specialist Periodical Report;* Davies, M. J., Gilbert, B. C., Murphy, D. M., Eds.; Royal Society of Chemistry: Cambridge, U.K., 2004; Vol. 19, Chapter 2, pp 33–81, and references therein.

Solid-State Spin Trapping

(a) Kuwabara, M.; Lion, Y.; Riesz, P. *Int. J. Radiat. Biol.* **1981**, *39*, 451; (b) Mossala, M. M.; Rosenthal, I.; Riesz, P. *Can. J. Chem.* **1982**, 60, 1493; (c) Ren, J.; Sakakibara, K.; Hirota, M. *React. Polym.* **1994**, *22*, 107; (d) Bhattacharjee, S.; Khan, Md. N.; Chandra, H.; Symons, M. C. R. *J. Chem. Soc. Perkin Trans. 2* **1996**, 2631.

Gas-Phase Spin Trapping

(a) Pryor, W. A.; Prier, G. G.; Church, D. F. *Environ. Health Perspect.* **1983**, *47*, 345; (b) Niki, E.; Minamisawa, S.; Oikawa, M.; Komuro, E. *Ann. N.Y. Acad. Sci.* **1993**, *686*, 29; (c) Kuno, N.; Sakakibara, K.; Hirota, M.; Kugane, T. *React. Funct. Polym.* **2000**, *43*, 43; (d) Baum, S. L.; Anderson, I. G. M.; Baker, P. R.; Murphy, D. M.; Rowlands, C. C. *Anal. Chim. Acta* **2003**, *481*, 1; (e) Culcasi, M.; Muller, A.; Mercier, A.; Clement, J. L.; Payet, O.; Rockenbauer, A.; Marchand, V.; Pietri, S. *Chem. Biol. Int.* **2006**, *164*, 215.

9 Radiation Produced Radicals

EINAR SAGSTUEN and ELI OLAUG HOLE
Department of Physics, University of Oslo, Oslo, Norway

9.1 INTRODUCTION

Absorption of ionizing radiation in matter results in a number of alterations of the electronic and geometric structures of the molecules constituting the matter. Ejection of electrons from their regular electronic states, capture of electrons into vacant electronic states, and excitation of electrons into states of higher energy than the ground state are the most common and important initial events following the radiation energy absorption process. The species resulting from these initial events subsequently develop into various secondary products that may have very different lifetimes, ranging from 10^{-13} s to many years.

Whereas most molecules in their normal state have all of their electrons coupled into pairs (forming chemical bonds or nonbonding states), the common feature of most of the radiation exposed species is that they exhibit one unpaired electron. Such molecules, which are charged or neutral, are commonly called free radicals. As described in many chapters in this book, free radicals are most conveniently studied by electron magnetic resonance (EMR) spectroscopic techniques by virtue of their paramagnetic properties. The present chapter is devoted to the qualitative and quantitative descriptions of radiation action to primarily organic molecules by way of their detection using EMR techniques. For the qualitative part, emphasis is placed on the most important biological structure, the DNA molecule, and its complexes with proteins. For the quantitative part, emphasis is placed on the application of electron paramagnetic resonance (EPR) spectroscopy to dosimetry, which is the science of measuring the amount of radiation energy absorbed in matter or the radiation *dose*.

Rather than presenting lengthy updated reviews of the literature in these fields, only selected themes and procedures will be treated in some detail according to the general idea behind this book. There has been a large number of comprehensive

---

Electron Paramagnetic Resonance. Edited by Brustolon and Giamello
Copyright © 2009 John Wiley & Sons, Inc.

reviews and book chapters on the radiation physics and chemistry of DNA as well as the use of EPR spectroscopy for quantitative studies within radiation biology, dosimetry, and geology during the last few decades. Thus, there is only room to provide key references to some of the most central works in this field, which are listed in Section 9.6.

9.2 INTERACTION OF RADIATION WITH MATTER

9.2.1 Physical Radiation Effects

Ionizing radiation is photons or particles with sufficient energy to eject an electron from (usually) the valence shell of a molecule, leaving a hole and thus creating a pristine molecular cation, which is a positively charged free radical. The major part of the energy deposited by X or γ rays will be transferred to the ejected electrons. These in turn will cause further ionizations and excitations by interaction with valence electrons of other molecules until the electron kinetic energy is sufficiently low for the electrons to become trapped in the (usually) lowest empty electronic state of a given molecule. This will then be the pristine molecular anion, which is a negatively charged free radical. In the solid state and in solution, excited states are normally not of significant importance as deexcitations by release of energy to the thermal lattice are very effective.

For moderately small charged particles such as electrons and protons [so-called low low-energy-transfer (LET) radiation] [1], the most abundant energy deposition process is also ionization by ejection of a molecular valence electron. Therefore, the actions of photon and low LET particle radiations are very similar, although some differences in the spatial distribution of ionization events will be evident.

A specific target molecule in a complex molecular system may be affected by radiation in two basically different ways, which are designated indirect and direct effects. The *direct effect* is when the radiation energy is deposited in the target molecule itself. The *indirect effect* is when the radiation energy is deposited in a neighboring (nontarget) molecule, and the resulting free radical subsequently reacts with the target molecule. A typical example of the latter effect is radicals formed in a solvent, most commonly water. These created water radicals (e.g., •H, •OH) may easily diffuse and reach the target molecule (the solute), reacting with it and forming a target molecule free radical (the dot in, e.g., •OH symbolizes the unpaired electron of a free radical). Similar indirect effects may be envisioned in, for example, a DNA–protein complex, where an initially formed protein radical may transfer its spin to DNA by some physical or chemical process.

9.2.2 Temporal Stages of Radiation Action and Radical Trapping

There are complex sequences of processes induced by an energy absorption event in a given system. It has been common to categorize these processes according to the timescales within which they occur at ambient temperatures [2]. This is illustrated in Box 9.1.

BOX 9.1 TEMPORAL STAGES OF RADIATION ACTION

Stage	Time Range	Typical Processes
Physical	$<10^{-13}$ s	Energy absorption, ionization and excitation processes, migration of electrons and holes
Physical–chemical	$<10^{-10}$ s	Ionic radicals and diffusible water radicals form and react with target molecules giving neutral radicals; proton transfer reactions
Chemical	$<10^{-4}$ s	Secondary radical processes, often initiated by proton transfer reactions, resulting in different neutral free radicals
Biochemical/biological	$<10^{8}$ s	Combination reactions. Neutral radicals lead to diamagnetic, altered molecules. In biological systems, this may be represented by DNA strand breaks. Effects on cellular function and replication, genetic or metabolic functions, eventually to mutations, tumors, or death of organism

Most free radicals have a very short lifetime, but there are exceptions. Particularly in solid-state systems (e.g., in single crystals or polycrystalline systems), secondary radicals may be stably trapped for years. Examples of this are given in Section 9.5.

Conventional continuous wave EPR spectroscopy is not a time-resolved technique; hence, short radical lifetimes represent a challenge. Other chapters in this book describe various techniques like *in situ* steady-state analysis (Chapter 4) and spin trapping (Chapters 8 and 11) for overcoming such problems. Otherwise, different means must be provided to stabilize short-lived radicals. Thermal energy is usually the key factor in driving chemical reactions; therefore, removing thermal energy will slow or stop them. Thus, cooling a system during irradiation will stabilize specific radicals, depending on the kinetic properties of the processes occurring. Subsequent controlled warming will allow for the processes to proceed, and recooling will facilitate further EMR studies of reaction products to be made. By such procedures, mapping of entire radiation induced radical reaction sequences in a given irradiated system is made possible.

9.3 QUALITATIVE DETECTION OF DNA RADICALS

The focus of this section will mainly be free radical formation by low LET radiation (primarily X and γ rays) in DNA and DNA components in the solid state. The DNA macromolecule is the most important target for radiation in a living individual, cell, or chromosome. Depending on the radiation dose and the specific system studied, the ultimate radiation effects span positive effects, no observed effects, and critical damage to the DNA molecule that may lead to cell death or mutations by altering the genetic information of the molecule. To fully understand the processes that occur when biological matter is exposed to ionizing radiation and to predict the biological outcome of a specific radiation dose to a given system, the molecular mechanisms that are involved must be understood. Thus, understanding the radiation induced free radical processes in DNA is of essential importance. The detailed molecular structure of the radiation products as well as their concentrations is necessary knowledge for the assessment of the potential harm or benefit the radiation action might cause.

There has been a long-standing discussion on the relative importance of direct and indirect effects in biological systems. Today, the scientific community generally agrees that for low LET radiation damage to DNA in a cell the direct effects account for about 40% of the events and indirect effects for the remaining 60% [3–5]. A large variety of methods can be used for studying radicals induced by indirect effects [4, 6] whereas the techniques are more limited for the direct effects. Consequently, many questions remain unanswered regarding the direct effects of ionizing radiation on DNA.

EPR and related techniques are perfectly suited for the study of free radicals formed by direct effects. Single crystals, polycrystalline systems, aqueous glass, and frozen aqueous solution systems have all proved to be useful matrices to isolate the effects of direct radiation action in these systems. If fully detailed molecular and mechanistic information is sought, single crystals usually must be used. Because of the tight and rigid packing of the molecules in single crystals, products and processes occurring in such systems may well represent processes induced by radiation in the tightly packaged DNA molecule of a biological system. However, because the environment of the DNA molecule or DNA components is clearly different in a cell and a solid-state sample, not all products and reactions observed in small molecular solid-state systems should be expected in larger model systems or in DNA itself.

9.3.1 Basic Concepts of the Structure of DNA

For the purpose of the discussions on radiation damage to DNA, a brief description of the structure of DNA will be provided.

The DNA molecule is a linear polymeric molecule of nucleotides. A nucleotide is composed of a nitrogenous base, a deoxyribose sugar, and a phosphate group. The $C1'$ carbon of the deoxyribose sugar (see numbering system below) is joined to the base, and the $C3'$ and the $C5'$ positions in the sugar are connected to phosphate groups by ester bonds. Four different nucleotides predominantly occur naturally in DNA, differing only in the nature of the bases. Two of the bases are pyrimidines (thymine and cytosine) and the other two are purines (adenine and guanine). These bases are shown in Fig. 9.1. Figure 9.2 displays a cytosine monophosphate nucleotide and a short single-stranded polynucleotide formed by connecting nucleotides by $3',5'$ phosphate ester bonds.

The DNA molecule consists of two strands of long polynucleotides, which are twisted together into a double-stranded helix with the sugar–phosphate backbone on the outside of the helix and the bases on the inside. The two strands are

9.3 QUALITATIVE DETECTION OF DNA RADICALS

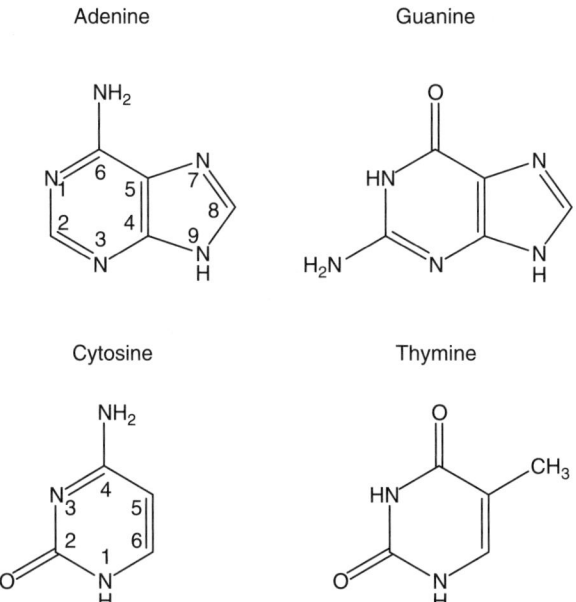

Fig. 9.1 The four most common bases in DNA. The numbering schemes used for the purines and pyrimidines are shown.

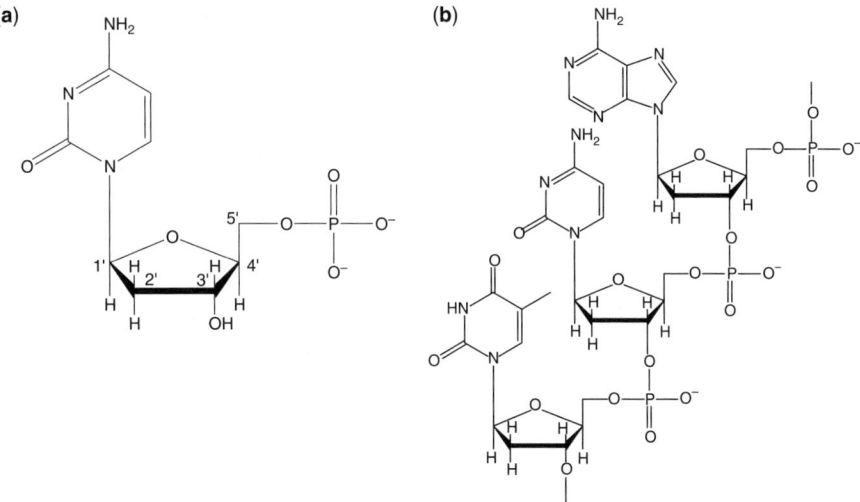

Fig. 9.2 (a) The nucleotide 2'-deoxycytidine 5'-monophosphate (5'-dCMP). The numbering scheme used for the deoxyribose sugar is shown. (b) A polynucleotide strand, with the backbone consisting of sugar–phosphate groups.

kept together by specific hydrogen bonds between the bases. These bonds are specific in that only a purine and a pyrimidine base may be combined into a hydrogen-bonded pair, and it is even more specific in that adenine only connects to thymine and guanine only connects to cytosine. This pairing scheme is depicted in Fig. 9.3a.

In the most common form of the DNA molecule, the so-called B-form shown in Fig. 9.3b, the (horizontal) base planes are almost perpendicular to the (vertical) helix axis. There are about 10 base pairs per period of the helix, the base plane distance is 0.34 nm, and the diameter of the double helix is about 1.8 nm.

The genetic information is contained in the base sequence along the strand, and the specific pairing scheme between the bases therefore makes each strand self-contained with respect to the genetic information. If one region is damaged (e.g., by ionizing radiation), the repair enzyme system of the cell nucleus can readily remove the damaged molecular fragments and insert new nucleotides because of the specific information contained on the complementary strand. For more detailed information on the structure and function of DNA, there are a number of excellent textbooks available [7].

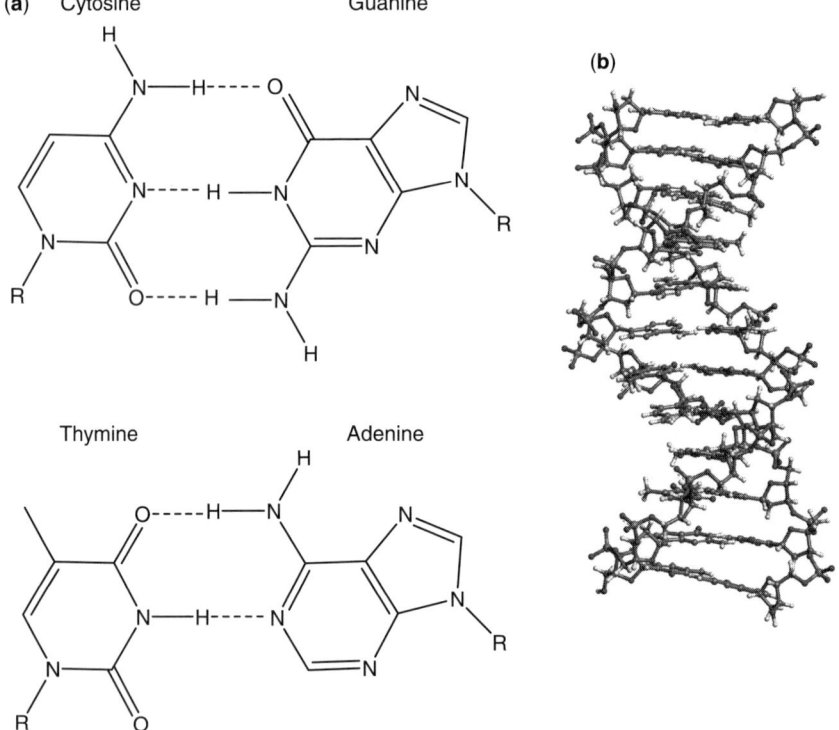

Fig. 9.3 (a) The specific hydrogen bonding scheme in DNA. (b) The DNA double helix structure (B form).

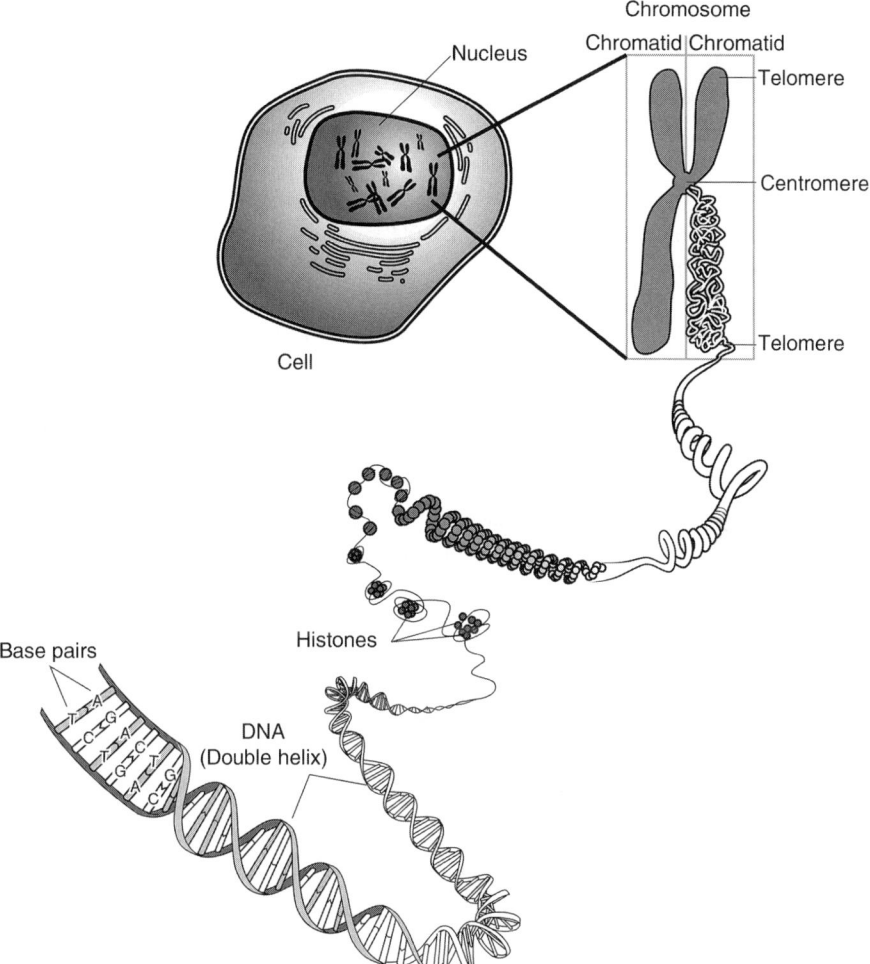

Fig. 9.4 The tight packaging of the DNA double helix into nucleosomes and chromosomes. Courtesy National Human Genome Research Institute.

Figure 9.4 is a schematic illustration of the packing of DNA in a cell. During most of the lifetime of the cell, DNA will be coiled around four different proteins (called histones) into small "balls," which are nucleosomes. These nucleosomes are in turn packed into the even more compact structures usually observed in a light microscope, which are chromosomes.

9.3.2 Initial Product Formation in DNA

The DNA molecule in a cellular environment usually is quite extensively hydrated (25–100 waters/nucleotide). Some of the water molecules are closely associated to the DNA double helix by strong associations (e.g., hydrogen bonding) and do not behave as bulk water [8, 9]. These water molecules comprise the first solvation shell and may count up to about 10 molecules/nucleotide ($\Gamma < 10$). Additional water molecules, typically up to about 22 waters/nucleotide ($10 < \Gamma < 22$), contribute to the solvation layer but are less tightly associated and may be considered like bulk water.

In fully hydrated DNA about 50% of the ionization events take place in the DNA molecule itself (direct effects) and about 50% occur in the bulk water and the first solvation shell. The probability for an ionization to take place in a specific part of the DNA molecule is roughly proportional to the number of valence electrons in that region. On average, a base contains about the same number of valence electrons as the sugar–phosphate region. Thus, there will be an almost equal initial population of oxidation products in the sugar–phosphate region and in the bases.

The fate of ionizations taking place in the solvation shell will vary according to the level of DNA hydration. Holes in the innermost water layer ($\Gamma < 10$) may transfer to the DNA quite effectively (and be counted as direct effect), whereas for the more remote water molecules ($\Gamma > 10$) the water cation will normally deprotonate faster and the created •OH radicals contribute to the indirect effect.

Whenever an ionization event takes place, an electron is ejected. This electron will cause further ionizations and more low energy electrons until it becomes trapped at the most energetically favorable site of the system. The nucleobases have the highest *electron affinity* (EA) among all DNA constituents, and virtually 100% of the electrons ejected from DNA become trapped at the bases. Electrons formed in the solvation shell will partly transfer to the nucleobases and partly contribute to the radiation chemistry taking place in the solvation layer itself, producing a variety of products. Such products (e.g., •H, •OH) will eventually contribute to the indirect effect.

A well-known and very important feature of ionizing events, even at low LET values, is that ionization almost always appears in clusters, which means that there is a high probability for multiple ionizations (ion pair formations) to occur in the near vicinity of each other [2, 10]. Even if some immediate recombination of ions will occur during thermalization of the system, there will still be a nonhomogenous distribution of ion pairs in the system at very short times after the radiation exposure. This is illustrated in Fig. 9.5, which shows clusters of ion pairs in the bulk and solvation water layers around the DNA molecule and on the DNA strands.

It has been shown repeatedly that the seriousness of the biological effects depends on the nonhomogenous distribution of trapped products (so-called multiple damaged sites) [11, 12]. Furthermore, the conductivity of electrons and holes along the DNA chain is the key factor in the progression of the initial damage to these nonhomogenous distributions. As clustered products develop from clustered ionizations, electron and hole transfer processes must exhibit a limited range. If the DNA molecule behaved as a molecular wire [13, 14], extensive recombination would occur and the yield of radical products in DNA would be negligible compared to that observed experimentally. Thus, even though charges may migrate along the axis of the DNA helix [15], DNA is a poor long-range electron and hole conductor [16].

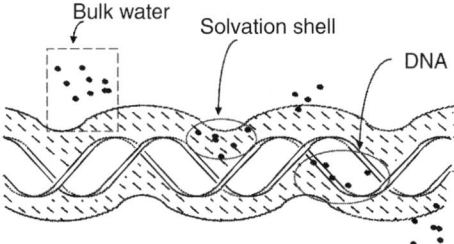

Fig. 9.5 Clustering of ionization events after exposure of solvated DNA to ionizing radiation. Courtesy Professor William A. Bernhard.

After thermalization of the system, there is only limited electron and hole mobility, sufficient to allow for some preferential trapping sites of the holes and electrons but not sufficient to obliterate all charges induced in DNA. The actual range of electron and hole transfers in systems irradiated at low temperatures appears to be less than 10 base pairs [16–18]. An example of such clustered damage is a double-strand break, commonly considered as two single-strand breaks localized close to each other on opposite strands of the DNA molecule [19, 20].

For high LET radiation the ionization density and the cluster structure is more pronounced than for that following low LET irradiation, and the initial recombination becomes more effective as the ionization density increases. As a consequence, the yield of stabilized DNA radicals at a given dose is often lower for high LET radiations [21].

9.3.3 Stabilization of Free Radicals in DNA

There is a large body of literature on radiation induced charge transfer processes in DNA, but here only a few central issues of the overall picture of damage progression will be discussed.

Two factors characterizing the electronic structure of the DNA constituents are important for the further discussion. The EA is a measure of the propensity for a given constituent to capture an excess thermal electron, that is, a measure for the energy reduction upon electron capture. Available experimental and theoretical results indicate that the pyrimidine bases cytosine (C) and thymine (T) have a larger EA than the purine bases guanine (G) and adenine (A), as well as the sugar–phosphate (SP) group: $C \approx T > A > G > SP$. The *ionization potential* (IP) is a measure of how hard it is to ionize a given constituent, that is, how much energy is required for an ionization event. Guanine is the constituent with the lowest IP (easiest to ionize), followed by adenine, thymine, cytosine, and the sugar–phosphate group: $G < A < T < C < SP$. As a consequence, guanine is predicted to be the site of localization of the hole in DNA and cytosine and/or thymine similarly are predicted to be the sites of electron localization.

The number of ionization processes in the sugar–phosphate region and in the DNA bases is about equal. The holes created in the bases will transfer (mainly) to guanine whereas the holes initially created in the sugar–phosphate region will partly remain in the sugar–phosphate region and partly transfer to the base stack. It is estimated that about 25% of all initial sugar–phosphate holes will remain trapped at the sugar–phosphate group. Most of the ejected electrons will be localized to the DNA pyrimidine bases (C and T).

A localized charge in an extended molecular system is a vulnerable entity, which is easily accessible for charge recombination during charge migration. There is, however, strong evidence that not only EA and IP determine the stabilization of both electron and hole trapping but also the hydrogen-bonding network is a crucial factor. The bases in DNA participate in hydrogen bonding to its neighboring base with a complementary pairing such that thymine always binds to adenine and cytosine always binds to guanine (Figs. 9.3 and 9.6). The proton positions in these hydrogen bonds are determined by the acid–base equilibrium between the two partners of the base pair. If one of the two partners either lose or gain an electron, the acid–base equilibrium will change and a proton transfer across a hydrogen bond may occur. It has been pointed out [22–24] that when cytosine captures an electron, the change in cytosine basicity is so large that a proton will readily transfer to cytosine from the guanine partner of the pair. A similar transfer will probably *not* take place from adenine to a thymine anion, making thymine a less suitable site for electron trapping and stabilization than cytosine. Similarly, if guanine loses an

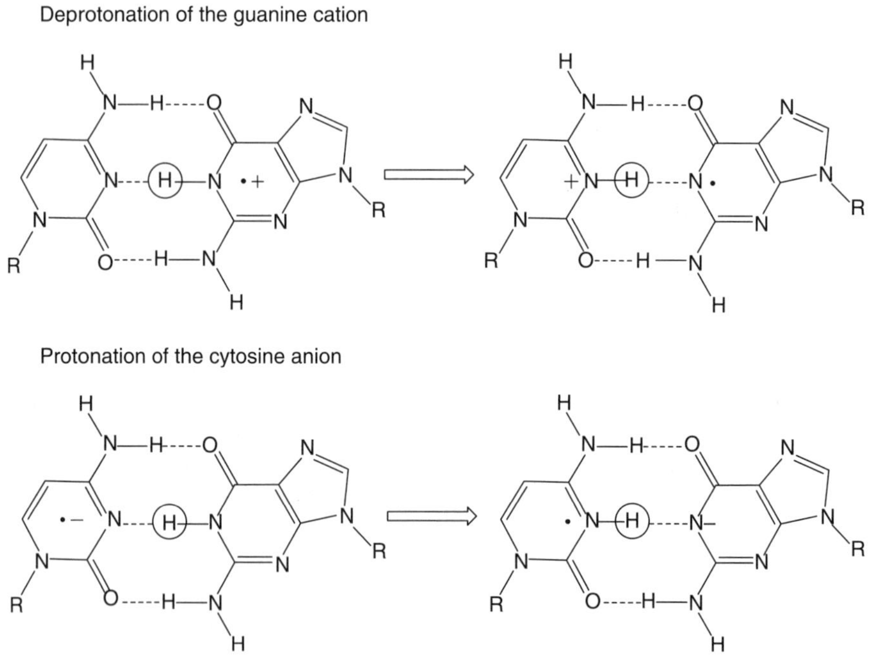

Fig. 9.6 Guanine/cytosine protonation and deprotonation reactions upon pristine ionization of guanine and electron capture of cytosine.

electron, the change in guanine acidity is so large that a proton will immediately transfer from the cytosine partner to guanine. Conversely, the adenine cation (or C^+ or T^+) will not have a similar driving force to become protonated by its complementary base.

With reference to Fig. 9.6, the reader will realize that such proton transfers in base pair hydrogen bonds will separate the excess charge from the spin, thus protecting the spin center from immediate charge recombination. The combination of EAs, IPs, and acid–base equilibria provide mechanisms for stabilizing two neutral products (protonated cytosine anions and deprotonated guanine cations) as the primary radiation defects in the DNA bases. These primary radicals are, in fact, observed in DNA and in components of DNA in the solid state when the samples are irradiated and detected by EPR or electron nuclear double resonance (ENDOR) at low temperatures (4–77K). Warming the samples at higher temperatures may induce detrapping of the primary neutral free radicals by backtransfer of the proton and induction of further charge migration along the DNA stack. Thus, initial base protonation and deprotonation processes are reversible and a variety of reactions may occur at elevated temperatures, leading partly to charge recombination and partly to a number of different stabilized free radical products including deprotonated cations, protonated anions, H-atom adduct radicals, and various fragmentation products [25, 26].

Less is known of the fate of the sugar–phosphate holes that initially become trapped at the sugar–phosphate group. However, recent studies have demonstrated indirectly that a significant fraction of these holes are irreversibly stabilized on the sugar. The stabilization apparently occurs by deprotonation of the sugar cation from one of the sugar carbon atom positions, as shown in Fig. 9.7 [27–29].

It is difficult to characterize early sugar radicals in DNA samples by EMR spectroscopy because of broad and featureless resonances [30], but carbon-centered sugar radicals have been observed in a few DNA and crystalline DNA model systems. The number and nature of the radicals vary to a large extent among the systems studied, and it is still not fully understood which factors (electronic, geometrical, environmental) direct the stabilization of a given neutral radical. In crystals of deoxyguanosine-5′-monophosphate, sugar radicals centered at each of the sugar carbon atoms have been observed [31]. In other systems, radicals centered at one or a few carbon atoms have been characterized. Extended studies of free radical stabilization in crystalline sugars [32–37] may contribute to understanding the nature of the products and secondary radical reactions occurring in the DNA sugar–phosphate moiety.

Sugar radicals have been considered to be far more important than base radicals with respect to immediate or catalyzed fatal biological effects like free base release and single-strand breaks and, by way of the clustered damage concept, to double-strand breaks and cell death. However, recent research has shown that base damage clustering associated with a radiation induced double-strand break may contribute to DNA damage complexity and affect break repair [38]. A summary of the essential features of the early radical formation in DNA by direct effects is provided in Box 9.2.

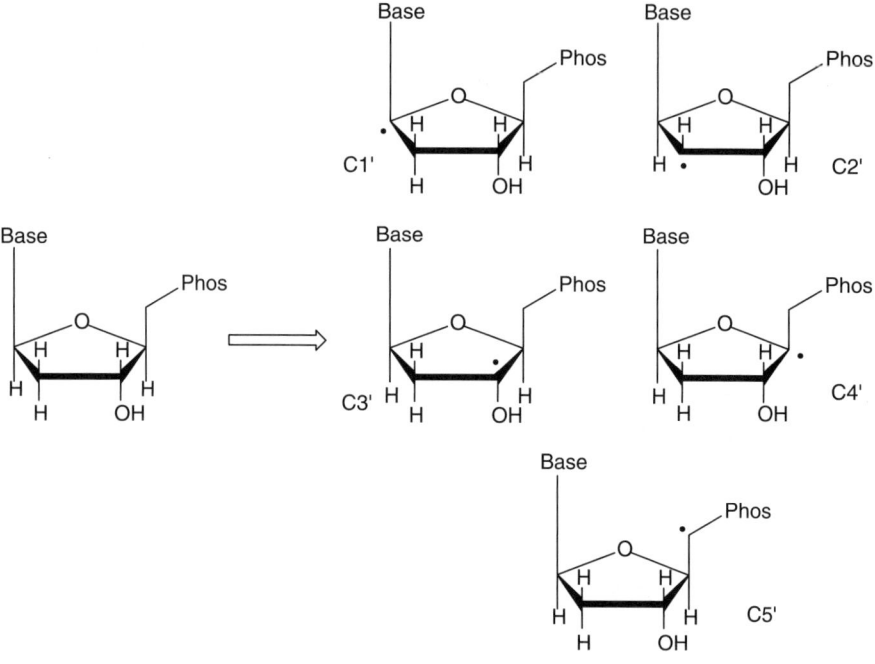

Fig. 9.7 Possible carbon-centered radicals formed in the DNA deoxyribose moiety by irreversible deprotonation from a pristine oxidation in the sugar–phosphate group.

BOX 9.2   EARLY RADICAL FORMATION IN DNA BY DIRECT EFFECTS

- ~50% of ionizations in the sugar–phosphate, ~50% in the bases
- All secondary electrons become trapped by the bases.
- About 50% of the holes in the sugar–phosphate transfers to the base stack.
- Electron or hole transfers in the base stack lead mainly to guanine cations and cytosine anions.
- Charged species are stabilized by protonation and deprotonation processes.
- The stabilized radical population at low temperatures consists of 80–90% base radicals (mainly deprotonated guanine cations and protonated cytosine anions) and 10–20% sugar-centered radicals.

9.3.4   Radical Structures in DNA and DNA Constituents

A large number of radiation induced free radicals from DNA and DNA constituents have been observed in crystalline and polycrystalline systems as well as in frozen aqueous solutions. Here, a short review of the essential radical structures from

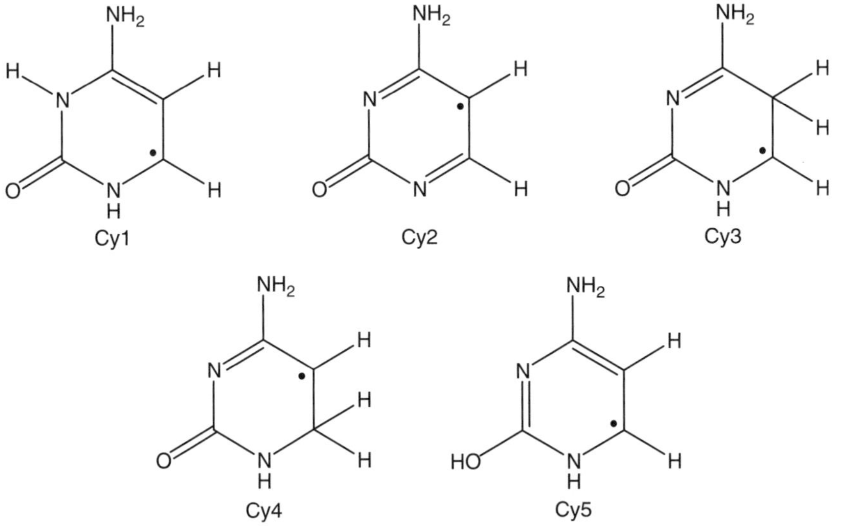

Fig. 9.8 Commonly observed cytosine-derived radicals observed after irradiation of solid-state samples.

bases, nucleosides, nucleotides, and DNA is provided. The structures of the various base radicals are given in Figs. 9.8–9.11 for cytosine, thymine, guanine, and adenine, respectively. The numbering schemes for the purines and pyrimidines are given in Fig. 9.1. For more detailed discussions of the radicals observed in DNA and DNA constituents, the reader should consult texts referenced in Section 9.6 [9, 25, 26, 39–41].

9.3.4.1 Cytosine. The N3-protonated cytosine anion (Cy1) and the N1-deprotonated cytosine cation (Cy2) were observed in single crystals of cytosine monohydrate subsequent to low temperature irradiation (Fig. 9.8) [42]. In the 5′-nucleotide, oxidative deprotonation at the N1 position is not possible (because N1 is bonded to C1′), and apparently the native cation is stabilized at low temperatures [43]. At higher temperatures, the C5 and C6 H-adduct radicals (Cy3 and Cy4, respectively) are commonly observed. An additional product is observed at room temperature. The molecular structure of this defect is not yet unambiguously determined, but it has been suggested to be the result of a net H addition to the O2 position (Cy5) [44, 45]. Sugar-centered radicals (alkoxy radicals) have also been observed in cytosine nucleotides [46, 47].

9.3.4.2 Thymine. The O4-protonated thymine anion (Th1) has been observed in crystalline systems (Fig. 9.9) [48, 49]. The pristine thymine cation might be expected to deprotonate at N3; however, that product has not been identified in the solid state. Rather, a product (Th2) formed by deprotonation of the thymine cation at the methyl group is present in all thymine derivatives studied [50–52]. The C6 H-addition radical (Th3) and, to a lesser extent, the C5 H-addition radical (Th4)

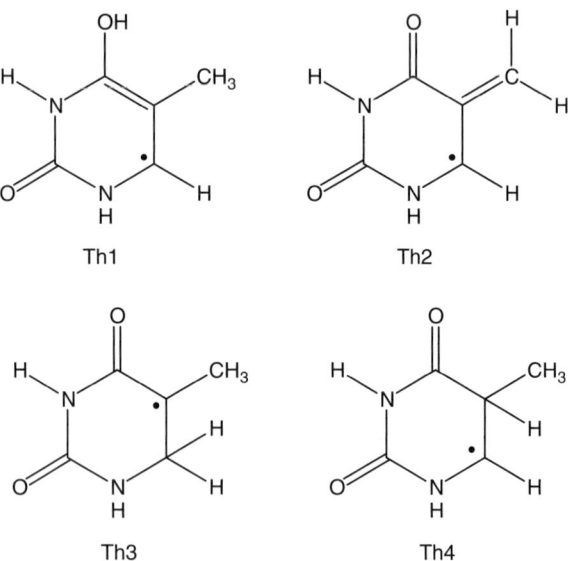

Fig. 9.9 Commonly observed thymine-derived radicals observed after irradiation of solid-state samples.

are commonly observed after annealing samples at higher temperatures [25]. There is evidence that the thymine anion is the precursor of the C6 H-addition radical [53], but this conversion has not been observed in the solid state. The C6 H-addition radical (Th3) is an important defect in irradiated DNA, and it was the very first stable free radical identified in irradiated DNA [54, 55].

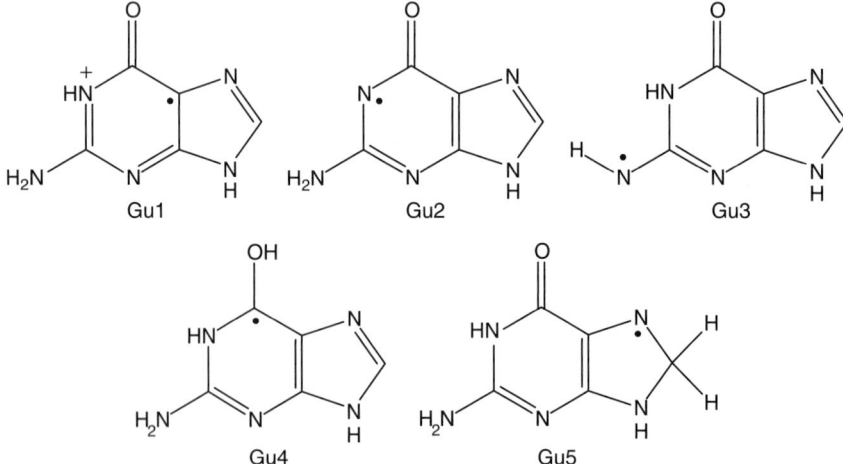

Fig. 9.10 Commonly observed guanine-derived radicals observed after irradiation of solid-state samples.

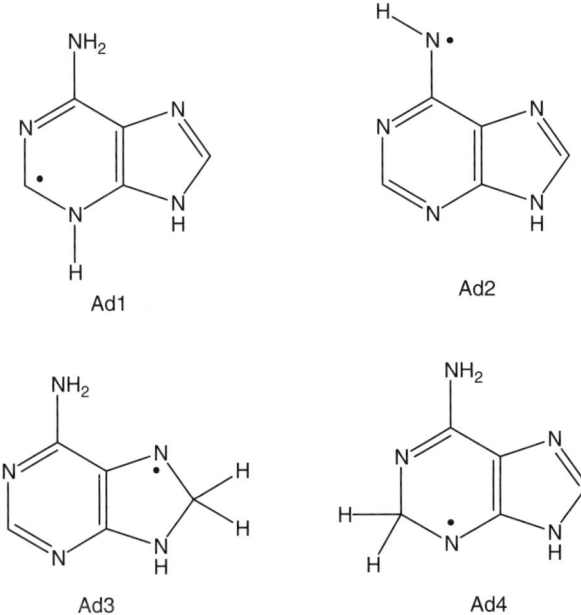

Fig. 9.11 Commonly observed adenine-derived radicals observed after irradiation of solid-state samples.

There is only weak evidence for a carbon-centered and an alkoxy radical in the sugar moiety of thymidine at low temperatures [48, 56].

9.3.4.3 Guanine. The purine base guanine has been a focus of attention since the early proposal that the primary oxidation of DNA almost exclusively takes place at guanine (Fig. 9.10) [57]. In solid-state model systems, the pristine guanine oxidation product has not been identified. However, in guanine·HCl, where the base is originally protonated at N7, the primary oxidation product (the dication) deprotonates at N7 and yields the structure Gu1 [58, 59], which would be analogous to the regular pristine guanine cation. In solid-state systems where the base is not protonated by the matrix, the pristine cations usually deprotonate at the exocyclic amino group [31], giving a product with structure Gu3. This structure exhibits resonance parameters that differ from those of the deprotonated guanine cation observed in frozen solutions and in frozen aqueous DNA. According to quantum chemical calculations, the guanine cation prefers to deprotonate at N1 (forming Gu2) in hydrated environments and at the exocyclic amino group (forming Gu3) in dry environments (e.g., crystalline systems) [60]. The most common reduction product observed in irradiated guanine derivatives, regardless of the initial protonation state of the base, is the O6-protonated guanine anion Gu4. At elevated temperatures, the most common radical product is the C8 H-addition radical Gu5.

In most guanine nucleosides and nucleotides, very few sugar-centered radicals have been observed with one noticeable exception: in 2′-deoxyguanosine-5′-monophosphate, radicals centered at each carbon atom of the deoxyribose moiety have been identified [31]. The origin of these differences is presently not understood.

9.3.4.4 Adenine. The N3-protonated anion radical (Ad1) is the most common reduction product in adenine derivatives (Fig. 9.11) [61, 62]. In an initially protonated adenine base, oxidation followed by deprotonation yielded a charged adenine radical cation that would be analogous to the native adenine cation. However, in an unprotonated base, the native cation deprotonates at the exocyclic amino group (Ad2) [61, 62]. The H-addition radicals at C2 and C8 (Ad3 and Ad4, respectively) are also commonly observed in the EPR spectra, even after irradiation at very low temperatures [63, 64].

In adenine derivatives, observations of sugar radicals vary considerably. In the nucleotide adenosine, no sugar radicals have been observed [61]. In contrast, no base oxidation product has been observed in deoxyadenosine; it appears as if oxidation is shifted to the deoxyribose moiety, yielding two carbon-centered radicals at C1′ and C5′ [64].

9.3.4.5 DNA. A vast number of studies of irradiated DNA and DNA constituents have been performed and different interpretations have dominated [65, 66]. The current dominant interpretation is that the cytosine anion (probably protonated at N3, i.e., Cy1) and the guanine cation (probably deprotonated at N1, i.e., Gu2) are the two major radicals initially formed. These products may become stabilized and studied at low temperatures. Additional minor radicals have been observed in various studies of irradiated DNA. These include thymine anions (10–20% of the reduction products) and adenine cations (about 10% of the base oxidation products) [67, 68], H abstraction from the thymine methyl group (Th2) [69], and C1′ and C3′ carbon-centered radicals in the deoxyribose moiety [69].

As mentioned above, the thymine C6 H-addition radical (Th3) was observed as a major radical in DNA after irradiation at room temperature. Its presence as a dominating radical is explained by the following and is based on experimental evidence. First, the fairly small amount of thymine anions initially formed will protonate irreversibly to form the thymine C6 H-adduct radical (Th3). Second, there is evidence that thermal annealing will promote detrapping of the initially formed cytosine anions [41] and that the released electrons will migrate to the thymine bases at which an irreversible protonation at C6 occurs, forming Th3 [53, 70].

The mechanisms for DNA strand breaks by direct effects are still not fully understood. Apparently, base damage is of little importance. It has been proposed that the initial precursors for strand breaks by direct-type effects are products formed by net H abstraction from carbon atoms at the deoxyribose sugar [69, 71]. Studies of irradiated crystalline DNA and plasmid DNA indicate that both singly and doubly oxidized sugars contribute [27–29, 72–74].

9.3.4.6 Nucleosomes. In irradiated nucleoproteins, electron transfer takes place between the protein and the DNA constituents [75]. Electrons transfer from the proteins to DNA, but the holes do not transfer [76, 77]. Thus, an increase in the DNA anion radicals and a loss of DNA cation radicals by recombinations is expected. The excess capture of electrons by thymine and cytosine is suggested to be the key for the spin transfer mechanism [78]. A recent single crystal study of a complex between a dipeptide and a nucleic acid base (as a model system for a protein–DNA complex) demonstrated that oxidation took place selectively at the dipeptide (protein) part of the complex and that electron capture selectively occurred at the nucleic acid base (DNA) [79]. This selectivity agreed with differences in the IPs and EAs of the different constituents of the complex.

If it is the oxidation pathway that leads to strand breaks, the above results all suggest that the nucleoproteins provide a radioprotective effect to DNA by providing electrons that can combine with the one-electron oxidation radicals of DNA.

9.4 TOOLS AND PROCEDURES FOR RADICAL STRUCTURE DETERMINATIONS

Section 9.3 presented a large number of free radicals formed by ionizing radiation in DNA and DNA without commenting on the actual methods used to arrive at these molecular structures. Three major tools are often combined when radical identifications are made by EMR techniques: the actual magnetic resonance study, quantum chemical calculations, and spectrum simulations. In the following, these three tools will be discussed in somewhat more detail using a single crystal study of *cytosine monohydrate* as a practical example [42].

9.4.1 Choice of System

The EMR studies of radicals in aqueous, frozen, polycrystalline, or single crystal samples may all provide some kind of information regarding a radical structure. The level of detail, however, strongly depends on the sample phase. Complete **g** tensors and hyperfine coupling tensors [characterized by both eigenvalues (principal values) and corresponding eigenvectors] can only be obtained from single crystal studies. Randomly oriented solid-state systems like polycrystalline samples and frozen solutions may provide information about principal values for the tensors (Chapters 1 and 6), but they usually do not give information about the eigenvectors (which often are crucial in determining the actual molecular structure). Aqueous solution studies are characterized by the complete averaging of the **g** tensor and the hyperfine coupling tensor and do only give information about the isotropic values, that is, no information about the important electron–nuclear dipolar coupling interactions. Thus, for structural determination of a given free radical structure, a single crystal is by far the best system, because of the wealth of information embedded in the dipolar component of the hyperfine interaction term of the spin Hamiltonian

(Chapter 1). As the level of detailed information is reduced and the credibility of the interpretation is decreased, the importance of additional methods like spectrum simulations and quantum chemical calculations increase. Even when single crystals are studied, these additional methods are sometimes a prerequisite for a complete and credible identification of the radical structures.

In some cases, however, frozen solutions are preferable. An excellent example of this is the long-standing efforts by Sevilla and Becker [80]. This group has studied radiation effects of DNA constituents and DNA using frozen solutions. By carefully varying the chemical properties of the solvent and methods for radical generation, they have been able to characterize radical products by using chemical and/or radiation physical arguments.

Almost all of the structures discussed and shown in Section 9.3 were identified using single crystal analysis techniques. A full single crystal study using EPR, ENDOR, and related techniques is an extensive and time consuming project. In the following, the reader will be guided through the steps involved in such a study, starting with the process of growing crystals and ending with the actual identification of the radiation induced radicals. The previously published study of single crystals of cytosine monohydrate [42] will be used to illustrate the various steps.

9.4.2 Crystallography

Good single crystals for X-band EPR/ENDOR work should exhibit well-developed faces and be of good quality without too many visible flaws. A crystal extending 5 mm in each direction would be considered as a very large crystal. Usually crystals are needle shaped or plate shaped. For higher frequency bands (K-, Q-, W-, etc.), correspondingly smaller samples must be chosen.

Many organic molecules crystallize into single crystals suitable for EMR studies. The major conditions determining the process of making a single crystal of a given compound are the solubility in different solvents and the temperature dependency of the crystallization process [81]. Many compounds are quite easily dissolved in water, and letting an almost saturated aqueous solution evaporate slowly at room temperature often may give nice crystals. Saturating the solution at a higher temperature and letting the sealed solution slowly cool to room temperature (or even below) is another common method. If the compound dissolves well in one solvent and poorly in another and the two solvents are fully comiscible, a common method is to let the vapor of the poor solvent slowly diffuse into the solution of the good solvent. Hints on the proper procedure for a given compound are usually found in the crystal structure analysis literature [81].

There are thousands of published crystal structures. The major crystallographic journal is *Acta Crystallographica*. For a full benefit of the single crystal analysis, the crystal and molecular structure of the compound in question should be known. Sometimes more than one crystal structure analysis is published for the same compound. A neutron diffraction analysis (not available for all systems) is preferable because the X—H (X=C, N, O) bond lengths are more precisely determined in such studies.

9.4 TOOLS AND PROCEDURES FOR RADICAL STRUCTURE DETERMINATIONS   343

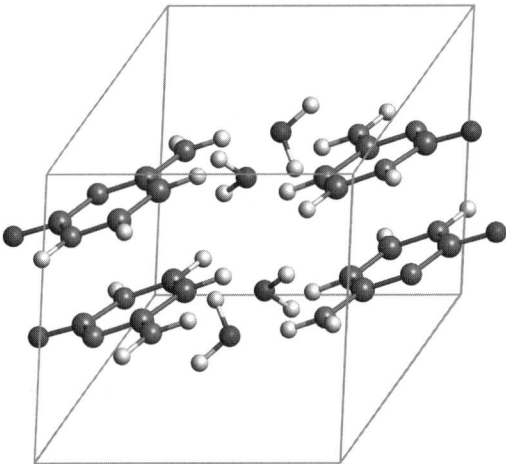

Fig. 9.12 The contents of one unit cell of cytosine monohydrate single crystals.

Most organic crystals grow in one of three crystal systems: orthorhombic, monoclinic, or triclinic. These are characterized by the length of the three unit cell axes, the angles between them, and the symmetry operations between each of the molecules contained within the unit cell. The position of each atom in the unit cell can be calculated from the crystal structure, and all atom positions in the crystal may be determined by using suitable translations [82]. By knowing the atomic positions, the directions of each bond as well as any inter- and intramolecular direction can be obtained.

In the case of cytosine monohydrate, single crystals were made by slow evaporation of nearly saturated aqueous solutions at room temperature. The crystals are monoclinic with four asymmetric units (each consisting of a cytosine molecule and a water molecule of crystallization) in the unit cell. Several crystal structure analyses were available [83–85], and the one involving neutron diffraction was chosen for the later analysis [85]. The unit cell of cytosine monohydrate is shown in Fig. 9.12.

9.4.3 Crystal Mounting and Irradiation

To obtain sufficient information to establish complete **g** and hyperfine coupling tensors, the single crystals must be rotated in the magnetic field of an EPR spectrometer around two or more axes of rotation. For the tensor analysis it is common to choose an orthogonal reference system consisting of direct crystallographic axes and their perpendiculars, such as $\langle a^*, b, c \rangle$, for a monoclinic system where $\mathbf{a}^* = \mathbf{b} \times \mathbf{c}$. It is also common to rotate the crystals about the three orthogonal axes, although other rotation axes may be chosen (see below).

The orientation sensitivity of ENDOR in particular is very high and the actual axis of rotation should be determined within about 1°. For EPR studies alone a precision of about 5° is sufficient. If the crystal faces are well developed it may be possible to achieve this precision optically, but usually assistance from X-ray diffraction will be necessary. A specific axis may be found by mounting the crystal to an X-ray diffraction goniometer, and using a coaxial transfer system the crystal may be transferred and glued to the EPR sample holder without loss of alignment.

If the crystals need to be irradiated at room temperature, it is most convenient to first irradiate the crystal, then orient it by X-ray diffraction, and finally transfer it to the sample holder (metal rod, quartz rod, or perspex rod). Thus, radicals induced in the glue or in the sample holder by the radiation may be avoided. The sample holder may then be inserted into a suitable EPR or ENDOR cavity for measurements. It is necessary to keep track of the angle of rotation in the cavity to within 0.5–1°. This requires that an angular scale is connected to the sample holder and a pointer fixed to the spectrometer (or vice versa). Following irradiation at room temperature, samples may easily be cooled down and observed at temperatures below room temperature by using liquid nitrogen or liquid helium insertion type dewars or by using commercial cryostats (e.g., flow cryostats like the Bruker ER 4131VT for temperatures down to about 110K or the Oxford Nanosciences ESR900 for temperatures down to 4K).

Irradiating samples at low temperature, however, requires much more attention because the crystal must be mounted to the sample holder and cooled to the desired

Fig. 9.13 An EPR/ENDOR X-band cavity system developed for use with a cold-finger liquid helium cryostat. The vacuum-sealed side walls of the system are removed showing the cavity with the EPR modulation coils, the iris adjustment system, optical and X-ray irradiation windows, electrical connections for the EPR modulation and the (internal) ENDOR coils, the waveguide connection, and the vacuum pump-out port.

temperature prior to irradiation. To avoid disturbing background signals from radicals induced in the glue or the sample holder, metallic sample holders and electrically conducting glue (e.g., a metal epoxy glue) may be used. There are no commercially available cavity systems allowing for X-ray irradiation at low temperatures, but there are several designs of cryostat/cavity constructions for this purpose in the literature. Flow cryostats [86], cold-finger cryostats [87], and cryogen bath cryostats [88] have been used. The most simple procedure for low temperature irradiation would be to irradiate the sample embedded in the suitable cryogen (liquid nitrogen, liquid helium) through the top part of an insertion dewar, lower the irradiated crystal to the bottom of the dewar, and insert the dewar into a suitable cavity.

Figure 9.13 shows the design used in the authors' laboratory, which was originally devised by W. H. Nelson [87]. A metal sample rod is fixed to the end of a cold-finger cryostat and inserted into the telescoping cryostat holder shown at the top of the system. The sample is lowered to the irradiation position, cooled to the desired temperature, and irradiated. After irradiation, the sample is lowered into the center of the microwave cavity by means of the telescoping cryostat holder.

9.4.4 Measurement and Analysis Procedures

Once the crystal has been irradiated and positioned in the microwave cavity, measurements may proceed. It is common to rotate the crystal about the crystallographic axes with the axes of rotation coinciding with the three reference axes. When rotating about orthogonal axes where (at least) two are crystallographic axes, three different experiments must be made. If one or more rotation axes are different from the crystallographic axes, sufficient information for a tensor analysis may in principle be obtained from two experiments or even only one experiment (depending on the choice of rotation axis), but in practice the analysis become more complex.

The anisotropy of the **g** tensor and the hyperfine coupling tensors will result in angular dependent EPR spectra from which the **g** tensor may be determined. In simple cases, the hyperfine interaction tensors may also be determined from the EPR alone. Normally, however, the spectra are so complex that the much better resolved ENDOR technique must be used. The resonance frequencies of the ENDOR lines will exhibit an anisotropic behavior characteristic for the interaction involved. ENDOR experiments give no information regarding **g** tensors.

When rotating a crystal about three orthogonal axes, each plane of rotation will provide three tensor elements. These are commonly determined by doing a least squares fitting of the observed field values (EPR, **g** tensor), splitting values (EPR, hyperfine coupling tensor), or ENDOR frequencies (ENDOR tensor) to the relation (first-order perturbation expansion, see Chapters 1 and 6):

$$t_k^2(\theta) = T_{ii} \cos^2\theta + T_{jj} \sin^2\theta + 2T_{ij} \sin\theta \cos\theta \qquad (9.1)$$

where t is the measured quantity; T_{ii} is the ii element of the squared tensor; θ is the angle of rotation from the i axis; and i, j, k are the reference axes in cyclic order. Thus, one rotation plane gives three tensor elements, and rotations about three orthogonal axes are necessary for all off-diagonal elements to be determined. In fact, the diagonal elements are determined twice, providing a good check of the internal consistency of the data. Once all tensor elements are determined, the tensor is diagonalized. This gives the three eigenvalues (T_l) of the squared tensor and their corresponding eigenvectors. The square root of these eigenvalues T_l corresponds to the principal tensor values for either the **g** tensor or the hyperfine coupling tensor. The eigenvectors remain the same as for the squared tensor.

If the measured quantities t are ENDOR frequencies, a slightly modified procedure must be followed, because the relation between the ENDOR tensor eigenvalues T_l and the hyperfine coupling principal values A_l is given by

$$A_l = 2(\mp \sqrt{T_l} + \nu_N) \tag{9.2}$$

for the electron spin quantum number $m_s = \pm(1/2)$ when assuming a positive hyperfine coupling constant $a < 2\nu_N$ and a fairly isotropic g value (an approximation that is valid for g variations within a range of 0.5% around the free electron g value of 2.0023). Here, ν_N is the nuclear Zeemann term in frequency units $g_N|\mu_B|/h$, which is about 14.5 MHz at X-band microwave frequencies ($B = 340$ mT, see Chapter 1).

Modern analysis procedures usually employ nonlinear regression methods to all available data simultaneously according to the more generalized version of Equation 9.1:

$$t^2(\vec{r}) = T_{xx}l_x^2 + T_{yy}l_y^2 + T_{zz}l_z^2 + T_{xy}l_xl_y + T_{xz}l_xl_z + T_{yz}l_yl_z \tag{9.3}$$

where $\vec{r}$ represents the direction of the magnetic field in the reference frame, and l_x is the direction cosine of the field direction with the x axis. This requires that a specification of the orientation of the magnetic field with respect to the reference system should be given to each measurement. Specifying the orientation of the rotation axis and the rotation angle of the field in the plane of rotation by polar or Euler angles usually fulfills this requirement. This procedure permits the use of nonorthogonal rotation axes and fewer experimental planes of rotation if the rotation axes are carefully selected, but the tensor analysis in often more complex. A fuller discussion of this, with reference to many available analysis programs, is given elsewhere [89].

A problem often encountered in single crystal analyses is called the Schonland ambiguity [90]. It refers to the uncertainty of the signs of the nondiagonal tensor elements T_{ij}, as discussed in some detail by Atherton (p 135) [91]. Because the unit cell of a crystal usually contains more than one molecule that is differently oriented in space according to the symmetry rules of that particular unit cell, the EMR experiment will at a general orientation of the magnetic field detect all of these differently oriented molecules simultaneously. The spectra and subsequent tensor analyses may become very complex. However, in the common cases of orthorhombic and monoclinic symmetries, the nature of the unit cell symmetry operations reduces the complexity considerably.

As a rule, in *monoclinic* crystals, all molecules degenerate into the same orientation upon rotation about the unique axis ⟨b⟩. Rotation around any axis perpendicular to ⟨b⟩ will group the molecules into two orientationally different sets, and these will both be observed in the EMR experiment. These are called two "sites." The value of the EMR parameter from one of the sites (e.g., the ENDOR frequency) at a rotation angle α will be identical to the value of that same parameter for the other site at the angle $180° - \alpha$. Thus, a kind of mirror symmetry at $\alpha = 90°$ is observed. The two sites coincide when the field is oriented along a crystallographic axis or a direction perpendicular to two crystallographic axes (e.g., **a***). It may be shown that one site corresponds to one sign of the corresponding off-diagonal element whereas the other site corresponds to the opposite sign. Because a given EMR parameter must be related to one unique molecule, it is a challenge to combine sites for several different experiments (rotation planes). There are eight different possible combinations, and upon diagonalization the solutions will be grouped into two subsets of solutions (four in each subset). One of these subsets is unphysical and must be rejected. The remaining subset contains the correct solution; but if there are only two molecules in the unit cell, two of these solutions are unphysical as well. The challenge is how to choose between the physical and unphysical sets. When performing a standard set of experiments, that is, using only three orthogonal rotation axes of which usually at least two coincide with crystallographic axes, there is no direct way to distinguish between the physical and unphysical solutions; thus, implicit methods must be used to solve this ambiguity. If, however, one additional rotation plane is recorded with the rotation axis at a skewed (and known) orientation with respect to the reference axis system [92], by using the two sets of tensors for calculating the relevant EMR parameter (g, a, ν) in the skewed plane, a comparison with the experimental data will distinguish between them and point out the correct solution. This may be realized from Equation 9.3, because such a calculation employs more than one off-diagonal tensor element [93]. Advanced analysis programs may take such skewed rotation planes into consideration from the beginning [89].

In *orthorhombic* crystals, rotation around any crystal axis (also ⟨b⟩) will group the molecules into two orientationally different sets that will both be observed in the EMR experiment as two (α, $180° - \alpha$) symmetry related sites. The two sites coincide when the field is oriented along a crystallographic axis. At a general orientation four sites are observed: two and two related by the (α, $180° - \alpha$) type of symmetry.

In *triclinic* crystals, no site splitting is observed at any orientation.

9.4.5 Results and Analysis

To obtain sufficient data for tensor analysis, (normally) three rotation planes of the irradiated samples must be obtained. The EPR (or ENDOR) spectra should be recorded with typically 5° intervals between each recording, and the crystal should be rotated a total of 90° or 180° (90° may be sufficient when the sites have mirror symmetry). Figure 9.14a provides an experimental EPR spectrum for a cytosine monohydrate crystal. The crystal was X-ray irradiated and measured at about 10K

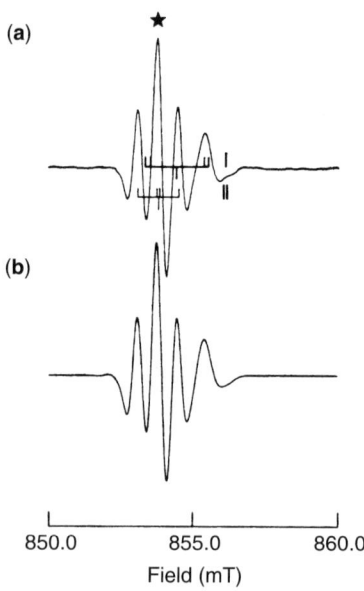

Fig. 9.14 (a) A second-derivative K-band EPR spectrum of a cytosine monohydrate crystal X-ray irradiated and measured at 10K. The magnetic field is perpendicular to the crystallographic $\langle b \rangle$ axis and in the plane of the cytosine ring, about 155° from $\langle a^* \rangle$. The microwave frequency was 23955.3 MHz. The stick spectra represent calculated line positions for radicals I and II. (b) Simulated spectra (see §9.4.8).

and recorded at K-band microwave frequencies (∼24 GHz). This specific orientation was deliberately chosen for illustration purposes for two reasons.

First, no site splitting complicated the spectra at this orientation. Second, at this orientation the magnetic field is *in* the plane of the cytosine base and most nitrogen couplings in a π-type radical are expected to be small (see below) and within the linewidth of the EPR resonance line, thus making the EPR spectrum as simple as possible. The available data at this and at other orientations indicated that at least three free radical species were present and that the g anisoptropy for all of these was quite small, in the range of $\langle 20035, 20020 \rangle$. Even though the spectrum in Fig 9.14a appears simple, the overall EPR resonance patterns were complex; for this reason, ENDOR was performed.

Figure 9.15 shows an ENDOR spectrum recorded with the magnetic field locked at the starred line in Fig. 9.14a. A number of ENDOR lines are denoted with Roman numerals in correspondence with the designation of the different radicals trapped in this crystal at 10K.

In the present cytosine monohydrate study, the crystals were rotated about two axes: the $\langle b \rangle$ axis (where no site splitting is expected) and a skewed axis with polar angles $\theta = 69°$ and $\varphi = 90°$ with respect to the $\langle a^*bc \rangle$ reference axis system. With this choice of skew rotation axis, these two rotation planes provided sufficient information to obtain complete hyperfine coupling tensors.

9.4 TOOLS AND PROCEDURES FOR RADICAL STRUCTURE DETERMINATIONS 349

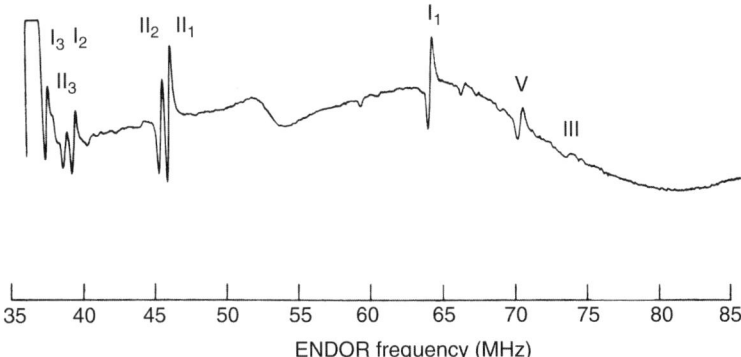

Fig. 9.15 An ENDOR spectrum obtained from the starred line in Fig. 9.14a. The different lines have been designated with roman numerals corresponding to the radical to which they have been assigned based on EPR, ENDOR, and EIE experiments [42]. The free proton frequency was 36.4 MHz.

Quite often an ENDOR spectrum will contain resonance lines due to several free radicals simultaneously present in the sample. There are various techniques to help assigning these to different radicals. The most common method is to record several ENDOR spectra at the same orientation but with the magnetic field locked at different positions in the EPR spectrum. If the EPR spectrum attributable to a given radical does not contribute at that field position, the corresponding ENDOR lines will not show up in the ENDOR spectrum. If the radicals present have significantly different microwave saturation behaviors, recording of the spectra at different microwave power levels may help the "grouping" of the ENDOR lines. Different temperature annealing behaviors may also help assign the ENDOR lines to different radicals. ENDOR induced EPR (EIE) is perhaps the most powerful procedure to assign specific ENDOR lines to specific radicals. In EIE, the radiofrequency (RF) field is locked to a partly saturated ENDOR transition (peak of the ENDOR line), and the magnetic field is swept. When the magnetic field coincides with an EPR transition, the population difference between the two energy levels constituting an ENDOR line will change. The resulting EIE spectrum will look like the EPR spectrum of the radical associated with the specific ENDOR line, and all ENDOR lines associated with the same radical will give similar EIE spectra. Because no magnetic field modulation is commonly used for the EIE (only RF modulation), the EIE spectra will look like EPR absorption patterns. It should be stressed that the relative intensities of the lines in an EIE spectrum are not representative for the relative line intensities in corresponding EPR spectrum. This is due to the number of different relaxation processes involved with the EIE mechanism being sensitive for both the electron and the nuclear spin quantum numbers and nuclear dynamics. Still, the pattern and total width of the EIE response most often provide a unique fingerprint of the responsible EPR pattern.

Figure 9.16 shows the EIE patterns obtained for cytosine monohydrate by locking the RF to the lines designated I and II in Fig. 9.15. The EIE spectrum attributable to

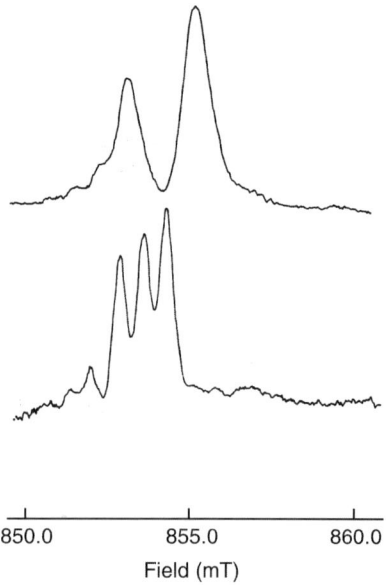

Fig. 9.16 (Top) EIE spectra from ENDOR lines designated I in Fig. 9.15. (Bottom) EIE spectra from ENDOR lines designated II in Fig. 9.15. The spectra are directly comparable with those in Fig. 9.14a and the corresponding stick spectra. Reprinted with permission [42]. Copyright 1996 American Chemical Society.

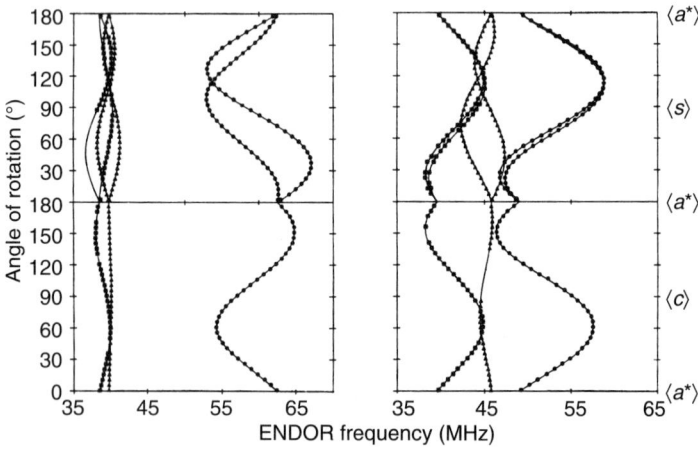

Fig. 9.17 The angular variation of the ENDOR lines due to (left) radical I and (right) radical II in two planes of rotation. For one rotation plane, a skewed axis was used, and the symbol $\langle s \rangle$ corresponds to a direction of the magnetic field defined by polar angles $\theta = 111°$ and $\varphi = 180°$. The fully drawn curves are calculated from the tensor data in Table 9.1.

9.4 TOOLS AND PROCEDURES FOR RADICAL STRUCTURE DETERMINATIONS

ENDOR lines I shows a doublet pattern with a splitting corresponding to the larger ENDOR frequency (64 MHz) in Fig. 9.15. The EIE due to ENDOR lines II shows a triplet pattern corresponding to the two nearly coinciding ENDOR lines at about 46 MHz in Fig. 9.15.

Focusing on the lines designated I and II in Fig. 9.15, complete plots showing the ENDOR line positions as a function of the rotation angle throughout the two experimental planes are displayed in Fig. 9.17. These data represent a sufficient set of data to allow for a full determination of the hyperfine coupling tensors, as described above. The program MagRes, written by W. H. Nelson, was used for the tensor determination [94, 95]. The actual hyperfine coupling tensors obtained are listed in Table 9.1.

TABLE 9.1 Magnetic Parameters for the RI and RII Resonances Observed in Single Crystals of Cytosine Monohydrate X-Ray Irradiated at 10K

Tensor	Isotropic value	Principal values	Eigenvectors		
			$\langle a \rangle$	$\langle b \rangle$	$\langle c^* \rangle$
I_1 (H-C6)	−37.8 (2)	−62.5 (4)	0.814 (5)	0.387 (5)	−0.434 (10)
		−35.5 (5)	0.458 (9)	0.035 (16)	0.888 (4)
		−15.3 (3)	0.359 (7)	−0.922 (2)	−0.149 (15)
I_2 (H-N3)	−5.5 (2)	−11.3 (3)	−0.719 (26)	−0.619 (16)	0.315 (69)
		−7.0 (4)	0.377 (48)	0.033 (63)	0.926 (21)
		1.7 (3)	−0.584 (15)	0.785 (11)	0.209 (37)
I_3 (H_b-N4)	−4.6 (2)	−9.2 (2)	−0.580 (41)	−0.751 (13)	0.315 (83)
		−6.7 (4)	0.492 (52)	−0.015 (75)	0.871 (28)
		2.3 (3)	−0.649 (13)	0.660 (15)	0.378 (27)
II_1 (H-C5)	−41.4 (2)	−62.4 (1)	0.040 (4)	0.999 (0)	−0.017 (8)
		−42.2 (4)	0.476 (12)	0.004 (9)	0.880 (7)
		−19.6 (6)	0.879 (7)	−0.043 (4)	−0.476 (12)
II_2 (H_b-N4)	−14.3 (2)	−23.6 (4)	−0.789 (17)	−0.508 (10)	0.345 (38)
		−16.1 (4)	0.429 (31)	−0.054 (31)	0.902 (14)
		−3.2 (3)	−0.440 (12)	0.859 (7)	0.261 (20)
II_3 (H_a-N4)	−13.0 (2)	−19.2 (1)	0.007 (30)	0.989 (9)	−0.148 (63)
		−16.6 (3)	0.445 (20)	0.129 (68)	0.886 (15)
		−3.3 (5)	0.896 (10)	−0.072 (10)	−0.439 (21)
Directions from the Crystal Structure					
Base perpendicular			0.4743	0.0238	0.8800
C6—H bond			−0.4349	0.8538	0.2860
C2—N3—C4 bisector			0.3839	−0.9021	−0.1972
C5—H bond			−0.4487	0.8661	0.2203
N4—H_b bond			0.8674	0.0067	−0.4975
N4—H_a bond			0.8808	−0.0047	−0.4734

The splitting parameters are in Megahertz. The numbers in parentheses represent the standard deviation in the last quoted digits.

9.4.6 Structural Identification

The tensor data in Table 9.1 represent the major part of the experimental data to be used for identification of the radical structures. What remains is to correlate these tensors, in particular their eigenvectors, with the known structure of the undamaged molecule in the single crystal. Theoretical and semiempirical rules of thumb for doing this have been formulated by many authors during the last 50 years. To enable a basic understanding of these, attention should be paid to Fig. 9.18.

Most of the radical structures discussed in this chapter are planar structures. In such structures, the unpaired electron is often delocalized over several nuclei of the planar molecular backbone in an extended π-molecular orbital. A planar structure is often an energetically more preferred conformation. Thus, even if the undamaged molecule is nonplanar, a planar radical conformation is frequently attained.

It is usual to make a broad classification of the most common π-type radical fragments into three groups. These are shown in Fig. 9.18 and are called an α-proton coupling, a β-proton coupling, and a central atom coupling. As depicted in Fig. 9.18a, an α-proton coupling is a coupling between an unpaired electron in a 2pπ-type atomic orbital and a proton (the α-proton) bonded to the same atom in a planar configuration. This atom is often, but not necessarily, a carbon atom. McConnell and colleagues showed that the isotropic coupling for a carbon-bonded α proton is given by the relation

$$a_H = Q_H^{CH} \rho^\pi \tag{9.4}$$

where ρ^π is the π-spin density, which is the fraction of the electron density residing in the 2pπ orbital [96]. For a carbon atom, the proportionality constant Q_H^{CH} is typically about -72 MHz or 2.6 mT for a carbon atom, but it may vary 10–15% at each side of this number, depending on the radical charge and geometrical strain [97]. If the α proton is bonded to a nitrogen atom, the value of the proportionality constant Q_H^{NH} is typically -80 MHz [98].

The McConnell constant Q is negative, implying that the α-proton hyperfine coupling is negative. The absolute sign of the isotropic value can usually not be decided

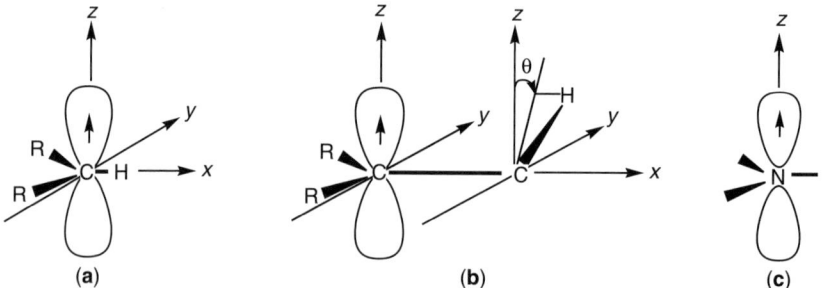

Fig. 9.18 The three major types of π-radical hyperfine interactions: (a) an α-proton coupling, (b) a β-proton coupling, and (c) the central atom coupling. (a, b) The C represents a carbon atom, but other atoms may also apply; (c) N represents a nitrogen atom, but any atom with $I \neq 0$ applies. The single, short arrows represent the unpaired electron.

9.4 TOOLS AND PROCEDURES FOR RADICAL STRUCTURE DETERMINATIONS 353

from EPR or ENDOR measurements alone, but it is assigned based on the theoretical knowledge of the nature of typical α-proton coupling tensors.

Because of the proximity of the main electron density and the proton, the electron–nuclear hyperfine interaction exhibits a strongly anisotropic component [99]. Due to the symmetry of the system, the x, y, z axes will be the principal axis system for this interaction. In the x-axis direction the dipolar coupling is positive and about 50% of the isotropic value. In the y-axis direction the dipolar coupling is negative, and it is also about 50% of the isotropic value. In the z-axis direction the dipolar coupling is very small. Thus, all three principal values are negative (being the sum of the isotropic and anisotropic component). The smallest negative value is along the C—H bond direction (x axis), the intermediate negative value is along the orbital of the unpaired spin (z axis), and the largest negative value is perpendicular to the other two (y axis). In the case of cytosine monohydrate, tensors $\mathbf{I}_1$ and $\mathbf{II}_1$ in Table 9.1 are typical carbon-centered α-proton couplings.

The relative magnitudes of the dipolar couplings are also dependent on the central nucleus. Therefore, the principal values of an N-centered α-proton interaction are somewhat different from that of a C-centered α-proton interaction. In the case of cytosine monohydrate, tensors $\mathbf{II}_2$ and $\mathbf{II}_3$ in Table 9.1 are typical N-centered α-proton couplings [98].

The spin density at the α-carbon atom can also be determined from the dipolar contribution to the hyperfine coupling tensor [100, 101], using a relation similar to the McConnell relation,

$$a_H^x = Q_H^{dip} \rho^\pi \quad (9.5)$$

where a_H^x is the largest positive dipolar coupling. When the Q_H^{dip} value is 38.7 MHz for carbon-bonded protons, a quite reliable value for the spin density is obtained. If the spin densities determined from the isotropic value (Equation 9.4) and the dipolar coupling (Equation 9.5) deviate significantly, the radical fragment is probably bent and not planar [102].

Figure 9.18b shows a β-proton interaction, an interaction between the unpaired electron and a proton bonded to an atom next to that carrying the unpaired electron. The isotropic value of the a_β coupling is commonly given by the form

$$a_\beta = (B_0 + B_2 \cos^2 \theta) r^\pi \quad (9.6)$$

where B_0 and B_2 are constants and θ (Fig. 9.18b) is the dihedral angle of the C—H bond with respect to the z axis [103]. Constants B_0 and B_2 will depend on the nature of the central atom and that of the intervening atom [25] and if both are C atoms, commonly used values for B_0 and B_2 are about 0 and 126 MHz (4.5 mT), respectively. The isotropic coupling constant a_β is usually positive [97].

The distance between the unpaired electron and the coupled proton is much larger in the case of β protons compared to α protons. Thus, the dipolar interaction is much smaller for β protons. In the limit of the so-called point–dipole approximation, the maximum dipolar coupling is positive and pointing from the central atom (with the unpaired electron) toward the β proton. Perpendicular to this direction, the dipolar interaction is negative. It exhibits the almost the same value in all of the perpendicular directions, and the magnitude is one-half of the maximum value. The β-proton coupling is thus nearly axially symmetric with its (positive) maximum value close to the C$\cdots$H$_\beta$ direction [104].

For a central atom coupling (Fig. 9.18c), the isotropic contribution depends mainly on the polarization of the unpaired electron density in the in-plane bonds between the atom and its neighbors. This dependence is very sensitive to small variations in the bonding geometries and configuration, and the isotropic hyperfine coupling is therefore quite unreliable for estimating the spin density at the central atom [91]. The spin density may, however, be estimated from the dipolar coupling tensor. This tensor should be close to axially symmetric with a maximum value $2a_d$ along the z axis (Fig. 9.18c) and the value $-a_d$ perpendicular to z. For π-type nitrogen central couplings, a_d and a_{iso} are often comparable in magnitude, resulting in small perpendicular principal values (because a_d and a_{iso} exhibit opposite signs). The value of a_d is fairly well defined for the different nuclei and may be expressed as

$$a_d = a_{d,0}^{nuc} \rho^\pi \tag{9.7}$$

where $a_{d,0}^{nuc}$ depends on the nature of the central atom. The $a_{d,0}^{nuc}$ values are tabulated in the literature. The most precise values to date were given by Fitzpatrick and coworkers [105]. These values appear to give a better fit to a large number of experimental data. They are close to an older set of calculated parameters [106] and are somewhat smaller than those given by Morton and Preston [107] and reproduced in the textbook by Atherton (p 197) [91].

In the case of cytosine monohydrate, all hyperfine coupling tensors presented in Table 9.1 exhibit principal values that are characteristic for α-proton interactions. All couplings have the intermediate principal value occurring close to the cytosine ring perpendicular, as expected for planar π-type radicals (the z direction in Fig. 9.18a). The two largest couplings ($\mathbf{I}_1$ and $\mathbf{II}_1$) have their smallest negative principal values close to the crystallographic C6-H and the C5-H directions, respectively (comparable with the x direction in Fig 9.18a). For radical I, the spin density at C6 can be estimated to 0.52 from both the isotropic and dipolar components of the coupling tensor. The two other couplings associated with radical I were assigned to a proton bonded to N3 and one of the two protons of the amino group. These tensors ($\mathbf{I}_2$ and $\mathbf{I}_3$) have dipolar values more typical of N–H α couplings, and their smallest negative principal values were close to the crystallographic N3-H and the N4-H$_b$ directions, respectively. Because cytosine monohydrate originally does not have a proton at N3, the observation of the N3-H α coupling ($\mathbf{I}_2$) shows that radical I has become protonated at N3. For radical II, the spin density at C5 was estimated to 0.57 (using elements of tensor $\mathbf{II}_1$), and the additional smaller couplings were assigned to the two amino protons, with a spin density of about 0.17 at the amino nitrogen. Again, the symmetry of the principal values (tensors $\mathbf{II}_2$ and $\mathbf{II}_3$) was typical of nitrogen-bonded α protons.

9.4 TOOLS AND PROCEDURES FOR RADICAL STRUCTURE DETERMINATIONS

With due precautions and care, the single crystal EMR analyses yields information about which nuclei interacts with the unpaired electron, and thereby a map of the unpaired electron density over the molecular skeleton. This information is in most cases sufficient for suggesting a model for the actual radical structure. This model may be further tested using a series of experimental variations like isotopic substitution, variable frequency measurements, variable temperature studies, and so on. Theoretical tools may also be used to test and support a given structural assignment. Two common tools will be described in the following sections on quantum chemistry calculations and spectrum simulations.

9.4.7 Quantum Chemistry Calculations

Ever since the early days of quantum mechanics, scientists have tried to calculate the electronic properties of molecules. From the 45-year-old calculations of Pullman and Pullman [108] to the current advanced quantum chemistry calculations on supercomputers, the goal has remained the same: to be able to precisely predict specific properties of molecules with accuracy comparable to that of experimental investigations (in particular, molecules of biological significance). It is beyond the scope of this book to review all of the work done on DNA and DNA components in this respect; instead some references to a few recent review articles are provided [109–111]. Even many years back, quite precise calculations of the electron spin density distributions in the DNA constituent free radicals could be made using semiempirical Hartree–Fock type calculations like CNDO and INDO [112, 113]. These results have not been significantly altered by the application of more advanced methods, for instance, by density functional theory (DFT), even if the level of detailed knowledge has increased tremendously. At present, Gaussian03 is one of the most commonly used computer programs for quantum chemistry calculations implementing a large number of different approaches and methods [114]. Chapter 7 gives an introduction to DFT methods and examples of how these may be integrated with EPR spectroscopy.

A number of calculations of the spin density distributions in cytosine anions have been published. Typical values for the N3-protonated anion of cytosine are about 0.50 π-spin density at C6; about 0.40 at C4; and much smaller π-spin densities (about 0.1) at the amino nitrogen, N3, and N1 (for an illustration, see Fig. 9.19) [115]. These calculations fit excellently with the experimental values obtained for radical I (Table 9.1) formed in cytosine monohydrate, X-ray irradiated at low temperature. Based on the experimental hyperfine couplings (Table 9.1), only the radical structure of an N3-protonated cytosine anion (Cy1 in Fig. 9.8) could be envisaged. Because the theoretical calculations on this radical structure are in good agreement with the experimental values, radical I can be assigned with a high degree of credibility to the N3-protonated cytosine anion. Similarly, a number of previous experimental results on cytosine nucleotides [116] combined with quantum chemistry calculations provide very strong evidence that radical II is the N1-deprotonated cation of cytosine (Cy2, Fig. 9.8). The expected (theoretical) π-spin density is about 0.60 at C5 and about

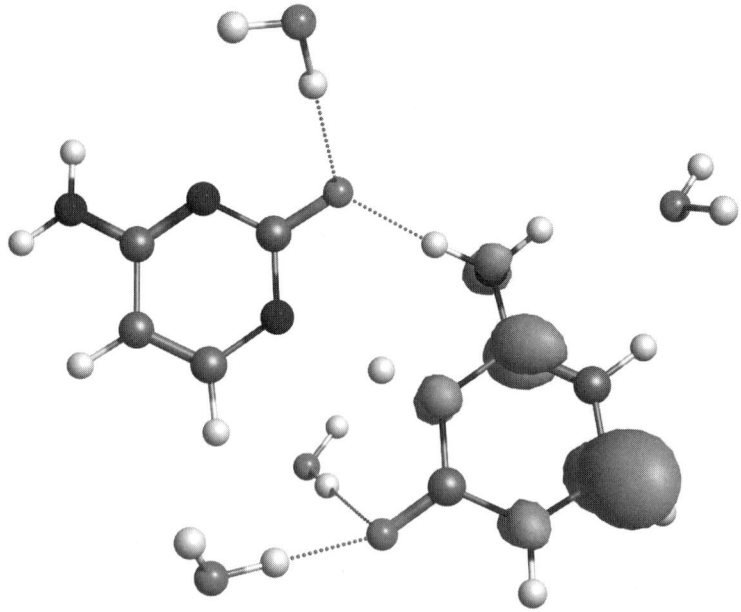

Fig. 9.19 The DFT-calculated unpaired spin density distribution of the cytosine monohydrate anion radical, protonated at N3 (Cy1). The proton at N3 was partly transferred through the (original) H bond from N1 of the neighboring cytosine molecule. The calculation was performed on a small cluster of molecules, consisting of two H-bond connected cytosine molecules and four water molecules. The figure clearly illustrates the larger π-spin densities at C6 and C4 and the smaller π-spin densities at the nitrogen atoms.

40% at N1 and the amino nitrogen together, which fit closely with the structural evidence extracted from the experimental data for radical II (Table 9.1).

Modern quantum chemistry calculations, for example, of the DFT type, are very important tools for predicting electron density distributions, hyperfine coupling tensors, and **g** tensors of free radicals. Whenever possible, such calculations should be used together with the structural information obtained from EMR experiments. There are many examples where DFT calculations of isolated radicals (i.e., in the gas phase) have reproduced the experimental tensors excellently. However, due caution should be taken with respect to the validity of any given model calculation. Within a rigid matrix like a single crystal it is very difficult to predict how the matrix will affect and hinder reorientations of the radical fragments produced [117]. Thus, in certain cases the theoretically calculated EPR parameters may be far from correct, and computationally more extended methods must be used [36, 118].

9.4.8 Computer Spectrum Simulations

Spectral parameters obtained from EPR and ENDOR analyses of a free radical system (most importantly the hyperfine coupling tensors and **g** tensors) may be used to

9.4 TOOLS AND PROCEDURES FOR RADICAL STRUCTURE DETERMINATIONS

reconstruct the EPR spectra by theoretical simulations. This is a very useful procedure both to validate the interpretation of the experimental data, and also to extract data not readily available from the experiments alone. A variety of simulation programs are available. *Win-EPR* is frequently used in the literature; this program is available from Bruker BioSpin GmbH. For the more recent version of Bruker's data acquisition system (*XEpr/XEprView*), an advanced simulation program called *XSophe-Sophe* is available. This program was originally written by Graeme Hanson and coworkers [119]. Many years ago, Lefebvre and Maruani published a simulation program based on second-order perturbation theory that also included a very effective polycrystalline (powder) system simulation algorithm [120]. Their program remains the core of a suite of simulation programs written by Erickson and Lund [121, 122]. One of these programs is called *KVASAT*, which can be used for $S = 1/2$ systems including nuclear quadrupolar couplings, microwave saturation effects, and modulation broadening effects [123]. *EasySpin* is a MatLab routine based on an exact solution of the spin Hamiltonian by energy matrix diagonalization [124]. *WinSim* is a first-order isotropic simulation program with a Windows interface that is simple in practical use [125].

In the case of cytosine monohydrate crystals X-ray irradiated at 10K, Lefebvre and Maruani's spectral simulation program was used in almost its original version. The separate radical spectra (radicals I and II) were simulated using the following experimental data: measured g values along the reference axes, the hyperfine coupling tensors for the given radical from Table 9.1, calculated nitrogen hyperfine coupling tensors based on the spin density estimates (Table 9.1 and §9.4.6), and finally the linewidths (Gaussian line shapes) as measured directly from the EPR spectra. The simulated spectra for radicals I and II at the specific orientation shown in Fig. 9.14a were added in a relative amount of 0.55/0.45, respectively, and the summed simulation is provided in Fig. 9.14b. The agreement with the experimental spectrum is excellent, and it provides an independent check for the interpretation that the ENDOR lines are mainly due to two different radicals (there are, however, additional minority radical species present).

Another illustrative example on the use of simulation programs to estimate the relative amounts of different radicals contributing to a given EPR spectrum is provided in the studies on γ-ray irradiated polycrystalline L-α-alanine at room temperature [126–129]. It was shown that there are predominantly three different radicals present at room temperature. These are designated R1, R2, and R3 in Fig. 9.20.

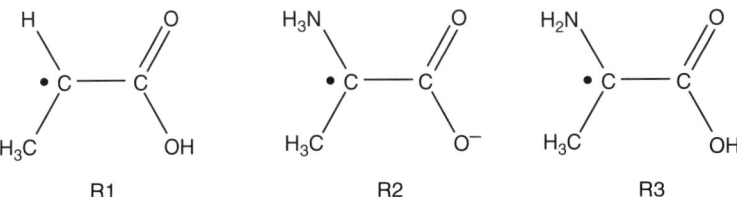

Fig. 9.20 The three major radicals in single crystals of L-α-alanine after irradiation at room temperature.

The R1 radical is the well-known stable alanine radical studied by several groups in the early 1960s [130–132]. Radical R2 is a secondary radical formed by net H abstraction from the C_α position of the amino acid and was recently thoroughly characterized using EPR, ENDOR, and EIE techniques [126]. Radical R3 was tentatively assigned to a radical similar to R2 but in a different tautomeric structure (deprotonated at the amino group and protonated at the oxygen). Based on the experimental EPR parameters obtained from the single crystal study [126], simulated polycrystalline spectra (so-called benchmark spectra) were prepared for each of the three components using KVASAT. The experimental polycrystalline spectrum is presented in Fig. 9.21 together with the three simulated benchmark spectra.

In the recent work [127] a small script was written within *MS Excel* for performing a linear fit of the experimental spectrum to the sum of the three benchmark spectra using the weighing factors of the three spectra as fitting parameters. With this simulation and fitting procedure the relative radical amounts R1/R2/R3 were estimated to be roughly 6:3:1. Furthermore, when combining the spectral fittings with EPR

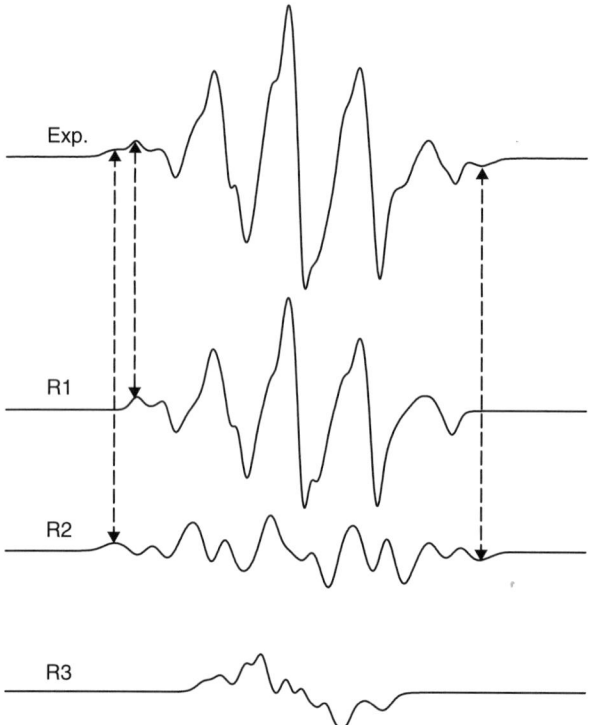

Fig. 9.21 The top trace shows the X-band EPR spectrum of polycrystalline L-α-alanine X-ray irradiated at 296K to a dose of 50 kGy. The three lower traces show the simulated polycrystalline EPR spectra for the three radicals R1, R2, and R3. The parameters used for these simulations are described in Reference 127. Vertical lines indicate characteristic features of the experimental and simulated spectra. All EPR magnetic field sweep widths are 20 mT. Reproduced with permission [128]. Copyright 2003 Radiation Research Society.

9.4 TOOLS AND PROCEDURES FOR RADICAL STRUCTURE DETERMINATIONS   359

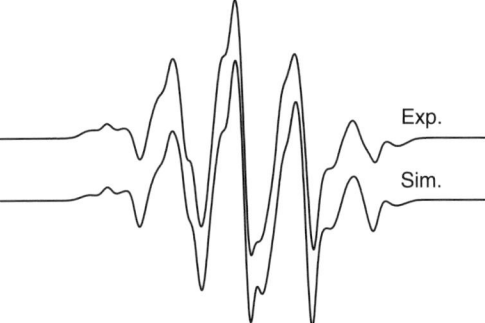

Fig. 9.22 Experimental and simulated X-band EPR spectra of polycrystalline L-α-alanine obtained after X-ray irradiation (60-kGy dose) at room temperature. The final weighing factors were 0.589, 0.335, and 0.076 for R1, R2, and R3, respectively. The goodness of fit (r^2) was 0.980.

experiments recorded under different conditions, it was possible to characterize the temperature behaviors and the dose–response curves for the three radical components separately. An experimental EPR spectrum of polycrystalline alanine at room temperature is shown in Fig. 9.22 together with the corresponding simulated spectrum.

9.4.9 Polycrystalline Spectra

The distinguishing feature of single crystal studies is the possibility to obtain the eigenvectors of the **g** tensors and coupling tensors. With the magnetic field directed along these (so-called canonical) orientations, the magnitude of the corresponding principal values can be measured directly.

In polycrystalline samples, this combined principal value and eigenvector information is lost because of the random orientation of all of the radicals in the sample. However, the principal values and the nature of the coupling nucleus may sometimes be obtained from the spectra. The theory and interpretations of polycrystalline EPR spectra have been treated in Chapters 1 and 6 and will not be discussed further here. Note, however, that multicomponent powder spectra are difficult to analyze unless the components have significantly different g anisotropies. At higher microwave frequencies small differences in the **g** tensors are magnified, which may enhance component separation of multicomponent spectra [37, 133].

Powder ENDOR, that is, ENDOR of polycrystalline samples, may sometimes be helpful when trying to obtain the hyperfine coupling values for powder samples with composite and overlapping radical spectra [134], even if the g value variations are small. During the study of the nature of the radicals induced in the amino acid L-α-alanine mentioned above, specific attention was paid to the radical designated R2 (see Figs. 9.20 and 9.21). Single crystal experiments indicated strongly that this radical is the only contributor to the outer flanking resonance lines of the polycrystalline alanine EPR spectrum (Fig. 9.21). To investigate this independently, a powder ENDOR experiment was performed. The powder ENDOR spectrum is shown in Fig. 9.23b [127]. The magnetic field was locked to the very high field

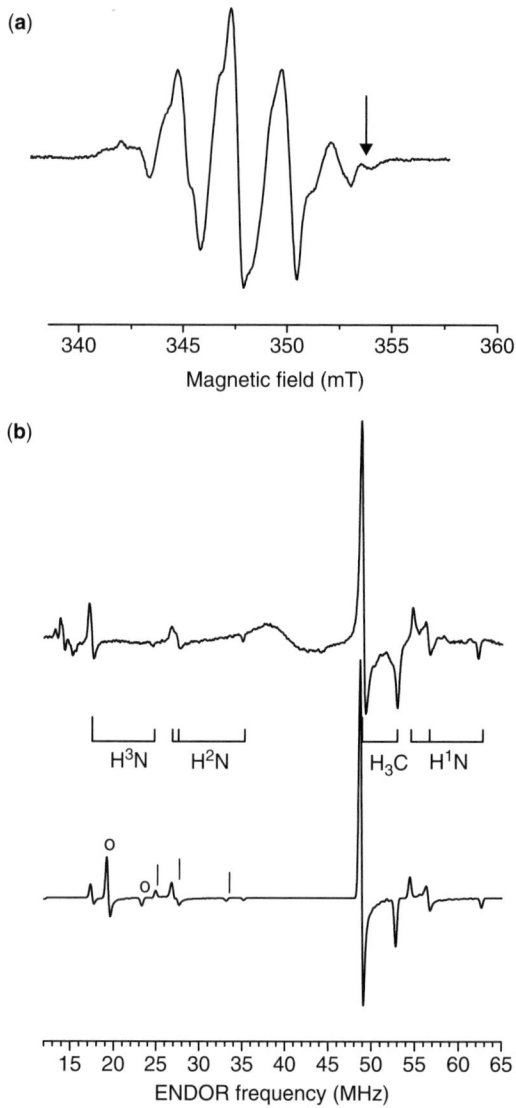

Fig. 9.23 (a) The first derivative X-band EPR spectrum from a polycrystalline sample of L-α-alanine (63 kGy dose) irradiated at 295K and measured at 221K. The downward arrow indicates the magnetic field position for the ENDOR spectrum. (b) The top spectrum is the experimental powder ENDOR spectrum recorded off the marked position in (a). Stick spectra indicate expected line positions from the single crystal study [126]. The bottom spectrum is the simulated powder ENDOR spectrum for alanine radical R2, using results from the single crystal study [126]. The lines denoted with symbols (o, |) were not detected experimentally because of relaxation effects. They represent low frequency branches of the methyl coupling and one of the amino proton couplings. Reprinted with permission [127]. Copyright 2002 American Chemical Society.

EPR resonance line of the powder alanine spectrum (downward arrow, Fig 9.23a). If R2 was the only radical contributing at this field position, only ENDOR lines associated with R2 should be observed and these ENDOR lines should occur at positions corresponding to the principal values of the associated hyperfine coupling tensors (known from single crystal studies). The strong resonance features near 48–53 MHz occur at the values expected for the nearly axially symmetric methyl group couplings. The weaker features also correspond perfectly with the hyperfine coupling tensor data extracted from the single crystal analysis [126]. This study demonstrated two important conclusions. First, R2 is the only significant radical contributing at the outer flanks of the polycrystalline EPR spectrum. Second, in this particular system the principal values for at least one of the radicals (R2) could have been obtained without performing a complete single crystal study.

A powder ENDOR spectrum can be computer simulated to various degrees of accuracy, depending on the knowledge of the **g** tensor, various coupling tensors, and the relaxation behavior of the system. The simulated spectrum of radical R2 in Fig. 9.23b was obtained using the published single crystal data for R2 and a simulation program written by Erickson [135]. This program includes the nuclear spin, hyperfine, and quadrupolar couplings in a second-order perturbation formalism together with appropriate expressions for the ENDOR line intensities, but relaxation effects have not been taken into account.

9.5 QUANTITATIVE DETECTION OF RADICALS

9.5.1 A Note on EPR Metrology

EPR spectroscopy is a very useful technique for studying the structure of free radicals and for following the radical function through various physically or chemically modified pathways, particularly when using single crystals. In addition to qualitative measurements, the EPR technique is also applicable for quantitative measurements. The absorption of microwave energy in a sample is directly proportional to the number of spins present. Thus, the total number of spins in a sample (or the specific radical density in a sample) may be measured by measuring the EPR energy absorption spectrum. This can be illustrated by looking at the Boltzmann distribution of spins in a two-level system. Using first-order expansions of exponential functions, the rate of microwave energy absorption by a sample may be expressed as

$$\frac{dE}{dt} = N\frac{(\Delta E)^2}{kT}\frac{P}{1+2PT_1} \quad (9.8)$$

where N is the total number of spins; P is the transition probability; T_1 is the spin–lattice relaxation time; k is the Boltzmann constant; T is the temperature; and ΔE is the spin level energy separation, $\Delta E = g_N|\mu_B|B$ [91]. In addition to being a function of the number of spins, the EPR absorption signal is also a function of several physical and instrumental factors [136]. Therefore, the relation between the microwave energy absorption and the number of spins also contains a proportionality factor that must be determined for each specific instrumental setup. This calibration factor is often obtained using a sample with a known number of spins, but there are large inherent uncertainties involved in these measurements.

Because measurements of the accurate number of spins are often hampered with large uncertainties, quantitative EPR measurements of spins are normally performed by measuring *relative* amounts of spins. When care is taken with the measurement procedures (discussed in more detail below), relative amounts of spins (i.e., *relative amount of radicals*) can be measured very precisely (uncertainties on the order of $\pm 1\%$ should be achievable in many cases). Quantitative EPR measurements are usually performed on polycrystalline samples and are the bases for radiation dosimetry (measurements of absorbed doses of ionizing radiation, e.g., EPR alanine dosimetry), retrospective dosimetry (measurements of radiation doses at a prolonged period after the irradiation has taken place, e.g., using tooth enamel or bricks), accident dosimetry (measurements of radiation doses shortly after an accident, e.g., using teeth *in vivo*), or geological and paleontological dating. EPR spectroscopy is also used for identifying irradiated foodstuffs. The qualitative characteristics of the signal yield information about what has been irradiated, whereas the quantitative characteristics may give information about the irradiation dose [137]. Detailed principles of EPR dosimetry will be discussed in the present chapter. The other applications will only be briefly discussed.

9.5.2 Introduction to EPR Dosimetry

Radiation dosimetry is the measurement of the absorbed dose in a given material resulting from an exposure to ionizing radiation. The *dosimeter reading* is the response produced by the dosimeter or by the radiosensitive material, and it should be proportional to the absorbed dose in that material. Dosimetry is necessary for most practical applications of ionizing radiation, but the requirements for accuracy and precision vary considerably. A number of different physical and chemical dosimetry methods have been developed. Ion chambers, thermoluminescense dosimetry, diode dosimetry, and various kinds of chemical dosimetry (e.g., Fricke dosimetry) are used for high accuracies [1].

EPR dosimetry is a method for measuring radiation doses in the gray to kilogray dose range with high precision. As discussed in Sections 9.2 and 9.3, the lifetimes of radiation induced radicals are often very short. However, some radicals may be very stable, in particular when trapped in dry materials. Such radicals may in some cases be used as probes for radiation doses. Because the area under the EPR absorption spectrum is proportional to the number of free radicals (spins) in the sample and the number of free radicals is proportional to the radiation energy absorbed in the sample, the area under the EPR absorption spectrum will be proportional to the absorbed dose. Thus, the radiation dose may be determined by measuring the relative concentration of radiation-generated free radicals in a given material (i.e., in the *dosimeter*) using EPR spectroscopy. In practice, the height of an EPR line is used instead of the area as a measure for the radical concentration. The readout process is *nondestructive*, meaning that the dosimeter is not influenced by the measurement procedure, and the dosimeter may be measured and remeasured as many times as wanted. The accurate dose is obtained by comparing the readout with relevant *reference dosimeters* (identical dosimeters irradiated to known doses).

Some radiation physical and chemical requirements *must* be met for a material to be suitable as a sensitive EPR dosimeter material, and other requirements are *preferable* but not absolutely necessary. Naturally, the number of detectable radiation induced radicals must reflect the amount of energy absorbed during the radiation exposure, and the relationship between the absorbed dose and the EPR response must be known (the mathematical expression). It is preferable that this dose–response relationship is linear, but it is not a necessity.

Other important properties have been summarized in Box 9.3. It is preferable that the dose–response relationship depends very little on factors like the irradiation temperature, radiation quality, dose rate, dose level, radiation energy, and radiation LET value. If there is any significant dependence on any of these factors, the relevant relationships must be established if high precisions are sought in the dose measurements. Furthermore, the radiation induced radicals must be stable for a sufficiently long time period to allow for precise dose measurements. It is important that the radical decay rate is known, and it is preferable that the decay rate depends very little on external factors like the storage temperature, light, and humidity conditions. If the radical stability is high, a dosimeter can be stored for a long time and serve as retrospective documentation for the dose deposited. It is preferable that the radical yield is high and the number of EPR lines is low because the signal intensity and the signal to noise ratio will increase as the radical yield increases and as the number of EPR lines decreases. The optimum EPR spectrum will consist of one narrow line, and there should be no disturbing zero-dose signal in the dosimeter material. The microwave power saturation properties of the radiation induced radicals are also important because high microwave power values will increase the intensity of the EPR resonance (i.e., increase the sensitivity of the dosimeter material). For medical applications, it is also preferable that the dosimeter material responds to the ionizing

---

BOX 9.3 PROFITABLE PROPERTIES OF EPR DOSIMETERS

- An EPR spectrum with few and narrow lines, and no zero-dose signal
- Stable radiation induced radicals under realistic environmental conditions
- A radiation quality (type, energy) dependence similar to that of water
- Low dose and dose rate and irradiation temperature dependence
- High radiation–chemical yield of radicals
- Suitable microwave power saturation properties allowing for high power values for increased sensitivity
- Robustness of dosimeter with regard to mechanical stress

radiation in a manner very similar to tissue and water. This means that the material should be close to water equivalent because water is the accepted reference material for clinical dosimetry. The robustness of dosimeters with regard to mechanical stress is also a factor that should be considered; the requirements will naturally depend on the intended use, but generally a robust dosimeter is preferable. Strategies for finding suitable dosimeter materials have also been discussed by Ikeya and coworkers [138].

When using EPR spectroscopy for quantitative measurements, it is important to be aware of and to control the various factors that may influence the EPR readout of a given sample. The double integral of the (first derivative) EPR spectrum is proportional to the number of free radicals and thus also to the absorbed dose in the sample. In practice, however, it is the peak–peak (PP) amplitude of the major EPR line that is used as the dose parameter. The dose precision at low doses is often better using PP amplitudes, it is easier to measure PP amplitudes, and integration problems associated with spectrum baseline and zero-dose (background) signals are reduced or avoided. The dose value of a sample given an unknown dose is found by comparing its PP value with a calibration curve made from identical samples irradiated with known doses. Figure 9.24b shows a typical calibration curve. Note that the calibration curves and the measured PP value must be given as *amplitude per mass*. This follows because the absorbed dose D is defined as the energy absorbed (ΔE) per mass m of an absorbant ($D = \Delta E/m$). Thus, for a given dose the amount of radicals (PP value) will increase with increasing sample mass.

The EPR amplitude (PP or integral) depends on the dose and sample mass; but it may also depend upon experimental conditions like the humidity in the laboratory (and *in* the EPR cavity), the sample position in the EPR cavity, the spectrometer parameters (field modulation, spectrometer gain, microwave power), as well as the overall stability of the EPR spectrometer. Even though the modulation and gain settings of the spectrometer may appear to be linear, this is often not exactly correct. It is thus important that *all* samples are measured under the same experimental conditions. Care must be taken that the samples are in identical positions (using specifically designed sample holders) and that the same spectrometer settings are used for all dosimeters. Because no EPR spectrometers are exactly calibrated, the correct parameters for modulation, power, and gain (in addition to microwave phase) need to be established for each separate instrument. If very high accuracy is sought, the dosimeters should all be measured on the same day. It is recommended to turn on the EPR spectrometer and let it warm up for 2 h prior to measurements.

Additional caution must be applied to the selection of the optimum microwave power. Increasing the power will within certain limits increase the EPR amplitude. However, when more than one radical type is stabilized in the dosimeter material, these radical types may exhibit different power saturation behaviors and therefore introduce power dependent changes in the EPR spectrum and in the PP amplitude. Power distortion of the EPR spectrum should be avoided, and *all* dosimeters to be compared should be recorded at identical power conditions. It is important to note that the actual power at the sample position depends both on the spectrometer settings and on the cavity used. For example, when using the Super-X cavity from Bruker

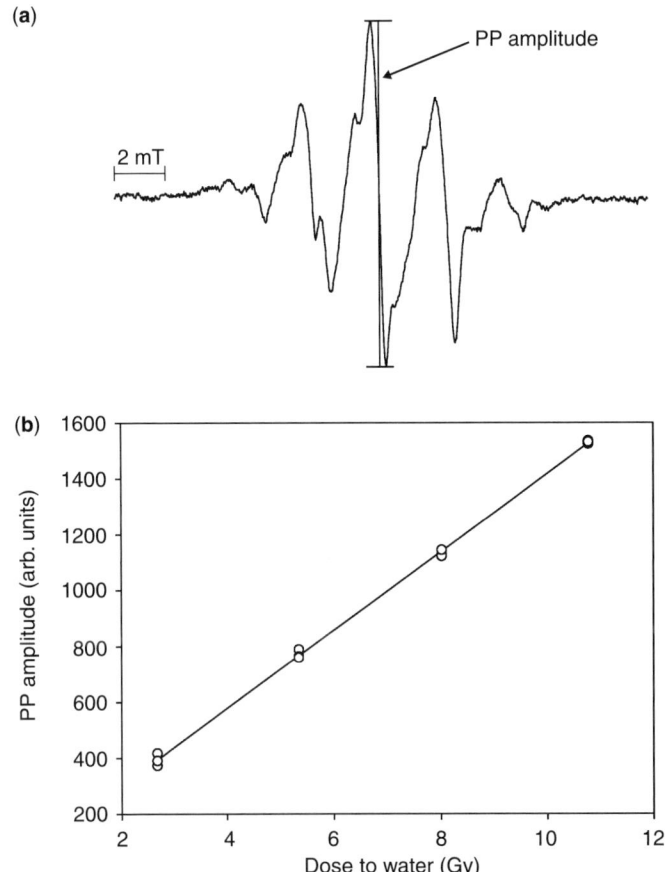

Fig. 9.24 (a) The EPR spectrum of an alanine dosimeter (80% L-α-alanine, 20% polyethylene, Bruker GmbH) irradiated to a dose of about 10 Gy using ^{60}Co γ rays. The PP amplitude of the central line, which is used to monitor absorbed dose levels, is indicated. (b) The PP amplitude is plotted against the absorbed dose for 12 alanine dosimeters (3 at each dose point) exposed to ^{60}Co γ rays.

Biospin, the actual power at the sample is increased by a factor of roughly 2.5 compared to that for a standard rectangular cavity for the same input (nominal) power.

9.5.3 EPR/Alanine Dosimetry

More than 40 years ago it was suggested that the amino acid L-α-alanine in polycrystalline form exhibits properties that are very suitable for EPR dosimetry [139]. This potential was investigated further by Regulla and Deffner [140] and other groups [141]. Today, EPR/alanine dosimetry is an accepted dosimetry system for high dose applications approved by IAEA as a secondary high dose standard [142] and with a specific internationally approved procedure and instrument settings for the measurements [143].

The popularity of alanine as a radiation dosimeter is attributable to a combination of very favorable properties with respect to most of the physical factors commented on above. It has an atomic composition that makes it near tissue (and water) equivalent with respect to energy absorption. The radical yield is high, and the radicals produced are quite stable (the fading is <1% per year) [140, 144, 145]. The dose–response relationship is linear in the region between $\sim$2 Gy and 5–10 kGy; and it has a very low dependency on dose rates, radiation type, and radiation energy above 150 keV. Correction factors for influences like the irradiation temperature, radical fading, beam quality, and light exposure after irradiation have been published. Alanine is very accurate when measuring high radiation doses (kGy), but it is not well suited for determining doses below a few greys unless very cumbersome precautions and procedures are followed [146]. The increased uncertainty at low doses is partly due to a variable background signal, but the presence of several radical types may also be of relevance [143, 147–149].

Figure 9.24a shows a first derivative EPR spectrum of X-irradiated polycrystalline alanine. As described in Section 9.4, this EPR spectrum is a complex, multiline spectrum that is due to at least three different radical types. The PP amplitude (per mass) of the major EPR line is used as the dose parameter. Figure 9.24b contains a typical alanine calibration curve.

A common procedure for EPR/alanine dosimetry is first to determine what kind of dosimeters to use, a choice that is mainly dictated by the actual application. The most typical dosimeters are shaped as *pellets* with a diameter of roughly 5 mm and a height of 3–5 mm, but other pellet dimensions are also in use. Alanine *film* dosimeters with a thickness of 100–200 μm are also available. Commercially available pellets (Gamma Service, Kodak/Bruker, Harwell Dosimeters) and films (Kodak/Bruker, Gamma Service) contain a binder material in addition to the alanine. Several laboratories use homemade dosimeters of various shapes and with various kinds of binder material, alternatively with no binder at all. Paraffin is a common binder material, and a typical amount is about 20%.

Suitable holders for the dosimeters must be used to preserve identical positions in the cavity. Quite advanced holder systems are being used in some reference laboratories [150, 151], but fairly straightforward combinations of Teflon pedestals and quartz tubing may also serve the purpose. It has become customary to include a reference sample permanently positioned in the cavity [151, 152]. Point samples of Mn^{2+}/MgO, Cr^{3+}/MgO, and ruby are the most commonly used standard samples. The reference sample is either mounted in the second cavity of a double cavity and recorded separately from the dosimeter or mounted within the standard cavity (e.g., in the Teflon support) and measured simultaneously with the dosimeter.

The unknown dose is determined by comparing the PP value of the (unknown) dosimeter with the PP values of dosimeters that have received known doses (reference dosimeters). The reference dosimeters naturally must have compositions identical to the (unknown) dosimeter, the reference doses should preferably be in the same dose range as the dose to be determined (the unknown dose), and mass corrections should be done. Normally, ^{60}Co γ radiation is used as the reference radiation. The irradiation conditions for the reference dosimeters should be as similar as possible to those of the unknown dosimeters; if not, relevant correction factors, for example, for different irradiation temperatures and different radiation quality, must be used when determining the dose from the calibration curve. The reference set of dosimeters is utilized to construct a calibration curve (see Fig. 9.24b), and the dose corresponding to the PP value of the unknown dose may be obtained from the calibration curve.

It is highly recommended to use more than one dosimeter at each dose point and to use at least four different dose points. Generally, the precision of the calibration curve will increase when the number of reference dosimeters increase. For a dose of 5 Gy, a precision of an unknown dose should quite easily be made better than $\pm 2\%$; for very careful measurements, a precision down to $\pm 0.5\%$ should be achievable [150, 151, 153, 154].

9.5.3.1 Self-Calibrated Alanine Dosimeters. Recently, a variant of the standard alanine dosimeter was introduced, which is a so-called *self-calibrated* dosimeter containing the radiation sensitive material (alanine), a binder, and an EPR active substance that is not influenced by radiation. The latter substance is used as an internal standard, which is recorded simultaneously with the dosimetric signal. Yordanov and Gancheva used Mn^{2+} ions in MgO in parts per million quantities for this internal standard [155]. Extensive testing and international intercomparisons have shown this to be a promising method [155, 156]. The dosimetric measure in this case is not the PP value of the radiation sensitive material alone, but it is instead the ratio of the PP amplitudes of the radiation sensitive material (I_{RS}) and of the reference material (I_{ref}). The advantages of using self-calibrated dosimeters are the independence of the cavity used, the independence of the positioning of the dosimeter in the cavity, and the independence of the actual EPR instrument used. In addition, a full set of reference dosimeters should not be necessary because it suffices to give a few dosimeters a known dose D and from that calculate the "calibration coefficient" $(I_{RS}/I_{ref})/D$. This coefficient may then be assumed to be valid for all dosimeters from the same production series. Note, however, that the microwave power and modulation properties must be known for both the dosimeter material and the internal standard to ensure that the calibration coefficient do not exhibit unknown variations with the instrumental parameter values. Furthermore, the dose precision will depend on the homogeneity of the dosimeters and on the dose span used for establishing the calibration coefficient. Small errors in the calibration coefficient obtained at one dose may multiply if the calibration dose is much higher or much lower than the dose to be determined. In addition, when there is a zero-dose signal (which is often the case for alanine), the uncertainties will increase more at low doses when using the self-calibrated system than when using a standard calibration curve.

9.5.4 Other Dosimeter Materials

In practical use, alanine dosimeters have a lower detection limit of a few grays when a precision as high as $\pm 2\%$ is required (this is partly due to a sample–sample variable nonlinear background signal [157]). As a consequence, alanine is not suited for clinical applications where doses down to 0.5 Gy should be easily detectable with high precision. This has motivated researchers to search for new sensitive EPR dosimeter materials that can do the job when alanine is inadequate.

Formates (salts of formic acid, HCOOH) were suggested as alternatives to alanine a few years ago [158] and have received attention from some EPR laboratories. Ammonium and lithium formate both have properties similar to water with respect to absorption and scattering of radiation, and they show promising potential for accurate dose estimation at low radiation doses [159–161]. Lithium formate has a sensitivity that is about 6 times as high as L-α-alanine (comparing pp values), and it has the propensity to measure radiation doses down to at least 50 mGy without unduly long spectrum recording or processing times [146]. Under normal temperature and humidity conditions the radicals are stable for many months, and the dose–response shows almost no dependence on radiation qualities and energies at low LET [160, 161]. No significant zero-dose signal is observed, and the dose–response relationship is linear up to at least 1 kGy. A dose–response curve for the 0–1.0 Gy region is displayed in Fig. 9.25.

Other materials that have been used for standard EPR dosimeters and for accident or retrospective EPR dosimetry (see below) include calcified tissues, various sugars, quartz in rocks, and sulfates [162]. The EPR dosimetric properties of some of these are summarized in Table 9.2 [163–190].

9.5.5 Food Irradiation

Irradiation of food is a widespread, although in some countries a controversial, method for increasing food safety. Depending on the dose used, the irradiation

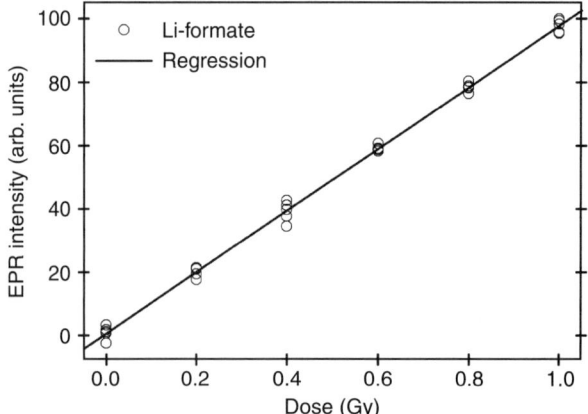

Fig. 9.25 The dose–response of the lithium formate EPR dosimeter in the dose range of 0–1.0 Gy.

TABLE 9.2 Dosimetric Properties of Different Candidates for EPR Dosimetric Materials

Compound	EPR Spectrum	Stabil. of Radiat. Induced Radicals	Effect. Z	Dose Depend.	Radiat. Chem. Yield, Radicals/ 100 eV	Power Sat. Prop.	Robustness	Ref.
Calcified tissue								
HA [Ca$_5$(OH)(PO$_4$)$_3$]	+	+	16.2					[163–165]
Tooth enamel (95–97% HA)	+	+	—					
Dentine (70–75% HA)	+	+	—					
Bone (60–70% HA)	−	+	~13					
Alanine								
Alanine (C$_3$H$_7$NO$_2$)	−	+	6.8	+	4.0	0	0	[166]
Sugars								
Sucrose (C$_{12}$H$_{22}$O$_{11}$)	−	−	7.0					[167]
Fructose (C$_6$H$_{12}$O$_6$)	−	−	7.1					
Glucose (C$_6$H$_{12}$O$_6$)	−	−	7.1					
Lactates								
Lithium lactate (C$_3$H$_5$O$_3$Li)	−	−	6.96		0.46	—		[138, 165, 168–171]
Magnesium lactate (C$_6$H$_{10}$O$_6$Mg)	−	0	9.10		1.15			
Li(Mg)–lactate (C$_3$H$_5$O$_3$Li)	−	0			0.78	—		
Mg(Li)–lactate (C$_6$H$_{10}$O$_6$Mg)	−	+			1.35	—		
Sulfates								
Lithium sulfate (Li$_2$SO$_4$)	0		11.8					[138, 165, 171–173]
Magnesium sulfate (MgSO$_4$)	+		12.20					
Barium sulfate (BaSO$_4$)	−	−	47		0.25			

(*Continued*)

TABLE 9.2 Continued

Compound	EPR Spectrum	Stabil. of Radiat. Induced Radicals	Effect. Z	Dose Depend.	Radiat. Chem. Yield, Radicals/ 100 eV	Power Sat. Prop.	Robustness	Ref.
Formates								
Ammonium formate (CHO_2NH_4)	+	0	7.12					[158–161, 174–178]
Lithium formate monohydrate ($HCO_2Li \cdot H_2O$)	+	+	7.31	+	3.2	+	+	
Magnesium formate dihydrate ($C_2H_2O_4Mg \cdot 2H_2O$)	–	–	9.64		3.7	+	0	
Sodium formate (CHO_2Na)	–		9.02		1.9		0	
Calcium formate ($C_2H_2O_4Ca$)	+	–	14.5		2.8	+	0	
Potassium formate (CHO_2K)	+	–	15.3		1.4		–	
Dithionates								
Dithionates (NH_4, Li, Ba, Cs, Rb)			>12					[158, 179–181]
Potassium dithionate ($K_2S_2O_6$)	+	0	15.5	+		+		
Materials with fair or good water equivalence								
Methylalanine ($C_4H_9NO_2$)	–	0	6.64			0		[182]
Tris(hydroxymethyl)aminomethane (($HOCH_2)_3CNH_2$)	+	–	6.79					[183]
Lithium acetate dihydrate ($C_2H_3O_2Li \cdot 2H_2O$)	0	–	7.15		0.4	+		[170]
Ammonium tartrate ($H_4NOOCHOH)_2$	+		7.26			–		[184–187]
Lithium carbonate (Li_2CO_3)	0		7.32					[188]
Bio-G (mix of SiO_2, CaO, Na_2O, P_2O_5)	0		–					[189]
Poly(tetrafluoroethylene)	0	0						[190]
Lithium phosphate (Li_3PO_4)	0	0	10.9		1.02	+		[170]

HA, hydroxyapatite; (+) favorable; (0) neither good nor bad; (−) not favorable.

may disinfect foodstuffs, provide insect control and extend shelf life (<1 kGy), and reduce pathogenic microorganisms in perishable food products (1–10 kGy); it may also be used for sterilization and the reduction of microbes (>10 kGy). In a 160-page report published in 1999, a joint FAO, IAEA, and WHO Study Group presented conclusions and recommendations for food irradiation based on more than 500 studies [191]. The Study Group found no justified reason to prohibit or restrict the use of food irradiation. A comprehensive review of food irradiation principles and applications is presented in the book by Molins [192].

Identification of irradiated food is based on detection of free radicals being formed by irradiation or molecules that have become altered by irradiation. There is at present no detection method that can be used for all groups of food being irradiated. EPR spectroscopy has turned out to be a quite successful method for identification of irradiation of three major food groups.

The European Committee for Standardization (CEN) lists three EPR-based standards for detection of irradiated food: one for foodstuffs containing bone, one for foodstuffs containing cellulose, and one for foodstuffs containing crystalline sugar [193–195]. The identification is based on the observation of irradiation induced radicals trapped in solid and/or dry parts of the food. The stability, as well as the dose detection limit, varies.

In foodstuffs containing meat or fish bones it is a quite stable radical ($\cdot CO_2^-$) formed in hydroxyapatite that is used as a probe for irradiation. An example of an EPR spectrum of irradiated bone is shown in Fig. 9.26. The detection limit and stabilization depends on the mineralization of the bones and the crystallinity of the hydroxyapatite. Normally radiation doses of approximately 0.5 kGy and above create a easily detectable amount of radicals, and even boiling of the samples shows little effect on the stability of the radicals formed. The method has been successfully tested with bones containing products of beef, trout, and chicken; and it is expected that it can be extended for all fish and meat products containing bones because hydroxyapatite is the principal component of bones. Because of the stability of the bone EPR signal, quantification of the radiation dose is often possible by using the additive dose method described below (see also Fig. 9.27).

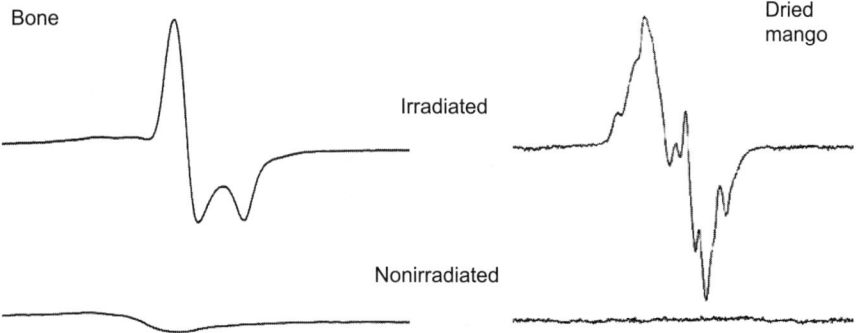

Fig. 9.26 The EPR signals from nonirradiated and irradiated samples from (left) chicken bone and (right) dried mango (containing fruit sugars).

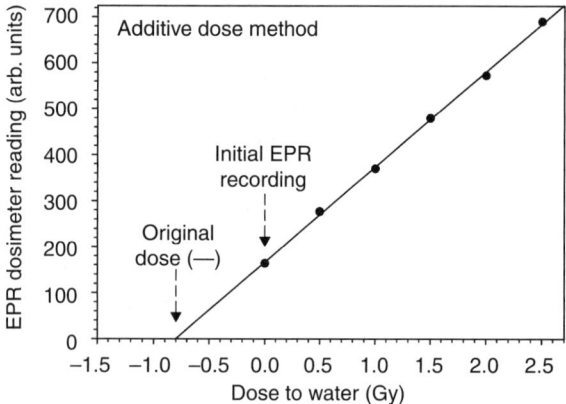

Fig. 9.27 The dose–response curve (not authentic) obtained by reirradiating a sample in the laboratory after the initial EPR recording. The unknown original dose in the sample is obtained by extrapolating the regression line back to the x axis. The regression line is given by $y = ax + b$ (a is the slope, x is the "laboratory dose," and b is the intersection with the y axis for $x = 0$). If there is negligible fading and no zero-dose signal, the original dose is simply given by the ratio b/a.

Dried fruits that have been irradiated have been successfully identified by observing the EPR spectrum from various sugar radicals formed in dried figs, mangoes (shown in Fig. 9.26), papayas, and raisins. The chemical structures of the actual radical products are not yet known, but they may be connected to the fruit sugars because irradiated samples show no EPR unless sugar is present. The dose detection limit depends on the amount of crystalline sugar present in the samples and the stability is at least several months, provided that the dried fruits have not been rehydrated prior to investigation [156].

The detection of stable radicals formed in cellulose is a method that is successfully used to identify irradiated strawberries, pistachio nut shells, and paprika powder. Doses as low as 0.5 Gy are detectable for strawberries. The stability of the formed radicals depends on the content of crystalline cellulose and the moisture of the sample from years for nuts to 3 weeks for fresh berries.

The legal regulations as well as common practice vary considerably between continents and countries, and even within the European Community the regulations are far from uniform. These variations are one reason why it is important to be able to detect whether foodstuffs have been irradiated. One other reason is to ensure that the customers who buy irradiated food for safety reasons in fact are getting irradiated items. EPR spectroscopy is now routinely implemented in many national food study and control programs.

9.5.6 Retrospective and Accident Dosimetry

In accident or emergency situations where no monitors have been present or the monitors have been overexposed and therefore are of limited use, it may be necessary to determine radiation doses some time (hours to months) after the actual radiation exposure. Examples of such situations are occupational exposure from nuclear facilities, nuclear power stations, accelerators, medical X-ray and radiation equipment, military installations, and space-based vehicles. These are situations where EPR spectroscopy has proven to be an important tool for dose reconstruction. Many different materials can be considered as dosimeter materials. Fingernails, hair, teeth, bone, household sugar, building bricks, pottery, fabrics, eggshell, shell buttons, and pharmaceuticals have all been tried with variable degrees of success. The main point is that the material should be dry, in the solid state, and preferably with some degree of crystallinity. Sugars, teeth, bone, and shell buttons so far exhibit the best potential for retrospective dosimetry.

One of the early studies in which EPR dosimetry was used for retrospective dose estimates following a fatal irradiation accident was performed in 1982 [196]. A radiation worker at a γ-irradiation facility was accidentally exposed to a Co-60 source and died about 2 weeks later. Using EPR dosimetry, the person's whole-body dose was estimated to be about 22.5 Gy. The dosimeter material that was used, which was found in the person's pocket, was heart disease tablets that consisted of more than 95% sugars (in addition to the medically active chemicals). Radiation induced radicals in dry sugars are quite stable and the radical yield is often high. Tablets from the same production lots were obtained from the pharmaceutical companies, and a calibration curve (similar to that in Fig. 9.24b) was established by irradiating them with well-known doses. The dose absorbed to the tablets carried by the radiation worker could then easily be established using EPR spectroscopy.

In situations where identical samples are not available for the generation of a calibration curve like it was in the accident described above, the absorbed radiation dose in the unique sample is commonly determined by the so-called additive dose method [193]. This involves a stepwise irradiation of the sample in the laboratory and recording the EPR spectrum after every dose increment. As illustrated in Fig. 9.27, the increasing EPR dosimeter reading is plotted against increasing (laboratory) radiation doses and the regression line is extrapolated back to the x-axis intersection for the determination of the initial dose. If the regression line is given by $y = ax + b$, then the original dose may simply be given by the ratio a/b. If the sample signal is fading in an unidentified manner or if it contains a radiation-irrelevant background signal, the method will be more complicated.

Quite reliable dose estimates may be provided by EPR-based biological dosimetry based on measurements of carbonate radicals in the enamel of irradiated teeth. Much of the current research in retrospective and accident EPR dosimetry is focused on teeth [137, 197, 198]. Tooth enamel has been used for the determination of radiation doses of victims of the atom bombs at Nagasaki and Hiroshima, in the Chernobyl accident, and for nuclear workers in the former USSR. There have been several recent international intercomparison studies [199], and IAEA has published a recommendation on measurement procedures for tooth dosimetry [200].

A major problem with standard tooth dosimetry is that it is an invasive method; teeth have to be extracted and treated before measurements may proceed. Instrumental designs and measurement procedures are, however, under development that allow for *in vivo* measurements of the EPR signal from teeth without actually extracting the teeth [201, 202].

9.5.7 EPR Dating

EPR spectroscopy has enjoyed various types of interdisciplinary research collaborations among physics, chemistry, geology, biology, and medicine, many of which are exemplified in the present book. Another interdisciplinary collaboration among these fields is *dating* [203]. Natural radiation from ^{238}U, ^{232}Th, and their daughters and from ^{40}K in the environment or inside a given specimen introduces radicals into the material. These radicals can be very stable and will accumulate over the years. Because the EPR signal is proportional to the number of radicals and hence to the accumulated dose, the time elapsed after the formation of the specimen may be estimated, provided the dose rate is known.

The first serious application of EPR on dating was Ikeya's work on dating stalagmites from the Akiyoshi cavern in Japan [204]. Since then, a huge number of materials of different kinds have been dated using EPR and cross-checked with other dating methods [203]. The time periods covered by EPR dating range from a few thousand years to several million years, far beyond the range of ^{14}C dating. The most useful dating materials are minerals like carbonate deposits (shells, corals), calcites, silicates (quartz crystals), phosphates, and carbonate radicals formed in the hydroxyapatites of fossil bones and teeth. Hydroxyapatite in teeth appears to be the only material applicable for ages as high as 10 million years. Both the accumulated dose and the environmental dose rate must be known in order to convert the estimated accumulated dose into time (age). The accumulated dose is usually obtained using the additive dose method (with the necessary refinements). The biggest challenge of EPR dating is normally the determination of the sample's radiation history [205]. A detailed practical and theoretical description of the EPR dating method is given in the EPR book by Ikeya [203]. Problems and potentials with the dating of humanoid fossil teeth were recently reviewed [206].

9.6 HIGHLIGHTED READING

The early books by Box [92], Hüttermann et al. [207] and von Sonntag [4] together provide an excellent overview of the radiation effects field. Detailed reviews by Bernhard [25], Becker and Sevilla [9], Close [26], Hüttermann [39], and Sagstuen and Sanderud [132] cover additional details of the many facets of the research that has been done. An important literature survey is found in Volumes 1–19 in the book series from The Royal Society of Chemistry, *EPR Specialist Periodical Report*, that contains reviews not only of radiation induced free radicals but also most other topics studied by EMR techniques [40]. Poole and Farach's *Handbook of Electron Spin Resonance* contains a useful introduction to EPR theory and instrumentation and a compilation of ENDOR data from a large number of systems [208]. In addition, Chapter 1 in this book contains an elaborate list of reviews and books up to about 1992. The most recent and very thorough review on radiation induced DNA damage was written by Bernhard and Close [41]. Lund and Shiotano recently published a new book on applications of EPR [89]. Finally, von Sonntag published an updated and greatly extended version of his first book, bringing the chemical aspect of radiation induced DNA damage up to date [6].

Ikeya has written a reference volume on quantitative applications of EPR with a main emphasis on dating [203]. There have been seven major international conferences devoted to EPR dosimetry and applications, the proceedings from the first six international symposia have been published [209–214]. A book [164] and a very detailed review on food irradiation [215] have also been published recently.

ACKNOWLEDGMENTS

We are grateful to many colleagues for their collaboration and discussions in connection with some of the work presented in this chapter, in particular, William H. Nelson, David M. Close, William A. Bernhard, Janko N. Herak, Eva Lund, Anders Lund, Bartlomiej Ciesielski, Eva S. Bergstrand, Eirik Malinen, Tor Arne Vestad, and Efim Brondz. Financial support from the U.S. National Institutes of Health and the Norwegian Research Council are acknowledged.

REFERENCES

1. Attix, F. H. *Introduction to Radiological Physics and Radiation Dosimetry*; Wiley-VCH Verlag & Co. KGaA: Weinheim, 2004.
2. Dertinger, H.; Jung, H. *Molecular Radiation Biology*; Springer-Verlag: Berlin, 1970.
3. Michaels, H. B.; Hunt, J. W. *Radiat. Res.* **1978**, *74*, 23.
4. von Sonntag, C. *The Chemical Basis of Radiation Biology*; Taylor & Francis: London, 1987.
5. Krisch, R. E.; Flick, M. B.; Trumbore, C. N. *Radiat. Res.* **1991**, *126*, 251.
6. von Sonntag, C. *Free-Radical-Induced DNA Damage and Its Repair. A Chemical Perspective*; Springer-Verlag: Heidelberg, 2005.
7. Alberts, B.; Johnson, A.; Lewis, J.; Raff, M.; Roberts, K.; Walther, R. *Molecular Biology of the Cell*, 4th ed.; Garland Science: New York, 2002.
8. Symons, M. C. R. The Role of Radiation Induced Charge Migration with DNA: ESR Studies. In *The Early Effects of Radiation on DNA*; Fielden, E. M., O'Neill, P., Eds.; Springer-Verlag: Berlin, 1991; Vol. 54, p 111.
9. Becker, D.; Sevilla, M. D. *Adv. Radiat. Biol.* **1993**, *17*, 121.
10. Pimblott, S. M.; LaVerne, J. A.; Mozumder, A.; Green, N. J. B. *J. Phys. Chem.* **1990**, *94*, 488.
11. Ward, J. F. *J. Chem. Ed.* **1981**, *58*, 135.
12. Goodhead, D. *Can. J. Phys.* **1990**, *68*, 872.
13. Murphy, C. J.; Arkin, M. R.; Jenkins, Y.; Ghatlia, N. D.; Bossmann, S. H.; Turro, N. J.; Barton, J. K. *Science* **1993**, *262*, 1025.

14. Wan, C.; Fiebig, T.; Schiemann, O.; Barton, J. K.; Zewail, A. H. *Proc. Natl. Acad. Sci. U.S.A.* **2000**, *97*, 14052.
15. Sevilla, M. D. Mechanisms for Radiation Damage in DNA. In *Excited States in Organic and Biochemistry*; Pullman, B., Goldblum, N., Eds.; D. Reidel: Boston, 1977; p 15.
16. Debije, M. G.; Milano, M. T.; Bernhard, W. H. *Angew. Chem. Int. Ed.* **1999**, *38*, 2752.
17. Spalleta, R. A.; Bernhard, W. A. *Radiat. Res.* **1992**, *130*, 7.
18. Razskazovskii, Y.; Swarts, S. G.; Falcone, J. M.; Taylor, C.; Sevilla, M. D. *J. Phys. Chem.* **1997**, *B101*, 1460.
19. Ito, T.; Baker, S. C.; Stickley, C. D.; Peak, J. G.; Peak, M. J. *Int. J. Radiat. Biol.* **1993**, *63*, 289.
20. Prise, K. M.; Davies, S.; Michael, B. D. *Radiat. Res.* **1993**, *134*, 102.
21. Goodhead, D. T. *Int. J. Radiat. Biol.* **1994**, *65*, 7.
22. Steenken, S. *Chem. Rev.* **1989**, *89*, 503.
23. Steenken, S. Radical Chemistry in Crystals, Matrices and in Aqueous Solution. Redox Properties and Proton Transfer Processes. In *The Early Effects of Radiation on DNA*; Fielden, E. M., O'Neill, P., Eds.; Springer-Verlag: Berlin, 1991; p 269.
24. Steenken, S. *Biol. Chem.* **1997**, *378*, 1293.
25. Bernhard, W. A. *Adv. Radiat. Biol.* **1981**, *9*, 199.
26. Close, D. M. *Radiat. Res.* **1993**, *135*, 1.
27. Swarts, S. G.; Sevilla, M. D.; Becker, D.; Tokar, C. J.; Wheeler, K. T. *Radiat. Res.* **1992**, *129*, 333.
28. Purkayastha, S.; Bernhard, W. A. *J. Phys. Chem.* **2004**, *B108*, 18377.
29. Purkayastha, S.; Milligan, J. R.; Bernhard, W. A. *J. Phys. Chem.* **2005**, *B109*, 16967.
30. Close, D. M. *Radiat. Res.* **1997**, *147*, 663.
31. Hole, E. O.; Nelson, W. H.; Sagstuen, E.; Close, D. M. *Radiat. Res.* **1992**, *129*, 119.
32. Sagstuen, E.; Lund, A.; Awadelkarim, O.; Lindgren, M. *J. Phys. Chem.* **1986**, *90*, 5584.
33. Sagstuen, E.; Lindgren, M.; Lund, A. *Radiat. Res.* **1991**, *128*, 235.
34. Vanhaelewyn, G. C. A. M.; Jansen, B.; Callens, F. J.; Sagstuen, E. *Radiat. Res.* **2004**, *162*, 96.
35. Vanhaelewyn, G. C. A. M.; Jansen, B.; Pauwels, E.; Sagstuen, E.; Waroquier, M.; Callens, F. J. *J. Phys. Chem.* **2004**, *A108*, 3308.
36. Vanhaelewyn, G. C. A. M.; Pauwels, E.; Callens, F. J.; Waroquier, M.; Sagstuen, E.; Matthys, P. F. A. E. *J. Phys. Chem.* **2006**, *A110*, 2147.
37. Georgieva, E.; Pardi, L. A.; Jeschke, G.; Gatteschi, D.; Sorace, L.; Yordanov, N. D. *Free Radical Res.* **2006**, *40*, 553.
38. Datta, K.; Jaruga, P.; Dizdaroglu, M.; Neumann, R. D.; Winters, T. A. *Radiat. Res.* **2006**, *166*, 767.
39. Hüttermann, J. Radical Ions and Their Reactions in DNA and Its Constituents. In *Radical Ionic Systems: Properties in Condensed Phases*; Lund, A., Shiotani, M., Eds.; Kluwer Academic: Dordrecht, 1991; Vol. 6, p 435.
40. Gilbert, B. C.; Davies, M. J.; Murphy, D. M. *Electron Paramagnetic Resonance*; Royal Society of Chemistry: Cambridge, U.K., 2003; Vol. 18.
41. Bernhard, W. A.; Close, D. M. DNA Damage Dictates the Biological Consequences of Ionizing Irradiation: The Chemical Pathways. In *Charged Particle and Photon*

Interactions with Matter: Chemical, Physicochemical, and Biological Consequences with Applications; Mozumber, A., Hatano, Y., Eds.; Marcel Dekker: New York, 2003; p 431.

42. Sagstuen, E.; Hole, E. O.; Nelson, W. H.; Close, D. M. *J. Phys. Chem.* **1992**, *96*, 8269.
43. Close, D. M.; Hole, E. O.; Sagstuen, E.; Nelson, W. H. *J. Phys. Chem.* **1998**, *A102*, 6737.
44. Herak, J. N.; Lenard, D. R.; McDowell, C. A. *J. Magn. Reson.* **1977**, *26*, 189.
45. Dertinger, H. *Z. Naturforsch* **1967**, *B22*, 1266.
46. Box, H. C.; Budzinski, E. E.; Potienko, G. *J. Chem. Phys.* **1980**, *73*, 2052.
47. Bernhard, W. A.; Close, D. M.; Hüttermann, J.; Zehner, H. *J. Chem. Phys.* **1977**, *67*, 1211.
48. Sagstuen, E.; Hole, E. O.; Nelson, W. H.; Close, D. M. *J. Phys. Chem.* **1989**, *93*, 5974.
49. Sagstuen, E.; Hole, E. O.; Nelson, W. H.; Close, D. M. *J. Phys. Chem.* **1992**, *96*, 1121.
50. Hüttermann, J. *Int. J. Radiat. Biol.* **1970**, *17*, 249.
51. Sevilla, M. D. *J. Phys. Chem.* **1971**, *75*, 626.
52. Hole, E. O.; Sagstuen, E.; Nelson, W. H.; Close, D. M. *J. Phys. Chem.* **1991**, *95*, 1494.
53. Wang, W.; Sevilla, M. D. *Radiat. Res.* **1994**, *138*, 9.
54. Salovey, R.; Shulman, R. G.; Walsh, W. M. *J. Chem. Phys.* **1963**, *39*, 839.
55. Ehrenberg, A.; Ehrenberg, L.; Löfroth, G. *Nature (Lond.)* **1963**, *200*, 376.
56. Box, H. C.; Freund, H. G. *Ann. N.Y. Acad. Sci.* **1973**, *222*, 446.
57. Gräslund, A.; Ehrenberg, A.; Rüpprecht, A.; Ström, G. *Biochim. Biophys. Acta* **1971**, *254*, 172.
58. Close, D. M.; Sagstuen, E.; Nelson, W. H. *J. Chem. Phys.* **1985**, *82*, 4386.
59. Close, D. M.; Nelson, W. H.; Sagstuen, E. *Radiat. Res.* **1987**, *112*, 283.
60. Adhikary, A.; Kumar, A.; Becker, D.; Sevilla, M. D. *J. Phys. Chem.* **2006**, *B110*, 24171.
61. Close, D. M.; Nelson, W. H. *Radiat. Res.* **1989**, *117*, 367.
62. Kar, L.; Bernhard, W. A. *Radiat. Res.* **1983**, *93*, 232.
63. Nelson, W. H.; Sagstuen, E.; Hole, E. O.; Close, D. M. *Radiat. Res.* **1992**, *131*, 272.
64. Close, D. M.; Nelson, W. H.; Sagstuen, E.; Hole, E. O. *Radiat. Res.* **1994**, *137*, 300.
65. Gräslund, A.; Ehrenberg, A.; Rupprecht, A.; Ström, G.; Crespi, H. *Int. J. Radiat. Biol.* **1975**, *28*, 313.
66. Hüttermann, J.; Voit, K.; Oloff, H.; Koehnlein, W.; Gräslund, A.; Rupprecht, A. *J. Chem. Soc. Faraday Trans.* **1984**, *78*, 135.
67. Bernhard, W. A. *J. Phys. Chem.* **1989**, *93*, 2187.
68. Sevilla, M. D.; Becker, D.; Yan, M.; Summerfield, S. R. *J. Phys. Chem.* **1991**, *95*, 3409.
69. Weiland, B.; Hüttermann, J. *Int. J. Radiat. Biol.* **1998**, *74*, 341.
70. Yan, M.; Becker, D.; Summerfield, S.; Renke, P.; Sevilla, M. D. *J. Phys. Chem.* **1992**, *96*, 1983.
71. Becker, D.; Bryant-Friedrich, A.; Trzasko, C.; Sevilla, M. D. *Radiat. Res.* **2003**, *160*, 174.
72. Purkayastha, S.; Milligan, J. R.; Bernhard, W. A. *J. Phys. Chem.* **2006**, *B110*, 26286.
73. Debije, M. G.; Bernhard, W. A. *Radiat. Res.* **2001**, *155*, 687.
74. Razskazovskiy, Y.; Debije, M. G.; Bernhard, W. A. *Radiat. Res.* **2000**, *153*, 436.

75. Alexander, P.; Lett, J. T.; Ormerod, M. G. *Biochim. Biophys. Acta* **1961**, *51*, 207.
76. Cullis, P. M.; Jones, G. D. D.; Symons, M. C. R.; Lea, J. S. *Nature (Lond.)* **1987**, *330*, 773.
77. Weiland, B.; Hüttermann, J. *Int. J. Radiat. Biol.* **2000**, *76*, 1075.
78. Faucitano, A.; Buttafava, A.; Martinotti, F.; Pedraly-Noy, G. *Radiat. Phys. Chem.* **1992**, *40*, 357.
79. Sagstuen, E.; Close, D. M.; Vågane, R.; Hole, E. O.; Nelson, W. H. *J. Phys. Chem.* **2006**, *A110*, 8653.
80. Sevilla, B. D.; Becker, D. ESR Studies of Radiation Damage to DNA and Related Biomolecules. In *Specialist Periodical Reports*; Gilbert, B. C., Davies, M. J., Eds.; Royal Society of Chemistry: Cambridge, U.K., 2004; Vol. 19, p 243.
81. Clegg, W., Ed. *Crystal Structure Analysis. Principles and Practice*; Oxford University Press: Oxford, U.K., 2002; Vol. 6.
82. International Union of Crystallography. *International Tables for X-Ray Crystallography, Space-Group Symmetry*, 5th ed.; Springer: Heidelberg, 2005; Vol. A.
83. Jeffrey, G. A.; Kinoshita, Y. *Acta Crystallogr.* **1963**, *16*, 20.
84. McClure, R. J.; Craven, B. M. *Acta Crystallogr.* **1973**, *B29*, 1234.
85. Weber, H. P.; Craven, B. M.; McMullan, R. K. *Acta Crystallogr.* **1980**, *B36*, 645.
86. Hüttermann, J.; Schmidt, G.; Weymann, D. *J. Magn. Reson.* **1976**, *21*, 221.
87. Nelson, W. H. *J. Magn. Reson.* **1980**, *37*, 205.
88. Mercer, K. R.; Bernhard, W. A. *J. Magn. Reson.* **1987**, *74*, 66.
89. Lund, A.; Shiotani, M. *EPR of Free Radicals in Solids*; Kluwer Academic: Dordrecht, 2003.
90. Schonland, D. S. *Proc. Phys. Soc* **1959**, *73*, 788.
91. Atherton, N. M. *Principles of Electron Spin Resonance*; Ellis Horwood PTR Prentice-Hall: London, 1993.
92. Box, H. C. *Radiation Effects: ESR and ENDOR Analysis*; Academic Press: New York, 1977.
93. Theisen, H.; Sagstuen, E. *J. Chem. Phys.* **1981**, *74*, 2319.
94. Nelson, W. H.; Nave, C. R. *J. Chem. Phys.* **1981**, *74*, 2710.
95. Nelson, W. H. *J. Magn. Reson.* **1980**, *38*, 71.
96. McConnell, H. M.; Chesnut, D. B. *J. Chem. Phys.* **1958**, *28*, 107.
97. Fessenden, R. W.; Schuler, R. H. *J. Chem. Phys.* **1963**, *39*, 2147.
98. Nelson, W. H.; Atwater, F. M.; Gordy, W. *J. Chem. Phys.* **1974**, *61*, 4726.
99. McConnell, H. M.; Strathdee, J. *Mol. Phys.* **1959**, *2*, 129.
100. Bernhard, W. A. *J. Chem. Phys.* **1984**, *81*, 5928.
101. Gordy, W. *Theory and Applications of Electron Spin Resonance*; Wiley: New York, 1980.
102. Erling, P. A.; Nelson, W. H. *J. Phys. Chem.* **2004**, *A108*, 7591.
103. Heller, C.; McConnell, H. M. *J. Chem. Phys.* **1960**, *32*, 1535.
104. Derbyshire, W. *Mol. Phys.* **1962**, *5*, 225.
105. Fitzpatrick, J. A. J.; Manby, F. R.; Western, C. M. *J. Chem. Phys.* **2005**, *122*, 084312.

106. Morton, J. R.; Rowlands, J. R.; Whiffen, D. H. *Atomic Properties for Reporting ESR Data. National Physical Laboratory BPR Report 13*; National Physical Laboratory: Teddington, U.K., 1962.
107. Morton, J. R.; Preston, K. F. *J. Magn. Reson.* **1978**, *30*, 577.
108. Pullman, B.; Pullman, A. *Quantum Biochemistry*; Wiley: New York, 1963.
109. Ban, F.; Gauld, J. W.; Wetmore, S. D.; Boyd, R. J. The Calculation of the Hyperfine Coupling Tensors of Biological Radicals. In *Progress in Theoretical Chemistry and Physics*; Lund, A., Shiotani, M., Eds.; Kluwer: Dordrecht, 2003; Vol. 10, p 239.
110. Close, D. M. Model Calculations of Radiation Induced Damage in DNA Constituents Using Density Functional Theory. In *Computational Chemistry, Review of Current Trends*; Leszczynski, J., Ed.; World Scientific: Singapore, 2004; Vol. 8, p 209.
111. Wetmore, S. D.; Eriksson, L. A.; Boyd, R. J. *Theor. Comput. Chem.* **2001**, *9*, 409.
112. Pople, J. A.; Beveridge, D. L. *Approximate Molecular Orbital Theory*; McGraw-Hill: New York, 1970.
113. Oloff, H.; Hüttermann, J. *J. Magn. Reson.* **1980**, *40*, 415.
114. Frisch, M. J.; Trucks, G. W.; Schlegel, H. B.; Scuseria, G. E.; Robb, M. A.; Cheeseman, J. R.; Montgomery, J. A., Jr.; Vreven, T.; Kudin, K. N.; Burant, J. C.; Millam, J. M.; Iyengar, S. S.; Tomasi, J.; Barone, V.; Mennucci, B.; Cossi, M.; Scalmani, G.; Rega, N.; Petersson, G. A.; Nakatsuji, H.; Hada, M.; Ehara, M.; Toyota, K.; Fukuda, R.; Hasegawa, J.; Ishida, M.; Nakajima, T.; Honda, Y.; Kitao, O.; Nakai, H.; Klene, M.; Li, X.; Knox, J. E.; Hratchian, H. P.; Cross, J. B.; Adamo, C.; Jaramillo, J.; Gomperts, R.; Stratmann, R. E.; Yazyev, O.; Austin, A. J.; Cammi, R.; Pomelli, C.; Ochterski, J. W.; Ayala, P. Y.; Morokuma, K.; Voth, G. A.; Salvador, P.; Dannenberg, J. J.; Zakrzewski, V. G.; Dapprich, S.; Daniels, A. D.; Strain, M. C.; Farkas, O.; Malick, D. K.; Rabuck, A. D.; Raghavachari, K.; Foresman, J. B.; Ortiz, J. V.; Cui, Q.; Baboul, A. G.; Clifford, S.; Cioslowski, J.; Stefanov, B. B.; Liu, G.; Liashenko, A.; Piskorz, P.; Komaromi, I.; Martin, R. L.; Fox, D. J.; Keith, T.; Al-Laham, M. A.; Peng, C. Y.; Nanayakkara, A.; Challacombe, M.; Gill, P. M. W.; Johnson, B.; Chen, W.; Wong, M. W.; Gonzalez, C.; Pople, J. A. *Gaussian 03*, Rev. B.04 ed.; Gaussian Inc.: Pittsburgh, PA, 2003.
115. Close, D. M.; Bernhard, W. A. *J. Chem. Phys.* **1977**, *66*, 5244.
116. Close, D. M.; Bernhard, W. A. *J. Chem. Phys.* **1979**, *70*, 210.
117. Wetmore, S. D.; Himo, F.; Boyd, R. J.; Eriksson, L. A. *J. Phys. Chem.* **1998**, *B102*, 7484.
118. Pauwels, E.; Van Speybroeck, V.; Waroquier, M. *Int. J. Quantum Chem.* **2003**, *91*, 511.
119. Hanson, G. R.; Gates, K. E.; Noble, C. J.; Mitchell, A.; Benson, S. W.; Griffin, M.; Burrage, K. X-Sophe-Sophe-XEprView: A Computer Simulation Software Suite for the Analysis of Continuous Wave EPR Spectra. In *EPR of Free Radicals in Solids*; Lund, A., Shiotani, M., Eds.; Kluwer Academic: Dordrecht, 2003; Vol. 10.
120. Lefebvre, R.; Maruani, J. *J. Chem. Phys.* **1965**, *42*, 1480.
121. Erickson, R.; Lund, A. *J. Magn. Reson.* **1991**, *92*, 146.
122. Lund, A.; Erickson, R. *Acta Chem. Scand.* **1998**, *52*, 261.
123. Sagstuen, E.; Lund, A.; Itagaki, Y.; Maruani, J. *J. Phys. Chem.* **2000**, *A104*, 6362.
124. Stoll, S.; Schweiger, A. *J. Magn. Reson.* **2006**, *178*, 42.
125. Duling, D. R. *J. Magn. Reson.* **1994**, *B104*, 105.

126. Sagstuen, E.; Hole, E. O.; Haugedal, S. R.; Nelson, W. H. *J. Phys. Chem.* **1997**, *A101*, 9763.
127. Heydari, M. Z.; Malinen, E.; Hole, E. O.; Sagstuen, E. *J. Phys. Chem.* **2002**, *A106*, 8971.
128. Malinen, E.; Heydari, M. Z.; Sagstuen, E.; Hole, E. O. *Radiat. Res.* **2003**, *159*, 23.
129. Malinen, E.; Hult, E. A.; Hole, E. O.; Sagstuen, E. *Radiat. Res.* **2003**, *159*, 149.
130. Miyagawa, I.; Gordy, W. *J. Chem. Phys.* **1960**, *32*, 255.
131. Morton, J. R.; Horsfield, A. *J. Chem. Phys.* **1961**, *35*, 1142.
132. Sagstuen, E.; Sanderud, A.; Hole, E. O. *Radiat. Res.* **2004**, *162*, 112.
133. Weiland, B.; Hüttermann, J.; Van Tol, J. *Acta Chem. Scand.* **1997**, *51*, 585.
134. Becker, D.; Kwiram, A. L. *Chem. Phys. Lett.* **1976**, *39*, 180.
135. Erickson, R. *Chem. Phys.* **1996**, *202*, 263.
136. Weil, J. A.; Bolton, J. R.; Wertz, J. E. *Electron Paramagnetic Resonance. Elementary Theory and Practical Applications*, 2nd ed.; Wiley: New York, 1994.
137. Regulla, D. F. *Appl. Radiat. Isot.* **2005**, *62*, 117.
138. Ikeya, M.; Hassan, G. M.; Sasaoka, H.; Kinoshita, Y.; Takaki, T.; Yamanaka, C. *Appl. Radiat. Isot.* **2000**, *52*, 1209.
139. Bradshaw, W. W.; Cadena, D. G.; Crawford, G. W.; Spetzler, H. A. W. *Radiat. Res.* **1962**, *17*, 11.
140. Regulla, D. F.; Deffner, U. *Appl. Radiat. Isot.* **1982**, *33*, 1101.
141. Olsen, K. J.; Hansen, J. W.; Waligorski, M. P. R. *Appl. Radiat. Isot.* **1989**, *40*, 985.
142. Mehta, K.; Girzikowsky, R. *Appl. Radiat. Isot.* **2000**, *52*, 1179.
143. ASTM International. *Nuclear Energy (II), Solar, and Geothermal Energy*; ASTM International: West Conshohocken, PA, 2004; Vol. 12.02, p 798.
144. Hansen, J. W.; Olsen, K. J.; Wille, M. *Radiat. Prot. Dosim.* **1987**, *19*, 43.
145. Sleptchonok, O. F.; Nagy, V.; Desrosiers, M. F. *Radiat. Phys. Chem.* **2000**, *57*, 115.
146. Hayes, R. B.; Haskell, E. H.; Wieser, A.; Romanykha, A. A.; Hardy, B. L.; Barrus, J. K. *Nucl. Instrum. Methods Phys. Res.* **2000**, *A440*, 453.
147. Wieser, A.; Siegele, R.; Regulla, D. F. *Appl. Radiat. Isot.* **1989**, *40*, 957.
148. Nagy, V.; Puhl, J. M.; Desrosiers, M. F. *Radiat. Phys. Chem.* **2000**, *57*, 1.
149. Sleptchonok, O. F.; Nagy, V.; Desrosiers, M. F. *Radiat. Phys. Chem.* **2000**, *57*, 115.
150. Sharpe, P. H. G.; Rajendran, K.; Sephton, J. P. *Appl. Radiat. Isot.* **1996**, *47*, 1171.
151. Anton, M. *Phys. Med. Biol.* **2006**, *51*, 5419.
152. Nagy, V.; Sleptchonok, O. F.; Desrosiers, M. F.; Weber, R. T.; Heiss, A. H. *Radiat. Phys. Chem.* **2000**, *59*, 429.
153. Nagy, V.; Sholom, S. V.; Chumak, V. V.; Desrosiers, M. F. *Appl. Radiat. Isot.* **2002**, *56*, 917.
154. Onori, S.; Bortolin, E.; Calicchia, A.; Carosi, A.; De Angelis, C.; Grande, S. *Radiat. Prot. Dosim.* **2006**, *120*, 226.
155. Yordanov, N. D.; Gancheva, V. *J. Radioanal. Nucl. Chem.* **2000**, *245*, 323.
156. Yordanov, N. D.; Gancheva, V. Recent Development of EPR Dosimetry. In *EPR of Free Radicals in Solids. Trends in Methods and Applications*; Lund, A., Shiotani, M., Eds.; Kluwer Academic: Dordrecht, 2003.
157. Ruckerbauer, F.; Sprunck, M.; Regulla, D. F. *Appl. Radiat. Isot.* **1996**, *47*, 1263.

158. Lund, A.; Olsson, S.; Bonora, M.; Lund, E.; Gustafsson, H. *Spectrochim Acta A* **2002**, *58*, 1301.
159. Gustafsson, H.; Olsson, S.; Lund, A.; Lund, E. *Radiat. Res.* **2003**, *161*, 464.
160. Vestad, T. A.; Malinen, E.; Lund, A.; Hole, E. O.; Sagstuen, E. *Appl. Radiat. Isot.* **2003**, *59*, 181.
161. Vestad, T. A.; Malinen, E.; Olsen, D. R.; Hole, E. O.; Sagstuen, E. *Phys. Med. Biol.* **2004**, *49*, 4701.
162. Ikeya, M. *Appl. Radiat. Isot.* **1996**, *47*, 1479.
163. Brady, J. M.; Aarestad, N. O.; Swartz, H. M. *Health Phys.* **1968**, *15*, 43.
164. Desrosiers, M. F.; Schauer, D. A. *Nucl. Instrum. Methods Phys. Res.* **2001**, *B184*, 219.
165. Da Costa, Z. M.; Pontuschka, W. M.; Campos, L. L. *Nucl. Instrum. Methods Phys. Res.* **2004**, *B218*, 283.
166. Regulla, D. F.; Deffner, U. *Appl. Radiat. Isot.* **1982**, *33*, 1101.
167. Yordanov, N. D.; Georgieva, E. *Spectrochim Acta A* **2004**, *60*, 1307.
168. Hassan, G. M.; Ikeya, M.; Toyoda, S. *Appl. Radiat. Isot.* **1998**, *49*, 823.
169. Hassan, G. M.; Ikeya, M.; Takaki, T. *Radiat. Meas.* **1999**, *30*, 189.
170. Hassan, G. M.; Ikeya, M. *Appl. Radiat. Isot.* **2000**, *52*, 1247.
171. Sharaf, M. A.; Hassan, G. M. *Nucl. Instrum. Methods Phys. Res.* **2004**, *B225*, 521.
172. Morton, J. R.; Ahlers, F. J.; Schneider, C. C. *Radiat. Prot. Dosim.* **1993**, *47*, 263.
173. Ohta, M.; Sakaguchi, M. *Radiat. Eff. Defects Solids* **1994**, *132*, 249.
174. Ovenall, D. W.; Whiffen, D. H. *Mol. Phys.* **1961**, *4*, 135.
175. Supe, A. A.; Zubarev, V. E.; Bugaenko, L. T. *Latv. PSR Zinat. Akad. Vestis* **1986**, *4*, 433.
176. Keizer, P. N.; Morton, J. R.; Preston, K. F. *J. Chem. Soc. Faraday Trans.* **1991**, *87*, 3147.
177. Köksal, F.; Karamustafaoglu, O.; Karabulut, B. *Solid State Ionics* **1999**, *123*, 325.
178. Komaguchi, K.; Matsubara, Y.; Shiotani, M.; Gustafsson, H.; Lund, E.; Lund, A. *Spectrochim. Acta* **2007**, *A66*, 754.
179. Bogushevich, S. E.; Ugolev, I. I. *Appl. Radiat. Isot.* **2000**, *52*, 1217.
180. Lund, E.; Gustafsson, H.; Danilczuk, M.; Sastry, M. D.; Lund, A. *Spectrochim. Acta* **2004**, *A60*, 1319.
181. Lund, E.; Gustafsson, H.; Danilczuk, M.; Sastry, M. D.; Lund, A.; Vestad, T. A.; Malinen, E.; Hole, E. O.; Sagstuen, E. *Appl. Radiat. Isot.* **2005**, *62*, 317.
182. Olsson, S.; Sagstuen, E.; Bonora, M.; Lund, A. *Radiat. Res.* **2002**, *157*, 113.
183. Azorin, J.; Rivera, T.; Solis, J. *Appl. Radiat. Isot.* **2000**, *47*, 1539.
184. Brustolon, M.; Zoleo, A.; Lund, A. *J. Magn. Reson.* **1999**, *187*, 137.
185. Olsson, S. K.; Lund, E.; Lund, A. *Appl. Radiat. Isot.* **2000**, *52*, 1235.
186. Yordanov, N. D.; Gancheva, V. *Radiat. Phys. Chem.* **2004**, *69*, 249.
187. Marrale, M.; Brai, M.; Triolo, A.; Bartolotta, A.; D'Oca, M. C. *Radiat. Res.* **2006**, *165*, 802.
188. Murali, S.; Natarajan, V.; Venkataramani, R.; Pushparaja; Sastry, M. D. *Appl. Radiat. Isot.* **2001**, *55*, 253.
189. Hassan, G. M.; Sharaf, M. A.; Desouky, O. S. *Radiat. Meas.* **2004**, *38*, 311.
190. Azorin, J.; Rivera, T.; Solis, J. *Appl. Radiat. Isot.* **2000**, *52*, 1243.
191. World Health Organisation. *Report of a Joint FAO/IAEA/WHO Study Group. Technical Report Series No. 890*; World Health Organisation: Geneva, 1999.

192. Molins, R. A., Ed. *Food Irradiation: Principles and Applications*; Wiley-Interscience: New York, 2001.
193. CEN. *EN 1786, Foodstuffs—Detection of Irradiated Food Containing Bone—Method by ESR Spectroscopy*; CEN: Brussels, 1997.
194. CEN. *EN 1787, Foodstuffs—Detection of Irradiated Food Containing Cellulose—Method by ESR Spectroscopy*; CEN: Brussels, 1997.
195. CEN. *EN 13708, Foodstuffs—Detection of Irradiated Food Containing Crystalline Sugar by ESR Spectroscopy*; CEN: Brussels, 2001.
196. Sagstuen, E.; Theisen, H.; Henriksen, T. *Health Phys.* **1983**, *45*, 961.
197. Callens, F.; Vanhaelewyn, G.; Matthys, P.; Boesman, E. *Appl. Magn. Reson.* **1998**, *14*, 235.
198. Romanyukha, A. A.; Desrosiers, M. F.; Regulla, D. F. *Appl. Radiat. Isot.* **2000**, *52*, 1265.
199. Wieser, A.; Debuyst, R.; Fattibene, P.; Meghzifene, A.; Onori, S.; Bayankin, S. N.; Blackwell, B.; Brik, A.; Bugay, A.; Chumak, V.; Ciesielski, B.; Hoshi, M.; Imata, H.; Ivannikov, A.; Ivanov, D.; Junczewska, M.; Miyazawa, C.; Pass, B.; Penkowski, M.; Pivovarov, S.; Romanyukha, A.; Romanyukha, L.; Schauer, D.; Scherbina, O.; Schultka, K.; Shames, A.; Sholom, S.; Skinner, A.; Skvortsov, V.; Stepanenko, V.; Tielewuhan, E.; Toyoda, S.; Trompier, F. *Appl. Radiat. Isot.* **2005**, *62*, 163.
200. IAEA. *IAEA-TECDOC-1331*; IAEA: Vienna, 2002.
201. Ikeya, M.; Yamamoto, M.; Ishii, H. *Rev. Sci. Instrum.* **1994**, *65*, 3670.
202. Iwasaki, A.; Walczak, T.; Grinberg, O.; Swartz, H. M. *Appl. Radiat. Isot.* **2005**, *62*, 133.
203. Ikeya, M. *New Applications of Electron Spin Resonance. Dating, Dosimetry and Microscopy*; World Scientific: Singapore, 1993.
204. Ikeya, M. *Nature* **1975**, *255*, 48.
205. Skinner, A. R. *Appl. Radiat. Isot.* **2000**, *52*, 1311.
206. Skinner, A. R.; Blackwell, B. A. B.; Chasteen, N. D.; Brassard, P. *Adv. ESR Applic.* **2002**, *18*, 77.
207. Hüttermann, J.; Kohnlein, W.; Teoule, R.; Bertinchamps, A. J. *Effects of Ionizing Radiation on DNA. Physical, Chemical and Biological Aspects*; Springer-Verlag: Berlin, 1978; Vol. 27.
208. Poole, C. P.; Farach, H. C. *Handbook of Electron Spin Resonance*; American Institute of Physics Press: New York, 1994.
209. Ikeya, M.; Miki, T. *ESR Dating and Dosimetry. Proceedings of the 1st International Symposium*; Ionics: Tokyo, 1985.
210. Regulla, D. F.; Deffner, U. *Appl. Radiat. Isot.* **1989**, *40*, 1039.
211. Skinner, A. F.; Desrosiers, M. F. *Appl. Radiat. Isot.* **1993**, *44*, 1.
212. Desrosiers, M. F.; Skinner, A. F.; Regulla, D. F.; Nagy, V.; McLauchlin, W. L.; Eaton, G.; Eaton, S. S. *Appl. Radiat. Isot.* **1996**, *47*, 1151.
213. Desrosiers, M. F.; Regulla, D. F.; Skinner, A. F.; Skøvortsov, V.; Fill, U. A.; Stepanenko, V. *Appl. Radiat. Isot.* **2000**, *52*, 1019.
214. Skinner, A. F.; Fill, U. A.; Baffa, O.; Eaton, S. S. *Appl. Radiat. Isot.* **2005**, *62*, 115.
215. Farkas, J. Food Irradiation. In *Charged Particle and Photon Interactions with Matter: Chemical, Physicochemical, and Biological Consequences with Applications*; Mozumber, A., Hatano, Y., Eds.; Marcel Dekker: New York, 2003; p 785.

10 Electron Paramagnetic Resonance in Biochemistry and Biophysics

PART I: SPIN LABELS, PARAMAGNETIC IONS, AND OXIMETRY

MICHAEL K. BOWMAN

Department of Chemistry, University of Alabama, Tuscaloosa, AL 35487-0336

10.1 INTRODUCTION

10.1.1 Paramagnetic Species in Biology

Electron paramagnetic resonance (EPR) has been used in many different ways in biochemical and biophysical studies. Many of the biological processes that involve free radicals were identified as such because EPR revealed the presence of free radicals. All one-electron redox reactions must involve at least one paramagnetic species in either the reactants or products. A large group of enzymes contain paramagnetic metals, such as Cu(II), Fe(III), and Mn(II), as active participants around the active site. Many more proteins contain metal ions that play a structural role and, even when the metal ions are not paramagnetic, it is often possible to substitute a paramagnetic metal ion as a spin probe with retention of structure and occasionally enzymatic activity. Even DNA binds paramagnetic metal ions that can serve as an EPR-active probe for both structure and dynamics.

Analogs of a biologically active molecule find use as spin probes to report on the normal environment of the native molecule. The spin probe, typically containing a nitroxide group, is designed to bind to a protein in place of a substrate, product, inhibitor, or cofactor, or to enter a particular structure, for example, a nitroxide-labeled fatty acid probing a biological membrane. Once in place, the spin probe is

---

Electron Paramagnetic Resonance. Edited by Brustolon and Giamello
Copyright © 2009 John Wiley & Sons, Inc.

used to probe the structure, dynamics, or other properties through the influence of the surroundings on the EPR signal of the probe [1].

10.1.2 Spin Labels

Another approach for the application of EPR to biochemical and biophysical problems involves the labeling of the protein or biomolecule with a paramagnetic tag or spin label. Although the spin label is often a stable nitroxide group, virtually any free radical or paramagnetic metal ion can be used or even a group with a long-lived triplet state. There is a wide variety of nitroxide spin labels that can be covalently attached to protein through functional groups, such as methane thiosulfonate (MTSL), iodoacetamide, N-ethyl maleimide, and fluorodinitrobenzene. The amino acid that is labeled varies with both the label and the protein; but typical targets include serine, lysine, and cysteine. The labels based on MTSL are specific for exposed cysteines on the protein. This feature led to the popular technique known as site-directed spin labeling (SDSL) [2].

In SDSL, all cysteines in the protein are changed to some other amino acid by site-directed mutagenesis. Then an amino acid is changed to cysteine, again by site-directed mutagenesis, and the protein is expressed and purified. Treatment with MTSL results in the formation of a disulfide bond linking the nitroxide label to the cysteine residue in the protein. The labeled protein can then be studied with a variety of EPR or nuclear magnetic resonance (NMR) techniques. In many, but not all cases, the SDSL protein has similar structure, dynamics, and even activity as the native protein. The labeled cysteine is certainly not one of the natural amino acids, but it is an α-amino acid. The SDSL protein is just another mutant protein, although with a very unnatural amino acid. One of the drawbacks of MTSL is that the adduct is susceptible to disulfide exchange. If there are any free —SH groups available, the nitroxide group can be released or exchanged.

10.1.3 Scopes of EPR Studies in Biology

The EPR of biological systems is, in one sense, no different than any other form of EPR. The electron spins interact with their surroundings and with the magnetic and microwave fields in the same way. Yet, a major purpose in EPR measurements on biological systems is to obtain information that is valid in the context of the living organism. This places limits on how the sample can be mistreated and the conditions under which measurements can be made. For instance, the large dielectric losses at microwave frequencies from water are often avoided in chemical applications of EPR through the use of nonaqueous, low-loss solvents. Water is an obligatory component in most biological samples and must be accommodated.

The information desired from biochemical and biophysical EPR measurements is sometimes different than that sought in other applications of EPR. Significant applications of EPR to biological samples include oximetry, which is the quantitative measurement of O_2 activity; measurement of surface potential or solvent accessibility at specific sites in proteins or membranes; accurate measurement of distances in the

1–10 nm range between specific sites in biological structures; measurement of transport properties and dynamics, such as local viscosity, librational amplitudes and frequencies, and diffusion tensors; and determination of the location and orientation of species in the myriad compartments, structures, and organelles that biological systems form.

10.1.4 Spectroscopic Applications

A major biochemical and biophysical use of EPR is for spectroscopic study of free radicals and metal ion centers at intrinsic binding sites of proteins and other biomolecules. Typically the object of such studies is initially the identification and quantitation of those paramagnetic species, followed by determination of their physical and electronic structure, and finally by the characterization of all interactions with their surroundings.

These spectroscopic studies proceed along a path similar to EPR spectroscopic measurements of any other sample. The first step is the measurement of good spectra using appropriate EPR techniques, for example, continuous wave EPR (CW-EPR), High–Field EPR, electron spin echo envelope modulation (ESEEM), electron nuclear double resonance (ENDOR), and so forth. The second step is the extraction of spectral parameters, such as the number of spins, **g** tensors, and hyperfine interactions. The third and final step is the interpretation of those spectral parameters in terms of the chemical, physical, and electronic structure of the paramagnetic species and the interactions with the surroundings, whether that is protein, DNA, lipid, or solvent. This procedure is identical to that for EPR study of simple free radicals and metal ion complexes and is described in detail in other chapters in this book. One aspect in which the EPR of paramagnetic metal centers is somewhat different in biological systems lies in the fact that biomolecules can create a highly specific ligand environment for the metal ion and can assemble complex and unique structures that are quite different from simple chemical species. As a result, there may be no good chemical model for a biological center and no compound with a similar EPR spectrum. A case in point is the Cu(II) in the protein azurin [3, 4]. No simple Cu(II) complex has the same ligands to Cu(II), geometric arrangement, and EPR spectrum. Thus, new and unique types of EPR spectra can be encountered in biological systems because the ligation can be controlled by the protein to a very high degree of precision.

Whenever starting on the study of a new biological system, it is extremely important to search the literature on that organism; that protein, cofactor, and so forth in all other organisms; and the expected free radical or paramagnetic center in any protein, biological, or chemical system. Even distantly related studies can save months of work in developing suitable protocols for sample preparation, conditions for EPR measurement, and approaches for analysis and interpretation. Although each protein is unique, it is nevertheless related to homologous proteins in other organisms and to unrelated proteins carrying out similar functions with the same cofactors or intermediates. In the postgenomic world, it is important to build on what others have done and to generalize findings to a broad range of other species; but this is possible only by knowing the literature.

10.2 EXPERIMENTAL CONSIDERATIONS

Several factors must be considered when contemplating EPR studies of biological systems. Biological systems are complex, and that complexity impacts EPR measurements. Some of the major experimental considerations will be discussed in three subsections that focus on the large dielectric losses from water, the wide range of timescales in the dynamics of the samples, and the heterogeneity of the sample.

10.2.1 Experimental Relevance: Water

Water and ions are components of all biological systems. Yet liquid water has high dielectric losses at microwave frequencies, and losses increase when ions from salts or buffers are present. These dielectric losses (see §2.1.1) cause nonresonant absorption of microwave energy that reduces or destroys EPR sensitivity and heats the sample. A 40-μL aqueous sample with 100 mM salt in a 4-mm EPR tube can absorb virtually all of the microwave power in an X-band EPR cavity. Just 100 mW of microwave power absorbed by the sample will heat it at a rate of $1\,°\text{C s}^{-1}$. EPR spectrometers are not very efficient microwave ovens, but they have boiled more than one aqueous protein sample.

Dielectric losses are much smaller at low frequencies, and EPR measurements have been made below 2 GHz in biological systems to take advantage of the lower losses. Low frequency EPR is not widely used because of its need for larger samples, the scarcity of low frequency spectrometers until recently, lower sensitivity, and lower g-factor dispersion.

The sample shape and position in the resonator have a large influence on nonresonant microwave absorption by the sample. Confining the sample to a microwave electric field node reduces microwave absorption and places the sample at a maximum of the microwave magnetic field where the EPR signal is most intense. A simple example of this strategy is the flat cell for the rectangular EPR cavity. Some sample geometries, such as a bundle of fine capillaries or a flat cell rotated 90° around its axis, also give low nonresonant losses in a rectangular cavity despite having some water in the microwave electric field.

A second strategy for minimizing the losses from water is to freeze the sample. Once the water is frozen and the water molecules completely immobilized, the dielectric losses drop dramatically. Freezing the sample also changes the motional dynamics of the proteins and other molecules in the sample; changes the relaxation times of the paramagnetic species; changes any chemical equilibria or kinetics in the sample; and requires the use of some sort of cryostat inside or around the resonator. Provided that temperatures of about 100K or below are used, the relaxation times generally become longer, potentially increasing spectral resolution for ESEEM and ENDOR if not in the EPR spectrum. Sensitivity often increases because of narrower lines and because of the more favorable Boltzmann factor. For many paramagnetic metal ions, low temperatures are necessary to increase EPR relaxation times to the extent that the EPR lines are narrow enough to be detectable.

Freezing a sample can be traumatic to it. Cells can be ruptured and membranes shredded by formation of ice crystals upon freezing. It is well known that proteins can be partially denatured by freezing. Research groups that freeze proteins for later use often add a "cryoprotectant" to their solutions. These are often molecules like dimethylsulfoxide, ethylene glycol, sugars, or glycerol. Proteins differ considerably in regard to what molecules and concentrations work best for cryoprotection. Biological activity, conformation, and crystallizability can often be maintained by the use of an optimal cryoprotectant. For EPR measurements, a good starting point is the cryoprotectant used by groups freezing that protein for crystallography or enzymology. Freezing aqueous samples is also hazardous to the quartz sample tube because water has a significant volume change near its freezing point. This change can shatter the sample tube either during freezing or on thawing. High concentrations of cryoprotectant can help the solution contract rather than expand on freezing and avoid shattering the tube.

Water tends to crystallize as it freezes. The crystals are generally purer than the solution, so that the proteins, buffers, and salts are squeezed into the interstices between water crystals. The ionic strength, pH, and protein concentration can be quite different from the starting solution. This will often be manifested in EPR measurements as poor resolution, short relaxation times, loss of signal intensity, and sample denaturation. High concentrations of most cryoprotectants prevent the formation of crystalline ice as the sample is cooled and increase the viscosity sufficiently so that the protein or small molecules simply cannot move far from their initial positions during freezing. For many protein solutions, addition of $\sim 30\%$ by volume of glycerol is sufficient to prevent aggregation upon freezing. The glycerol can be added directly or the solution can be dialyzed briefly against pure glycerol. An aqueous sample that freezes into a polycrystalline mass typically is an opaque white or nearly white solid because of the strong scattering of light. As soon as crystal formation is prevented, the sample often is highly cracked so that it is not transparent but appears more translucent and has more of the color of the starting solution. This stage is often adequate for EPR measurements. As the amount of cryoprotectant increases further, there is less cracking until the sample resembles a transparent glass rod with a few cracks in it.

The speed with which a sample is frozen can also affect the EPR spectrum. Proteins, DNA, and membranes are dynamic structures that are always in motion at room temperature and have a variety of closely related structures of similar energy. As a sample is cooled (annealed) and frozen, the higher-energy structures convert into lower-energy structures until the sample temperature drops below the activation energy to change conformation. At that point, the conformations are frozen in and persist through the EPR measurement. Proteins control a significant portion of the ligand fields for the paramagnetic centers they contain. Consequently, any heterogeneity or distribution in protein conformation will produce heterogeneity in the ligand fields that become, in turn, heterogeneity in EPR parameters such as **g** tensors and hyperfine couplings (hfcs). This can produce a few distinct spectra, analogous to site splitting in high-resolution optical spectra, to a continuous distribution of spectral parameters known as g-strain or A-strain. It is not uncommon for a rapidly-frozen sample to have less resolution and broader lines than a slowly-frozen sample of the same starting solution. Unfortunately, slow cooling also encourages crystallization or precipitation of proteins and salts in the sample and should not be carried to extremes.

A final consequence of freezing concerns the effect of temperature on the supposedly "inert" components of the sample. Buffers generally are temperature dependent, and the pH of their solutions changes with temperature. For instance, 100 mM Tris buffer changes about 0.5 pH units between 5 and 25°C. Sodium phosphate buffer is quite nasty in this regard when it freezes. The pH can jump enough to completely denature proteins. Yet, potassium phosphate buffer is free of this problem. Lipids can also undergo dramatic changes with temperature. The phase diagram for a single or binary mixture of lipids and water can be quite complex with many drastic phase changes between room temperature and the freezing point of water. Thus, a sample of well-behaved, planar lipid bilayers at room temperature may change into a three-dimensional gel-like network as the temperature drops, so that a membrane protein that starts out in a synthetic lipid bilayer can be in quite a different environment after freezing.

10.2.2 Experimental Relevance: Dynamics

Biological systems are dynamic and constantly changing with characteristic timescales that range from that of molecular vibrations to the life span of the organism. The dynamic processes are varied and range from the rotation or libration of a protein or the diffusion of lipids in a membrane to the consumption of O_2 in a respiring cell. The consequences of dynamic processes in biological samples are no different from those of dynamics in other applications of EPR. When the dynamic timescale is fast compared to the spin interaction it affects ($\tau_c \cdot \Delta\omega \ll 1$, see Chapter 4), that interaction is averaged out in the EPR measurement; whereas with a slower timescale, that interaction appears in the EPR spectrum. When the dynamic timescale and the interaction are comparable, strong "motional effects" occur.

In most chemical solutions, there are only a few dynamic processes that need to be considered, such as exchange between conformers, molecular rotation, and chemical exchange. Their dynamic timescales are usually well within the fast or slow dynamic regimes. In biological systems, biomolecules and aggregates have a wide range of sizes and consequently a wide range of timescales that often spans the region of strong motional effects. For instance, a free radical in solution typically tumbles rapidly enough that anisotropic hyperfine and Zeeman interactions are averaged, producing an isotropic spectrum with sharp lines and long relaxation times. That same free radical in a biological membrane may undergo rapid rotation around one axis, averaging anisotropy perpendicular to that axis. The free radical will undergo motion about its other axes caused by its own anisotropic rotation in the membrane, the rotation of the membrane itself, and local fluctuations or "ripples" in the membrane. The additional dynamics often include the regime of motional effects and cause a major loss in resolution.

A small variation in sample temperature or composition is usually ineffective in biological samples for addressing dynamic timescale problems. A small change may move one dynamic process out of the motional effects regime, only to be replaced by another dynamic process with similar effects on the spectrum. Freezing the sample to 10K or lower will usually move all dynamic processes into either the fast or slow dynamic regimes but with the full **g** tensor and hyperfine anisotropy appearing in the spectrum and with all the potential problems discussed in the previous subsection associated with freezing.

The impact of dynamics needs to be considered during the interpretation of changes in EPR spectra measured near room temperature. If the EPR signal intensity or spectral shape changes when the sample is altered, changes in dynamics need to be considered as well as changes in the numbers of spins or spin Hamiltonian parameters. However, the very fact that biological systems encompass such a wide and dense range of dynamic processes means that EPR can measure quite a range of dynamic properties in biological systems. Some of these will be described in 10.3.

10.2.3 Experimental Relevance: Heterogeneity

Biological samples have several sources of significant heterogeneity, both from sample to sample and within the same sample. The heterogeneity arises from impurities, variations in the protein or biomolecule, differences in conformation or cofactors, and differences in environment. Sample heterogeneity may be obvious from the EPR spectrum or it may have no impact at all, depending on the type of heterogeneity and EPR measurement.

Biological samples rarely reach the purity of a typical chemical. A "purified" protein may be 90% of the desired protein and 10% other proteins. Fortunately, EPR is exquisitely selective for paramagnetic species, so that large amounts of diamagnetic proteins will not affect many EPR measurements if those proteins do not interact with the protein of interest. For instance, the photosynthetic reaction center from photosynthetic bacteria has been studied extensively by EPR in crude chromatophore preparations and even whole cells because it is the only protein with a light-inducible free radical with a large quantum yield. In contrast, a minor flavoprotein impurity can produce the most intense EPR line in a cytochrome sample because the differences in the widths of the respective spectra (10 vs. 10^3 G) are greater than their relative numbers.

Proteins, DNA, and other biomolecules can be heterogeneous because of mutations, degradation, or modification. Different strains or isolates of the same organism can have slightly different copies of a gene. During isolation or purification, chemicals and enzymes, such as proteases and nucleases, can partially degrade proteins, DNA, and other biomolecules, producing a mixture of species derived from the biomolecule of interest. Cells can extensively modify biomolecules over their lifetime. Posttranslational modification of proteins can involve removal of residues targeting the protein to specific organelles; methylation, phosphorylation, acetylation, and glycosylation; modification of specific residues, as in the green fluorescent protein; and covalent attachment of cofactors such as *c*-type hemes.

Proteins, DNA, and other large biomolecules are in a constant equilibrium involving different conformations; the binding of ions, cofactors, and small molecules; and the binding of other proteins. The timescales for these equilibria can extend from the fast dynamic limit to the slow, resulting, in an EPR measurement, in several distinct spectra, dynamic broadening, or in the *g*-strain and hyperfine strain mentioned in the previous subsection. Many of these equilibria can be shifted by adjusting the pH or concentrations of the various species. However, solubility and protein stability restrict the extent to which equilibria can be shifted and it is by no means certain that a single species in the slow dynamic limit can be obtained in the sample.

The environment or surroundings of the paramagnetic center under study is yet another source of heterogeneity for EPR measurements. Environmental differences are more of a concern in highly-ordered or structured samples, such as whole cells. The compositions, ionic strength, and pH can be quite different in the cytoplasmic, periplasmic, or extracellular regions and affect measurements on species in more than one of those environments. Cellular, nuclear, mitochondrial, and other membranes differ from each other in terms of thickness, fluidity, composition, charge, and polarization. Even in the same membrane, the composition can change from one region to another, between folded and flat regions and from one side to the other. Even simple detergent-solubilized proteins differ in terms of the number of associated detergent and lipid molecules.

This brief discussion of experimental considerations is by no means exhaustive. In many measurements, they will have no impact on the results or interpretation. However, the discussion suggests a few of the reasons why one sample may have more than a single EPR spectrum, why two samples may have different spectra, and why two labs may report different results. Consistency and accuracy in sample preparation are quite important any time that two samples will be compared. Hypotheses are often posed and tested with respect to an "ideal" sample that is pure and homogeneous. It is necessary to recognize and deal with the full complexity and heterogeneity of "real" experimental samples before rejecting a hypothesis. This discussion of heterogeneity is relevant not only to EPR spectroscopic measurements mentioned earlier in this section but also to measurements of dynamics and distances in the following sections.

10.3 DYNAMICS

Several types of dynamics can be measured in biological samples. Perhaps the most familiar is chemical reaction kinetics. For the most part, the same experimental EPR techniques are applied to biological systems and to chemical systems. These include measurements of chemical exchange from line broadening and relaxation and measurements of chemical kinetics from the change in EPR intensity with time. This latter type of kinetic measurement is made by initiating the reaction by mixing reagents, by light excitation, by ionizing radiation, and by temperature jumps and then following the time course by monitoring the CW or pulsed EPR signal in real time either by chemical quenching or by freeze-quenching techniques. Such measurements are extensions of standard techniques used on chemical systems and will not be discussed in this section. The focus here is on application of EPR to dynamics that are strongly shaped by the biochemistry and biophysics.

One strong theme that has run through the field of biophysical EPR is the desire to know and understand the motion of proteins and the transport properties of membranes. The cell's membranes both define the cell and control how it interacts and responds to its environment and to other cells. The membrane has a significant amount of protein embedded or associated with it. Those proteins, along with small molecules such as cholesterol, help control the fluidity and flexibility of the membrane. The membrane, in turn, controls what surface of the protein is presented to the interior and exterior of the cell; how the protein rotates and diffuses in the membrane; and how small cofactors, such as lipophilic quinones, approach and leave the protein.

10.3.1 Dynamics and Nitroxide Spin Probes

Nitroxide radicals were recognized rather quickly as being well suited for the study of motion in biological systems [1]. Both the **g** tensor of the electron spin and the nitrogen hyperfine tensor are anisotropic. That is, both the center of the EPR spectrum and the splitting of the spectrum into three lines depend on the orientation of the nitroxide radical relative to the magnetic field of the spectrometer. As the nitroxide rotates, the spectrum changes and the correlation time τ_c for the rotation can be extracted from the spectrum. As nitroxides were first introduced into lipid bilayers as models of biological membranes, it became apparent that the rotation in such an anisotropic medium was itself quite anisotropic and could be characterized by a rotational diffusion tensor that depended on the shape of the nitroxide molecule and on the properties of the bilayer. EPR studies of nitroxides made significant contributions to the understanding of ordered phases, such as bilayers and liquid crystals, and helped bring about the high-performance liquid-crystal displays for computers and flat panel televisions. Our understanding of the motion of proteins and properties of membranes has advanced as a result of EPR studies. The questions investigated by EPR have become more complex and subtle as EPR techniques improved and as the appreciation of the complexities of proteins and membranes developed. The evolution of nitroxide EPR for the study of proteins and membranes is chronicled over the years in the *Biological Magnetic Resonance Series* edited by Berliner and colleagues (see Further Reading).

10.3.1.1 Dynamics and Nitroxide Spin Probes: Line Shapes. Nitroxide spin labels were covalently attached to proteins in order to study protein motion as soon as they became available. The dynamics of the nitroxide were greatly altered by the protein, but it soon became clear that the nitroxide was not reporting the dynamics of the protein as a whole. The spin label actually undergoes three types of motion when attached to the protein. The first is the motion of the nitroxide about the chemical bonds linking it to the amino acid residue (usually cysteine or lysine). This internal motion of the label is restricted by the protein, depending on the extent to which the protein engulfs the label. The second component of label dynamics is the motion of the site of attachment relative to the rest of the protein. This is an internal motion or conformational flexibility of the protein. The third component of dynamics is the motion of the protein as a whole. When spin-labeled proteins are cross-linked or immobilized in a solid matrix so that they are unable to rotate, the EPR spectrum usually shows considerable motion of the spin label because of the internal motion of the label and conformational flexibility of the protein.

These three dynamic components greatly complicate the analysis of the spectrum. The internal motion of the spin label is usually in the rapid dynamic limit and averages some (but not all if the internal motion is not isotropic) of the anisotropic Zeeman and hyperfine interactions. The conformational flexibility of the protein is often in the rapid to intermediate dynamic range and produces varying amounts of line broadening. The rotation of the protein, which was the original object of such studies, is the slowest and often falls into the intermediate or slow dynamic regime. As a result, the spectra usually have features indicative of both fast and slow motions. A spectrum often can be approximated by a simple motional model to provide some apparent rotational correlation time. However, when spectra measured at different microwave frequencies, for example, X- and Q-band, are examined, quite different rotational correlation times are obtained. It was clear that a CW-EPR spectrum, measured at a single EPR frequency, lacked the detail to allow complete description of the spin label motion.

Similar measurements were attempted in membranes when the first nitroxides became available. Nitroxide spin probes in bilayers or vesicles made from pure lipid showed partial alignment and anisotropic rotation of the nitroxide probe that strongly depended on the temperature and lipid composition. The dynamics of the nitroxide probe were reflecting the anisotropic environment of the lipid. The phase transitions of the lipids were clearly seen through the linewidths, line shapes, and hyperfine splittings in the CW-EPR spectrum. Studies of nitroxides in biological membranes showed similar behavior, but the phase transitions were less distinct and more gradual because of the many components that make up biological membranes.

Membrane measurements suffered from the same challenges as spin-labeled protein measurements [5]. There is rapid anisotropic motion of the nitroxide spin probe in the potential of the anisotropic "solvent cage" consisting of the membrane. The "cage" and the potential field it creates then changes on a slower timescale as the result of motion of the membrane constituents. In addition, there is the slower motion of the membrane as a whole.

It takes a complex model of the dynamics to accurately simulate EPR spectra at several EPR frequencies for labeled proteins or probes in membranes. In the early 1970s the capabilities for numerical simulations and for multifrequency EPR were not sufficient for making reliable studies of protein or membrane dynamics except for a limited range of very favorable conditions. As a result, nonlinear forms of CW-EPR spectroscopy were used to obtain spectra that were mainly sensitive to the slower motions of proteins and membranes.

10.4 SATURATION TRANSFER

The quest for an EPR method to measure the slow rotational dynamics of proteins and membranes led to the development of a method that Hyde and Dalton named saturation-transfer EPR [6]. The concept of saturation-transfer EPR is quite elegant and insightful. It relies on the fact that the T_1 of a nitroxide label or probe is within a couple of orders of magnitude of the rotational correlation time for slow motion in proteins and membranes. Consequently, if a spin packet in the EPR spectrum is saturated by high microwave power, rotation of the protein will result in spectral diffusion (because of the anisotropic hyperfine and Zeeman interactions) that spreads or transfers that saturation to other parts of the spectrum. The electron T_1 determines how broadly rotational diffusion will spread the saturated spin packet and how long that saturation will persist. The magnetic field modulation used to record the CW-EPR spectrum allows the spectrometer to probe how far that saturation spreads through the spectrum. The saturation transfer appears in the amplitude and phase of the magnetic field modulated EPR spectrum.

Saturation-transfer EPR produces a family of spectra measured in absorption or dispersion mode, in-phase or out-of-phase with the magnetic field modulation, and depending strongly on the microwave power, modulation frequency, and amplitude. Saturation-transfer spectra are relatively easy to measure. They are primarily determined by the slowest motion of a protein and are quite sensitive to small changes in the motion.

10.5 TWO-DIMENSIONAL PULSED EPR

In the 1980s pulsed EPR technology reached the point that spin-labeled proteins and membranes could be measured routinely. It was also realized that the fast motions of nitroxides contribute much less to electron and nuclear relaxation than does slow motion or protein rotation. This prompted Freed and coworkers [7] (see Further Reading) to examine the two-pulse electron spin echo decay at a series of points across the EPR spectrum. The echo decay is caused by spectral diffusion resulting from rotational motion of the nitroxide. The echo decay varies across the spectrum because of anisotropy of the Zeeman and hyperfine interactions coupled to the possibly anisotropic rotation. These decays were presented as a function of the interpulse delay time t and the position in the EPR spectrum in a two-dimensional surface or a contour plot. Such presentations are quite effective at revealing anisotropic motion and at picking out subtle changes in the dynamics of proteins and membranes.

This pulsed EPR approach from the Freed group has grown into a family of pulse sequences that highlight different types of motion. For example, one variant measures the collision rate between nitroxides in membranes based on the transfer of magnetization from one radical to another. Those measurements probe the lateral diffusion rate of the nitroxide spin probes in the membranes and provide a view into membrane fluidity that is complementary to that provided by rotational diffusion measurements. These measurements are extremely demanding and require large bandwidths and high microwave powers in order to achieve uniform coverage in both excitation and detection.

10.6 PROTEIN TOPOLOGY AND SDSL

Although the initial focus of EPR studies of spin-labeled proteins was the overall motion of the protein, there was still a lot to be learned from the more localized motions of the spin label. It was quickly evident that labels attached to amino acid residues on the surface of a protein were able to move rather freely and quite rapidly, resulting in sharp, motionally-narrowed EPR spectra. Labels attached to residues buried inside the protein have broad spectra resulting from motion that is strongly hindered by the protein. These differences are easily seen in CW-EPR spectra and provide some information about the location of a particular amino acid in the folded protein. Changes in the CW-EPR spectrum were also useful in following the folding and unfolding of a protein, denaturation, or the formation of protein complexes.

A real quantum leap was made by Hubbell and Khorana [8] in their EPR studies on rhodopsin. They began to systematically move a cysteine residue along rhodopsin by making site-directed mutants at each position. After expressing, purifying, and spin labeling each of these mutants, they measured the CW-EPR spectrum. Rhodopsin is a membrane protein with a bundle of membrane-spanning helices linked by short flexible loops. The EPR spectra showed which amino acids were on the surface of the protein exposed to water or lipid, which were partially exposed and completely buried in the protein. The massive amount of data they collected allowed them to see periodic changes in mobility that corresponded to the pitch of the protein α helix, which side of the helix faced out toward the membrane and which side faced other helices, and where the connecting loops emerged from the membrane and where they reentered.

Subsequent advances in molecular biology and the MTSL spin label have made SDSL routine. Systematic studies on several proteins, particularly lysozyme, have established several empirical correlations between secondary protein structure and the EPR spectrum of the label, so that secondary and tertiary structural features can be determined by systematic SDSL studies. An important aspect of this SDSL approach to protein structure is that the protein can be studied under a wide range of conditions, unlike structural studies using diffraction or NMR where the conditions required for crystallization or NMR measurement may demand unrealistic temperatures, concentrations, counterions, pH, or lipids. EPR/SDSL structural studies can be made with the protein poised in different physiological states throughout its catalytic or functional cycle and the structural changes in progressing from step to step followed. This is being done quite elegantly with rhodopsin.

10.7 SURFACE POTENTIALS/ACCESSIBILITY AND SDSL

Another application of SDSL is the measurement of the surface potential of proteins. It is an outgrowth of the use of paramagnetic broadening agents to shorten the relaxation times of a nitroxide, typically by broadening the EPR spectrum of the nitroxide. High concentrations of a water-soluble or membrane-soluble paramagnetic species whose spectra do not interfere with the nitroxide are added to a sample. If the nitroxide is accessible to the added paramagnetic species (sometimes referred to as a relaxing or broadening agent), the EPR spectrum will broaden and relax more quickly. Chromium(III) oxalate is frequently used as an aqueous-phase broadening agent and O_2 for lipid phases.

The extent of broadening depends on how close the broadening agent approaches the nitroxide; on the concentration, magnetic properties, and diffusion constant of the broadening agent; and on the viscosity around the broadening agent. The broadening depends on so many different parameters that most measurements are primarily qualitative, for example, one residue of the protein is more exposed to the aqueous phase than another, or the diffusion of O_2 in a lipid bilayer increases with the addition of cholesterol. One major application is to distinguish the inside of a cell or vesicle from the outside. If a broadening agent that cannot cross the cell membrane is added to a cell suspension, it is often possible to broaden the extracellular nitroxide beyond detectability and thus only measure intracellular nitroxide or spin trap adduct.

Broadening agents have been used to make quantitative measurements of surface potential by normalizing against a suitable reference broadening agent [9]. If two reagents have the same diffusion coefficient, relaxivity, and size but differ in charge, then the ratio of the relaxation rate changes they cause in a spin label attached to a protein measures their chemical activities or, equivalently, the surface potential of the protein at the spin label. The potential of the entire protein can be measured by scanning the spin label through the protein because the variations that are caused, for example, by changes in the exposed surface area of the spin label at the different positions, are effectively removed by taking ratios.

10.8 OXIMETRY

Oxygen plays a central role in most living organisms and the concentration, or more accurately, the activity of O_2 is important in physiology and pathology. The measurement of O_2 is problematic. Most oximetric methods consume O_2, destroy the sample, and/or are difficult to calibrate. The advantages of EPR oximetry lie in being relatively noninvasive, sensitive at physiological levels, and easily calibrated.

EPR oximetry is based on the broadening of EPR spectra by O_2. For most applications of EPR, O_2 is a nuisance because it is paramagnetic and broadens and relaxes any electron spins with which it collies. To achieve maximum resolution in liquids or some solids, it is necessary to rigorously exclude O_2 during measurement. The O_2 is a ground state triplet molecule with a very short T_1, and virtually every collision between O_2 and a free radical results in electron spin relaxation of the free radical. The O_2 induced increase in linewidth gives the rate at which O_2 collides with the free radical. The rate of O_2 consumption by 100 cells was measured in an early EPR oximetry study by measuring the EPR linewidth of a concentrated nitroxide solution decrease as the O_2 was consumed [10]. That measurement was far from ideal because the high concentration of nitroxide had the potential to interfere with normal physiology and was not easily extended to measurements *in vivo*.

Hyde and Subczynski [11] realized that EPR oximetry was possible under tissue culture conditions or even *in vivo* using low frequency EPR to reduce dielectric losses, low concentrations of nitroxide to minimize its effect on physiology, and the partially resolved proton hyperfine splittings in the CW-EPR spectrum as an internal standard against which to measure the O_2 induced broadening (see Further Reading). EPR oximetry has developed rapidly on several fronts. There are now a number of spin probes for use as the EPR active media for oximetry, including soluble nitroxides and trityl radicals and insoluble carbon chars, lithium phthalocyanine, and India ink. EPR instrumentation for oximetry extends down to almost 200 MHz and includes CW and pulsed EPR detection and even three-dimensional spatial imaging of the O_2 concentration.

In EPR oximetry, the spin packet linewidth, $1/T_2$, or their equivalent is determined in the presence and absence of O_2. The increase in $1/T_2$ caused by O_2, that is, $1/T_2(O_2) - 1/T_2(0)$, is equal to $k[O_2]$. Because O_2 diffuses fairly rapidly in aqueous media, the constant k is large enough to more than double the spin packet linewidth at physiological O_2 concentrations with some of the free radical probes used for EPR oximetry. EPR oximetry is a rapidly advancing field at present because of its potential use in clinical medicine [12]. A number of groups are developing instrumentation and protocols to determine the value of rapid tissue oximetry in guiding medical treatment.

10.9 NANOSCALE DISTANCE MEASUREMENT

The measurement of nanoscale distances between 1 and 10 nm has always been difficult, particularly in noncrystalline materials. The interaction between unpaired electron spins is appreciable for such distances. A recent convergence between pulsed EPR spectroscopy and molecular biology now provides an accurate method for measuring distances in the 1 – 10 nm range.

EPR spectroscopy in solids quickly found that spectra were severely broadened by high concentrations of radicals or paramagnetic centers. The origin of the broadening comes from two types of interactions between radicals. One is the exchange interaction $JS_1 \cdot S_2$, mediated by overlap between the wavefunctions of the two unpaired electron spins. Beyond a few tenths of a nanometer, the exchange interaction falls off exponentially with a characteristic length on the order of 0.1 nm, so that for two free radicals separated by 1 nm, the exchange interaction is usually negligible. However, in some enzymes with multiple paramagnetic centers, extended conjugation between paramagnetic centers can result in unexpectedly large exchange interactions. The second and more common interaction between unpaired spins is the magnetic dipolar interaction, which contains a term

$$\frac{\gamma_e^2 \hbar^2}{r^3} \left(S_{Z1} S_{Z2} - \tfrac{1}{4}(S_1^+ S_2^- + S_1^- S_2^+) \right)(1 - 3\cos^2 \theta)$$

that shifts the EPR resonance and that converges to a more familiar form

$$\frac{\gamma_e^2 \hbar^2}{r^3} S_{Z1} S_{Z2}(1 - 3\cos^2 \theta)$$

when the dipolar interaction is smaller than the difference in resonant frequencies of the isolated spins. This interaction amounts to a splitting of 81.32 MHz at 1 nm separation between free radicals. Other terms in the dipolar interaction cause spin relaxation and have a similar r^{-3} or r^{-6} distance dependence.

10.9.1 Nanoscale Distance Measurement: Relaxation

Kevan and Voevodsky each independently proposed in the 1960s that EPR relaxation be used to measure the distribution of free radicals in radiation-produced tracks and spurs. Voevodsky [13] proposed the use of electron spin echo spectroscopy to directly measure the relaxation times of the ionizing radiation produced radicals, whereas Kevan [14, 15] used measurements of EPR in the rotating frame (now called multiquantum EPR) to determine something akin to the T_{1r} of NMR. Kevan and Tsvetkov (following Voevodsky's death) had some success with their approaches, but the relatively crude state of instrumentation and computer technology limited the quality of data that could be obtained and the sophistication with which it could be processed and analyzed.

The Likhtenshtein group pursued the use of CW-EPR relaxation and line broadening measurements for determination of distances between centers in several proteins during the 1970–1980s with some notable success [16], whereas the Norris group used electron spin echo measurements of T_1 to probe the distance between Fe and the special pair in photosynthetic reaction centers [17]. These approaches have been extended by several groups over the years, with varying degrees of reported success, depending on how closely the outcomes agreed with the current models.

The use of relaxation or line broadening to determine distance involves simple steps.

1. The spectrum or relaxation rate of one center in the protein or biological sample is measured (often as a function of temperature).

2. That portion of the linewidth or relaxation that is caused by center B in the protein is extracted. Ideally, center B would be removed or converted into a diamagnetic form without affecting center A in any way, so that the difference in the measured linewidths or relaxation of center A is caused only by dipole–dipole interactions between the two centers and not by the sample treatment.

3. One or more of the six terms in the dipole–dipole interaction between centers A and B are selected as being responsible for the linewidth or relaxation difference.

4. The line broadening or relaxation caused by center B is calculated (using the appropriate T_1 and T_2 for center B) for different values of the distance between centers A and B and compared to the measurements. The dipole–dipole interaction is anisotropic, which makes the calculated relaxation anisotropic and almost always nonexponential, so that simple comparisons with experimental results are difficult, but nevertheless made.

Three problems may be lurking in this simple distance determination recipe.

1. Center B is often a metal center with an anisotropic **g** tensor. The effect of such an anisotropic metal then depends on distance, on the principal values of the **g** tensor, and on the three Euler angles relating the vector between centers A and B to the **g** tensor axes.
2. Center B is often a fast relaxing center with T_1 and T_2 poorly known and possibly anisotropic. Selecting the correct terms in the dipole–dipole interaction and then calculating their contribution to the relaxation of center A is problematic.
3. The formulae for calculating the linewidth or relaxation contribution of center B on center A are available in the absence of zero-field splittings. However, center B, in many important cases, is an $S > 1/2$ ion whose spin eigenfunctions are not even close to those of the Zeeman interaction. Consequently, it is impossible to evaluate the terms in the dipole–dipole interaction producing the linewidth or relaxation changes without first determining the complete spin Hamiltonian and eigenfunctions for center B at all orientations. It does not appear to be correct to approximate an $S > 1/2$ B spin as an effective $S = 1/2$ for dipolar interactions.

For many systems, several of the parameters needed to extract a distance are so poorly known that there is the danger of selecting an incorrect set of parameters that predicts a distance that agrees quite well with one's expectations unless rigorous care is taken.

10.9.2 Nanoscale Distance Measurement: Line Broadening

The development of SDSL methods makes it possible to perform very careful and controlled studies of line broadening with doubly spin-labeled proteins. SDSL is used to express three spin-labeled proteins: a protein doubly labeled at two selected sites and two proteins singly labeled at each of those sites. Assuming that the mutations and the labels have not caused major changes in the protein structure, the CW-EPR spectrum of the doubly labeled sample will be equal to the sum of the spectra from the other two proteins plus the broadening caused by spin–spin interactions between the two labels.

In the limit that the spin–spin interactions are weak and that the protein does not strongly orient the spin labels, the EPR spectrum is uniformly broadened and the broadening function can be obtained by deconvoluting the doubly-labeled spectrum by the sum of the singly-labeled spectra. The broadening function is composed of a distribution of Pake-like doublets modified by any exchange interaction. This broadening function is equivalent to the Fourier conjugate of the double electron–electron resonance (DEER) signal and can be analyzed using any of the approaches used for DEER.

This line broadening approach to distance measurement relies on accurate deconvolution of the EPR spectra and thus works best when the line broadening causes substantial changes in the appearance of the sharpest spectral features, which is distances of less than ~ 2 nm. Fortunately, this is the regime where DEER becomes complicated so the two approaches are complementary. The deconvolution also requires spectra with excellent signal to noise ratios and flat baselines. Analysis of the broadening function in terms of a distribution of distances is the same ill-conditioned problem faced by DEER. The use of line broadening for distance determination is limited to a rather narrow window of about 1–2 nm.

10.9.3 Nanoscale Distance Measurement: DEER

One of the most significant recent developments in biochemical and biophysical EPR is DEER [also known as pulsed electron–electron double resonance (PELDOR)] for the measurement of distances and distance distributions. DEER is the analog of spin echo double resonance techniques in NMR but has a very different development. The evolution of pulsed EPR techniques that eventually produced DEER began back in the early 1960s with the study of phase memory relaxation in ionic crystals at low temperature. It was noticed that the two-pulse electron spin echo decay depended on the concentration of paramagnetic ions in the crystal and on the amplitude and width of the second microwave pulse.

The effect was explained by Klauder and Anderson [18] as due to the change in the dipole field of the spins being observed (called the "A" spins) that was caused by the flipping of other spins (called the "B" spins) by the second microwave pulse. In short, there was spectral diffusion occurring instantaneously with the second pulse, resulting in incomplete refocusing of the echo. As the time between the pulses increases, the impact of instantaneous spectral diffusion increases and the echo decreases as mentioned in Chapter 5. The decay caused by this "instantaneous diffusion" depends on the concentration of spins (i.e., dipolar field) and the turning angle of the second microwave pulse and has been used over the years as an accurate measure of the local dipolar field. Its major drawback is that it measures the root mean square average dipolar interaction, making it difficult, if not impossible, to extract a dominant dipolar interaction.

The next step toward DEER was the report in 1969 of ESEEM in biradicals caused by the exchange and dipolar interaction between the spins in the biradical [19]. This modulation was perhaps closer to double quantum EPR measurements than to DEER, but it demonstrated that distance information did appear in pulsed EPR. Recovering that information from the two-pulse echo signal is difficult because the modulation from the dipole–dipole interaction is mixed together with the ESEEM from weakly interacting nuclei and with the decay from instantaneous diffusion and phase memory relaxation.

In 1984 Milov, Ponomarev, and Tsvetkov applied a new PELDOR sequence to a biradical [20, 21], observing only the ESEEM caused by a combination of exchange and dipolar interactions. A two-pulse electron spin echo sequence with fixed separation τ between pulses was used to generate an echo. Another pulse at a different frequency was swept in time between the two pulses, producing a periodic modulation of the echo intensity. If we speak of the biradical as two weakly interacting radicals (Klauder and Anderson's A and B spins), that additional pulse flipped the electron spin of the partner (B spin) of the radical (A spin), producing the detected echo signal. The advantage of this sequence, which is also known as the three-pulse DEER sequence, is that with τ held constant, the nuclear ESEEM and phase memory decay do not appear in the signal as the position of the additional pulse is swept. This sequence was used by Larsen and Singel [22] in 1993 to measure the dipolar interaction in a weakly coupled nitroxide biradical and to determine the distance between the two nitroxide groups. They proposed this method for measuring the separation distance and relative orientations of the nitroxide groups.

A variant of this pulsed ELDOR sequence was described by Kurshev et al. [23] in 1989, in which all three pulses have the same frequency. This was named the "2 + 1" pulse sequence to differentiate it from the three-pulse stimulated echo sequence that it resembles. By careful selection of the widths and intensities of the pulses, the dipolar interaction between spins can be measured. Raitsimring et al. [24] used the 2 + 1 sequence in 1992 to measure a distance of 3.5 nm between two nitroxide spin labels attached to hemoglobin at cysteines separated by 3.4 nm. The 2 + 1 sequence has not had much use in distance measurement because of the careful adjustment of pulse widths needed for optimal results.

The first application of DEER for distance measurement was in photosynthetic proteins [25] that contain a number of paramagnetic species, both free radicals and metal centers. The paramagnetic centers in the photosynthetic proteins are highly ordered and highly oriented and make ideal subjects for distance determinations because the distribution of distances is basically a δ-function. Those initial measurements used a free radical as one center and a metal center with large anisotropy as the other center. Consequently, there was a high degree of orientation selection involved with the metal center and only part of the Pake pattern was measured. It was still possible to analyze the data to determine distances and some orientation information.

By the late 1990s, the Tsvetkov group had established a good theoretical and experimental understanding of DEER on model systems and started to apply DEER to biological molecules [26, 28]. They studied a group of small peptide-like antibiotics called peptiabols. These molecules were small enough that they could be chemically synthesized using automated protein synthesis techniques. This allowed the incorporation of 2,2,6,6-tetramethyl-piperidine-1-oxyl-4-amino-4-carboxyl (TOAC) spin labels that are an analog of the natural but rare amino butyric acid found in these antibiotics. The nitroxide group in TOAC is firmly anchored to the peptiabol backbone, resulting in a narrow distribution of distances from the restricted range of conformations. The spin-labeled peptiabols were studied using DEER in model membranes and a range of solvents. Aggregates and helical structures were found, depending on the environment and concentration, and detailed structures based on DEER data and molecular modeling were proposed. This success and that of double-quantum EPR [29] inspired other groups to use DEER to determine distances in spin-labeled proteins, RNAs, and DNAs.

The adoption of DEER methods was advanced by two major developments. One was the introduction by Gunnar Jeschke of an improved four-pulse DEER sequence [30] that largely eliminated the dead time and missing data associated with the initial three-pulse DEER sequence and that really made it possible to obtain the distribution of distances in a sample. The second advance was the development of methods to analyze the DEER decays in terms of distributions of distances [31–35].

Most DEER measurements on spin-labeled proteins produce modulation that decays much faster than expected. That decay results from the interference of the modulation from a distribution of distances and was recognized in the original measurement by Raitsimring et al. [24]. Because the modulation pattern at any distance is readily calculated, it should be possible to reconstruct the distance distribution function. However, this turns out to be one of those ill-conditioned mathematical problems. There is literally an infinite number of distance distributions that will give the same experimental DEER signal to within the experimental noise level. The question is how to pick a reasonable or most probable distribution.

One approach is to assume a particular distribution function, such as a Gaussian distribution of distances or one based on molecular modeling of nitroxide conformations. Then a small number of parameters, for example, mean distance and standard deviation, are varied until the best fit of the predicted DEER signal to the experimental signal is obtained. The results depend, of course, on how accurate the selected distribution model is. The alternate approach is to regularize the ill-conditioned problem in some way. Various criteria have been used with excellent results, for example, distributions should be nonnegative or smooth or have maximum entropy.

DEER is still a rapidly developing technique in both data acquisition and data analysis. X-band measurements have almost become routine for doubly-labeled samples between 2 and 8 nm with a distance distribution narrower than about 0.5 nm. It appears that DEER measurements of distances should be possible for samples with four to six labels, although analysis will be much more difficult as the number of labels increases. DEER at Q-band seems to have some advantages relative to X-band, but it has not been extensively explored. DEER at W-band and higher fields seems to have inherent difficulties arising from orientation selection produced by the relatively weak (compared to the anisotropy) microwave fields. At one extreme are samples with highly oriented labels, so that each measurement observes only a limited set of orientations of the protein. This situation is similar to the measurements made at X-band in the early 1990s on photosynthetic proteins. Analysis is extremely difficult, but it yields considerable detail about the distance and the relative orientations of the labels. At the other extreme are labels that are totally unoriented by the protein yet separated by a well-defined distribution of distances. In that case, there are no orientation selection problems and the standard X-band analysis should be applicable. The difficulty is that, with all combinations of pump and observed frequencies, the amplitude of the DEER modulation is small relative to the total echo, making sensitivity an issue. Between these two extremes only time and the scientific literature will reveal what can be achieved.

REFERENCES

1. *Spin Labeling Theory and Applications*; Berliner, L. J., Ed.; Academic Press: New York, 1976.
2. Hubbell, W. L.; McHaourab, H. S.; Altenbach, C.; Lietzow, M. A. *Structure* **1996**, *4*, 779–783.
3. Coremans, J. W. A.; van Gastel, M.; Poluektov, O. G.; Groenen, E. J. J.; den Blaauwen, T.; van Pouderoyen, G.; Canters, G. W.; Nar, H.; Hammann, C.; Messerschmidt, A. *Chem. Phys. Lett.* **1995**, *235*, 202–210.
4. Goldfarb, D.; Arieli, D. *Annual Review of Biophysics and Biomolecular Structure* **2004**, *33*, 441–468.
5. Barnes, J. P.; Liang, Z.; McHaourab, H. S.; Freed, J. H.; Hubbell, W. L. *Biophys J* **1999**, *76*, 3298–3306.
6. Hyde, J. S.; Thomas, D. D. *Annual Review of Physical Chemistry* **1980**, *31*, 293–317.
7. Gorcester, J.; Millhauser, G. L.; Freed, J. H. *Two-Dimensional Electron Spin Resonance* In *Modern Pulsed and Continuous-Wave Electron Spin Resonance*; Kevan, L., Bowman, M. K., Eds.; Wiley: New York, 1990; Vol. 1st, pp. 119–194.
8. Hubbell, W. L.; Altenbach, C.; Hubbell, C. M.; Khorana, H. G. *Advances in Protein Chemistry* **2003**, *63*, 243–290.
9. Franklin, J. C.; Cafiso, D. S.; Flewelling, R. F.; Hubbell, W. L. *Biophys J* **1993**, *64*, 642–653.
10. Backer, J. M.; Budker, V. G.; Eremenko, S. I.; Molin, Y. N. *Biochim Biophys Acta* **1977**, *460*, 152–156.
11. Subczynski, W. K.; Hyde, J. S. *Biophys. J.* **1983**, *41*, 283–286.
12. *Biomedical EPR, Part A: FreeRadicals, Metals, Medicine, and Physiology*; Eaton, S. S.; Eaton, G. R.; Berliner, L. J., Eds.; Kluwer Academic: New York, 2005.
13. Tsvetkov, Y. D.; Raitsimring, A. M.; Zhidomirov, G. M.; Salikhov, K. M.; Voevodsky, V. V. *High Energy Chemistry* **1968**, *2*, 529–535.
14. Zimbrick, J.; Kevan, L. *The Journal of Chemical Physics* **1967**, *47*, 2364–2371.
15. Zimbrick, J.; Kevan, L. *The Journal of Chemical Physics* **1967**, *47*, 5000–5008.
16. Likhtenshtein, G. I. *Depth of Immersion of Paramagnetic Centers in Biological Systems* In *Distance Measurement in Biological Systems by EPR*; Berliner, L. J., Eaton, G. R., Eaton, S. S., Eds.; Kluwer Academic: New York, 2000, pp. 309–345.
17. Norris, J. R.; Thurnauer, M. C.; Bowman, M. K. *Adv Biol Med Phys* **1980**, *17*, 365–416.
18. Klauder, J. R.; Anderson, P. W. *Physical Review* **1962**, *125*, 912–932.
19. Yudanov, V. F.; Salikhov, K. M.; Zhidomirov, G. M.; Tsvetkov, Y. D. *Theoretical and Experimental Chemistry* **1969**, *5*, 663–668.
20. Milov, A. D.; Ponomarev, A. B.; Tsvetkov, Y. D. *Zhurnal Strukturnoi Khimii* **1984**, *25*, 710–713.
21. Milov, A. D.; Ponomarev, A. B.; Tsvetkov, Y. D. *Chemical Physics Letters* **1984**, *110*, 67–72.
22. Larsen, R. G.; Singel, D. J. *Journal of Chemical Physics* **1993**, *98*, 5134–5146.
23. Kurshev, V. V.; Raitsimring, A. M.; Tsvetkov, Y. D. *Journal of Magnetic Resonance* **1989**, *81*, 441–454.

24. Raitsimring, A.; Peisach, J.; Lee, H. C.; Chen, X. *Journal of Physical Chemistry* **1992**, *96*, 3526–3531.
25. Hara, H.; Kawamori, A.; Astashkin, A. V.; Ono, T. *Biochimica et Biophysica Acta-Bioenergetics* **1996**, *1276*, 140–146.
26. Milov, A. D.; Maryasov, A. G.; Tsvetkov, Y. D. *Applied Magnetic Resonance* **1998**, *15*, 107–143.
27. Milov, A. D.; Tsvetkov, Y. D.; Formaggio, F.; Crisma, M.; Toniolo, C.; Raap, J. *J Am Chem Soc* **2001**, *123*, 3784–3789.
28. Maryasov, A. G.; Tsvetkov, Y. D.; Raap, J. *Applied Magnetic Resonance* **1998**, *14*, 101–113.
29. Saxena, S.; Freed, J. H. *Journal of Chemical Physics* **1997**, *107*, 1317–1340.
30. Pannier, M.; Veit, S.; Godt, A.; Jeschke, G.; Spiess, H. W. *Journal of Magnetic Resonance* **2000**, *142*, 331–340.
31. Jeschke, G.; Panek, G.; Godt, A.; Bender, A.; Paulsen, H. *Applied Magnetic Resonance* **2004**, *26*, 223–244.
32. Jeschke, G.; Koch, A.; Jonas, U.; Godt, A. *Journal of Magnetic Resonance* **2002**, *155*, 72–82.
33. Bowman, M. K.; Maryasov, A. G.; Kim, N. K.; DeRose, V. J. *Applied Magnetic Resonance* **2004**, *26*, 23–39.
34. Chiang, Y. W.; Borbat, P. P.; Freed, J. H. *Journal of Magnetic Resonance* **2005**, *177*, 184–196.
35. Chiang, Y. W.; Borbat, P. P.; Freed, J. H. *Journal of Magnetic Resonance* **2005**, *172*, 279–295.

PART II: PHOTOSYNTHESIS

DONATELLA CARBONERA

Dipartimento di Scienze Chimiche, Università di Padova, Via Marzolo 1, 35131, Padova, Italy

10.10 INTRODUCTION

Photosynthesis is the natural process that converts the energy of solar photons to chemical forms of energy that can be used by biological systems. Many different organisms, ranging from plants to bacteria, perform the photosynthetic energy conversion. The best-known form of photosynthesis is the one carried out by higher plants, algae, and cyanobacteria, which convert CO_2 to carbohydrates in several steps of enzymatic reactions. Electrons for this reduction reaction ultimately come from water, which is converted to oxygen and protons. Energy for this process is provided by light, which is absorbed by pigments (phycobilines, chlorophylls, and carotenoids) bound to "light-harvesting proteins." Light energy absorbed by the pigments is transferred to "special" chlorophylls that are bound to a membrane protein complex, called a reaction center, where the primary energy conversion event occurs: the light energy is used to transfer an electron to a neighboring pigment creating a charge separated state. Many redox (reduction–oxidation) reactions follow this first event, along a chain of cofactors (such as chlorophyll, pheophytin, and quinone molecules) in the reaction centers and can be described by the following general scheme of photoinduced electron transfer including several donor (D) and acceptor (A) molecules:

$$h\nu \longrightarrow \cdots D_2 D_1^* A_1 A_2 \cdots \longrightarrow \cdots D_2 D_1^+ A_1^- A_2 \cdots \longrightarrow \cdots D_2^+ D_1 A_1 A_2^- \cdots$$

All photosynthetic organisms that produce oxygen have two types of reaction centers, which are named photosystem II (PS II) and photosystem I (PS I), both of which are pigment–protein complexes that are located in specialized membranes called thylakoids. PS II is the complex where water oxidation occurs. In PS I electrons are transferred to nicotine adenine dinucleotide phosphate ($NADP^+$). Therefore, electron transfer through PS II and PS I results in water oxidation and production of reduced NADP (NADPH), which has a high reduction potential. The light-induced electron flow from water to $NADP^+$ that requires light is also coupled to generation of a proton gradient across the thylakoid membrane. This proton gradient is used for synthesis of adenosine 5'-triphosphate (ATP), a high energy molecule, by the enzyme ATP-ase, another membrane bound protein complex. ATP and NADPH are used for CO_2 fixation in a process that is independent of light (dark reactions in the Calvin–Benson cycle).

The first charge separation produced by light generates the radicals $D_1^{\cdot +}, A_1^{\cdot -}$. The further migration of the charges leads to the formation of other secondary radicals. It is immediately clear that EPR is a potentially powerful tool to investigate the light reactions of photosynthesis because of the presence of unpaired electrons in the donor and acceptor molecules of the electron-transfer chain.

Since the first detection by EPR spectroscopy of paramagnetic species in leaves by Commoner et al. [1] in 1956, many applications of all of the basic and advanced EPR techniques in the field of photosynthesis have been reported in the literature (for recent reviews, see Refs. 2 and 3). On the other hand, photosynthesis has driven the development of EPR in a number of areas, particularly in the spectroscopy of excited states, interacting radicals, chemically induced electron polarization, and nonequilibrium systems. Some EPR experiments have been performed for the first time in photosynthesis research and have contributed to the development of the technique.

In photosynthesis, the results derived from EPR have contributed to the knowledge of the light-induced processes and the coupled enzymatic reactions. Actually, all aspects of EPR are well represented in this area. The paramagnetic species produced by light include not only radicals and radical pairs but also paramagnetic metal centers in the reaction centers and triplet states in the light harvesting complexes and in the reaction centers.

In the last 20 years the progress obtained in the fields of molecular biology, semiempirical density functional theory, and *ab initio* calculations together with the resolution of the most significant X-ray structures of the photosystems has opened new possibilities for the application of EPR in photosynthesis that is well beyond the identification of the paramagnetic species as in the early applications. The structural X-ray data of the photosynthetic reaction centers provide information on the geometric structure but only little information regarding the electronic structure determined from the wavefunctions of the valence electrons of the cofactors. The electronic structure, by contrast, plays a key role in the electron-transfer process. EPR techniques have been major tools to determine the electronic structure of paramagnetic centers in photosynthetic reaction centers. Much of the information that EPR provides about the electronic structure of paramagnetic centers is obtained by analysis of the **g** tensor, the hfcs, and when possible, the quadrupole couplings. A crucial question is now how the electronic structure of the cofactors, which is related to their finely tuned redox properties, is influenced by the interaction with other cofactors and protein environment and by the dynamics of the protein itself.

Because it is impossible to cover the whole subject of the applications of EPR in photosynthesis in this contribution, only some examples are dealt with here. For this reason, although bacterial photosynthesis has been investigated much more in the past, because of the simpler assembly of the photosynthetic apparatus and the earlier determination of the X-ray structure at atomic resolution, only the recent EPR applications in oxygenic photosynthesis are described. As an example of methodology, it will be shown how detailed information on the structure and function of the paramagnetic species involving the primary donors, in the early events of the electron-transfer process, can be obtained starting from the EPR experiments and how this information can be exploited to gain insights into biological function.

10.11 OXYGENIC PHOTOSYNTHESIS

During the early events of photosynthesis the light quanta are harvested from "antenna" pigment–protein complexes and funneled to the reaction center complexes where electron transfer reactions between protein bound donor and acceptor pigments across the membrane are initiated. The architecture of all of the natural photosystems shows many common features concerning the reaction center structure and the photochemical reactions that involve donor and acceptor molecules arranged in two branches. Two types of reaction centers have been recognized: *type I reaction centers* are found in PS I of plants, algae, and cyanobacteria and in green sulfur bacteria; *type II reaction centers* are found in PS II of plants, algae, and cyanobacteria and in purple bacteria. Only plants, algae, and cyanobacteria possess both types of reaction centers.

The primary donors are a closely related pair of chlorophyll (Chl) or bacteriochlorophyll molecules; the acceptors comprise monomeric chlorophylls, pheophytins (Ph), and quinones (Q) in type II reaction centers and chlorophylls, quinones, and iron–sulfur centers in type I reaction centers.

Cytochromes, iron–sulfur centers (in PS I and green bacteria), and the water splitting manganese cluster (Mn4) in PS II are the main electron-transfer metal centers involved in the early events of photosynthesis. Figure 10.1 shows the cofactor arrangements of reaction centers of PS I and PS II as determined by the recent X-ray structures [4–7].

After the first photoinduced charge separation several other electron-transfer steps occur between the various redox partners in the transmembrane with very different reaction rates ranging from <1 ps to >1 ms [8]. These events are described in the

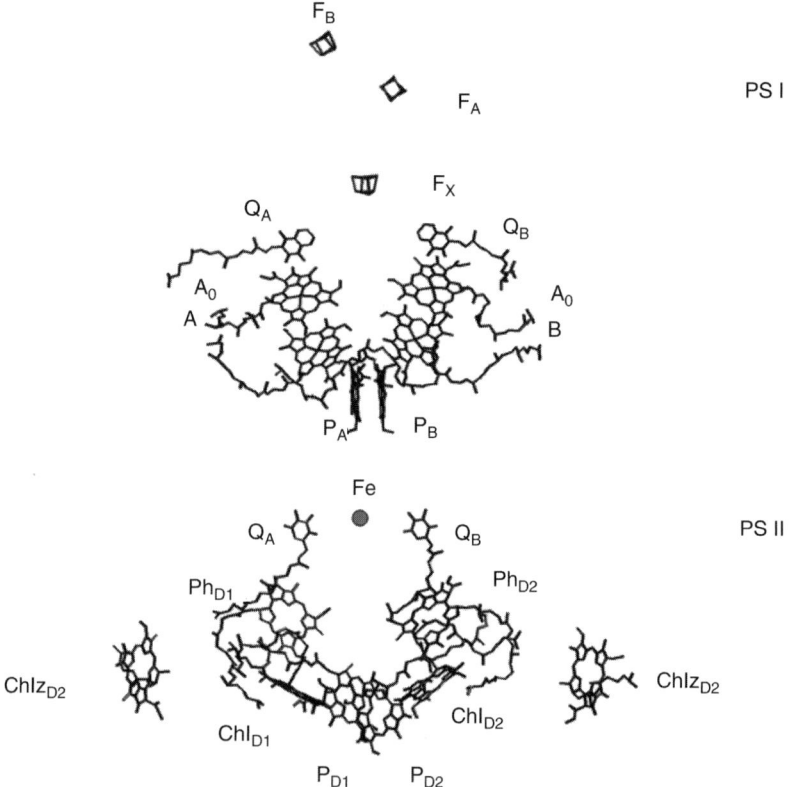

Fig. 10.1 The cofactor arrangement derived from the X-ray structure of PS I [4] and PS II [7].

so-called Z scheme reported in Fig. 10.2, which shows the cofactors of the two light reactions associated with PS I and PS II reaction centers in the sequential electron transport.

Light excitation of the primary donors associated with PS I and PS II, which are chlorophyll dimers P700 (P_A/P_B in Fig. 10.1) and P680 (P_{D1}/P_{D2} in Fig. 10.1) that absorb predominantly at 700 and 680 nm, respectively, leads to their first excited singlet states P*. An electron is transferred from the primary donor *P700 (reduction midpoint potential $E_m + 430$ mV) to the primary acceptor, Chl a (A_0), then to a phylloquinone (A_1, Q_A/Q_B in Fig. 10.1), and sequentially to the three iron–sulfur centers (so-called F_x, F_A, and F_B). The electrons on the PS I acceptor side finally reach the NADP$^+$. In turn, P700$^{·+}$ is reduced by the electrons coming from PS II. On the donor side of PS II, electrons from water are extracted by the water splitting enzyme (OEC) to reduce P680$^{·+}$ after donation of one electron from *P680 to Pheo a, the primary acceptor of PS II. Thus, in the whole process PS II catalyzes the oxidation of water and the rereduction of P700$^{·+}$ through a cascade of electron acceptors: Pheo a, the bound quinones Q_A and Q_B, the plastoquinone (PQ) lipo-soluble pool, the complex cytochrome b$_6$f (cyt b/f) that is also involved in the H$^+$ translocation

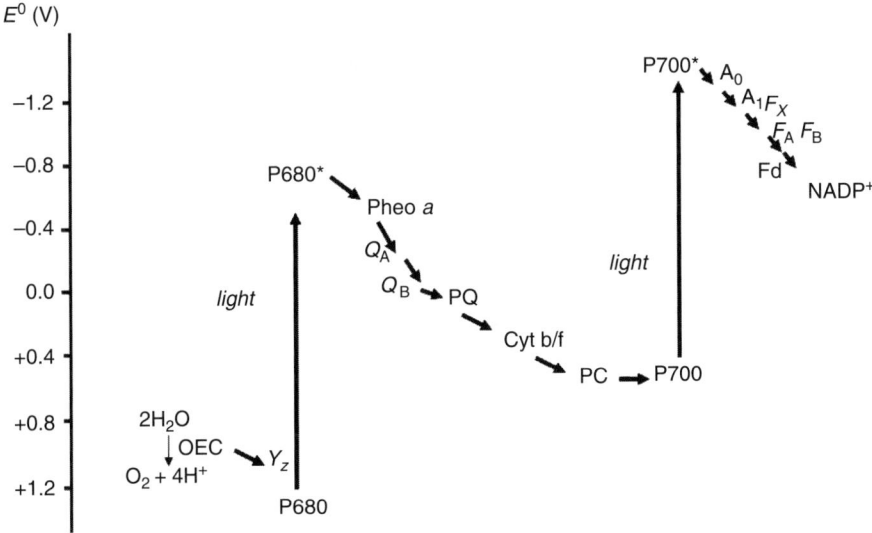

Fig. 10.2 The Z scheme for electron transfer in the "light reactions" of plant photosynthesis showing the redox potential of the main cofactors.

associated with the ATP synthesis, and finally the plastocyanin (PC). The production of ATP and reducing equivalents by PS I during the light reactions is finally coupled to the dark reactions for CO_2 fixation.

10.11.1 EPR of Primary Donors

EPR techniques have been widely applied to study the radical ions, radical pairs, and paramagnetic metal centers present in PS I and PS II. Here, the description is restricted to some of the representative results obtained relative to the primary donors. The knowledge of the electronic structure of the primary donors is an important point to correlate the special redox properties of these molecules to their function in the electron-transfer chain. The electronic structure can be studied by EPR techniques applied to the paramagnetic states involving the primary donors, such as $P^{\cdot+}$ (cation radical), $P^{\cdot+}A^{\cdot-}$ (radical pairs), and eventually ^3P (triplet state), by determining the electronic **g** tensor and the electron–nuclear hfc and nuclear quadrupole coupling tensors.

10.11.1.1 $P700^{\cdot+}$. According to the recently published X-ray crystallographic structure at 2.5-Å resolution of PS I from the cyanobacterium *Thermosynechococcus elongatus* [4], the cofactors of the electron-transfer chain are arranged in two branches related by a pseudo-C2 axis as shown in Fig. 10.1. The orientation of electron-transfer components and transmembrane helices found in cyanobacterial PS I are very similar to those of plants. P700 is composed of two Chl *a* molecules that are not identical and therefore deviate from perfect symmetry. A pair of Chl *a* molecules situated symmetrically at about 16 Å from P700 were assigned to the spectroscopically characterized primary acceptor A_0. However, another pair of accessory Chl *a* monomers is located approximately halfway between P700 and A_0. These chlorophylls may participate in excitation and/or electron transfer. From A_0, the electron is transferred to A_1. Two clearly resolved quinones are placed on the pseudo-twofold symmetry axis, but their angles and interactions with the protein are clearly not identical. The high resolution raised questions about the mode of action of the single cofactors, and the presence of two cofactor branches in PS I raised the question of whether one or both are active in electron transport. Recently, some evidence has been obtained for the use of both potential electron-transfer branches because the electron transfer to F_X is biphasic with the time constants of 20 and 200 ns, which is possibly due to the reoxidation of the two quinones [9]. Many EPR studies performed on PS I are aimed to answer the question of the electronic nature of the primary donor and to assign the cation species to specific molecules in the complex structure. This is relevant to define at which level the branches are involved in the electron-transfer process and if asymmetry plays an important role in the electron transfer through PS I.

The $P700^+$ radical cation can be easily produced by photoaccumulation or chemical oxidation, with ferricyanide, of PS I preparations. The EPR signal of the radical $P700^{·+}$ shows a structure-less Gaussian shape, 7.0-G broad, which is about $2^{-1/2}$ times the linewidth of the Chl $a^{·+}$ in organic solvent *in vitro* (for a review on the historical experiments, see Ref. 10). This was taken in the early studies on PS I as an indication of the dimeric nature of the primary donor. If an unpaired electron is shared by two identical molecules, this leads to a halving of the hyperfine interactions (the unpaired electron would spend only half of the time on each nucleus) that corresponds to a reduction of the linewidth by a factor of $\sqrt{2}$. The same kind of experimental results lead to the conclusion that the primary donor of purple bacteria is a dimer of bacteriochlorophyll molecules. However, it has been demonstrated from the *in vitro* and *in vivo* studies of Chl *a* radical cations that the EPR linewidth of these large conjugated molecules are strongly dependent on the solvent and molecular environment. Although the X-ray crystallographic structure has shown that P700 is indeed a Chl '*a*−Chl *a* heterodimer, where Chl '*a* is an epimer of Chl *a* (P_A and P_B in Fig. 10.1), it must still be clarified whether the electronic structure of this species is better described by a wavefunction extending over both dimer halves or by a localized, monomeric electronic structure. This is relevant for the function of P700. In principle, the **g** tensor is a property of the paramagnetic center, probing the average valence electron density distribution. Conventional CW-EPR provides *g* values; however, to fully resolve the **g** tensor components of $P700^{·+}$, it is necessary to perform EPR measurements at very high microwave frequency because the *g* anisotropy of $P700^{·+}$ is too small to be resolved at the conventional EPR frequency. The EPR experiments performed at 300 GHz on $P700^{·+}$ and monomeric $Chl^{·+}a$ in organic solvent showed small differences between the *g* values of the two species [11, 12]. These differences were tentatively interpreted as resulting from dimerization. Moreover, the observed temperature dependence of the *g* values could also be attributable to a redistribution of the spin density in a dimeric species [13]. A recent powerful tool to investigate the spin density localization in P700$^{\cdot+}$ comes from site-directed mutagenesis. A series of mutants in which altered amino acids in the surrounding of P700 are introduced may be produced and the effects on EPR spectra of the cation monitored. Petrenko et al. [11] measured an effect on the **g** tensor values of P700$^{\cdot+}$ when the original ligand on the P$_B$ side was replaced; but they also found that the replacement of a threonine residue, which forms a hydrogen bond to the 13-keto group of P$_A$, changed one g component [14]. This indicates that the P dimer shows an electronic coupling between the two halves.

Additional information can be obtained from studies on single crystals. Zech et al. obtained the orientation of the **g** tensor, which is also related to the electronic structure, from the W-band EPR spectra of P700$^{\cdot+}$ measured in PS I single crystals [15]. The important result obtained from the analysis of the experimental data is a clear deviation of the **g** tensor components with respect to the molecular axes of P700. A possible reason for this deviation can be the influence of the second dimer half, even though the effect of neighboring pigments and/or coordinating amino acids cannot be excluded.

Together with the measurements of the **g** tensor values, information on the electronic structure may be obtained by means of the electron–nuclear hyperfine interactions. Hyperfine tensors are local probes that reflect the distribution of the unpaired spin density at individual nuclei of the paramagnetic centers and their immediate environment. In principle, the hyperfine couplings are manifested as splittings in the EPR spectra but, with the exception of single crystal studies, nuclear hyperfine splittings are not resolved in EPR spectra of biological samples, especially at low temperature; consequently, the associated information is lost. As expected, hyperfine interactions are not resolved in the EPR spectrum of P700$^{\cdot+}$; however, they can be measured by using ENDOR spectroscopy (Chapter 1). The same ENDOR approach unraveled the electronic structure of the primary donor of the bacterial reaction centers, where the unpaired electron resulted in an asymmetrical distribution (2 : 1) in the two dimer halves. ENDOR experiments in frozen solution and in PS I single crystals of *T. elongatus* have been performed by several groups (see Ref. 3 and references therein). The cation radical can be easily generated chemically or by light in the single crystals of *T. elongatus*. The angular dependence of the ENDOR spectra was obtained by Käss et al. [16] at a temperature of 100K. The principal values and corresponding axes orientations were determined for three different proton hfc tensors. On the basis of the tensor magnitudes, their symmetry, and the relative orientations of their axes, they were assigned to the protons of three methyl groups of a single Chl a molecule. Assignment of other observed small hfcs to a putative second dimer half was also suggested. Thus, the EPR/ENDOR experiments show that the second chlorophyll half is involved in determining the electronic structure of P700 but carries very little spin density. Moreover, because the molecular plane of the spin-carrying Chl a molecule results to be parallel to the crystallographic c axis, the assignment to P$_B$ in the X-ray structure is possible.

A complementary technique to ENDOR is ESEEM, which is suited for measuring weak hyperfine couplings. In the ESEEM experiment the nuclear transition frequencies are monitored indirectly through EPR transitions and are observed because of the mixing of the frequencies of the semiforbidden and allowed EPR transitions, which can be coherently excited using short, intense microwave pulses. In the presence of nuclear spins weakly coupled with the electron spin, the echo intensity generated by a sequence of resonant microwave pulses separated by evolution times is modulated at the nuclear transition frequencies of the interacting nuclei that contain the information about the hyperfine and quadrupole couplings. ESEEM and ENDOR may both be used to measure the hyperfine tensors of nitrogen from ^{14}N and ^{15}N nuclei in P700$^{\cdot +}$. This information is important because the presence of more than the four nitrogen couplings arising from the tetrapyrrol nitrogens of a single chlorophyll could be taken as good evidence for the spin density being delocalized on an additional chlorophyll. For a largely asymmetric dimer the couplings on the half of P700 with the minor fraction of the delocalized spin are expected to be very weak. Käss et al. [17, 18] performed these kinds of experiments and provided evidence for four nitrogen couplings in addition to a weak fifth nitrogen coupling, and they assigned the fifth nitrogen coupling to the second half of the putative P700 dimer. The assignment is, however, under discussion in the literature because axial ligands, such as histidines, may produce the same effect of coupling (for a recent review, see Ref. 19). In any case, the absence of eight different comparable couplings excludes the possibility of a dimer with an equal distribution of the spin density.

More recent ENDOR studies of mutants of the axial ligands to P700 in the P_A and P_B moieties (see Fig. 10.3, His A676 and His B656) in the green alga *Chlamydomons reinhardtii* have been performed. It was concluded that spin and positive charge mainly resides on P_B, and P_A has only a minor effect on determining the spin density [20] because the electron spin density distribution of P700$^{\cdot +}$ determined by ENDOR spectroscopy was only changed when His B656 was replaced whereas the ENDOR spectra of P700$^{\cdot +}$ of the A branch mutants HS A676 and HQ A676 were all unchanged compared with those of the wild type.

Witt et al. [21] demonstrated that the replacement of Thr A739 in *C. reinhardtii*, an amino acid that is hydrogen bonded to the 13^1-keto group of P_A, induces changes in the ENDOR spectra of the major hyperfine constants assigned to the methyl protons at position 12 of the spin-carrying chlorophyll in P_B. In contrast to the histidine mutants, the hyperfine constants are decreased.

This gives evidence that, although the spin is mainly localized in the P_B moiety, P_A is electronically coupled to P_B and is thus involved in determining the electronic structure of P700. The decrease in spin density on P_B in the threonine mutant can be interpreted as a spin density shift from P_B toward P_A. Figure 10.3 shows the ENDOR spectra of the wild type and mutants.[1]

10.11.1.2 P680$^{\cdot +}$.
Recently, the structure of the PS II core particles was resolved by X-ray crystallography, and the number and geometry of chlorophyll molecules and

---

[1] Note that FTIR studies on PS I suggested the delocalization of unpaired electron over the two chlorophyll molecules in P700$^+$ at a ratio of 1–2 : 1. The discrepancy between the EPR/ENDOR spectroscopy and FTIR data is still unresolved.

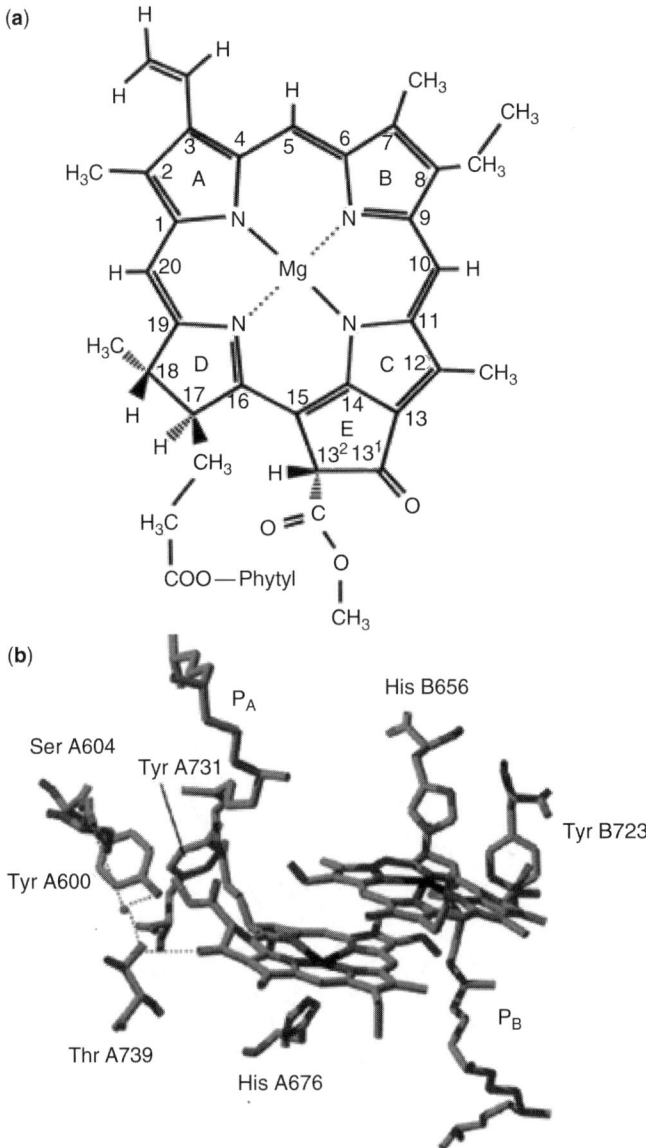

Fig. 10.3 (a) The molecular structure and numbering scheme for Chl a; Chl a' is the C_{13}^2 epimer of Chl a. (b) The primary donor P700 with His A676 and His B656 as the coordinating ligands to P700 and Thr A739 as a donor of a putative hydrogen bond to the 13^1-keto group of P_A.

cofactors were well defined [6, 7]. The heterodimer of D1 and D2 proteins of the PS II complex binds almost all of the redox components of PS II. P680 is a dimer of Chl a (P_{D1} and P_{D2} in Fig. 10.1). The oxidized P680 is reduced by the secondary donor Y_Z (Tyr 161 in D1), and the oxidized $Y_Z^{\cdot}$ is in turn reduced by the oxygen evolving complex. There are also additional redox components in PS II, including cytochrome (Cyt) b559,

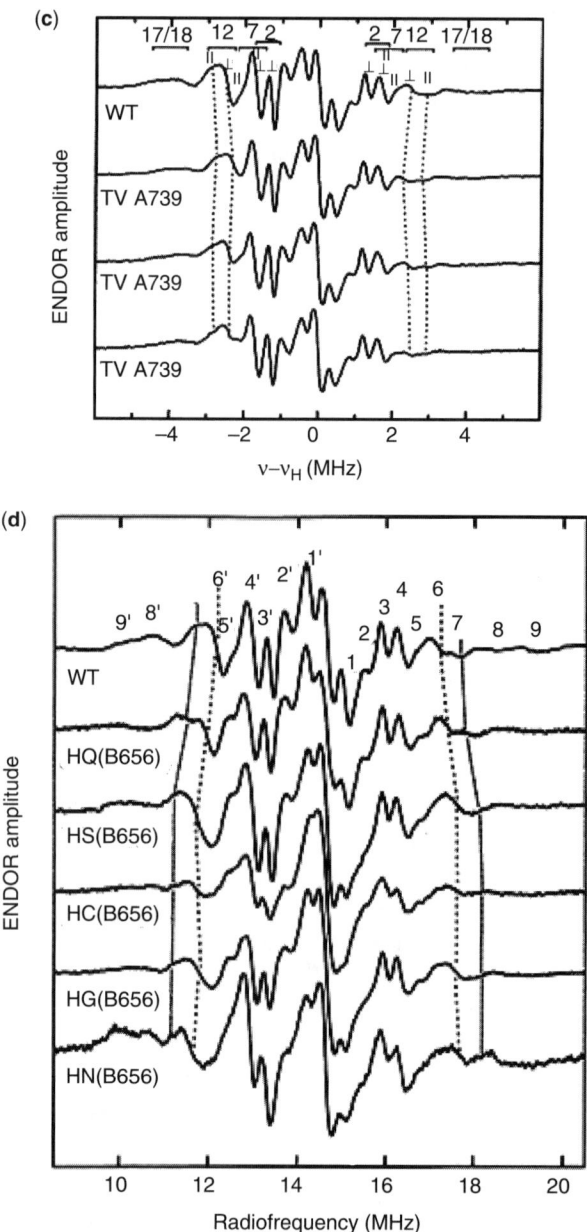

Fig. 10.3 (*Continued*) (c) ^{1}H ENDOR spectra of P700$^+$ in PS I from *C. reinhardtii* wild type (WT) and PsaA mutants in frozen solution. The principal components of the hfc tensors for the different protons are indicated: 12,12-methyl group; 7,7-methyl group; 2,2-methyl group; 17 and 18, α protons. Adapted with permission [21]. Copyright 2002 American Chemical Society. (d) ^{1}H ENDOR spectra of P700$^+$ in PS I from *C. reinhardtii* wild type (WT) and PsaB mutants in frozen solution. Adapted with permission [20]. Copyright 2000 American Chemical Society.

chlorophyll Z (Chl$_Z$), and another redox-active tyrosine residue Y_D in D2. Chl$_Z$ functions as an electron donor to the oxidized P680, is reduced by the nearby Cyt b559 [22], and is known to play a role under photoinhibitory conditions. The identification in the structure of the functional Chl$_Z$ defined by spectroscopy is still not clear.

The primary donor of PS II is, in its excited state, a highly oxidizing species that is able to oxidize water. For this reason it has a midpoint potential that is much more positive than *P700. A recent reevaluation of the P680 reduction potential suggests an even more positive potential (+1.27 V) with respect to the previous estimates. It seems unlikely that this extreme property can be played by a dimer, because dimerization lowers the oxidation potential. The recent publication of the X-ray structure shows a larger distance (8.3 Å) between the putative chlorophylls forming a dimer (P$_{D1}$, P$_{D2}$) compared to the dimer halves of PS I. The longer distance is in accord with the proposal that, in PS II, the entire array of reaction center pigments might be able to function as a multimeric reaction center, even though a localized chlorophyll cation radical could be present in the charge separated state. In agreement with this remark, ENDOR spectroscopy has shown that the hyperfine constants of P680$^{\cdot+}$ are compatible with those expected for monomeric Chl a^+ (see Refs. 23 and 24 and references therein).

10.11.2 Radical Pairs and Spin Polarization

The photosynthetic process generates a series of sequential correlated pairs of radicals $D_i^{\cdot+}A_i^{\cdot-}$. These radical pairs are short-living species and suitable techniques for studying them are time-resolved EPR (TR EPR, §2.2.3) and pulsed EPR methods (Chapter 5). Many studies have been performed on photoinduced radical pairs in bacteria and higher plant reaction centers (for reviews, see Refs. 25–30) and in artificial reaction centers [31–34]. In reaction centers, light-induced spin polarization occurs in the separated radicals because the electron is transferred initially from the excited singlet state of the donor *P, a state in which the electrons are spin paired ($S = 0$), to the primary acceptor. The singlet precursor generates a pair of separated but spin-correlated radicals with selective population of their spin states. This spin polarization manifests itself as very intense, short-lived EPR signals that can be either absorptive or emissive. In addition, coherence effects such as transient nutations, quantum beats, and nuclear modulations can be observed.

The theoretical model of "spin-correlated radical pairs" provides the basis for the interpretation of EPR data on radical pairs in photosynthesis. The radical pairs are generated starting from the excited singlet state of the primary donors. In the separated state the relative motion of the spins leads to a deviation from a pure singlet character, and a well-polarized EPR-detectable state of the pair is generated. The spins are far apart in the two radicals, even in the first photoinduced radical pair, and the interaction between them is weak. In the formalism of Hore for non-diffusing and weakly interacting radical pairs generated from a singlet state [35–37], for each orientation of the molecule with respect to the external magnetic field **B** and, without taking into account explicit hyperfine interactions, which can be often included in the inhomogeneous broadening of the absorption lines, the spin Hamiltonian of the radical pair may be written as

$$\hat{H} = \omega_A \hat{S}_{AZ} + \omega_B \hat{S}_{BZ} - J(\hat{S}^2 - 1) + 1/2 D_{ZZ}(3\hat{S}_Z^2 - \hat{S}^2)$$

where $\hat{S}_{AZ}$ and $\hat{S}_{BZ}$ are the Z components of the electron spin angular momentum of the two radicals; $\hat{S}$ and $\hat{S}_Z$ are the total electron spin angular momentum operator and its component along Z, respectively; J is the isotropic exchange interaction; and ω_A and ω_B are the EPR frequencies of the two radicals in the absence of magnetic spin–spin interactions $\omega_{A,B} = g_{A,B}(\varphi, \theta), \mu_B \hbar^{-1} B_0$, where φ and θ denote the angles defining the orientation of the magnetic field **B** with respect to the axis system of the **g** tensors. The term for the dipolar interaction between unpaired electrons is $D_{ZZ} = D(\cos^2 \theta - 1/3)$, where θ is the angle between the dipolar axis and the magnetic field direction **B**. In this framework, the two electron spins in the magnetic field have four energy levels, as shown in Fig. 10.4. States 1 and 4 correspond to the states with both electrons having the projection of the spin antiparallel and parallel to the magnetic field direction, respectively. These states are equal to the $m_S = +1$ and -1 pure triplet states, and they are not populated starting from a singlet precursor in the presence of the magnetic field if J is small as expected for radicals at distances of the order of tens of angstroms. States 2 and 3 correspond to admixtures of the two possible states of one spin having a parallel projection and the other having an antiparallel projection on the magnetic field direction. They can be considered as admixtures of the $m_S = 0$ of a triplet state (T_0) and a singlet state $m_S = 0$. For radical pairs created in a singlet spin state, the population of the energy levels is proportional to their respective singlet character and all population is found in the two states corresponding to 2 and 3. Four EPR transitions are possible between a couple of states. All four transitions have equal intensity, which is due to a compensation of the population difference between levels 2 and 3 by different transition probabilities.

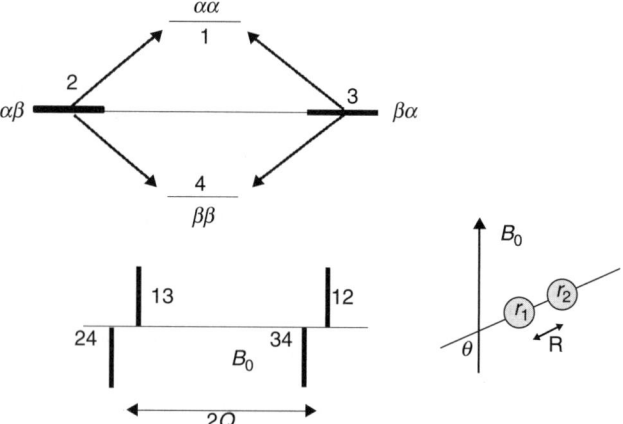

Fig. 10.4 The energy level diagram of a radical pair ($r_1 - r_2$) with a singlet precursor (top) and a schematic EPR initial spectrum, $J > 0$ and $Q > 0$.

In this framework, at short times after its generation from a singlet precursor, the spin-correlated radical pair spectrum consists of two doublets in antiphase of equal amplitude, centered at the Larmor frequencies of the individual radicals forming the pair, as shown in Fig. 10.4. The splitting of each doublet is due to the exchange contribution J and the magnetic dipolar contribution D_{ZZ}. The frequencies of the four spin-polarized EPR lines are

$$\omega_{12} = \omega_0 - \Omega - J + D_{ZZ}$$
$$\omega_{34} = \omega_0 - \Omega + J - D_{ZZ}$$
$$\omega_{13} = \omega_0 + \Omega - J + D_{ZZ}$$
$$\omega_{24} = \omega_0 + \Omega + J - D_{ZZ}$$

where $\omega_0 = (\omega_A + \omega_B)/2$, $\Omega^2 = (J + D_{ZZ}/2)^2 + Q^2$, and $Q = (\omega_A - \omega_B)/2$. The antiphase doublet splitting is equal to $2J + 2D(\cos^2\theta - 1/3)$.

Anisotropic g values and hyperfine interactions may be introduced explicitly in the Hamiltonian:

$$\hat{H} = \beta B_0 \mathbf{g_1}\hat{S}_1 + \beta B_0 \mathbf{g_2}\hat{S}_2 + \hat{S}_1\sum \mathbf{A_{1j}}I_j + \hat{S}_2 \sum \mathbf{A_{2j}}I_j - J\left(\hat{S}^2 - 1\right)$$
$$+ 1/2 D_{ZZ}\left(3\hat{S}_Z^2 - \hat{S}^2\right)$$

The EPR spectra in frozen solution reflect the powder average over all possible orientations.

In the reaction centers the first light-generated radical pair is too short-lived to be detected by TR EPR techniques, and secondary radical pairs are the first radical pairs observed after laser illumination. A theory has been developed that takes into account the amount of additional polarization that can be produced by the motion of the spins in a sequential radical pairs electron-transfer chain. However, the polarization of the secondary radical pairs predominately inherits the singlet character of the first donor and the above description is a good approximation for the interpretation of the EPR spectra of observed radical pairs in photosynthesis [30] (Box 10.1).

10.11.2.1 TR EPR of P700$^{\cdot +}$A$_1^{\cdot -}$. The TR EPR technique is described in Chapter 2 (§2.2.3). Here, the application of this technique to the study of P700$^{\cdot +}$A$_1^{\cdot -}$ radical pair will be reported. Special pulsed EPR techniques and their application to the photosynthetic radical pairs may be found in Appendix A.10.1.

The first observation of spin polarization in PS I was reported by Blankenship et al. [38] in 1975. Many other studies were performed after that, and recently the subject of light-induced spin polarization in PS I was reviewed by van der Est [28]. The application of the spin-correlated radical pairs model to interpret the transient X- and K-band (24 GHz) EPR spectra of the P700$^{\cdot +}$A$_1^{\cdot -}$ radical pair state in PS I obtained by Stehlik et al. [39] yielded structural information on PS I well before the X-ray structure determination of the protein complex.

> **BOX 10.1 EPR AND ELECTRON TRANSFER KINETICS**
>
> In photosynthesis a relevant point concerns the evaluation of the factors that are important for the efficiency of the electron-transfer reactions. The classical theory describing the rate of electron transfer is attributable to Marcus and comprises two terms: the actual transfer of the electron, which needs the overlap (V) of the electronic wavefunctions of the initial and final state, and the coupling of the nuclear motions in the two states, the Franck–Condon factor (FC), which is related to the activation energy (E_a).
>
> $$k_{ET} = \frac{2\pi}{\hbar} V^2 FC$$
>
> $$k_{ET} = \frac{2\pi}{\hbar} V^2 \frac{1}{\sqrt{4\pi\lambda k_B T}} \exp\left[\frac{-E_a}{k_B T}\right]$$
>
> where k is the Boltzmann constant and ET is the electron transfer.
>
> The electron coupling V is a parameter that is quite difficult to calculate with *ab initio* methods. The EPR technique in some cases can be a powerful technique for the experimental evaluation of the electronic coupling by allowing the measure of the exchange parameter J in the charge-transfer state of radical pairs [2]. This is directly correlated to V:
>
> $$J = -2V^2(\Delta G^0 + \lambda)$$
>
> where ΔG^0 is the free energy of the electron-transfer process and λ is the reorganization energy corresponding to the difference between the energy of the final state with its nuclei in the transition configuration and in equilibrium.

The light-induced spin polarization of P700$^{\cdot+}$A$_1^{\cdot-}$ in PS I can be used to determine the orientation of P700 and A$_1$ because, in the description of the charge separated state as a correlated coupled radical pair with a singlet precursor, the polarization pattern depends on the angles describing the relative orientation of the **g** tensors of the two radical partners and the distance vector between them. By measuring the spectrum of P700$^{\cdot+}$A$_1^{\cdot-}$ in fully deuterated whole cells of *Synechococcus sp.* at K-band, Stehlik et al. obtained greatly improved spectral resolution with respect to the X-band spectra that were characterized by the overlap of the **g** tensor components and line broadening caused by unresolved hyperfine splittings. Stehlik and coworkers showed that the axis connecting the two carbonyl oxygens of phylloquinone A$_1$ is parallel to the axis connecting the donor P700 and the acceptor A$_1$. Moreover, the spectra of P700$^{\cdot+}$A$_1^{\cdot-}$ were compared to those of P865$^{\cdot+}$Q$_A^{\cdot-}$ from the bacterial reaction center at X- and K-band [40] and at W-band (95 GHz) [41]. An important conclusion from these studies is that the orientation of A_1 is quite different from that of Q_A in purple bacterial reaction centers. In particular, the differences in the polarization patterns are mainly due to the different quinone orientations in the respective reaction centers. In addition, it has been shown that the anisotropy in the **g** tensor of phylloquinone $A_1^{\cdot-}$ in its binding site is considerably different from that in frozen 2-propanol solution. These differences are attributable to the interactions of $A_1^{\cdot-}$ with its surroundings in the protein matrix, showing that the protein environment is functional for the modulation of the electronic and redox properties of the cofactor.

Another effect that can be observed in the time dependence of the spin-polarized EPR signals of radical pairs is due to the correlation between the electron spins in the initial singlet state. The initial states of the radical pair are not necessarily stationary states or eigenstates of the spin Hamiltonian of the radical pair but are instead some coherent superposition of states. The probability of finding the system in one of the eigenstates oscillates in time. This can occur for electron and nuclear spin states. The result is a spin state that evolves in time once it is formed, producing periodic oscillations or quantum beats in the EPR signal.

Quantum beats involving the electron spins carry information about the dipolar and exchange interactions within the radical pair whereas quantum beats involving nuclear spin states occur at the nuclear resonance or ENDOR frequencies. These effects also depend on the strength of the dipolar coupling and are orientation dependent. Kothe and colleagues [42] were able to detect quantum beats from $P700^{\cdot+}Q_A^{\cdot-}$ and obtained a prediction of the geometry of the radical pair. The analysis is very complex and yields the same angles that are more easily obtainable from the spin polarization patterns.

10.11.3 Recombination Triplet States

Photosynthesis proceeds via excited singlet states. Electronic triplet states of the cofactor molecules may be formed occasionally, especially in the light harvesting complexes, during light stress conditions. Triplet states may also be formed in the reaction centers by charge recombination of the primary radical pairs, when overreduction of the acceptors occurs [2]. In some cases, for example, for the primary donor of PS II, the recombination triplet state $^{T}P680$ is involved in the so-called photoinhibition phenomenon, which is light-induced damage of the reaction centers. In any case, even though the triplet is not a functional state in the electron-transfer process, its formation can be induced and exploited for the investigation of the excited state electronic properties of the primary donors. The two unpaired electrons in the triplet states belong to the highest occupied molecular orbital and to the lowest unoccupied molecular orbital. In principle, the direct information on the electronic distribution of the lowest unoccupied molecular orbital obtained by the characterization of the triplet state by TR EPR can be used to attain information on the first excited singlet state that is involved in the charge separation.

As mentioned above, the differences in resonant frequencies of the two radicals generated by illumination in the radical pair produce a loss of the initial spin correlation and some triplet character appears [35–37]. If the distance between the two radicals is large, the exchange interaction J between the spins is very small compared to the Zeeman energy. In this case only the T_0 state can interact with the singlet state (see Fig. 10.5). States 2 and 3 (Fig. 10.4) correspond to admixtures of the two possible states of one spin having a parallel and the other having an antiparallel projection on the magnetic field direction. They can be considered as admixtures of the $m_S = 0$ of a triplet state (T_0) and a singlet state $m_S = 0$. If recombination of this pair of radicals is induced, for example, by removing the secondary acceptors, a triplet in the T_0 state (but not in the T_{+1} or T_{-1} states) will be formed. This gives rise to a characteristic EPR spectrum with a polarization pattern that is now well known as $S-T_0$ polarization (for a chlorophyll triplet state this is AEEAAE, as shown in Fig. 10.5). The strong polarization gives population differences between the spin levels in the magnetic field that are far from the thermal population differences.

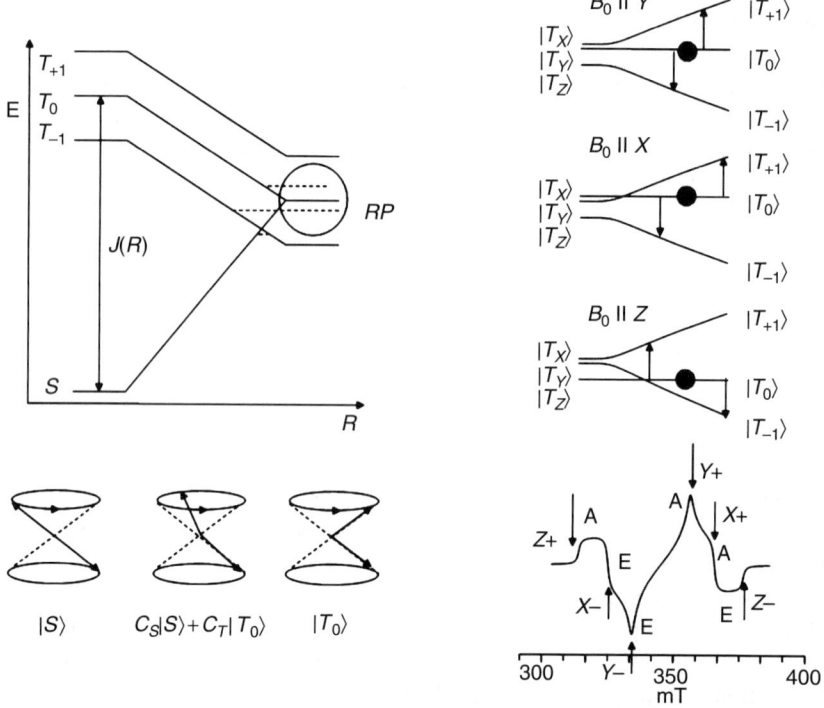

Fig. 10.5 (Left) The singlet and triplet energy level of a radical pair in an external magnetic field as a function of the distance R between the two radicals. (Left bottom) The vector model for $S-T_0$ mixing in a radical pair with a singlet precursor. (Right) The spin energy levels of a triplet state molecule for the three zero-field splitting axes parallel to the magnetic field. The T_0 states are selectively populated by charge recombination of a radical pair with a singlet precursor. (Right bottom) The corresponding TR EPR spectrum for a chlorophyll triplet in frozen solution is also shown.

In all of the reaction centers of the investigated photosystems in which electron transfer is blocked at the level of secondary acceptors, recombination triplet states become finally localized on the primary donors according to the following scheme:

$$PA \longrightarrow P^*A \longrightarrow (P^+A_0^-) \longrightarrow {}^T(P^+A_0^-) \longrightarrow {}^TPA_0$$

The results obtained for the recombination triplet states localized in the primary donors of the reaction centers of the two photosystems of higher plants, namely, $^TP700$ and $^TP680$, are described in detail in Appendix A.10.2.

APPENDIX A.10.1: PULSE EPR EXPERIMENTS ON RADICAL PAIRS

A.10.1.1 Out-of-Phase ESEEM (OOPESEEM) of $P700^{\cdot+}A_1^{\cdot-}$ and $P680^{\cdot+}Q_A^{\cdot-}$

Pulsed EPR experiments on radical pairs are also possible and very informative. For a Hahn echo-type pulsed EPR experiment, which is obtained by applying short microwave pulses that cover the total spectral width of the spectrum, the in-phase echo amplitude vanishes because of the net intensity of the spectrum. However, because of the relative shift of the two absorptive lines with respect to the two emissive lines in the radical pair spectrum (see Fig. 10.4), an out-of-phase echo may be observed. Tang et al. [43] presented a theory of ESEEM for spin-polarized radical pairs showing that the spin echo for a spin-polarized radical pair is phase shifted compared to the echo of the same spin system at thermal equilibrium. The amplitude of the out-of-phase echo is modulated as a function of the interpulse time in a two-pulse EPR experiment, with a frequency that is proportional to the dipolar and exchange spin–spin interactions between the two unpaired electron spins. This kind of ESEEM is known as OOPESEEM.

The magnetization along the x and y axes for a pulse scheme as shown in the Fig. A.10.1 are given by

$$M_x = \frac{\Delta\omega^2(2J + D_{ZZ})^2}{4\Omega^4} \sin(\Gamma\tau)[1 - \cos(2\Omega\tau)]$$

$$M_y = 0$$

where

$$\Delta\omega = \frac{1}{2}\beta B_0(g_1 - g_2) + \frac{1}{2}\sum_j (A_{1j} - A_{2j})m_j$$

$$\Omega^2 = \Delta\omega^2 + (J + D_{ZZ}/2)^2$$

$$\Gamma = 2(J - D_{ZZ})$$

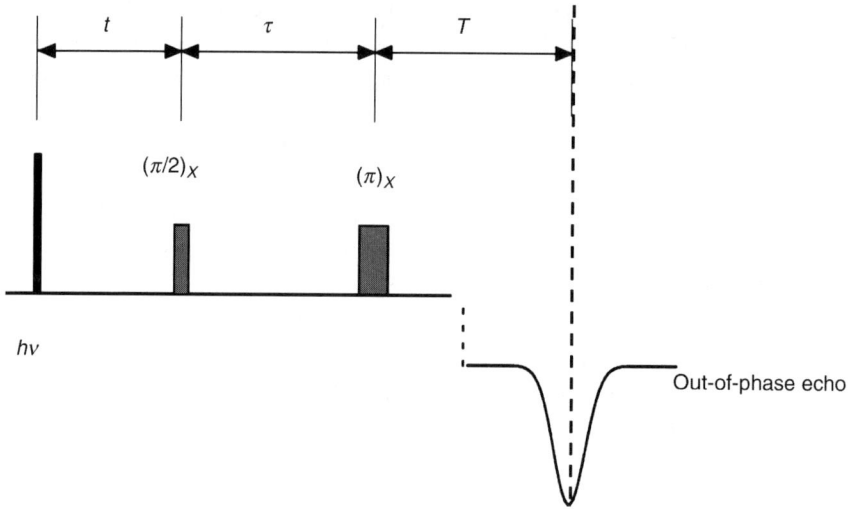

Fig. A.10.1 The pulse scheme for the out-of-phase electron spin echo experiment. The laser pulse generates the radical pair. The delay time t is kept constant while the time τ between the microwave pulses is varied.

where $\Delta\omega$ is the difference in the Larmor frequency for the two radicals, which is due to different g factors and hyperfine interactions, and Ω is related to the zero-quantum coherence. The most important term for the modulation of the out-of-phase echo is the term in $\sin(\Gamma\tau)$ that is due to the spin–spin interactions, whereas the term in $[1 - \cos(2\Omega\tau)]$ can be neglected for a broad excitation because of the wide distribution of the zero-quantum frequency. ESEEM due to hyperfine couplings gives only a minor contribution to the out-of-phase echo modulation on radical pairs [44] if short microwave pulses are used, because the hfcs with large anisotropy contribute to the modulation only as a "second-order effect." Thus, the OOPESEEM spectrum is dominated by the electron–electron spin couplings.

An important condition for the observation of modulations induced by spin–spin interactions is the full excitation of both radicals in the pair by the microwave pulse.

In a randomly oriented sample the contributions of the isotropic exchange interaction and of the dipolar interaction can be separated. From the dipolar contribution one can obtain information about the distance between the unpaired electron spin (R in Fig. 10.4), and thus about the distance between the cofactors, being $D/\text{Gauss} = [-2.78 \times 10^4]/[R/\text{Å}]$. The modulation pattern observed in OOPESEEM is very sensitive to the distance of the two radicals but not very sensitive to their relative orientation. In this respect the method is complementary to TR EPR.

A.10.1.2 OOPESEEM of Radical Pair of PS I: $P700^{\cdot+}A_1^{\cdot-}$

After the first experiment on the $P_{865}^{\cdot+}Q_A^{\cdot-}$ state of bacterial reaction centers, OOPESEEM was also successfully performed for the study of radical pairs of higher plants. Pulsed EPR applied to radical pairs in PS I was reviewed by Bittl and Zech [45]. For $P700^{\cdot+}A_1^{\cdot-}$ in PS I, a distance of 25.4 Å was deduced from the dipolar spin–spin coupling. OOPESEEM experiments performed on PS I single crystals allowed the determination of the dipolar coupling axis with respect to the crystallographic axes, from the angular dependence of the observed echo modulation. From the analysis of the experiments, a position of the quinone $A_1^{\cdot-}$ of the radical pair in the A branch was deduced [45].

A.10.1.3 OOPESEEM of Radical Pair of PS II: $P680^{\cdot+}Q_A^{\cdot-}$

The analysis of the OOPESEEM spectra of $P680^{\cdot+}Q_A^{\cdot-}$ obtained in preparations of PS II reaction centers, both frozen and liquid solutions, using a continuous flow system on a microsecond timescale, yielded a distance of 27.4 Å for the two radicals. This allows the assignment of the $P680^{\cdot+}$ species in the X-ray structure of PS II to one of the two chlorophylls (P_{D1} or P_{D2} in Fig. 10.1), P_{D1} being the most likely because of the proximity to Y_Z (for a recent review, see Ref. 24).

APPENDIX A.10.2: RECOMBINATION TRIPLET STATES OF THE PRIMARY DONORS

A.10.2.1 Recombination Triplet States of the Primary Donors

A.10.2.1.1 $^T P700$. Treatment of PS I preparations with a strong reductant reduces the iron–sulfur centers (F) and the phylloquinone (A_1). In such conditions the triplet state $^T P700$ is obtained by radical pair recombination of $P700^{\cdot+}A_0^{\cdot-}$. The TR EPR spectrum of $^T P700$, a short-lived species, taken at an early time after laser excitation, shows the AEEAAE polarization pattern, which confirms the mechanism of triplet recombination from a charge separated state with a singlet precursor. From the simulation of the spectrum, the zero-field splitting parameters D and E for $^T P700$ can be easily extracted and compared to those of $^T Chl\ a$ in organic solvent [46–48]. No significant difference was found at low temperature. A reduction of parameter E of $^T P700$ by increasing the temperature was reported by Sieckmann et al. [49] and interpreted in terms of triplet localization at low temperature and triplet delocalization, over two coplanar Chl a molecules, at elevated temperatures. The orientation dependence of the triplet state spectra shows that the plane of the chlorophyll, which carries the triplet at low temperature, is oriented perpendicular to the membrane [50]; that is, the triplet state must be located on P_A or P_B (see Fig. 10.1) as expected. The D and E values for $^T P700$ in PS I mutants of *C. reinhardtii*, in which the histidines coordinating the central Mg atoms of P_A and P_B and the hydrogen bond to P_A were altered, confirmed that the triplet state is trapped on a single chlorophyll at cryogenic temperatures [20, 21]. The TR EPR spectra at D-band (130 GHz) and 75K show that the ordering of the g_x and g_y components is switched between $^T P700$ and $^T Chl\ a\ in\ vitro$ [51]. The different orientation of the zero-field axes with respect to the **g** tensor main axes in $^T P700$ compared to $^T Chl\ a$ means that the spin densities are very different. All of these results suggest that the molecular environment of P700 is important in determining the electronic properties of the primary donor.

A.10.2.1.2 ^{T}P680. The TR EPR spectrum taken early after laser excitation in prereduced PS II particles presents the characteristic AEEAAE polarization pattern. Surprisingly, the detection of recombination triplet states in oriented PS II particles shows that the molecule carrying the triplet state forms an angle of about 30° with the membrane normal [52]. The more recent TR EPR data in PS II single crystals at low temperature [53] confirmed these data; by comparison with the structural data, the localization of the triplet state can be assigned not to the dimer P_D but to one accessory chlorophyll (Chl_D). TR ENDOR [54] and pulsed ENDOR [24] combined with repetitive laser excitation at cryogenic temperatures were also done to obtain the hyperfine structure of the triplet state. The TR spectrum is provided in Fig. A.10.2. A comparison with the ENDOR data for ^{T}Chl *a* confirms that ^{T}P680 is a monomeric species at low temperature (10K). The observed shifts of the hyperfine couplings of individual nuclei indicate specific protein interactions.

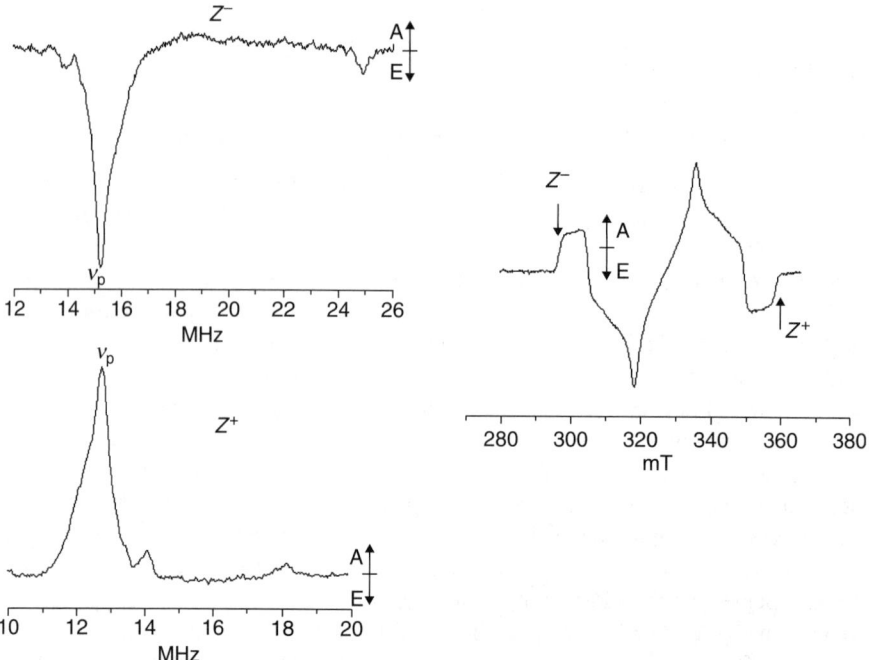

Fig. A.10.2 (Left) The ENDOR spectra of ^{T}P680 in frozen solution PS II ($T = 15$K) for the EPR transition, Z^- and Z^+; A, absorption; E, emission. (Right) The TR EPR spectrum of ^{T}P680 in frozen solution PS II ($T = 15$K) and spin energy levels of a triplet state molecule ($S = 1$) when the triplet Z axis is parallel to the B_0 direction. The first-order electron Zeeman, nuclear Zeeman, and hyperfine splittings ($D > 0$, $A_{ZZ} > 0$) are shown on the right together with the expected ENDOR transitions. Modified with permission [54]. Copyright 1996 Elsevier.

Pashenko et al. studied the triplet state by high field/high frequency (130 GHz) pulsed EPR in a 50–90K temperature range. They found that at 50K the zero-field splitting parameters correspond well to those of a chlorophyll monomer, but at higher temperatures the shape of the spectra changes significantly [55]. Triplet excitation hopping involving the accessory chlorophyll Chl_D and P_{D1} or P_{D2} can partially explain those changes, but the most prominent features indicate that the primary electron acceptor (Pheo a) must also be involved.

REFERENCES

1. Commoner, B.; Heise, J. J.; Townsend, J. *Proc. Natl. Acad. Sci. U.S.A.* **1956**, *42*, 710.
2. Hoff, A.; Deisenhofer, J. J. *Phys. Rep.* **1997**, *287*, 1.
3. Lubitz, W. EPR in Photosynthesis. In *Electron Paramagnetic Resonance. A Specialist Periodical Report*; Gilbert, B., Davies, M., Murphy, D., Eds.; Royal Society of Chemistry: Cambridge, U.K., 2004; Vol. 19, pp 174–242.
4. Jordan, R.; Fromme, P.; Witt, H. T.; Klukas, O.; Saenger, W.; Krauß, N. *Nature* **2001**, *411*, 909.
5. Ben-Shem, A.; Frolow, F.; Nelson, N. *Nature* **2003**, *426*, 630.
6. Ferreira, K. N.; Iverson, T.; Maghlaoui, M.; Barber, K.; Iwata, J. S. *Science* **2004**, *303*, 1831.
7. (a) Zouni, A.;Witt, H.-T.; Kern, J.; Fromme, P.; Krauss, N.; Saenger, W.; Orth, P. *Nature* **2001**, *409*, 739; (b) Loll, B.; Kern, J.; Saenger, W.; Zouni, A.; Biesiadka, J. *Nature* **2005**, *438*, 1040.
8. Ort, D. R., Yocum, C. F., Eds. *Oxygenic Photosynthesis: The Light Reactions;* Kluwer Academic: Dordrecht, 1996; p 503.
9. Guergova-Kuras, M.; Boudreaux, A.; Joliot, A.; Joliot, P.; Redding, K. *Proc. Natl. Acad. Sci. U.S.A.* **2001**, *98*, 4437.
10. Hoff, A. J. *Phys. Rep.* **1979**, *54*, 75.
11. Petrenko, A.; Maniero, A. L.; van Tol, J.; MacMillan, F.; Li, Y.; Brunel, L.-C.; Redding, K. *Biochemistry* **2004**, *43*, 1781.
12. Bratt, P. J.; Poluektov, O. G.; Thurnauer, M. C.; Krzystek, J.; Brunel, L.-C.; Schrier, J.; Hsiao, Y.-W.; Zerner, M.; Angerhofer, A. *J. Phys. Chem.* **2000**, *B104*, 6973.
13. Bratt, P. J.; Rohrer, M.; Krzystek, J.; Evans, M. C. W.; Brunel, L.-C.; Angerhofer, A. *J. Phys. Chem.* **1997**, *101*, 9686.
14. Li, Y.; Lucas, M.-G.; Konovalova, T.; Abbott, B.; MacMillan, F.; Petrenko, A.; Sivakumar, V.; Wang, R.; Hastings, G.; Gu, F.; van Tol, J.; Brunel, L.-C.; Timkovich, R.; Rappaport, F.; Redding, K. *Biochemistry* **2004**, *43*, 12634.
15. Zech, S.; Hofbauer, W.; Kamlowski, A.; Fromme, P.; Stehlik, D.; Lubitz, W.; Bittl, R. *J. Phys. Chem.* **2000**, *B104*, 9728.
16. Käss, H.; Fromme, P.; Witt, H. T.; Lubitz, W. *J. Phys. Chem.* **2001**, *B105*, 1225.
17. Käss, H.; Fromme, P.; Lubitz, W. *Chem. Phys. Lett.* **1996**, *257*, 197.
18. Käss, H.; Lubitz, W. *Chem. Phys. Lett.* **1996**, *251*, 193.
19. Deligiannakis Y.; Rutherford A. W. *Biochim. Biophys. Acta* **2001**, *1507*, 226.
20. Krabben, L.; Schlodder, E.; Jordan, R.; Carbonera, D.; Giacometti, G.; Lee, H.; Webber A. N.; Lubitz, W. *Biochemistry* **2000**, *39*, 13012.

21. Witt, H.; Schlodder, E.; Teutloff, C.; Niklas, J.; Bordignon, E.; Carbonera, D.; Kohler, S.; Labahn, A.; Lubitz, W. *Biochemistry* **2002**, *41*, 8557.
22. Koulougliotis, D.; Innes, J. B.; Brudvig, G. W. *Biochemistry* **1994**, *33*, 11814.
23. Rigby, S. E. J.; Nugent, J. H. A.; O'Malley, P. J. *Biochemistry* **1994**, *33*, 10043.
24. Lubitz, W. *Phys. Chem. Chem. Phys.* **2002**, *4*, 5539.
25. Angerhofer, A.; Bittl, R. *Photochem. Photobiol.* **1996**, *63*, 11.
26. Stehlik, D.; Moebius, K. *Annu. Rev. Phys. Chem.* **1997**, *48*, 745.
27. Moebius, K. *Chem. Soc. Rev.* **2000**, *29*, 129.
28. van der Est, A. *Biochim. Biophys. Acta* **2001**, *1507*, 212.
29. Bittl, R.; Weber, S. *Biochim. Biophys. Acta* **2005**, *1707*, 117.
30. Kandrashkin, Y. E.; van der Est, A. *Appl. Magn. Reson.* **2007**, *31*, 105.
31. Levanon, H.; Hasharoni, K. *Prog. React. Kinet.* **1995**, *20*, 309.
32. Levanon, H.; Möbius, K. *Annu. Rev. Biophys. Biomol. Struct.* **1997**, *26*, 495.
33. Di Valentin, M.; Bisol, A.; Agostini, G.; Fuhs, M.; Liddell, P. A.; Moore, A. L.; Moore, T. A.; Gust, D.; Carbonera, D. *J. Am. Chem. Soc.* **2004**, *126*, 17074.
34. Carbonera, D.; Di Valentin, M.; Corvaja, C.; Agostini, G.; Giacometti, G.; Liddell, P. A.; Kuciauskas, D.; Moore, A. L.; Moore, T. A.; Gust, D. *J. Am. Chem. Soc.* **1998**, *120*, 4398.
35. Hore, P. J. Analysis of Polarized Electron Paramagnetic Resonance Spectra. In *Advanced EPR: Applications in Biology and Biochemistry;* Hoff, A. J., Ed.; Elsevier: Amsterdam, 1989; pp 405–440.
36. Till, U.; Hore, P. J. *Mol. Phys.* **1997**, *90*, 289.
37. Till, U.; Klenina, I. B.; Proskuryakov, I. I.; Hoff, A. J.; Hore, P. J. *J. Phys. Chem.* **1997**, *B101*, 10939.
38. Blankenship, R.; McGuire, A.; Sauer, K. *Proc. Natl. Acad. Sci. U.S.A.* **1975**, *72*, 4943.
39. Stehlik, D.; Bock, C. H.; Petersen, J. *J. Phys. Chem.* **1989**, *93*, 1612.
40. Füchsle, G.; Bittl, R.; van der Est, A.; Lubitz, W.; Stehlik, D. *Biochim. Biophys. Acta* **1993**, *1142*, 23.
41. van der Est, A.; Prisner, T.; Bittl, R.; Fromme, P.; Lubitz, W.; Möbius, K.; Stehlik, D. *J. Phys. Chem.* **1997**, *B101*, 1437.
42. Kothe, G.; Weber, S.; Ohmes, E.; Thurnauer, M. C.; Norris, J. R. *J. Phys. Chem.* **1994**, *98*, 2706.
43. Tang, J.; Thurnauer, M. C.; Norris, J. R. *Chem. Phys. Lett.* **1994**, *219*, 283.
44. Dzuba, S. A. *Chem. Phys. Lett.* **1978**, *278*, 333.
45. Bittl, R.; Zech, S. G. *Biochim. Biophys. Acta* **2001**, *1507*, 194.
46. Angerhofer, A. Chlorophyll Triplets and Radical Pairs. In *Chlorophylls;* Scheer, H., Ed.; CRC Press: Boca Raton, FL, 1991; p 945.
47. Budil, D. E.; Thurnauer, M. C. *Biochim. Biophys. Acta* **1991**, *1057*, 1.
48. Carbonera, D.; Collareta, P.; Giacometti, G. *Biochim. Biophys. Acta* **1997**, *1322*, 115.
49. Sieckmann, I.; Brettel, K.; Bock, H.; van der Est, A.; Stehlik, D. *Biochemistry* **1993**, *32*, 4842.
50. Rutherford, A. W.; Sétif, P. *Biochim. Biophys. Acta* **1990**, *1019*, 128.
51. Poluektov, O. G.; Utschig, L. M.; Schlesselman, S. L.; Lakshmi, K. V.; Brudvig, G. W.; Kothe, G.; Thurnauer, M. C. *J. Phys. Chem.* **2002**, *B106*, 8911.
52. van Mieghem, F. J. E.; Satoh, K.; Rutherford, A. W. *Biochim. Biophys. Acta* **1991**, *1058*, 379.
53. Kammel, M.; Kern, J.; Lubitz, W.; Bittl, R. *Biochim. Biophys. Acta* **2003**, *1605*, 47.

54. Di Valentin, M.; Kay, C. W. M.; Giacometti, G.; Moebius, K. *Chem. Phys. Lett.* **1996**, *248*, 434.
55. Pashenko, S. V.; Proskuryakov, I. I.; Germano, M.; van Gorkom, H. J.; Gast, P. *Chem. Phys.* **2003**, *294*, 439.

FURTHER READING

Nitroxide Spin Labels

Berliner, L. J., Ed. *Spin Labeling: Theory and Applications*; Academic Press: New York, 1976.

Berliner, L. J., Reuben, J., Eds. Spin Labeling: Theory and Applications. In *Biological Magnetic Resonance*; Plenum: New York, 1989; Vol. 8.

Berliner L. J., Ed. Spin Labeling: The Next Millennium. In *Biological Magnetic Resonance*; Plenum: New York, 1998; Vol. 14.

Eaton, G. R.; Eaton, S. S. Distance Measurements in Biological Systems by EPR. In *Biological Magnetic Resonance*; Berliner, L. J., Ed.; Plenum: New York, 2001; Vol. 19.

EPR and Biological Membranes

Borbat, P. P.; Costa-Filho, A. J.; Earle, K. A.; Moscicki, J. K.; Freed, J. H. *Science* **2001**, *291*, 266.

Hemminga, M. A., Berliner, L. J., Eds. *ESR Spectroscopy in Membrane Biophysics*; Springer: New York, 2007.

Protein Topology

Hubbell, W.; Altenbach, C.; Hubbell, C.; Khorana, H. G. *Adv. Prot. Chem.* **2003**, *63*, 243.

Oximetry

Subczynski, W. K.; Hyde, J. S.; Kusumi, A. *Proc. Natl. Acad. Sci. U.S.A.* **1989**, *86*, 4474.

Ligeza, A. A.; Tikhonov, A. N.; Hyde, J. S.; Subczynski, W. K. *Biochim. Biophys. Acta Bioenergetics* **1998**, *1365*, 453.

Photosynthesis

Hoff, A. J., Ed. *Advanced EPR: Applications in Biology and Biochemistry*; Elsevier: New York, 1989.

Amesz, J., Hoff, A. J., Eds. *Biophysical Techniques in Photosynthesis. Advances in Photosynthesis*; Kluwer Academic: New York, 1996; Vol. 3.

11 Electron Paramagnetic Resonance Detection of Radicals in Biology and Medicine

MICHAEL J. DAVIES

The Heart Research Institute, 114 Pyrmont Bridge Road, Camperdown, Sydney, NSW 2050, Australia

11.1 FREE RADICALS IN DISEASE PROCESSES

There is now considerable evidence for, and interest in, the role that free radicals play in a number of human diseases. Although in some cases the formation of radicals is likely to turn out to be a secondary event and irrelevant to disease formation and progression, there are now a number of examples where it is clear that radicals are a major cause of damage and/or play a major role in disease development [1]. The need for definitive information as to whether and how radicals are involved has sparked considerable interest in the detection of radicals in biological and medical systems. Because only the most extreme pathologies give rise to massive and acute fluxes of radicals, the detection of radicals is often fraught with problems and the literature abounds with erroneous data. As one of the few methods that allow direct detection of radicals in complex systems such as intact cells and tissue samples, electron paramagnetic resonance (EPR) has, and will hopefully continue to, play a vital role in this process. However, the low rates and extents of radical formation in intact biological systems make radical detection especially challenging, and this is compounded by the knowledge that many secondary downstream processes that are irrelevant to the disease progression often result in radical formation (e.g., cell necrosis). Given the complex nature of these systems, a number of different approaches have been developed to obtain the maximal amount of information from a given system, and consistent data obtained from multiple methods is often

---

Electron Paramagnetic Resonance. Edited by Brustolon and Giamello
Copyright © 2009 John Wiley & Sons, Inc.

required to obtain a complete picture. The information reviewed here is designed to provide guidance and approaches to tackling this complex and challenging area and will not attempt to cover all of the detailed experimental information required to tackle particular problems. However, consideration of these overriding principles and examination of the cited references should hopefully allow educated decisions to be made as to the methods that are most likely to yield positive and informative results.

11.2 NATURE OF FREE RADICALS INVOLVED IN DISEASE PROCESSES AND POTENTIAL CATALYSTS FOR RADICAL FORMATION

It is now well established that there are multiple endogenous and exogenous processes that can give rise to radical formation in biological systems (see, e.g., Ref. 1). Some of the radicals that are formed are intentionally generated, for example, in enzyme catalytic cycles, in the killing of invading pathogens, or as signaling molecules (e.g., NO·). Other endogenous sources appear to be a necessary evil associated with electron transport and the use of oxygen as the terminal electron acceptor in mitochondria. Thus, electron leakage to oxygen from mitochondrial and endoplasmic reticulum electron transport chains has been reported to occur at levels of up to a few percent; the exact level of superoxide and hydrogen peroxide formation via this pathway has been the subject of some dispute. Other potential endogenous sources of radicals include a large number of redox-active enzymes that may oxidize inappropriate substrates, including surrounding protein structures, to yield reactive intermediates, and include activated white cells (neutrophils, monocytes, macrophages, eosinophils) that use oxidative chemistry to tackle bacterial, parasitic, and fungal infections. Indeed, it is now relatively well accepted that chronic, inappropriate inflammation is a major factor in generating oxidative damage in many human diseases ranging from heart disease to rheumatoid arthritis, asthma, cystic fibrosis, and a significant number of cancers [1].

A wide range of external factors can also stimulate radical generation. These sources range from various types of high energy radiation (e.g., X-ray and γ-, β-, and α-particle radiation, neutrons, UV light, visible light in the presence of sensitizers) to various atmospheric pollutants (sulfur and nitrogen oxides, particulate matter, ozone), inappropriate metabolism of various drugs, solvents (e.g., chlorinated hydrocarbons), industrial and household chemicals (e.g., the weedkiller Paraquat), and exposure to various industrial compounds (e.g., asbestos, chromium, arsenic). A full discussion of the extent of such sources is beyond the scope of this chapter; a number of review articles and books have covered this subject in depth (see, e.g., [1]).

11.3 DIRECT EPR DETECTION OF REACTIVE RADICALS *IN VIVO* AND *EX VIVO*

Because many of the processes involved in radical formation in biological systems generate radicals at relatively slow rates, the steady-state concentration of many radicals *in vivo* is rather low, with low micromolar concentrations often being an absolute maximum. Because the sensitivity limit of most commonly available spectrometers is in the range of $10^{-7}-10^{-8}$ mol dm^3, direct EPR studies of radicals are limited to a relatively small number of cases, primarily involving long-lived radicals where the low rates of radical formation are compensated for by a slow rate of decay. Examples of such systems include the following:

1. the ascorbyl radical, which is generated on oxidation of the key cellular and plasma antioxidant ascorbate (vitamin C);
2. the α-tocopheroxyl radical, which is generated on oxidation of the key membrane and lipoprotein antioxidant α-tocopherol (a component of vitamin E);
3. phenoxyl radicals generated from hindered phenols, structures that are present in a number of synthetic antioxidants (e.g., butylated hydroxytoluene and butylated hydroxyanisole), a large number of natural products (e.g., flavonoids) that have antioxidant properties, and tyrosine residues on proteins;
4. nitroaromatics that yield long-lived radical anions on reduction; and
5. semiquinone radicals (e.g., those generated from the important membrane electron transport component, coenzyme Q, and a number of important drugs, such as adriamycin and doxorubicin).

A significant number of biologically important metal ions can also be detected directly by EPR in biological systems, and these include a number of widely distributed enzyme cofactors (e.g., Fe^{3+}, Cu^{2+}, Mn, Mo) and carcinogenic species such as Cr salts. Because of the much greater amount of information that can be obtained from direct studies (hyperfine couplings, g values, line shapes and linewidths, absolute concentrations, and more definitive, although not always easier, spectral analysis), direct detection should always be considered as the first option.

In cases where high yields of radicals are generated in a transient fashion (e.g., via enzyme burst kinetics, on addition of high concentrations of an oxidant, during reperfusion injury after ischemia, etc.), it is possible to obtain valuable information from low temperature EPR studies on rapidly frozen samples. This can be achieved by freeze-clamping tissue samples with tongs cooled in liquid nitrogen or by use of a conventional freeze-quenching apparatus for enzyme reactions. Once the radicals are trapped in the frozen matrix, they typically exhibit long lifetimes that allow the acquisition of multiple EPR spectra over an extended period of time with various spectrometer conditions. However, the interpretation and assignment of the resultant EPR spectra to particular radicals can be highly problematic, which is due to the anisotropy induced by freezing. This is not necessarily a problem in well-defined systems with single or few components (cf. the large number of highly successful, low temperature studies carried out on isolated enzymes), but it can be a huge problem in complex biological systems where radical formation may be occurring at (literally) thousands of different sites. There are other drawbacks associated with carrying out low temperature studies, not the least of which are the increased number of samples that are needed to allow the time course to be constructed (particularly a problem with large animal models), problems associated with introducing tissue samples into EPR tubes while keeping the material frozen, and the natural bias toward the successful detection of longer lived species resulting from the finite time required for freezing. It is now well established that manipulation of frozen tissue samples is undesirable and should be avoided if at all possible; it is known that numerous artifacts are introduced during cutting, grinding, or other sample processing.

Continuous generation of radicals has also been used as a means of circumventing the short lifetimes of many radical species. Although rapid flow systems have been used very successfully to detect radicals in isolated chemical systems, and in a few cases with isolated enzymes, the large volumes of reagents that are required can be prohibitively expensive. Stopped flow systems have also been employed with some success, but again such studies are limited to systems where relatively high reagent concentrations are achievable and to systems that can withstand the stresses imposed by the conditions employed (e.g., many cells are lysed by rapid pressure changes).

11.4 SPIN TRAPPING OF REACTIVE RADICALS IN VIVO AND EX VIVO

11.4.1 General Principles

The technique of spin trapping was developed in the late 1960s to facilitate the detection of reactive free radicals in chemical systems (reviewed in Refs. 2–5). This method involves the addition of a spin trap, typically a nitrone or nitroso compound (Fig. 11.1), to the system under study at a concentration sufficient to ensure significant extents of trapping of any radicals present to give stable, detectable, nitroxide radical adducts (Fig. 11.2). Nitroso spin traps, of which 2-methyl-2-nitrosopropane (MNP) and 3,5-dibromo-4-nitrosobenzene sulfonic acid (DBNBS) are the most commonly employed, have the advantage that the reactive radical attaches directly to the nitroso nitrogen atom and is therefore in proximity to the unpaired electron that is localized primarily on the nitroxide function. This usually results in the detection of additional distinctive hyperfine couplings from magnetic nuclei present in the added radical. The size and nature of these couplings helps identify the added radical, and a number of compilations of radical adduct data are available (e.g., Ref. 6 and the NIEHS website at http://EPR.niehs.nih.gov).

A number of studies have determined kinetic data for the rates of addition of radicals to various common spin traps; in general, the rate of addition of radicals to nitroso traps is more rapid than to nitrones (reviewed in Ref. 7). These traps do, however, have the disadvantage that they form long-lived readily detectable adducts with a more limited range of radicals (usually limited to carbon-centered species) than nitrone traps (reviewed in Refs. 8 and 9). Many nitroso traps are also often thermally and photochemically unstable, and this results in higher concentrations of artifactual EPR signals that complicate spectral analysis.

In contrast, nitrone spin traps, including 5,5-dimethyl-1-pyrroline *N*-oxide (DMPO), 5-diethoxyphosphoryl-5-methyl-1-pyrroline *N*-oxide (DEPMPO), *N-tert*-butyl-α-phenylnitrone (PBN), and α-(4-pyridyl-1-oxide)-*N-tert*-butylnitrone (POBN),

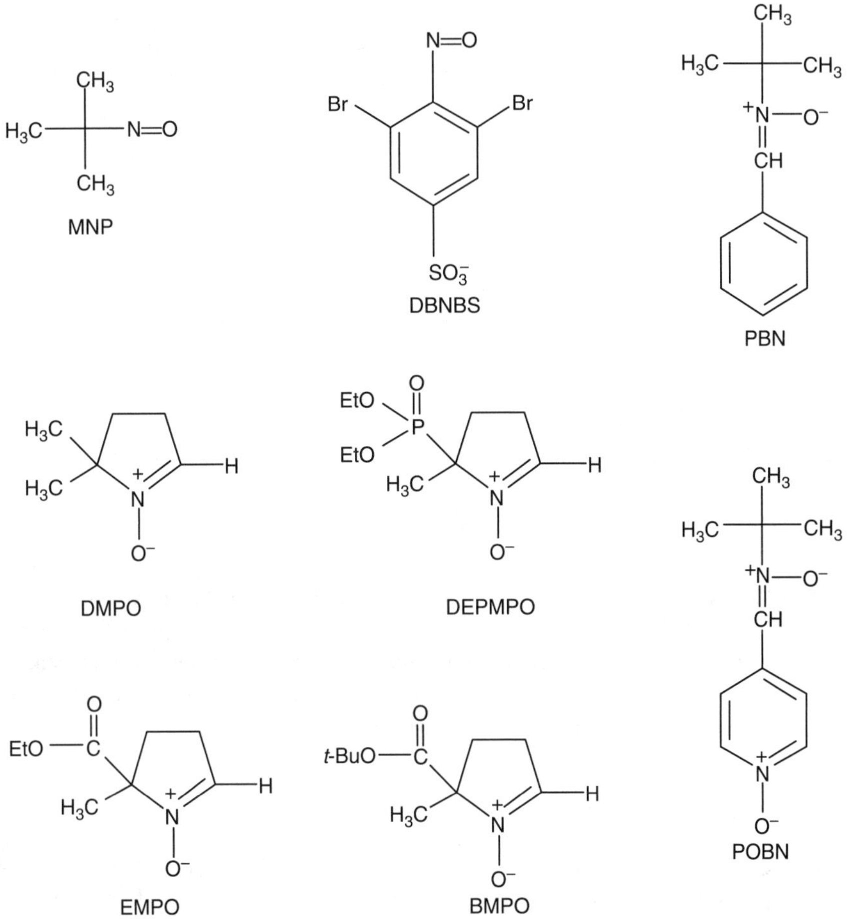

Fig. 11.1 The structures of commonly used spin traps.

often form long-lived adducts with a wider range of radical types (e.g., carbon-, oxygen-, sulfur-, and nitrogen-centered species; reviewed in Refs. 8 and 9). However, with nitrone spin traps the reactive radical adds to the carbon atom adjacent to the incipient nitroxide group and is therefore more distant from the orbitals containing the unpaired electron. As a result, it is often not possible to resolve hyperfine couplings from the added radical; this makes the definitive assignment of the observed spectral lines to a particular species problematic. However, the magnitude of the hyperfine couplings arising from the spin trap derived nitroxide nitrogen and (especially) the β hydrogen are markedly dependent, particularly with the cyclic nitrones, on the nature and structure of the added radical as a result of the influence of this species on the conformation of the nitrone. The size of these couplings can therefore provide valuable information on the nature of the trapped radical [3, 4, 6]. The use of spin traps to detect radicals in biological systems has been reviewed extensively (see e.g., the specialist periodical series *Electron Spin Resonance*, later

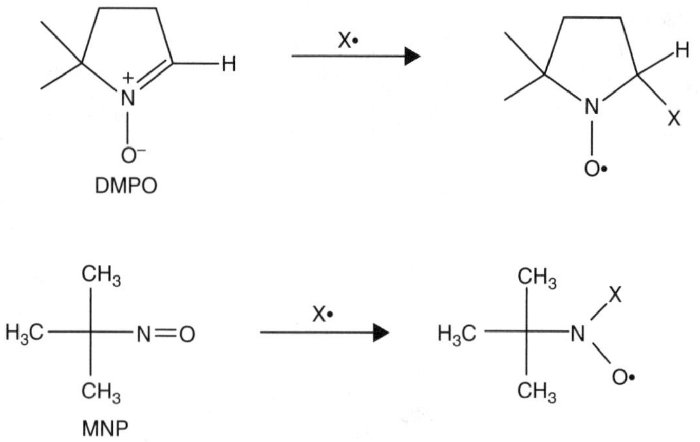

Fig. 11.2 The formation of radical adducts from nitrone (DMPO) and nitroso (MNP) spin traps.

changed to *Electron Paramagnetic Resonance*, published by the Royal Society of Chemistry [8]). Although spin trapping was developed initially to examine reactive radicals formed from low molecular mass compounds [2–5], an increasing number of studies have used this methodology to examine radical formation on proteins and other macromolecules in both isolated and complex biological systems [10–13].

11.4.2 Choice of Spin Trap

Given the number of spin traps now available (reviewed in Refs. 8 and 9), choosing the correct spin trap for the studies to be undertaken is a very important factor. Having some idea of the nature of the species that might be formed or having a clear idea of what information is required can aid such choices. Thus, nitrone traps are likely to be the traps of choice for heteroatom-centered species (oxygen, nitrogen, sulfur); if detailed information is required as to the nature of a carbon-centered species, nitroso traps are the most sensible choice (reviewed in Refs. 8 and 9). There are also numerous other factors that need to be taken into consideration for biological studies, which are discussed below.

11.4.2.1 Kinetics of Adduct Formation and Decay. The radicals under study should react rapidly with the spin trap to give long-lived adduct species; some rate constants have been measured for the reaction of a range of radicals with spin traps, and these data are compiled in Table 11.1 [7, 14]. Information is less readily available on the rates of decay of adduct radicals once formed, and what data are available is likely to be situation specific. Thus, many of the decay processes are very dependent on the exact reaction conditions; the presence of metal ions and reductants, pH, temperature, and solvent are all known to play a role in determining the rate of decay (reviewed in Refs. 8 and 9).

TABLE 11.1 Selected Rate Constants for the Addition of Radicals to Spin Traps

Spin Trap	Radical	k, M^{-1} s^{-1}	Spin Trap	Radical	k, M^{-1} s^{-1}
MNP	·CH$_3$	1.7×10^7	DMPO	·C(OH)(CH$_3$)$_2$	5.4×10^7
	·CH$_2$CH$_3$	5.3×10^7		PhC(O)O·	8.0×10^{7a}
	·CH$_2$COO$^-$	7×10^6		·OH	2.7×10^9
	·CH(CH$_3$)$_2$	4.6×10^7		·OOH	6.6×10^3
	·CH$_2$OH	1.4×10^8		O$_2^-$·	~10
	·CH(OH)CH$_3$	3.2×10^8		·SPh	5.2×10^{7b}
	·C(CH$_3$)$_3$	3.3×10^{6a}		Glutathionyl-S·	2.6×10^8
	·CCl$_3$	1.9×10^{6b}		Cysteinyl-S·	2.1×10^8
	·SO$_3^-$	4.3×10^7		·OC(CH$_3$)$_3$	9.0×10^6
	·CO$_2^-$	1.7×10^9		·OOC(CH$_3$)$_3$	$<10^6$
	e$^-$	6.2×10^9	PBN	·CH$_3$	4.0×10^{6b}
	·H	9.1×10^8		·CH$_2$(CH$_2$)$_3$R	$1.3 \times 10^{5a,c}$
	·OH	2.5×10^9		·C(CH$_3$)$_3$	$<10^a$
	·OC(CH$_3$)$_3$	1.3×10^8		·CH$_2$Ph	2.0×10^{7a}
	·OCH$_3$	1.3×10^{8d}		·Ph	1.2×10^{7e}
	·SPh	1.7×10^{8b}		·CCl$_3$	7.0×10^{4b}
DBNBS	·CH$_3$	1.5×10^9		·CH$_2$OH	4.3×10^7
	·CH$_2$OH	2.5×10^8		·CH(OH)CH$_3$	1.6×10^7
	·CH(OH)CH$_3$	6.3×10^7		·C(OH)(CH$_3$)$_2$	1.0×10^7
	·OH	3.9×10^9		·C(O)Ph	8.1×10^5
	·CO$_2^-$	1.5×10^9		·H	5.5×10^8
	·N$_3$	2.4×10^8		·OH	8.5×10^9
	O$_2^-$·	4.3×10^7		·OOH	50
DMPO	·CH$_3$	$<10^7$		·OC(CH$_3$)$_3$	7.9×10^7
	·CH$_2$(CH$_2$)$_3$R	$2.6 \times 10^{6a,c}$		·OCH$_3$	1.2×10^{8f}
	·C(CH$_3$)$_3$	5.0×10^{7b}		PhC(O)O	5.0×10^{6a}
	Ph·	7.0×10^{7b}		·SPh	1.1×10^{7g}
	·CH$_2$OH	2.3×10^7		Glutathionyl-S·	7.5×10^7
	·CH(OH)CH$_3$	6.1×10^7		Et$_3$Si·	7.1×10^{7h}

All conditions are room temperature and in aqueous solution, unless stated otherwise by the italic footnotes. Data are taken from References 7 and 14.
[a]At 40°C in benzene.
[b]At room temperature in organic solvents (usually benzene).
[c]R = —CH=CH$_2$.
[d]At −45°C in ethanol.
[e]At 25°C in methanol.
[f]At 20°C, solvent not given.
[g]At 60°C in chlorobenzene.
[h]At 27°C in di-*tert*-butylperoxide/triethylsilane.

11.4.2.2 Distinctiveness of Adduct Signals.

The adduct species should preferably give rise to highly characteristic hyperfine coupling constants. In this respect, the nitroso traps (e.g., MNP) and cyclic nitrones (e.g., DMPO, DEPMPO) tend to provide more distinctive data than the acyclic nitrones such as PBN and POBN. For the nitroso traps this is because of the potential for distinctive g shifts and additional hyperfine coupling constants from the added radical, and for the cyclic nitrones it is because of the large variation in the magnitude of the β-hydrogen couplings.

11.4.2.3 Purity of Spin Trap. The spin trap should be as pure as possible! Even though spin traps may be obtained from commercial suppliers at a purity of >99%, the high concentrations of traps used in many EPR experiments (often up to hundreds of millimoles) result in impurities also being present at high levels (e.g., up to millimolar levels). If any of these species are themselves radicals or species that can be readily oxidized (e.g., hydroxylamines) or reduced to radicals, then these can give rise to artifact signals. It is therefore essential that adequate controls are carried out for all studies.

11.4.2.4 Potential Transformations of Spin Trap and Adducts. The potential transformation of both the spin trap and/or its adducts to other species can be an important factor. Rapid reduction of the spin adducts to the corresponding hydroxylamines is a particularly important route to loss of adduct signals in intact biological systems [15]. Oxidation to the nitrone can also occur, but this is far less likely. This route is, however, of some considerable significance because it can result in the formation of long-lived products that can be detected by several non-EPR approaches [e.g., mass spectroscopy (MS) and Western blotting; see below]. Addition of a nucleophile to the β position of nitrone traps (often called the Forrester–Hepburn mechanism), oxidation of the spin trap and subsequent reaction with a nucleophile (inverted spin trapping), or "ene" addition of a nitroso group to a double bond and subsequent oxidation of the resulting hydroxylamines to nitroxide radicals are all potential sources of artifacts, which are sometimes very difficult to distinguish from "real" spin trapping [5, 16–18].

11.4.2.5 Pharmacokinetics of Trap and Adducts. The pharmacokinetics of both the trap and its adducts need to be appropriate for the planned studies. Partition coefficient data, which give an estimate of how a spin trap might distribute in complex systems, are available for many common spin traps (Table 11.2) [19]. These data, although useful, need to be treated with care, because many other factors (e.g., binding to proteins) may modulate these effects. Penetration across specific membranes (e.g., across the blood–brain barrier) is also not readily predicted from this data and requires experimental verification. Lack of spin adduct signals may merely be a consequence of poor localization of the spin trap at the site of radical formation. Relatively little is known about the partitioning of spin adducts once formed (much more is known about stable nitroxides that have been extensively studied as magnetic resonance imaging contrast agents and oxygen probes [15]), and the rapid translocation of such species from their site of generation may result in erroneous conclusions about the site of radical generation. Such mobility after radical adduct formation can, however, be of benefit as well. Thus, a significant number of spin adducts end up in bile for excretion, and this has been shown to be a powerful tool in investigating spin adduct formation *in vivo* (see, e.g., Refs. 20 and 21).

TABLE 11.2 Partition Coefficients for Spin Traps in 1-Octanol/Phosphate Buffer (10 mM, pH 7.5)

Spin Trap	Partition Coefficient
DMPO	0.08
POBN	0.09
DBNBS	0.15
MNP	8.2
PBN	10.4
Nitrosobenzene	73.0

Data are taken from Reference 19.

11.4.2.6 Toxicity. The toxicity of a spin trap may have an important bearing on the veracity of the information obtained. Although some of the more hydrophilic nitrone traps such as DMPO appear to be relatively nontoxic and have been given to animals at concentrations of up to high millimolar amounts in acute experiments, most of the hydrophobic materials, particularly some nitroso traps, have significant toxic effects at the sort of concentrations that are typically employed in spin trapping experiments [19, 22, 23]. This may arise via multiple pathways including membrane perturbation effects and the inhibition of key enzymes [24]. These effects obviously makes interpretation of any resulting radical adduct data problematic and of doubtful veracity. It is therefore of great importance that careful analysis is undertaken of the direct effects of the spin trap on the system under study before analysis of any spectra is undertaken; it would be surprising if high concentrations of spin traps did not perturb the system under study to some extent. Unfortunately, relatively little is known on this topic and extrapolation of the (few) data available to other systems is likely to be fraught with dangers. Detailed information on the effects of nitroso and nitrone compounds on enzyme inhibition, antioxidant status, and induction of cell necrosis or apoptosis would be of great benefit to the EPR community, although each system is likely to be unique.

11.4.2.7 Method of Spin Trap Introduction. The method by which the spin trap is introduced into the studied system needs to be appropriate. Although it is obvious that the spin trap must be added to the system before or during the period when radicals are formed, the method by which this is done needs some consideration. For cell culture experiments, simple addition of a short period before the insult is applied is usually sufficient. However, it should be borne in mind that penetration of some of the spin traps into particular subcellular locations may not be instantaneous, with this likely to be spin trap dependent. However, a very long preincubation period of cells with spin traps is also potentially problematic because this may result in greater toxicity and or metabolism of the trap [24]. For poorly soluble traps the use of cosolvents needs to be carefully considered, because of their potential toxicity and radical scavenging actions; in some cases cellular loading can be achieved by use of carrier materials such as proteins (cf. the well characterized action of serum albumins in binding hydrophobic compounds). Such protein binding may be problematic because it may reduce the effective concentration of the spin trap. Similar considerations apply to tissue experiments. For animal studies, a number of approaches have been employed successfully including oral dosing via a stomach tube (oral gavage), intraperitoneal injection (a highly efficacious route for getting spin traps into the liver and hence the circulation), and intravenous administration (which ensures rapid systemic dosing; reviewed in Ref. 24). Microdialysis has also been employed in some animal studies where the spin trap is introduced via a semipermeable membrane on a probe, with the resulting spin adducts diffusing out via the same route for sampling [25]. After introduction of the spin trap into an animal, a large number of avenues are available for sampling of the spin adducts including urine, bile, and blood collection and tissue biopsies; the cost of animals and the principle of reduction of animal usage makes the former methods preferable, although not always possible. With regard to (intact) human studies, *ex vivo* addition of a spin trap (e.g., to blood samples) is currently the only method that can be readily employed [26–28]. The interpretation of data from such studies is fraught with problems; even with the fastest sampling procedure (drawing blood directly into a syringe containing a solution of the spin trap) the time frame is too great to detect any radicals that were originally present in the circulation. Thus, any signals that are detected are likely to arise from secondary reactions (e.g., decomposition of preexisting peroxides present on lipids [26]). Unfortunately, in most, if not all, cases the process of drawing blood results in some cell lysis and the resulting release of hemoglobin and other cellular materials that can confound data analysis.

11.4.3 Aids to Spectral Analysis

In most biological studies the likelihood of the formation of multiple radicals is high, which can make analysis of the resulting spectra problematic. Given the variable kinetics of radical trapping and spin adduct decay, the most intense EPR signal may not necessarily arise from the radical generated in the highest yield nor be from the most important radical in terms of biological action or effect. Thus, a minor species that reacts rapidly with a spin trap and that gives a persistent adduct may give a much more intense EPR spin adduct signal than a major radical that reacts slowly with a trap and/or gives a short-lived (or undetectable) adduct. Thus, it is always important to try and determine whether the radicals that have been detected are central to the observed biological effect. This may require complimentary functional and endpoint studies to determine whether radical scavenging by the spin trap correlates with a diminution of the biological endpoint. Analysis of mixtures of spin adducts can be challenging, and the possibility of making definitive assignments is not always great. However, there are a number of methods that have been developed that can aid this process. These are discussed below.

11.4.3.1 Chemical Modification of the Suspected Site of Radical Formation. Modification or blocking of particular residues that are believed to be the site of radical formation can be a powerful tool to investigate the nature of trapped radicals, although it should be noted that such methods are not always as specific as might be wished. This method has found particular utility in examining the sites of radical formation on macromolecules such as proteins and enzymes, but this sort of approach has also been used to examine low molecular mass targets (e.g., by blocking of reduced glutathione). It should be borne in mind that many chemical agents alter all residues of a particular type instead of individual residues and are often not completely specific for particular amino acids (reviewed in Refs. 29 and 30). However, chemical modification is often relatively easy to achieve, and a number of reagents have been developed for the (relatively) specific modification of residues on proteins (reviewed in Refs. 29 and 30). Examples of the use of this technique include confirmation of the generation of thiyl radicals on proteins where data from native proteins are compared to that from proteins treated with alkylating agents, such as *N*-ethylmaleimide that modifies cysteine residues [31, 32]. Other amino acid residues that have been modified to facilitate spectral assignments include lysine (via reductive methylation [33]), tyrosine (by iodination [34, 35]), tryptophan (treatment with *N*-bromosuccinimide [34]), and histidine (treatment with diethylpyrocarbonate [36]). Negative data obtained using this approach do not always eliminate particular residues as sites of radical formation because there are multiple reasons why such modification reactions may not be effective including steric and electronic interactions that prevent modification of inaccessible residues (e.g., those buried within proteins). Care should always be taken to use the minimum amount of reactive agent necessary to achieve modification because high concentrations of many of these species can alter other sites in addition to those planned [29, 30], and confirmation of the nature and extent of modification (e.g., by loss of parent amino acid or generation of products) is always a useful addendum to convince manuscript referees.

11.4.3.2 Isotopic Modification of the Suspected Site of Radical Formation.

This approach relies on the fact that the presence of particular isotopes alters the observed spectrum of the radical adduct by removing an existing splitting (by substitution of the desired atom with a nonmagnetic, $I = 0$, isotope), by reducing or altering an existing coupling (e.g., by substitution of a hydrogen atom with a deuterium, which results in the conversion of a large doublet coupling into a much smaller 1:1:1 triplet, or substitution of a ^{14}N nucleus with ^{15}N, which results in the conversion of a 1:1:1 triplet pattern into a larger 1:1 doublet pattern), or by generating an additional coupling (e.g., substitution of a ^{12}C atom with a ^{13}C atom, $I = 0.5$, which results in an additional doublet coupling). Although this type of approach is relatively easy to achieve with low molecular mass compounds such as drugs, solvents, and other chemicals, the introduction of a suitable isotope label into macromolecules such as proteins, DNA, lipids, or complex carbohydrates is challenging. Many of the examples of the latter approach have been via isolated well-characterized enzymes with the label introduced at a particular residue by use of a labeled amino acid into a bacterial or cell-free expression system [37, 38]. This approach results in the labeling of all residues of a particular type and hence does not provide exact identification of the site of the radical, but it does yield data on the *type* of amino acid involved. Of the various possible isotope substitutions, those that give rise to additional large couplings (e.g., ^{12}C to ^{13}C) are usually the most advantageous, because these give rise to pronounced spectral changes. A related approach, site-directed mutagenesis, in which the suspected site of radical formation is mutated to a different residue that either does not undergo the same chemistry or results in a spectrally distinct species, has been used with great success for some isolated proteins with well-defined structures. Thus, radical formation on myoglobin and various peroxidases treated with H_2O_2 has been examined by use of single and multiple mutants of the tyrosine, tryptophan, and cysteine residues on which these species have been postulated to be formed (see, e.g., Refs. 37–41). Such studies have provided important information on the role of particular residues as "sinks" for oxidizing species and electron (or positive hole) transfer within three-dimensional structures. As with chemical modification, such changes need to be carried out with care and the effects of substitution assessed, because mutations can sometimes lead to changes in structure and stability, which may interfere with the rate or extent of the process under study. It is also worth noting that this methodology may merely change the site of radical formation instead of preventing it (e.g., it may merely move the site from one residue to another [42]); hence, radical adduct signals may still be detected.

11.4.3.3 Isotope Substitution in the Spin Trap. The resolution of the additional couplings from an added radical can provide very valuable data on the nature of these species. The detection of such (often very small) couplings can be aided by using isotope-substituted spin traps, in particular deuterated materials [e.g., wholly (d_{14}) or partly (d_9) deuterated PBN, wholly (d_9) deuterated MNP, wholly deuterated (d_2) DBNBS, deuterated DEPMPO], because this results in smaller intrinsic linewidths of the adducts derived from these traps as a result of the decreased size of the couplings to the deuterium atoms compared with the original hydrogen atoms (see, e.g., Ref. 43). ^{15}N-Labeled spin traps have also been employed because this isotope substitution results in a much larger (1:1) doublet coupling from the nitroxide nitrogen compared to the 1:1:1 triplet coupling observed with the ^{14}N-labeled trap; this results is an increased spectral width and hence a decreased potential overlap of the spectral lines [44].

11.4.3.4 Spectroscopic Simulations. Simulation of experimental spectra has proved to be a powerful tool to confirm that a proposed analysis is consistent with experimental data. It cannot *prove* that an assignment is correct, so it is important that this approach is used carefully and that the input parameters make chemical sense. In addition to confirming that a proposed assignment is consistent with the experimental data, this methodology can provide valuable data on the relative concentrations of different adducts when multiple species are present. The data need to be interpreted with care as they only provides data on the abundance of the *adducts* and not the parent radicals from which they were formed (the rate of trapping and the stability of the adducts formed play an important role in this process). Nonetheless, such data can provide valuable information as to the relative concentration of different species and changes in these parameters. This approach has been widely used for isotropic spectra from low molecular mass, rapidly tumbling species, but it has also been recently extended to the anisotropic spectra observed from the protein–radical adducts such as those formed from the cyclic nitrone traps DMPO and DEPMPO [45, 46]. The greater amount of data that need to be input to simulate such anisotropic spectra limits this approach to relatively clean systems (e.g., those where a single adduct is formed, such as on some proteins), because the analytical challenges that are inherent in deconvoluting anisotropic spectral species from multiple species has yet to be fully solved. However, considerable progress has been made in this area; it is now possible to distinguish the trapping of oxygen-, sulfur-, or carbon-centered species with some confidence [45, 46].

11.4.3.5 Use of Novel Spin Traps. A number of spin traps have recently been developed that have specific features that aid the detection of particular types of radicals (e.g., superoxide radicals) or radical formation in particular locales (reviewed in Refs. 8 and 9). Thus, considerable effort has been expended in the design of new traps that react rapidly with superoxide radicals to give adducts with much longer half-lives (and hence allow more ready detection of this species in complex biological systems). Remarkable progress has been made in this area, and several recently developed traps have orders of magnitude longer lifetimes (reviewed in Refs. 8 and 9). A few of these have entered commercial production [e.g., DEPMPO, 5-alkoxycarbonyl-5-methyl-1-pyrroline *N*-oxide (EMPO), 5-*tert*-butoxycarbonyl-5-methyl-1-pyrroline-*N*-oxide (BMPO)]. One added advantage of some of these new traps is the decreased propensity of the superoxide radical adduct to undergo decay to a species that is identical with the hydroxyl radical adduct. This has been a major problem (e.g., with DMPO) in determining whether superoxide radical formation occurs exclusively with some enzymes or whether both superoxide and hydroxyl radicals are formed concurrently. Although many of the newer traps do not undergo such spontaneous decay, some enzymes can or may catalyze this type of interconversion (e.g., glutathione peroxidases that reduce peroxides to alcohols [47]); hence, this problem may only have been lessened rather than eliminated.

A range of spin traps have also been developed that localize to particular regions within cells. Thus, spin traps have been created that localize in cell membranes [48, 49] and allow trapping of radicals at particular depths in membranes [50], bind to specific targets [50], and localize to particular subcellular compartments (e.g., derivatives of PBN, such as mito-PBN, and BMPO, which selectively localize to mitochondria as a result of the presence of a triphenylphosphonium substituent [53–55]). The design of such site-specific spin traps is obviously of great potential significance, but the limited market for these materials may limit the widespread commercial availability of these materials.

11.5 SPIN SCAVENGING OF REACTIVE RADICALS *IN VIVO* AND *EX VIVO*

Spin scavenging is related to spin trapping and comes in two forms: trapping of a radical with another radical, typically a nitroxide, and subsequent product analysis, and use of hydroxylamines that are readily oxidized to nitroxide radicals, thereby providing (limited) information as to the presence of radical species.

In the former approach the initial radicals are "trapped" using a stable nitroxide, such as 2,2,6,6-tetramethylpiperidine 1-oxyl (TEMPO), to give nonradical products and the resulting adduct species are characterized by MS. This approach has been used successfully to examine radical formation on a number of proteins exposed to oxidants such as H_2O_2 [56]. This method has some advantages over conventional spin trapping, in that the rate constants for the reactions of the initial radical with the trapping agent are often very large (cf. rate constant data for the reaction of a number of low molecular mass radicals with nitroxides [57]) and in many cases greater than for traditional trapping reactions (cf. data in Refs. 7 and 14). This arises from the fact that these are radical–radical rather than a radical–molecule reactions and hence occur at, or near, diffusion-controlled rates. Although many of the resulting TEMPO–adduct species are stable under the conditions used for MS, thereby eliminating a potential source of artifacts [56], it is unclear whether all of the species originally trapped survive the processing period. In particular, it is likely that adducts formed with heteroatom-centered radicals will rapidly decompose, because these contain weak N–O–X bonds (X = S, N, OR, OOR, OH, and OOH, when sulfur, nitrogen, alkoxyl, peroxyl, hydroxyl, and superoxide radicals are trapped). This has not been explored in detail. This technique may therefore be limited to, although very effective for, carbon-centered species.

In the second approach oxidation of an EPR-silent hydroxylamine to an EPR-detectable nitroxide is used as a method for assessing radical formation (e.g., Ref. 58). Because oxidation can be achieved by a wide range of radicals and is not limited to radicals (e.g., oxidation may be mediated by peroxidase enzymes), this approach has obvious limitations and drawbacks. However, some of these hydroxylamines (which are typically highly substituted five- or six-membered ring species, as the resulting nitroxide radicals are persistent [58]) react with radicals such as superoxide at appreciably greater rates than most traditional spin traps [59, 60] and hence may be more sensitive markers of the formation of this radical [59–61]. Experiments with specific protective enzymes are employed to address the specificity issue (i.e., which radical is causing the observed oxidation); a diminution of the nitroxide signal in the presence of added superoxide dismutase implicates superoxide radicals (or species derived therefrom) [61]. Although this is an attractive, and potentially very sensitive, approach, it is not always clear whether the enzymes (such as superoxide dismutase) required to elucidate the source of oxidation can be readily introduced into the systems under study (e.g., into cells, tissues, animals).

11.6 SPIN TRAPPING OF NITRIC OXIDE

Nitric oxide in now recognized as a key intracellular signaling molecule and an important regulator of a wide range of biological processes (e.g., vascular dilation and blood pressure regulation, platelet aggregation, neurotransmission). There is therefore widespread interest in the detection and quantification of this radical in biological systems ranging from humans to plants. Because of the delocalized nature of the unpaired electron on this species, it cannot be readily trapped by traditional EPR spin trapping as discussed above. As a consequence, a number of other methodologies have been developed to allow detection of this important radical; these are outlined briefly in following subsections.

11.6.1 Detection by Use of Heme Proteins

NO· binds avidly to Fe(II) heme moieties, with high association constants. The resultant nitrosyl-heme species can be readily detected by EPR with the spectra having highly distinctive triplet 1:1:1 couplings arising from the interaction of the unpaired electron with the nitrogen atom. The origin of this species can be readily checked by using the ^{15}N-labeled precursors, which results in the expected conversion to a large doublet coupling [62]. The ease of detection and quantification of these species has resulted in the widespread use of endogenous Fe(II) heme proteins as detection agents for this radical, in particular hemoglobin (Hb) and myoglobin. The use of these endogenous species allows ready measurements to be made on humans and the prospect of *in vivo* quantification via noninvasive methods using L-band spectrometers [63, 64]. It has been reported that the detection limit of NOHb in human blood is on the order of 200 nM, and the basal levels of this species in both arterial and venous blood are below this limit. Exposure of subjects to NO· results in dramatic elevations in the detected levels to 0.25–2.5 μM, and these levels remain elevated for considerable periods (half-life estimated to be ∼40 min) [65]. These levels are comparable to those detected by some other methods (e.g., triiodide chemiluminescence) but lower than others (e.g., photolysis chemiluminescence) [65]. The use of endogenous heme proteins as trapping agents has allowed this methodology to be extended to plants where, for example, NO· has been detected using the endogenous hemoprotein leghemoglobin in the symbiotic root nodules on leguminous plants [66].

11.6.2 Detection Using Nitronyl Nitroxides

A number of nitronyl nitroxides such as 2-(4-carboxyphenyl)-4,4,5,5-tetramethylimidazoline-1-oxyl-3-oxide (carboxy-PTIO) have been shown to react with NO· to yield iminonitroxides [e.g., 2-(4-carboxyphenyl)-4,4,5,5-tetramethylimidazoline-1-oxyl from carboxy-PTIO]. The latter have distinctly different EPR spectra from the former, allowing the interconversion of these species, and hence the formation of NO· to be examined [67–69]. The stoichiometry of the reaction of NO· with carboxy-PTIO was originally reported as 1:1 [68], but later work revised this figure to 0.63:1 as a result of competing reactions [69]. This stoichiometry also appears to vary with the rate of NO· formation, making quantification by this method inexact [69]. Later studies on cell systems have indicated that nitronyl nitroxides react rapidly with superoxide radicals ($k \sim 8 \times 10^5$ mol^{-1} s^{-1} [70]) and that this reaction does not generate the corresponding iminonitroxide [70]. Furthermore, a range of other reducing agents including reduced glutathione and ascorbate also remove the nitronyl nitroxide [70], indicating that this technique is unlikely to give accurate quantification of NO· formation in biological systems.

11.6.3 Detection Using Iron–Dithiocarbamate Complexes and Related Species

A range of water-soluble Fe(II) dithiocarbamate complexes [e.g., complexes with diethyldithiocarbamate, N-methyl-D-glucamine (MGD), N-(dithiocarboxy)sarcosine (DTCS)] have been shown to bind NO· and thereby give rise to EPR-detectable Fe complexes at ambient temperatures (reviewed in Ref. 63). With $(MGD)_2$–Fe(II), the parameters of the resulting complex are $a_N \sim 1.25$ mT and $g \sim 2.04$ [63, 71]; use of ^{15}N-labeled materials resulted in the expected changes in the coupling pattern and hyperfine coupling constants [63]. The choice of ligand is dictated by a number of factors including the need for long-lived NO· adducts to facilitate detection, particularly when the rate of NO· generation is slow; a requirement for the Fe center to be kept in the Fe(II) state (or to be readily reduced to this oxidation state [72, 73]); significant water solubility to allow ready addition to cells, plant, or animal systems [73]; and a requirement that the complex (and its resulting adduct) is minimally toxic. Although these complexes have been used successfully in a number of studies, the need for relatively high concentrations of complexes (to ensure rapid and efficient trapping of the NO·) has to be counterbalanced by the effects of the complexes on the system under study. It is well established that iron complexes can participate in radical generation (e.g., by autoxidation, or reaction with peroxides) and show significant toxicity. The $(MGD)_2$–Fe(II) and $(DTCS)_2$–Fe(II) species fulfill many of the above criteria (e.g., they have very long lifetimes of up to 12 h in blood [63]); can be used in animals and plants [73]; can be detected noninvasively using L-band EPR [71]; and can be detected in multiple biological fluids including blood, tissues (e.g., brain, liver, kidney, heart), and urine. However, they do have significant drawbacks including their potential toxicity (although no toxicity has been reported at levels of up to 2.5 g kg^{-1} for MGD itself [74], subtle perturbations may occur at lower levels for the Fe complex), the instability of the complexes in water (particularly at a pH of <7), and the poor water solubility of the diethyldithiocarbamate species.

11.6.4 Use of NO· Sensitive Chars

Broadening of the EPR linewidth of solid carbon-based paramagnetic species (chars) by O_2 has been widely used as a method for assessing the concentration of O_2 in biological systems [75]. It has been demonstrated that similar line broadening occurs with NO· and hence that these materials can potentially be used to measure NO· levels in biological systems [76]. However, the sensitivity of this approach is in the pharmacological range rather than the physiological; thus, this approach is unlikely to be sensitive enough to measure *in vivo* concentrations. Furthermore, line broadening by O_2 may confound measurements [76].

11.7 VERIFICATION OF THE OCCURRENCE OF RADICAL-MEDIATED PROCESSES

Although EPR is a wonderfully specific and hugely powerful method for detecting radical formation in biological and medical systems, recent advances in sensitivity have created a major problem for the interpretation of biological data. Therefore, the very low detection limits of modern spectrometers (nanomolar) have resulted in strong EPR signals sometimes being detected from processes that are not primarily radical in nature. Thus, 99% of the metabolism of a particular drug, solvent, or chemical may be occurring via nonradical reactions, but the 1% that proceeds via radical intermediates may appear to be the major process as a result of the detection of strong EPR signals. It has therefore become increasingly necessary to pair EPR measurements with other methods (e.g., product analysis) in order to provide a true picture of the process being studied. A wide range of quantitative techniques has been developed that allow such verification to be untaken. These include the quantification of the loss of parent materials; the formation of a specific protein (e.g., side-chain oxidation products [77, 78]), DNA (base oxidation products; see, e.g., Refs. 79 and 80), and lipid oxidation products (isoprostanes, hydroperoxides, alcohols, epoxides, hydrocarbons, aldehydes [81, 82]); and detection of generic markers of such reactions (e.g., protein carbonyls [83, 84], DNA strand breaks [85]). These methods rely heavily on the dramatic increase in sensitivity and resolving power of techniques such as high performance liquid chromatography (HPLC) and gas chromatography, with detection of materials by UV, visible, and fluorescence spectroscopy; electrochemistry; and various MS techniques. The latter in particular has become a very powerful tool in the search for, and quantification of, low levels of oxidation products in biological samples.

A wide range of other assays has also been developed to examine radical reactions in biological systems. Although quantification is possible with some of these, in many cases semiquantification is probably the best that can be reasonably expected. These include the use of dye molecules that become, lose, or change fluorescence on oxidation (e.g., Refs. 86–90; note that many of these are not wholly specific to radical processes, with some enzymes also inducing such changes); specific electrodes (e.g., for NO·, O_2, H_2O_2); and ELISA and Western blotting methods using antibodies raised against specific oxidation products (e.g., protein carbonyls, 3-nitrotyrosine, isoprostanes).

A number of these techniques have been coupled with EPR to aid the identification of radical intermediates and the products arising from these reactions. Direct MS analysis has been attempted for spin adducts, and these studies have been primarily carried out with well-characterized systems in which the identity of the major radical adducts had already been established. Thus, MS in conjunction with MNP spin trapping has been employed to detect radicals generated from halocarbons by rat liver microsomes [91, 92]. Although unambiguous assignments were possible in these systems, the situation with complex materials, in which many other species are also present and the nature of the species being sought is unknown, is much less promising. In more recent studies HPLC has been used to resolve materials before examination by EPR and MS. HPLC is probably the separation method of choice for such studies because of the widespread use of water-based eluents and the low (ambient) temperatures that can be used, which minimize degradation of the adduct materials. This approach has been used extensively and successfully by Iwahashi and colleagues [93–101] and Qian and colleagues [102–104] to examine a range of (relatively) well-characterized systems. Although very promising data have been obtained in these studies, the situation is far less clear when it comes to the analysis of real biological samples. In many of these studies the species detected by MS are not the radical adducts themselves, but species derived from them, either by oxidation or reduction (e.g., Refs. 105 and 106). This is not necessarily a disadvantage, as long as the chemistry that results in the formation of these species is unique to radical adducts and well understood. Indeed, the ability of a number of techniques to detect the nitrone species resulting from oxidation or disproportionation of the initial radical adducts has resulted in the development of some novel approaches to the detection of radicals in biological systems.

One of the first examples of this approach was a method developed by Kagan and colleagues to assay for glutathione thiyl radicals in cells, where the nitrone formed from decay of the initial (short-lived) GS· adduct to DMPO was detected by liquid chromatography and MS [107]. Although this approach can yield valuable information, note that absolute quantification of GS· by this methodology may be limited by the presence of alternative removal pathways for the initial adduct (and specifically the rapid reduction of the nitroxide radical to the hydroxylamine). Because the ratio of these two routes may vary from system to system, the yield of GS-DMPO nitrone may vary significantly and hence give misleading data.

This approach of detecting the nitrone arising from adduct decay has been developed into a powerful new methodology (immuno-spin trapping) by Mason and colleagues [108]. In this (non-EPR) technique, antibodies raised against a DMPO nitrone derivative are used to probe for the presence of DMPO adducts to macromolecules. These antibodies (which can be used in either ELISA or Western blotting experiments [108]) detect the nitrone structure present on the macromolecules as a result of initial adduct decay; by linking this recognition (which is highly specific because cyclic nitrones appear to be very rare structures in nature) to highly sensitive detection systems (e.g., enhanced chemiluminescence), the presence of radical adduct species can be detected with great sensitivity and specificity even in highly complex systems. This approach has been used with great success to detect radicals formed on a range of proteins [108–114] and DNA [115] in both isolated systems and in cells (reviewed in Ref. 116). Furthermore, use of a fluorescently tagged secondary antibody and confocal microscopy has allowed the *site* of radical adduct formation to be determined in intact cells [114]. This methodology has the potential to be a major advance in free radical research, although the exact nature of the adducts that are detected by this approach cannot be determined as yet because the antibody recognized only the nitrone function and not the group through which the nitrone is linked to the target. Thus, this approach should be seen as a highly sensitive and specific complimentary technique to more traditional EPR studies in which information is obtained on the nature of the intermediate species.

11.8 CONCLUSIONS

EPR has been and continues to be one of the most powerful methods of examining radical formation in biological systems. It is one of the few, if not the only, major method that allows specific and sensitive detection of such intermediates. Recent advances in technology have pushed the detection limit into the region where reactive radicals present at physiologically and pathologically relevant concentrations can now be detected.

However, the technique also has some disadvantages to counteract these major plus points. The greater sensitivity of modern spectrometers has resulted in significant problems in a number of areas, including the determination of whether any detected radical species are major intermediates in a biological pathway, as opposed to a minor process, and problems with the detection of minor impurities in reagents and artifacts arising from side reactions (or introduced as components of supposedly pure spin traps!).

A number of problems still remain to be overcome including the following:

1. a lack of knowledge of the rate constants for spin trapping reactions and the decay kinetics of adducts that prevent true quantification of spin trapping data;
2. the analysis of complex mixtures of radical adducts (although this can be tackled relatively readily for species that give isotropic spectra, deconvolution of mixed anisotropic spectra resulting from damage to biological macromolecules is still very much in its infancy);
3. the never-ending problem of deciphering which signals are true adduct species and which are artifacts;
4. the short-lived nature of some spin adduct signals (e.g., those of superoxide and thiyl radicals), although major advances are being made in this area, and an inability to trap other (often delocalized) radicals at a sufficiently rapid rate to allow ready detection;
5. a lack of compartmentalization and spatial resolution of radical formation in intact systems (although immuno-spin trapping offers promise in this area);
6. problems arising from the toxicity of spin traps and the perturbation that these species produce on the systems under study; and, finally,
7. our current inability to readily examine many radical processes in humans as a result of the inability to use spin traps.

Although this list is still long and somewhat dispiriting, rapid progress is being made in this area. Moreover, EPR is, and is likely to remain, the "gold standard" for the detection and identification of radicals in biological systems.

ACKNOWLEDGMENTS

The author is grateful to the Australian Research Council (through the Discovery and Centres of Excellence programs) and the National Health and Medical Research Council of Australia for financial support.

REFERENCES

1. Halliwell, B.; Gutteridge, J. M. C. *Free Radicals in Biology and Medicine*, 3rd ed.; Oxford University Press: Oxford, U.K., 1999.
2. Lagercrantz, C. *J. Phys. Chem.* **1971**, *75*, 3466.
3. Janzen, E. G. *Acc. Chem. Res.* **1971**, *4*, 31.
4. Janzen, E. G.; Haire, D. L. *Adv. Free Radical Chem.* **1990**, *1*, 253.
5. Perkins, M. J. *Adv. Phys. Org. Chem.* **1980**, *17*, 1.
6. Buettner, G. R. *Free Radical Biol. Med.* **1987**, *3*, 259.
7. Davies, M. J.; Timmins, G. S. EPR Spectroscopy of Biologically Relevant Free Radicals in Cellular, *Ex Vivo*, and *In Vivo* Systems. In *Biomedical Applications of Spectroscopy*; Clark, R. J. H., Hester, R. E., Eds.; Wiley: New York, 1996.
8. Tordo, P. Spin-Trapping: Recent Developments and Applications. In *Electron Paramagnetic Resonance;* Gilbert, B. C., Atherton, N. M., Davies, M. J., Eds.; Royal Society of Chemistry: Cambridge, U.K., 1998; Vol. 16, pp 116–144.
9. Davies, M. J. Recent Developments in Spin Trapping. In *Electron Paramagnetic Resonance;* Gilbert, B. C., Davies, M. J., Murphy, D. M., Eds.; Royal Society of Chemistry: Cambridge, U.K., 2002; Vol. 18, pp 47–73.
10. Davies, M. J. *Res. Chem. Intermed.* **1993**, *19*, 669.
11. Davies, M. J.; Hawkins, C. L. *Free Radical Biol. Med.* **2004**, *36*, 1072.
12. Rees, M. D.; Hawkins, C. L.; Davies, M. J. *J. Am. Chem. Soc.* **2003**, *125*, 13719.
13. Rees, M. D.; Davies, M. J. *J. Am. Chem. Soc.* **2006**, *128*, 3085.
14. Madden, K. P.; Taniguchi, H. *J. Am. Chem. Soc.* **1991**, *113*, 5541.
15. Kocherginsky, N.; Swartz, H. M. *Nitroxide Spin Labels. Reactions in Biology and Chemistry;* CRC Press: Boca Raton, FL, Chapter 3, pp. 25–66, 1995.
16. Mason, R. P.; Kalyanaraman, B.; Tainer, B. E.; Eling, T. E. *J. Biol. Chem.* **1980**, *255*, 5019.
17. Eberson, L. *J. Chem. Soc. Perkin Trans. 2* **1992**, 1807.
18. Eberson, L.; Persson, O. *J. Chem. Soc.* **1997**, 893.
19. Konorev, E. A.; Baker, J. E.; Joseph, J. E.; Kalyanaraman, B. *Free Radical Biol. Med.* **1993**, *14*, 127.
20. Sentjurc, M.; Mason, R. P. *Free Radical Biol. Med.* **1992**, *13*, 151.
21. Knecht, K. T.; Degray, J. A.; Mason, R. P. *Mol. Pharmacol.* **1992**, *41*, 943.
22. Kocherginsky, N.; Swartz, H. M. *Nitroxide Spin Labels. Reactions in Biology and Chemistry;* CRC Press: Boca Raton, FL, Chapter 10, pp. 199–206, 1995.
23. Rohr-Udilova, N.; Stolze, K.; Marian, B.; Nohl, H. *Bioorg. Med. Chem. Lett.* **2006**, *16*, 541.
24. Knecht, K. T.; Mason, R. P. *Arch. Biochem. Biophys.* **1993**, *303*, 185.
25. Zini, I.; Tomasi, A.; Grimaldi, R.; Vannini, V.; Agnati, L. F. *Neurosci. Lett.* **1992**, *138*, 279.
26. Coghlan, J. G.; Flitter, W. D.; Holley, A. E.; Norell, M.; Mitchell, A. G.; Ilsley, C. D.; Slater, T. F. *Free Radical Res. Commun.* **1991**, *14*, 409.
27. Flitter, W. D. *Br. Med. Bull.* **1993**, *49*, 545.

28. Ashton, T.; Young, I. S.; Peters, J. R.; Jones, E.; Jackson, S. K.; Davies, B.; Rowlands, C. C. *J. Appl. Physiol.* **1999**, *87*, 2032.
29. Means, G. E.; Feeney, R. E. *Chemical Modification of Proteins;* Holden-Day: San Francisco, CA, 1971.
30. Lundblad, R. L. *Techniques in Protein Modification;* CRC Press: Boca Raton, FL, 1995.
31. Graceffa, P. *Arch. Biochem. Biophys.* **1983**, *225*, 802.
32. Davies, M. J.; Gilbert, B. C.; Haywood, R. M. *Free Radical Res. Commun.* **1993**, *18*, 353.
33. Hawkins, C. L.; Davies, M. J. *Biochem. J.* **1998**, *332*, 617.
34. Chen, Y. R.; Mason, R. P. *Biochem. J.* **2002**, *365*, 461.
35. Davies, M. J. *Biochim. Biophys. Acta* **1991**, *1077*, 86.
36. Gunther, M. R.; Peters, A.; Sivaneri, M. K. *J. Biol. Chem.* **2002**, *277*, 9160.
37. DeGray, J. A.; Gunther, M. R.; Tschirret-Guth, R.; Ortiz de Montellano, P. R.; Mason, R. P. *J. Biol. Chem.* **1997**, *272*, 2359.
38. Gunther, M. R.; Tschirret-Guth, R. A.; Lardinois, O. M.; Ortiz de Montellano, P. R. *Chem. Res. Toxicol.* **2003**, *16*, 652.
39. Gunther, M. R.; Tschirret-Guth, R. A.; Witkowska, H. E.; Fann, Y. C.; Barr, D. P.; Ortiz de Montellano, P. R.; Mason, R. P. *Biochem. J.* **1998**, *330*, 1293.
40. Witting, P. K.; Douglas, D. J.; Mauk, A. G. *J. Biol. Chem.* **2000**, *275*, 20391.
41. Miller, V. P.; Goodin, D. B.; Friedman, A. E.; Hartmann, C.; Ortiz de Montellano P. R. *J. Biol. Chem.* **1995**, *270*, 18413.
42. Ostdal, H.; Andersen, H. J.; Davies, M. J. *Arch. Biochem. Biophys.* **1999**, *362*, 105.
43. Haire, D. L.; Janze, E. G. *Magn. Reson. Chem.* **1994**, *32*, 151.
44. Timmins, G. S.; Wei, X.; Hawkins, C. L.; Taylor, R. J. K.; Davies, M. J. *Redox Rep.* **1996**, *2*, 407.
45. Clement, J.-L.; Gilbert, B. C.; Rockenbauer, A.; Tordo, P. *J. Chem. Soc. Perkin Trans. 2* **2001**, 1463.
46. Clement, J.-L.; Gilbert, B. C.; Rockenbauer, A.; Tordo, P.; Whitwood, A. C. *Free Radical Res.* **2002**, *36*, 883.
47. Olive, G.; Mercier, A.; Le Moigne, F.; Rockenbauer, A.; Tordo, P. *Free Radical Biol. Med.* **2000**, *28*, 403.
48. Stolze, K.; Udilova, N.; Nohl, K. *Free Radical Biol. Med.* **2000**, *29*, 1005.
49. Stolze, K.; Udilova, N.; Nohl, H. *Acta Biochem. Pol.* **2000**, *47*, 923.
50. Hay, A.; Burkitt, M. J.; Jones, C. M.; Hartley, R. C. *Arch. Biochem. Biophys.* **2005**, *435*, 336.
51. Ouari, O.; Polidori, A.; Pucci, B.; Tordo, P.; Chalier, F. *J. Org. Chem.* **1999**, *64*, 3554.
52. Ouari, O.; Chalier, F.; Bonaly, R.; Pucci, B.; Tordo, P. *J. Chem. Soc. Perkin Trans. 2* **1998**, 2299.
53. Murphy, M. P.; Echtay, K. S.; Blaikie, F. H.; Asin-Cayuela, J.; Cocheme, H. M.; Green, K.; Buckingham, J. A.; Taylor, E. R.; Hurrell, F.; Hughes, G.; Miwa, S.; Cooper, C. E.; Svistunenko, D. A.; Smith, R. A.; Brand, M. D. *J. Biol. Chem.* **2003**, *278*, 48534.
54. Reddy, P. H. *J. Biomed. Biotechnol.* **2006**, 31372.
55. Xu, Y.; Kalyanaraman, B. *Free Radical Res.* **2007**, *41*, 1.

56. Wright, P. J.; English, A. M. *J. Am. Chem. Soc.* **2003**, *125*, 8655.
57. Hill, R. P.; Fielden, E. M.; Lillicrap, S. C.; Stanley, J. A. *Int. J. Radiat. Biol.* **1975**, *27*, 499.
58. Fink, B.; Dikalov, S.; Bassenge, E. *Free Radical Biol. Med.* **2000**, *28*, 121.
59. Dikalov, S.; Skatchkov, M.; Bassenge, E. *Biochem. Biophys. Res. Commun.* **1997**, *230*, 54.
60. Dikalov, S.; Skatchkov, M.; Bassenge, E. *Biochem. Biophys. Res. Commun.* **1997**, *231*, 701.
61. Dikalov, S.; Skatchkov, M.; Fink, B.; Bassenge, E. *Nitric Oxide* **1997**, *1*, 423.
62. Jiang, J.; Jordan, S. J.; Barr, D. P.; Gunther, M. R.; Maeda, H.; Mason, R. P. *Mol. Pharmacol.* **1997**, *52*, 1081.
63. Fujii, H.; Berliner, L. J. *Phys. Med. Biol.* **1998**, *43*, 1949.
64. Tsai, P.; Porasuphatana, S.; Halpern, H. J.; Barth, E. D.; Rosen, G. M. *Methods Mol. Biol.* **2002**, *196*, 227.
65. Piknova, B.; Gladwin, M. T.; Schechter, A. N.; Hogg, N. *J. Biol. Chem.* **2005**, *280*, 40583.
66. Mathieu, C.; Moreau, S.; Frendo, P.; Puppo, A.; Davies, M. J. *Free Radical Biol. Med.* **1998**, *24*, 1242.
67. Joseph, J.; Kalyanaraman, B.; Hyde, J. S. *Biochem. Biophys. Res. Commun.* **1993**, *192*, 926.
68. Akaike, T.; Yoshida, M.; Miyamoto, Y.; Sato, K.; Kohno, M.; Sasamoto, K.; Miyazaki, K.; Ueda, S.; Maeda, H. *Biochemistry* **1993**, *32*, 827.
69. Hogg, N.; Singh, R. J.; Joseph, J.; Neese, F.; Kalyanaraman, B. *Free Radical Res.* **1995**, *22*, 47.
70. Haseloff, R. F.; Zollner, S.; Kirilyuk, I. A.; Grigor'ev, I. A.; Reszka, R.; Bernhardt, R.; Mertsch, K.; Roloff, B.; Blasig, I. E. *Free Radical Res.* **1997**, *26*, 7.
71. Fujii, H.; Berliner, L. J. *Biol. Magn. Reson.* **2003**, *18*, 381.
72. Vanin, A. F.; Poltorakov, A. P.; Mikoyan, V. D.; Kubrina, L. N.; van Faassen, E. *Nitric Oxide* **2006**, *15*, 295.
73. Vanin, A. F.; Bevers, L. M.; Mikoyan, V. D.; Poltorakov, A. P.; Kubrina, L. N.; van Faassen, E. *Nitric Oxide* **2007**, *16*, 71.
74. Shinobu, L. A.; Jones, S. G.; Jones, M. M. *Acta Pharmacol. Toxicol.* **1984**, *54*, 189.
75. Khan, N.; Hou, H.; Hein, P.; Comi, R. J.; Buckey, J. C.; Grinberg, O.; Salikhov, I.; Lu, S. Y.; Wallach, H.; Swartz, H. M. *Adv. Exp. Med. Biol.* **2005**, *566*, 119.
76. Zweier, J. L.; Samouilov, A.; Chzhan, M. *J. Magn. Reson.* **1995**, *B109*, 259.
77. Hawkins, C. L.; Davies, M. J. *Biochim. Biophys. Acta* **2001**, *1504*, 196.
78. Davies, M. J.; Fu, S.; Wang, H.; Dean, R. T. *Free Radical Biol. Med.* **1999**, *27*, 1151.
79. Cadet, J.; Douki, T.; Gasparutto, D.; Ravanat, J. L. *Mutat. Res.* **2003**, *531*, 5.
80. Dizdaroglu, M. *Free Radical Biol. Med.* **1991**, *10*, 225.
81. Porter, N. A.; Wagner, C. R. *Adv. Free Radical Biol. Med.* **1986**, *2*, 283.
82. Yin, H.; Porter, N. A. *Antioxid. Redox Signal.* **2005**, *7*, 170.
83. Levine, R. L.; Williams, J. A.; Stadtman, E. R.; Shacter, E. *Methods Enzymol.* **1994**, *233*, 346.

84. Levine, R. L.; Wehr, N.; Williams, J. A.; Stadtman, E. R.; Shacter, E. *Methods Mol. Biol.* **2000**, *99*, 15.
85. Moller, P. *Mutat. Res.* **2006**, *612*, 84.
86. Borisenko, G. G.; Martin, I.; Zhao, Q.; Amoscato, A. A.; Tyurina, Y. Y.; Kagan, V. E. *J. Biol. Chem.* **2004**, *279*, 23453.
87. Borisenko, G. G.; Martin, I.; Zhao, Q.; Amoscato, A. A.; Kagan, V. E. *J. Am. Chem. Soc.* **2004**, *126*, 9221.
88. Zhao, H.; Joseph, J.; Fales, H. M.; Sokoloski, E. A.; Levine, R. L.; Vasquez-Vivar, J.; Kalyanaraman, B. *Proc. Natl. Acad. Sci. U.S.A.* **2005**, *102*, 5727.
89. Robinson, J. P.; Bruner, L. H.; Bassoe, C. F.; Hudson, J. L.; Ward, P. A.; Phan, S. H. *J. Leukoc. Biol.* **1988**, *43*, 304.
90. Bonini, M. G.; Rota, C.; Tomasi, A.; Mason, R. P. *Free Radical Biol. Med.* **2006**, *40*, 968.
91. Sang, H.; Janzen, E. G.; Poyer, J. L.; McCay, P. B. *Free Radical Biol. Med.* **1997**, *22*, 843.
92. Janzen, E. G.; Towner, R. A.; Krygsman, P. H. *Free Radical Res. Commun.* **1990**, *9*, 353.
93. Iwahashi, H.; Ikeda, A.; Negoro, Y.; Kido, R. *Biochem. J.* **1986**, *236*, 509.
94. Iwahashi, H.; Parker, C. E.; Mason, R. P.; Tomer, K. B. *Rapid Commun. Mass Spectrom.* **1990**, *4*, 352.
95. Iwahashi, H.; Albro, P. W.; McGown, S. R.; Tomer, K. B.; Mason, R. P. *Arch. Biochem. Biophys.* **1991**, *285*, 172.
96. Iwahashi, H.; Parker, C. E.; Mason, R. P.; Tomer, K. B. *Biochem. J.* **1991**, *276*, 447.
97. Iwahashi, H.; Parker, C. E.; Mason, R. P.; Tomer, K. B. *Anal. Chem.* **1992**, *64*, 2244.
98. Iwahashi, H.; Parker, C. E.; Tomer, K. B.; Mason, R. P. *Free Radical Res. Commun.* **1992**, *16*, 295.
99. Iwahashi, H.; Deterding, L. J.; Parker, C. E.; Mason, R. P.; Tomer, K. B. *Free Radical Res.* **1996**, *25*, 255.
100. Kumamoto, K.; Hirai, T.; Kishioka, S.; Iwahashi, H. *Toxicol. Lett.* **2004**, *154*, 235.
101. Iwahashi, H.; Hirai, T.; Kumamoto, K. *J. Chromatogr.* **2006**, *A1132*, 67.
102. Qian, S. Y.; Chen, Y. R.; Deterding, L. J.; Fann, Y. C.; Chignell, C. F.; Tomer, K. B.; Mason, R. P. *Biochem. J.* **2002**, *363*, 281.
103. Qian, S. Y.; Guo, Q.; Yue, G. H.; Tomer, K. B.; Mason, R. P. *Free Radical Biol. Med.* **2001**, *31* (Suppl. 1), S14.
104. Qian, S. Y.; Mason, R. P. *Free Radical Biol. Med.* **2001**, *31* (Suppl. 1), S13.
105. Dage, J. L.; Ackermann, B. L.; Barbuch, R. J.; Bernotas, R. C.; Ohlweiler, D. F.; Haegele, K. D.; Thomas, C. E. *Free Radical Biol. Med.* **1997**, *22*, 807.
106. Domingues, P.; Domingues, M. R.; Amado, F. M.; Ferrer-Correia, A. J. *J. Am. Soc. Mass Spectrom.* **2001**, *12*, 1214.
107. Goldman, R.; Claycamp, G. H.; Sweetland, M. A.; Sedlov, A. V.; Tyurin, V. A.; Kisin, E. R.; Tyurina, Y. Y.; Ritov, V. B.; Wenger, S. L.; Grant, S. G.; Kagan, V. E. *Free Radical Biol. Med.* **1999**, *27*, 1050.
108. Detweiler, C. D.; Deterding, L. J.; Tomer, K. B.; Chignell, C. F.; Germolec, D.; Mason, R. P. *Free Radical Biol. Med.* **2002**, *33*, 364.
109. Guo, Q.; Detweiler, C. D.; Mason, R. P. *J. Biol. Chem.* **2004**, *279*, 13272.

110. He, Y.-Y.; Ramirez, D. C.; Detweiler, C. D.; Mason, R. P.; Chignell, C. F. *Photochem. Photobiol.* **2003**, *77*, 585.
111. Deterding, L. J.; Ramirez, D. C.; Dubin, J. R.; Mason, R. P.; Tomer, K. B. *J. Biol. Chem.* **2004**, *279*, 11600.
112. Chen, Y. R.; Chen, C. L.; Chen, W.; Zweier, J. L.; Augusto, O.; Radi, R.; Mason, R. P. *J. Biol. Chem.* **2004**, *279*, 18054.
113. Nakai, K.; Mason, R. P. *Free Radical Biol. Med.* **2005**, *39*, 1050.
114. Bonini, M. G.; Siraki, A. G.; Atanassov, B. S.; Mason, R. P. *Free Radical Biol. Med.* **2007**, *42*, 530.
115. Ramirez, D. C.; Mejiba, S. E.; Mason, R. P. *Nat. Methods* **2006**, *3*, 123.
116. Mason, R. P. *Free Radical Biol. Med.* **2004**, *36*, 1214.

12 Electron Paramagnetic Resonance Applications to Catalytic and Porous Materials

DANIELLA GOLDFARB

Department of Chemical Physics, Weizmann Institute of Science, Rehovot, Israel

12.1 INTRODUCTION

Electron paramagnetic resonance (EPR) spectroscopy is an excellent tool for obtaining information about the geometric and electronic structures of paramagnetic centers. Therefore, it is a valuable characterization tool for heterogeneous catalysts when paramagnetic transition-metal ions (TMIs) and radical reaction intermediates are of interest. EPR can be applied to explore the properties of active sites comprising paramagnetic TMIs introduced into a variety of high surface area solid supports. It can identify and quantify the amount of different TMI binding sites and determine special coordination and oxidation states that the TMI adapts upon preparation and activation. Relating these properties with the catalytic performance of the material leads to a better understanding of the catalysis mechanism. A long-term challenge is the observation of the active site in action, which implies carrying out EPR measurements under extreme conditions of high temperatures and pressure with a high time resolution. This is beyond the current state of the art of high resolution EPR techniques, but some efforts are currently being directed toward this goal using continuous wave EPR (CW-EPR) within the framework of Operando spectroscopy [1].

The catalytic activity of TM-based heterogeneous catalysts is often a result of the synergy between the nature of the TMI or complex and the solid support. Therefore, the properties of the surface (internal or external) of the support are also of interest. For this purpose one can employ spin probes that can be either TMI complexes or stable organic radicals like nitroxides.

Electron Paramagnetic Resonance. Edited by Brustolon and Giamello
Copyright © 2009 John Wiley & Sons, Inc.

EPR spectroscopy encompasses several families of techniques designed to highlight different types of information, as manifested by different magnetic interactions:

1. CW-EPR, which is the routine, fastest and easiest way to record an EPR spectrum;
2. pulse EPR techniques that are used mainly for measurements of relaxation rates, chemical exchange, and distance measurements between electron spins using double electron–electron resonance;
3. CW and pulse electron–nuclear double resonance (ENDOR); and
4. electron spin echo envelope modulation (ESEEM).

The last two groups of techniques (see Chapter 5) are high resolution in nature and resolve weak magnetic interactions, which are usually not resolved in the EPR spectrum. These methods are often referred to as hyperfine spectroscopy because they provide the nuclear magnetic resonance (NMR) frequencies (often also called ENDOR frequencies) of nuclei coupled to unpaired electrons. The hyperfine and nuclear quadrupole interactions of these nuclei can be derived from these frequencies.

In principle, all EPR experiments can be carried out on spectrometers operating at different frequencies. The majority of commercial spectrometers operate at X-band frequencies ($\sim$9.5 GHz); Q-band ($\sim$34 MHz) is also a rather popular frequency, mostly for CW-EPR. Recently Q-band pulse spectrometers have been shown to be highly valuable. Lower frequency spectrometers, such as those operating at 4–7 GHz (S- and C-band) are also available but are not as common. Spectrometers operating at frequencies higher than 70 GHz are considered as high field/high frequency spectrometers. The most common high frequency spectrometers operate at W-band (95 GHz), and are commercially available. Higher frequency spectrometers are less abundant; they are homemade spectrometers and thus far are available only in highly specialized laboratories. It is not possible to define an optimum spectrometer frequency because such a frequency is system dependent and a function of the relative magnitude of its different magnetic interactions with respect to the electron–Zeeman interaction. Therefore, in some cases low frequencies are better than high frequencies, but for others it is the other way around. Often a complete characterization will require several spectrometer frequencies, and this is usually referred to as the multifrequency approach. Pulse EPR experiments, particularly those applied to TMIs, are usually carried out at low temperatures in order to overcome fast spin relaxation that prevents echo detection beyond the spectrometer dead time (usually around 100 ns).

For many decades CW-EPR spectroscopy has been applied successfully to problems in heterogeneous catalysis and for characterization of solid supports, and details can be found in recent reviews [2, 3]. This chapter describes only a few CW-EPR applications and concentrates primarily on the application of pulse ENDOR and ESEEM techniques. These should be applied only after the CW-EPR or the echo detected EPR (ED-EPR) spectrum has been recorded and analyzed. Although most of these techniques were developed more than two decades ago, their applications have been limited to expert laboratories, mainly because of unavailability of pulse spectrometers. This situation has changed considerably with the introduction of commercial pulse spectrometers, and therefore the potential of these techniques can now be fully realized. The purpose of this chapter is to highlight in a number of examples how such methods can be applied and what type of information they can yield. All examples involve orientationally disordered samples, which are typical of heterogeneous catalysts and porous materials. These can be highly crystalline like zeolites, partially ordered like mesoporous materials in which the ordered elements are the pores, or completely amorphous solids. Some of the examples concern characterization of the catalyst prior to its use, which is targeted at the identification and characterization of what are believed to be catalytically active sites. Others describe the use of spin probes with well-defined properties, which are designed to interact with the surface in a specific way and thereby explore the surface properties. The different examples are grouped according to the type of material or problem investigated, rather than the techniques used, to emphasize that often several techniques are needed for a complete characterization. In each case, a brief introduction on the material involved is presented, followed by a description of the problem and the EPR experiment used to address it. This chapter ends with practical information regarding sample concentration, preparation, and handling. The details of the different EPR techniques have been described in earlier chapters and therefore the pulse sequence and spin physics underlying the experiments will not be repeated.

12.2 PARAMAGNETIC TMIs

12.2.1 Zeolites and Zeotypes

Zeolites are aluminosilicate microporous crystalline solids with well-defined structures [4]. Their frameworks are made up of tetrahedra, with a silicon or aluminum atom in the center and oxygen atoms at the corners. These tetrahedra can then link together by their corners (see Fig. 12.1) to form a rich variety of structures. These may contain linked cages, cavities, or channels, the size of which allows small molecules to enter. Zeolites are characterized by their structure and the Si/Al ratio. The presence of a trivalent Al results in a negatively charged framework, and this charge is compensated by the presence of extra framework ions like Na^+. In addition to having silicon or aluminum as the tetrahedral atom, other compositions have also been synthesized, among them the aluminophosphates zeotypes known as AlPOs [5]. Because of their pore dimension, zeolites and their related materials are also referred to as *molecular sieves*. Zeolites are used in a variety of commercial applications like catalysis, mainly in petrochemical cracking, ion exchange (water softening and purification), and the separation and removal of gases and solvents.

The loosely bound nature of extraframework ions means that they can be readily exchanged for other types of metal ions when in aqueous solution. One class of catalytic activity concerns redox reactions where TMIs are introduced by cation exchange, such as copper zeolites for NOx decomposition, or by framework substitution, as in titanium in ZSM-5 in the production of caprolactam [6, 7]. Underlying

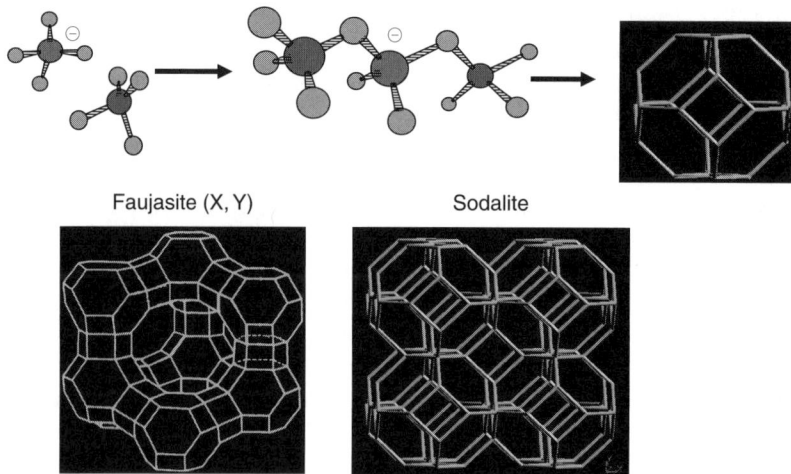

Fig. 12.1 A schematic representation of a zeolite structure built of connected SiO_4 and AlO_4^- tetrahedra that form cages and channels. The structures (top right) of a sodalite cage and (bottom) the zeolites sodalite and faujasite. Reprinted from International Zeolite Association (http://www.iza-online.org). Copyright International Zeolite Association.

all of these types of reactions is the unique microporous nature of zeolites, where the shape and size of a particular pore system exerts a steric influence on the reaction, controlling the access of reactants and products. Thus, zeolites are often said to act as *shape-selective catalysts*. The following subsections present examples related to extraframework TMIs and then framework substituted TMIs.

12.2.1.1 Extraframework Paramagnetic Metal Ions. The catalytic and absorptive properties of zeolites are influenced by the type, amount, and distribution of exchangeable cations. Consequently, the location and immediate environment of the cations are highly relevant to the function of the zeolite and need to be determined. Naturally, when the cation involved is paramagnetic, EPR is the method of choice. Examples of such cations are Cu^{2+} (d^9, $S = 1/2$), V^{4+} (d^1, $S = 1/2$), Mn^{2+} (d^5, $S = 5/2$), Fe^{3+} (d^5, $S = 5/2$), Co^{2+} (d^7, $S = 1/2, 3/2$), Ni^{3+} (d^7, $S = 1/2, 3/2$), and Ni^{1+} (d^9, $S = 1/2$).

12.2.1.1.1 The EPR Spectrum. The first step in the characterization of a sample of interest is recording the CW-EPR spectrum, usually using an X-band ($\sim$9.5 GHz) spectrometer. For some TMIs, such as Cu^{2+}, Mn^{2+}, and V^{4+}, it is possible to observe the EPR spectrum already at room temperature. For lower temperatures (77–150K), liquid N_2 cooling may be required, because of too fast spin lattice relaxation rates at room temperature. In some cases, such as Co^{2+}, even lower temperatures are required ($\sim$10K, liquid He cooling). Because the zeolite samples used are powders, the EPR spectrum is inhomogenously broadened and governed by anisotropic interactions (as described in detail in Chapter 6). The EPR spectrum is highly sensitive to the local environment of the paramagnetic metal ions and therefore can be used as a fingerprint for a specific location of the TMI within the zeolite. EPR is also a quantitative method; if a proper standard is used, it can determine the absolute amount of the paramagnetic TMIs in the sample.

Copper exchanged zeolites have been of interest since the late 1970s because of their catalytic properties [8]. Extensive investigations of the location of the Cu ions within the zeolite structure, its accessibility to reactants, and the nature of its coordination shell to rationalize the catalytic activity were carried out [9–13]. In order to resolve the ^{63,65}Cu hyperfine couplings, which serve as "fingerprinting" identification (along with the g values), the concentration of Cu^{2+} in the sample has to be kept low (a few weight percent) to avoid Heisenberg exchange that averages out the hyperfine interaction. This also minimizes electron spin dipole–dipole interactions that broaden the lines and reduce resolution.

The following example reveals how the presence of several Cu^{2+} sites is manifested in the EPR spectrum and how the technique can follow the variations of the coordination shell and location of the Cu^{2+} ions induced by dehydration. Figure 12.2 shows the X-band EPR spectra of Cu^{2+} exchanged into the NaY zeolite (Si/Al = 12, termed CuY12) at three hydration states. The EPR spectrum is determined by the principal components $g_\perp$, $g_\parallel$, $A_\perp(Cu)$, and $A_\parallel(Cu)$.

Usually for Cu^{2+}, $A_\parallel > A_\perp$ and the corresponding splitting are usually well resolved at X-band; therefore, $A_\parallel$ and $g_\parallel$ can be read out directly from the spectrum, as illustrated in the inset of Fig. 12.2. For each Cu^{2+} type, four equidistant (to first order) peaks, corresponding to the $A_\parallel$ feature of each of the four m_I states arising from the nuclear spin of 3/2, are expected. The center of these four features corresponds to the $g_\parallel$ values. The $A_\parallel$ and $g_\parallel$ values can be used as identity cards because they are very sensitive to the environment of the Cu^{2+} ions. The $A_\perp$ value is usually small (1–2 mT) and thus the hyperfine splitting in the $g_\perp$ region is not always resolved, but the position of $g_\perp$ can usually be well estimated, as marked in Fig. 12.2. The $A_\perp$ and $g_\perp$ values can be determined more accurately using simulations. An excellent software package that can be used to simulate EPR spectra and explore the effect of the various magnetic interactions on the spectrum is EasySpin, which can be downloaded free of charge from the web [14].

The sensitivity of the $A_\parallel$ and $g_\parallel$ values to the hydration degree of the zeolites is illustrated next. The spectrum of a sample obtained after Cu^{2+} exchange and mild evacuation at room temperature (E-CuY12) is provided in the middle trace of Fig. 12.2 [15]. It shows four resolved features in the low field region from which $A_\parallel = 435$ MHz (13 mT) and $g_\parallel = 2.39$ were obtained, and $g_\perp = 2.08$ was estimated from the intense high field peak. This particular Cu^{2+} site is named $Cu^{2+}(A)$. The spectrum of the hydrated sample (H-CuY12) recorded prior to evacuation is shown in the bottom of Fig. 12.2. Here, more than four $A_\parallel$ peaks are resolved in the low field region, immediately revealing the presence of two Cu^{2+} types: one with a parameters similar to $Cu^{2+}(A)$ and a new one, called $Cu^{2+}(B)$, the parameters of which are listed in Table 12.1. This shows that the evacuation transformed $Cu^{2+}(B)$ into $Cu^{2+}(A)$. This sample was also dehydrated at 400°C under evacuation at 10^{-3} torr [15]. The spectrum of this dehydrated sample (D-CuY12) is displayed in the top of

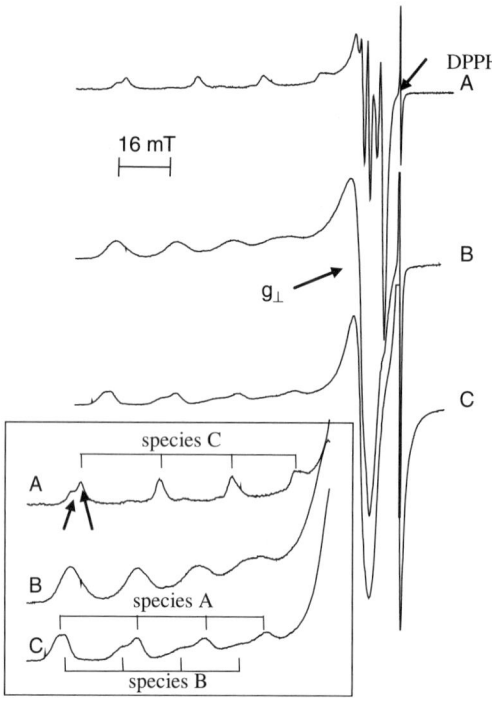

Fig. 12.2 Low temperature (160K) X band CW-EPR spectra of (spectrum A) D-CuY12, (spectrum B) E-CuY12, and (spectrum C) H-CuY12. (Inset) The expansion of the parallel region from 214 mT to 312 mT, indicating the $A_\parallel$ features of the various species. The center of this quartet gives the position of the corresponding $g_\parallel$. The arrows mark splitting due to ^{63,65}Cu isotopes. Reproduced from Reference 15. Copyright 2002 American Chemical Society.

Fig. 12.2. The peaks are significantly sharper and some resolution is observed in the $g_\perp$ region. This line narrowing shows that when water is present there is a distribution of Cu^{2+} sites, each with slightly different EPR parameters, probably induced by freezing a slightly different arrangement of water ligands. In the dehydrated sample, the Cu^{2+} site is well defined and its local geometry is dictated by the zeolite

TABLE 12.1 EPR Parameters of Cu^{2+} Exchanged Zeolite Y (Si/Al = 12) and Faujasite Determined from X-Band CW-EPR Spectra

Sample	Type	$g_\parallel \pm 0.01$	$A_\parallel$/MHz, ±25 ($A_\parallel$/mT, ±0.7)	$g_\perp \pm 0.01$
H-CuY12	Cu^{2+}(A)	2.39	481 (14.4)	2.08
	Cu^{2+}(B)	2.42	420 (12.4)	
E-CuY12	Cu^{2+}(A)	2.39	435 (13.0)	2.08
D-CuY12	Cu^{2+}(C)	2.34	505 (15.4)	2.06
D-CuFAU1.0	Cu^{2+}(D)	2.38	410 (12.6)	2.08

oxygens, which serve as ligands. In this spectrum the signals of the ^{63}Cu and ^{65}Cu isotopes are well resolved in the lowest field feature as indicated by the arrows in the inset of Fig. 12.2. The EPR parameters of this Cu^{2+} site are different than the other two (see Table 12.1), and it is termed Cu^{2+}(C). This example demonstrates that different Cu^{2+} sites can be distinguished on the basis of their $A_{\parallel}$ and $g_{\parallel}$. An example of the correlation existing between $g_{\parallel}$ and $A_{\parallel}$ in a variety of copper exchanged zeolites having different Si/Al ratio is provided in Fig. 12.3.

Another way to measure an EPR spectrum is in the pulse mode by recording the echo intensity as a function of the magnetic field. This referred to as ED-EPR. It yields a spectrum that is similar to the absorption spectrum as opposed to the CW-EPR spectrum, which is the first derivative of the absorption spectrum. The ED-EPR spectrum differs from the integrated CW-EPR spectrum when strong nuclear modulations are present [16], when the relaxation times are anisotropic, or when several species with different relaxation times are present. The advantage of measuring the spectrum this way is its higher sensitivity for broad lines, which are difficult to detect by CW-EPR. The ED-EPR spectra of the hydrated and evacuated samples compared to that of a frozen solution of 2 mM Cu(NO$_3$)$_2$ are shown in Fig. 12.4a. These ED-EPR spectra were recorded at 95 GHz (W-band), which is 10 times higher than the spectrometer frequency at which the CW-EPR spectra shown in Fig. 12.2 were acquired. The spectra are characteristic of an axially symmetric powder pattern due to g anisotropy. The hyperfine coupling is not resolved because of the extra broadening from the g strain discussed above. Although the X-band spectrum of E-CuY12 revealed only one type of Cu^{2+}, the W-band spectrum resolves in the $g_{\perp}$ region the presence of an additional minor Cu^{2+} site. This example shows that hyperfine resolution is best at low frequencies whereas g resolution increases at high

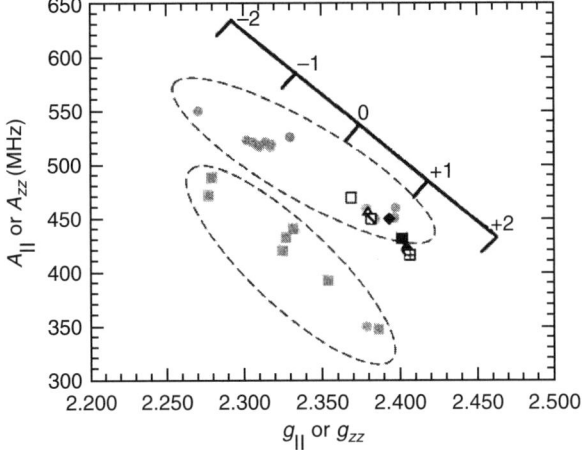

Fig. 12.3 A correlation plot of Cu^{2+} $g_{\parallel}$ and the absolute value of $A_{\parallel}$ showing group 1 (gray circles) and group 2 (gray squares) from a variety of hydrated Cu exchanged zeolites with varying Si/Al ratios. The bar reflects the charge of the complex. Reproduced from Reference 13. Copyright 2000 American Chemical Society.

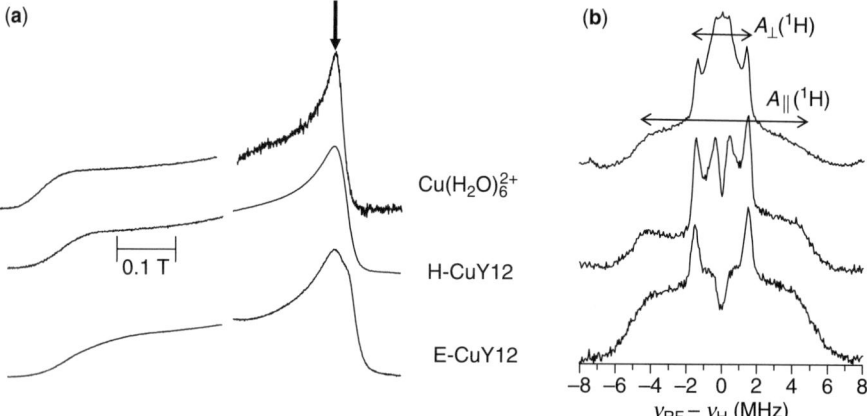

Fig. 12.4 (a) The W-band ED-EPR of a 2 mM Cu(NO$_3$)$_2$ solution, H-CuY12, and E-CuY12 and the corresponding Davies ^{1}H ENDOR spectra recorded at (b) $g_\perp$ as indicated by the arrow. (The gap in the ED-EPR spectrum is due to an instrument problem.)

frequencies, emphasizing the notion that an optimum spectrometer frequency does not exist and a multifrequency approach is often necessary.

Peisach and Blumberg [17] have shown that through correlations between $g_\|$ and $A_\|$ the number of nitrogen and oxygen ligands can be determined. This is mostly relevant to biological systems where nitrogen ligands prevail. A similar approach can be adapted to Cu^{2+} sites in zeolites where the goal is to find a $A_\|$–$g_\|$ correlation that will provide insight into the structure of different types of Cu^{2+} binding sites in zeolites [13]. Figure 12.3 shows that for hydrated zeolites the correlation plots reveal two groups. Group 1 zeolites are copper exchanged ZSM-5, beta, and ferrierite, all having Si/Al ratios of >10, and exhibiting $A_\|$ values approximately 20% higher than group 2 zeolites. Group 2 zeolites consists of copper exchanged X, Y, and mordenite with Si/Al ratios of <10. This shows that the differentiation into the two groups is attributable to the different Si/Al ratios. Within each group the spread correlated with the total local charge of the Cu^{2+}.

12.2.1.1.2 The First Few Coordination Shells. Table 12.1 lists three major different Cu^{2+} sites in CuY12 that have been identified and characterized by their different g and A values. However, more information regarding the close environment of these Cu^{2+} types is required. For example, in which cage is the Cu^{2+} located (α cage or sodalite)? How accessible is it? Does it have water ligands? Is it coordinated to zeolite oxygens? To answer these questions one has to apply high resolution EPR techniques, which can identify the nuclei near the Cu^{2+} ion and determine their location relative to the Cu^{2+} ion. This type of information can be obtained from the hyperfine interactions of such nuclei. Unlike the ^{63,65}Cu^{2+} hyperfine interaction, these so-called ligand hyperfine interactions are weak (1–10 MHz) and are not resolved in the broad EPR powder line shapes and therefore cannot be determined by CW-EPR. ESEEM and ENDOR can obtain them. The following example will show how pulsed ENDOR can provide some answers to the above questions. The ENDOR measurements identified the Cu^{2+} ligands and gave some information about their arrangement around the Cu^{2+}.

Figure 12.4b depicts W-band ^{1}H ENDOR of H-CuY12 and E-CuY12 recorded with the Davies ENDOR sequence at a magnetic field corresponding to the $g_\perp$ of a 2 mM frozen aqueous solution [15]. The $Cu(NO_3)_2$ solution was used as a model for a fully hydrated Cu^{2+}, which is $Cu(H_2O)_6^{2+}$. All three samples exhibit similar spectral features, showing two powder patterns centered about the ^{1}H Larmor frequency ν_H. The first, with small splittings (<1 MHz), is due to remote protons from solvation water. The intensity of these peaks decreases upon evacuation of the hydrated sample. The second powder pattern shows singularities with splittings of ~3 and ~9 MHz corresponding to $A_\perp(^1H)$ and $A_\parallel(^1H)$, respectively. From these, assuming that the point–dipole approximation is valid such that $A_\perp(^1H)$ is taken as negative, an isotropic hyperfine coupling a_{iso} of $(-3 - 3 + 9)/3 = 1$ MHz and an anisotropic component $T_\perp = 4$ MHz are obtained. This yields a Cu–H distance of 0.27 nm, in agreement with the distance of a proton of a water ligand from the Cu^{2+} ion.

The similarity in the ^{1}H splittings of the large coupling for $Cu(NO_3)_2$ and H- and E-CuY-12 shows that the Cu^{2+} has water ligands, with a similar orientation of the protons with respect to the Cu^{2+} local symmetry axes in all samples. The Cu^{2+}(A) and Cu^{2+}(B) species found in low Al content zeolite Y (Si/Al = 12) are therefore assigned to $Cu(H_2O)_n^{2+}$ complexes located in the large α cage (see Fig. 12.1). The number of water ligands n cannot be determined from the ENDOR spectrum without a proper standard. The two complexes, however, show different strengths of interaction with the zeolite framework, where the more weakly interacting complex has a larger hydration shell about the Cu^{2+} center. By carrying out spectral simulations of the ^{1}H ENDOR spectra, it is possible to show that the water ligands are orientationally disordered with respect to the Cu—O bond [15]. The ENDOR spectrum of Cu^{2+}(C) in D-CuY12 did not reveal any protons, thus indicating that it has no water ligands and it is coordinated to the zeolite oxygens. To obtain information on the interaction of the Cu^{2+} ions with the zeolites, one would like to detect hyperfine couplings with a framework nucleus such as ^{27}Al. This can be achieved by X-band hyperfine sublevel correlation (HYSCORE) spectroscopy. Here, the choice was X-band because ^{27}Al modulations appear when ν_{Al} and $A(Al)$ are of the same order. Because Al is not expected to be directly coordinated to Cu^{2+}, $A(Al)$ is expected to be small and at high field nuclear modulations will be suppressed.

The HYSCORE experiment is a two-dimensional experiment, which correlates NMR transitions of one electron spin manifold with those of the other electron spin manifold (see Chapter 5). Briefly recall that for $S = 1/2$ and $I = 1/2$, the $m_I = +1/2 \leftrightarrow -1/2$ NMR transition in the $m_S = \alpha$ (1/2) manifold ν_α correlates with the $m_I = +1/2 \leftrightarrow -1/2$ transition in the $m_S = \beta$ (−1/2) manifold ν_β. These correlations appear in the spectrum as off-diagonal cross-peaks between basic frequencies (ν_α, ν_β) and (ν_β, ν_α) in the (+, +) and/or (−, +) quadrants. For weak couplings $(\nu_I > |A/2|)$ the basic frequencies are centered about the nuclear Larmor frequency v_I, such that $(v_\alpha + v_\beta) \sim 2v_I$. These frequencies provide the identity of the nucleus through v_I whereas the hyperfine coupling of the nucleus is estimated by $|A| \sim (v_\alpha - v_\beta)$. Peaks along the diagonal in the experimental HYSCORE spectrum are often a result of incomplete inversion of the electron spin echo by the π pulse and from weakly coupled nuclei for which $v_\alpha \sim v_\beta$.

For nuclei with $I \geq 1$ (^{2}H, ^{27}Al, ^{14}N), the spectrum becomes more complicated because the number of nuclear states m_I increases and the nuclear quadrupole effect becomes important. With more nuclear states present, higher order nuclear frequencies, corresponding to transitions with $\Delta m_I = 2, 3, \ldots$, become possible; consequently, the number of cross-peaks increases as well. In the case of a half-integer, high spin nucleus like ^{27}Al ($I = 5/2$), the intensity of these peaks decrease as Δm_I increases, so typically only cross-peaks corresponding to single–single, single–double, and double–double quantum transitions are observed in the experimental spectrum.

The X-band HYSCORE spectra of H-CuY12 and E-CuY12 exhibited ^{1}H cross-peaks corresponding to the water ligand, consistent with the ENDOR spectra shown in Fig. 12.4b. In both samples ^{27}Al signals were not detected. Upon dehydration, the ^{1}H signals disappear and ^{27}Al signals with major intensities at 3.0 and 4.5 MHz appear. The latter corresponds to the basic frequencies $(v_\alpha, v_\beta)_{Al}$ with an estimated hyperfine coupling of 1.5 MHz (Fig. 12.5b). Additional signals at 3.0 and 9.9 MHz correspond to $(2v_\alpha, v_\beta)_{Al}$ and at 4.9 and 6.1 MHz to $(v_\alpha, 2v_\beta)_{Al}$. The appearance of the Al cross-peaks clearly provides direct experimental evidence for the coordination of Cu^{2+} to framework oxygens that are bonded to Al in the dehydrated state and all water ligands have been removed [15]. Although the HYSCORE spectrum of D-CuY12 clearly shows the presence of ^{27}Al hyperfine couplings, its interpretation is complex because of the presence of a large quadrupole interaction and the hyperfine and quadrupole parameters could not be determined uniquely.

When the Si/Al ratio is higher, there is an increased affinity of the Cu^{2+} ion to the zeolite oxygens and coordination to the latter ones can already be observed after mild evacuation. The HYSCORE spectrum of a sample of a Cu^{2+} exchanged Y zeolite with Si/Al = 5 recorded after evacuation (E-CuY5) is shown in Fig. 12.5a. The spectrum shows ^{1}H ridges centered at (13.3, 13.3) MHz with a width of 8 MHz perpendicular to the diagonal, which is typical for water ligands. In addition, there are clear ^{27}Al signals at 1.8 and 4.8 MHz with an estimated $A(^{27}\text{Al})$ of 3.0 MHz. This dependence on the Si/Al ratio of the zeolite shows that the zeolite oxygens become competitive ligands for Cu^{2+} after a specific framework charge density is achieved.

An example of W-band ENDOR is added to end this section, which has concentrated on the interaction of the Cu^{2+} ions with the zeolite framework. The W-band ENDOR spectra should be easier to interpret because only $\Delta m_I = \pm 1$ transitions appear. The concept of orientation selectivity is introduced here. For an orientationally disordered sample, information on the orientation of the ligand hyperfine tensor $\mathbf{A}(L)$ with respect to the $\mathbf{g}$ tensor can be obtained using orientation-selective ENDOR [18, 19]. In such measurements the magnetic field where the ENDOR spectrum is collected is varied within the EPR spectrum, thereby selecting different orientations of the paramagnetic center with respect to the external magnetic field. For a system

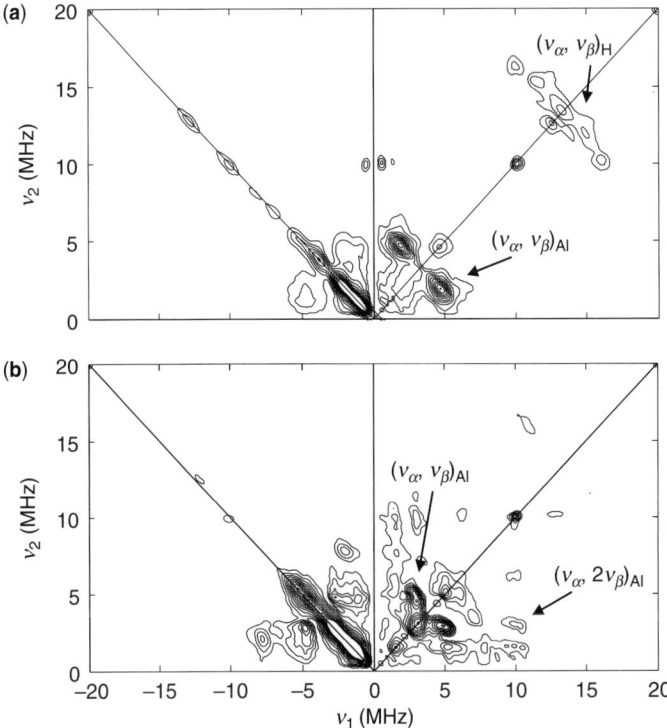

Fig. 12.5 The X-band HYSCORE spectra recorded at $g_\perp$ of (a) E-CuY5 and (b) D-CuY12, $\tau = 0.24$ μs. The peaks on the high end of the $(-, +)$ quadrant are noise. Reproduced from Reference 15. Copyright 2002 American Chemical Society.

with an axial symmetry, setting the magnetic field at $g_\perp$ will select paramagnetic centers with their symmetry axis perpendicular to the field direction whereas choosing $g_\parallel$ will select paramagnetic centers with their symmetry axis parallel to the field. These different field selections will generate different ENDOR spectra. The ENDOR line shape variation with the selected field within the EPR powder pattern provides the orientation of the **A** tensor with respect to the **g** tensor. Because $g_\parallel$ is usually oriented along the local symmetry axis, orientation of the **A** tensor can then be translated into bond angles.

A better orientation selection is achieved at W-band because of the larger spread of the g anisotropy (in terms of magnetic field units). This approach is demonstrated on a sample of dehydrated Cu^{2+} exchanged X zeolite with Si/Al = 1, which is termed D-CuFAU1.0. The EPR parameters of this Cu^{2+} are also listed in Table 12.1. The ENDOR spectra recorded at several magnetic fields within the listed EPR powder patterns [20] are presented in Fig. 12.6. The spectrum recorded closest to $g_\parallel$ (2.8799 T) is "single crystal-like" and is therefore highly resolved, showing a superposition of two equivalent quintets, with a splitting of ~1.6 MHz. This splitting is due to the quadrupolar interaction of the ^{27}Al nucleus [20]. The distance between the centers of the two

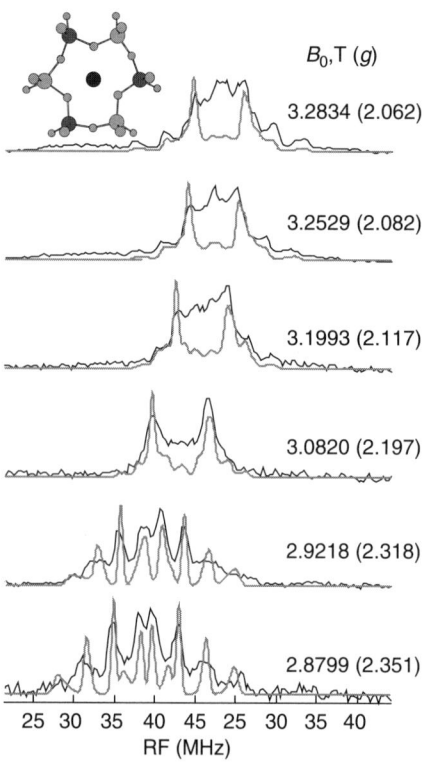

Fig. 12.6 (—) Orientation selective W-band Davies ^{27}Al ENDOR spectra at 5K of D-Cu-FAU1.0 recorded at the indicated field positions and g values (in parentheses), along with (—) simulations with the parameters listed in Table 12.2; RF, radiofrequency. Reproduced from Reference 20. Copyright 2006 American Chemical Society.

quintets yields a hyperfine coupling of 4.2 MHz. As the field increases toward $g_\perp$ the lines broaden and the splitting changes, revealing a clear orientation dependence that is due to the large spread of the g anisotropy, which at W-band exceeds the ^{63,65}Cu hyperfine anisotropy. Simulations using one type of ^{27}Al nucleus yielded the hyperfine and quadrupole interaction parameters (principal values and orientation relative to **g**) as listed in Table 12.2 [20]. The experimental and calculated spectra are compared in Fig. 12.6. The signals near the ^{27}Al Larmor frequency (not reproduced by the simulations) are attributed to distant Al nuclei. The orientation of the tensor relative to the **g** tensor principal axis system, which is given by Euler angles α, β, and γ, shows that the ^{27}Al is situated in a plane perpendicular to $g_\parallel$. Only a highly symmetric geometry where the Cu^{2+} located in the six-membered ring with three equal Cu–Al distances and $g_\parallel$ perpendicular to the Cu–Al axis could be described by a single ^{27}Al with the parameters listed in Table 12.2. An example of such a site is the Cu^{2+} ion situated at the center of a six-membered ring shown in the inset of Fig. 12.6 (see also Fig. 12.1), where the Cu^{2+} ion is coordinated to three oxygen atoms only (the direction of $g_\parallel$ is perpendicular to the six-membered ring). This last example is

TABLE 12.2 Best fit ^{27}Al Isotropic Hyperfine Coupling, Anisotropic Hyperfine Components, Quadrupole Coupling Constant, Asymmetry Parameter, and Orientation of Hyperfine and Quadrupole Tensors Relative to g Tensor Principal Axis System (MHz) of D-CuFAU1.0

a_{iso}	T_{xx}, T_{yy}, T_{zz} of D-CuFAU1.0	α, β, γ	e^2qQ/h	η	α', β', γ'
−3.43	−0.66, −0.07, 0.73	0°, 90°, 0°	14.8	0.7	90°, 90°, 0°

Key: a_{iso}, isotropic hyperfine coupling; T_{xx}, T_{yy}, T_{zz}, anisotropic hyperfine components; e^2qQ/h, quadrupole coupling constant; η, asymmetry parameter; α, β, γ, orientation of the hyperfine tensor; α', β', γ', orientation of the quadrupole tensor with respect to the g-tensor.

exceptional in terms of the spectral resolution that is observed. It shows that the geometry of the Cu^{2+} site can be determined using the orientation selective approach.

The application of HYSCORE and high field ENDOR techniques are rather recent, and prior to their introduction the method of choice for characterization of the environment of paramagnetic cations in zeolites has been two- and three-pulse ESEEM. The one-dimensional ESEEM experiments are simple to carry out; when the hyperfine coupling is very small such that ν_α and ν_β are not resolved, ESEEM is the method of choice. Otherwise, the HYSCORE experiment is more attractive due to its better resolutions because the spectrum is spread into two dimensions and the correlations it provides help in signal assignment. However, because recording HYSCORE spectra is time consuming, ESEEM may be the only choice for samples with a poor signal to noise ratios.

The next example is of an ESEEM application to Cu^{2+} doped NaH-ZSM-5 zeolites [11], where information on the Cu^{2+} ion location, mobility, and accessibility were obtained. This is a particularly interesting system because Cu/ZSM is highly active for decomposition of NO, and therefore the study of the Cu^{2+} location after various thermal treatments and the interaction with adsorbates is highly relevant. Unfortunately, EPR cannot probe Cu^{1+}, which plays an important role in catalysis, because it is diamagnetic. However, its Cu^{2+} precursors and their reduction can be easily followed. The X-band EPR spectrum distinguishes different Cu^{2+} species according to their $g_\parallel$ and $A_\parallel$ as described earlier. The room temperature EPR spectrum of Cu^{2+} in the hydrated zeolite is characteristic of a highly mobile species; therefore, Cu^{2+} is assigned to the intersection of the ZSM-5 channels. In addition, three-pulse ESEEM of a sample rehydrated with D_2O exhibited very deep modulation, as shown in Fig. 12.7a. The time domain trace can be simulated with 12 coupled ^2H nuclei, at a distance of 0.28 nm, which is in agreement with six water ligands. This is similar to the distance determined by the W-band ENDOR as described earlier.

Evacuation at room temperature reduces the ^2H modulation depth, and simulations revealed that only six ^2H nuclei remained coupled. Complete dehydration is achieved after activation at 400°C, and the Cu^{2+} resides in the small channel and is inaccessible to the large channel. This is deduced from the slow change of the EPR spectrum following adsorption of ethylene compared to polar adsorbates. This tardiness is due

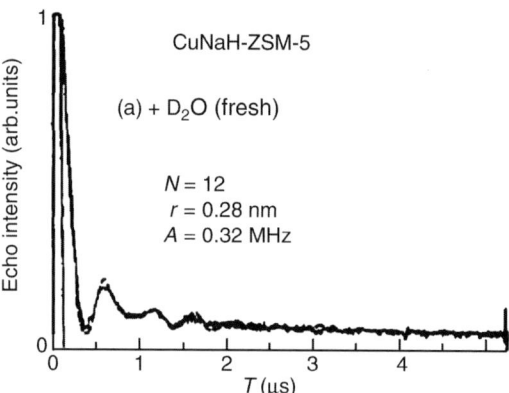

Fig. 12.7 (—) Experimental and (- - -) simulated three-pulse ESEEM spectra (4K) of CuNaH-ZSM-5 with adsorbed D_2O; N, number of interacting nuclei; A, isotropic hyperfine coupling; r, Cu–D distance. Reproduced from Reference 11. Copyright 1987 American Chemical Society.

a slow migration of the Cu^{2+} into an accessible site. Adsorption of polar molecules such as water, alcohol, pyridine, and ammonia draws the Cu^{2+} back to the channel intersection position where it forms complexes with the adsorbed molecules as evident again from ESEEM experiments showing ^2H modulation from specifically deuterated asdorbates. Now such ESEEM experiments can be replaced by X-band ^2H HYSCORE or Q- or W-band ^2H ENDOR, showing frequency domain spectra that are more intuitively analyzed.

To summarize, by combining routine CW-EPR measurements with a variety of ENDOR, ESEEM, and HYSCORE experiments carried out at different frequencies, this section showed how a detailed characterization of Cu^{2+} sites in zeolite can be obtained though its interaction with nuclei such as ^1H, ^2H, and ^{27}Al. This set of examples has conveyed the strategy for the application of the various methods, and it has become clear that these can be applied to paramagnetic TMIs in different types of systems.

12.2.1.2 Clusters Trapped in Zeolite Cages. Section 12.2.1.1 presented examples of cation exchanged zeolites, particularly Cu^{2+}. Another interesting group of paramagnetic species in zeolites is constituted by small paramagnetic metal clusters, which often have enhanced catalytic activity and specific selectivity. An example is Pt supported on KL zeolite, which was found to be a highly active catalyst for the hydrogenation of aromatics [21]. This is a useful reaction to transform carcinogenic aromatics in car fuel into less poisonous compounds of reasonably high octane number. Platinum clusters can be formed by exchanging $[Pt(NH_3)_4]^{2+}$ ions in aqueous solution against K^+ ions, followed by calcination to destroy the amine ligands [21]. After reduction with hydrogen a well-defined paramagnetic cluster is obtained, $Pt_{13}H_{12}^{n+}$ ($n = 1$ or 3) with $S = 1/2$, with the hydrogen atoms chemisorbed to its surface. The following example focuses on how the composition and structure of the cluster were determined by EPR. The number of Pt atoms in the cluster is determined from the hyperfine structure of the EPR spectrum. The natural abundance of ^{195}Pt ($I = 1/2$) is 33.8%; therefore, the spectrum is a superposition of contributions from ^{194}Pt, which does not give hyperfine splitting, and ^{195}Pt, which does. To facilitate the interpretation of the spectra, samples with enriched ^{195}Pt and ^{194}Pt were prepared. The X-band EPR spectra of these samples are provided in Fig. 12.8a. To better resolve the g anisotropy, the EPR spectrum was also recorded at a higher frequency, Q-band (34 GHz, ~1.2 T). Although the increased linewidth (g strain) suppressed the hyperfine splitting (as observed earlier for the Cu^{2+} spectrum at W-band), g values were easily determined ($g_\perp = 2.354$, $g_\parallel = 2.426$). Using these g values, the X-band EPR spectra were simulated with $A_\perp(^{195}Pt) = 226$ MHz and $A_\parallel(^{195}Pt) = 186$ MHz for 12 identical Pt nuclei. This shows that the unpaired electron is delocalized equally over the whole cluster, with the thirteenth Pt in the center having no spin density.

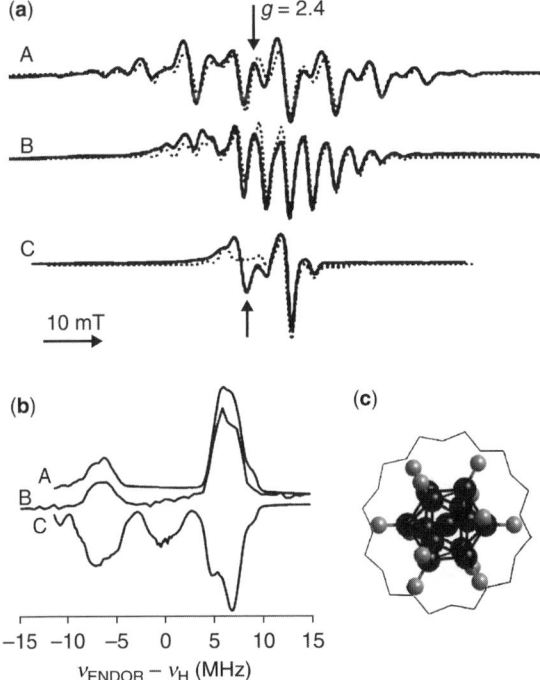

Fig. 12.8 (a) The X-Band spectra of 10% Pt/KL after (—) D_2 reduction and (- - -) simulation (12 nuclei): (spectrum A) enriched ^{195}Pt, (spectrum B) natural abundance Pt, and (spectrum C) enriched ^{194}Pt. The arrow denotes an unidentified additional species. (b) (Spectrum A) The X-band ^{1}H Davies ENDOR spectrum recorded at 291.9 mT, (spectrum B) the Q-band Davies ENDOR spectrum recorded at 1045.4 mT, and (spectrum C) the X-band Mims ENDOR spectrum recorded at 291.9 mT. (c) The structure of the Pt cluster. Reproduced from Reference 21. Copyright 2003 RSC Publishing.

When the sample was reduced with D_2 rather than H_2, a narrowing of the EPR lines was observed. This is due to a reduction in the unresolved hyperfine couplings, which is $\gamma_H/\gamma_D = 6.5$ times smaller for ^2H than for ^1H. The hyperfine couplings of hydrogen, which are not resolved in the EPR spectrum, can be obtained from ENDOR measurements, as shown in Fig. 12.8b. The X-band Davies ENDOR spectrum shows a doublet centered at the ^1H Larmor frequency (12.5 MHz at the field of measurements) with a splitting of about 12 MHz. Each of the doublet components has a line shape characteristic of an axial hyperfine interaction, yielding $|A_\perp(^1H)| = 9.0$ MHz and $|A_\parallel(^1H)| = 18.5$ MHz. Similar measurements were also carried out at Q-band (Fig. 12.8b) and the ENDOR spectrum shows the same doublet, but this time centered at 44.5 MHz, which is the corresponding ^1H Larmor frequency. Unfortunately, the ENDOR spectra cannot give information on the number of coupled protons. In contrast, HYSCORE measurements can provide information on the number of equivalent nuclei through the appearance of combination lines of the type $(2\nu_\alpha, \nu_\beta)$ and $(\nu_\alpha, 2\nu_\beta)$. The HYSCORE spectrum of the sample reduced with D_2 showed a large number of combination peaks, revealing that the number of coupled deuterons is larger than one, but the exact number could not be determined. The HYSCORE spectrum also showed a peak at the ^{27}Al Larmor frequency, confirming the proximity of the cluster to the zeolite framework [21].

The number of adsorbed hydrogen nuclei was estimated from the broadening of the EPR linewidth; it exceeded 8 and was probably close to 12. Considering further that the 12 observed Pt atoms are equivalent, it was concluded that each of them must be bound to the same number of hydrogen atoms; that is, the number of hydrogen atoms in the cluster is indeed exactly 12. Obviously, the observed cluster must be particularly stable. It is well known that cluster sizes matching a magic number are often much more stable than others. For platinum, Pt_{13} is one of the smallest magic clusters, and it has approximately icosahedral symmetry. Therefore, the prominent EPR active species is assigned to a $Pt_{13}H_{12}^{n+}$ close to icosahedral symmetry, which is shown in Fig. 12.8c.

12.2.1.3 Encapsulation of Complexes in Zeolite Cages. One of the approaches for the preparation of redox-active zeolite catalysts is the encapsulation of TM complexes in the zeolite channels where the general idea is to combine solution-like activity with the shape-selective control induced by the zeolite. In addition to the space constraints imposed by the zeolite, the net negative charge of the zeolite framework and the distribution of the positive charges of the cations can lead to specific interactions with the zeolite framework that, in turn, induce structural and functional modifications compared to solution activities [22–24]. One such example is the complex of Cu^{2+}–histidine (CuHis) exchanged into Y zeolites, which exhibits catalytic activity in the epoxidation of alkenes with peroxides at moderate temperatures [25, 26]. Depending on the concentration of CuHis (holding the Cu^{2+}/histidine molar ratio constant) and on the pH of the exchange medium, the EPR spectra reveal that two types of complexes, referred to as complex A and complex B, with different $g_\parallel$ and $A_\parallel$ (Cu) values are formed [27]. To rationalize the structural basis for these differences and to understand what the effect of the zeolite environment on the structure and chemical nature of the encapsulated complexes is, the structure of these complexes and their interaction with the zeolite framework were probed using W-band ENDOR and X-band HYSCORE experiments applying a strategy similar to that described in Section 12.2.1.1. The combination of all of these methods was necessary because of the spread of the hyperfine couplings of the various nuclei involved.

Here there is an additional complication because histidine is a multidentate ligand and can coordinate in several modes, depending on the pH and the [Cu]/[histidine] ratio [28]. The coordination sites of histidine are the amino (N_a) and imidazole imido (N_δ) or N_ε nitrogen atoms and the carboxylate oxygen (see Fig. 12.9a). The strategy was to identify ^{14}N and ^{1}H hyperfine couplings that can be used as indicators for the binding of a particular atom using CuHis complexes in aqueous solution at different pH as references for potential zeolite complexes [29]. This comparison highlighted the differences between the two types of complexes and thereby revealed the effect of the zeolite framework. The evidence for the coordination of N_δ is obtained from the characteristic ENDOR frequencies of the remote N_δ, which was extensively studied in the past in many biological systems using ESEEM and HYSCORE at X-band frequencies [30]. ESEEM and HYSCORE spectra also detect the interaction of the complex with the zeolite framework through the appearance of a ^{27}Al signal at its Larmor frequency, just as observed for the Pt cluster (§12.2.1.2).

The identification of the N_a binding and of the bidentate coordination mode of N_a and N_ε was obtained through the ^{1}H hyperfine couplings of the amino proton

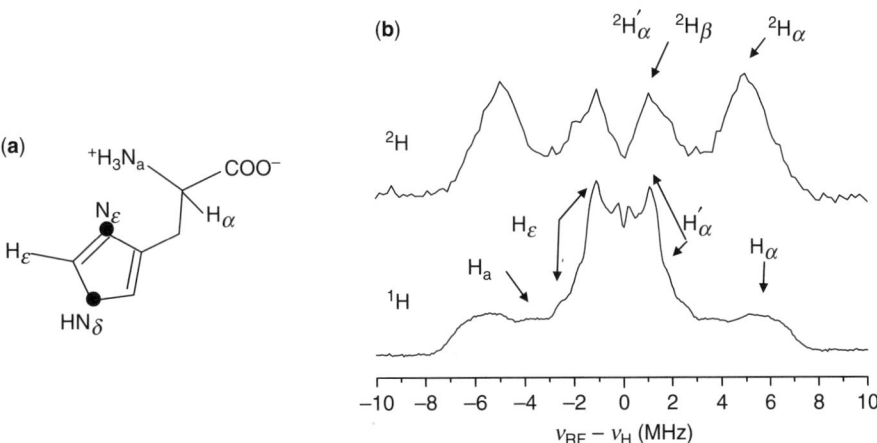

Fig. 12.9 (a) The histidine molecule and the labeling of the different protons. (b) (Bottom) The W-band ^{1}H Davies ENDOR spectrum of CuHis in zeolite Y and (top) the ^{2}H Mims ENDOR spectrum of CuHis-α,β-d$_3$ in zeolite Y. The spectra were recorded at 4.5K and $g_\perp$. The frequency scale of the ^{2}H spectrum was multiplied by γ_H/γ_D and the intensity scale by -1 for a convenient comparison with the ^{1}H Davies ENDOR spectra.

H_a and the α-proton H_α, respectively, which were determined by W-band ENDOR [28, 29]. Here it was essential to carry out the experiment at high frequency to resolve the $^{14}N_a$ and ^1H signals. Proton ENDOR signals are usually centered around ν_H because they behave like weakly coupled nuclei, whereas directly coordinated ^{14}N nuclei are within the strong coupling regime and the signals are centered about $A/2$. Hence, typical ^{14}N hyperfine couplings of 20–40 MHz will appear at X-band in the region of 10–20 MHz, which is also the region of the proton signals ($\nu_H \sim 15$ MHz).

The assignment of the various signals is carried out using specific ^2H labeling and comparing ^1H and ^2H spectra, which is a common practice in ENDOR. The signals of the exchangeable protons H_a are identified by comparing the spectra of CuHisY prepared in H_2O and D_2O solutions. Fortunately, the relatively small ^2H quadrupole coupling introduces only limited additional broadening, such that line positions in the ^1H and ^2H spectra can be matched by multiplying the ^2H frequency scale by $\gamma_H/\gamma_D = 6.5144$. The spectra shown in Fig. 12.9b are of a CuHisY sample with ~ 0.5 Cu/unit cell, containing mostly complex B. The bottom spectrum is the ^1H Davies ENDOR of CuHisY with natural abundance histidine, whereas the top spectrum is the ^2H spectrum of CuHisY prepared with histidine specifically deuterated at positions α and β. The spectra were recorded with the magnetic field set to $g_\perp$, and the assignment of the peaks is noted on the figure. The frequency axis is shifted by ν_H, such that the frequency separation between each of the two symmetric features gives the hyperfine coupling directly. These spectra show that H_a, H_ε, and H_α exhibit characteristic hyperfine couplings that can be easily distinguished. Although the former two have a large anisotropic contribution, H_α exhibits a large $a_{iso} = 10.9$ MHz and a small anisotropic component $T_\perp = 1.3$ MHz (determined from simulations of a series of orientation-selective spectra) [25]. Moreover, this large a_{iso} value was found to be a signature of the chelating coordination mode where both N_a and N_ε are coordinated [29].

The ^1H and ^2H spectra of zeolite complex B are similar to those of the CuHis complex in a pH 7.3 solution, with the difference that an extra feature ($H_{\alpha'}$ in Fig. 12.9b) is found in the zeolite spectrum. This feature was assigned to an H_α in a histidine ligand that is coordinated through only one nitrogen (monodentate). Therefore, the first coordination shell of Cu^{2+} ions in complex B in the zeolite was found to constitute N_a, N_ε of one histidine molecule and N_ε, O_c from a second molecule. The coordination of the O_c was further confirmed through the ^{13}C hyperfine interaction of the carboxylic carbon [31]. This coordination mode implies that the complex is positively charged, as expected, because the complex exchanged into the zeolite must compensate the negative framework charges. This selective exchange of positively charged complexes is one of the reasons for the differences in the structure between CuHis in solution and encapsulated in the zeolite. Similar ENDOR and HYSCORE measurements applied to complex A obtained with lower CuHis loadings assigned it to a bis-histidine complex, with an $N_aN_\varepsilon;O_c;O_z$ coordination mode, where O_z corresponds to a zeolite oxygen [31]. To summarize, this example shows how the coordination mode of the encapsulated complexes can be determined by a combination of high resolution EPR techniques.

12.2.1.4 Framework Substitution. In the previous sections all of the examples dealt with zeolites with TMIs or complexes introduced into the zeolite cavities by cation exchange. A limited amount of TM elements can also be introduced into tetrahedral (T) framework sites during synthesis, usually substituting an Al or Si atom. The prospect of TM incorporation into the framework of aluminosilicate and aluminophosphate molecular sieves [32] is of interest because of the new catalytic properties of the modified materials. These can be manifested in different acidities and in new catalytic activities, depending on the nature of the incorporated TMI. In the case of aluminophosphate zeotypes this substitution is particularly important because it introduces negative charges to the framework, which can be further used for cation exchange and the generation of acid sites. A major problem in such materials is that the metal ions do not necessarily occupy exclusively framework sites but can also exist as extraframework cations balancing the negative charge of the framework or as an interstitial phase of small particles located either within the molecular sieve cavities or in between the crystallites. Moreover, even if the metal ion has assumed a tetrahedral position during synthesis, it may pop out during calcination. Here too, EPR spectroscopy and its related high resolution techniques can be highly instrumental in determining the local environment of paramagnetic metal ions and determining unambiguously their incorporation into the framework.

Although previous sections gave examples of $S = 1/2$ TMIs, here the example will be of a high spin system, Mn(II), with $S = 5/2$. When dealing with half-integer, high spin systems in the solid state, the dominant interaction is the crystal field interaction that results in a zero-field splitting, which is characterized by the D and E parameters. Out of the $2S$ EPR transitions, the frequencies of $2S - 1$ EPR transitions depend to first order on D and E and on the orientation of the magnetic field relative to the local symmetry axis of the paramagnetic center. Accordingly, in polycrystalline powders, such as zeolites, these transitions appear as a broad unresolved line because of the superposition of the spectra of all crystallites, each having a different orientation relative to the magnetic field. In contrast, the central transition, corresponding to the $m_S = |-1/2\rangle \rightarrow |1/2\rangle$ transition, depends on D and E according to D^2/ν_0, for $D \ll \nu_0$, and has a limited orientational dependence (ν_0 is the EPR spectrometer frequency). Therefore, the higher the field is, the narrower the central transition [33].

The next example shows how high field EPR can resolve different types of Mn(II) substituted into the framework of AlPOs. Figure 12.10 compares the X-band (9.2 GHz), Q-band (35 GHz), W-band (95 GHz), and G-band (180 GHz) CW-EPR spectra of as-synthesized Mn(II) substituted UCSB-10 Mg [34]. The latter is a unidimensional cage, Mg-containing, AlPO. The spectra show the region of the central transition, which is split into a sextet because of the ^{55}Mn hyperfine coupling ($I = 5/2$). The resolution of the X-band spectrum is low because of the inhomogeneous broadening and the presence of forbidden transitions ($\Delta m_S = \pm 1$, $\Delta m_I = \pm 1$). This prevents resolving different Mn(II) species. The Q-band spectrum is somewhat better resolved; yet, at this frequency the six hyperfine lines are still broad. At W- and G-bands the resolution increases and a line splitting in the high field component is clear, resolving two different species with different hyperfine couplings of -248 and -264 MHz. Additional resolution can be obtained from ^{55}Mn ENDOR, because the ENDOR

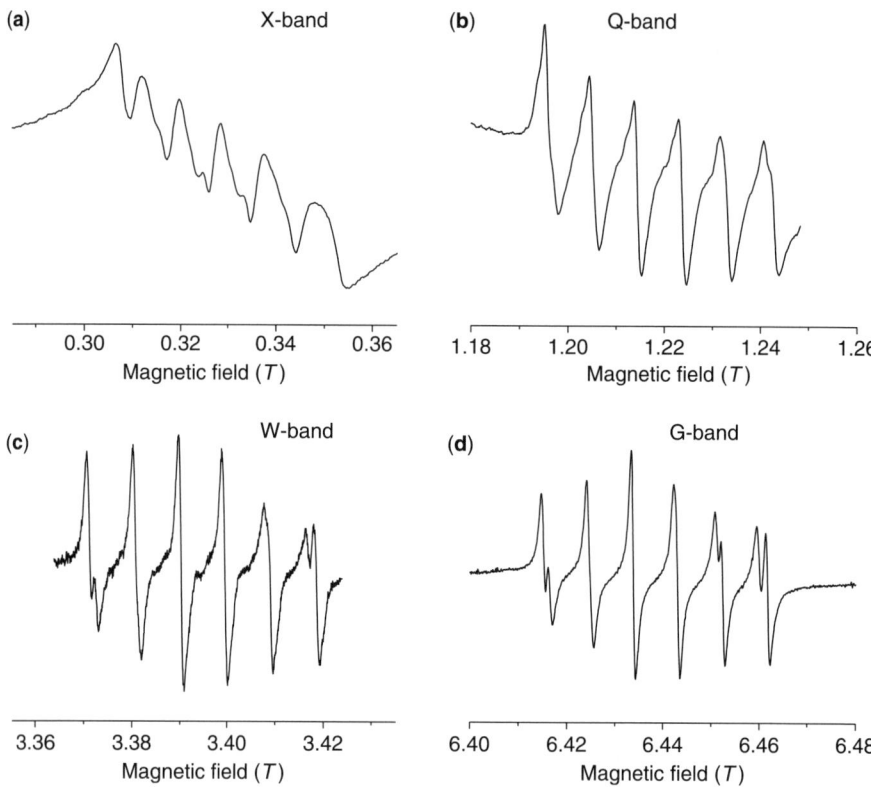

Fig. 12.10 CW-EPR spectra of as-synthesized Mn-UCSB-10 Mg: (a) X-band, 160K, $v_0 = 9.2$ GHz; (b) Q-band, 298K, $v_0 = 35$ GHz; (c) W-band, 298K, $v_0 = 95$ GHz; and (d) G-band, 298K, $v_0 = 180$ GHz.

spectra are better resolved than the EPR spectrum. Indeed, the ENDOR spectrum of as-synthesized Mn-UCSB-10 Mg disclosed the presence of a third Mn(II) type. The difference between the species was associated with the degree of local hydration and not different T sites [34].

Next is demonstrated how evidence for framework substitution can be obtained from high field ENDOR. The AlPO-n frameworks contain two magnetic nuclei, ^{27}Al and ^{31}P, the hyperfine interaction of which can be used to distinguish between framework and extraframework substitution. ENDOR measurements on Mn-AlPO$_{20}$, which has a structure similar to sodalite, show $A(^{31}P) \approx 8$ MHz and $A(^{27}Al) \sim 0$. This indicates framework substitution because Mn(II) located in a T site should have four ^{31}P nuclei as second shell neighbors (oxygens are the first) and ^{27}Al as a fourth shell. In the case of the extraframework location, both ^{27}Al and ^{31}P are in the second coordination shell and both should have finite hyperfine couplings. A substantial $A(^{31}P)$ was found also in Mn(II) incorporated into AlPO-5, AlPO-11, UCSB-6 Mg, and UCSB-10 Mg, as shown by their ENDOR spectra, compared with that of

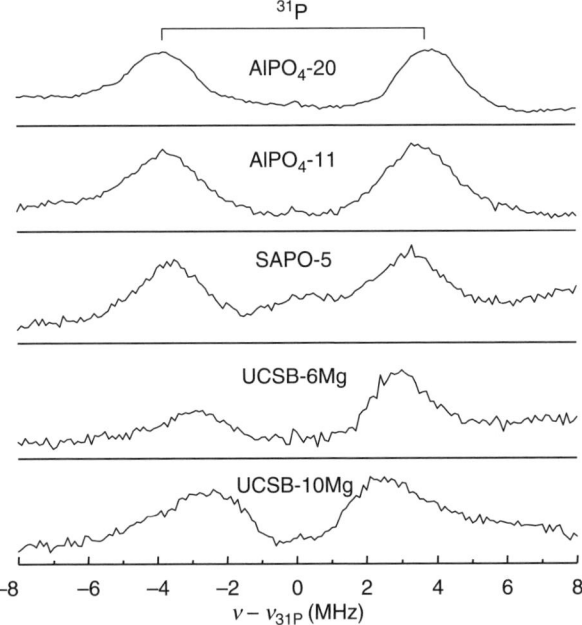

Fig. 12.11 The W-band Davies ^{31}P ENDOR spectra of several Mn(II) containing zeotypes measured at 5K and recorded with the magnetic field set to one of the $|-1/2\rangle$, $m_I \rightarrow |1/2\rangle$, m_I transitions.

AlPO$_4$-20 in Fig. 12.11 [35, 36]. All spectra show a ^{31}P doublet with 5–8 MHz splitting centered at about ν_P. The relatively large linewidth and the lack of powder pattern singularities indicate a distribution of the $a_{iso}(^{31}P)$ values [36]. The different $a_{iso}(^{31}P)$ values of the different structures show that $a_{iso}(^{31}P)$ is sensitive to the local environment and bonding topology of these substituted framework sites. However, although in some of the materials the EPR spectra resolved two types of Mn(II), only one type of ^{31}P coupling was detected. In principle, X-band ESEEM or HYSCORE experiments could also be carried out to detect the ^{31}P hyperfine coupling. However, because of the broad lines, they may escape detection of the doublet.

In the above, the focus was mostly on structural aspects of the metal ion environment in the final product. Another interesting basic question deals with the mechanistic details by which metal elements are incorporated into the framework during the synthesis of aluminophosphate zeotypes. The realization that $a_{iso}(^{31}P)$ is a clear signature of framework substitution can be used to follow the process of Mn(II) incorporation into AlPO$_4$-n materials during synthesis [37]. This can be accomplished by subjecting samples recovered from reaction mixtures quenched at different times to ENDOR measurements and examining the variation of the ^{31}P and ^{1}H hyperfine couplings, which are sensitive probes to the Mn–P interaction and the Mn(II) hydration states, respectively. The intensity of the ^{1}H ENDOR signal decreased with reaction time, indicating that the number of water ligands and solvent water molecules in the vicinity of the Mn(II) decreased. An a_{iso} value of 4–5 MHz was detected in the ENDOR spectra of samples quenched at the early stages of the reaction, but in the final material it was 7 MHz. The former was assigned to Mn(II) incorporated into a network of disordered aluminophosphate precursors. These precursors formed prior to the detection of any X-ray diffraction peaks.

12.2.2 Metal Ions in Mesoporous Materials

Ordered mesoporous materials comprise a rather new family of materials that are synthesized with surfactant molecules as templates [38]. The final material is a hybrid mesostructural material with the organic aggregates trapped within the inorganic solid, with structures that are similar to that exhibited by lyotropic liquid crystals with hexagonal, cubic, or lamellar structures. Removal of the organic molecules yields materials with ordered arrays of pores of hexagonal or cubic symmetries and disordered silica walls [39–41]. These materials are characterized by a large surface area ($\sim$500–1000 m^2) and a narrow distribution of pore volumes. The final structure and the pore diameters can be controlled by the choice of the template, the composition of the reaction mixture (inorganic/organic ratio and pH), and the addition of salts and cosurfactants [42]. These materials have attracted considerable attention that is due to their potential industrial applications in a wide range of areas such as catalysis [43], electrochemistry, adsorption, and separation technology. The majority of the mesoporous materials are silica based, but many others have been synthesized [44]. In addition, many different types of surfactant molecules, both charged (anionic and cationic) and neutral molecules, are used as templates [40]. Just as in zeolites, TMIs can be introduced during and after synthesis and can be characterized by EPR spectroscopy. One such example is presented: VO^{2+} in SBA-15.

One of the most popular mesoporous materials is the hexagonal SBA-15 material, prepared with Pluronic block copolymers [45]. One of the suggested ways to increase the mechanical stability of SBA-15 is to generate microporous amorphous nanoparticles (plugs) inside of the mesoporous channels of SBA-15 [46]. Such plugs can be zeolite nanoparticles and, if they have catalytic activities, then the plugs and the catalytically active elements are combined. This is the case for nanoparticles of vanadium silicalite-1 (VS-1), deposited in the mesoporous channels of SBA-15 by postsynthesis [46]. The VS-1 particles were prepared using tetrapropylammoniumhydroxide (TPAOH) as a template, and X-band CW-EPR and HYSCORE measurements were used to characterize the final material and learn about the specific interactions between the VO^{2+} and the TPAOH template.

The EPR spectrum of the as-prepared sample showed the presence of clusters of VO^{2+}. After dehydration at 373K, the presence of isolated VO^{2+} became clear from the well-resolved ^{51}V hyperfine splitting. Yet, rehydration did not restore the original spectra, yielding a VO^{2+} with different ^{51}V hyperfine coupling and motional characteristics. To rationalize the effects of the dehydration and rehydration on the local coordination environment of the VO^{2+} ions, HYSCORE measurements were carried out. In the hydrated state the HYSCORE spectrum (Fig. 12.12) showed only proton signals

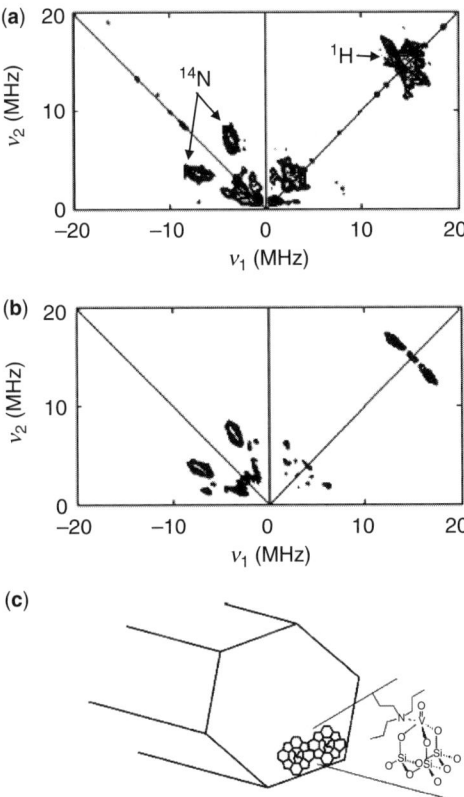

Fig. 12.12 (a) The HYSCORE spectrum of SBA-VS dehydrated under a dynamic vacuum at 373K and (b) a computer simulation of (a). All spectra were taken at the same observer position (346.5 mT) with $\tau = 176$ ns. (c) The local VO^{2+} environment for non-calcined VS-1 nanoparticles deposited on the walls of SBA-15 after dehydration. Reproduced from Reference 46. Copyright 2006 American Chemical Society.

with significant hyperfine couplings, indicating that in the hydrated material the VO^{2+} is at least partially hydrated. However, in the dehydrated sample cross-peaks in the $(-, +)$ quadrant at $(-3.5, 6.9)$ and $(-6.9, 3.5)$ MHz are clear evidence for the coordination to nitrogen atoms. These cross-peaks are assigned to the ^{14}N double–double quantum transitions. The assignment was corroborated by computer simulation of the spectrum (Fig. 12.12b), which provided the ^{14}N hyperfine and quadrupole couplings. The ^{14}N nucleus was assigned to an equatorially bound ^{14}N from the TPAOH employed as a template in the synthesis. Rehydration of SBA-VS removed these ^{14}N signals, showing that admission of water displaces the nitrogen ligand. Interestingly, the HYSCORE spectrum of full-grown VS-1 does not show ^{14}N signals. These results indicate that upon dehydration one of the propyl ligands is removed from the template and is replaced by vanadyl [46]. A schematic representation of the bonding of vanadyl to the zeolite nanoparticle after dehydration is shown in

Fig. 12.12c. This example shows that EPR can provide mechanistic details regarding the formation of composite solids of mesoporous materials and nanoparticles involving paramagnetic TMIs.

12.2.3 Metal Ions in Oxides

Often solid supports in heterogeneous catalysis are crystalline or amorphous oxides with a large surface area such as SiO_2, Al_2O_3, $SiO_2 \cdot Al_2O_3$, TiO_2, ZrO_2, MgO, La_2O_3, SnO_2, HfO_2, ZnO_2, and CeO_2. The metal ion is usually introduced via impregnation of a corresponding salt solution, and anchoring to the oxide occurs following a heat treatment. Other methods include coprecipitation of the active metal ion and the support, grafting of metal complexes to specific functional groups, and so forth.

The examples of EPR applications described thus far in this chapter were limited to the catalyst before or after activation, where the goal was to determine the structure of the TMI binding site, which is essential for the rationalization of the structural basis of the specific function. However, monitoring events taking place during catalysis is crucial for understanding the reaction mechanisms of many important chemical processes. This includes the observation of the active sites, the quantification of unusual oxidation states and coordination environments of metal ions during action, as well as the migration and mobility of species at the catalyst surface. Such information can be obtained from measurements carried out under real reaction conditions, namely, *in situ* spectroscopy [1]. This presents a considerable challenge because it requires measurements under extreme conditions of temperature and pressure and requires instrumental adaptation of the EPR spectrometer, which is not commercially available. Consequently, the number of such EPR studies is relatively limited and restricted to X-band CW-EPR.

Such an example is the *in situ* investigation of the dehydroaromatization of *n*-octane over supported chromium oxide catalysts using a special *in situ* EPR flow reactor with on-line gas chromatography analysis [47]. A representative set of *in situ* EPR spectra obtained with this setup is shown in Fig. 12.13 for a 1 wt% CrO_3/ZrO_2 catalyst. After calcinations the catalysts contains several types of Cr ions. The majority is Cr^{6+} (d^0), which is EPR inactive, along with some small amounts of Cr^{5+} (d^1) and Cr^{3+} (d^3) species, which are EPR active. The isolated Cr^{5+} surface species, referred to as γ, have $g_{\|\gamma} = 1.959$ and $g_{\perp\gamma} = 1.978$. The Cr^{3+} ions exhibit broad lines assigned to Cr_2O_3-like clusters, called β, with $g_\beta = 2$, and isolated coordinatively unsaturated Cr^{3+} ions on the zirconia surface, called δ, with $g_\delta = 4.3$. Upon heating the catalyst in a gas mixture of 0.6 mol% *n*-octane in N_2, the γ signal disappears above 260°C but the signal intensity of both β and δ increase. This indicates the formation of Cr^{3+} species by reduction of surface Cr^{6+} and Cr^{5+} species. Moreover, the intensity of the β signal passes through a maximum at the very beginning of the dehydroaromatization reaction, whereas on-line gas chromatography results indicate a gradual decrease of the catalyst activity. The partially deactivated catalyst is black and shows a narrow EPR signal at $g_\alpha = 2.003$, which is characteristic for coke radicals.

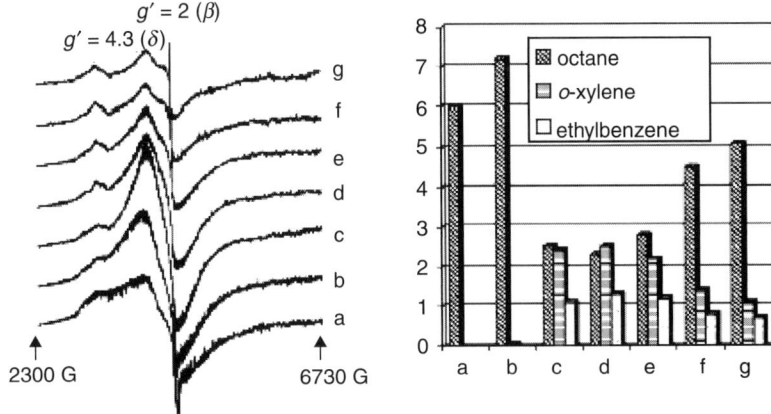

Fig. 12.13 The *in situ* EPR spectra of a 1 wt% CrO$_3$/ZrO$_2$ catalyst during the catalytic reaction in a 0.6 mol% octane/N$_2$ mixture and online gas chromatography signal areas of the product mixture (unreacted octane and formed *o*-xylene and ethylbenzene): a, $T = 300°C$; b, $T = 350°C$; c, $T = 400°C$, $t = 0$ min; d, $T = 400°C$, $t = 10$ min; e, $T = 400°C$, $t = 20$ min; f, $T = 400°C$, $t = 50$ min; g, $T = 400°C$, $t = 75$ min. Reproduced from Reference 47. Copyright 2002 RSC Publishing.

The changes in the intensity of the various EPR signals show that the initial reduction stage is governed by the formation of magnetically isolated Cr^{3+} signals (γ signal), and further reduction results in an increase of clustering of the Cr^{3+} species at the surface (β signal). From Fig. 12.13 it is clear that the β signal decreases faster than the δ signals, suggesting that the coke deposition occurred preferentially at the surface of the Cr$_2$O$_3$ clusters. Thus, the remaining activity of the CrO$_3$/ZrO$_2$ catalyst after 75 min on stream at 400°C must be due to the presence of the still accessible and active isolated Cr^{3+} species.

Unfortunately, this *in situ* EPR setup cannot be used also for ESEEM, HYSCORE, or ENDOR type experiments because these are carried out at low temperatures. This, in principle, can be achieved by freeze quench experiments but will require a rather complicated sample preparation setup.

12.3 SPIN PROBES

The previous sections described the application of EPR spectroscopy to study systems where the paramagnetic center comprised an intrinsic part of the system, and often it was an essential component to the material's function. EPR can be applied to diamagnetic systems through the incorporation of paramagnetic probes such as nitroxides. Nitroxide spin labels are extensively used in biological applications, in characterization of soft matter like liquid crystals and micellar solutions [48], and in probing the properties of pores and surfaces. The pioneering works on this context date back to the mid-1970s [49–51]. It is possible to use spin probes with different properties, such as polarity, size, and charge, or to specifically label constituent molecules at different positions, thereby allowing different properties and locations within the system to be explored. The amount of spin probe introduced is minute, and the high resolution EPR techniques allow the targeting of different types of information as demonstrated in the previous sections. The CW-EPR spectrum provides the degree of mobility of the spin label and its ordering. The polarity of the environment can be sensed by the ^{14}N hyperfine coupling $A(^{14}N)$, which increases when the polarity of the environment increases.

12.3.1 Probing Lewis Acidity

A nitroxide acting as a Lewis base can be used to probe Lewis acid sites. It can coordinate to an electron acceptor site through the nonbonding electrons of the nitroxide oxygen and thereby change the ^{14}N A_{zz} value according to the following resonance structures:

$$R_2N^{\bullet+}-\bar{\underset{\bullet\bullet}{O}}:A \longleftrightarrow R_2\overset{\bullet}{N}-\underset{\bullet\bullet}{\overset{\bullet}{O}}:A \quad (12.1)$$

Hence, the charge density shift toward the oxygen means a spin density shift to the nitrogen and a higher A_{zz}. This was shown in 1975 by Lozos and Hoffman [49] on di-*tert*-butyl nitroxide adsorbed on silica, silica-alumina, alumina, and zeolites X and Y. Two types of adsorption sites were resolved. One arises from di-*tert*-butyl nitroxide that is hydrogen bonded to surface hydroxyls and one from complexation to surface Al^{3+}. In this case ^{27}Al hyperfine splittings were clearly resolved in the EPR spectrum.

Another version of this idea is the use of NO as a paramagnetic probe molecule to characterize the structure, concentration, and acid strength of Lewis acid sites [52, 53]. Such Lewis acid sites in zeolites can be cations or aluminum defect centers $(Al_xO_y)^{n+}$, both of which form adsorption sites for NO molecules. The particular adsorption sites and the coordination geometry of the NO adsorption complexes can be obtained from the hyperfine interaction between the unpaired electron spin of the NO molecules and the nuclear spin of the metal ions, which form the Lewis acid sites. Unfortunately, metal ion hyperfine couplings are not always resolved in CW-EPR spectra as in the case of the Na$^+$—NO adsorption complexes where ENDOR had to be applied [54]. The ENDOR experiments were carried out at W-band where signals of ^{27}Al and ^{23}Na can be distinguished and where the orientation selection is good. A series of orientation-selective W-band Davies ENDOR spectra of the Na$^+$—NO adsorption complex in zeolite NaA is displayed in Fig. 12.14a. The various magnetic field positions at which the ENDOR spectra were taken are indicated by the arrows in Fig. 12.14b. The Mims ENDOR spectrum (Fig. 12.14f), which is more sensitive to small couplings than the Davies ENDOR, displays an intense triplet with a total splitting of about 0.74 MHz, which is shifted from ν_{Na}. These are assigned to weakly coupled ^{27}Al. In addition, all spectra reveal two groups of signals that are symmetrically situated with respect to ν_{Na} and assigned to Na$^+$ cations bound to the NO. The ^{23}Na ENDOR signals have an average splitting of about 6–9 MHz, indicating a

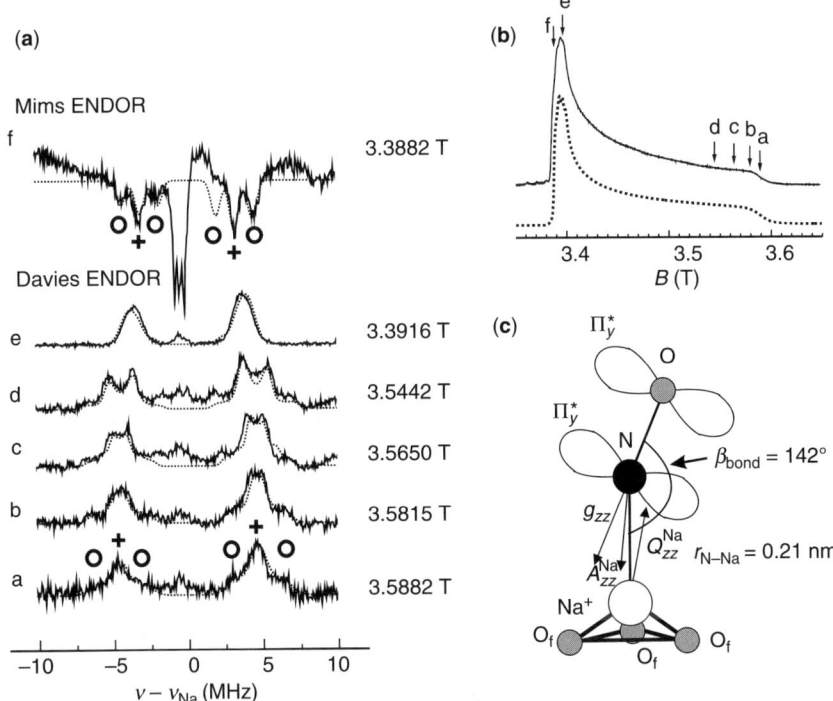

Fig. 12.14 (a) (—) Selected experimental and (- - -) simulated W-band orientation-selective (spectra a–e) Davies and (spectrum f) Mims ENDOR spectra of the Na$^+$–NO adsorption complex in zeolite NaA, demonstrating the ^{23}Na quadrupole splittings. The (+) $m_I = 1/2 \leftrightarrow -1/2$ and (O) $m_I = \pm 1/2 \leftrightarrow \pm 3/2$ ^{23}Na ENDOR transitions. (b) The ED-EPR spectrum with the field position at which the ENDOR measurements were carried out along with a (···) simulated trace. (c) A schematic drawing of the bent structure of the Na$^+$–NO adsorption complex in zeolite NaA. The z principal axes of the **g**, **A**(Na), and **Q**(Na) tensors lie within the Na–N–O complex plane and form angles with the Na$^+$–NO bond direction of 38°, 3°, and 8°, respectively. The unpaired electron is localized mainly in the antibonding $^2\Pi_y^*$ molecular orbital of the NO molecule, which is also within the complex plane. The cation at site S2 is coordinated to the framework oxygens O$_f$ in the six-membered rings in a trigonal symmetry. Only the three oxygens in the first coordination sphere are shown. Reproduced from Reference 54. Copyright 2000 American Chemical Society.

substantial isotropic ^{23}Na hyperfine coupling. The spectra recorded near g_{zz} exhibit additional splitting in each ^{23}Na ENDOR line, which is attributed to ^{23}Na ($I = 3/2$) quadrupolar splitting. Simulations of these spectra yielded the ^{23}Na hyperfine and quadrupole tensors [54]. This, together with the analysis of the ^{14}N hyperfine coupling determined from the CW-EPR and X-band Davies ENDOR spectrum, led to a detailed characterization of the formed Na$^+$–NO adsorption complexes and the determination of their geometrical and electronic structure of the complex, as shown in Fig. 12.14c. It was found that the complex formation is associated with a shift of the NO electron density toward the nitrogen atom, which is due to the electron pair acceptor property of the Na^+ ion (Eq. 12.1). The geometrical and electronic structure of the Na^+–NO complex was obtained from the comparison of the experimentally determined ^{23}Na and ^{14}N hyperfine coupling, but information about the specific cation adsorption site of the NO molecules within the zeolite was obtained from the ^{23}Na nuclear quadruple interaction based on a comparison with earlier NMR results [55]. These determined the ^{23}Na quadrupole tensor of Na^+ in the different sites of zeolite A. The Na^+–NO adsorption complexes formed most likely with sodium cations at the six-membered rings (see Fig. 12.1) [54].

Nitric oxide forms an important EPR active complex with Cu(I) in zeolites, which is a crucial intermediate of the direct decomposition of NO into N_2 and O_2 catalyzed by Cu^+/ZSM-5 zeolites [56]. Also in this case the unpaired electron originates from the NO because Cu(I) has a d^{10} configuration. The X-band EPR spectrum resolved both the ^{63,65}Cu and ^{14}N hyperfine couplings [57–59]. Analysis of the EPR spectrum showed that the Cu^+NO complex has an η^1 bent structure with the unpaired electron residing mainly on the coordinated NO molecule. To examine the motional characteristics of the complex, the multifrequency approach was applied and temperature dependent CW X-, Q-, and W-band measurements were used to analyze the dynamic behavior of the complex and confirm that NO complexes are formed at two different Cu(I) cationic sites in ZSM-5 [60].

12.3.2 Exploring Pore and Surface Properties

The motional characteristics of the spin label, which are clearly manifested in the spectral line shape, can also be used to differentiate types of pores and identify their source. Examples are the micro- and mesopores in SBA-15 mentioned in Section 12.2.2. SBA-15 is prepared with Pluronic triblock copolymers consisting of poly(ethylene oxide) (PEO_x)–poly(propylene oxide) (PPO_y)–PEO blocks (PEO_x–PPO_y–PEO_x) as templates. In aqueous solutions the Pluronics form micelles, where the PEO part comprises the hydrophobic core and the PEO blocks the hydrophilic corona [61]. SBA-15 has a two-dimensional hexagonal arrangement of channels, which after synthesis are filled with the template molecules. Calcination removes the template molecules, and the mesoporous, silica based material can be used for various applications. It was found that the calcined material contains, in addition to the ordered mesopores, micropores that provide connectivities between the mesopores [62]. These can be nicely distinguished using a series of spin-labeled Pluronics block copolymers, where a nitroxide group is added to the ends of the PEO chains (see Fig. 12.15a) [63]. The spectra of as-synthesized SBA-15 with three Pluronic spin probes (F127-NO, P123-NO, and L62-NO) are provided in Fig. 12.15c. The different spin probes have different PEO block lengths, and therefore they sense different regions in the PEO part. The averaged location of their nitroxide spin label with respect to the corona is also shown in Fig. 12.15b. The spectrum of F127-NO reveals the presence of a single species, with a rigid limit powder pattern showing that the spin label is immobilized on the EPR timescale (at room temperature). The spectra of the other two show, in addition to the immobilized species, a mobile species; and the closer the label is to the corona–water interface, the larger

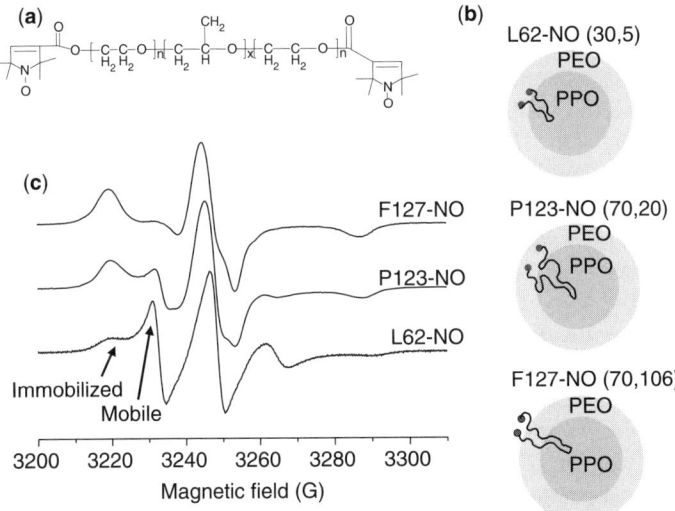

Fig. 12.15 (a) The structure of the Pluronics spin labels. (b) A schematic representation of the Pluronics micelles with the core and corona represented by light and dark grey, respectively. The numbers in parentheses correspond to the number of segments in the PPO and PEO blocks, respectively. The approximate averaged location of the spin labels are indicated for micelles of P123. (c) The EPR spectra of the as-synthesized SBA-15 with the three spin probes indicated. The features of the immobile and mobile species are also indicated.

is the contribution of the rigid species, which are trapped in the silica layer. The more mobile label is situated in the mesopores, where there is no silica. Calcination removes the chains trapped in the silica, generating micropores.

Spin probes of different character can be used as models for the adsorption of different kinds of organic molecules on surfaces. This approach was used in the study of the host properties of templated mesoporous materials [64]. Here, spin probes with a different polarity and charge are used to examine differently prepared surfaces or treated materials. The EPR spectrum of the guest usually exhibits a superposition of different spectra, differing in their motional characteristics; namely, the strongly adsorbed ones exhibit a slow motion spectrum, whereas the weakly bound ones show high mobility. The correlation times of the various species and their relative amounts can be obtained from computer simulations, and these are further used to elucidate the surface properties.

Spin probes can also be used to characterize the external surface of zeolites [65]. Here the challenge is to distinguish between the internal and external surfaces, where the internal surface is larger. This was demonstrated on a series of monodispersed crystallites of the zeolite silicalite. A spin probe, such as 2,2,6,6-tetramethylpiperidine 1-oxyl (TEMPO), is first absorbed and at low loadings. Its EPR spectrum is characteristic of an immobilized species because of binding at strong binding sites. Then, a large diamagnetic probe molecule, *ortho*-methyldibenzyl ketone, is coabsorbed. This probe is too large to enter the channels of the zeolite and will adsorb on the external surface. Because of the small amount of the EPR probe, it will first occupy the empty binding sites on the external surface; as its amount increases, at a certain stage it will replace the EPR probe. The latter then binds to weaker binding sites and its spectrum will change to that characteristic of fast motion, superimposed on the broad line of the strongly bound nitroxides. The onset of this narrow line spectrum is used as an indication of the saturation of the binding sites on the external surface. A rather good correlation of the external surface was obtained by this method compared with other methods (12% error).

Another example is the study of the internal surface of zeolites, as demonstrated on zeolite X and Y [66]. Here a cationic spin probe, 4-trimethylammonium-2,2,6,6-tetramethyl piperidinyloxy (TempTMA$^+$), was exchanged into the zeolite. Stronger binding, expressed in lower correlation times, was found in zeolite Y compared to zeolite X with the higher Al/Si ratio. Additional information on the close environment of the spin probe can be obtained from ESEEM experiments. In these experiments the adsorption of the probe was carried out in a D_2O solution, and ESEEM provided the ^2H modulation pattern. It was found that in zeolite X the NO of TempTMA$^+$ is hydrogen bonded to a water molecule, but in zeolite Y the water molecules are further away. This is consistent with the EPR results, showing that in zeolite Y the spin probe is bound to the surface.

12.3.3 Study of the Formation Mechanism of Mesoporous Materials

The high sensitivity of nitroxide spin probes to their microenvironment and the short time (a few seconds) it takes to record a CW-EPR spectrum makes it an excellent tool for the investigation of the formation mechanism of various templated mesoporous materials by *in situ* measurements. These provide the temporal evolution of the local viscosity and polarity during the formation of mesoporous materials; by using both "surfactant-like" and "silica-like" probes, it is possible to follow different compartments of the system, namely, the organic assemblies and the forming inorganic wall [63, 67–73]. In such experiments one follows $a_{iso}(^{14}N)$ and the correlation time τ_c when the molecular motion is in the fast motion regime. In the intermediate and slow motion regimes it is the anisotropic rotation diffusions constants that dictate the shape of the spectrum. The measurements, naturally, can be carried out at different frequencies, opening different dynamic windows (ranges). Thus far such experiments have been carried out only at X-band frequencies; but, in principle, they can be carried out at higher frequencies, although the sample handling may be more difficult.

The CW-EPR experiments can be further complemented by ESEEM experiments. Because these experiments are carried out at low temperatures, samples have to be rapidly freeze quenched at different reaction times. A number of studies have shown that upon fast freezing (plunging into isopentane cooled in liquid N_2) the micellar structure is generally preserved [71, 74]. Specific isotopic labeling of different components of the reaction mixture allows the targeting of specific regions. For example, using D_2O as a solvent, a surfactant-like spin probe gives information on the degree of

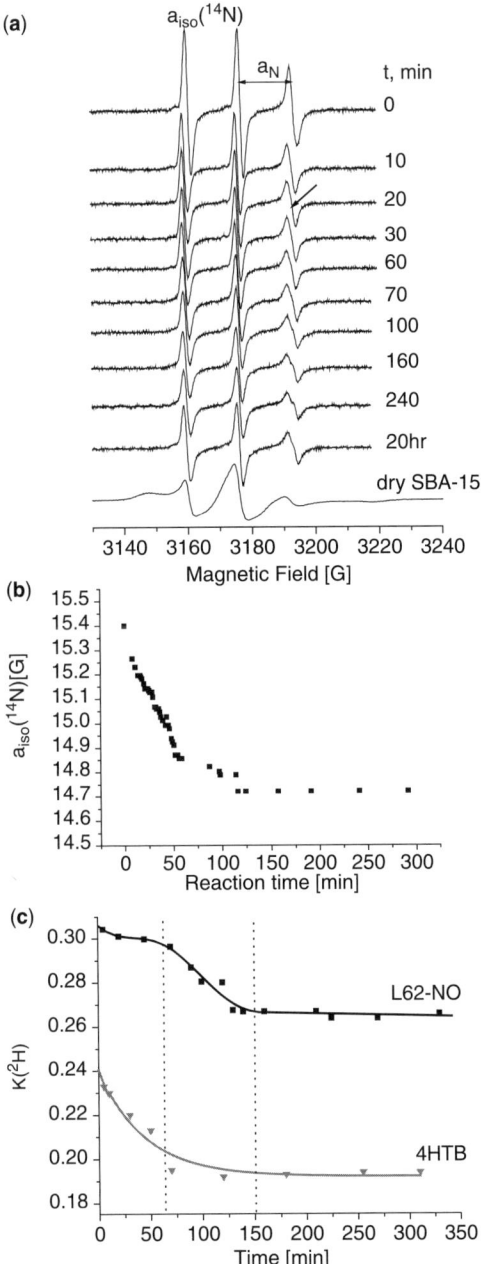

Fig. 12.16 (a) The time evolution of the EPR spectrum of L62-NO during the formation of SBA-15 (Si/P123 = 59) at 50°C and the dry product recorded at room temperature. (b) The time evolution of $a_{iso}(^{14}N)$ of the more hydrophobic species, first resolved at $t = 20$ min as indicated by the arrow. (c) The time evolution of $k(^2H)$ for L62-NO and 4HTB in the SBA-15 reaction mixture. (The vertical lines are drawn to guide the eye.)

water penetration into specific regions in the micelles [70] whereas a silane-based spin probe that copolymerizes with the silica source will report on the properties of the forming silica wall [67]. Deuteration of the surfactant molecules, in contrast, provides information on the packing of the molecules [71] and isotope labeling of promoter molecules and anions ($^{15}NO_3^-$) allows the probing of their location with respect to the spin probe [75], rationalizing their role in the determination of the final structure.

The combined application of CW-EPR and ESEEM to study the formation of SBA-15 is demonstrated next [63, 70]. Figure 12.16a provides the EPR spectra of L62-NO recorded during the reaction carried out at 50°C, as a function of time. The addition of the silica source, tetraethoxy-*ortho*-silane (TEOS), is taken as time $t = 0$. This spin probe is situated at the core–corona interface, and prior to the addition of TEOS the spectrum is characteristic of a single species with $a_{iso}(^{14}N) = 15.5$ G and a correlation time $\tau_c = 2.0 \times 10^{-10}$ s. As the time increases the high field triplet component splits, indicating the formation of a second environment for the spin label. With time the $a_{iso}(^{14}N)$ of this second species becomes smaller, showing that the environment becomes more hydrophobic and τ_c increases to 2.8×10^{-10} s. Changes are no longer observed after 120 min (Fig. 12.16b). The new species is assigned to spin labels occupying the hydrophobic core of the micelles, the size of which increases with time because of water depletion induced by the silica polymerization at the micelle's/solvent interface and within the corona. This dehydration is detected by the ESEEM experiments of a reaction carried out on a D_2O reaction mixture (Fig. 12.16c). Two types of spin probes were used: L62-NO and 4-hydroxy-TEMPO-benzoate (4HTB). The latter is highly hydrophobic and therefore it is located primarily in the core of the micelles. The ^2H modulation depth, $k(^2H)$, exhibits a reduction with time. This reduction is sensed earlier for 4HTB than for L62-NO, showing that the water depletion occurs from the core outward to the corona.

12.4 REACTION INTERMEDIATES AND TRAPPED RADICALS

Another interesting application of EPR to porous materials is related to characterization of trapped radicals. The lifetime of many organic radicals in solution is often too short to allow their characterization. However, if such radicals can be formed and trapped within the zeolite pores, where their diffusion is considerably limited such that the probability of encountering other radicals or reactants is very low, their lifetime can be significantly prolonged allowing their characterization by EPR techniques. One such classical example is the "blue" sodalite, which is also known as ultramarine blue. There, the color originates from an S_3^- radical trapped in the sodalite cage (see Fig. 12.1). This radical has been studied extensively by EPR at X- and W-band [76] and by W-band ENDOR [77], focusing on its interaction with the sodalite framework and with Na^+ ions in the cage. Another recent example is the characterization of radical products of the photolysis of dibenzyl ketones adsorbed on ZSM-5. Radicals formed on the external surface can diffuse fast and react with other radicals and cannot be detected. In contrast, the radicals that are inside the channels persist for a long enough time (hours and days) and can be studied by regular CW-EPR. They could also react with oxygen to generate peroxy persistent radicals and with NO to form diamagnetic nitroso compounds. The structure of the radicals was determined using specific deuteration and by computer simulations [78].

12.5 SAMPLE PREPARATION CONSIDERATIONS

Unlike NMR, where the sample concentration scales linearly with the signal intensity (and therefore concentrated samples are welcome in terms of the signal to noise ratio), in EPR a high concentration of paramagnetic centers is problematic because of spin–spin interactions that affect resolution and relaxation times. In CW-EPR the Heisenberg exchange removes the hyperfine splitting that is a very valuable source of information. Hence, concentrations in the millimolar range are preferred for frozen solutions and the amounts of paramagnetic centers should be in the range of a few weight percents for solids. For pulse EPR, where one would like to extend the phase memory time and limit spectral diffusion, which leads to faster echo decay, the concentrations for frozen solutions should be even lower (<2 mM) and for solids should be around $0.1-0.5$ wt%. The amount of sample required is $50-100$ μL of frozen solution or ~ 50 mg of solids. Naturally, the amount needed is lower if the spectrum is narrower.

In solids the distribution of the paramagnetic centers may not be homogeneous, especially in the case of metal ions impregnated onto oxides. In this case, although the total sample concentration may be low, the local concentration is high because of the formation of clusters. In such a situation one may have nice and strong EPR signals, but in the pulse experiments the echo decays too fast and cannot be detected. The disappearance of the echo signal that is due to a particular treatment, such as calcination, without a noticeable change in the EPR spectrum can be used as an indication for clustering.

The high sensitivity of pulse experiments to relaxation times requires low temperature measurements. For TMs the temperature range is usually between 4 and 25K whereas for nitroxides a higher range of $50-77$K can be used. For X-band measurements the sample tubes should be of high quality quartz (usually $3-4$ mm o.d.) because glass tubes usually have paramagnetic metal ion impurities that give signals at low temperatures. This tube size is also convenient and easy to use when samples should be prepared under vacuum because a connection to a vacuum line is rather easy. Teflon tubes can be used as well, particularly when frozen aqueous solutions are used, such as in ESEEM measurements on reaction mixtures of mesoporous materials. This prevents the sample tubes from breaking. Sample handling becomes less convenient as the spectrometer frequency increases because the sample tube size decreases. For example, the tube size for W-band is 0.8-mm o.d. This does not present a problem for samples that are not airtight. However, preparation under vacuum is more involved but can be worked out [15]. The amount of sample is minute, $1-2$ μL or 1–2 mg. The concentration range again depends on the width of the spectrum and can go down to 0.1 mM for nitroxide and even less for Mn^{2+}.

12.6 SUMMARY AND OUTLOOK

This chapter presented a series of examples of a variety of EPR methods applied to a range of problems in porous materials and high surface area catalytic materials. This involved paramagnetic TMIs or nitroxide spin labels. All of these are paramagnetic in the ground state. One can also think about similar application to excited triplet states. The tool boxes offered by EPR spectroscopy are very rich and consist of many different experimental techniques that can be carried out on a variety of spectrometers operating at different frequencies. This chapter did not describe applications of pulse doubled electron–electron resonance experiments designed for distance measurements. This is an emerging technique that has been found to be most effective for biological systems. It can be very useful also for applications to porous materials for determining the spatial distribution of radicals and TMIs. One major drawback of the high resolution EPR methods is that they have to be carried out at low temperatures, and they are applicable only to dilute systems. Systems where the paramagnetic ions constitute the majority of the sample are not amenable to high resolution techniques. Nonetheless, they still can be probed by CW-EPR at conventional and high frequencies.

These high resolution methods are widely applied to problems in biophysics and structural biology, but their applications in the field of material science has thus far been more scarce and their potential is far from being fully exploited. This is attributed to the high cost of the spectrometers, primarily the pulse spectrometers, and more so to the expertise needed for the spectral interpretation. Therefore, it is recommended to perform the first steps of the application of advanced EPR through collaboration with one of the many "expert" laboratories around the world. Finally, although the paramagnetic centers that comprise the active site itself or its precursor can now be characterized in detail, the link with the catalytic activity is often missing. This is mainly attributable to the lack of coordination of the expertise in catalysis and spectroscopy, requiring close collaboration between two such laboratories.

REFERENCES

1. Weckhuysen, B. M. *Phys. Chem. Chem. Phys.* **2003**, *5*, 4351.
2. Dyrek, K.; Che, M. *Chem. Rev.* **1997**, *97*, 305.
3. Sojka, Z.; Che, M. *Colloid Surf.* **1999**, *A158*, 165.
4. Breck, D. W. *Zeolite Molecular Sieves*; Wiley: New York, 1974.
5. Wilson, S. T.; Lok, B. M.; Messina, C. A.; Cannan, T. R.; Flanigen, E. M. *J. Am. Chem. Soc.* **1982**, *104*, 1146.
6. Reddy, J. S.; Ravishankar, R.; Sivasanker, S.; Ratnasamy, P. *Catal. Lett.* **1993**, *17*, 139.

7. Iwamoto, M.; Yahiro, H.; Tanda, K.; Mizuno, N.; Mine, Y.; Kagawa, S. *J. Phys. Chem.* **1991**, *95*, 3727.
8. Herman, R. G.; Lunsford, J. H.; Beyer, H.; Jacobs, P. A.; Uytterhoeven, J. B. *J. Phys. Chem.* **1975**, *79*, 2388.
9. Dedecek, J.; Sobalik, Z.; Tvaruzkova, Z.; Kaucky, D.; Wichterlova, B. *J. Phys. Chem.* **1995**, *99*, 16327.
10. Attfield, M. P.; Weigel, S. J.; Cheetham, A. K. *J. Catal.* **1997**, *172*, 274.
11. Anderson, M. W.; Kevan, L. *J. Phys. Chem.* **1987**, *91*, 4174.
12. Larsen, S. C.; Aylor, A.; Bell, A. T.; Reimer, J. A. *J. Phys. Chem.* **1994**, *98*, 11533.
13. Carl, P. J.; Larsen, S. C. *J. Phys. Chem.* **2000**, *B104*, 6568.
14. Stoll, S.; Schweiger, A. *J. Magn. Res.* **2006**, *178*, 42.
15. Carl, P. J.; Vaughan, D. E. W.; Goldfarb, D. *J. Phys. Chem.* **2002**, *B106*, 5428.
16. Goldfarb, D.; Kevan, L. *J. Magn. Res.* **1988**, *76*, 276.
17. Peisach, J.; Blumberg, W. E. *Arch. Biochem. Biophys.* **1974**, *165*, 691.
18. Hurst, G. C.; Henderson, T. A.; Kreilick, R. W. *J. Am. Chem. Soc.* **1985**, *107*, 7294.
19. Rist, G. H.; Hyde, J. S. *J. Chem. Phys.* **1968**, *49*, 2449.
20. Carl, P. J.; Vaughan, D. E. W.; Goldfarb, D. *J. Am. Chem. Soc.* **2006**, *128*, 7160.
21. Schmauke, T.; Eichel, R. A.; Schweiger, A.; Roduner, E. *Phys. Chem. Chem. Phys.* **2003**, *5*, 3076.
22. DeVos, D. E.; KnopsGerrits, P. P.; Vanoppen, D. L.; Jacobs, P. A. *Supramol. Chem.* **1995**, *6*, 49.
23. Devos, D. E.; Knopsgerrits, P. P.; Parton, R. F.; Weckhuysen, B. M.; Jacobs, P. A.; Schoonheydt, R. A. *J. Inclusion Phenom. Mol. Recognit. Chem.* **1995**, *21*, 185.
24. KnopsGerrits, P. P.; Trujillo, C. A.; Zhan, B. Z.; Li, X. Y.; Jacobs, P. A. *Top. Catal.* **1996**, *3*, 437.
25. Weckhuysen, B. M.; Verberckmoes, A. A.; Fu, L. J.; Schoonheydt, R. A. *J. Phys. Chem.* **1996**, *100*, 9456.
26. Weckhuysen, B. M.; Verberckmoes, A. A.; Vannijvel, I. P.; Pelgrims, J. A.; Buskens, P. L.; Jacobs, P. A.; Schoonheydt, R. A. *Angew. Chem. Int. Ed. Engl.* **1996**, *34*, 2652.
27. Grommen, R.; Manikandan, P.; Gao, Y.; Shane, T.; Shane, J. J.; Schoonheydt, R. A.; Weckhuysen, B. M.; Goldfarb, D. *J. Am. Chem. Soc.* **2000**, *122*, 11488.
28. Deschamps, P.; Kulkarni, P. P.; Gautam-Basak, M.; Sarkar, B. *Coord. Chem. Rev.* **2005**, *249*, 895.
29. Manikandan, P.; Epel, B.; Goldfarb, D. *Inorg. Chem.* **2001**, *40*, 781.
30. Deligiannakis, Y.; Louloudi, M.; Hadjiliadis, N. *Coord. Chem. Rev.* **2000**, *204*, 1.
31. Baute, D.; Arieli, D.; Neese, F.; Zimmermann, H.; Weckhuysen, B. M.; Goldfarb, D. *J. Am. Chem. Soc.* **2004**, *126*, 11733.
32. Hartmann, M.; Kevan, L. *Chem. Rev.* **1999**, *99*, 635.
33. Markham, G. D.; Rao, B. D. N.; Reed, G. H. *J. Magn. Res.* **1979**, *33*, 595.
34. Arieli, D.; Prisner, T. F.; Hertel, M.; Goldfarb, D. *Phys. Chem. Chem. Phys.* **2004**, *6*, 172.
35. Arieli, D.; Vaughan, D. E. W.; Strohmaier, K. G.; Goldfarb, D. *J. Am. Chem. Soc.* **1999**, *121*, 6028.

36. Arieli, D.; Delabie, A.; Vaughan, D. E. W.; Strohmaier, K. G.; Goldfarb, D. *J. Phys. Chem.* **2002**, *B106*, 7509.
37. Arieli, D.; Delabie, A.; Groothaert, M.; Pierloot, K.; Goldfarb, D. *J. Phys. Chem.* **2002**, *B106*, 9086.
38. Beck, J. S.; Vartuli, J. C.; Roth, W. J.; Leonowicz, M. E.; Kresge, C. T.; Schmitt, K. D.; Chu, C. T. W.; Olson, D. H.; Sheppard, E. W.; McCullen, S. B.; Higgins, J. B.; Schlenker, J. L. *J. Am. Chem. Soc.* **1992**, *114*, 10834.
39. Ying, J. Y.; Mehnert, C. P.; Wong, M. S. *Angew. Chem. Int. Ed.* **1999**, *38*, 56.
40. Wan, Y.; Zhao, D. *Chem. Rev.* **2007**, *207*, 2821.
41. Epping, J. D.; Chmelka, B. F. *Curr. Opin. Colloid Interface Sci.* **2006**, *11*, 81.
42. Schuth, F. *Annu. Rev. Mater. Res.* **2005**, *35*, 209.
43. Taguchi, A.; Schuth, F. *Micropor. Mesopor. Mater.* **2005**, *77*, 1.
44. Schuth, F. *Chem. Mater.* **2001**, *13*, 3184.
45. Zhao, D. Y.; Feng, J. L.; Huo, Q. S.; Melosh, N.; Fredrickson, G. H.; Chmelka, B. F.; Stucky, G. D. *Science* **1998**, *279*, 548.
46. Chiesa, M.; Meynen, V.; Van Doorslaer, S.; Cool, P.; Vansant, E. F. *J. Am. Chem. Soc.* **2006**, *128*, 8955.
47. Weckhuysen, B. M. *Chem. Commun.* **2002**, 97.
48. Berliner, L. J., Reuben, J., Eds. *Spin Labeling Theory and Applications*; Plenum: New York, 1989; Vol. 8.
49. Lozos, G. P.; Hoffman, B. M. *J. Phys. Chem.* **1974**, *78*, 2110.
50. Rigo, A.; Viglino, P.; Ranieri, G.; Orsega, E. F.; Sotgiu, A. *Inorg. Chim. Acta* **1977**, *21*, 81.
51. Evreinov, V. I.; Golubev, V. B.; Lunina, E. V. *Zh. Fiz. Khim.* **1975**, *49*, 966.
52. Lunsford, J. H. *J. Phys. Chem.* **1968**, *72*, 4163.
53. Rudolf, T.; Pöppl, A.; Brunner, W.; Michel, D. *Magn. Res. Chem.* **1999**, *37*, S93.
54. Pöppl, A.; Rudolf, T.; Manikandan, P.; Goldfarb, D. *J. Am. Chem. Soc.* **2000**, *122*, 10194.
55. Tijink, G. A. H.; Janssen, R.; Veeman, W. S. *J. Am. Chem. Soc.* **1987**, *109*, 7301.
56. Giamello, E.; Murphy, D.; Magnacca, G.; Morterra, C.; Shioya, Y.; Nomura, T.; Anpo, M. *J. Catal.* **1992**, *136*, 510.
57. Sojka, Z.; Che, M.; Giamello, E. *J. Phys. Chem.* **1997**, *B101*, 4831.
58. Umamaheswari, V.; Hartmann, M.; Pöppl, A. *J. Phys. Chem.* **2005**, *B109*, 1537.
59. Umamaheswari, V.; Hartmann, M.; Pöppl, A. *J. Phys. Chem.* **2005**, *B109*, 10842.
60. Umamaheswari, V.; Hartmann, M.; Pöppl, A. *J. Phys. Chem.* **2005**, *B109*, 19723.
61. Alexandridis, P.; Hatton, T. A. *Colloids Surf.* **1995**, *A96*, 1.
62. Kruk, M.; Jaroniec, M.; Ko, C. H.; Ryoo, R. *Chem. Mater.* **2000**, *12*, 1961.
63. Ruthstein, S.; Frydman, V.; Kababya, S.; Landau, M.; Goldfarb, D. *J. Phys. Chem.* **2003**, *B107*, 1739.
64. Moscatelli, A.; Galarneau, A.; Di Renzo, F.; Ottaviani, M. F. *J. Phys. Chem.* **2004**, *B108*, 18580.
65. Liu, Z. Q.; Ottaviani, M. F.; Abrams, L.; Lei, X. G.; Turro, N. J. *J. Phys. Chem.* **2004**, *A10*, 8040.
66. Martini, G.; Ristori, S.; Romanelli, M.; Kevan, L. *J. Phys. Chem.* **1990**, *94*, 7607.

67. Baute, D.; Frydman, V.; Zimmermann, H.; Kababya, S.; Goldfarb, D. *J. Phys. Chem.* **2005**, *B109*, 7807.
68. Galarneau, A.; Di Renzo, F.; Fajula, F.; Mollo, L.; Fubini, B.; Ottaviani, M. F. *J. Colloid Interface Sci.* **1998**, *201*, 105.
69. Ottaviani, M. F.; Galarneau, A.; Desplantier-Giscard, D.; Di Renzo, F.; Fajula, F. *Micropor. Mesopor. Mater.* **2001**, *44*, 1.
70. Ruthstein, S.; Frydman, V.; Goldfarb, D. *J. Phys. Chem.* **2004**, *B108*, 9016.
71. Zhang, J. Y.; Carl, P. J.; Zimmermann, H.; Goldfarb, D. *J. Phys. Chem.* **2002**, *B106*, 5382.
72. Zhang, J. Y.; Luz, Z.; Goldfarb, D. *J. Phys. Chem.* **1997**, *B101*, 7087.
73. Zhang, J. Y.; Luz, Z.; Zimmermann, H.; Goldfarb, D. *J. Phys. Chem.* **2000**, *B104*, 279.
74. Ruthstein, S.; Potapov, A.; Raitsimring, A. M.; Goldfarb, D. *J. Phys. Chem.* **2005**, *B109*, 22843.
75. Baute, D.; Goldfarb, D. *J. Phys. Chem.* **2007**, *C111*, 10931.
76. Eaton, G. R.; Eaton, S. S.; Stoner, J. W.; Quine, R. W.; Rinard, G. A.; Smirnov, A. I.; Weber, R. T.; Krzystek, J.; Hassan, A. K.; Brunel, L. C.; Demortier, A. *Appl. Magn. Res.* **2001**, *21*, 563.
77. Arieli, D.; Vaughan, D. E. W.; Goldfarb, D. *J. Am. Chem. Soc.* **2004**, *126*, 5776.
78. Turro, N. J.; Lei, X. G.; Jockusch, S.; Li, W.; Liu, Z. Q.; Abrams, L.; Ottaviani, M. F. *J. Org. Chem.* **2002**, *67*, 2606.

13 Electron Paramagnetic Resonance of Charge Carriers in Solids

MARIO CHIESA and ELIO GIAMELLO
Department of Chemistry IFM, University of Torino, Via P. Giuria 7, 10125 Torino, Italy

13.1 INTRODUCTION

The aim of this chapter is to illustrate the considerable potential and applicability of electron paramagnetic resonance (EPR) as a powerful probe of the microscopic electronic structure of paramagnetic species in different materials from insulating solids to metallic conductors. The intention is not to produce a comprehensive in-depth survey of the vast literature in the field but to attempt a fairly concise description of some important aspects of the magnetic resonance phenomenon in solid materials. Particular emphasis is placed on experimental results and theoretical descriptions of charge carriers associated to point defects in insulating and semiconducting solids that are characterized by either ionic or covalent bonds. In particular, attention is focused on the detailed structural information available from the spectrum of a defect (symmetry from its angular dependence, atomic and lattice structure from its hyperfine interactions), which makes EPR uniquely able to identify a defect, to map out its wavefunction in the lattice, and to determine its microscopic structure. In doing so, the choice of examples and associated literature has been rather selective, focusing the attention on conventional continuous wave EPR and electron nuclear double resonance (ENDOR) results of classical and particularly tutorial cases.

The end of the chapter also includes a cursory discussion of conduction electron spin resonance (CESR) in metallic materials to illustrate the wealth of useful information that can be obtained on systems where the unpaired electrons are delocalized throughout the entire sample.

Electron Paramagnetic Resonance. Edited by Brustolon and Giamello
Copyright © 2009 John Wiley & Sons, Inc.

13.2 POINT DEFECTS, CHARGE CARRIERS, AND EPR

Real solids are never perfect. The defects they contain are not, however, a simple curiosity of the solid-state world. Conversely, the fundamental properties of a solid, such as mechanical, optical, electrical, and chemical properties, are very often connected to the presence of defects. In particular, the threshold concentration of defects causing the onset of a given property is extremely low in many cases. Defects are usually classified as intrinsic defects when the composition of a solid is not altered by the presence of the defect itself and as extrinsic defects when a chemical impurity is present in the solid. Another kind of classification concerns the "dimension" of the defect. Point defects have "zero dimension" (in contrast with "one dimension" or line defects like dislocations) and are actually a modification of the regular crystal lattice that concerns a single site. These defects are the most important ones as far as optical, electric, and magnetic properties are concerned. Three typical examples of point defects in an ionic solid are the following:

1. ion vacancies, which are usually found as pairs of independent cation–anion vacancies called Schottky defects (Fig. 13.1);
2. interstitial ions formed in parallel with an ion vacancy (Frankel defects); and
3. single-atom chemical impurities.

Figure 13.1a illustrates a two-dimensional structure of NaCl, which is the most common structure of ionic systems (the majority of alkali halides and MeO type binary oxides exhibit this simple structure). The Schottky defect is sketched in Fig 13.1b.

For the reasons mentioned above, solid-state physics, solid-state chemistry, and materials science have a strategic interest in methods allowing the understanding of the nature, location, and electronic structure of point defects. Among all physical

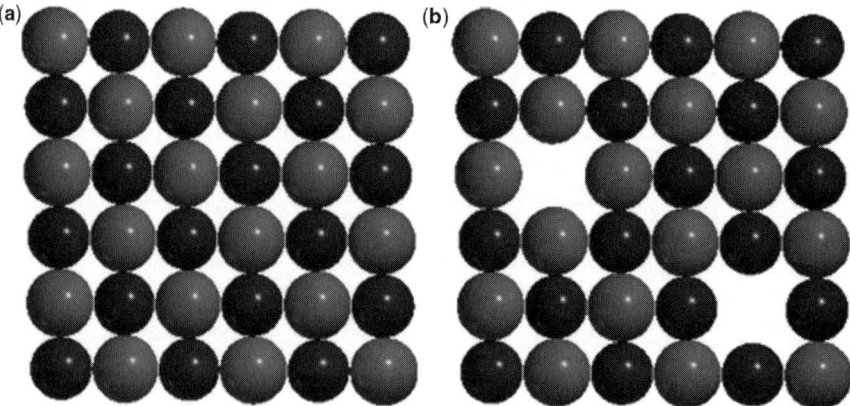

Fig. 13.1 A two-dimensional scheme of (a) a perfect ionic crystal (rock salt or NaCl structure) and (b) a Schottky defect.

13.2 POINT DEFECTS, CHARGE CARRIERS, AND EPR

methods, EPR techniques are the most powerful tools for the description of defects in solids. This is due to two basic factors: the intrinsic accuracy of EPR techniques in describing the structure and electron spin density distribution of a species and the high sensitivity of the techniques that allows detection of very tiny concentrations of paramagnetic centers. The intrinsic limitation of EPR that cannot be applied to diamagnetic systems is, in this case, attenuated by the fact that many highly important point defects in solid materials are paramagnetic.

EPR detectable charge carriers (electrons and electron holes) are usually related to the presence of point defects in the solid at least in the case of insulators and semiconductors. They can be either localized or mobile and these two opposite properties depend, as a first approximation, on the nature of the solid matrix and the temperature.

The concept of the electron hole is not typical of molecular chemistry and cannot be disjoined by a view of the electronic structure of a solid in terms of band theory. Removal of an electron from a filled band leaves behind a net positive charge, which is a positive hole or more simply a hole. This can typically be observed when irradiating a solid with suitable radiation capable of promoting an electron from the valence band to the conduction band (in this case the photon energy $h\nu$ is higher than the energy of the band gap; Fig. 13.2).

Once separated by the photon energy, the charge carriers tend to recombine unless they find a suitable site for their stabilization. These sites can be chemical impurities, intrinsic defects (ion vacancies), and even surface sites whose energy is usually placed in the forbidden gap (intra band gap states). When stabilized in a localized site, both charge carriers can be observed, in principle, by EPR.

Figure 13.3 shows the EPR powder spectrum of a trapped electron–hole pair formed by UV excitation of MgO, an insulator having a band gap of 7.8 eV. Both charge carriers are stabilized in particular locations of the surface of the crystals. The trapped electron signal is observed close to the free electron g value (2.0023), the perpendicular component of the axial **g** tensor of the trapped hole falls at $g = 2.0357$.

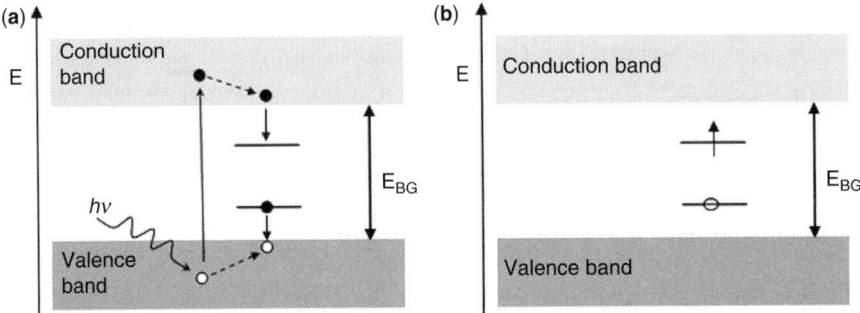

Fig. 13.2 (a) A schematic of the photonic excitation of an electron from the valence band to the conduction band. (b) The electron and hole are then stabilized in localized states in the band gap.

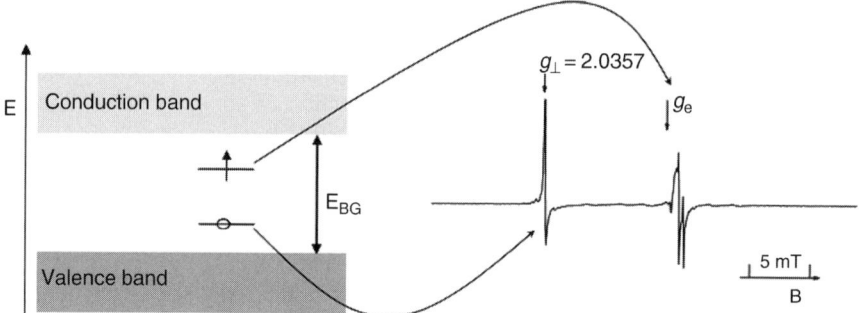

Fig. 13.3 An X-band EPR spectrum of an trapped electron–hole pair stabilized at the surface of MgO.

13.3 LOCALIZED ELECTRONS: COLOR CENTERS IN IONIC SOLIDS

Ionic insulating solids, which are formed by regular arrays of ions having alternating electric charge and generating strong electric potential within the matrix, are suited to host localized charge carriers whereas metallic solids typically contains mobile carriers. It will be shown in the following that semiconductors, and in particular those based on doped silicon, have a somewhat intermediate behavior, the localization–delocalization of the charge carrier strongly depending on the temperature.

The simplest systems in the solid state containing one unpaired electron are the color centers formed in ionic solids. These systems basically consist of one electron trapped in the volume of an anion vacancy (Fig. 13.4). Color centers can be found in naturally occurring ionic crystals or can be generated in unperturbed crystals by high energy irradiation or chemical treatments (metal addition).

The presence of such centers originate the energy states in the band gap of the insulating solid that are able to absorb visible light, thus coloring the colorless (when without defects) crystals. For this reason they were called color centers or F centers from *Farbe*, the German word for color. Using the classic symbology of Kröger and Vink, an F center, in the case of an alkali halide such as KCl, can be written as V_{Cl}^x, where V stands for vacancy, the subscript indicates the missing ion, and the superscript is the charge of the center that is neutral in the present example (x indicates neutrality).

Alkali halides and alkali-earth oxides are the solids in which color centers have been most frequently found and investigated. The model for the color centers described above and sketched in Fig. 13.4 is attributable to De Boer (1937) [1]. In spite of the very intense experimental investigation by various physical methods, it was confirmed about 20 years later, only after the advent of electron magnetic resonance methods (EPR and ENDOR). Because the De Boer picture for F centers is very close to the ideal model of a particle in a box, the F centers became an important reference for solid-state physics. As convincingly said by Wood and Joy in a paper published in *Physical Review* [2], "... this defect is one of the simplest which

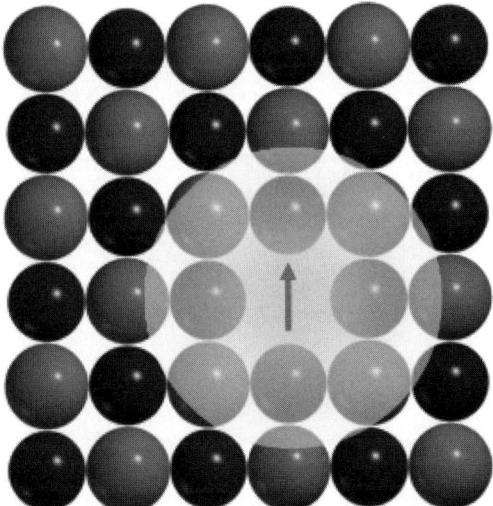

Fig. 13.4 A two-dimensional scheme of a color center or F center consisting of an electron trapped in an anion vacancy. The shadowed region schematically indicates the unpaired electron density distribution.

can occur in ionic crystals and, in the physics and chemistry of lattice defects it occupies a position of importance roughly comparable to that of the hydrogen atom in ordinary chemistry."

Among the various cases of electrons observed by EPR in the solid state, color centers differ from all other types of localized and mobile electrons. In fact, although the major fraction of the electron spin density is found within the volume of the oxygen vacancy (*vide infra*), the ground state wavefunction of the electron exhibits a tail that causes the electron to interact with a surprisingly high number of ionic shells surrounding the vacancy [3]. From the EPR point of view, this fact confirms the formation, for F centers in alkali halides, of a rich and very complex superhyperfine structure[1] because all alkali metal and halogen nuclei have nonzero nuclear spin.

The superhyperfine structure of F centers was monitored in the early 1960s by EPR and ENDOR, which were essential for building up an accurate model for this kind of defects. The F centers in alkali halides have g values smaller but very close to the free electron value. In a few cases (LiF, NaF, RbCl, CsCl) the superhyperfine structure of the EPR spectrum is partially resolved (Fig. 13.5), and in all other cases (LiCl, NaCl, NaBr, KCl, KI, etc.) the spectrum consists of a broad unstructured Gaussian line the width of which ranges from a few tenths to hundreds of gauss.

---

[1]The term superhyperfine is usually adopted instead of hyperfine to emphasize that the interacting nuclei surround the unpaired electron in analogy to the case of transition-metal ion compounds, in which the central ion is surrounded by ligands. In such a case the structure produced by the interaction of the unpaired electron with the transition ion nucleus is called a hyperfine structure and that due to interaction with the ligands is called superhyperfine.

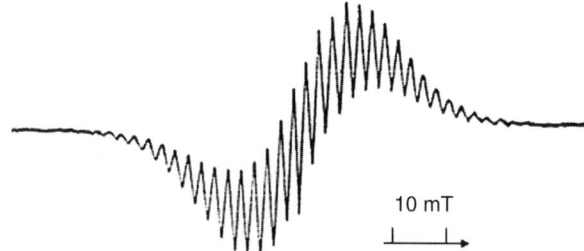

Fig. 13.5 The EPR spectrum of F centers in LiF. The spectrum was recorded on a single crystal after neutron irradiation. Adapted from Reference 3. Copyright 1963 American Physical Society.

The presence of a high number of lines in the spectra of F centers in alkali halides is because the single trapped electron interacts with the magnetic nuclear moments of a high number of ions. The F centers thus represent the most valuable tutorial example in the inorganic world to illustrate complex hyperfine structures, and it is analogous to the case of complex multiline organic radicals illustrated in Chapter 4. The complexity of the hyperfine structure is best understood with an example.

Let us consider in detail the origin of the EPR spectrum of the F center in KCl. In this example the unpaired electron is trapped into an anion vacancy, surrounded by 6 K^+ cations that form the first coordination sphere around the vacancy at a distance r ($r = 0.315$ nm) from the center of the vacancy. Figure 13.6 illustrates the geometry of the crystal and the various ionic shells that surround the vacancy. Because of the spherical symmetry of the center in the ground state, all of the ions belonging to a

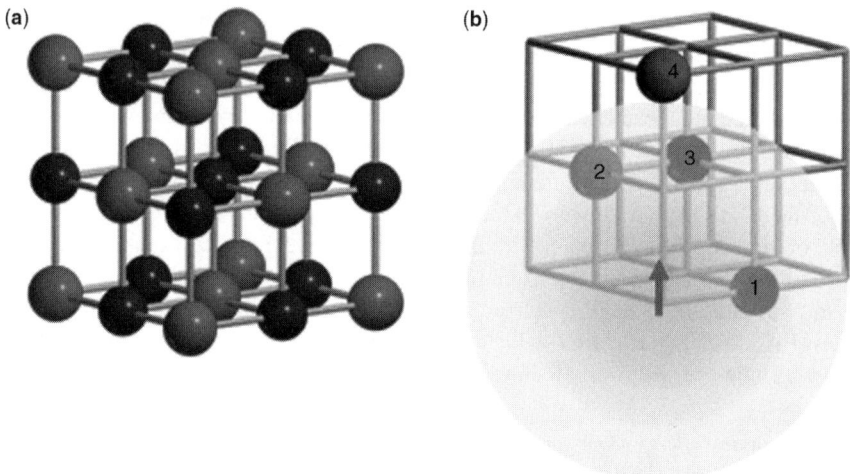

Fig. 13.6 The structure of KCl. (a) The trapped electron center is placed at the corner of the cube. (b) The various shells surrounding the defect are labeled with progressive numbers.

13.3 LOCALIZED ELECTRONS: COLOR CENTERS IN IONIC SOLIDS

given shell interact in the same way with the unpaired electron and thus originate the same hyperfine coupling. In other words, all K^+ ions of the first shell are magnetically equivalent. The interaction is mainly based on the isotropic Fermi contact term; for the sake of simplicity, the minor contributions of dipolar anisotropic hyperfine terms will be neglected in this example.

The most naturally abundant (93%) potassium isotope is ^{39}K, and it has a nuclear spin of $I = 3/2$. For chlorine, both ^{35}Cl (75.5% natural abundance) and ^{37}Cl (24.5%) have $I = 3/2$ and a slightly different nuclear magnetic moment. For simplicity, the presence of ^{41}K (~7%) and the differences of the magnetic moments of ^{35}Cl and ^{37}Cl can be neglected. Let us now take into account the number of equivalent nuclei coupled to the electron, starting from the first coordination cell. Because there are 6 equivalent potassium nuclei (n_1), one expects $2n_1 I + 1 = 19$ nuclear sublevels (for each electron spin state) corresponding to 19 different M_I, where M_I is the total nuclear spin quantum number of a given nuclear state that assumes the values $(nI, nI-1, nI-2, \ldots, -nI)$, which in the present case is $(9, 8, \ldots, 0, \ldots, -9)$. This corresponds to 19 allowed transitions and therefore to 19 equally spaced hyperfine lines, the intensity of which is proportional to the statistical weight of each state of a given M_I. The intensities can be easily calculated and are reported in the stick diagram of Fig. 13.7.

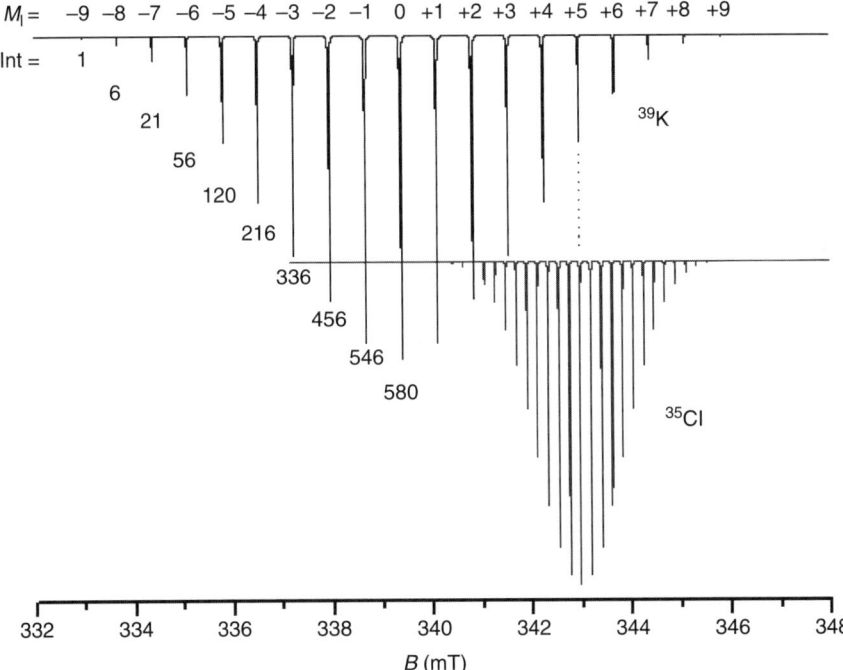

Fig. 13.7 A stick diagram of the hyperfine lines due to the first (6 equivalent K nuclei) and second (12 equivalent Cl nuclei) shells of an F center in KCl. Only the ^{39}K and ^{35}Cl isotopes are considered.

The separation between two consecutive lines corresponds to the hyperfine coupling that is 20.7 MHz in the present case. The situation due to the described interaction is graphically sketched in Fig. 13.7 using the typical stick diagram approach.

A real EPR spectrum based on this collection of lines is intrinsically complex, but the situation in the system analyzed here is far more intricate. In the typical rock salt structure of KCl, a second shell (Fig. 13.6) is found around the vacancy constituted by 12 (n_2) Cl$^-$ anions at $\sqrt{2}r$ from the center of the vacancy. Each Cl nucleus of this second shell ($I_2 = 3/2$) interacts with the unpaired electron, although at lesser extent than the first potassium sphere, producing a further contribution to the hyperfine structure (∼6.90 MHz in the present case). Drawing the nuclear sublevel formed by the addition of this second contribution is rather complex. It is sufficient to think that the EPR spectrum will be based on the splitting of *each* of the 19 K hyperfine lines in $2n_2 I + 1$ new lines that is due to Cl hyperfine interaction. Because $I = 3/2$ and $n_2 = 12$, there are 37 chlorine lines for each line that are due to K (Fig. 13.7). This means that the EPR spectrum involving the first and second coordination shells is constitutes 703 lines. Simulated spectra of the F centers in KCl, limited to the hyperfine interaction with the first and second coordination spheres, are reported in Fig. 13.8. Using a realistic linewidth

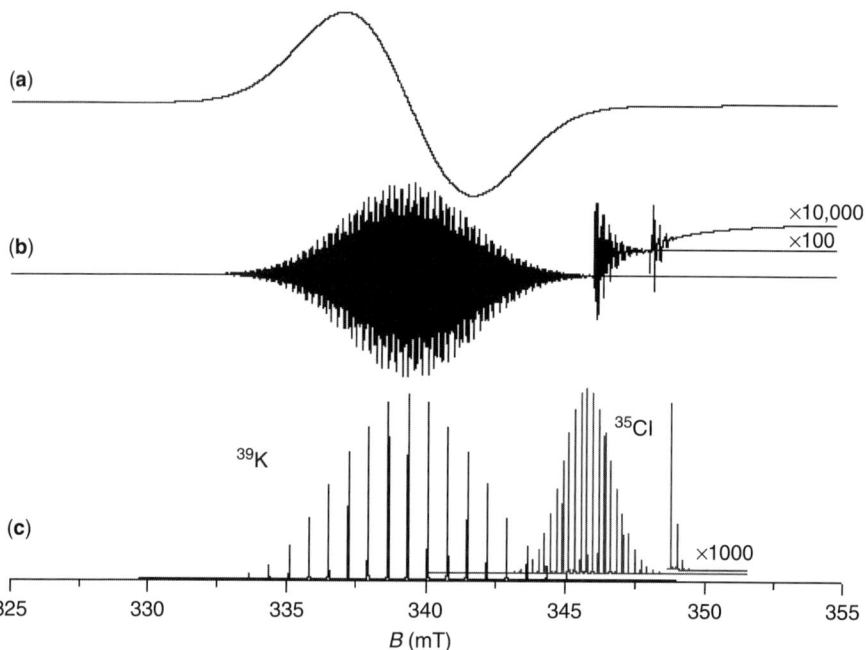

Fig. 13.8 Simulations of the EPR spectrum of the F center in KCl considering the first and second coordination spheres only. (a) $\Delta B = 0.6$ mT, (b) $\Delta B = 0.06$ mT, and (c) a stick diagram. Simulation (a) fits the experimental spectrum.

($\Delta B = 0.6$) for each hyperfine line, the spectrum (Fig. 13.8a) appears as a unique unstructured broad line with a Gaussian shape. This is also the profile of the experimental spectrum [4].

Using a narrower linewidth ($\Delta B = 0.06$ mT), the 703 line spectrum in Fig. 13.8b still shows resolution. Spectra were simulated using the EasySpin program [5]. This example illustrates a typical case of heterogeneous broadening. The investigation of the system is limited by the high number of nuclear sublevels and, consequently, of EPR allowed transitions. In general, the expected number of transitions and thus hyperfine lines can be calculated using the following equation:

$$\text{number of lines} = \prod_i (2n_i I_i + 1) \tag{13.1}$$

where i indicates the shell and n_i is the number of ions belonging to the ith shell.

Extending the analysis to the fifth shell surrounding the F center, the total number of expected lines thus becomes more than 12×10^6!

As shown, the hyperfine coupling constants of the trapped electron interacting with the surrounding nuclei cannot be directly measured by EPR but become measurable by ENDOR. Such a measurement was one of the first successes of this double resonance technique a short time after its discovery. Because a description of the ENDOR principles has already been provided in Chapters 2 and 6, it is sufficient here to recall that the during ENDOR experiments an EPR transition is saturated (α and β electron spin populations are equalized by high microwave power) and the nuclear spin resonance of the nuclei coupled to the unpaired electron can be induced by a suitable radiofrequency that desaturates the EPR line. The number of nuclear spin transitions observed in the experiment according to the nuclear magnetic resonance selection rule ($\Delta m_S = 0$, $\Delta m_I = \pm 1$) is dramatically lower than that of the EPR ones. This fact induces a tremendous simplification of the experimental datum. Figure 13.9 reports the classic ENDOR experiment performed by Siedel on F centers of KBr [6]. A series of lines is observed in a large range of frequencies and can be assigned either to K^+ or Br^- ions belonging to progressive outer coordination spheres. In the particular case of KBr, a detectable interaction was observed until the eighth coordination sphere of Br^- ions. The whole set of EPR (when resolved spectra are observed) and ENDOR results on F centers in alkaline halides represents a confirmation of the de Boer model and a spectacularly detailed determination of their wavefunction in the ground state. No computational method of similar accuracy based on the resolution of the Schrödinger equation was available in the early 1960s to build a theoretical model of these solid-state defects having approximations comparable to that reached by electron magnetic resonance experiments.

Things are different now, and modern quantum chemistry computational techniques in the solid state allow accurate calculations on solid systems. The theoretical investigation at the Hartree–Fock level of an F center in LiF appeared in 1998 [7]. A calculated hyperfine structure was computed involving several coordination shells around the defect whose values, depending on the basis set adopted, were in

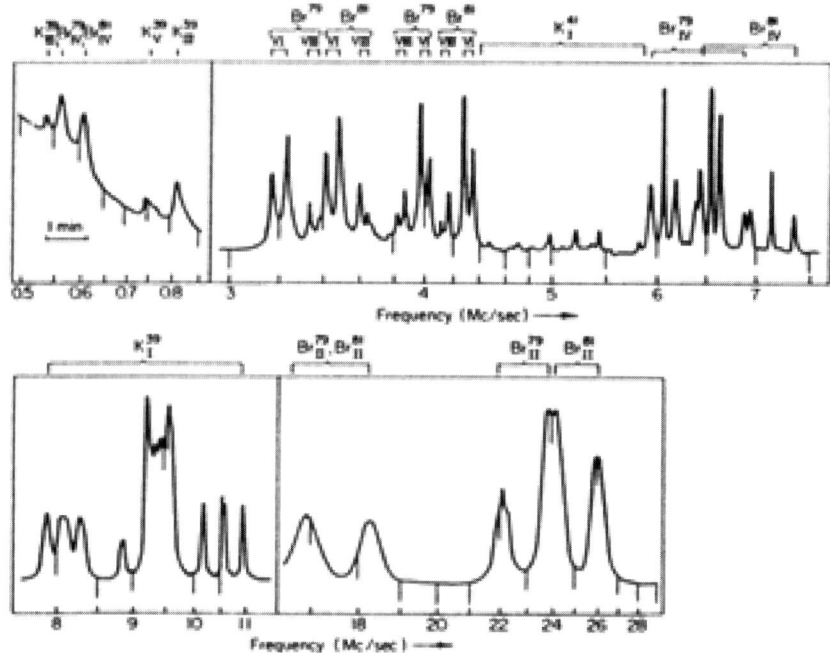

Fig. 13.9 The ENDOR spectra of F centers in KBr. The data are from Reference 6. Copyright 1961.

quite good agreement with the experiments. This is shown in Table 13.1, which compares experimental and calculated data for LiF.

The discrepancies before calculated and experimental data are apparently not negligible. However, taking into account the delicate calculations related to the evaluation of the Fermi contact term for the various nuclei, the agreement between experimental and computed data and the description of the whole wavefunction is satisfactory. Figure 13.10 reports the calculated profiles of the spin density that graphically illustrates the results in Table 13.1 and confirms the first intuition of de Boer about the nature of color centers.

TABLE 13.1 Calculated (Hartree–Fock Hamiltonian) and Experimental Isotropic Hyperfine Coupling Constants (MHz) of the First Seven Coordination Shells for F Centers in LiF

Hamiltonian	Shell I Li	Shell II F	Shell III Li	Shell IV F	Shell V Li	Shell VI F	Shell VII Li
Hartee–Fock	39.17	76.40	0.11	0.89	0.07	0.46	0.83
Experimental	39.06	105.94	0.50	0.48	0.27	0.88	1.34

Data adapted from References 6 and 7. Copyright 1961 and 1998 APS.

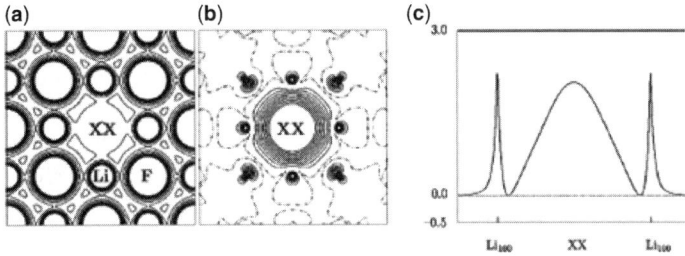

Fig. 13.10 (a) The electron charge density and (b) electron spin density for an F center in LiF; XX indicates the center of the anion vacancy V. (c) The spin density profile along the direction Li–V–Li is reported. Adapted from Reference 7. Copyright 1998 American Physical Society.

13.4 AGGREGATE COLOR CENTERS

Irradiation of crystals containing F centers leads to the partial formation of aggregate centers having specific optical and magnetic properties. In particular, as far as EPR is concerned, examples of $S > 1/2$ systems can be observed. The F_2 or M center is based on two neighboring anion vacancies, each containing one electron, along a [001] direction (the symmetry of the center is D_{2h}). In the ground state the two electrons interact to form an EPR silent singlet state ($S = 0$), but under irradiation a triplet state ($S = 1$) is observed that has a rhombic fine structure characterized by $D/g\mu_B = -161$ G and $E/g\mu_B = 54$ G. A similar system is that formed by three anion vacancies, each containing one electron, the so-called R center, lying in a (111) plane in C_{3v} symmetry (Fig. 13.11).

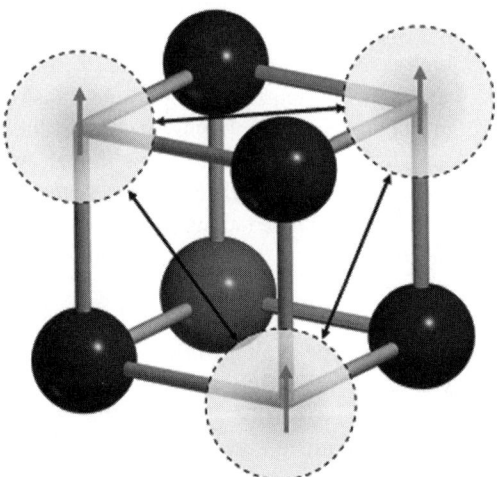

Fig. 13.11 A model of the aggregate R center in alkali halides.

The ground state of this center is a twofold orbitally degenerated doublet ($S = 1/2$) that, because of its degeneracy, can be observed only at very low temperature and under mechanical stress. Under optical irradiation of the crystal containing R centers, however, a photoexcited transient quartet state ($S = 3/2$) is populated. This signal is attributable to a state having $S = 3/2$ (quartet state), which is the first excited state of the R center having the three electron spins parallel to each other. The EPR spectrum of this center can be studied by varying the orientation of the crystal in the magnetic field. The observed parameters are $g = 1.996$ and $D/g\mu_B = +168.5$ G. Because of the axial symmetry of the center, the E term is zero (see Chapter 1).

13.5 LOCALIZED HOLES IN IONIC SOLIDS

Trapped hole centers are electron deficient centers formed by removal of electrons in a band. Removal of an electron leaves behind a net positive charge, that is, a positive electron hole, or more simply a hole. Electron holes are common defects of the solid state. They can be formed by ionizing radiation or can be induced by chemical modification of a crystal. In the former case irradiation usually promotes an electron from the valence band to the conduction band leaving a hole in the valence band. In chemical terms, and referring to an ionic solid, an electron is ionized from an anion site (Cl^-, O^{2-}) and leaves a hole on this site. Several ways can be found to stabilize the hole and, if stabilization occurs, the hole center is usually observable by EPR. A rich family of such defects, called V centers, has been identified in ionic solids. As opposed to the localized electron, which can be stabilized in the volume of an anion vacancy (F center), holes formed in ionic solids cannot stabilize in a corresponding manner (that is in the volume of a cation vacancy), tending instead to localize in an atomic (or molecular) orbital of the system. In other words, the antimorph of an F center does not exist.

Typical behavior of irradiated alkaline halides (e.g., KCl) after generation of a hole is the relaxation of the crystal around the hole site with formation of a center that is molecular in nature, which is called a self-trapped hole.

Adopting chemical terms, a Cl^- ion localizing a hole transforms into a Cl atom that binds to one of the neighboring Cl^- [along the (110) direction] and forming the molecular radical anion Cl_2^- (Fig. 13.12). This center, which is called V_K in the original nomenclature [8] and is related to the optical properties of point defects, is described as a σ radical trapped along the [110] crystallographic direction and it has the unpaired electron in the σ_u antibonding orbital (Fig. 13.13). The Cl–Cl distance in this center is intermediate between the Cl^-–Cl^- distance in the crystal (4.44 Å) and that typical of the chlorine molecule (2.00 Å). The ionic environment around the Cl_2^- center is not the same in all directions. This causes a certain distortion of the axial magnetic symmetry expected for the free anion radical that results in the rhombic symmetry of the **g** tensor that has $g_{xx} = 2.0428$, $g_{yy} = 2.0447$, and $g_{zz} = 2.0010$. The unpaired electron is equivalently shared by the two Cl nuclei as indicated by the hyperfine structure. The **A** tensor was found to be axial, which is different from the **g** tensor. The case of Cl_2^- is a particularly interesting example of a

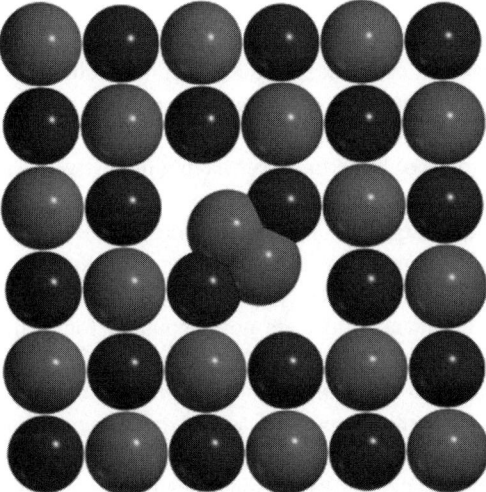

Fig. 13.12 A schematic of a V_K center in KCl.

hyperfine pattern, the features of which are determined by the presence of different isotopomers. Chlorine has two magnetic nuclei ^{35}Cl and ^{37}Cl with natural abundance of $x_{35} = 0.75$ and $x_{37} = 0.25$, respectively. Both have nuclear spin $I = 3/2$ but different nuclear magnetic moments ($\gamma_{37}/\gamma_{35} = 0.83$) and thus originate distinct hyperfine structures.

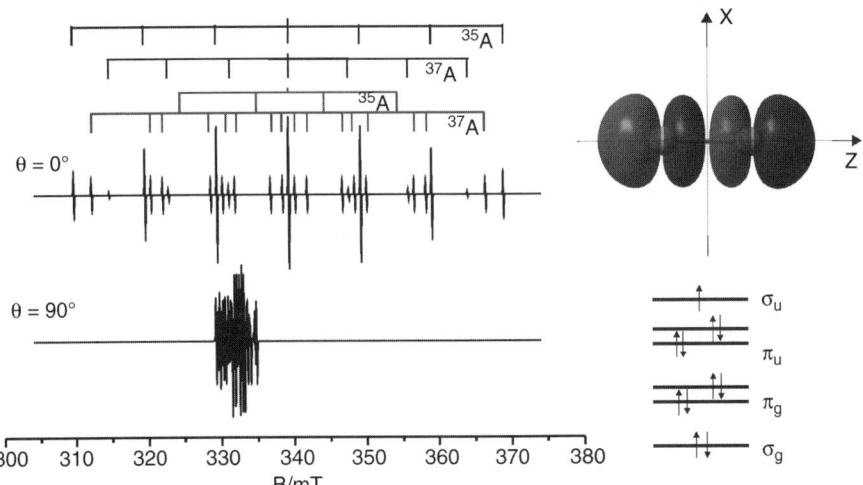

Fig. 13.13 The V_K center (Cl$_2^-$). Simulations of the EPR spectrum in the direction corresponding to the molecular z and x axes and a plot of the singly occupied molecular orbital and molecular orbital scheme.

To understand this complex structure, one has to consider the presence of three isotopomers $(^{35}\text{Cl}-^{35}\text{Cl})^-$, $(^{37}\text{Cl}-^{37}\text{Cl})^-$, and $(^{35}\text{Cl}-^{37}\text{Cl})^-$, the abundance of which are x_{35}^2, $x_{37}^2 = (1-x_{35})^2$, and $2x_{35}(1-x_{35})$, respectively. Figure 13.13 shows the simulation of such a spectrum referring to the species in a single crystal oriented in two different directions with respect to the molecular axes. The spectrum is based on two distinct septuplets of hyperfine lines having intensity $1:2:3:4:3:2:1$ that is due to $^{35}\text{Cl}_2^-$ and $^{37}\text{Cl}_2^-$ (stick diagram in Fig. 13.13). The third isotopomer $(^{35}\text{Cl}-^{37}\text{Cl})^-$ originates a quartet of quartets where all lines have the same intensity. The real spectra of V_K centers are even more complex than those simulated in the figure because the diatomic Cl_2^- species in the crystal can assume various orientations with respect to the crystal axis that are not accounted for in the simplified model of Fig. 13.12.

The above example of a self-trapped hole is typical of alkali halides. In the case of ionic oxides, which contain divalent O^{2-} anions (alkaline earth oxides are the most investigated materials of this class), the formation of a hole does not give rise to self-trapping. In an otherwise perfect lattice that does not relax so as to trap a hole, the latter will, in fact, migrate freely until it becomes trapped at a suitable defective site.

The latter can be a preexisting cation vacancy (in that case a center called V^- is formed, Fig. 13.14), an impurity like an Li^+ ion ($[Li]^0$ center), or a low surface coordinated site. An electron hole center associated with a cation vacancy in a ionic solid, as in the first of the three examples mentioned previously, is known as a V center. Few words of warning are necessary concerning the nomenclature of such defects that appears contradictory at first glance. The negative charge does not indicate the charge of the carrier but instead indicates the whole charge of the defect because a cation vacancy locally generates a double negative charge (only partially compensated by the incoming positive hole). The same center in the more exhaustive Kröger and Vink notation is indicated (considering, e.g., the case of MgO) as V'_{Mg}.

As introduced at the beginning of this section, the positive hole in ionic oxides localizes in the orbitals of one of the oxygen anions surrounding the defect. In

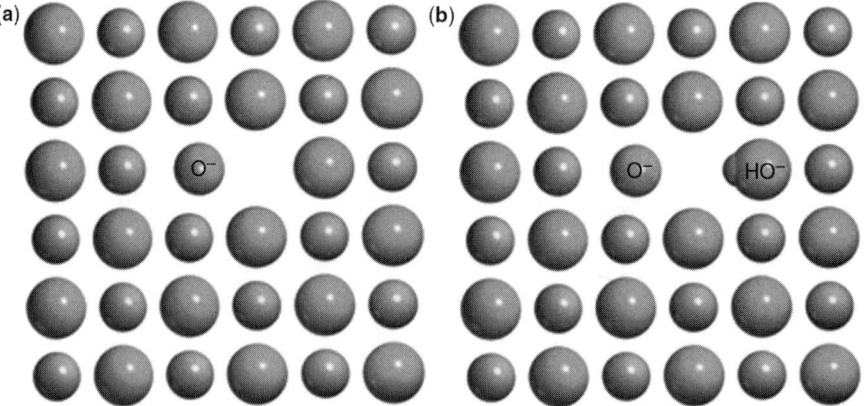

Fig. 13.14 A schematic of (a) V^- and (b) V_{OH} centers in MgO.

chemical terms, a diamagnetic O^{2-} ion transforms into a paramagnetic O^- one. This description is fully supported by EPR data. The g values measured for V in MgO ranges from 2.038 to 2.0003 according to the orientation of the crystal [9], and the high deviation from g_e that is measured in a particular crystallographic direction is attributable to the oxygen spin–orbit coupling constant value that is quite high.

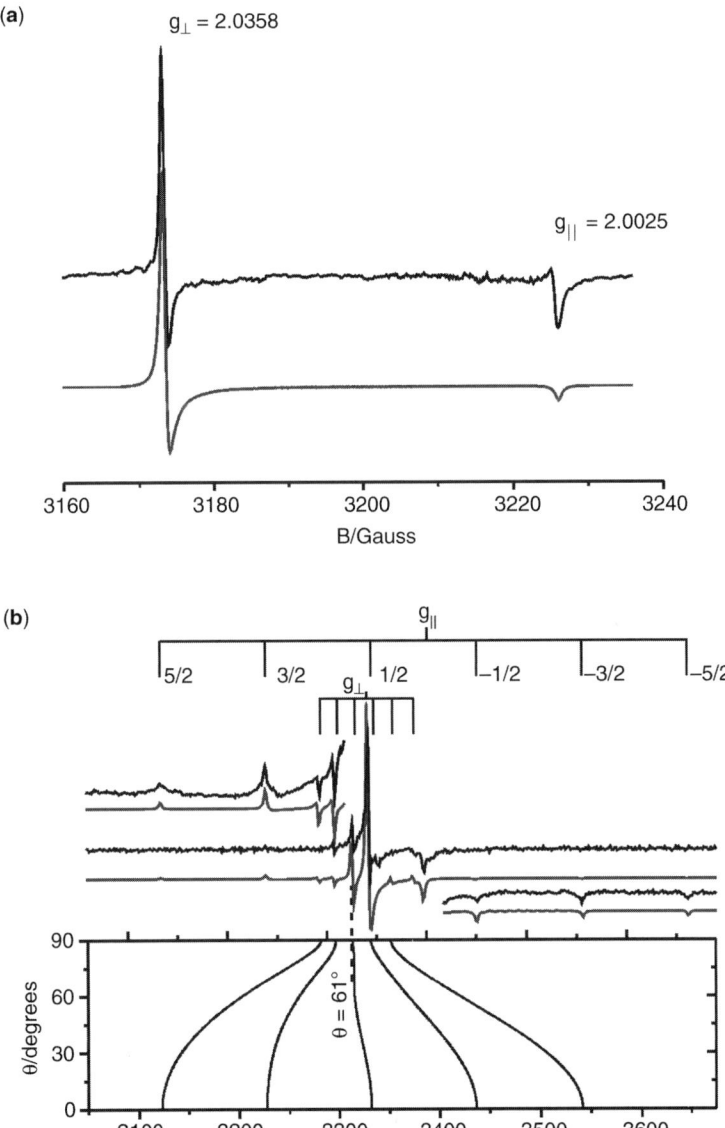

Fig. 13.15 The EPR spectra of a surface O^- center (a) on MgO and (b) on ^{17}O enriched MgO.

More explicit is the result obtained by labeling the oxygen ions of an oxide with ^{17}O. This isotope has nuclear spin $I = 5/2$, and it generates a hyperfine structure.

Figure 13.15 compares two EPR spectra of hole centers formed in polycrystalline MgO and localized at the surface of the solid where the ions are less coordinated than into the bulk and can function as a hole trap [10]. Both spectra in the figure are thus powder spectra (see Chapter 6). The former one is obtained in "normal" MgO and shows the expected profile of an axial signal with $g_\perp > g_\parallel \cong g_e$ ($g_\perp = 2.0358$, $g_\parallel = 2.0025$). The second spectrum is obtained after enriching the surface with ^{17}O, and it shows two distinct sextets in correspondence with the two principal g values with separation $A_\perp = 19.2\,G$ and $A_\parallel = 105.6\,G$. The hyperfine matrix can be

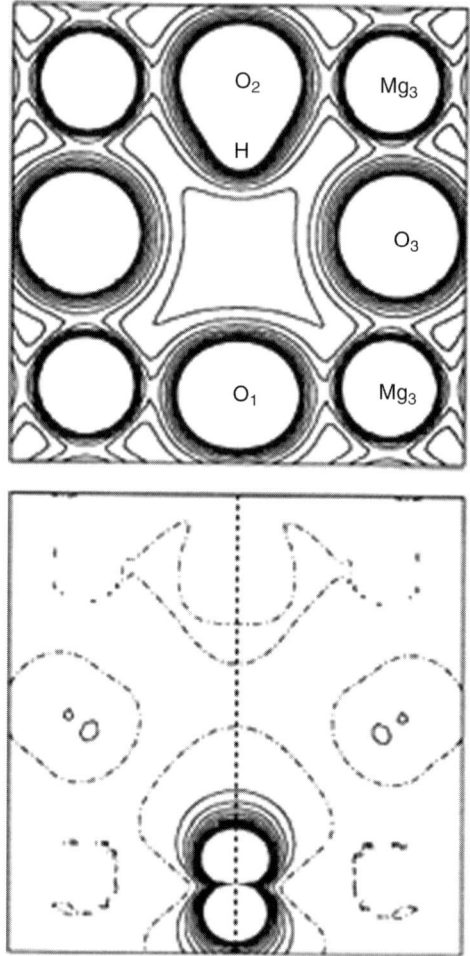

Fig. 13.16 The electron charge and electron spin density of a V_{OH} center in MgO. Adapted from Reference 11. Copyright 2000 Elsevier.

analyzed by deriving the isotropic a_{iso}(−22.5 G) and anisotropic dipolar (−82.3 G) terms. The spin density ρ on the oxygen orbitals, obtained by comparing the above values with the tabulated atomic values for the ^{17}O isotopes, gives $\rho^{2s} = 0.012$ and $\rho^{2p} = 0.820$, respectively. The total spin density on the O$^-$ ion can then be assessed to be $\rho^{tot} = 0.832$. The spin density values indicate that the unpaired electron wavefunction has a small s character, probably due to a small degree of spin polarization, and the unpaired electron spin density is mainly localized in a p orbital of oxygen as expected for an O$^-$ radical ion.

Available theoretical calculations of hole centers trapped on an oxide O^{2-} ion are in line with the described observations. Figure 13.16 reports the computed electron charge and electron spin density of a V$_{OH}$ center (which is similar to a V$^-$ center but contains an OH group opposite to the trapped hole, Fig. 13.14b) in MgO [11]. The calculations confirm the above EPR analysis results with the electron hole fully localized on one oxygen atom, namely, on the p orbital protruding through the vacancy.

13.6 CHARGE CARRIERS IN SEMICONDUCTORS

Electron spin resonance (ESR) in semiconductors is a huge and ever growing field of research, and the allotted space will not suffice to cover it in any detail. Our aim here is to provide some qualitative understanding of the information that can be obtained from ESR concerning these important systems. For a deeper discussion of the numerical results and formal derivations, the reader is referred to the detailed books and review articles that are available (see Further Readings). In the selective and arbitrary choice of examples here the intention is to focus in particular on results from doped silicon semiconductors. Silicon-based semiconductors have represented an ideal system to study resonances since the early days of EPR and provide a very nice tutorial example. They can be prepared with a very high degree of purity, and chemical impurities can be introduced in a controlled way. They possess a simple diamondlike structure (Fig. 13.17) that facilitates the interpretation of the spectra. Silicon-based semiconductors, however, not only provided an ideal system for EPR studies as pointed out by Ludwig [12] in a *Science* paper in 1962 but also ESR "...is leading to a new understanding of impurity centers in semiconductors such as silicon." Indeed, the use of ESR was (and still is) instrumental for the comprehension of the electronic properties of these solids.

Elemental silicon has a $3s^2 3p^2$ configuration with four electrons in the valence shell that can give rise to chemical bonding. As shown in Fig. 13.17, the basic arrangement of atoms in a silicon crystal is tetrahedral with each silicon atom forming electron pair bonds with four neighbors. In the language of solid-state physics, the allowed energy levels fall into discrete bands. The highest filled band is termed the valence band. It is separated from the highest unoccupied band, which is termed the conduction band, by an energy gap that is schematically shown in Fig. 13.18.

In general terms a semiconductor can be described as an insulator in which in thermal equilibrium some charge carriers are mobile. At 0K a pure perfect crystal of a

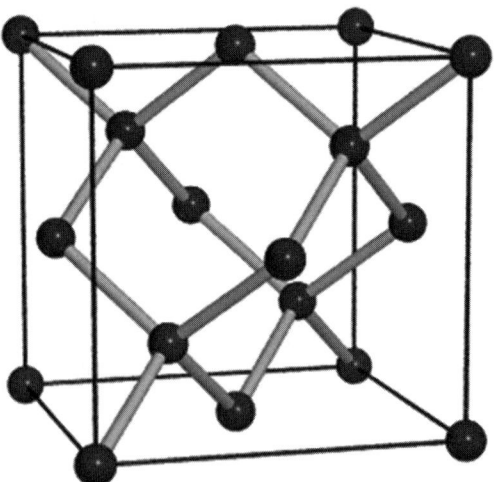

Fig. 13.17 The basic arrangement of atoms in the silicon lattice.

semiconductor has a zero conductivity because all states in the valence band are filled and all states in the conduction band are empty. In the same way at low temperature a silicon crystal containing no impurities is diamagnetic because filled bands, like closed shells, contain no unpaired electrons. As the temperature is raised electrons are thermally excited in the conduction band where they become mobile and contribute to both conductivity and spin susceptibility.

For the reason discussed above, ESR is typically associated with the presence of defects, which can be intrinsic or deliberately introduced in the matrix. It is this last case that is our focus. Let us consider an impurity atom of valence five such as phosphorus, which has been introduced in the lattice in the place of a normal atom. There

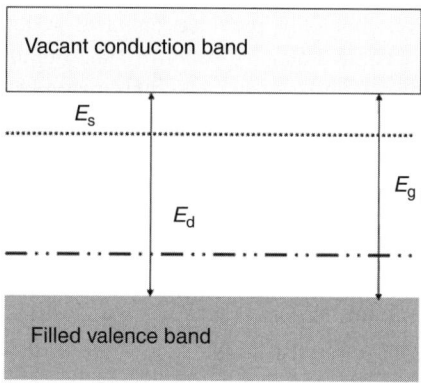

Fig. 13.18 The band scheme for a doped semiconductor. The energy levels for shallow (E_s) and deep (E_d) donor impurities are indicated.

will be one excess electron from the impurity atom left over after the four covalent bonds are formed. Impurity atoms that may be ionized to give up an extra electron are called donors. The extra electrons move in the Coulomb attractive potential of the parent donor ion (e.g., P^+) in a manner reminiscent of a 1s electron orbiting around a proton in an isolated H atom. However, because the parent phosphorus cation is embedded within the silicon dielectric host, the Coulomb attractive field between the extra electron and the parent P cation, which for a hydrogen atom is given by $V(r) = -e^2/r$, will be reduced to $V(r) = -e^2/\varepsilon r$, where ε is the dielectric constant of the medium (ε is ~13 for Si). The factor $1/\varepsilon$ accounts for the fact that the Coulomb energy interaction between the extra electron and the P^+ nucleus is screened by the high dielectric constant of the medium. As a result, the fifth phosphorus valence electron will move in a highly expanded Bohr orbit whose Bohr radius is of the order of 20 Å, see Fig. 13.19. For the same reason the extra valence electron will be only loosely bound to the parent cation, with the ionization energy on the order of 0.045 eV, which is well below the energy gap. In the language of semiconductors, impurities are usually indicated as "shallow or deep," which refers to the energy required to ionize the impurity. The P in silicon thus represents a paradigmatic example of a shallow donor. Note that as the unpaired electron travels in an orbit that encompasses many atoms (Fig 13.19) the resonance properties will be largely determined by the band structure of the matrix instead of the particular impurity to which it is loosely bound.

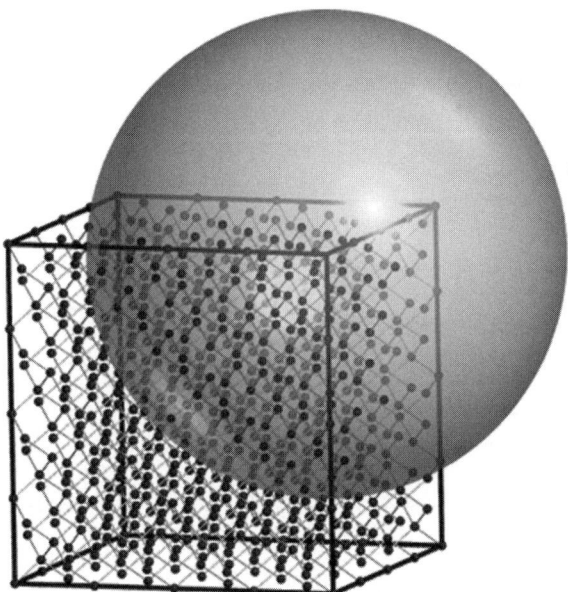

Fig. 13.19 An illustration of P doped silicon. The dopant P atom is located at the upper right corner of the cell. The extension of the unpaired electron wavefunction localized around the dopant atom is represented by the shadowed volume.

Conversely, in the case of deep impurities, the unpaired electron is tightly bound to the impurity atom and the energy required to ionize the extra electron in the conduction band is of the order of the forbidden gap. This situation is typical of transition-metal ions, and the impurity can be treated as a free ion whose spin Hamiltonian includes energy terms associated with the crystal field of the lattice (see Chapter 6). The attention in the following will be restricted to shallow impurities in Si and highlighting the relevant questions that can be answered with EPR concerning the electronic structure of these systems.

The resonance spectrum of P doped silicon recorded at different temperatures is provided in Fig. 13.20 [13]. At 3.6K the extra valence electron is localized over the parent nucleus, and this situation is clearly monitored by the EPR spectrum via the hyperfine interaction. The donor ionization energy of 0.045 eV may be compared with kT at room temperature (0.026 eV), indicating that low temperatures are needed to localize the extra electron whereas considerable thermal ionization is expected at room temperature. At temperatures on the order of a few kelvins, the EPR spectrum is dominated by two phosphorus hyperfine lines that are due to the hyperfine interaction of the unpaired electron with the ^{31}P nucleus ($I = 1/2$), indicating that the electrons are localized on the parent nuclei. The situation is schematically represented in Fig. 13.20. The magnitude of this hypefine splitting, which is of about 117.5 MHz, is directly related to $|\psi_0|^2$, the unpaired electron spin density at the donor nucleus. The experimental value may be compared to the theoretical value calculated for a gas phase P atom, which is 13,306 MHz. This drastic reduction in the Fermi contact term is the direct consequence of the large Bohr orbit described by the unpaired electron, with consequently lowered $|\psi_0|^2$ at the P nucleus. This agrees with the theory of Khon and Luttinger [14] that the donor wavefunction is spread over a large portion of the crystal rather than concentrated on the donor nucleus, as shown in Fig. 13.19, where the P$^+$ cation coincides with the upper right corner of the cell and the extension

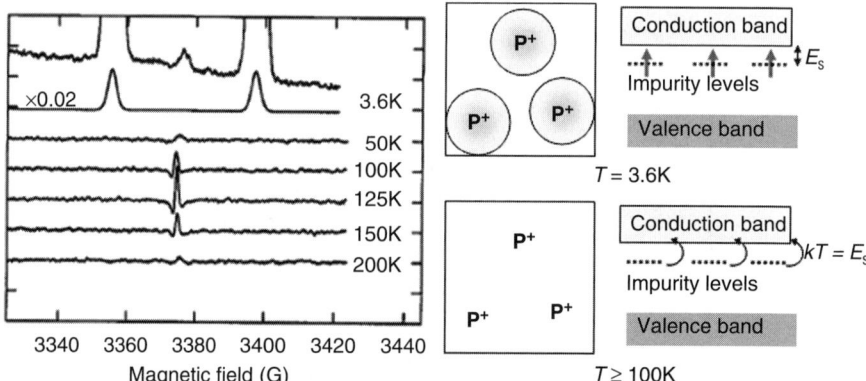

Fig. 13.20 The EPR spectra of P doped silicon with a P concentration of 5×10^{15} P cm^{-3} measured at different temperatures [13]. Spectra are recorded using second harmonic detection. The right-hand side of the figure shows a schematic of the two extreme cases of localized ($T = 3.6$K) and thermally excited ($T \geq 100$K) electrons.

of the s-like expanded orbital is represented by the shadowed volume. A direct measure of the extent of the donor wavefunction was provided by Feher's [15] classical ENDOR experiment, which allowed the measurement of the hyperfine interaction of the unpaired electron with ^{29}Si nuclei situated at different lattice points in the vicinity of the donor atom. The strength of the ^{29}Si interaction (i.e., the ENDOR frequency) is a direct measure of the probability of finding the extra valence electron at a given lattice point. Thus, a direct mapping of the donor wavefunction is made possible by the ENDOR technique establishing in beautiful detail the correctness of the Khon and Luttinger theory.

As the temperature is raised more and more the electrons are thermally excited in the conduction band and thus delocalized over the entire sample. Under these circumstances the electron is no longer localized but moves freely throughout the sample, averaging out the local hyperfine fields. At 100K about 70% of the donor electrons are excited into the conduction band and a single sharp resonance line is observed that is due to conduction band electrons. This situation is schematically described in Fig. 13.20.

The complementary experiment, which is monitoring the fate of electron spin delocalization (at constant temperature) as the concentration of donors is increased, is just as interesting. The relevant experiment [15] that shows the transition from isolated impurities to nonlocalized electrons is reported in Fig. 13.21. At low temperature (<4K) and low P doping the unpaired electron is bound to the donor level as previously described. When the donor concentration is increased, the distance between the impurity P atoms (indicated as d in Fig. 13.21) is reduced and the electron orbits overlap. At sufficiently high concentrations ($\cong 10^{18}$ P cm^3) this overlap is

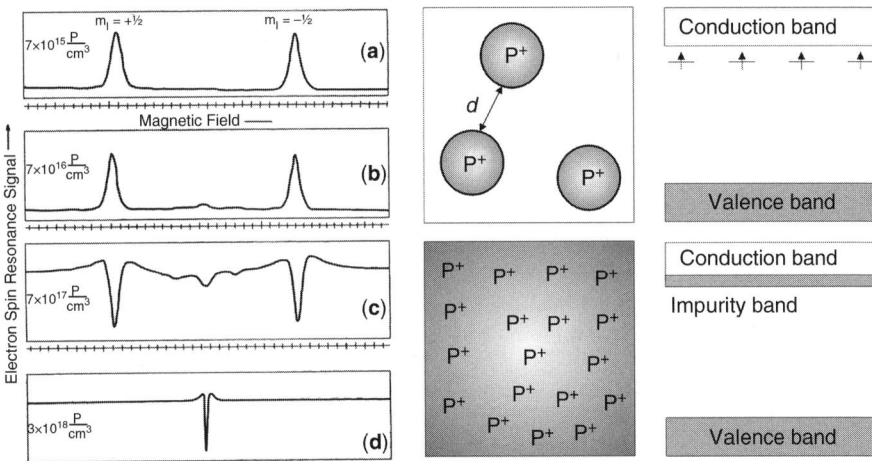

Fig. 13.21 (Left) The ESR signal from silicon doped with different amounts of phosphorus. The range covers the transition from (spectrum a) localized centers to itinerant electrons. Spectra are recorded at 1.25K [15]. The different line shapes are a result of the change in relaxation times that accompanies the transition to the metallic state. (Right) A schematic representation of the two limiting situations corresponding to low doping (spectrum a) and high doping (spectrum d) levels.

so large that the electron is no longer localized but moves freely in a so-called impurity band throughout the sample. The situation is schematically described in the right-hand side of Fig. 13.21 and has profound effects on the EPR spectrum. When this happens the situation is analogous to that of a metallic conductor and the electrons' behavior is metallic, averaging out the local hyperfine fields. It is important to stress, however, that the onset of spin delocalization occurs well before that of electron (charge) delocalization because spin delocalization does not require the formation of charge carriers. The characteristics of ESR in metallic systems are described in the next paragraph. It is interesting that at intermediate concentrations (Fig. 13.21b and 13.21c) there are extra lines in the middle of the spectrum. These are due to impurity atoms that are sufficiently close together that their wavefunctions overlap to form pairs whose electron spins are strongly coupled together by the exchange interaction. This problem was theoretically analyzed by Slichter [16], who showed that in the case of strong coupling between centers each electron effectively samples the nucleus of all of the coupled centers, not just one. Applied to two P impurities in silicon, this means that each spin will effectively experience the hyperfine field of two P nuclei and the magnitude of the interaction with each nucleus will be reduced by the same factor of 2. In general, the pattern for pairs consists of $4I + 1$ hyperfine lines spaced at half the interval of the isolated impurities but extending over the same total range. Moreover, the intensity will not be uniform but will depend on the number of ways of getting the same sum $m_1 + m_2$ (where m_1 and m_2 are the nuclear quantum numbers for the two centers), which in the case of $m = \pm 1/2$ as for P, leads to a pattern with intensity $1:2:1$. Only the central line will be seen in the spectrum because the external transitions will fall at the same field as that of isolated impurities. This embryonic spin delocalization was regarded by Slichter "...as simply the first step in the process of turning an insulator into a metal" [16]. In fact, as the concentration of P impurity atoms grows in the sample, small P clusters will gradually develop, leading to a situation in which the exchange coupling is so strong that the system behaves like a metal with completely nonlocalized electrons, even at <4K, and giving rise to the singlet EPR resonance line shown in Fig. 13.21d.

13.7 CESR IN METALS

13.7.1 The CESR Phenomenon

The overwhelming majority of EPR measurements are made on systems where the unpaired electron spins are localized on individual atoms or relatively small groups of atoms. In this chapter the attention is on a more specialized area of EPR and the case of CESR measurements on metallic materials is considered. In these systems, as observed in the previous section for highly P doped silicon, the unpaired electrons are delocalized throughout the entire sample and this deeply affects the nature of the resonance signal. The electrons in this so-called conduction band exhibit paramagnetism but their ESR absorption is quite different from that of unpaired electrons in localized systems and determined by a number of factors, which are discussed in an illuminating review by Edmonds et al. [17] on which this account has drawn extensively.

When the electrons are localized on individual atoms, one of the most important sources of information is represented by the hyperfine interaction with nuclei bearing a nuclear magnetic moment. Under favorable circumstances the hyperfine interaction not only enables a paramagnetic species to be identified but also its local environment may often be described in detail.

A fundamentally different situation occurs in the case of a metal when the valence electrons can move freely as a sort of gas throughout the entire material. Under these circumstances the electron spins move so rapidly from site to site that they are able to experience the complete range of orientational configurations of the nuclear moments and each electron spin produces an identical signal. The effect of this "motional narrowing" is then to "wash out" any hyperfine interaction, leaving a single, unresolved resonance signal. This situation may be illustrated by comparing the spectra of isolated K atoms, small K clusters, and metallic potassium as revealed in Fig. 13.22.

Figure 13.18a shows the spectrum of atomic potassium in a supersonic beam of atoms [18]. The spectrum is characterized by two sets of hyperfine lines arising from the interaction of the unpaired 4s valence electron to ^{39}K and ^{41}K nuclei that both have $I = 3/2$ but different magnetic moments and natural abundance; hence, the two quartets show different coupling constants and intensities. In the case of a

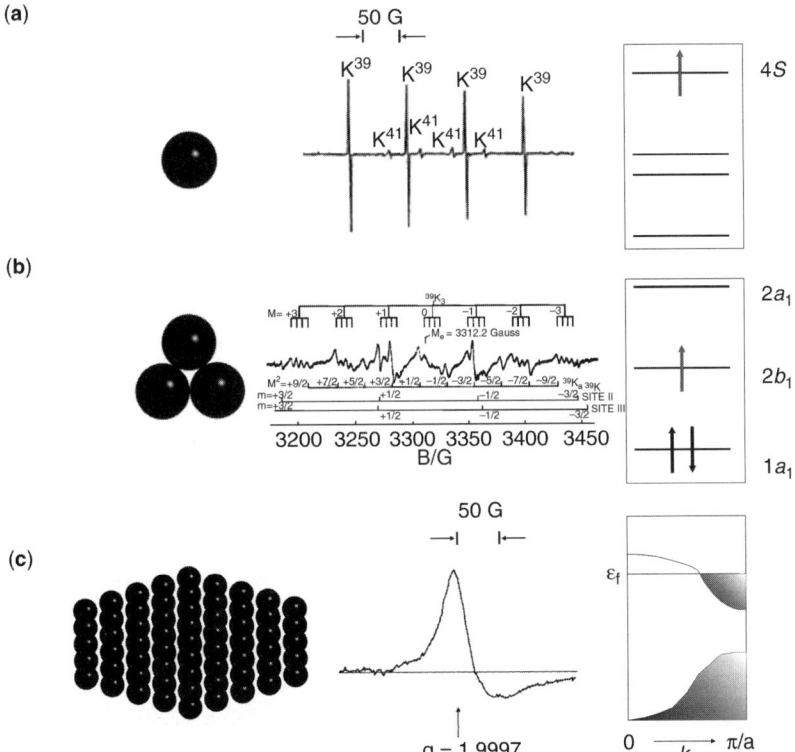

Fig. 13.22 The EPR spectra and energy levels for (a) isolated K atoms [18], (b) trimeric K clusters [19], and (c) bulk K metal [17].

trimeric neutral cluster, such as the one reported in Fig. 13.22b and obtained by Lindsay [19], the EPR spectrum is composed of a complex hyperfine pattern that is attributable to two equivalent and one inequivalent K, reflecting the molecular orbitals associated with the unpaired electron. A completely different situation occurs in a sample of potassium metal. In this case the 4s valence electrons are no longer associated with the parent nuclear cores and move freely from one ion to another in the conduction band. In solid-state physics the orbitals of an infinite chain of K atoms are written by making use of translational symmetry, leading to so-called Bloch functions. Thus, what chemists usually draw as a block (containing roughly an Avogadro number of levels) are depicted in the solid-state trade as an $E(k)$ versus k diagram as shown in Fig. 13.22c, where k represents the wave vector [20]. The EPR spectrum in this case appears as a single resonance signal, which is representative of the overall form of CESR in bulk metallic conductors (Fig. 13.22c). The resonance shows no evidence of any hyperfine interaction and is characterized by a distinct asymmetric shape. Moreover, the signal is rather weak considering that in a typical sample of bulk metal the number of potassium atoms is on the order of 10^{22}! The factors that contribute to determine this signal will now be examined. In particular, the elements that determine the intensity, position, width, and peculiar line shape of the CESR resonance signal are schematically considered in Fig.13.23.

13.7.1.1 Signal Intensity. In order to understand the CESR phenomenon it is important to realize that the conduction electrons are delocalized over the entire sample and can be treated as a Fermi gas. Contrary to the case of isolated, independent spins (such as in molecules or discrete levels localized within the band gap of insulators or semiconductors), the alignment of the spin magnetic moments along an applied magnetic field is not only contrasted by the thermal energy but also governed by the Pauli principle. In other words, an itinerant gas of electrons will obey Fermi–Dirac statistics. Consequently, the only electrons that can be observed in a CESR experiment are those located in a narrow band of levels close to the Fermi energy whereas the other electrons can be treated as silent. This fact explains why CESR signals are always relatively weak, despite the large number of electrons in a metal sample.[2]

13.7.1.2 Linewidth and Position. The CESR resonance linewidths (indicated as ΔH in Fig. 13.23) is markedly affected by the spin relaxation rate. The electron spin–lattice and electron spin–spin relaxation times (T_1 and T_2) in metals are extremely short and of the same order of magnitude. Spin–lattice interactions are strong because of the coupling of the oscillatory and translational motions of

---

[2]Another major consequence of this state of affairs is the differences in magnetic susceptibilities between localized and itinerant electrons. An assembly of noninteracting spins, governed by Maxwell–Boltzmann statistics, will display Curie-like behavior. In the case of a metal, only those electrons located within an interval of the order of kT at the top of the Fermi level will be free of changing orientation with the applied field. Thus, only a fraction T/T_F of the total number of electrons, where T_F is the temperature of the Fermi level, contribute to the susceptibility. Because T_F is a constant, the susceptibility will be temperature independent. This behavior is known as Pauli-like paramagnetism.

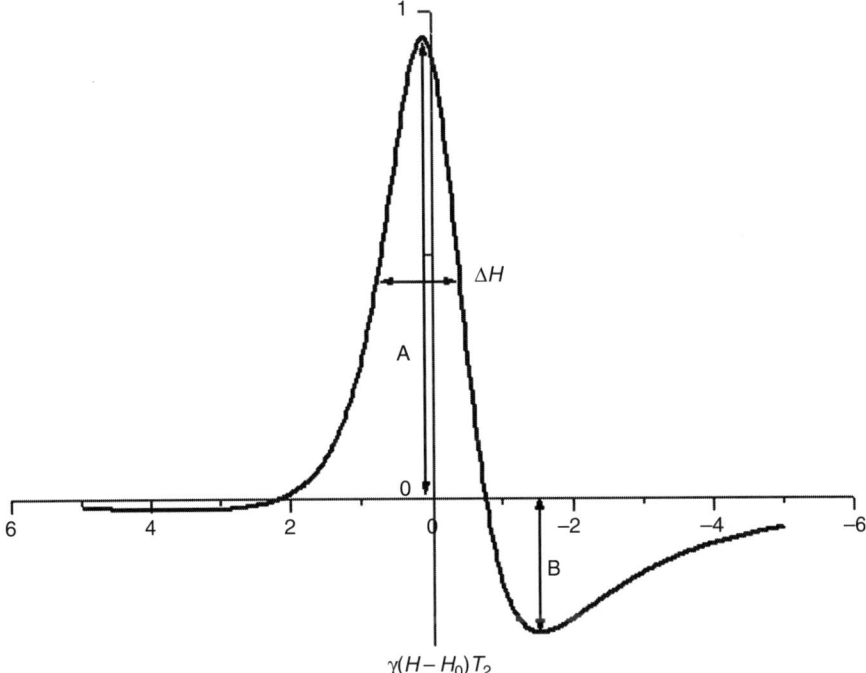

Fig. 13.23 A computer simulation of a CESR signal in a metal; A/B is the asymmetry parameter, and ΔH is the linewidth of the signal.

the free electrons with the lattice vibrations. As an example, in sodium metal $T_1 = T_2 \cong 10^{-6}$ s at 20K. Scattering mechanisms, which are the same as that giving rise to the electrical resistivity of a metal, determine the linewidth of the CESR resonance signal by reducing the lifetime of a given spin state. The mechanism of electron spin relaxation in metals was first considered by Overhauser [21] and later by Elliott [22], who showed that the relaxation occurs by the combined effects of electron–phonon and spin–orbit couplings. The spin–orbit coupling causes admixtures of the spin-up and spin-down Bloch wavefunctions of the conduction electrons. Thus, the electron–phonon scattering process, which causes transitions between k states giving rise to electrical resistivity, may simultaneously cause an electron spin–flip transition with a probability of approximately $(\Delta g)^2$, where Δg is the deviation of the electronic g factor from the free spin value (2.0023). Elliott's relation between the spin–lattice relaxation rate and the resistivity rate τ_R (which is proportional to the electrical resistivity) is

$$T_2^{-1} = T_1^{-1} \simeq \alpha(\Delta g)^2 \tau_R^{-1} \qquad (13.2)$$

where α is a numerical factor.

Spin–orbit coupling also determines the magnitude of the g shifts observed experimentally. As in the case of localized unpaired electrons, this g shift can be calculated by perturbation theory leading to values on the order of $\Delta g = \lambda/\Delta E$, where λ is the spin–orbit coupling constant and ΔE is the energy difference between the relevant Bloch states.

13.7.1.3 Line Shape.
The main factor that determines the shape of a CESR signal is that microwaves penetrate only a short distance into a metallic sample. Thus, contrary to classic EPR, the electron spins of the sample do not experience the same microwave field. The extent that microwaves penetrate into a metal is known as the skin depth, which is given by

$$\delta = \left(\frac{c^2 \varepsilon_0 \rho}{\pi v}\right)^{1/2} \tag{13.3}$$

where c is the velocity of light, ε_0 is the vacuum permittivity, ρ is the electrical resistivity of the metal, and v is the frequency of the microwave radiation. The skin depth can be very small; for example, at 9.5 GHz the experimental skin depth for bulk potassium derived from electrical resistivity measurements is on the order of 10^3 Å at 20K. Only a fraction of the metal sample within the skin depth of the metal is sampled by CESR experiments; therefore, for bulky samples a rather small number of conduction electrons is actually observed. A further complication arises from the motion of the conduction electrons in and out of the skin depth region and a further parameter that describes this motion has to be introduced in order to explain CESR line shapes. Therefore, T_D is termed the time it takes an electron to diffuse through the skin depth δ.

The theory of the CESR line shape was developed by Dyson [23] and is briefly recapitulated here following the treatment of Feher and Kip [24]. The main parameters in Dyson's theory are the time it takes an electron to diffuse through the skin depth (T_D), the electron–spin relaxation time (T_2), and the time it takes the electron to traverse the sample (T_T). Two limiting situations may be considered:

1. A metal particle of dimensions small in comparison to the skin depth. In this case $T_T \ll T_D$ and the derivative of the absorbed power is

$$\frac{\partial P}{\partial \omega} = -\frac{\omega H_1^2}{4} V \omega_0 \chi T_2 \frac{2(\omega - \omega_0)T_2^2}{[1 + (\omega - \omega_0)T_2^2]^2} \tag{13.4}$$

   where V is the volume of the sample, H_1 is the magnetic field associated to microwave radiation, χ is the paramagnetic susceptibility, and ω_0 is the resonant frequency satisfying the resonance condition. Under these circumstances the microwave field is uniform throughout the sample; thus, the result is entirely independent of the diffusion and gives a symmetrical Lorentzian line with the natural half-width $1/T_2$.

2. A metal particle whose dimensions are of the same order of or greater than the skin depth. Under these circumstances the microwave field experienced by a conduction electron changes as it moves through and in and out of the skin

depth region. The resonance signal will be characterized by an asymmetric shape whose asymmetry parameter A/B (see Fig. 13.23) is dictated by the ratio T_d/T_2. For a fixed g value and T_2, both the line shape and linewidth are dependent on the sample thickness.

The Dyson formalism can thus be used to calculate the CESR resonance line shape under different conditions and, based on the A/B asymmetry ratio, the particle dimension can be obtained.

13.7.2 The Effect of Metal Particle Size

The CESR phenomenon in bulk metals has been considered here within the framework of band theory, which depicts a metal in terms of an infinite macroscopic system in which the itinerant conduction electrons are shared among all of the constituents. A particularly interesting situation arises when the metal is fragmented into smaller and smaller pieces, leading to the fundamental and important question, "How many atoms are needed to maketh a metal?" as posed by Edwards and Sienko [25]. As the size of the sample is reduced new effects arise from the splitting of the energetic levels, which result in the changeover of the density of state function from quasicontinuous to discrete. This has profound consequences on the electronic and magnetic properties of the system and a metal to nonmetal transition must eventually occur, the lower limit for this critical size being a small cluster and eventually a single metal atom, as described in Fig. 13.18. There is great interest in characterizing the intermediate states between these two extremes because the properties of small metallic particles are becoming in the "nanochemistry" era increasingly important from both fundamental and applied research perspectives.

The differences in the electronic structure peculiar to finely divided metals are also of paramount practical importance in observing the CESR signal from certain metals. As mentioned in the previous paragraph, the typical asymmetric (Dysonian) line shape observed in bulky metal samples originates from the ratio between the metal particle dimension and the skin depth. In particular, symmetrical Lorentzian line shapes are observed when the particle dimension is of the same order or smaller than the skin depth. An example of a typical signal obtained from small potassium particles (colloids) is reported in Fig. 13.22 [26].

Figure 13.24 contains the CESR spectra for K metal particles of different size in a frozen solvent. The noticeable feature is the change in the line shape passing from small clusters (diameter of ~ 1000 Å) to larger ones, which is due to the decrease of the skin depth as the size of the metal particle increases.

The high surface to bulk ratio and the fact that the microwaves homogeneously permeate the entire metal particle not only produces a symmetric line shape but also greatly enhances the intensity of the resonance signal. As the size of the metal particle is further reduced, a limit is reached where the (quasi) continuous energy level spectrum characteristic of a metal is no longer appropriate and a number of interesting effects are expected to be observed in the CESR spectrum. One of these effects

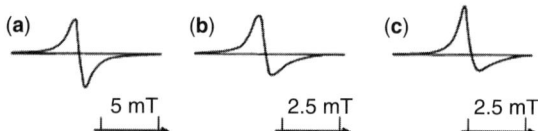

Fig. 13.24 The typical CESR spectra from samples of colloidal potassium recorded at 4K [26]. Three samples with different metal particle sizes are shown. Average particle sizes: (a) <1000 Å, (b) ≅3500 Å, and (c) ≅10,000 Å.

is concerned with the spin relaxation time that in this so-called quantum size regime is effectively hindered, leading to considerable narrowing of the spectral linewidth. The linewidth of the resonance is determined by the electron spin relaxation time (T_2) that is considerably affected by the increased surface area brought about by the reduction of the sample size. Kawabata [27] predicted an increase in the linewidth from the bulk value at a particular temperature as the particle diameter is reduced, and sample dimensions can be evaluated based on the width of the resonant line.

The above considerations pertain to clusters with dimensions ranging between 100 and 10,000 atoms and radii between approximately 2 and 9 nm. As the size of the metal particle is further reduced, a small metal cluster regime (<100 atoms) is encountered where the energy levels become discrete and new effects start to appear. It is these effects and their influence on the CESR signal that will be considered now.

One important aspect to be considered when dealing with small metal particles is the effect of the discretization of the electron energy levels on the metallic behavior, an effect generally known as the quantum size effect. Kubo [28] suggested a simple criterion for a finite metallic particle (large cluster) to exhibit metallic conductivity, which is based on the existence of a partially occupied band with level spacing sufficiently small near the Fermi level so that a small external energy (thermal activation) can create electron–hole pairs, allowing electrical conductivity. More synthetically, the Kubo criterion can be expressed by the following equation:

$$\Delta = E_f/N < kT \qquad (13.5)$$

where Δ is the band gap or the electronic energy level spacing at the Fermi level (i.e., the highest occupied orbital in a cluster). Based on Equation 13.5, the suppression of metallic behavior will appear when $\Delta > kT$. Because smaller particles will have larger Δ, a progressively higher temperature will be needed for metallic properties to show up.[3]

---

[3] According to the Kubo criterion, an Na particle of 16,000 atoms (d = 10 nm) will display metallic behavior at 5K, whereas a particle made up of about 125 atoms (d = 2 nm) will exhibit metallic conductivity at room temperature.

Another very important effect predicted by Kubo [28] is the so-called even–odd effect. Small metal clusters of monovalent metals such as potassium will have the highest energy levels occupied by either one or two electrons, depending on the number of constituent atoms. Kubo first showed that at high temperature both odd and even particles exhibit a Pauli-like magnetic susceptibility. At low temperatures the odd particles display a Curie-like behavior and the even particles are diamagnetic. Thus, CESR signals will only be observed from odd particles in the sample. The general magnetic behavior of an ensemble of particles will then be described by some appropriate probability distribution function, and the temperature behavior of the magnetic susceptibility will display a transition from a Pauli paramagnetism to a Curie-like behavior at a temperature corresponding to $kT/\Delta = 1$.

In bulk metals, especially in the heavier metals, the conduction–electron spin relaxation times are very short, giving rise to very broad CESR absorption lines, which are in some cases impossible to detect. Another consequence of the level discreteness encountered in the quantum size regime is that the usual relaxation mechanisms should be quenched, leading to narrowed CESR lines. Kawabata [27] has given various criteria for narrowing due to level discreteness:

$$g\beta B \ll \Delta \tag{13.6}$$
$$h/T_2 \ll \Delta \tag{13.7}$$

where Δ is the mean energy level spacing and the other terms have the usual meaning. At X-band frequency, particles in the size range below 100 Å are required for Kawabata's criteria to hold. Under these circumstances the CESR signal comes from particles with an odd number of electrons, its intensity will follow the Curie law, and sharp lines resonating at g values slightly higher than the bulk metal value are expected. A few words of caution should be injected at this stage regarding the analysis of CESR lines in the quantum size regime. The interpretation of small particle CESR data is complicated by a number of corollary effects such as the particle size distribution, surface interactions, nuclear hyperfine broadening, and the presence of impurity resonances that may obscure or be mistaken for CESR signals. To distinguish among these effects it is desirable to obtain data for a wide range of particle sizes from bulk to a few tens of angstroms. Such studies are in part available in the literature and the reader is directly referred to these studies for further details on the analysis of the experimental data.

REFERENCES

1. De Boer, J. H. *Recl. Trav. Chim.* **1937**, 301.
2. Wood, R. F.; Joy, H. W. *Phys. Rev.* **1946**, *136*, 451A.
3. Kaplan, R.; Bray, P. J. *Phys. Rev.* **1963**, *129*, 1919.
4. Seidel, H.; Wolf, H. C. EPR and ENDOR Spectroscopy of Color Centers in Alkali Halide Crystals. In *Physics of Color Centers*; Beall Fowler, W., Ed.; Academic: New York, 1968.
5. Stoll, S.; Schweiger, A. *J. Magn. Reson.* **2006**, *178*, 42.

6. Seidel, H. *Z. Phyzik* **1961**, *165*, 218.
7. Mallia, G.; Orlando, R.; Roetti, C.; Ugliengo, P.; Dovesi, R. *Phys. Rev.* **1998**, *B63*, 235102.
8. Castner, T. G.; Kanzig, W. *J. Phys. Chem. Solids* **1957**, *3*, 178.
9. Wertz, J. E.; Auzins, P.; Griffiths, J. H. E.; Orton, J. W. *Discuss. Faraday Soc.* **1959**, *28*, 136.
10. Chiesa, M.; Giamello, E.; DiValentin, C.; Pacchioni, G. *Chem. Phys. Lett.* **2005**, *403*, 124.
11. Lichanot, A.; Orlando, R.; Mallia, G.; Mérawa, M.; Dovesi, R. *Chem. Phys. Lett.* **2000**, *318*, 240.
12. Ludwig, G. W. *Science* **1962**, *135*, 899.
13. Young, C. F.; Poindexter, E. H.; Gerardi, G. J.; Warren, W. L.; Keeble, D. J. *Phys. Rev.* **1997**, *B55*, 16245.
14. (a) Khon, W.; Luttinger, J. M. *Phys. Rev.* **1955**, *97*, 1721; (b) Khon, W.; Luttinger, J. M. *Phys. Rev.* 1955, *98*, 915.
15. Feher, G. *Phys. Rev.* **1959**, *114*, 1219.
16. Slichter, C. P. *Phys. Rev.* **1955**, *99*, 479.
17. Edmonds, R. N.; Harrison, M. R.; Edwards, P. P. *Annu. Rep. Prog. Chem. C* **1985**, 265.
18. Knight, W. D. *Surf. Sci.* **1981**, *106*, 172.
19. Thompson, G. A.; Lindsay, D. M. *J. Chem. Phys.* **1981**, *74*, 959.
20. Kittel, C. *Introduction to Solid State Physics;* Wiley: New York, 2005.
21. Overhauser, A. W. *Phys. Rev.* **1953**, *89*, 689.
22. Elliott, R. J. *Phys. Rev.* **1954**, *96*, 266.
23. Dyson, F. J. *Phys. Rev.* **1955**, *98*, 349.
24. Feher, G.; Kip, A. F. *Phys. Rev.* **1955**, *98*, 337.
25. Edwards, P. P.; Sienko, M. *Int. Rev. Phys. Chem.* **1983**, *3*, 83.
26. Guy, S. C.; Edmonds, R. N.; Edwards, P. P. *J. Chem. Soc. Faraday Trans. 2* **1985**, *81*, 937.
27. Kawabata, A. *J. Phys. Soc. Jpn.* **1970**, *29*, 902.
28. Kubo, R. *J. Phys. Soc. Jpn.* **1962**, *17*, 975.

APPENDIX

Physical Tables

Fundamental Physical Constants (CODATA 2002)[a]

Quantity	Symbol	SI Value
Speed of Light (vacuum)	c_o	299 792 458 m s^{-1} (defined)
Permeability of vacuum	μ_o	$4\pi \times 10^{-7}$ H m^{-1} or N A^{-2} (defined)
Permittivity of vacuum	$\varepsilon_o = 1/(\mu_o c_o^2)$	$8.854\ 187\ 817 \times 10^{-12}$ F m^{-1} (defined)
Planck Constant	h	$6.626\ 0693\ (11) \times 10^{-34}$ J s
	$\hbar = h/2\pi$ (au)	$1.054\ 571\ 68\ (18) \times 10^{-34}$ J s
Elementary Charge (au)	e	$1.602\ 176\ 53\ (14) \times 10^{-19}$ C
Electron Rest Mass (au)	m_e	$9.109\ 3826\ (16) \times 10^{-31}$ kg
Proton Rest Mass	m_p	$1.672\ 621\ 71\ (29) \times 10^{-27}$ kg
Proton/Electron Mass Ratio	m_p / m_e	1836.152 672 61 (85)
Neutron Rest Mass	m_n	$1.674\ 927\ 28\ (29) \times 10^{-27}$ kg
Deuteron Rest Mass	m_d	$3.343\ 583\ 35\ (57) \times 10^{-27}$ kg
Atomic Mass Unit (^{12}C/12)	$m_u = 1\ u = 1$ Da	$1.660\ 538\ 86\ (28) \times 10^{-27}$ kg
Avogadro's Number	N_A	$6.022\ 1415\ (10) \times 10^{23}$ mol^{-1}
Boltzmann Constant	k	$1.380\ 6505\ (24) \times 10^{-23}$ J K^{-1}
Faraday Constant	F	$9.648\ 5336\ (16) \times 10^{4}$ C mol^{-1}
Gas Constant	R	8.314 472 (15) J mol^{-1} K^{-1}
Molar Volume of ideal gas[b]	$V_m = RT/p$	$22.413\ 996\ (39) \times 10^{-3}$ m^3 mol^{-1}
Standard Atmosphere	atm	101.325 kPa (defined)
Fine Structure Constant	$\alpha = \mu_o e^2 c_o / 2h$	$7.297\ 352\ 568\ (24) \times 10^{-3}$
Inverse Fine-Structure Constant	$1/\alpha$	137.035 999 11 (46)
Bohr Radius (au)	$a_o = 4\pi\varepsilon_o \hbar^2 / m_e e^2$	$0.529\ 177\ 2108\ (18) \times 10^{-10}$ m
Hartree Energy (au)	$E_h = \hbar^2 / m_e a_o^2$	$4.359\ 743\ 17\ (75) \times 10^{-18}$ J
Rydberg Constant	$R_\infty = E_h / 2hc_o$	$1.097\ 373\ 156\ 8525\ (73) \times 10^{7}$ m^{-1}
Compton Wavelength (Electron)	$\lambda_C = h / m_e c_o$	$2.426\ 310\ 238\ (16) \times 10^{-12}$ m
Bohr Magneton (β, β_e)	$\mu_B = e\hbar / 2m_e$	$9.274\ 009\ 49\ (80) \times 10^{-24}$ J T^{-1}
Electron Magnetic Moment	μ_e	$-9.284\ 764\ 12\ (80) \times 10^{-24}$ J T^{-1}
Electron Magnetogyric Ratio	$\gamma_e = 2\mu_e / \hbar$	$1.760\ 859\ 74\ (15) \times 10^{11}$ s^{-1} T^{-1}
	$\gamma_e / 2\pi$	28.024 9532 (24) GHz T^{-1}
Free Electron Landé g factor	$g_e = 2\mu_e / \mu_B$	$-2.002\ 319\ 304\ 3718\ (75)$
Nuclear Magneton (β_N)	$\mu_N = (m_e / m_p)\mu_B$	$5.050\ 783\ 43\ (43) \times 10^{-27}$ J T^{-1}
Proton Magnetic Moment (free)	μ_p	$1.410\ 606\ 71\ (12) \times 10^{-26}$ J T^{-1}
(shielded, H$_2$O sphere, 25°C)	μ_p'	$1.410\ 570\ 47\ (12) \times 10^{-26}$ J T^{-1}
Proton Magnetogyric Ratio (free)	γ_p	$2.675\ 222\ 05\ (23) \times 10^{8}$ s^{-1} T^{-1}
(shielded, H$_2$O sphere, 25°C)	γ_p'	$2.675\ 153\ 33\ (23) \times 10^{8}$ s^{-1} T^{-1}
Proton MR freq. in H$_2$O	$\gamma_p' / 2\pi$	42.576 3875 (37) MHz T^{-1}
Electron/Proton Magn. Mom. Ratio	μ_e / μ_p	$-658.210\ 6862\ (66)$
Deuteron Magnetic Moment	μ_d	$0.433\ 073\ 482\ (38) \times 10^{-26}$ J T^{-1}
Gravitation Constant (Newtonian)	G	$6.6742\ (10) \times 10^{-11}$ m^3 kg^{-1} s^{-2}
Standard Acceleration (Earth gravity)	g_n	9.806 65 m s^{-2} (defined)
π = 3.141 592 653 59	e = 2.718 281 828 46	ln 10 = 2.302 585 092 99

[a] au = atomic units; uncertainty of last digits shown in ( ); source: http://physics.nist.gov/constants
[b] at STP of 273.15 K and 101.325 kPa = 1 atm.

APPENDIX 521

Conversion Factors for Important Physical Units

Table entries represent multipliers for converting unit in row heading to unit in column heading.

Energy Equivalents

from \ to	joule	hertz	m^{-1}	kelvin	eV	E_h (Hartree)
joule	1	1.5091904 E+33	5.0341172 E+24	7.2429626 E+22	6.2415095 E+18	2.2937126 E+17
hertz	6.6260693 E−34	1	3.3356410 E−09	4.7992372 E−11	4.1356674 E−15	1.5198298 E−16
m^{-1}	1.9864456 E−25	2.9979246 E+08	1	1.4387751 E−02	1.2398419 E−06	4.5563353 E−08
kelvin	1.3806505 E−23	2.0836644 E+10	6.9503564 E+01	1	8.6173432 E−05	3.1668154 E−06
eV	1.6021765 E−19	2.4179894 E+14	8.0655444 E+05	1.1604505 E+04	1	3.6749325 E−02
E_h	4.3597442 E−18	6.5796839 E+15	2.1947463 E+07	3.1577464 E+05	2.7211385 E+01	1

based on CODATA 2002 with $E = mc_0^2 = hc_0/\lambda = h\nu = kT$; 1 eV = e*1 V; $E_h = 2R_\infty hc_0$; 1 m^{-1} = 100 cm^{-1}

Force Units: SI unit = Newton (N), cgs unit = dyne, Weight = mass × g_n

From \ to	N	kp (kilopond)	dyn	lbf (pound-force)
N	1	0.1019716	1.000 E+05	0.2248089
kp	9.806650	1	9.806650 E+05	2.204622
dyn	1.000 E−05	1.019716 E−06	1	2.248089 E−06
lbf	4.448222	0.4535924	4.448222 E+05	1

Energy, Heat and Work Units: SI unit = joule (J), cgs unit: 1 erg = 10^{-7} J

From \ to	J (N m)	kp m	kWh	kcal(th)	Btu(th)	ft lbf
J	1	1.019716 E−01	2.777778 E−07	2.390057 E−04	9.484517 E−04	7.375621 E−01
kp m	9.806650	1	2.724069 E−06	2.343846 E−03	9.301133 E−03	7.233014 E+00
kWh	3.600 E+06	3.670978 E+05	1	8.604207 E+02	3.414426 E+03	2.655224 E+06
kcal(th)	4184.000	4.266493 E+02	1.162222 E−03	1	3.968322 E+00	3.085960 E+03
Btu(th)	1054.350	1.075138 E+02	2.928750 E−04	2.519957 E−01	1	7.776486 E+02
ft lbf	1.355818	1.382550 E−01	3.766161 E−07	3.240483 E−04	1.285928 E−03	1

Power Units: SI unit = watt (W); cgs unit: 1 erg/s = 10^{-7} W

from \ to	W (J/s)	kp m / s	cal / s	kcal / h	ft lbf / s	horsepower
W	1	1.019716 E−01	2.390057 E−01	8.604207 E−01	7.375621 E−01	1.341022 E−03
kp m / s	9.806650	1	2.343846 E+00	8.437844 E+00	7.233014 E+00	1.315093 E−02
cal / s	4.184000	4.266493 E−01	1	3.600000 E+00	3.085960 E+00	5.610836 E−03
kcal / h	1.162222	1.185137 E−01	2.777778 E−01	1	8.572111 E−01	1.558566 E−03
ft lbf / s	1.355818	1.382550 E−01	3.240483 E−01	1.166574 E+00	1	1.818182 E−03
horsepower	745.6999	7.604023 E+01	1.782266 E+02	6.416156 E+02	5.500000 E+02	1

Pressure Units: SI unit = pascal (Pa); cgs unit: 1 dyne / cm^2 = 0.1 Pa; 1 psi = 6.894757 kPa

from \ to	Pa (N / m^2)	kp / m^2	atm (std)	bar	Torr (mmHg)	at (techn)
Pa	1	0.1019716	9.86923 E−06	1.000 E−05	7.500617 E−03	1.019716 E−05
kp / m^2	9.806650	1	9.678411 E−05	9.806650 E−05	7.355692 E−02	1.000000 E−04
atm (std)	1.013250 E+05	1.033227 E+04	1	1.013250 E+00	7.600000 E+02	1.033227 E+00
bar	1.000 E+05	1.019716 E+04	9.869233 E−01	1	7.500617 E+02	1.019716 E+00
Torr	133.3224	1.359510 E+01	1.315789 E−03	1.333224 E−03	1	1.359510 E−03
at	9.806650 E+04	1.000000 E+04	9.678411 E−01	9.806650 E−01	7.355692 E+02	1

Time Units : SI unit = second (s)

from \ to	s	min	h	d	week	year (Gregorian)
s	1	1.666667 E−02	2.777778 E−04	1.157407 E−05	1.653439 E−06	3.168874 E−08
min	60	1	1.666667 E−02	6.944444 E−04	9.920635 E−05	1.901324 E−06
h	3600	60	1	4.166667 E−02	5.952381 E−03	1.140795 E−04
d	86400	1440	24	1	1.428571 E−01	2.737907 E−03
week	604800	10080	168	7	1	1.916535 E−02
year (Greg.)	31556952	525949.2	8765.820	365.2425	52.17750	1

1 sidereal year (astron.) = 31558150 s

Temperature Conversion: SI unit = kelvin (K)

from \ to	Kelvin (K)	Celsius (°C)	Fahrenheit (°F)	Rankine (°R)
K	1	$T_C = T_K - 273.15$	$T_F = 1.8T_K - 459.67$	$T_R = 1.8T_K$
°C	$T_K = T_C + 273.15$	1	$T_F = 1.8T_C + 32$	$T_R = 1.8(T_C + 273.15)$
°F	$T_K = (T_F + 459.67)/1.8$	$T_C = (T_F − 32)/1.8$	1	$T_R = T_F + 459.67$
°R	$T_K = T_R/1.8$	$T_C = (T_R/1.8) − 273.15$	$T_F = T_R − 459.67$	1

EPR Tables

Useful Relationships for EPR

Magn. Moment of the electron $\mu_e = g_e \mu_B S = g_e \mu_B / 2$ (g_e defined as negative) alternatively: $\mu_e = -g_e \mu_B / 2$ (with g_e positive) $\mu_B = \beta_e$ = Bohr magneton	**Magn. Moment for nucleus n** with spin I_n $\mu_n = g_n \mu_N I_n = \gamma_n \eta I_n$ $\mu_N = \beta_N$ = nuclear magneton
Resonance Condition - EPR $v_e = \|g\| \mu_B B_0 / h$ v_e [GHz] = 13.996246 $\|g\| B_0$ [T] B_0 [T] = 0.071447730 v_e [GHz] / $\|g\|$ $\|g\|$ = 0.071447730 v_e [GHz] / B_0 [T] $\|g\|$ = 3.04198626 v_e [GHz] / v_{H2O} [MHz]	**Resonance Condition - NMR** $v_n = \|g_n\| \mu_N B_0 / h = \|\gamma_n\| B_0 / 2\pi$ for ^{1}H: v_{H2O} [MHz] = 42.5763875 B_0 [T] B_0 [T] = 0.0234871970 v_{H2O} [MHz]
Hyperfine Coupling A [MHz] = 2.99792458 ×10^4 A [cm^{-1}] $\phantom{A\ [\text{MHz}]}$ = 13.996246 $\|g\| A$ [mT] $\phantom{A\ [\text{MHz}]}$ = 1.3996246 $\|g\| A$ [G]	A [cm^{-1}] = 0.333564095 × 10^{-4} A [MHz] $\phantom{A\ [\text{cm}^{-1}]}$ = 4.6686451 × 10^{-4} $\|g\| A$ [mT] $\phantom{A\ [\text{cm}^{-1}]}$ = 0.46686451 × 10^{-4} $\|g\| A$ [G]
Magnetic Field (flux density) = B_0 [in Tesla]	1 T = 10^4 G = 10 kG; 1 mT = 10 G; 1 G = 0.1 mT

see Physical Tables for Fundamental Constants

EPR/ENDOR Frequency Table

Z	A	E	Spin I	Nat. Abund. [%] (Half-life)	calc. X-Band ENDOR Freq. [MHz at 0.350T] (free nucleus)	$g = \mu / (I \mu_N)$	$g \mu_N / g_e \mu_B$	Quadrupole Moment Q [fm² = 0.01 barn]
0	1	n	1/2		10.2076431	−3.8260854	1.040669 E−03	
1	1	H	1/2	99.9885	14.9021185	5.58569468	−1.519270 E−03	
	2	H	1	0.0115	2.28756615	0.857438228	−2.332173 E−04	0.286
	3	H	1/2	(12.32 y)	15.8951943	5.95792488	−1.620514 E−03	
2	3	He	1/2	0.000134	11.3519356	−4.25499544	1.157329 E−03	
3	6	Li	1	7.59	2.193146	0.8220473	−2.235912 E−04	−0.0808
	7	Li	3/2	92.41	5.791961	2.1709750	−5.904902 E−04	−4.01
4	9	Be	3/2	100.0	2.09429	−0.784993	2.13513 E−04	5.288
5	10	B	3	19.9	1.601318	0.600214927	−1.632543 E−04	8.459
	11	B	3/2	80.1	4.782045	1.792433	−4.875293 E−04	4.059
6	13	C	1/2	1.07	3.747940	1.404824	−3.821023 E−04	
7	14	N	1	99.636	1.077197	0.40376100	−1.098202 E−04	2.044
	15	N	1/2	0.364	1.511043	−0.56637768	1.540508 E−04	
8	17	O	5/2	0.038	2.02098	−0.757516	2.06039 E−04	−2.558
9	19	F	1/2	100.0	14.01648	5.253736	−1.428980 E−03	
10	21	Ne	3/2	0.27	1.177076	−0.441198	1.20003 E−04	10.155
11	22	Na	3	(2.6019 y)	1.5527	0.5820	−1.583 E−04	
	23	Na	3/2	100.0	3.944334	1.4784371	−4.021247 E−04	10.4
12	25	Mg	5/2	10.00	0.91290	−0.34218	9.3071 E−05	19.94
13	27	Al	5/2	100.0	3.886082	1.4566028	−3.961859 E−04	14.66
14	29	Si	1/2	4.685	2.96293	−1.11058	3.02070 E−04	
15	31	P	1/2	100.0	6.03801	2.26320	−6.15575 E−04	
16	33	S	3/2	0.75	1.145104	0.429214	−1.16743 E−04	−6.78
17	35	Cl	3/2	75.76	1.461790	0.5479162	−1.490294 E−04	−8.165
	36	Cl	2	(3.01 E5 y)	1.71476	0.642735	−1.748195 E−04	−1.80
	37	Cl	3/2	24.24	1.216786	0.4560824	−1.240513 E−04	−6.435
18	39	Ar	7/2	(269 y)	1.21	−0.454	1.234 E−04	
19	39	K	3/2	93.258	0.696337	0.2610049	−7.099152 E−05	5.85
	40	K	4	0.0117	0.865803	−0.324525	8.82686 E−05	−7.3
	41	K	3/2	6.730	0.382209	0.143261827	−3.896623 E−05	7.11
20	41	Ca	7/2	(1.02 E5 y)	1.215637	−0.4556517	1.239341 E−04	−6.7
	43	Ca	7/2	0.135	1.004386	−0.3764694	1.023971 E−04	−4.08
21	45	Sc	7/2	100.0	3.625677	1.358996	−3.696376 E−04	−22.0
22	47	Ti	5/2	7.44	0.84144	−0.31539	8.5784 E−05	30.2
	49	Ti	7/2	5.41	0.84166	−0.315477	8.58076 E−05	24.7
23	50	V	6	0.250	1.487665	0.5576148	−1.516674 E−04	21
	51	V	7/2	99.750	3.924649	1.4710588	−4.001178 E−04	−5.2
24	53	Cr	3/2	9.501	0.844019	−0.31636	8.6048 E−05	−15 or −2.8
25	53	Mn	7/2	(3.74 E6 y)	3.8296	1.4354	−3.9043 E−04	
	55	Mn	5/2	100.0	3.701688	1.38748716	−3.773869 E−04	33

EPR/ENDOR Frequency Table

Z	A	E	Spin	NA [%]	ENDOR Freq.	$g = \mu / (I \mu_N)$	$g \mu_N / g_e \mu_B$	Q [fm²]
26	57	Fe	1/2	2.119	0.483548	0.18124600	−4.92977 E−05	
27	59	Co	7/2	100.0	3.527	1.322	−3.596 E−04	42
	60	Co	5	(1925 d)	2.027	0.7598	−2.067 E−04	44
28	61	Ni	3/2	1.1399	1.33399	−0.50001	1.3600 E−04	16.2
29	63	Cu	3/2	69.15	3.961568	1.484897	−4.038817 E−04	−22.0
	65	Cu	3/2	30.85	4.2359	1.5877	−4.318525 E−04	−20.4
30	67	Zn	5/2	4.102	0.933986	0.35008196	−9.521988 E−05	15
31	69	Ga	3/2	60.108	3.58672	1.34439	−3.6567 E−04	17.1
	71	Ga	3/2	39.892	4.55726	1.70818	−4.6461 E−04	10.7
32	73	Ge	9/2	7.76	0.521409	−0.1954373	5.315759 E−05	−19.6
33	75	As	3/2	100.0	2.56025	0.959647	−2.61017 E−04	31.4
34	77	Se	1/2	7.63	2.855058	1.0701486	−2.910730 E−04	
	79	Se	7/2	(2.95 E5 y)	0.7760	−0.2909	7.911 E−05	80
35	79	Br	3/2	50.69	3.746454	1.404267	−3.819508 E−04	30.5
	81	Br	3/2	49.31	4.038433	1.513708	−4.117181 E−04	25.4
36	83	Kr	9/2	11.500	0.575479	−0.215704	5.86701 E−05	25.9
	85	Kr	9/2	(10.756 y)	0.596	−0.2233	6.075 E−05	43.3
37	85	Rb	5/2	72.17	1.44385	0.541192	−1.47200 E−04	27.6
	87	Rb	3/2	27.83	4.89349	1.83421	−4.98892 E−04	13.35
38	87	Sr	9/2	7.00	0.648363	−0.243023	6.61005 E−05	33.5
39	89	Y	1/2	100.0	0.733223	−0.2748308	7.475208 E−05	
40	91	Zr	5/2	11.22	1.39118	−0.521448	1.41830 E−04	−17.6
41	93	Nb	9/2	100.0	3.6583	1.37122	−3.72963 E−04	−32
42	95	Mo	5/2	15.90	0.9756	−0.3657	9.9462 E−05	−2.2
	97	Mo	5/2	9.56	0.9962	−0.3734	1.016 E−04	25.5
43	99	Tc	9/2	(2.111 E5 y)	3.3703	1.2633	−3.4360 E−04	−12.9
44	99	Ru	5/2	12.76	0.684	−0.256	6.97 E−05	7.9
	101	Ru	5/2	17.06	0.764	−0.286	7.79 E−05	45.7
45	103	Rh	1/2	100.0	0.47169	−0.17680	4.8088 E−05	
	105	Rh	7/2	(35.36 h)	3.39	1.27	−3.46 E−04	
46	105	Pd	5/2	22.33	0.685	−0.257	6.98 E−05	66
47	107	Ag	1/2	51.839	0.606574	−0.2273593	6.184016 E−05	
	109	Ag	1/2	48.161	0.69734	−0.26138	7.1094 E−05	
48	111	Cd	1/2	12.80	3.174203	−1.1897722	3.236098 E−04	
	113	Cd	1/2	12.22	3.320483	−1.2446018	3.385231 E−04	
49	113	In	9/2	4.29	3.2779	1.2286	−3.3418 E−04	79.9
	115	In	9/2	95.71	3.2850	1.2313	−3.3490 E−04	81
50	115	Sn	1/2	0.34	4.9027	−1.8377	4.9983 E−04	
	117	Sn	1/2	7.68	5.34136	−2.00208	5.44552 E−04	
	119	Sn	1/2	8.59	5.5881	−2.09456	5.69706 E−04	
51	121	Sb	5/2	57.21	3.5893	1.34536	−3.65929 E−04	−36 or −45
	123	Sb	7/2	42.79	1.9436	0.72851	−1.9815 E−04	−49
	125	Sb	7/2	(2.75856 y)	2.00	0.75	−2.04 E−04	

EPR/ENDOR Frequency Table

Z	A	E	Spin	NA [%]	ENDOR Freq.	$g = \mu / (I\,\mu_N)$	$g\,\mu_N / g_e\,\mu_B$	Q [fm²]
52	123	Te	1/2	0.89	3.932218	−1.4738956	4.008894 E−04	
	125	Te	1/2	7.07	4.740899	−1.7770102	4.833345 E−04	
53	127	I	5/2	100.0	3.00222	1.125308	−3.060760 E−04	−71
	129	I	7/2	(1.57 E7 y)	1.9979	0.7489	−2.0368 E−04	−48
54	129	Xe	1/2	26.4006	4.15114	−1.555952	4.232082 E−04	
	131	Xe	3/2	21.2324	1.230549	0.461241	−1.254545 E−04	−11.4
55	133	Cs	7/2	100.0	1.968173	0.7377214	−2.006551 E−04	−0.343
	134	Cs	4	(2.0652 y)	1.9967	0.74843	−2.0357 E−04	38.9
	135	Cs	7/2	(2.3 E6 y)	2.0828	0.78069	−2.12341 E−04	5.0
	137	Cs	7/2	(30.07 y)	2.163	0.8109	−2.205 E−04	5.1
56	133	Ba	1/2	(10.51 y)	4.11749	−1.54334	4.19778 E−04	
	135	Ba	3/2	6.592	1.491586	0.559085	−1.520672 E−04	16.0
	137	Ba	3/2	11.232	1.66716	0.62489	−1.69967 E−04	24.5
57	137	La	7/2	(6 E4 y)	2.054	0.7700	−2.094 E−04	26
	138	La	5	0.090	1.981533	0.7427292	−2.020172 E−04	45
	139	La	7/2	99.910	2.121403	0.7951559	−2.1627690 E−04	20
59	141	Pr	5/2	100.0	4.5625	1.71016	−4.65152 E−04	−5.89
60	143	Nd	7/2	12.2	0.8118	−0.3043	8.2764 E−05	−63
	145	Nd	7/2	8.3	0.5000	−0.1874	5.0979 E−05	−33
61	147	Pm	7/2	(2.6234 y)	1.97	0.737	−2.00 E−04	74
62	147	Sm	7/2	14.99	0.6211	−0.2328	6.332 E−05	−25.9
	149	Sm	7/2	13.82	0.5120	−0.1919	5.220 E−05	7.5
	151	Sm	5/2	(90 y)	0.3874	−0.1452	3.949 E−05	71
63	151	Eu	5/2	47.81	3.70487	1.38868	−3.77711 E−04	90.3
	152	Eu	3	(13.537 y)	1.7253	0.6467	−1.759 E−04	271
	153	Eu	5/2	52.19	1.6353	0.6130	−1.667 E−04	241
	154	Eu	3	(8.593 y)	1.783	−0.6683	1.818 E−04	280
	155	Eu	5/2	(4.753 y)	1.62	0.608	−1.65 E−04	240
64	155	Gd	3/2	14.80	0.4575	−0.1715	4.664 E−05	127
	157	Gd	3/2	15.65	0.5999	−0.2249	6.116 E−05	135
65	157	Tb	3/2	(71 y)	3.57	1.34	−3.64 E−04	141
	158	Tb	3	(180 y)	1.563	0.5860	−1.594 E−04	270
	159	Tb	3/2	100.0	3.5821	1.3427	−3.6520 E−04	143.2
66	161	Dy	5/2	18.889	0.512	−0.192	5.22 E−05	251
	163	Dy	5/2	24.896	0.718	0.269	−7.32 E−05	265
67	165	Ho	7/2	100.0	3.150	1.181	−3.211 E−04	358
68	167	Er	7/2	22.869	0.4298	−0.1611	4.382 E−05	357
69	169	Tm	1/2	100.0	1.23	−0.462	1.257 E−04	
	171	Tm	1/2	(1.92 y)	1.22	−0.456	1.240 E−04	
70	171	Yb	1/2	14.28	2.6341	0.98734	−2.6855 E−04	
	173	Yb	5/2	16.13	0.72555	−0.27196	7.3970 E−05	280
71	173	Lu	7/2	(1.37 y)	1.74	0.651	−1.772 E−04	
	174	Lu	1	(3.31 y)	5.1	1.9	−5.2 E−04	
	175	Lu	7/2	97.41	1.7016	0.6378	−1.735 E−04	349
	176	Lu	7	2.59	1.208	0.4527	−1.231 E−04	497

EPR/ENDOR Frequency Table

Z	A	E	Spin	NA [%]	ENDOR Freq.	$g = \mu / (I \mu_N)$	$g \mu_N / g_e \mu_B$	Q [fm²]
72	177	Hf	7/2	18.60	0.6049	0.2267	−6.166 E−05	337
	179	Hf	9/2	13.62	0.380	−0.142	3.87 E−05	379
73	181	Ta	7/2	99.988	1.8069	0.67729	−1.8422 E−04	317
74	183	W	1/2	14.31	0.628478	0.2355695	−6.407328 E−05	
75	185	Re	5/2	37.40	3.4012	1.2748	−3.4675 E−04	218
	187	Re	5/2	62.60	3.4359	1.2879	−3.5029 E−04	207
76	187	Os	1/2	1.96	0.344971	0.12930378	−3.516974 E−05	
	189	Os	3/2	16.15	1.173760	0.439955	−1.196648 E−04	85.6
77	191	Ir	3/2	37.3	0.26804	0.10047	−2.7326 E−05	81.6
	193	Ir	3/2	62.7	0.2912	0.1091	−2.9684 E−05	75.1
78	195	Pt	1/2	33.832	3.25229	1.21904	−3.31570 E−04	
79	197	Au	3/2	100.0	0.26351	0.098772	−2.6865 E−05	54.7
80	199	Hg	1/2	16.87	2.699312	1.011771	−2.751947 E−04	
	201	Hg	3/2	13.18	0.996420	−0.3734838	1.015850 E−04	38.6
81	203	Tl	1/2	29.52	8.656069	3.24451580	−8.824859 E−04	
	204	Tl	2	(3.78 y)	0.12	0.045	−1.22 E−05	
	205	Tl	1/2	70.48	8.741211	3.2764292	−8.911661 E−04	
82	207	Pb	1/2	22.1	3.1065	1.1644	−3.1670 E−04	
83	207	Bi	9/2	(32.9 y)	2.419	0.9069	−2.4667 E−04	−58
	209	Bi	9/2	100.0	2.437	0.9134	−2.4844 E−04	−51.6
84	209	Po	1/2	(102 y)	3.63	1.36	−3.70 E−04	
86	211	Rn	1/2	(14.6 h)	3.207	1.202	−3.269 E−04	
87	212	Fr	5	(20 min)	2.47	0.924	−2.51 E−04	−10
88	225	Ra	1/2	(14.9 d)	3.916	−1.468	3.993 E−04	
89	227	Ac	3/2	(21.772 y)	2.0	0.73	−1.99 E−04	170
90	229	Th	5/2	(7.34 E3 y)	0.491	0.184	−5.00 E−05	430
91	231	Pa	3/2	100.0 (3.276 E4 y)	3.57	1.34	−3.64 E−04	−172
92	233	U	5/2	(1.592 E5 y)	0.630	0.236	−6.42 E−05	366.3
	235	U	7/2	0.7204 (7.04 E8 y)	0.290	−0.109	2.95 E−05	493.6
93	237	Np	5/2	(2.144 E6 y)	3.351	1.256	−3.416 E−04	386.6
94	239	Pu	1/2	(2.411 E4 y)	1.08	0.406	−1.104 E−04	
	241	Pu	5/2	(14.35 y)	0.726	−0.272	7.40 E−05	560
95	241	Am	5/2	(432.2 y)	1.69	0.632	−1.72 E−04	314
	243	Am	5/2	(7.37 E3 y)	1.60	0.60	−1.63 E−04	286
96	243	Cm	5/2	(29.1 y)	0.438	0.164	−4.46 E−05	
	245	Cm	7/2	(8.5 E3 y)	0.38	0.14	−3.89 E−05	
	247	Cm	9/2	(1.56 E7 y)	0.22	0.082	−2.24 E−05	
97	249	Bk	3.5	(320 d)	1.5	0.6	−1.6 E−04	
99	253	Es	3.5	(20.47 d)	3.13	1.17	−3.19 E−04	670

revision 2006 based on NMR Properties Table, W. E. Hull; no. of decimal places reflects precision.

SUBJECT INDEX

ADF, 140
Alanine Dosimeters, 368
Angular Anomalies, 219
Angular dependencies, 204
Angular momentum
   in Quantum Mechanics, 31–32
   of electron spin, 5, 6
Anisotropy in magnetic parameters, 12, 22–24, 195, 210, 221, 248
Antibodies, 441
Aqueous Solution, 39–40, 341, 386ff.
Artifact(s), 75, 176, 185, 430, 434, 440, 445
Asymmetry parameter, see Quadrupole
Axes
   Coincident, 216, 245
   Crystal, 202, 241
   Molecular, 203
   Principal, 24, 197, 199, 202, 212, 215, 216, 227, 241
   Transformations, 241–242

Base Radicals, see DNA
Biradicals, 93, 134, 254, 260, 262, 280, 282, 283
   exchange interaction, 134
   with TOAC, 275–278
Birch reduction, 110
Bloch equations, 16–17, 163, 165, 167, 175
Bloch functions, 512ff.
Bohr magneton, 6, 518
Boltzmann populations, 9, 25, 236, 258, 273, 361, 386
Born-Oppenheimer approximation, 255

C(13), 74, 113, 117, 130, 144, 148–150, 292, 468
   ENDOR, 224ff.
Carbohydrate(s), 435
Cell membranes, 437
Cellulose, 371, 372
Central Atom, 352, 354
CESR, Conduction Electron Spin Resonance, 508ff.
Charge separation, 100, 404
Chromosomes, 327, 331
CIDEP, 90, 152–153
CIDNP, 153–154
Cluster(s)
   of radicals, 332, 333
Coal, 101
Coenzyme Q, 427
Color centres
   in ionic solids, 492ff.
Computational chemistry, 252, 263
Conductivity, 332
Correlated Radical Pair, see radical pair, correlated
Cryoprotectant, 387
Cryostat, 344, 345
Crystal, see Single crystal
Cu(II)
   complexes, 196, 221, 227, 248
   EPR, 195, 220–221, 452–456
   EPR linewidths, 233
   g tensor, 231–232
   in zeolites and zeotypes, 453–462
   proteins, 99
CYCLOPS, see Phase Cycling

D, E

fine interaction parameters, 30–31, 235
Dating, 101, 102, 362, 374
d-block, 197
DEER, 182–183, 399–401
Density Functional Theory, 139, 251, 259, 272, 283, 355, 356
  functional
    B3LYP, 271
    PBE0, 271
Density Matrix, 4, 163, 165, 255, 270
Derivative signal amplitude, 49
Deuteration, 483
DFT, see Density Functional Theory
Diagonalization, 200, 202, 211, 222, 241
Dielectric loss, see Resonator Q factor
Diffusion tensor, 252, 253, 257–259, 261, 262, 266–269, 273
Dihedral Angle, 353
Dipolar Coupling, 4, 22ff., 209–210, 224, 341, 353, 354, 161
  with ^{17}O, 505
  electron-electron, 30, 68
    calculation, 260, 278
    in pulsed EPR 172, 182–183, 396ff.
    in radical pairs, 414ff.
Direct Effect
  of ionizing radiation, 326, 328, 340
Direction cosines, 242
Discrete-continuum solvent models, 253
Distance measurement, see DEER, 396ff.
DNA, 327, 328, 332, 335, 340, 435, 441
DNA
  base radicals, 335, 337
Dose-Response, 363, 368
Dosimeter(s)
  properties, 363, 369
  self-calibrated, 367
  Table, 369–370
Dosimetry
  accident, 362, 372
  dried fruits, 372
  EPR, 362, 365
  radiation, 362
  retrospective, 362, 372
Double Electron Electron Resonance, see DEER and ELDOR
DPPH, 45, 86
Dynamic models
  for spectrum calculation, 253, 282
Dynamics, see Spin Relaxation and Dynamics, 388, 391
Dysonian CESR line, 513

EA, see electron affinity
Easyspin, 135, 147, 261, 284, 357, 455, 497
Echo, see Electron Spin Echo
Echo-detected EPR, see ED-EPR
ED-EPR
  echo-detected EPR, 175, 176, 452, 457–458, 477
EIE, see ENDOR induced EPR
Eigenvalues
  of Spin Hamiltonian, 211
ELDOR, 181–183
Electrically Detected Magnetic Resonance, 53
Electrochemical generation of radicals, 88, 113–114, 115
Electron Affinity, 332
Electron Magnetic Resonance, see EPR
Electron Paramagnetic Resonance, see EPR
Electron spin, 5, 12
Electron Spin Echo, 169–172, 175
  2 + 1 sequence, 400
  decay, 171
  Stimulated, 175
  Two-pulse, Hahn, 175
Electron Transfer, 113–114, 151–152, 332, 341, 414
  reactivity, 100
Electron Zeeman, see Zeeman effect, 6, 32, 197
Electron-hole pair, 489
ELISA, 443
EMR, see Electron Magnetic Resonance
ENDOR
  CW, 27, 71
  CW in solution, 131–132
  in Photosynthesis, 409–410
  modulation techniques, 74
  of TP680, 420–421
  of P680+, 411
  of P700+, 408–411
  of single crystals, 343, 345–351
  orientation selection, 222
  principles, 25–28
  Pulsed

Davies, 180–181, 224, 225
Mims, 179, 224, 225
Resonator, 73, 344
TRIPLE, 27, 76
ENDOR Induced EPR, 76, 349–350
ENDOR of
^{14}N
quadrupole effects, 227–229
^{23}Na, 474
^{27}Al, 460
^{31}P, 469
^{13}C, 224
ENDOR Spectrum
pattern, 27
Enzymes, 99, 330, 383, 389, 396, 404, 406, 428ff., 435, 439, 440
EPR
CW in liquid phase, 109–150
CW instrumentation, 37–80
High Field, 8, 13, 28, 98, 104, 217, 218, 225, 423, 452, 469
in disordered solid phase, see Powder pattern
in Photosynthesis, 100, 404–405, 407–408
in solid state, 195–236
multifrequency, 67, 450
pulsed, 159–192
2D-sequences, 183–184
in biological applications, 393
quantitative, 66, 361
Time Resolved, 63, 90, 152, 236, 411, 414
Transient, see Time Resolved
radiation effects, 327, 328, 344, 361, 371
EPR spectrum
CW
modeling, 251–283
EPRFIT, 262
EPRLL, 261
ESE, see Electron Spin Echo
ESEEM, 172, 175ff., 209, 222, 227, 386, 399, 400, 409, 410, 463, 464, 467, 480ff.
2-D, 184–185
^{2}H, 480
data fitting, 186–187
E-SpiReS, 262–265, 269, 272, 274, 275, 279, 280

ESR, 3
Euler angles, 222, 227, 243–244, 273, 274
EWSim, 262
Ex vivo, 426, 428
Exchange interaction, see Biradicals

F centres (or color centres), 492ff.
Fe(III), 101, 112, 113
proteins, 99
Fenton reactions, 112, 304
Fermi contact interaction, 19, 116, 135, 140, 142, 249, 271, 498
calculation, 271
Fermi level, 516
Ferromagnetism, 102, 272
FID, see Free Induction Decay
Field modulation
amplitude, 47, 54ff.
frequency, 46, 58ff.
principle, 46, 48
scheme, 48
Fitting of the spectrum, 253, 254, 261, 269, 272, 280, 282, 283
Flow systems, 89
Fokker Planck
operator, 257
Food Irradiation, 368, 369
Free electron, 6
Free Induction Decay (FID), 165, 174
Freon matrices, 92
Friction tensor, 261, 268, 269
Frozen solution, 90, 195, 217, 221, 342
Fullerene
anions, 88, 89
endohedral, 58, 97, 98
C_{60} triplet state, 236

g factor, 6, 12, 120, 198, 200, 249
measurement, 44–46
g tensor, 12, 198, 199, 202–203, 207, 215, 246, 252, 260, 341, 345, 359
transition metal ions, 229–233, 246
GAMESS, 139
Gaussian
lineshape, 17, 53
Gaussian
program package, 139, 262, 269
Gaussian 03, 355
GaussView, 263

g-radiation, 426
Graphical interface, 262, 265
γ-rays, *see* Ionizing Radiation
Group theory, 246

H(2), 227–228
Half-life, 437
Hall sensor, 44
Heisenberg Exchange, 168
Heterogeneity, 389
Heterogeneous broadening, 495
HF-EPR, *see* EPR High Field
High field approximation, 20, 21, 68, 212, 236
Hole centres
   EPR spectra, 499, 501
   in ionic solids, 498
   produced by ionizing radiation, 326
   transfer along the DNA chain, 333
Homogeneous linewidth, 17
HPLC, 441, 442
Humic substances, 101
Hydrogen bond, 227, 256, 330
Hyperconjugation, 118–119
Hyperfine coupling constant(s), 19–20, 33
   for 100% s-Spin population, 116
Hyperfine Coupling Tensors, *see*
   Dipolar Coupling, 341, 345, 351, 354, 357
Hyperfine interaction, *see* Fermi
   interaction, 209, 212
   α-Proton, 118, 352
   β-Proton, 118–120, 146–147, 353
   second order effects, 130
HYSCORE, 184–185
HYSCORE of
   ^{27}Al, 457–459

ICS, *see* Integrated Computational Strategy
Imaging, 78
   resolution, 79, 80
Immuno spin trapping, 442
In vivo, 426, 428
Indirect Effect
   of ionizing radiation, 326, 328
Inhomogeneous linewidth, 17
in-situ
   EPR, 472–473, 478
Instantaneous Diffusion, 172, 176

Integrated Computational Strategy
   definitions, 253, 254, 255, 259, 262, 278, 280, 282, 283
   implementation, 253
Inversion Recovery, 168, 174
Inverted spin trapping, 432
Ion pairs, *see* Radical anions
Ion vacancy, 488
Ionization Potential, 333
Ionizing Radiation, 326
   generation of radicals in solids, 90
   X-rays, 426
IP, *see* ionization potential
Isotope substitution, 435, 436
Isotropic interaction, 195, 197, 200, 352

Jahn-Teller, 230

Kinetics, 430, 443
Kramers doublets, 235
KVASAT
   simulation program, 357

Laboratory frame, 215, 226, 241
Lanczos, 261, 263
Larmor nuclear frequency, 73–74, 459–460
   precession, 15
Lattice, 9
L-band, 40, 441, 442
Least Squares Fitting, 345
LET, *see* Linear Energy Transfer
Line of nodes, 246
Linear Energy Transfer, 326, 327, 363
Linewidths, *see* Spin Relaxation and EPR linewidths
Linux, 262, 263, 280
Liouville equation, 252, 255
Liquid crystals, 254, 282
Lorentzian line shapes, 17, 252

M centres, 497
Magnetic Field
   gradient, 79
   homogeneity, 43
   measurement, 45
   modulation, 46, 54, 58
      amplitude, 55
   precision, 44

sensors, 45
stability, 46
sweep, 46
types, 50
Magnetic materials, 103
Magnetic parameters
  and spectra calculation, 251, 253, 255, 256, 279
Magnetically dilute
  single crystal, 197
Magnetization
  equilibrium, 14
  time evolution, 15–17
Mass spectroscopy, 432, 441, 442
Matrices, 197, 199, 211–212
  Manipulations, 239–241
  Pauli, 241
  Transpose, 241
Metrology, see EPR, Quantitative, 361
Microdialysis, 436
Microwave
  automatic frequency control, 43
  bands, 8, 41, 67, 74
  bridge, 42, 188
  components, 42
  field, 35
  power, 42, 49, 55, 58, 64
Mitochondria, 428, 439
Mn(II)
  in proteins, 99, 224
  as spin probe, 100, 102
  ENDOR in zeolites and zeotypes, 469
  energy levels of spin states, 234
Modulation depth
  in ESEEM of ^{2}H, 480
Modulation
  Frequency, 58
Modulation schemes
  in ENDOR, 74–75
Molden, 263, 265
Monoclinic, 343, 347
Multifrequency EPR, see EPR multifrequency

Neutron Diffraction, 342
Nitric oxide
  spin trapping, 438–440
Nitronyl nitroxides, 84, 103, 272, 274, 275, 276, 439

Nitroxides
  as spin labels, 86, 216, 384
  as spin probes, 86, 391–392, 473–474, 476–480
  di-tert-butyl nitroxide, 473
  EPR powder spectra, 216
  TEMPO, 12, 438
NLSL, 261
NMR, 13, 21, 25, 26, 27, 28, 35
Norrish I cleavage, 113
Nuclear spins, 18
Nuclear spins frequencies, 74
Nuclear spins frequencies
  Table, 523–526
Nucleosomes, 331, 341
Nucleotides, 328
Number of spins, 47, 66, 361

O$^-$ in MgO, 502
ODMR, see Optically Detected Magnetic Resonance
Off axes extrema, 219
OOPESEEM
  out-of-phase ESEEM, 417–419
Optically Detected Magnetic
  Resonance, 39
Orbital, 229, 248
ORCA, 140
Orientation
  Field, 198, 199, 200
  Sample, 196
  Selection, 222, 227
Orthorhombic, 343, 347
Oxidation, 332, 341
Oximetry, 395–396, 440
Oxygen (O$_2$), 97
Oxygen concentration, see Oximetry
Oxygen Evolving Complex (OEC), 101, 405
Oxygen radicals, 97

P doped Silicon, 506
P680, 405, 411, 419, 421
P700, 405, 406–411, 414–415, 417–420
Pake Pattern, 214
Partition coefficient, 433
Pauli
  Spin operator, 199, 200

PELDOR, *see* DEER and ELDOR
Phase Cycling, 190–191
Phase Memory Time, 172, 176
Photolysis, 87, 90, 95, 96, 113, 114, 152–154, 292, 295ff., 302, 310, 482
Photosynthesis, *see* EPR in Photosynthesis
Polar angles, 199, 241
Polarizable continuum
 model, 254
 solvent described as a, 254
Polycrystalline, 197, 341, 359, 366
Polymerization via radical, 104, 145, 295, 313
Polymers
 conformation, 145–146
Polyradicals, 96, 103
Post-translational modification, 389
Powder pattern, 13, 196, 203, 205, 210, 213, 216
Powder spectra
 simulation program, 357
Power Saturation, *see* Spectrum saturation
Principal axes, 24, 29, 202, 241
Principal values, 24, 29, 197, 199, 203
Protein, 62, 99–100, 182, 224, 227, 320, 331, 341, 383–426, 436–445
Proton Transfer, 334
Pulsed EPR
 2D, 393
 see EPR pulsed

Q factor, *see* Resonator Q factor
Q-band, *see* Microwave bands
Quadrupole, 197, 225, 228
 asymmetry parameter, 227
Quadrupole coupling
 of ^{14}N, 227–229
 of ^{27}Al, 461
Quantitative EPR, *see* EPR quantitatve
Quantum Chemistry, 259, 355
Quartet state, 96, 98
Quenching
 of the orbital angular momentum, 12

R centres, 499
Radiation
 ionizing, 325, 326, 362, 371

Radical pairs
 correlated, 93, 411
Radical Trapping, 335
Radical anions, 100
 by ionizing radiation, 326, 355
 generation, 89, 115
 ion pairs, 148–149
Radical cations
 amines, 141–144
 by ionizing radiation, 326, 355
 generation, 89, 114
Radical(s)
 in biological systems
  agents of formation
  in solids, 90
  in solution, 85, 110–131
 inorganic, 96, 208
 neutral
  generation, 111
 nitroxides, *see* Nitroxides
 organic, 208
 primary, 90
 radiation produced, 325–375
 secondary, 91
 phenoxyl, 427
 in disease
 π-type, 348, 352, 354
 σ-character, 140–142
Rate constant, 431, 443
Recombination, 332, 334, 335
Recombination
 triplet, 416, 419–421
Redfield method, 128, 252, 281
Reduction, 340
Reference Sample, 366
Relaxation, *see* Spin relaxation
Reperfusion injury, 427
Resonator, 37, 191–192
 conversion factor, 38, 40, 60
 ENDOR, 73
 field distribution, 39, 40, 67
 filling factor, 39, 49
 matching, 41
 Q-factor
  in transient EPR, 63
 Q-factor, 37–39, 41, 49, 52, 191–192, 386

sensitivity, 39, 49
temperature, 52
types, 40
Rotation, 343, 345, 346

Sample holder, 365, 366
Satellite lines
  C(13), 130
Saturation, *see* Spectrum saturation
Saturation Recovery, 168
Saturation Transfer, 393
Schonland ambiguity, 346
Schottky defects, 490
SDSL, *see* Site directed spin labelling
Second order effect, 69, 70, 130, 133, 142, 185, 233, 420
Secular
  constant, 200, 239, 242
  terms, 211
Selection rules, 21, 35, 172, 177–178, 212, 226, 497
Semiconductors
  charge carriers in, 503
Sensitivity, 39, 361, 363
  noise, 49
  number of spins, 49, 66
Signaling, 440
SimFonia, 134, 262
Single crystal, 195, 201–203, 223, 341, 342, 344, 359
Single Molecule Magnet, 103, 104, 220
Singlet state, 28
Singularities, 204, 214–215, 224, 235
Site directed spin labeling, 62, 384
  and protein topology, 394–395
Site Splitting, 347, 348
Solid state, 195ff.
Solvent effects
  in spectra simulation, 254
SOMO, 144–145, 208, 247, 248
Spatial resolution, 78, 445
Spectrometer
  CW, 37–49
  Pulsed, 187–193
Spectrum
  absorption/dispersion, 43
  amplitude, 48, 49

detection, 48
g-factor, 44, 68
reference sample, 45, 66
saturation, 9, 55, 58–62, 363, 364
hyperfine pattern, 120–124
Spectrum analysis
  flow chart, 137
Spectrum Simulation, 71, 125–130, 134–138, 161, 215, 224, 227, 251–284, 355, 356–359, 392, 438
  benchmark spectra, 358
Spin Density, 352, 353, 355
  Hückel MO, 138
Spin diffusion, 168
Spin Echo, *see* Electron Spin Echo
Spin Hamiltonian, 195, 210, 222, 227, 251, 255, 257, 266
Spin orbit coupling
  and g tensor, 12, 230, 246
Spin Packet, 163
Spin polarization, 411
  $\sigma - \pi$, 117
Spin populations
  non equilibrium, 236
Spin probe(s), *see* Nitroxides, spin probes
  silane based, 478
Spin relaxation, 4, 9, 16ff.
  and Cr(III), 395
  and distance measurements, 397–399
  and dynamics, 124–130, 251ff., 388–389, 391–392
  and EPR linewidths, 126–130, 397
  and EPR linewidths
    for transition metal ions, 233
  and O_2, 395, 440
  and saturation, 58–62
  cross, 26
  in CESR, 511–512, 515
  in EIE, 349
  in ENDOR, 25–27, 72, 77
  in pulsed EPR, 163, 168, 170, 172–173, 393, 481
  of transition metal ions, 132, 235
Spin scavenging, 437
Spin trapping, 63, 287–321, 428, 443
  in biological systems, 320

534 SUBJECT INDEX

Spin-lattice relaxation time, 9–10, 16–17, 53, 58, 59ff., 163, 164, 168
  and O_2, 395, 397, 398
  measure, 174, 189
  in metals, 512ff.
Spin-Orbit Coupling, 9, 11–12, 99, 120, 197, 207–208, 231, 233
  and g values in Transition Metal Ions, 246, 248
  in CESR, 513, 514
  operators, 270
Spin-spin relaxation time, 16–17, 53, 58, 59ff., 64, 132, 163ff.
  and O_2, 396
  as measured by Pulsed EPR, 171
  in metals, 512ff.
Stalagmites, 374
Sterilization, 369
Stimulated Echo, 175–177
Stochastic
  effects, 253
  models, 258
Stochastic Liouville Equation, 252, 253, 257, 259, 261, 269, 273, 275, 279
Strand Break, 333, 340, 341
Sugar Radicals, 335, 340
Sugar-Phosphate Backbone, 328, 332, 334
Superoxide dismutase, 438
Superoxide radical, 307, 318, 320, 439, 445
Surface potential, 394
Symmetry, 198
  Axial, 204
  Point, 230, 247
  Rhombic, 204–206
System bath, 254

T_1, see Spin-lattice relaxation time
T_2, see Spin-spin relaxation time
TABLES
  Conversion factors, 521
  EPR Tables, 522
  EPR-ENDOR Tables, 523–526
  Physical Tables, 520
Temperature
  as experimental parameter, 51
Tensor
  diagonalization, 202, 211, 222, 241
  dipolar interaction
    electron-electron, 29–30

  dipolar interaction
    hyperfine, 22, 210
  hyperfine interaction, 212, 215
  quadrupolar, 227
  transformations, 241
Tensor Elements, 345, 346
Tensor g, see g tensor
Tensor(s)
  rotation, 241–243
  magnetic, 196, 197
Time resolution
  direct detection, 63
  methods, 63
  transient, 63, 65
Timescale, 326
Time-scale separation, 255
Torsional angle, 258, 268
Toxicity of spin traps, 433, 443
Transition metal ions, 98–100, 132–133, 197, 229, 230, 246
  electron configuration, 230–233
  in mesoporous materials, 470
  in oxides, 472
  in solution, 132–133
  in zeolites and zeotypes
    by cation exchange, 451–466
    by framework substitution
Transverse relaxation time, see Spin-spin relaxation time
TR-EPR, see EPR, Time Resolved
Triclinic crystals, 343, 347
Triplet(s), 28, 93–96, 235, 416, 419–421
  ground state, 95
  photoexcited, 63, 96, 236
TURBOMOL, 139
Tyrosyl Radical, 216ff., 227

V centres, 498
Vector Model, 162ff., 418
Vibrational averaging effects, 256, 260

Water
  in biological systems, 386
Water Equivalent, 364, 366
Water Radicals, 326, 332
Water solutions
  precautions in biological applications, 385–388
W-band, see Microwave bands

Western blotting, 434, 443, 444
Wigner matrix functions, 274

X-band, *see* Microwave bands
X-Diffraction, 342, 344
X-rays, *see* Ionizing Radiation
Xsophe, 262, 357

Z scheme, 405
Zavoisky, 3
Zeeman effect, 6–7, 32–33, 257, 258, 259, 262, 270
Zeolites, 92, 451–470
Zero Field Splitting (ZFS), 99, 197, 233–235, 420

CHEMICAL INDEX

2,2-diphenyl-1-picrylhydrazyl (DPPH), 45, 86
2,3-di(tert-butyl)buta-1,3-diene radical anion, 147
2,3-Dicyano-5,6-dichlorobenzoquinone, 114
4-hydroxy-TEMPO-benzoate 4HTB, 482
4PyPN, *see* Spin traps
4-trimethylammonium-2,2,6,6, tetraethyl piperidinyloxy (TempTMA+), 480

Adenine, 340
Al(27)
    ENDOR of, 462
    HYSCORE of, 459–464
Alanine, 357–362, 365
Alpha-tocopherol, 429
AlPO, 453, 469–471
Aluminosilicates, 453
Arsenic, 428
Asbestos, 428
Ascorbyl radical, 429
α-tocopheroxyl radical, 429

Barium sulfate
    as dosimeter, 369
Butylated hydroxytoluene, 429

C(13) (*see also* Subject Index), 224
Chlorinated hydrocarbons, 428
Chromium, 395, 428, 474
Chromium oxide
    catalysts, 474
Cl_2^-, 97
Cl_2^- in KCl, 500

Cr^{3+}
    in MgO, 45, 366
    multifrequency spectra, 70
$Cr^{6+}, Cr^{5+}, Cr^{3+}$
    in CrO_3/ZnO_2 catalysts, 474
CrO_3/ZnO_2
    catalysts, 474
Cu
    isotopes, 132
Cu(I)/ZSM5, 478
Cu(II) (*see also* Subject Index)
    EPR, 99, 110, 132, 133, 196, 221, 230–232, 248
    exchange zeolites, 454–464
    in azurine, 385
    Histidine complex in zeolite Y, 466–468
$Cu(NO_3)_2$, 457–459
$CuCl_2$, 3, 133
Cysteine, 435
Cytosine, 337–338
Cytosine Monohydrate, 341, 343, 347, 354, 357

DBNBS, *see* Spin traps
DEPMPO, *see* Spin traps
DEPNP, *see* Spin traps
Dibenzo[b,h]biphenylene
    radical ions, 143
dibenzyl ketone
    on ZSM-5, 480, 483
dibenzylmercury, 316
di-tert-butyl nitroxide, 86, 256, 257, 292, 296, 476
di-tert-butyl peroxide, 87, 316
Dithiazolyl radicals, 219–221

Electron Paramagnetic Resonance. Edited by Brustolon and Giamello
Copyright © 2009 John Wiley & Sons, Inc.

CHEMICAL INDEX

DMPO, *see also* spin traps, 308, 318–322, 430
DNA, 327–341
DPPH, *see* 2,2-diphenyl-1-picrylhydrazyl
DTBN, *see* di-tert-butyl nitroxide

EMPO, *see* Spin traps

F127-NO, 478–479
Faujasite, 454, 456
Fe(II)-dithiocarbamate, 442
Ferrierite
  Cu2+ in, 458
Flavonoids, 429
Fmoc-(Aib-Aib-TOAC)2-Aib-OMe, 275, 277
Formate
  ammonium, 368
  lithium, 368

GDP, *see* guanosine-5′-diphosphate
Glutathione, 439
Glycerol
  as additive to water solutions, 387
GTP, *see* guanosine-5′-triphosphate
Guanine, 328–330, 333, 339
Guanosine-5′-diphosphate, 224

H(2)
  modulation depth, 480
histidine
  Cu(II) ligand, 187, 410, 421, 466–468
Hydrocarbons
  non Kekulé, 95
Hydrogen peroxide, 112, 305, 428
Hydroperoxides, 112, 443
Hydroxyapatite, 371, 374
Hydroxyl radical, 295, 307, 319, 320, 439
Hydroxylamine, 292, 293, 320, 434, 440

Isoprostanes, 443

Ketone(s), 113, 313, 482
Ketyl radical, 116

L62-NO, 478–479,481
LAMPBN, *see* Spin traps
LAMPPN, *see* Spin traps
Lewis acid
  sites, 114, 476
Lipids, 388, 392, 394, 436, 437

Lithium formate
  as dosimeter, 368
Lithium phthalocyanine, 396

MBN, *see* Spin traps
Methane, 92
Methanethiosulfonate, 216–217, 384
Mn(II), 45, 99–102, 110, 224
  extraframework, 454
  in AlPO$_4$-11
    ENDOR of, 469
  in AlPO$_4$-20
    ENDOR of, 469
  in UCSB-10 Mg
    ENDOR of, 469
  in MgO, 366
  in SAPO-5
    ENDOR of, 469
  in UCSB-6 Mg
    ENDOR of, 467–469
  in CaO, 55–57
  in UCSB-10 Mg
    EPR of, 467–469
Mn$_{12}$O$_{12}$, 104
MNP, *see* Spin traps
MnSO$_4$, 133
Molecular sieves, 453
MTPNN, *see* p-(Methylthio)phenyl Nitronyl Nitroxide
MTSL, *see* methanethiosulfonate
Myoglobin, 438

N(14), 187, 216, 220, 227–229, 410, 437
Na(23), 115, 116, 147, 476–478
Naphtalene radical anion
  EPR spectrum, 136
NB, *see* Nitrosobenzene
ND, *see* Nitrosodurene
Nitric oxide, 97, 296, 321, 426, 440–442, 463, 476–480
Nitroaromatics, 429
Nitrone(s), *see also* spin traps, 289, 430, 442
Nitroso compounds, *see* Spin Trap(s)
Nitrosobenzene, 296ff.
Nitrosodurene, 297ff.
Nitroxide(s), *see* Spin Traps and Subject index
NNP, *see* Spin traps
NO, *see* Nitric oxide

OH, *see* hydroxyl radical
Ozone, 428

P(31), 74, 75, 116, 121, 153, 187, 224–225, 302, 306–313, 470, 471, 508
p-(Methylthio)phenyl Nitronyl Nitroxide spectrum calculation, 272–275
P123-NO, 478–479
PBN, *see* Phenylbutylnitrone and Spin traps
Pentacene in p-terphenyl
    triplet, 63–65
Peroxidase, 438–440
Phenoxyl radicals(s), 429
Phenylbutylnitrone, 63, 64
Phosphoryl Dithioesters, *see also* Spin traps, 213
Phosphoryldithioformates, 290
Phylloquinone, 406, 417, 421
Plastoquinone, 407
POBN, *see* Spin traps
PPN, *see* Spin traps
Pt
    cluster, 464–465
    isotopes, 465
PyOPN, *see* Spin traps

Quinone(s), 101, 114, 115, 290, 309, 310, 315, 391, 403, 406

SBA-15, 472
Si(29)
    satellites, 130, 311, 312, 149–150, 509
Silyl radicals, 112, 127, 148, 309, 311, 315
Sodalite, 454, 459, 470, 482
Spin Trap(s)
    Cyclic nitrones
        DEPMPO, 308, 309, 319, 320, 321, 430, 433, 438, 439
        DMPO, 305–308, 318–321, 431ff.
        EMPO, 308, 321, 431, 439
    Nitroso aliphatic
        DEPNP, 296
        MNP, 291ff., 430ff.
    Nitroso aromatic
        DBNBS, 298ff., 430ff.
        TBNB, 298

Open chain Nitrones
    4PyPN, 300, 301
    LAMPBN, 300
    LAMPPN, 300
    MBN, 300
    PBN, 300ff., 431
    POBN, 301ff., 431
    PPN, 300
    PyOPN, 300ff.
Phosphoryl dithioesters, 313, 314
Polyfunctional
    NNP, 316ff.
Thioketones
    TBTPS, 311
    Thiobenzophenone, 311
Stannyl radicals, 112
Superoxide radical anion, 307, 318, 320, 428, 439–441, 445

TBNB, *see* Spin traps
TBTPS, *see* Spin traps
Tempo, 12, 85, 440, 480
Tempone, 86, 263, 265
Tempo-palmitate, 279, 280, 281
TEOS
    tetraethyl-ortho-silane, 482
Tetrachlorobenzoquinone, 114
Tetrapropylammoniumhydroxide, 472
Thioketones, *see* Spin Trap(s)
Thiyl radical, 291, 306, 437, 444, 445
Thymine, 328ff.
TPAOH, *see* tetrapropylammoniumhydroxide

UCSB-10 Mg, 469
Ultramarine blue, 483

Vitamin C, 429
Vitamin E, 429
VO^{2+}, 99, 110, 132, 472

zeolite(s), 92, 100
    and TMI, 453ff.
    KL, 464
    NaY, 455
    ZSM-5, 453, 458ff.